Symbiotic Microbiomes of Coral Reefs Sponges and Corals

Zhiyong Li
Editor

Symbiotic Microbiomes of Coral Reefs Sponges and Corals

Springer

Editor
Zhiyong Li
Marine Biotechnology Laboratory,
State Key Laboratory of Microbial Metabolism,
School of Life Sciences and Biotechnology
Shanghai Jiao Tong University
Shanghai, People's Republic of China

ISBN 978-94-024-1610-7 ISBN 978-94-024-1612-1 (eBook)
https://doi.org/10.1007/978-94-024-1612-1

Library of Congress Control Number: 2019935528

This Springer imprint is published by the registered company Springer Nature B.V.
The registered company address is: Van Godewijckstraat 30, 3311 GX Dordrecht, The Netherlands

Preface

Coral reefs, which are often called "rainforests of the sea," form some of the most diverse ecosystems on Earth, i.e., they occupy less than 0.1% of the world's ocean surface and provide a home for 25% of all marine species. Paradoxically, coral reefs flourish even though they are surrounded by ocean waters that provide few nutrients. The exploration to explain how can the world's most productive and diverse ecosystems thrive in the marine equivalent of a desert has made the invertebrates in coral reefs an important object of study.

"Microbiome" means the entire microbial community and genes that reside in an environmental niche. Coral reefs provide habitat for an array of marine invertebrates that host symbiotic microbiomes that perform important ecosystem functions such as chemical defense, nutrient and energy transformation, and waste and toxic compound degradation. It was proved that microbiomes have numerous beneficial functions relevant to supporting the host's life. In some aspects, invertebrate-symbiotic microbes play a key role in the biological diversity of coral reefs ecosystem. The significance of the associations and interactions between symbionts and their invertebrate hosts is becoming increasingly apparent.

As two main predominant invertebrates in the coral reefs, the microbial diversity and putative symbiotic functions have been well explored in corals and sponges. Particularly recently, omics technology has been successfully used in the function revelation of sponge and coral microbiomes. The relationship between sponges/corals and their microbial/algal associates represents a valuable model that can be applied to the broader discipline of invertebrate–symbionts. Thus, this book focuses on the symbiotic microbiomes of sponges and corals in coral reefs, both applied and basic in ecology, chemistry, and biotechnology, providing in-depth and up-to-date reviews on the microbial structure and diversity, metabolism and function, environment and adaption, and bioactive potentials. Meanwhile, the future perspectives will be discussed according to the existing problems and the development trend.

Chapters 1 and 2 provide a general overview of coral reefs ecosystem and sponge/coral microbiome. The structure and diversity of prokaryotic and eukaryotic symbionts in sponges and corals including the core microbiome and host-specific symbionts are covered in Chaps. 3–6. To assist in the understanding of sponge/coral microbiome, Chap. 7 covers the sponge holobionts in health and disease. Chapter 8–10 introduces the metabolism and function of sponge/coral microbiomes with emphasis on biogeochemical cycles and community metabolic characteristics. The response of sponge microbiomes to environmental variations is discussed in Chap. 11. We designate Chaps. 12–18 to introduce the potential and utilization of sponge/coral microbiomes in the biosynthesis of antibiotics, natural products, and enzymes from sponge/coral holobionts.

We try to use an integrative approach to relate new and important topics to the broader scientific field of symbiotic microbiomes of coral reef sponges and corals. This book will be of particular interest to the professionals in marine ecology, marine biotechnology, as well as medicinal chemists and molecular biologists.

Marine Biotechnology Laboratory, State Key Laboratory of Microbial Metabolism, School of Life Sciences and Biotechnology
Shanghai Jiao Tong University
Shanghai, People's Republic of China

Zhiyong Li

Acknowledgement

We gratefully acknowledge the financial supports for the sponge/coral microbiology and biotechnology research in Professor Zhiyong Li's group at Shanghai Jiao Tong University from the Natural Science Foundation of China (NSFC) (31861143020, 41776138, 41742002, U1301131, 41176127, 41076077, 31000062, 31300104, 81102417), the National Major Scientific Research Program of China (2013CB956103), and the High-Tech Research and Development Program of China (national 863 program) (2013AA092901, 2011AA090702, 2007AA09Z447, 2004AA628060, 2002AA608080).

Contents

Chapter 1
Coral Reef Ecosystem

Baolin Liao, Baohua Xiao, and Zhiyong Li

Contents

Abstract Coral reefs construct the most bustling metropolis in the oceans. Known as the "rainforests of the sea," coral reefs form some of the most diverse ecosystems. They provide many complicated marine habitats to support a wide range of other living beings. There are diverse coral reef organisms, such as sponge, sea anemone, coral algae, etc. besides corals. A coral reef is one of the most vital ecosystems on Earth and plays an irreplaceable role in maintaining the biodiversity and ecological balance of the oceans.

B. Liao (✉)
Shenzhen Institute of Guangdong Ocean University, Shenzhen, People's Republic of China

B. Xiao
Shenzhen Institute of Guangdong Ocean University, Shenzhen, People's Republic of China

Guangdong Ocean University, Zhanjiang, People's Republic of China

Z. Li
Marine Biotechnology Laboratory, State Key Laboratory of Microbial Metabolism,
School of Life Sciences and Biotechnology, Shanghai Jiao Tong University,
Shanghai, People's Republic of China

Z. Li (ed.), *Symbiotic Microbiomes of Coral Reefs Sponges and Corals*,
https://doi.org/10.1007/978-94-024-1612-1_1

Keywords Coral reefs · Coral · Sponge · Zooxanthellae · Ecological function

1.1 Definition and Classification of Coral Reefs

Coral reefs are varied underwater ecosystems held together by calcium carbonate structures that are excreted by corals, which construct the most bustling metropolis in the oceans. Coral reefs are one of the main types of reefs in the ocean [1], which are constructed by stony corals, through the gradual accumulation of calcium carbonate skeleton (Fig. 1.1). Based on the topographic features, coral reefs are generally classified in marginal reef, barrier reef, and atolls [2, 3]. Marginal reefs are close to the land, e.g., the Red Sea coast and the Hainan Island's coastal zone, China. Barrier reefs, like a dike, are not connected with the land, for example, the Great Barrier Reef in Australia, which includes more than 2800 independent coral reefs, is the world's biggest coral reef with a depth of about 30 m and a total length of about 2000 km; its widest point is about 72 km. Atoll reefs are distributed in zonal shape around the lagoon, with a diameter from a few hundred meters to tens of kilometers. The known atoll reefs are mainly distributed in the Western Pacific trade winds and tropical waters of the Indian Ocean. On the base of relationship between coral reefs and sea level, coral reefs can be distinguished as uplift reef and drowning reef.

1.2 Ecological Value and Biodiversity of Coral Reef Ecosystem

Known as the "blue desert oasis" [4], or the "rainforests of the sea," coral reefs make which the most diverse ecosystems on Earth come into being. They provide complex and manifold marine habitats that support a large number of other biologies, e.g., at least 25% of all halobios including fish, mollusks, worms, crustaceans, echinoderms, sponges, scales, and other cnidarians inhabit the coral reef ecosystem [5] (Fig. 1.2).

Fig. 1.1 Coral reef ecosystem. (Photos by Ran Li)

Fig. 1.2 Reef corals in the South China Sea. (Photos by Li Ran and Bing Wang)

For example, the Great Barrier Reef ecosystem contains more than 400 kinds of corals, more than 1500 kinds of fish, more than 4000 kinds of invertebrates, and more than 400 kinds of sponges. Therefore, the coral reef is one of the most crucial biogeocenose on Earth, which is very important for maintaining biodiversity and ecological balance of the oceans.

The coral reef ecosystem's function also includes the energy flow, material cycle (e.g., C, N, and P), and productivity. For instance, coral reefs play an important part in maintaining the steady state of carbon dioxide; the dissolved carbon dioxide will be used to build the coral skeleton of calcium carbonate.

1.3 Corals

1.3.1 Physiological Characteristics of Corals

Corals are one of the main component organisms in the coral reef ecosystem, which are oceanic invertebrates in the class Anthozoa of phylum Cnidaria. The organization includes coral reef creators who live in tropical seas and excrete calcium carbonate to form rigid skeletons.

In appearance, the coral body has a simple structure with a cylindrical body, an opening at the top, and one or more rings of tentacles used for predation. The lower part of the mouth is a pouch-shaped intestinal cavity, which is usually separated by a membrane to increase the area of digestion and absorption. There are some tubular membrane filaments in the body, which are connected to the intestinal wall and secrete digestive enzymes in coral endoderm, and zooxanthellae can be found. Most scleractinian corals have symbiotic zooxanthellae; they can promote the corals to secrete calcium carbonate and provide carbon for the coral host through photosynthesis. There are multiple shapes for coral colony, such as branching, leafy, massive, laminar, etc. (Figs. 1.3 and 1.4). Coral can reproduce through asexual and sexual reproduction [6] (Fig. 1.5).

(1) Asexual Reproduction

Asexual reproduction of corals includes fission and budding. The common point in all asexual reproduction is that the offspring inherit all the genetic traits of the mother and are actually the exact copies or clones [7]. The main way of hermatypic corals is to reproduce asexually, while stony corals have two ways: the first is gemmatio exogenous, and the second is endogenous budding.

(2) Sexual Propagation

Sexual reproduction is an important strategy to maintain the scleractinian coral's population evolution [7, 8]. Most scleractinian corals are hermaphroditic, few dioecious. The same strain of coral can produce sperm and eggs; only about 1/5 of the corals are male and female, which can't be distinguished in appearance [8, 9]. The sexual reproduction of corals can be divided into two types: emission type and incubation type [1]. About 80% of the coral species belong to the discharge type; the insemination is finished in seawater, and the fertilized eggs develop into larvae. In addition, there are about 20% species of corals that are incubation type; only sperm is released, along with the water flow into the cavity in the coral or egg to form a fertilized egg, and then released into the seawater [10] (Fig. 1.6).

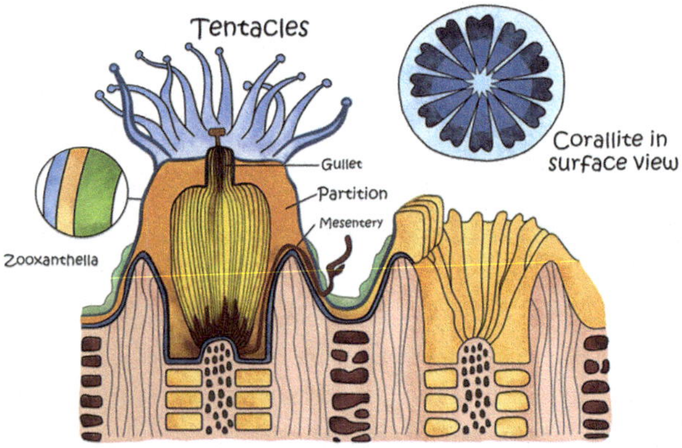

Fig. 1.3 Structure of scleractinian coral. (Photo by Baolin Liao)

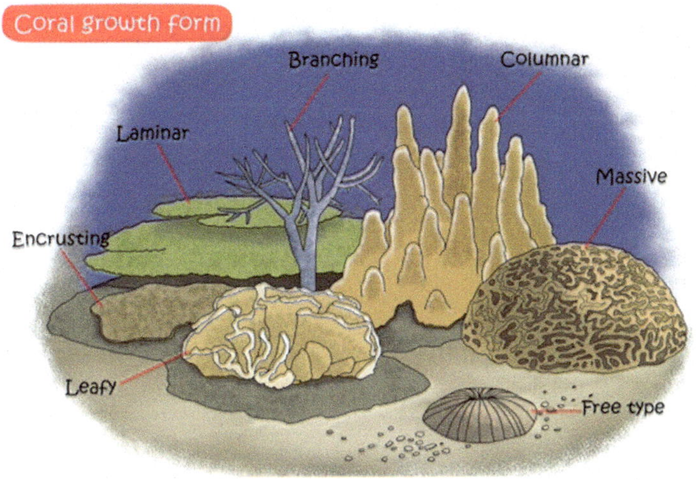

Fig. 1.4 Shapes of the coral colony [1]. (Photo by Changfeng Dai)

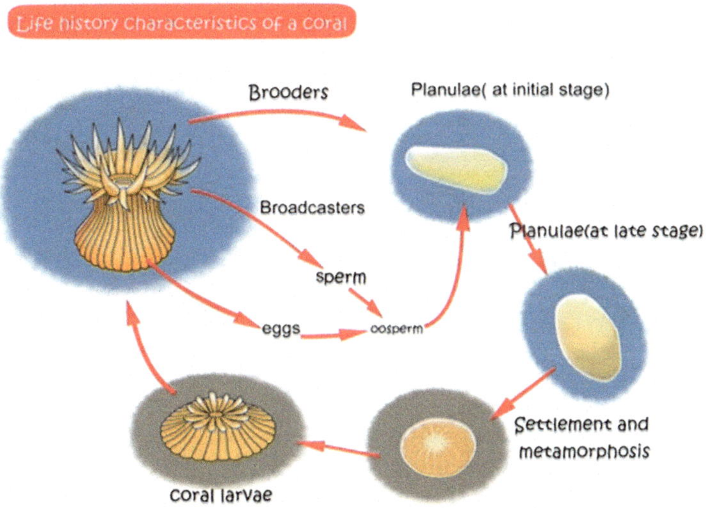

Fig. 1.5 Life history of a coral [1]. (Photo by Changfeng Dai)

1.3.2 The Geographic Distribution of Reef Corals

Reef corals are mainly distributed in the tropical waters, and coral species decrease gradually with the increase of latitude. Warm water will make the distribution range of reef corals extend to high latitude, for example, in the Western Pacific Ocean, influenced by the Kuroshio Current, some corals inhabit even the Tokyo Bay, Japan;

a

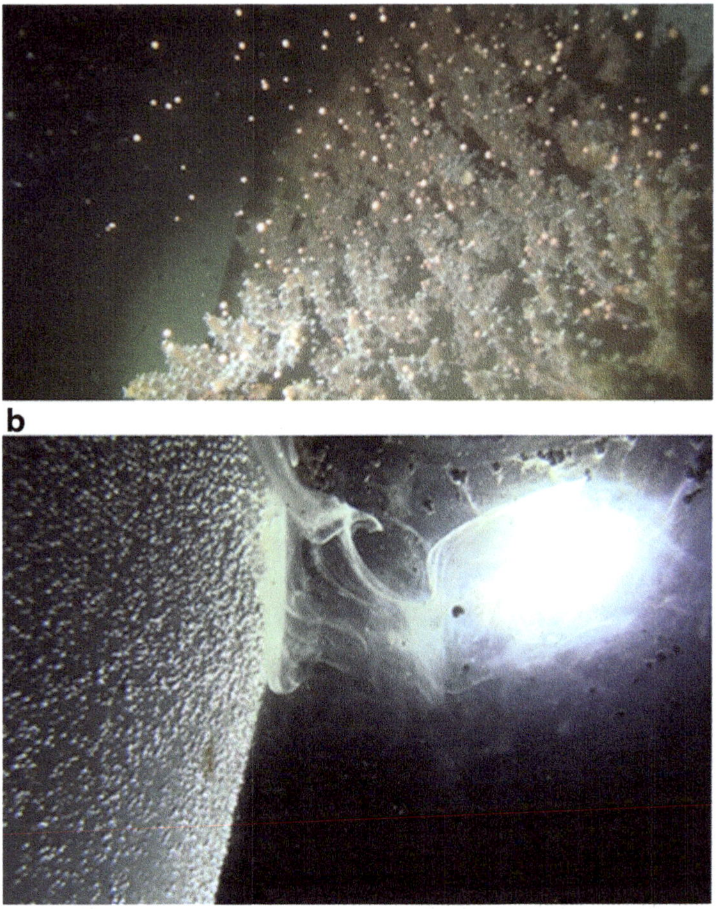

b

Fig. 1.6 The release of coral eggs (a) and sperm (b). (Photos by Baolin Liao)

on the contrary, in the east of the Galapagos Islands, Pacific Ocean, due to the impact of cold water upwelling, even in the tropics, coral species grow slowly [1].

The global distribution of reef corals can be divided into two biogeographic provinces: India and the Atlantic – the pacific fauna and flora of the Caribbean. There are about 800 coral species in the former, with the highest diversity in the triangle area of the Philippines, Indonesia, and Malaysia, namely, "coral triangle," with ca. 600 kinds of corals, accounting for 3/4 of the total number of species of corals in the world. As for the global distribution of a hermatypic corals which have no obvious relationship and coral reefs, there is no obvious trend; these corals usually have their symbiotic algae, and the distribution is not limited to tropical and shallow water, which can be found even in the region with the depth of over 1000 m. Chinese coral reef areas, roughly 38,405 km^2, account for 13.48% of the total area of the world's coral reefs. There are about 174 species of scleractinian corals, mainly distributed in Nansha, Xisha, Hainan, Taiwan, and Dongsha, Southern China coast [11] (Fig. 1.7).

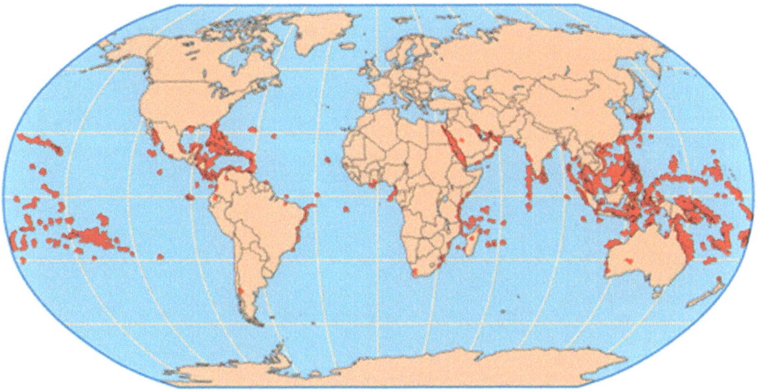

Fig. 1.7 Global distribution of reef corals. (Photo by Baolin Liao)

1.3.3 Effect of Anthropic and Environment Factors on Coral Reefs

In recent years, human activities increased pressures on the destruction of coral reefs, some area even face the danger of disappearing [12, 13]. There are many factors influencing the coral reef ecosystem, including the global space scale pressure, such as overfishing, disease, poaching, eutrophication, and global climate change (e.g., seawater acidification, global warming), resulting in declining the coverage of coral reefs [14, 15]. Many of these changes are directly or indirectly related to human activities. Reclamation, mining, filling, and dumping of wastes in the coastal areas make a lot of silt into the seawater, and the seawater turbidity increases, which reduces the light into coral reefs, and meanwhile may result in hypoxia [16–19]. In addition, human activities will generate a lot of wastewater, causing water eutrophication and pollution, which will promote the outbreak of red tide, causing the overgrowth of algae or predators, and consequently causing the death of corals [15]. Destructive fishing activities, e.g., using explosives and toxic substances, will cause irreversible devastating damage to coral reefs. Marine aquarium trade (MAT) and coral arts and crafts are also listed as the main threats to the protection of coral reefs [20], since thousands of species of invertebrates, e.g., corals, are collected or smuggled illegally.

Coral bleaching events in 1998 led to a loss of 16% of the world's coral reefs. Normal healthy coral is showing a colorful variety of different colors, mainly because of coral symbiotic algae in vivo [21]. The association of scleractinian corals and their symbiotic algae is the foundation of the coral reef ecosystem. Under dramatic changes in the external environment, e.g., high temperature or low temperature, strong ultraviolet, seawater acidification, and heavy metal pollution, etc., the symbiotic relationship could be disrupted; corals will lose the symbiotic algae and bleach, i.e., "coral albino," and the corals will eventually completely die (Fig. 1.8).

Fig. 1.8 Long spine starfish erosion coral. (Photos by Baolin Liao)

1.4 Sponges

Sponges are the oldest multicellular animals of the phylum Porifera. And their fossils can be traced back to 630 million years ago. As one of the most primitive animals, they lack true tissues and organs. Sponge's size ranges from a few millimeters to more than 1–2 m. Most of adult sponges are rich in pores and channels that allow water to circulate through them to get food and oxygen and to filter waste by pumping a large amount of water from the body, filtering bacteria, microeukaryotes, and particles. On the other hand, few sponges, carnivorous sponges in the deep sea, are not filter feeders; instead, they capture prey on hooks on the body's surface. Despite the presence of freshwater species, the vast majority are marine species, living in a wide range of marine habitats, from the polar to the tropical, tidal zones to depths exceeding 8800 m. The majority of sponges are much richer in temperate waters than in tropical waters. Most sponges usually appear on hard surfaces such as rocks or coral reefs, but some sponges can attach themselves to soft sediment with the help of a root-like base.

Sponge's calcareous or siliceous skeletal morphology is more complex, called spicules or bone. Spicules are one of the important bases for sponge classification since different sponge species have different spicule shapes. Sponges are traditionally grouped in four types: calcareous sponges (Calcarea), glass sponges (Hexactinellida), demosponges (Demospongiae), and Homoscleromorpha. Currently, the World Porifera Database lists 8370 valid sponge species within 680 genera in the 4 classes above [22]. Among the four classes, Demospongiae is the largest class and accounts for ca. 83% of the total valid species [23].

Similar as corals, sponges are able to reproduce by both sexual and asexual strategies. Most spongy species use sexual reproduction to release sperm cells into water, fertilize the eggs, and be released in some species, while others are preserved by the "mother." Fertilized eggs form larvae, swimming in search of habitat. Sexual reproduction will produce larvae with flagella, and larvae will find new habitable place for attachment. In addition, a few species reproduce asexually by budding, i.e., produce gemmules. The bud will eventually fall off and adsorb in an appropriate place [24]. Sponges have a strong regenerative capacity; each small part of a sponge is able to grow into an individual.

Sponges are important members of benthic communities and play a vital function in marine nutrient cycle [25], e.g., dissolved inorganic nitrogen (DIN). They are important sources of dissolved organic carbon (DOC), particulate organic carbon (POC), and dissolved organic nitrogen (DON) [26]. Some sponges also have a function of reef construction instead of reef corals, e.g., in deep water below 80 m where almost no living corals use skeletal aragonite, siliceous spicules, and organic fibers. For example, in Jamaica and the Virgin Islands, the deep slope of 70–150 m in depth, a hard sponge is an important reef-building organism instead of scleractinian coral (Fig. 1.9).

1.5 Other Coral Reef Organisms

1.5.1 Coral Algae

Coralline alga is a kind of algae in Rhodophyta and is another large family of coral reef organisms. It plays a key role in the coral reef-building process [27], because coral algal cells can secrete calcareous skeletons, also called the calcareous sheath. Coralgal not only helps to build coral reefs; coral algae are frequently attached to the reef surface or parcel to make the coral reefs more firmly. The lateral edge atolls in the Pacific Ocean and Indian Ocean are built by the coral reefs of the intertidal algal ridge which can buffer the impact of the waves (Fig. 1.10).

Zooxanthella is one of the most fatal primary producers of the coral reef ecosystem, providing nutrients from photosynthesis for corals, e.g., to produce organic compounds by carbon dioxide fixation and promote the building of coral calcium carbonate skeleton. Thus, coral algae and other important primary producers play an important role in the element cycle in coral reefs (Fig. 1.11).

Fig. 1.9 Demosponges in the South China Sea. (Photos taken by Bing Wang)

Fig. 1.10 Coral algae. (Photos by Bing Wang)

1.5.2 Sea Anemone

Sea anemone is a kind of coral. The sea anemone's body looks like a cylinder at both ends. The upper end is composed of a dish, a ring of tentacles around the entrance. The groove in the back of the mouth keeps flowing into the seawater to supply oxygen to the inner tissues. Sea anemones use chassis to fix on a hard surface. Some sea anemones can also secrete mucus to help the fixation. Although sea

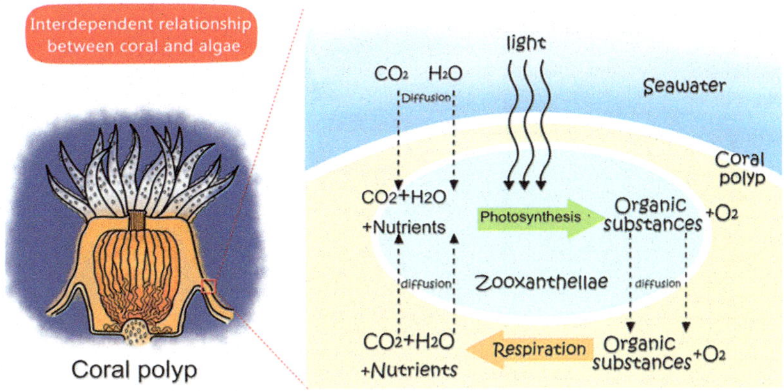

Fig. 1.11 Interdependent relationship between coral and algae [1]. (Photo by Changfeng Dai)

anemones are classified as attaching organisms, they can sometimes move very slowly. One way is to use their base plate to move on the surface of coral reefs, which is very similar to a snail. Another way is to use the antennae to somersault. Sometimes sea anemones can float up to the surface of the sea [28].

1.6 Coral Reef Conservation

Coral reef ecosystem is an extremely important marine ecosystem, which plays an irreplaceable role in maintaining biodiversity and maintaining ecological balance, which is related to the life of human beings. The coral reef ecosystem is extremely fragile and very sensitive to the change of the marine environment. The formation of coral reefs needs hundreds of thousands of years; it is difficult to recover after destruction.

Coral reef health survey (Reef Check) is the world's largest coral reef monitoring program jointly participated by divers and marine scientists. The UN Global Coral Reef Monitoring Network has decided to make the Reef Check as a "public" coral reef monitoring program. Since 2007, China has carried out the census of coral reefs in Guangdong Province along the coast; more than 300 diving volunteers have participated in protecting coral reefs (Fig. 1.12).

Meanwhile, in order to promote the ecological restoration of coral reefs, China has set up the Coral Reef Conservation Center in Shenzhen, Dapeng, and Guangdong Ocean University, through coral artificial cultivation of reef corals and propagation by artificial reef construction and reef coral's transplantation (Fig. 1.13).

Artificial reef is a kind of multidimensional spatial structure based on the principle of bionics, which is suitable for the ecological restoration of coastal coral reef area. The coral reef can fix and grow on the inner surface (Fig. 1.14).

Fig. 1.12 Reef check in China. (Photos by Guangdong Reef Check)

Fig. 1.13 Coral reef conservation center of Shenzhen Research Institute of Guangdong Ocean University. (Photo by Baolin Liao)

Fig. 1.14 Artificial reefs and coral's transplantation. (Photos by Baolin Liao)

The artificial cultivation coral seedlings can be transplanted to the natural area, making the damaged coral reef area to be fast and effective recovery.

Acknowledgments We gratefully acknowledge financial supports from the Natural Science Foundation of China (NSFC) (31861143020, 41776138, 41742002, U1301131, 41176127, 41076077) and the National Major Scientific Research Program of China (2013CB956103).

Ethical Statement The corals mentioned in this chapter for transplant and restoration are approved by the Guangdong Science and Technology Bureau and Shenzhen Science and Technology Innovation Commission. We have obtained the license of the People's Republic of China for aquatic wild animals and the license of the People's Republic of China for domestication and breeding of aquatic wild animals; all of the studies were carried out in compliance with the law and regulations under the supervision and approval of the government.

References

1. Dai C. Map of Taiwan's coral reefs. Margate: Commonwealth Publishing; 2011.
2. Veron JEN. Corals in space and time: the biogeography and evolution of the Scleractinia. Sydney: University of New South Wales Press; 1995.
3. Veron JEN. Corals of the world, vol. 3. Townsville: Australian Institute of Marine Sciences and CRR QLD Pty Ltd; 2000. p. 11–57.
4. Hughes TP, Baird AH, Bellwood DR, Card M, Connolly SR, Folke C, et al. Climate change, human impacts, and the resilience of coral reefs. Science. 2003;301:929–33.
5. Barnes RSK, Mann KH. Fundamentals of aquatic ecology. Oxford: Blackwell Publishing; 1991. p. 217–27.
6. Harriott VJ. Reproductive seasonality, settlement, and post-settlement mortality of *Pocillopora damicornis* (Linaeus), at Lizard Island, Great Barrier Reef. Coral Reefs. 1983;2:151–7.
7. Rinkevich B. Restoration strategies for coral reefs damaged by recreational activities: the use of sexual and asexual recruits. Restor Ecol. 1995;3:241–51.
8. Fan TY, Dai CF. Sexual reproduction of the scleractinian coral *Merulina ampliata* in southern Taiwan. Bull Mar Sci. 1998;62:897–904.
9. Kojis BL. Sexual reproduction in *Acropora* (Isopora) (Coelenterata: Scleractinia) I. A. cuneata and A. palifera on Heron Island reef, Great Barrier Reef. Mar Biol. 1986;91:311–8.
10. Soong K. Sexual reproductive patterns of shallow-water reef corals in Panama. Bull Mar Sci. 1991;49:832–46.
11. Zhou R. Fauna Sinica—Scleractinia coral. Beijing: Science Press; 2001.
12. McCorry, D. Hong Kong's scleractinian coral communities: status, threats and proposals for management. Unpublished PhD thesis. The University of Hong Kong; 2002.
13. Harrison PL, Booth DJ. Coral reefs: naturally dynamic and increasingly disturbed ecosystems. In: Connell SD, Gillanders B, editors. Marine ecology. Melbourne: Oxford University; 2007. p. 316–77.
14. Fox HE, Pet JS, Dahuri R, Caldwell RL. Recovery in rubble fields: long-term impacts of blast fishing. Mari Pollut Bull. 2003;46:1024–31.
15. Richmond RH. Coral-reefs: present problems and future concerns resulting from anthropogenic disturbance. Am Zool. 1993;33:524–36.
16. Jokiel PL, Coles SL. Effects of temperature on the mortality and growth of Hawaiian reef corals. Mar Biol. 1977;43:201–8.
17. Houlbreque F, Tambutte E, Richard C, Ferrier-Pagès C. Importance of a micro diet for scleractinian corals. Mar Ecol. 2004;282:151–60.

18. Alutoin S, Boberg J, Nystrom M, Tedengren M. Effects of the multiple stressors copper and reduced salinity on the metabolism of the hermatypic coral *Porites lutea*. Mar Environ Res. 2001;52:289–99.
19. Anthony KRN. A tank system for studying benthic aquatic organisms at predictable levels of turbidity and sedimentation: case study examining coral growth. Limnol Oceanogr. 1999;44:1415–22.
20. Rhyne AL, Tlusty MF, Kaufman L. Long-term trends of coral imports into the United States indicate future opportunities for ecosystem and societal benefits. Conserv Lett. 2012;5:478–85.
21. Nystroum M, Nordemar I, Tedengren M. Simultaneous and sequential stress from increased temperature and copper on the metabolism of the hermatypic coral *Porites cylindrica*. Mar Biol. 2001;138:1225–31.
22. Van Soest RW, Boury-Esnault N, Vacelet J, Dohrmann M, Erpenbek D, De Voogd NJ, et al. Global diversity of sponges (Porifera). PLoS One. 2012;7:e35105.
23. Jack SA, Kennedy J, Margassery M, Flemer B, O'Leary N, Morrissey JP, et al. Marine sponges-molecular biology and biotechnology. In: Kim S, editor. Handbook of marine biotechnology. New York: Springer; 2015. p. 219–54.
24. Walker P, Wood E. Life in the sea: the coral reef. Shanghai: Shanghai Scientific & Technological Literature Publishing House; 2011.
25. De Goeij JM, van Oevelen D, Vermeij MJA, Osinga R, Middelburg JJ, de Goeij AFPM, et al. Surviving in a marine desert: the sponge loop retains resources within coral reefs. Science. 2013;342:108.
26. Diaz MC, Ward BB. Sponge-mediated nitrification in tropical benthic communities. Mar Ecol Prog Ser. 1997;156:97–107.
27. Pfister CA. Seaweed ecology and physiology. Cambridge: Cambridge University Press; 2014.
28. García-Ortega L, Alegre-Cebollada J, García-Linares S, Bruix M, Martínez-Del-Pozo A, Gavilanes JG. The behavior of sea anemone actinoporins at the water-membrane interface. Biochim Biophys Acta. 2011;1808:2275–88.

Chapter 2
Sponge and Coral Microbiomes

Zhiyong Li

Contents

Abstract Coral/sponge holobiont is the stable assemblage of the host and its symbiotic bionts, e.g., microalgae, bacteria, archaea, virus, fungi, and protists. Coral/sponge microbiome means the entire microbial community and genes that reside within a coral/sponge. Sponges host abundant and diverse microbes including bacteria, archaea, and fungi. Corals form a close mutualistic relationship with photosynthetic, endosymbiotic dinoflagellates of the genus *Symbiodinium*, along with microorganisms including bacteria, archaea, fungi, and viruses. These microbiota and algae are thought to have various symbiotic relationships with coral/sponge host including mutualism, commensalism, and parasitism.

Keywords Sponge · Coral · Holobiont · Microbiome · Structure · Function

Z. Li (✉)
Marine Biotechnology Laboratory, State Key Laboratory of Microbial Metabolism,
School of Life Sciences and Biotechnology, Shanghai Jiao Tong University,
Shanghai, People's Republic of China
e-mail: zyli@sjtu.edu.cn

Z. Li (ed.), *Symbiotic Microbiomes of Coral Reefs Sponges and Corals*,
https://doi.org/10.1007/978-94-024-1612-1_2

2.1 Introduction

Strictly speaking, "microbiota" means a collection or community of microbes, while "microbiome" refers to the full collection of genes of all the microbes in a community. However, now in many cases, "microbiome" is also used to mean all the microbes in a community. Meanwhile, although originally coined to specifically refer to host-associated microbial communities [1], the word "microbiome" is now utilized broadly to refer to any habitats. Therefore, the term "microbiome" used in this book includes two meanings: all microbes and their genes in a community.

"Holobiont" was first used to describe the assemblage of different species that form ecological units, typically symbiosis. According to Lynn Margulis, all individuals who participate in a particular symbiosis are bionts, and the entire organism that is comprised of these bionts is a holobiont [2]. By this definition, nearly every macrospecies is a holobiont because it always lives in symbiosis with some other species. For example, all plants and animals, from lower organisms, e.g., invertebrates to humans, live in close association with microbial organisms.

Coral/sponge holobiont means the stable assemblage of the host and its symbiotic bionts, e.g., microalgae, bacteria, archaea, virus, fungi, and protists. The entire assemblage of genomes in the holobiont is termed a "hologenome" which includes the host's genome and its microbiome. The concept of the holobiont was first used to understand corals' components, ecological functions, and their evolution over time and then expanded to other species, e.g., sponges. There is a similar concept "metaorganism" (an entity formed by the aggregation of a number of individual organisms) or "superorganism" (an organism consisting of many organisms) to mean the association of one macrospecies and its bionts.

In 2007, the Human Microbiome Project (HMP) was launched by the NIH (National Institutes of Health, USA) with the aim to characterize the human microbiota (microbiome) and analyze their role in human health. In 2015, the USA announced a new National Microbiome Initiative (NMI) to foster the integrated study of microbiomes across different ecosystems including marine microbiomes. Microbiome represents a frontier in the microbiology field. In the field of marine microbiology, *Science* journal published a special issue on the marine microbiome in 2015. Coral/sponge microbiome is one of the hotspots of marine microbiome because of its important value in ecology, evolution, and biotechnology. In 2016, using next-generation sequencing, the World Sponge Microbiome Project has been achieved, finding phylogenetically diverse microbes on a global scale [3]. Soon afterward, the dataset of the Sponge Microbiome Project was announced in 2017, which represented a comprehensive resource of sponge-associated microbial communities [4].

2.2 Sponge Microbiome

2.2.1 Sponge Microbial Community

Sponges are complex holobionts or metaorganisms because they host abundant and diverse microbes including at least 46 bacterial phyla, 3 archaeal phyla, 3 fungal phyla, and phylogenetically diverse algae [3, 5–10]. These bionts are thought to have various symbiotic relationships with sponge host including mutualism, commensalism, and parasitism.

The presence of bacteria in the mesohyl of sponges was first confirmed in the early 1960s by the use of electron microscopy (EM) [11]. Scanning electron microscopy (SEM) was subsequently employed to detect cyanobacteria in sponges [12]. Traditional culture-dependent approach has been used to investigate the diversity of microbes in sponge holobionts. For instance, a lot of novel fungi and actinobacteria from sponges, e.g., *Marihabitans*, *Polymorphospora*, and *Streptomonospora*, were isolated from marine sponges for the first time in my group [13–15]. Even now, it is still an effective strategy to study the function of microbes derived from sponges, e.g., natural products with medical potentials for marine drug development. In order to recover novel species, some innovative culture methods have been employed, e.g., through antibiotic administration [16] or floating filter cultivation [17].

Because of the medium and condition limitation in the simulation of in situ environment, only a small percentage of microbial populations, i.e.,$<1\%$, could be isolated from sponges in the laboratory. Therefore, culture-independent methods have become the main strategies for the revelation of community structure of sponge microbial symbionts. For example, polymerase chain reaction (PCR)-based cloning library was first successfully employed to detect the diversity of unculturable sponge-associated bacteria. Denaturing gradient gel electrophoresis (DGGE) fingerprint was successfully used to compare the bacterial components among different species of sponges [18]. The subsequent fluorescence in situ hybridization (FISH) allows to reveal the bacterial spatial distribution within sponge tissues [19]. The great advances in the bacterial symbionts' diversity evaluation come from pyrosequencing which makes the comprehensive descriptions of bacterial community structures possible especially the rare biosphere [20].

Sponges could get their microbial symbionts through horizontal transmission from the environmental seawater and vertical transmission from parents [21]. Sponges are suggested to be capable of differentiating food bacteria from symbionts. Wilkinson et al. [22] suggested that the chemical composition of the bacterial outer layer may play a role in sponge symbionts' recognition. However, the question of how sponges discriminate between food and symbionts remains unsolved. Ankyrin-repeat proteins (ARP) may interact with surrounding cells and proteins and might be involved in the recognition and protection from host phagocytosis that allows the host to discriminate between food and symbiont bacteria [23].

In 2002, Hentschel et al. revealed monophyletic clusters of sponge-derived sequences more closely related to each other than that from non-sponge sources and suggested the sponge-specific microbes [24]. Whereafter, by 16S ribosomal RNA gene amplicon pyrosequencing of 32 sponge species from eight locations around the world's oceans, only a minimal core bacterial community consisting of very few OTUs was found. In contrast, a large species-specific bacterial community, which is represented by OTUs present in only a single sponge species, was detected [25]. The sponge species-specific bacteria represent the unique association of sponge microbes which was in depth proved by the global analysis [3]. The species-specific bacteria are probably vertically transmitted which has been demonstrated in sponge larvae by FISH [26].

Cenarchaeum symbiosum was first reported in association with marine sponge *Axinella mexicana* in 1996 [27]. Till now, three archaeal phyla, i.e., *Crenarchaeota*, *Euryarchaeota*, and *Bathyarchaeota*, have been detected in sponges [6, 9, 10, 28]. Particularly, the vertical transmission of archaea in sponge larvae was demonstrated, suggesting a very close coevolutionary relationship of archaea with sponge host [26, 29].

Prokaryotic symbionts show different distribution characteristics in one sponge, for instance, a significant difference of bacterial phylotypes between the cortex and endosome was revealed in sponge *Astrosclera willeyana* [30]. *Bacteroidetes*, *Frankineae*, and *Propionibacterineae* were detected only in the endosome, whereas *Cyanobacteria*, *Planctomycetacia*, and *Micrococcineae* were only associated with the cortex. Some branches of a-*Proteobacteria*, c-*Proteobacteria*, *Corynebacterineae*, *Acidimicobidae*, *Crenarchaeota*, and *Euryarchaeota* also showed distribution difference in *Astrosclera willeyana*.

Compared with the knowledge of sponge-associated bacterial diversity, the diversity of eukaryotic symbionts in sponges remains largely unknown. Phylogenetically diverse eukaryotic symbionts were detected in the *N. huxleyi* metagenome [31]. Using 454 pyrosequencing of the V4 region of 18S ribosomal ribonucleic acid gene of eukaryota associated with 11 species of South China Sea sponges, 2 phyla of fungi (*Ascomycota* and *Basidiomycota*) and 9 phyla of protists including 5 algal phyla (*Chlorophyta*, *Haptophyta*, *Streptophyta*, *Rhodophyta*, and *Stramenopiles*) and 4 protozoal phyla (*Alveolata*, *Cercozoa*, *Haplosporidia*, and *Radiolaria*) including 47 orders (12 fungi, 35 protists). Entorrhizales of fungi and 18 orders of protists were detected in sponges, and sponge species-specific eukaryotic symbionts were suggested [5]. Particularly, the in situ active fungi in sponges *T. swinhoei* and *X. testudinaria* were revealed using 18S rRNA gene transcripts [32].

2.2.2 Sponge Microbial Function

Sponges probably represent one of the most complex symbioses on earth with a core microbial community and sponge-specific or sponge species-specific microbial lineages. After learning more about the diversity of sponge microbial symbionts, the function evaluation of the microbial symbionts represents the frontier and hot issue of sponge symbioses; however, to date, the function of sponge microbiomes lags behind the understanding of taxonomic affiliation. The primal strategy for the function investigation of sponge microbes is culture dependent. Particularly, the potentials in producing biologically active natural products have been carried out for sponge-derived microbes especially actinobacteria and fungi.

Compound isolation-based activity assay and functional gene-/genome-based evaluation are two main ways to analyze the microbial functions [31, 33–36]. In particular, omics provides a promising strategy for understanding the metabolism and function of sponge microbiomes and has revealed previously unknown diversity and functions of sponge symbionts [37–40]. In 2010, Thomas et al. first explored the functional genomic signature of bacteria associated with the sponge *Cymbastela concentrica* [37]. Thereafter, Liu et al. analyzed the bacterial functional proteins in the sponge *Cymbastela concentrica* using metaproteogenomic technique [38]. Fan et al. investigated the metabolisms of the bacterial communities of six sponges using metagenomics and suggested the functional equivalence and evolutionary convergence in complex microbial communities of sponge symbionts [39]. To date, the well-known function of sponge microbiomes mainly includes an element cycle for providing nutrients for the sponge hosts and removing metabolic wastes [41–43] and chemical defense by producing bioactive compounds [35, 36].

The nitrogen cycle is a critical biogeochemical process of the oceans. Marine sponges have been suggested to play an important role in the marine nitrogen cycle. Mohamed et al. [44] provided the first molecular evidence for the presence of potential anammox bacteria in sponges. Using functional genes (*amoA*, *nirS*, *nirK*, and *nxrA*) involved in ammonia oxidization and denitrification and 16S rRNA genes for specific bacterial groups as markers, phylogenetically diverse prokaryotes including bacteria and archaea, which may be involved in the ammonia oxidization and denitrification processes in sponges, were revealed in seven South China Sea sponge species [45, 46]. Sulfate-reducing bacteria (SRB) are known to play a key role in the cycling of marine elements. Phylogenetically diverse SRB, which mainly belonged to the genus *Desulfovibrio* in the class *Deltaproteobacteria*, in three sponges *Arenosclera heroni*, *Dysidea arenaria*, and *Astrosclera willeyana* from the South China Sea were detected [47].

In 2010, López-Legentil et al. detected the expression of ammonia monooxygenase genes in ammonia-oxidizing archaea associated with the barrel sponge *Xestospongia muta* [48]. Liu et al. proved the expression of the ammonia monooxygenase membrane-bound subunits (*AmoB* and *C*) and an ammonia transporter (*AmtB*) in the microbial community of *C. concentric* by metaproteogenomic analysis [38]. Nitrifying community with transcriptional activity in sponge microbiomes was observed in South China Sea sponges. For example, the expression of *ureC* genes from *Proteobacteria*, which were the predominant component in sponge *X. testudinaria*, suggested the function of bacterial symbionts in urea utilization [49]. In addition, the inhabitancy and transcriptional activity of *Nitrosopumilus*-like AOA (ammonia-oxidizing archaea) and *Nitrospira* NOB (nitrite-oxidizing bacteria) in this sponge *T. swinhoei* from the South China Sea were confirmed [50].

The metabolic analysis of sponge holobionts at the whole community level including prokaryotes and eukaryotes is helpful for understanding the biology and ecology of sponge symbioses. In 2014, phylogenetically diverse prokaryotes and eukaryotes were detected in deep-sea sponge *N. huxleyi* using metagenomics in my group. MEGAN and gene-enrichment analyses indicated different metabolic potentials of prokaryotic symbionts from eukaryotic symbionts, especially in nitrogen and carbon metabolisms [31].

As the oldest multicellular animals lack active defense ability and developed immune system, still sponges have survived in the complex sea environment for almost 630 million years, mainly because of their chemical defense against predator, other colonial organisms, and pathogenic microbes besides their strong regenerative capacity. The secondary metabolites produced by marine sponges include steroids, isoprenoids, non-isoprenoids, quinones, nitrogen and nitrogen-sulfur heterocyclic compounds, alkaloids, peptides, and terpenes, and most of them show higher biological activities, e.g., cytotoxicity, anti-pathogens, enzymic inhibition, etc. [51–54]. Sponges are currently the most important marine sources of biologically active natural products [55], since the number of natural products isolated from sponges, *ca.* 6000, accounts for nearly one-third of the total marine natural products. Among the seven marine drugs in the market before 2016, three are derived from sponges, e.g., anticancer drug cytarabine (Ara-C) and eribulin mesylate (E7389) and antivirus vidarabine (Ara-A). Thus, bioactive compounds isolated from marine sponges have become a starting point for developing new marine drugs.

It is worth mentioning that some of these compounds isolated from marine sponges are only synthesized in symbiotic relationships with fungi, microalgae, archaea, cyanobacteria, and bacteria [56, 57]. In 1996, macrolide swinholide A was limited to unicellular heterotrophic bacteria in sponge *Theonella swinhoei*, and an antifungal cyclic peptide was found to occur only in the filamentous heterotrophic bacteria [58], providing the first chemical evidence for the uncultured bacterial origin of sponge-derived compounds. Afterward, Piel et al. found the bacterial gene cluster which was responsible for biosynthesizing onnamides and proved the producer was uncultured *Entotheonella* spp., providing gene evidence for bacterial origin of sponge-derived compounds [59, 60].

2.3 Coral Microbiome

2.3.1 Coral Microbial Community

Corals are holobionts or "metaorganisms," e.g., in a mutualistic relationship with photosynthetic, endosymbiotic dinoflagellates of the genus *Symbiodinium*, which can provide >90% of a coral's nutritional requirements, along with microorganisms including bacteria, archaea, fungi, and viruses.

Coral microbiome means the entire microbial community (and associated genes) that resides on or within a coral. Extensive phylogenetic surveys of coral microbiomes have revealed that the dominant symbionts reside within the *Proteobacteria* (particularly *Gammaproteobacteria* and *Alphaproteobacteria*) as well as *Actinobacteria*, *Bacteroidetes* (especially *Flavobacteria*), and *Cyanobacteria* [61]. For example, in *Porites astreoide*, the most prominent bacterial groups were *Proteobacteria* (68%), *Firmicutes* (10%), *Cyanobacteria* (7%), and *Actinobacteria* (6%) [62]. In particular, Roder et al. showed that bacterial diversity of fungid host species *Ctenactis echinata* is primarily structured by one bacterial taxon (genus *Endozoicomonas*) representing more than 60% of all bacteria [63].

The coral microhabitats include coral mucus, tissues, skeleton, and gastric cavity. The surface mucopolysaccharide layer, produced by endosymbiotic *Symbiodinium* spp., is composed of glycoproteins and provides an ideal habitat for microbes, e.g., 10^6–10^8 microbial cells per milliliter [61]. However microbial community in the coral mucus is not very stable because of the environmental effects. In contrast, microbes inhabiting in the coral epithelium and gastrodermal tissues are always specific to coral host, though with a generally low microbial abundance. Coral skeleton provides a unique habitat for coral symbiotic microbes.

Similar with sponges, the establishment of coral-microbes symbioses includes two strategies, inheritance, i.e., vertical transmission, and acquisition from the surrounding environment, i.e., horizontal transmission. However, the molecular mechanisms that allow for the establishment, recognition, and maintenance of microbial symbionts are still unknown. Meanwhile, coral microbiome research that mainly focus on bacteria, archaea, virus, while fungi are rarely involved.

The coral holobiont is a dynamic assemblage of the coral host, zooxanthellae, endolithic algae, fungi, bacteria, archaea, and viruses. The coral animal can adapt to differing ecological niches by "switching" its microbial associates. Zooxanthellae and some bacteria form relatively stable and species-specific associations with coral hosts. Other associations are less specific, e.g., coral-associated archaea [61]. According to Roder et al. [63], the content and structure of the coral microbiome aligns with environmental differences and denotes habitat adequacy. On the other hand, an inflexible bacterial community under different environmental conditions was also suggested [64]. Compared to the changes in the *Symbiodinium* community, the associated bacterial community remains remarkably stable even under conditions of coral bleaching. Totally, coral holobionts might occupy structural landscapes ranging from a highly flexible to a rather inflexible composition with consequences for their ability to respond to environmental change [64].

2.3.2 *Coral Microbial Function*

Symbiodinium spp. fixes carbon by photosynthates and transfer nutrients to the coral host. The coral microbes play an important role in the element cycling, e.g., carbon, nitrogen, phosphorus, and sulfur [61].

It is known that coral microbiota is involved in the carbon fixation by the Calvin cycle, a reductive acetyl-CoA pathway, or reverse Krebs cycle, and carbon degradation. Coral microbiota undertakes nitrogen cycling via nitrogen fixation, nitrification, and denitrification. Metagenomics analysis suggested that the coral-associated bacteria possessed a large number of genes for the uptake and processing of sugars and proteins [62]. Ceh et al. revealed that coral larvae acquire nutrients previously taken up from the environment by bacteria, which may increase the survival rate and fitness of the developing coral and therefore contribute to the successful maintenance of coral reefs [65].

Corals are one of the largest producers of dimethylsulfonipropionate (DMSP) in the oceans. Some coral microbes, e.g., *Endozoicomonas* spp., have been proved to be capable of DMSP metabolism [66]. In addition, inorganic sulfur can also be cycled via sulfate-reducing bacteria (SRB). Dissolved organic phosphorus may be recycled by coral microbes, e.g., *Vibrio* spp. is capable of phosphorus degradation [67].

Take reef-building coral *Porites astreoide* as an example, functionally, the bacterial community is primarily heterotrophic and includes a number of pathways for the degradation of aromatic compounds, and the most abundant is the homogentisate pathway. Particularly, a wide diversity of fungal genes involved in carbon and nitrogen metabolism were detected, which suggested that endolithic fungi could be converting nitrate and nitrite to ammonia within the coral holobionts [62].

It was confirmed (both spatially and temporally) that a nitrogen fixer (*Prosthecochloris*, a green sulfur bacteria) in the green layers of coral skeletons, played an essential role in providing nutrients for the coral holobiont in the nutrient-limited reef ecosystem [68].

Archaea associated with the surface mucus of corals include marine group II, anaerobic methanotroph, anaerobic nitrate reducers (i.e., denitrification), and marine group III (8%). Coral-associated archaea may contribute to nitrogen recycling in the holobiont, presumably by acting as a nutritional sink for excess ammonium trapped in the mucus layer, through nitrification and denitrification processes [69].

Marine viral assemblages within the coral holobionts probably have important but currently unknown functions in the coral stress response, coral disease, and the adaptive potential of the coral holobionts with respect to climate change [70].

Using *Porites* spp. as a case study, Sogin et al. presented evidence that the relative abundance of different subclades of *Symbiodinium* and bacterial/archaeal families were linked to positive and negative metabolomic signatures. Consequently, coral partner choice likely influences cellular metabolic activities and, therefore, holobiont nutrition [71].

The shifts in the prokaryotic community composition during mucus aging may lead to the prevalence of opportunistic and potentially pathogenic bacteria. Particularly, microbe-depleted corals started exhibiting clear signs of bleaching and necrosis. Thus, it could be concluded that the natural prokaryotic community inhabiting the coral SML contributes to coral health [72].

Coral microbiome responds and quickly adapts to disturbance and has central roles in the coral reef ecosystem. Prosser [73] recently stated that quantitative information on the links between microbial community structure, populations, and activities will allow predictions on the impacts of climate change to ecosystem processes. Thus, theoretically, coral microbes may be used for predicting responses of reef ecosystems to climate changes, if important linkages occurring between the microbial communities and macroecological change. Ultimately, this microbial perspective will improve our ability to accurately predict the resilience of specific reefs and contribute to the conservation of these important ecosystems [74].

Acknowledgments We gratefully acknowledge financial supports from the National Natural Science Foundation of China (NSFC) (31861143020, 41776138, 41742002, U1301131, 41176127, 41076077), the National High-Tech Research and Development Program of China (2013AA092901, 2011AA090702, 2007AA09Z447, 2004AA628060, 2002AA608080), and the National Major Scientific Research Program of China (2013CB956103).

References

1. Lederberg J, McCray AT. Ome sweet omics – a genealogical treasury of words. Scientist. 2001;15:8.
2. Margulis L, Fester R. Symbiosis as a source of evolutionary innovation: speciation and morphogenesis. Cambridge, MA: MIT Press; 1991.
3. Thomas T, Moitinho-Silva L, Lurgi M, Björk JR, Easson C, Astudillo-García C, et al. Diversity, structure and convergent evolution of the global sponge microbiome. Nat Commun. 2016;7:11870.
4. Moitinho-Silva L, Nielsen S, Amir A, Gonzalez A, Ackermann GL, Cerrano C, et al. The sponge microbiome project. GigaScience. 2017;6:1–7.
5. He L, Liu F, Karuppiah V, Ren Y, Li Z. Comparisons of the fungal and protistan communities among different marine sponge holobionts by pyrosequencing. Microb Ecol. 2014;67:951–61.
6. Li Z, Wang Y, Li J, Liu F, He L, He Y, Wang S. Metagenomic analysis of genes encoding nutrient cycling pathways in the microbiota of deep-sea and shallow-water sponges. Mar Biotechnol. 2016;18:659–71.
7. Webster NS, Negri AP, Munro MM, Battershill CN. Diverse microbial communities inhabit Antarctic sponges. Environ Microbiol. 2004;6:288–300.
8. Lee OO, Wang Y, Yang J, Lafi FF, Al-Suwailem A, Qian PY. Pyrosequencing reveals highly diverse and species specific microbial communities in sponges from the Red Sea. ISME J. 2011;5:650–64.
9. Pape T, Hoffmann F, Quéric NV, Juterzenka JR, Michaelis W. Dense populations of Archaea associated with the demosponge *Tentorium semisuberites* Schmidt, 1870, from Arctic deepwaters. Polar Biol. 2006;29:662–7.
10. Holmes B, Blanch H. Genus-specific associations of marine sponges with group I crenarchaeotes. Mar Biotechnol. 2007;150:759–72.
11. Wilkinson CR. Microbial associations in sponges. I. Ecology, physiology and microbial populations of coral reef sponges. Mar Biol. 1978;49:161–7.
12. Magnino G, Sarà A, Lancioni T, Gaino E. Endobionts of the coral reef sponge *Theonella swinhoei* (Porifera, Demospongiae). Invertebr Biol. 1999;118:213–20.

13. Yu Z, Zhang B, Sun W, Zhang F, Li Z. Phylogenetically diverse endozoic fungi in the South China Sea sponges and their potential in synthesizing bioactive natural products suggested by PKS gene and cytotoxic activity analysis. Fungal Divers. 2013;58:127–41.
14. Zhou K, Zhang X, Zhang F, Li Z. Phylogenetically diverse cultivable fungal community and polyketide synthase (PKS), non-ribosomal peptide synthase (NRPS) genes associated with the South China Sea sponges. Microb Ecol. 2011;62:644–54.
15. Ding B, Yin Y, Zhang F, Li Z. Recovery and phylogenetic diversity of culturable fungi associated with marine sponges *Clathrina luteoculcitella* and *Holoxea* sp. in the South China Sea. Mar Biotechnol. 2011;13:713–21.
16. Richardson C, Hill M, Marks C, RunyenJanecky L, Hill A. Experimental manipulation of sponge/bacterial symbiont community composition with antibiotics: sponge cell aggregates as a unique tool to study animal/microorganism symbiosis. FEMS Microbiol Ecol. 2012;81:407–18.
17. Sipkema D, Schippers K, Maalcke WJ, Yang Y, Salim S, Blanch HW. Multiple approaches to enhance the cultivability of bacteria associated with the marine sponge *Haliclona* (gellius) sp. Appl Environ Microbiol. 2011;77:2130–40.
18. Li Z-Y, He L-M, Wu J, Jiang Q. Bacterial community diversity associated with four marine sponges from the South China Sea based on 16S rDNA-DGGE fingerprinting. J Exp Mar Biol Ecol. 2006;329:75–85.
19. Ridley CP, Faulkner DJ, Haygood MG. Investigation of Oscillatoria spongeliae-dominated bacterial communities in four dictyoceratid sponges. Appl Environ Microbiol. 2005;71:7366–75.
20. Sogin ML, Morrison HG, Huber JA, Welch DM, Huse SM, Neal PR, et al. Microbial diversity in the deep sea and the underexplored "rare biosphere". Proc Natl Acad Sci USA. 2006;103:12115–20.
21. Taylor MW, Radax R, Steger D, Wagner M. Sponge-associated microorganisms: evolution, ecology, and biotechnological potential. Microbiol Mol Biol Rev. 2007;71:295–347.
22. Wilkinson CR. Immunological evidence for the Precambrian origin of bacterial symbioses in marine sponges. Proc R Soc. 1984;220:509–17.
23. Thomas T, Rusch D, Demaere MZ, Yung PY, Lewis M, Halpern A, et al. Functional genomic signatures of sponge bacteria reveal unique and shared features of symbiosis. ISME J. 2010;4:1557–67.
24. Hentschel U, Hopke J, Horn M, Friedrich AB, Wagner M, Hacker J, et al. Molecular evidence for a uniform microbial community in sponges from different oceans. Appl Environ Microbiol. 2002;68:4431–40.
25. Schmitt S, Tsai P, Bell J, Fromont J, Ilan M, Lindquist N, et al. Assessing the complex sponge microbiota: core, variable and species-specific bacterial communities in marine sponges. ISME J. 2012;6:564–76.
26. Sharp KH, Eam B, Faulkner DJ, Haygood MG. Vertical transmission of diverse microbes in the tropical sponge Corticium sp. Appl Environ Microbiol. 2007;73:622–9.
27. Preston CM, Wu KY, Molinski TF, Delong EF. A psychrophilic crenarchaeon inhabits a marine sponge: *Cenarchaeum symbiosum* gen. nov., sp. nov. Proc Natl Acad Sci USA. 1996;93:6241–6.
28. Radax R, Hoffmann F, Rapp TR, Leninger S, Schleper C. Ammonia-oxidising Archaea as main drivers of nitrification in cold-water sponges. Environ Microbiol. 2012;14:909–23.
29. Steger D, Ettinger-Epstein P, Whalan S, Hentschel U, de Nys R, Wagner M, et al. Diversity and mode of transmission of ammonia oxidising Archaea in marine sponges. Environ Microbiol. 2008;10:1087–94.
30. Yang Z, Li Z. Spatial distribution of prokaryotic symbionts and ammoxidation, denitrifier bacteria in marine sponge *Astrosclera willeyana*. Sci Rep. 2012;2:528.
31. Li Z-Y, Wang Y-Z, He L-M, Zheng H-J. Metabolic profiles of prokaryotic and eukaryotic communities in deep-sea sponge *Neamphius huxleyi* indicated by metagenomics. Sci Rep. 2014;4:3895.
32. Jin L, Liu F, Sun W, Zhang F, Karuppiah V, Li Z. Pezizomycotina dominates the fungal communities of South China sea sponges *Theonella swinhoei* and *Xestospongia testudinaria*. FEMS Microbiol Ecol. 2014;90:935–45.

33. Karuppiah V, Li Y, Sun W, Feng G, Li Z. Functional gene-based discovery of phenazines from the Actinobacteria associated with marine sponges in the South China Sea. Appl Microbiol Biotechnol. 2015;99:5939–50.
34. Sun W, Zhang F, He L, Loganathan K, Li Z. Actinomycetes from the South China Sea sponges: isolation, diversity and potential for aromatic polyketides discovery. Front Microbiol. 2015;6:1048.
35. Zhao H-Y, Anbuchezhian R, Sun W, Shao C-L, Zhang F-L, Yin Y, et al. Cytotoxic nitrobenzoyloxy-substituted sesquiterpenes from sponge derived endozoic fungus *Aspergillus insulicola* MD10-2. Curr Pharm Biotechnol. 2016;17:271–4.
36. Li Z. Advances in marine microbial symbionts in the China Sea and related pharmaceutical metabolites. Mar Drugs. 2009;7:113–29.
37. Thomas T, Rusch D, DeMaere MZ, Yun PY, Lewis M, Halpern A, et al. Functional genomic signatures of sponge bacteria reveal unique and shared features of symbiosis. ISME J. 2010;4:1557–67.
38. Liu M, Fan L, Zhong L, Kjelleberg S, Thomas T. Metaproteogenomic analysis of a community of sponge symbionts. ISME J. 2012;6:1515–25.
39. Fan L, Reynolds D, Liu M, Stark M, Kjelleberg S, Webster NS, et al. Functional equivalence and evolutionary convergence in complex communities of microbial sponge symbionts. Proc Natl Acad Sci USA. 2012;109:E1878–87.
40. Trindade-Silva AE, Rua C, Silva GGZ, Dutilh BE, Moreira Ana PB, Edwards RA. Taxonomic and functional microbial signatures of the endemic marine sponge *Arenosclera brasiliensis*. PLoS One. 2012;7:e39905.
41. Hoffmann F, Radax R, Woebken D, Holtappels M, Lavik G, Rapp HT, et al. Complex nitrogen cycling in the sponge *Geodia barretti*. Environ Microbiol. 2009;11:2228–43.
42. Schlappy ML, Schottner SI, Lavik G, Kuypers MMM, de Beer D, Hoffmann F. Evidence of nitrification and denitrification in high and low microbial abundance sponges. Mar Biol. 2010;157:593–602.
43. Siegl A, Kamke J, Hochmuth T, Piel J, Richter M, Liang C, et al. Single cell genomics reveals the lifestyle of Poribacteria, a candidate phylum symbiotically associated with marine sponges. ISME J. 2011;5:61–70.
44. Mohamed NM, Saito K, Tal Y, Hill RT. Diversity of aerobic and anaerobic ammonia oxidizing bacteria in marine sponges. ISME J. 2010;4:38–48.
45. Han M, Li Z, Zhang F. The ammonia oxidizing and denitrifying prokaryotes associated with sponges from different sea areas. Microb Ecol. 2013;66:427–36.
46. Han M, He L, Li Z, Lin H. Bacterial and archaeal symbionts in the South China Sea sponge *Phakellia fusca*: community structure, relative abundance, and ammonia-oxidizing populations. Mar Biotechnol. 2012;14:701–13.
47. Zhang D, Sun W, Feng G, Zhang F, Anbuchezhian R, Li Z, et al. Phylogenetic diversity of sulfate-reducing desulfovibrio associated with three South China Sea sponges. Lett Appl Microbiol. 2015;60:504–12.
48. López-Legentil S, Erwin PM, Pawlik JR, Song B. Effects of sponge bleaching on ammonia-oxidizing Archaea: distribution and relative expression of ammonia monooxygenase genes associated with the barrel sponge *Xestospongia muta*. Microb Ecol. 2010;60:561–71.
49. Su J, Jin L, Jiang Q, Sun W, Zhang F, Li Z. Phylogenetically diverse *ure*C genes and their expression suggest the urea utilization by bacterial symbionts in marine sponge *Xestospongia testudinaria*. PLoS One. 2013;8:e64848.
50. Feng G, Sun W, Zhang F, Karthik L, Li Z. Inhabitancy of active Nitrosopumilus-like ammonia oxidizing archaea and *Nitrospira* nitrite-oxidizing bacteria in the sponge *Theonella swinhoei*. Sci Rep. 2016;6:24966.
51. Engel S, Jensen PR, Fenical W. Chemical ecology of marine microbial defense. J Chem Ecol. 2002;2(8):1971–85.
52. Walters KD, Pawlik JR. Is there a trade-off between wound-healing and chemical defenses among Caribbean reef sponges? Integr Comp Biol. 2005;45:352–8.

53. Green G. Ecology of toxicity in marine sponges. Mar Biol. 1977;40:207–15.
54. Thoms C, Ebel R, Proksch P. Activated chemical defense in *Aplysina* sponges revisited. J Chem Ecol. 2006;32:97–123.
55. Koopmans M, Martens D, Wijffels RH. Towards commercial production of sponge medicines. Mar Drugs. 2009;7:787–802.
56. Piel J. Metabolites from symbiotic bacteria. Nat Prod Rep. 2009;26:338–62.
57. Unson MD, Holland ND, Faulkner DJ. A brominated secondary metabolite synthesized by the cyanobacterial symbiont of a marine sponge and accumulation of the crystalline metabolite in the sponge tissue. Mar Biol. 1994;119:1–11.
58. Bewley CA, Holland ND, Faulkner DJ. Two classes of metabolites from *Theonella swinhoei* are localized in distinct populations of bacterial symbionts. Experientia. 1996;52:716–22.
59. Piel J, Hui D, Wen G, Butzke D, Platzer M, Fusetani N, et al. Antitumor polyketide biosynthesis by an uncultivated bacterial symbiont of the marine sponge *Theonella swinhoei*. Proc Natl Acad Sci U S A. 2004;101:16222–7.
60. Wilson MC, Mori T, Rückert C, Uria AR, Helf MJ, Takada K, et al. An environmental bacterial taxon with a large and distinct metabolic repertoire. Nature. 2014;506:58–62.
61. Wegley L, Edwards R, Rodriguez-Brito B, Liu H, Rohwer F. Metagenomic analysis of the microbial community associated with the coral *Porites astreoides*. Environ Microbiol. 2007;9:2707–19.
62. Bourne DG, Morrow KM, Webster NS. Insights into the coral microbiome: underpinning the health and resilience of reef ecosystems. Annu Rev Microbiol. 2016;70:317–40.
63. Roder C, Bayer T, Aranda M, Kruse M, Voolstra CR. Microbiome structure of the fungid coral *Ctenactis echinata* aligns with environmental differences. Mol Ecol. 2015;24:3501–11.
64. Pogoreutz C, Rädecker N, Cárdenas A, Gärdes A, Wild C, Voolstra CR. Dominance of endozoicomonas bacteria throughout coral bleaching and mortality suggests structural inflexibility of the *Pocillopora verrucosa* microbiome. Ecol Evol. 2018;8:2240–52.
65. Ceh J, Kilburn MR, Cliff JB, Raina J, van Keulen M, Bourne DG. Nutrient cycling in early coral life stages: *Pocillopora damicornis* larvae provide their algal symbiont (*Symbiodinium*) with nitrogen acquired from bacterial associates. Ecol Evol. 2013;3:2393–400.
66. Raina J-B, Dinsdale E, Willis BL, Bourne DG. Do organic sulphur compounds DMSP and DMS drive coral microbial associations? Trends Microbiol. 2010;18:101–8.
67. Gilbert JA, Thomas S, Cooley NA, Kulakova A, Field D, et al. Potential for phosphonoacetate utilization by marine bacteria in temperate coastal waters. Environ Microbiol. 2009;11:111–25.
68. Yang S-H, Lee STM, Huang C-R, Tseng C-H, Chiang P-W, Chen C-P, et al. Prevalence of potential nitrogen-fixing, green sulfur bacteria in the skeleton of reef-building coral *Isopora palifera*. Limnol Oceanogr. 2016;61:1078–86.
69. Siboni N, Ben-Dov E, Sivan A, Kushmaro A. Global distribution and diversity of coral-associated Archaea and their possible role in the coral holobiont nitrogen cycle. Environ Microbiol. 2008;10:2979–90.
70. Wood-Charlson EM, Weynberg KD, Suttle CA, Roux S, van Oppen MJH. Metagenomic characterization of viral communities in corals: mining biological signal from methodological noise. Environ Microbiol. 2015;17:3440–9.
71. Sogin EM, Putnam HM, Nelson CE, Anderson P, Gates RD. Correspondence of coral holobiont metabolome with symbiotic bacteria, archaea and *Symbiodinium* communities. Environ Microbiol Rep. 2017;9:310–5.
72. Glasl B, Herndl GJ, Frade PR. The microbiome of coral surface mucus has a key role in mediating holobiont health and survival upon disturbance. ISME J. 2016;10:2280–92.
73. Prosser JI, Bohannan BJM, Curtis TP, Ellis RJ, Firestone MK, Freckleton RP, et al. The role of ecological theory in microbial ecology. Nat Rev Microbiol. 2007;5:384–92.
74. Ainsworth TD, Thurber RV, Gates RD. The future of coral reefs: a microbial perspective. Trends Ecol Evol. 2009;25:233–44.

Chapter 3
Microbial Diversity of Sponge/Coral Microbiome

Sandi Orlić

Contents

Abstract The ocean is the largest habitat on our planet for microbes. These microorganisms play a key role in global biogeochemical cycles. Still, we are missing more detail on their phylogenetic, genomic, and metabolic diversity. Microbes play a key role in the sponge and coral biology. Sponges and corals can no longer be considered as autonomous entities but rather as holobionts. Microbes contribute to the nutrition, defense, immunity, and development of the host. Associated microbes can comprise as much as half of their tissue volume, with densities in excess of a billion cells of the sponge tissue, several orders of magnitude higher than those typical for seawater. Each of the three domains of life, i.e., *Bacteria*, *Archaea*, and *Eukarya* (single-celled eukaryotes: fungi and microalgae), is now known to reside within sponges and corals. This chapter focuses on domains of life that are present in sponges and corals and gives an overview of their biodiversity and significance: (1) microbes in sponge and (2) microbes in corals.

Keywords Bacteria · Fungi · Sponges · Corals

S. Orlić (✉)
Ruđer Bošković Institute, Zagreb, Croatia

Center of Excellence for Science and Technology Integrating Mediterranean Region, Microbial Ecology, Zagreb, Croatia
e-mail: sorlic@irb.hr

Z. Li (ed.), *Symbiotic Microbiomes of Coral Reefs Sponges and Corals*,
https://doi.org/10.1007/978-94-024-1612-1_3

3.1 Introduction

More than 70% of the Earth's surface is covered by oceans, and this huge environment is considered a great reservoir of natural resources. However, the extent of marine biodiversity, especially of microorganisms, is barely known. Also, the oceans house a wide array of symbioses, where different organisms depend on one another for survival. The word symbiosis literally means *living together*. In the context of marine biology, this is referring to an intimate relationship between two different species.

3.2 Microbes in Sponges

Sponges (phylum Porifera) with more than 8600 formally described and 15,000 estimated species are very important benthic aquatic animals of great ecological, commercial, and biopharmaceutical importance. They are ancient metazoan animals with Precambrian origins that date back more than 600 million years ago [1]. Sponges are distributed across different shallow and deepwater habitats from the hot to the cold regions, from tropics to the poles. They are living in almost all marine environments, providing the habitat for a wide range of fauna, coupling the benthic and pelagic species; they filter enormous quantities of seawater and mediate biogeochemical fluxes by respiring organic matter and facilitating the consumption and release of nutrients, including nitrate, nitrite, ammonium, and phosphate. Sponges are known for hosting dense, diverse, and highly specific microbial communities, and different interactions can occur as different microbes can represent food sources, pathogens/parasites, or mutualistic symbionts. Sponge-associated microbes can comprise as much as 40% of sponge volume, with densities in excess of 10^9 CFU of sponge tissue, several orders of magnitude higher than those typical for seawater. All the three domains of life, i.e., *Bacteria*, *Archaea*, and *Eukarya* (single-celled eukaryotes: fungi and microalgae), are part of the sponge microbiome, and they have played an important role in shaping sponge evolution. Several reviews can be consulted for a perspective on the understanding of the microbial/bacterial associations with sponges [2–6].

Prokaryotic microorganisms (*Bacteria* and *Archaea*) are the basis of all life on Earth. They are found in all viable habitats everywhere, from deep subsurface sediments and bedrocks to high up in the atmosphere, with a variety and total biomass that by far exceeds eukaryotic life [7]. Life on Earth has depended on the activity of microorganisms that harvest chemical and light energy from the environment to generate organic matter [8]. The biogeochemical processes they mediate are of vital importance for the whole biosphere. Nowadays, nearly all terrestrial and aquatic environments are based on energy and organic carbon that ultimately originated from photosynthesis, that is, on light energy converted into chemical energy used to fix carbon dioxide from the atmosphere. However, many prokaryotes, called chemotrophs, can obtain energy by oxidizing or reducing compounds such as hydrogen, ammonia, nitrate, sulfide, methane, sulfate, and metal ions in light-independent processes. Some microorganisms derive all their cellular carbon from carbon dioxide and are thus able to grow without any organic compounds. Such *Bacteria* and *Archaea* are called autotrophs. Others rely on organic compounds as carbon source and are called heterotrophs. Regardless of their lifestyle, all microorganisms interact with and directly affect the environment they inhabit. Many even engage into symbiotic, commensalist, or parasitic relationships with multicellular organisms and play important roles in nutrition and disease.

In the last four decades, numerous culture-independent techniques, which bypass the need for isolation and laboratory cultivation of individual microbial species, have been developed. This novel approach has fundamentally changed the field of environmental microbiology, as it has become possible to investigate microorganisms, and their interactions with the environment and other organisms in situ. Examples for such culture-independent approaches include the cataloging and phylogenetic analysis or rRNAs and other housekeeping or pathway-specific genes; the analysis of the whole-community DNA, RNA, or protein composition in meta-omic approaches; or the identification of in situ active microbial populations through stable and radioisotope labeling techniques [9, 10]. The sensitivity and accuracy of these techniques made it possible for environmental microbiologist and microbial ecologist to investigate even the most distant, inaccessible, and uncommon environments on Earth.

All three cellular domains of life—*Bacteria*, *Archaea*, and *Eukarya* [4]—are found in sponges. The *Eukarya* members include dinoflagellates and fungi (16 different sponge species). Sponge archaea include important actors in the nitrogen cycle, especially in the ammonia elimination [11, 12]. It is speculated that sponges have some of the most diverse microbial populations, most clearly at the phylum level of diversity (according to today's molecular tool resolution). Phylogenetic studies have shown that sponges contain more than 30 different bacteria phyla (*Proteobacteria* (e.g., classes *Alpha, Gamma-*, and *Deltaproteobacteria*), *Chloroflexi, Actinobacteria, Acidobacteria, Nitrospirae*) [7, 13, 14], including the candidate phylum '*Poribacteria*' that is almost exclusively found within sponges [15, 16]. Sponge symbionts are capable of diverse metabolic processes such as nitrification, nitrogen fixation, sulfate reduction, and photosynthesis [17–21] and probably contribute to sponge nutrition [22]. Additionally, certain sponge symbionts produce secondary metabolites that might be involved in the chemical defense of their hosts [23–25].

In the literature, we can recognize two types of sponges: (1) the so-called high-microbial abundance (HMA) sponges with a rich microflora and (2) low-microbial abundance (LMA) sponges with relatively low numbers in their microbial biomass [3]. LMA sponges have microbial communities that are less diverse than those in

HMA sponges and particular single clades dominate [26, 27]; they can however still harbor interesting diversity at least in particular divergent. In some LMA sponges, researchers were able to identify only one single phylotype, e.g., the sponge *Crambe crambe*, where a single *Betaproteobacterium* symbiont dominates the community [28] of phyla, e.g., the *Planctomycetes* [29], or the sponge *Axinella mexicana*, where a member of the *Thaumarchaeota*, *Cenarchaeum symbiosum*, dominates the microbial biomass [30, 31].

High similarity among sponge-associated microbes is a common feature [32]. In the literature, the term "sponge-specific" was introduced for all microbes that were repeatedly detected in sponges around the Earth but different from seawater communities [6, 32]. Schmitt et al. [33] identified the potential bacterial subpopulations in relation to sponge phylogeny and sampling sites. The authors proposed vertical transfer of species-specific bacterial community and that different sponges contain different bacterial species. However, these bacteria are still closely related to each other. They analyzed 16S rRNA gene amplicon pyrosequencing data from 32 sponges around the world's oceans and found six new phyla. The comparison of bacterial communities at broader taxonomic levels (phylum, order) in their study revealed a high overall similarity, but no correlation with sponge phylogeny was identified, except for tropical sponges. Comparison of bacterial communities identified a minimal core and a large host species-specific bacterial community. Different hypotheses were proposed: different sponges contain bacterial communities consisting of mainly different bacterial species (species-specific community) and share very few bacterial species (core community) [33].

The relationship of sponge and bacteria was demonstrated long before culture-independent molecular methods were applied [34]. Only with the advent of molecular tools the existence of such bacteria has only become clearer [26]. Using PCR-DGGE analyses of the total bacterial community, Hardoim et al. [35] revealed the bacterial population in *A. fulva*, which differed drastically from the seawater. The same group has constructed *Poribacterial* clone libraries for *Aplysina fulva* sponge specimens. They demonstrated the coexistence of several "intraspecific" poribacterial genotypes in a single-sponge host [36]. Still, our understanding of sponge-associated microbiota remains limited to a few host species found in restricted geographical localities. Hardoim et al. [37] analyzed bacterial abundance and diversity of two temperate marine sponges belonging to the Irciniidae family—*Sarcotragus spinosulus* and *Ircinia variabilis*—in the northeast Atlantic Ocean. Using two approaches (epifluorescence microscopy and PCR-DGGE), they revealed that *S. spinosulus* hosted significantly more and different prokaryotic cells than *I. variabilis*. Their results confirm the hypothesized host-specific composition of bacterial communities between phylogenetically and spatially close sponge species. More than 7500-member SSU rRNA sequence database of bacteria, archaea, and fungi [38] were analyzed on both culture-independent clones and denaturing gradient gel electrophoresis (DGGE) assays. Bacterial sponge-specific clusters were mainly dominated by members of the genus *Synechococcus*. Less abundant were the members of phyla *Chloroflexi*, *Proteobacteria*, *Acidobacteria*, and *Poribacteria*.

Massively parallel pyrosequencing of the 16S rRNA has been applied to three Australian sponge species. The results suggested that the diversity of bacterial symbionts of sponges may be comparable to that of the human gut microbiome—at least ca. 3000 symbiont taxa can be detected in a single-sponge species [39]. Using NGS Jackson et al. [40] have revealed diverse and distinct sponge-specific microbial communities in marine sponges *Raspailia ramosa* and *Stelligera stuposa* sampled from a single geographical location in Irish waters. The data showed that only 2.8% of classified reads from the sponge *R. ramose* and 26% of *S. stuposa* represent sponge-specific bacteria. Novel sponge-specific clusters were identified, whereas the majority of previously reported sponge-specific clusters (e.g., *Poribacteria*) were absent from these sponge species. Recently, 16S rRNA amplicon sequencing was applied to investigate the diversity, specificity, and transcriptional activity of microbes associated with an LMA sponge (*Stylissa carteri*), HMA sponge (*Xestospongia testudinaria*), and seawater collected from the central Saudi Arabia coast of the Red Sea [41]. The most abundant OTUs were shared, while rare OTUs were unique to any given source.

The majority of the research on sponge microbiome was performed on temperate regions. Some authors have studied the sponge microbiome in deep waters. Analysis of these deepwater sponges gives us some interesting insights into sponge symbiosis. Kennedy et al. [42] have analyzed the microbiota of three sponges, *Lissodendoryx diversichela*, *Poecillastra compressa*, and *Inflatella pellicula*, using pyrosequencing. The microbial communities of *L. diversichela*, *P. compressa*, and *I. pellicula* were typical of low-microbial abundance (LMA) sponges, while *S. normani* had a community more typical of high-microbial abundance (HMA) sponges. Their results showed that the three LMA-like sponges shared a set of abundant OTUs that were distinct from those associated with sponges from shallow waters. The same group have [40] determined that sponge-associated microbial communities were less diverse and less even than any other sponge-microbial community investigated to date. They used pyrosequencing of 16S rRNA genes to compare the bacterial and archaeal communities associated with two individuals of the marine sponge *Inflatella pellicula* from the deep sea, sampled from a depth of 2900 m, the deepest sampling for sponges. They noticed that archaea appear to dominate the microbiome of *Inflatella pellicula*. In the sponge *Polymastia* cf. *corticata* collected from 1127 m in the Caribbean Sea, 38 distinct microbial phylotypes were identified, and 53% of these fell within previously described sponge-specific sequence clusters (SSSC) [43]. Similarly, *Scleritoderma cyanea* collected at 242 m depth off the coast of Curacao and *Scleritoderma* sp. collected from 255 m depth off Bonaire hosted microbial communities most similar to uncultivated microbes retrieved from the shallow-water sponges *Theonella swinhoei* and *Aplysina aerophoba* [44]. In contrast, three specimens of *Geodia* collected at depths of 197–304 m were analyzed with group-specific fluorescence in situ hybridization (FISH) probes and found to host a microbial community which was highly similar to those in sediment samples from the same region, including an abundance of *Archaea*, *Gammaproteobacteria*, and *Firmicutes* [45]. A study of the deep-sea sponges *Characella* sp., collected at a hydrothermal vent off the coast of Japan (686 m depth) and *Pachastrella* sp. and an unidentified *Poecilosclerid* sponge from an oil seep in the Gulf of Mexico (572 m depth) led to the first report of thioautotrophic symbionts in sponges [46].

The bacterial communities associated with giant barrel sponges *Xestospongia muta* and *Xestospongia testudinaria* revealed remarkable similarity in the bacterial communities. The communities were species-specific, and sequences found in other sponge species were determined (*Chloroflexi*, *Acidobacteria*, and *Actinobacteria*). This research, comparing the bacterial communities associated with closely related but geographically distant sponge hosts, gives new insight into the relationships between marine sponges and some of their bacterial symbionts [47]. In Antarctica, marine sponges are abundant and important members of the benthos, structuring the Antarctic marine ecosystem. Recently, Rodriguez-Marconi et al. [48] have used NGS to investigate the microbial diversity from eight different Antarctic sponges (families Acarnidae, Chalinidae, Hymedesmiidae, Hymeniacidonidae, Leucettidae, Microcionidae, and Myxillidae). Their study indicates that there are different diversity and similarity patterns between bacterial/archaeal and eukaryote microbial symbionts from these Antarctic marine sponges.

A global Porifera microbiome survey set out to establish the ecological and evolutionary drivers of these host–microbe interactions was presented recently [49]. The authors provide a comprehensive analysis of microbial symbiont communities associated with 81 sponge species. A total of 804 sponge samples were collected from the waters of 20 countries bordering the Atlantic, Pacific, and Indian Oceans as well as the Mediterranean and Red Seas, primarily from shallow-water habitats; this study represents the largest sponge microbiome survey. Except for the sponge samples, the authors have simultaneously collected 133 seawater and 36 sediment samples as potential sources or sinks of microorganisms associated with sponges. Microbial community composition for each sample was determined using standardized DNA extraction and 16S rRNA gene-sequencing protocols established by the Earth Microbiome Project. The results have revealed that sponges are a reservoir of exceptional microbial diversity and a major contributor to the total microbial diversity found in the world's oceans. Microbial communities exhibit little similarity in species composition or structure although a number of emerging properties related to community organization are evident. They show that the core symbiont communities in sponges are strongly density dependent and have few and weak interactions, low connectance, and amensal and/or commensal interactions indicative of stable core symbionts within the Porifera. On the other hand, symbionts that appear to be phylogenetically unique (i.e., having previously been defined as "sponge-specific") did not contribute disproportionally to the core microbiome. These results indicate that symbiont communities have independently assembled or evolved across the Porifera and that convergent forces have resulted in the analogous community organization and interactions. The complexity rather than the composition of the symbiont community primarily impacts host phylogeny. These findings further support a model of convergent evolution in symbiont communities across the entire host phylum.

Fungi are often isolated from sponges, although the extent and ecological nature of fungal–sponge associations remain to be detailed and investigated. Eighty-five fungal taxa have been isolated from the Mediterranean Sea sponge *Ircinia variabilis*, with classical approaches [50]. Morrison-Gardiner [51] found different fungal species from tropical, subtropical, and temperate water sponges.

Like in bacterial diversity, there are few studies in evidence for specificity of fungal communities within different sponges [52, 53]. In the deep-sea sponge *Lamellomorpha* sp. using a metagenomic and gene enrichment analysis, Li et al. [54] proposed that eukaryotic symbionts, including fungi, along with their prokaryotic communities, can exhibit different metabolic potentials, especially in nitrogen and carbon metabolism.

3.3 Microbes in Corals

Corals are marine invertebrates in the class *Anthozoa* of the phylum *Cnidaria*. They typically live in compact colonies of many identical individual polyps. The group includes the important reef builders that inhabit tropical oceans and secrete calcium carbonate to form a hard skeleton. Coral reefs are among the most biodiverse marine ecosystems on the planet and provide substantial economic and ecological benefits to coastal communities. The phylum Cnidaria is comprised of an approximately 10,000 species [55] and is divided into four classes: Anthozoa, Cubozoa, Scyphozoa, and Hydrozoa. Huge biodiversity of microbial species involved in the coral symbiosis is similar to that found in sponges. Coral reefs are covered by phototrophic organisms through their obligate symbiosis with dinoflagellate endosymbionts in the genus *Symbiodinium*. *Symbiodinium* convert sunlight and carbon dioxide into organic carbon and oxygen to fuel coral growth and calcification, creating habitat for these diverse and productive ecosystems.

Symbiodinium putative species ("types") are commonly identified using different nuclear- and chloroplast-encoded ribosomal DNA regions. Population studies have provided some insights into *Symbiodinium* biogeography. Although these results are showing a complex population, they are complicated by (i) a lack of consensus criteria used to delineate inter- vs. intragenomic variation within species and (ii) the high density of *Symbiodinium* in host tissues, which results in single samples comprising thousands of individuals.

As in the sponge microbiome research advances in prokaryote, coral research was made following the introduction of culture-independent methods [56]. One of the major challenges during coral microbiome research is the sampling methods [57, 58]. Although healthy corals are crucial to the productivity and sustainability of reef ecosystems and surrounding human communities, a decline in coral reefs has been documented over the last decades. Global climate change, increasing ocean acidification, overfishing, and other anthropological influences have all been linked to a decrease in coral cover worldwide and/or a rapid structural change, often associated with the loss of biodiversity in reef ecosystems. Coral microbiome research is usually correlated with some anomalies in coral community growth like diseases. The microbiome (16S rRNA) of an almost extinct red coral *Corallium rubrum* was recently analyzed with Illumina MiSeq, and *Spirochaeta* was identified as the dominant population in the microbial community on a broad geographic scale across the Western Mediterranean Sea [59].

In the core microbiome, 12 bacterial species, which accounted for 94.6% of the overall bacterial community, were identified. The bacteria of the orders *Spirochaetales* and *Oceanospirillales* were particularly abundant in the core microbiome, orders important in nutrient cycling, including nitrogen, carbon, and sulfur. Cold water corals are usually not often investigated, and Lawler et al. [60] sampled 23 specimens of the family Anthothelidae from the Western Atlantic. The research was performed by a novel approach (454 pyrosequencing), and the coral host was identified as the primary driver of bacterial community composition. *Al. grandiflorum* were dominated by *Alteromonadales* and *Pirellulales* and had much higher species richness, and a distinct bacterial community compared to other *Anthothela* samples. *Anthothela* species (*A. grandiflora* and *Anthothela* sp.) had very similar bacterial communities, dominated by *Oceanospirillales* and *Spirochaetes*. The core microbiome in this investigation was composed of unclassified *Oceanospirillales*, *Kiloniellales*, *Campylobacterales*, and genus *Spirochaeta*. These bacterial orders were previously associated with functional capabilities in nitrogen cycling and suggest the possibility of a nearly complete nitrogen cycle within *Anthothela* species.

In scleractinian corals microorganisms form dynamic associations and exhibit substantial genetic and ecological diversity. The diversity of coral-associated bacteria within the surface mucosal layer (SML) of healthy corals (*Montastraea faveolata* and *Porites astreoides*) from a different location in the Caribbean Sea was detected using different molecular tools (DGGE and 454 pyrosequencing; [61]). In the coral species *M. faveolata*, bacterial diversity was more significant and exhibited higher specificity at the family level than *P. astreoides* assemblages (dominated by the *Endozoicomonas*, within the order *Oceanospirillales*).

One of the major questions in bacteria coral symbiosis is whether these associations are spatial–temporally stable or species-specific. In a recent publication, Hester et al. [62] have collected coral (*Acropora rosaria*, *Acropora hyacinthus*, and *Porites lutea*) and algae samples (crustose coralline algae and turf algae) from the Central Pacific, and the bacterial communities (determined by 454 pyrosequencing) were analyzed using a novel statistical approach termed the Abundance–Ubiquity (AU) test. Their results are showing that there are stable associations between the bacteria and corals, but these associations are not exclusive. The coral core microbiome of *Acropora granulose* was studied utilizing different approaches: coupling laser microdissection, fluorescence in situ hybridization (FISH), and NGS [63]. Their results reveal several globally stable coral–bacterial interactions across highly diverse reef habitats. They showed also that these symbiotic bacterial interactions are enriched from the rare biosphere of the coral host and are closely associated with the dinoflagellate endosymbionts. The coral core microbiome consists of a sustainably lower bacterial diversity, comprising only several hundred distinct phylotypes. Their study further provides strong evidence for niche habitat partitioning of the bacterial community within the coral host, an observation comparable to mutualistic bacterial symbioses in more complex metazoans.

Although there are some researchers of fungi presence in coral [64–66], insight as to the mechanistic nature of the interactions between the two is well documented. Amend et al. [67] have used in an in-depth SSU-based amplicon sequencing/transcriptomic analysis of the fungal community associated with the coral *Acropora*. Interestingly, their results are showing that the community was correlated with the host rather than differences in environment (e.g., water temperature) or the presence of *Symbiodinium* partners in the sample. In the analysis of bleaching of *A. millepora*, Littman et al. [68] revealed a threefold increase in fungi-like sequences which could be relevant in the nitrogen cycles [69].

Viruses and viruslike particles are abundant in coral reef environments and have been found in the water column adjacent to reefs, in the coral surface microlayer, and inside the tissues of corals and their dinoflagellate endosymbionts (zooxanthellae) [70–74].

3.4 Future Perspectives

In the last years, the application of next-generation sequencing technologies revealed that sponge and coral ecosystems host highly diversified bacterial, archaeal, and unicellular eukaryotic assemblages and contain novel taxa. The analysis of micro-eukaryotic diversity in the sponge and coral ecosystems represents a key step of future scientific research in this field. Another focal point should be the study of the genetic diversity of viruses. Their study is more difficult due to insufficient information on conserved genes in the viral genomes (such as the ribosomal genes of bacteria, archaea, and eukaryotes) and because most of the viral hosts cannot be cultivated. The metagenomic approaches could represent a novel opportunity for characterizing the viral diversity. The marine environment is changing rapidly [75], and for sure this will influence the relationships of microbes, sponges, and corals. The potential impact of these global changes (e.g., temperature increase, decrease of pH, deoxygenation of bottom waters) can co-occur in time and space with anthropogenic impacts, which can or will influence the sponges' and corals' lifestyles. Although, we cannot predict the influence, today this is visible in the mortality of corals. Certainly, high-throughput sequencing will help in the identification of functional signals contained within microbial metagenomes. New identified process and activities could help us to better understand the complicated relationship between sponges/corals and microbes.

Acknowledgment We would like to thank the 6th China–Croatia Science and Technology Cooperation Committee Program (No.6–13) for their support.

References

1. Bergquist PR. Sponges. London: Hutchinson & Co.; 1978.
2. Abdelmohsen UR, Bayer K, Hentschel U. Diversity, abundance and natural products of marine sponge-associated actinomycetes. Nat Prod Rep. 2014;**31**:381–99.
3. Hentschel U, Fieseler L, Wehrl M, Gernert C, Steinert M, Hacker J, et al. Microbial diversity of marine sponges. Prog Mol Subcell Biol. 2003;**37**:59–88.
4. Hentschel U, Piel J, Degnan SM, Taylor MW. Genomic insights into the marine sponge microbiome. Nat Rev Microbiol. 2012;**10**:641–54.
5. Imhoff JF, Stohr R. Sponge-associated bacteria: general overview and special aspects of bacteria associated with *Halichondria panicea*. Prog Mol Subcell Biol. 2003;**37**:35–57.
6. Taylor MW, Radax R, Steger D, Wagner M. Sponge-associated microorganisms: evolution, ecology, and biotechnological potential. Microbiol Mol Biol Rev. 2007;**71**:295–347.
7. Schleifer K. Microbial diversity: facts, problems and prospects. Syst Appl Microbiol. 2004;**27**:3–9.
8. Macalady JL, Hamilton TL, Grettenberger CL, Jones DS, Tsao LE, Burgos WD. Energy, ecology, and the distribution of microbial life. Philos Trans R Soc Lond Ser B Biol Sci. 2013;368:20120383.
9. Olsen CR, Larsen IL, Lowry PD, Cutshall NH, Nichols MM. Geochemistry and deposition of 7Be in river-estuarine and coastal waters. J Geophys Res. 1986;**91**:896–908.
10. Handelsman J, Rondon MR, Brady SF, Clardy J, Goodman RM. Molecular biological access to the chemistry of unknown soil microbes: a new frontier for natural products. Chem Biol. 1998;**5**:245–9.
11. Radax R, Hoffmann F, Rapp HT, Leininger S, Schleper C. Ammonia-oxidizing archaea as main drivers of nitrification in cold-water sponges. Environ Microbiol. 2012;**14**:909–23.
12. Turque AS, Batista D, Silveira CB, Cardoso AM, Vieira RP, Moraes FC, et al. Environmental shaping of sponge associated archaeal communities. PLoS One. 2010;**12**:e15774.
13. Webster NS, Cobb RE, Negri AP. Temperature thresholds for bacterial symbiosis with a sponge. ISME J. 2008;**2**:830–42.
14. Sipkema D, Holmes B, Nichols SA, Blanch HW. Biological characterisation of *Haliclona* (*?gellius*) sp.: sponge and associated microorganisms. Microb Ecol. 2009;**58**:903–20.
15. Fieseler L, Horn M, Wagner M, Hentschel U. Discovery of the novel candidate phylum "Poribacteria" in marine sponges. Appl Environ Microbiol. 2004;**70**:3724–32.
16. Lafi FF, Fuerst JA, Fieseler L, Engels C, Goh WWL, Hentschel U. Widespread distribution of Poribacteria in Demospongiae. Appl Environ Microbiol. 2009;**75**:5695–9.
17. Wilkinson CR. Nutrient translocation from symbiotic cyanobacteria to coral reef sponges. In: Levi C, Boury-Esnault N, editors. Biologie des Spongiaires. Paris: C.N.R.S.; 1979. p. 373–80.
18. Hoffmann F, Rapp HT, Reitner J. Monitoring microbial community composition by fluorescence *in situ* hybridization during cultivation of the marine cold-water sponge *Geodia barretti*. Mar Biotechnol. 2006;**8**:373–9.
19. Hoffmann F, Radax R, Woebken D, Holtappels M, Lavik G, Rapp HT, et al. Complex nitrogen cycling in the sponge *Geodia barretti*. Environ Microbiol. 2009;**11**:2228–43.
20. Bayer K, Schmitt S, Hentschel U. Physiology, phylogeny and *in situ* evidence for bacterial and archaeal nitrifiers in the marine sponge *Aplysina aerophoba*. Environ Microbiol. 2008;**10**:2942–55.
21. Mohamed NM, Saito K, Tal Y, Hill RT. Diversity of aerobic and anaerobic ammonia-oxidizing bacteria in marine sponges. ISME J. 2010;**4**:38–48.
22. Weisz JB, Hentschel U, Lindquist N, Martens CS. Linking abundance and diversity of sponge-associated microbial communities to metabolic differences in host sponges. Mar Biol. 2007;**152**:475–83.
23. Kennedy J, Marchesi JR, Dobson ADW. Metagenomic approaches to exploit the biotechnological potential of the microbial consortia of marine sponges. Appl Microbiol Biotechnol. 2007;**75**:11–20.

24. Siegl A, Hentschel U. PKS and NRPS gene clusters from microbial symbiont cells of marine sponges by whole genome amplification. Environ Microbiol Rep. 2010;**2**:507–13.
25. Thomas TRA, Kavlekar DP, LokaBharathi PA. Marine drugs from sponge-microbe association-a review. Mar Drugs. 2010;**8**:1417–68.
26. Taylor MW, Tsai P, Simister RL, Deines P, Botte E, Ericson G, et al. Sponge-specific bacteria are widespread (but rare) in diverse marine environments. ISME J. 2013;**7**:438–43.
27. Giles EC, Kamke J, Moitinho-Silva L, Taylor MW, Hentschel U, Ravasi T, et al. Bacterial community profiles in low microbial abundance sponges. FEMS Microbiol Ecol. 2013;**83**:232–41.
28. Croue J, West NJ, Escande ML, Intertaglia L, Lebaron P, Suzuki MT. A single betaproteobacterium dominates the microbial community of the crambescidine-containing sponge *Crambe crambe*. Sci Rep. 2013;**3**:2583.
29. Izumi H, Sagulenko E, Webb RI, Fuerst JA. Isolation and diversity of planctomycetes from the sponge *Niphates* sp., seawater, and sediment of Moreton Bay, Australia. Antonie Van Leeuwenhoek. 2013;**104**:533–46.
30. Preston CM, Wu KY, Molinski TF, DeLong EF. A psychrophilic crenarchaeon inhabits a marine sponge: *Cenarchaeum symbiosum* gen. nov., sp. nov. Proc Natl Acad Sci USA. 1996;**93**:6241–6.
31. Hallam SJ, Konstantinidis KT, Putnam N, Schleper C, Watanabe Y, Sugahara J, et al. Genomic analysis of the uncultivated marine crenarchaeote *Cenarchaeum symbiosum*. Proc Natl Acad Sci USA. 2006;**103**:18296–301.
32. Hentschel U, Hopke J, Horn M, Friedrich AB, Wagner M, Hacker J, et al. Molecular evidence for a uniform microbial community in sponges from different oceans. Appl Environ Microbiol. 2002;**68**:4431–40.
33. Schmitt S, Tsai P, Bell J, Fromont J, Ilan M, Lindquist N, Perez T, et al. Assessing the complex sponge microbiota: core, variable and species-specific bacterial communities in marine sponges. ISME J. 2012;**6**:564–76.
34. Wilkinson CR, Nowak M, Austin B, Colwell RR. Specificity of bacterial symbionts in Mediterranean and Great Barrier-Reef sponges. Microb Ecol. 1981;**7**:13–21.
35. Hardoim CCP, Costa R, Araújo FV, Hajdu E, Peixoto R, Lins U, et al. Diversity of bacteria in the marine sponge *Aplysina fulva* in Brazilian coastal waters. Appl Environ Microbiol. 2009;**75**:3331–43.
36. Hardoim CCP, Cox CJ, Peixoto RS, Rosado AS, Costa R, van Elsas JD. Diversity of the candidate phylum *Poribacteria* in the marine sponge *Aplysina fulva*. Braz J Microbiol. 2013;**44**:329–34.
37. Hardoim CC, Esteves AI, Pires FR, Gonçalves JM, Cox CJ, Xavier JR, et al. Phylogenetically and spatially close marine sponges harbour divergent bacterial communities. PLoS One. 2012;**7**:e53029.
38. Simister R, Taylor MW, Tsai P, Fan L, Bruxner TJ, Crowe ML, et al. Thermal stress responses in the bacterial biosphere of the Great Barrier Reef sponge, *Rhopaloeides odorabile*. Environ Microbiol. 2012;**14**:3232–46.
39. Webster NS, Taylor MW, Behnam F, Lücker S, Rattei T, Whalan S, et al. Deep sequencing reveals exceptional diversity and modes of transmission for bacterial sponge symbionts. Environ Microbiol. 2010;**12**:2070–82.
40. Jackson SA, Kennedy J, Morrissey JP, O'Gara F, Dobson AD. Pyrosequencing reveals diverse and distinct sponge-specific microbial communities in sponges from a single geographical location in Irish waters. Microb Ecol. 2012;*64*:*105*–16.
41. Moitinho-Silva L, Bayer K, Cannistraci CV, Giles EC, Ryu T, Seridi L, et al. Specificity and transcriptional activity of microbiota associated with low and high microbial abundance sponges from the Red Sea. Mol Ecol. *2014*;*23*:*1348*–63.
42. Kennedy J, Flemer B, Jackson SA, Morrissey JP, O'Gara F, Dobson ADW. Evidence of a putative deep sea specific microbiome in marine sponges. PLoS One. 2014;**9**:e91092.
43. Meyer B, Kuever J. Homology modeling of dissimilatory APS reductases (AprBA) of sulfur-oxidizing and sulfate-reducing prokaryotes. PLoS One. 2008;**3**:e1514.

44. Olson JB, Mccarthy PJ. Associated bacterial communities of two deep-water sponges. Aquat Microb Ecol. 2005;**39**:47–55.
45. Brück WM, Brück TB, Self WT, Reed JK, Nitecki SS, McCarthy PJ. Comparison of the anaerobic microbiota of deep-water *Geodia* spp. and sandy sediments in the Straits of Florida. ISME J. 2010;**4**:686–99.
46. Nishijima M, Lindsay DJ, Hata J, Nakamura A, Kasai H, Ise Y, et al. Association of thioautotrophic bacteria with deep-sea sponges. Mar Biotechnol. 2010;**12**:252–60.
47. Montalvo NF, Hill RT. Sponge-associated bacteria are strictly maintained in two closely related but geographically distant sponge hosts. *Appl Environ Microbiol.* 2011;**77**:7207–16.
48. Rodríguez-Marconi S, De Iglesia R, Díez B, Fonseca CA, Hajdu E, Trefault N. Characterization of bacterial, archaeal and eukaryote symbionts from Antarctic sponges reveals a high diversity at a three-domain level and a particular signature for this ecosystem. PLoS One. 2015;**10**:e0138837.
49. Thomas T, Moitinho-Silva L, Lurgi M, Björk JR, Easson C, Astudillo-García C, et al. Diversity, structure and convergent evolution of the global sponge microbiome. Nat Commun. 2016;**7**:11870.
50. Paz Z, Komon-Zelazowska M, Druzhinina IS, Aveskamp MM, Shnaiderman A, Aluma Y, et al. Diversity and potential antifungal properties of fungi associated with a Mediterranean sponge. Fungal Divers. 2010;**42**:17–26.
51. Morrison-Gardiner S. Dominant fungi from Australian coral reefs. Fungal Divers. 2002;**9**:105–21.
52. Gao Z, Li B, Zheng C, Wang G. Molecular detection of fungal communities in the Hawaiian marine sponges *Suberites zeteki* and *Mycale armata*. Appl Environ Microbiol. 2008;**74**:6091–101.
53. Li Q, Wang G. Diversity of fungal isolates from three Hawaiian marine sponges. Microbiol Res. 2009;**164**:233–41.
54. Li ZY, Wang YZ, He LM, Zheng HJ. Metabolic profiles of prokaryotic and eukaryotic communities in deep-sea sponge *Lamellomorpha* sp. indicated by metagenomics. Sci Rep. 2014;**4**:3895.
55. Zhang ZQ. Animal biodiversity: an introduction to higher-level classification and taxonomic richness. Zootaxa. 2011;**3148**:7–12.
56. Rohwer F, Breitbart M, Jara J, Azam F, Knowlton N. Diversity of bacteria associated with the Caribbean coral *Montastraea franksi*. Coral Reefs. 2001;**20**:85–91.
57. Sweet MJ, Croquer A, Bythell J. Bacterial assemblages differ between compartments within the coral holobiont. Coral Reefs. 2011;30:39–52.
58. Kellogg CA, Piceno YM, Tom LM, DeSantis TZ, Zawada DG, Andersen GL. PhyloChip™ microarray comparison of sampling methods used for coral microbial ecology. J Microbiol Methods. 2012;**88**:103–9.
59. van de Water JAJM, Melkonian R, Junca H, Voolstra CR, Reynaud S, Allemand D. Ferrier-Pagès, C. Spirochaetes dominate the microbial community associated with the red coral *Corallium rubrum* on a broad geographic scale. Sci Rep. 2016;**6**:27277.
60. Lawler SN, Kellogg CA, France SC, Clostio RW, Brooke SD, Ross SW. Coral-associated bacterial diversity is conserved across two deep-sea *Anthothela* species. Front Microbiol. 2016;**7**:458.
61. Morrow KM, Moss AG, Chadwick NE, Liles MR. Bacterial associates of two Caribbean coral species reveal species-specific distribution and geographic variability. Appl Environ Microbiol. 2012;**78**:6438–49.
62. Hester ER, Barott KL, Nulton J, Vermeij MJ, Rohwer FL. Stable and sporadic symbiotic communities of coral and algal holobionts. ISME J. 2016;**10**:1157–69.
63. Ainsworth TD, Heron SF, Ortiz JC, Mumby PJ, Grech A, Ogawa D, et al. Climate change disables coral bleaching protection on the Great Barrier Reef. Science. 2016;352:338–42.
64. Kendrick B, Risk MJ, Michaelides J, Bergman K. *Amphibious microborers*: bioeroding fungi isolated from live corals. Bull Mar Sci. 1982;**32**:862–7.

65. Bentis CJ, Kaufman L, Golubic S. Endolithic fungi in reef-building corals (order: Scleractinia) are common, cosmopolitan, and potentially pathogenic. Biol Bull. 2000;**198**:254–60.
66. Golubic S, Radke G, Le-Campoin Alsumard T. Endolithic fungi in marine ecosystems. Trends Microbiol. 2005;**12**:229–35.
67. Amend AS, Barshis DJ, Oliver TA. Coral-associated marine fungi form novel lineages and heterogeneous assemblages. ISME J. 2012;**6**:1291–301.
68. Littman R, Willis BL, Bourne DG. Metagenomic analysis of the coral holobiont during a natural bleaching event on the Great Barrier Reef. Environ Microbiol Rep. 2011;**3**:651–60.
69. Wegley L, Edwards RA, Rodriguez-Brito B, Liu H, Rohwer F. Metagenomic analysis of the microbial community associated with the coral *Porites astreoides*. Environ Microbiol. 2007;**9**:2707–19.
70. Wilson SK. Multiscale habitat associations of detritivorous blennies (Blenniidae: Salariini). Coral Reefs. 2001;**20**:245–51.
71. Seymour J, Patten N, Bourne D, Mitchell J. Spatial dynamics of virus-like particles and heterotrophic bacteria within a shallow coral reef system. Mar Ecol Prog Ser. 2005;**288**:1–8.
72. Patten NL, Harrison PL, Mitchell JG. Prevalence of virus-like particles within a staghorn scleractinian coral (*Acropora muricata*) from the Great Barrier Reef. Coral Reefs. 2008;**27**:569–80.
73. Davy J, Patten N. Morphological diversity of virus-like particles within the surface microlayer of scleractinian corals. Aquat Microb Ecol. 2007;**47**:37–44.
74. Lohr J, Munn CB, Wilson WH. Characterization of a latent virus-like infection of symbiotic Zooxanthellae. Appl Environ Microbiol. 2007;**73**:2976–81.
75. Doney SC, Ruckelshaus M, Duffy JE, Barry JP, Chan F, English CA, et al. Climate change impacts on marine ecosystems. Annu Rev Mar Sci. 2012;**4**:11–37.

Chapter 4
Endolithic Microbes in Coral Skeletons: Algae or Bacteria?

Shan-Hua Yang and Sen-Lin Tang

Contents

Abstract Microbial endoliths exist ubiquitously, in several terrestrial and marine environments, including inside pores of coral skeletons. Although coral skeletons provide a harsh environment due to limited light and circadian fluctuations in pH and oxygen, visible green, black, brown, and red bands comprised of cyanobacteria, fungi, bacteria, and red and green algae, respectively, are usually observed in the skeletons of corals. Based on microscopic observation and culture-based methods, many studies showed algae, fungi, and cyanobacteria as dominant microorganisms in coral skeleton and also suggested that the endolithic microbes may be nutrient source of their coral hosts. Recently, various bacteria in coral skeletons have been illustrated, thanks to culture-independent methods. This chapter focuses on (1) challenges of environment in coral skeleton to endolithic microbes, (2) the endolithic microorganism and potential roles in coral skeletons, and (3) perspectives.

Keywords Endolithic bacteria · Coral skeleton

S.-H. Yang · S.-L. Tang (✉)
Biodiversity Research Center, Academia Sinica, Taipei, Taiwan
e-mail: sltang@gate.sinica.edu.tw

Z. Li (ed.), *Symbiotic Microbiomes of Coral Reefs Sponges and Corals*,
https://doi.org/10.1007/978-94-024-1612-1_4

4.1 Introduction

In marine environments, including coral reef ecosystems, all biomes are fundamentally dependent on their microbial constituents [1]. Consequently, the importance of diverse microbial communities involves coral nutrition, health, and disease [2–4]. These microbes, such as bacteria, viruses, and eukaryotic, together with their scleractinian host, form an integrated holobiont [3, 5]. Although there are coral-associated microbes in coral skeletons, they have not been well characterized.

Endolithic communities, groups of organisms that live inside rocks or similar hard materials, have important roles in biogeochemical cycles in coral reef ecosystems. It has been reported that endolithic microorganisms live within hard substrates in coastal regions [6–10] and they may actively dissolve some of the carbonate substrates where they live [11], via a process termed bioerosion [12]. For instance, endolithic algae and cyanobacteria play a major role of removing dead carbonate substrates [6, 11]. It is noteworthy that endolithic microorganisms are considered major components in food chains [7, 13, 14], as coral tissue may be able to use the nutrients generated by coral-associated endolithic microorganisms [15–17] and some endolithic bacteria within coral skeletons contain genes of nitrogen fixation, making nitrogen available to the coral [18]. Hence, to elucidate the ecosystem functions, it is important to characterize endolithic microbial diversity in the coral skeletal and the roles they play in coral host-microbial interactions.

4.2 Coral Skeleton Conditions

4.2.1 Physical and Chemical Features of Coral Skeletons

From a microscale to a macroscale (e.g., microbial mats and meromictic lakes, respectively), biogeochemical characteristics of these environments can be represented vertically by steep gradients, due to various environmental factors [19, 20]. Among these gradient systems, oxygen concentration, which decreases with increasing water depth, has a key influence on various biochemical processes, including photosynthesis, N_2 fixation, denitrification, sulfate reduction, methanogenesis, iron, and metal reduction reactions, as well as microbe distribution [21]. Therefore, as this gradient system is the foundation for distribution of microorganisms, various microbes form vertical zones, corresponding to specific environmental factors [20, 21]. Generally, the upper layer is an oxidized layer, followed by a layer with an oxidized nitrogen compound that provides a niche for aerobic microbes, such as cyanobacteria and algae, and aerobic heterotrophs [21]. At greater depths, an anaerobic layer provides niches for anaerobic bacteria and anoxygenic photoautotrophs, with fermentative heterotrophs and sulfate reducers in the bottom region, which has high concentrations of hydrogen sulfide [22, 23].

4.2.2 Challenges for Organisms Living in Coral Skeletons

To adapt to a low-nutrient environment, there is obligate symbiosis between coral and their symbiotic partner zooxanthellae [24]. Furthermore, coral-associated microbes, including endolithic microbes in coral skeletons, might be essential for coral to access nutrients [25]. However, these regions provide a specialized niche in what is overall a harsh environment [26], for example, due to low light. In that regard, the morphology of coral can influence light capturing [27]. Moreover, even within a single coral species, there is phenotypic plasticity in morphology according to light conditions [28]. It is suggested that photosynthetically active radiation (PAR) within the coral skeleton is less than 2% because coral tissue and the inorganic calcium carbonate of the skeleton are able to absorb penetrating light [15, 29, 30]. To survive under diverse light conditions and enhance capture of available wavelengths, endolithic photosynthetic algae absorb the red and far red light of the light spectrum, which are not used by zooxanthellae in coral tissues [29, 31, 32]. For instance, by using chlorophyll *d* in lieu of chlorophyll *a* to absorb near-infrared light, endolithic *Acaryochloris*-like cyanobacteria may thrive in coral skeletons and other marine habitats [33].

In addition to limited light, endolithic microbes should adapt to a condition of drastic diurnal fluctuations in pH value and oxygen concentrations, which resulted from the photosynthesis and respiration reactions that are processed by zooxanthellae in the coral tissue [26, 34]. In the skeleton of *Porites compressa*, the pH can exceed 8.5 in light conditions but can be as low as 7.7 under shade or dark conditions [26]. There are also substantial fluctuations in oxygen concentration, ranging from 210% to 10% of ambient concentrations on a daily basis in the skeleton of *Porites compressa* [26], which might enhance production of hydrogen peroxide [35]. Although this substance is well known for its potential to damage cell membranes of endolithic algae and their DNA [35, 36], this threat is mitigated by high catalase activity in endolithic algae [26].

Although the coral skeleton is critical for supporting coral, it can also provide a refuge for endolithic microbes. Coral tissues of *Porites* can filter 93.98–99.5% of the ambient ultraviolet radiation that damages endolithic algae [37]. In addition, endolithic microbes, such as algae, are relieved from grazing pressure by fish [37, 38]. In addition, endolithic microbes might get at least some of their nutrients from other endolithic communities in the coral skeleton [18, 39].

4.3 Endolithic Microorganisms in Coral Skeletons

Approximately 98% of reef-building corals contain diverse endolithic microorganisms [40], containing algae, fungi, sponges, bacteria, viruses, and archaea living within the coral skeleton [3, 38]. Furthermore, endolithic communities differ between living and dead coral substrates [41]. In living coral, endolithic microbes are oligotrophic and phototrophic, have a fast growth rate, and follow the accretion rate of skeletons [41]. However, in dead substrates, there is succession of endolithic microbial communities, because of reduced light conditions and shade due to epilithic organisms [41]. The first report of endolithic microorganisms in coral skeletons was in the early twentieth century [42]. Now, it is known that endolithic microbes are abundant in coastal environments [43], with roles including microborer and primary producer.

Various colored bands, including green, black, gray-green, yellow-green, brown, pink, and gray-black, have been reported in coral skeletons [44–47] (Fig. 4.1). Furthermore, it is common to have distinct green bands (layers), formed by endolithic microflora, running parallel to the coral surface [40, 46, 48]. These green layers in coral skeleton can be single or multiple [40, 46]. In addition, there is a white region between the green layer and coral tissue; the lack of color is that algae in the outer region may be photoinhibited by higher PAR levels [49, 50]. The green layer can be one to four to five layers with interval layers [40], but it is usually closer to the coral surface when coral is located at greater depths [50].

4.3.1 Algae and Genus Ostreobium

Many studies on endolithic microbial composition mainly focused on aerobic microorganisms, such as algae, cyanobacteria, and fungi [45, 51–55], considered major components of conspicuous green layers in the skeleton. Although endolithic algae, cyanobacteria, and fungi exist in branch-type corals, such as *Acropora acuminata* [14, 56], mostly, they live in massive scleractinian-type corals, such as *Astreopora*, *Cyphastrea*, *Goniastrea*, *Platygyra*, and *Porites* [40, 46, 48, 50]. Coral skeleton-associated microbes are summarized in Table 4.1.

Fig. 4.1 Note the various colored layers in a coral skeleton

Table 4.1 Endolithic microbes present in coral skeletons

Microbes		Coral species	Reference
Fungi	Fungi	*Acropora cytherea, Porites lutea, Porites lobata, Pocillopora eydouxi*	[45, 54, 57, 74, 75]
Algae	Algae	*Acropora cervicornis, Acropora palmata, Acropora acuminata, Favia pallida, Goniastrea australensis, Madracis mirabilis, Millepora* sp., *Oculina patagonica, Porites compressa, Porites lobata, Porites* sp.	[6, 12, 14, 17, 26, 34, 55, 56, 76]
	Ostreobium quekettii, O. constrictum	*Acropora cytherea, Astreopora myriophthalma, Cyphastrea serailia, Favia pallida, Favia* sp., *Goniastrea australensis, Goniastrea retiformis, Montipora monasteriata, Platygyra lamellina, Pocillopora eydouxi, Porites cylindrical, Porites lobata, Porites lutea, Tubastrea micranthus*	[15, 29, 31–33, 40, 48–50, 57, 75]
Bacteria	Cyanobacteria	*Porites lobata, Porites* sp., *Goniastrea aspera*	[6, 12, 33, 40, 46, 52, 55]
	Other bacteria	*Porites lobata, Porites lutea, Goniastrea aspera, Isopora palifera*	[44, 47, 72, 73]

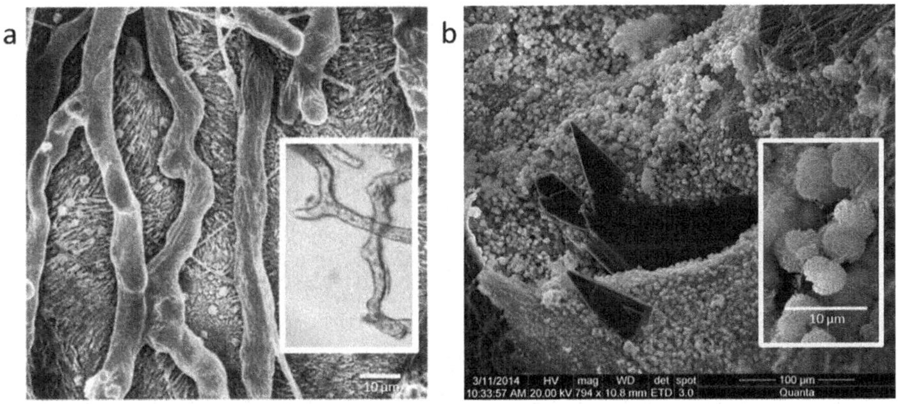

Fig. 4.2 Green algae and bacteria in coral skeletons. (**a**) Algae *Ostreobium quekettii* in live coral *Porites lobata* [40]; (**b**) endolithic bacteria in live coral *Isopora palifera* [47]

Among the endoliths, the algae *Ostreobium* (Fig. 4.2) are widely present in skeletons of living corals [48] such as *Favia* sp. [32], *Porites lobata* [40], and *Goniastrea aspera* [46]. *Ostreobium* is considered responsible for the formation of green layers beneath and parallel to the surface of coral tissue [40]. Two major species of *Ostreobium* are usually reported: *O. quekettii* which is common in reef corals from Atlantic and Pacific and *O. constrictum* which is in Caribbean corals [49]. They might offer their photosynthates as carbon resources to bleached corals that have lost symbiotic zooxanthellae [17].

4.3.2 Fungi

Both coral-associated endolithic green algae and fungi grow outward from coral skeletons [40, 49]. However, unlike endolithic algae which may be neutral or beneficial to coral hosts, fungi living in coral skeletons have been regarded as opportunistic pathogens [6, 45]. Also, endolithic fungi are usually associated with black or grayish bands [57]. In coral, *Porites* spp., *Halosphaeriaceae* (*Ascomycetes*), and *Oospora halophila* v. Beyma were isolated [57]. Furthermore, in coral *Porites lobata*, *Aspergillus*, and similar fungi attack endolithic algae *Ostreobium quekettii* in the coral skeleton, transforming the green band formed by *O. quekettii* into black by depositing pigments containing polysaccharides [45].

4.3.3 Cyanobacteria and Other Bacteria

The most common genera of autotrophic endoliths (in addition to algae) are filamentous cyanobacteria [6, 8, 12, 40, 46, 52, 53, 55, 58], including genera of *Hyella*, *Mastigocoleus*, *Plectonema*, *Solentia*, and *Acaryochloris* [33, 41], which belong to the diverse Gram-negative bacterial group and are considered the major primary producer for coral tissue [17] and other endolithic communities [18, 39]. Cyanobacteria have an important role in the carbonate cycle, due to their capacity for photosynthesis, which converts CO_2 into organic carbon. In addition, cyanobacteria support carbonate mineralization, another mode of the carbon fixation, which converts CO_2 into inorganic carbon [59]. Furthermore, given the extracellular polymeric substances produced by cyanobacteria, adherent heterotrophic bacteria contribute to various carbonate structures and defunctionalization of calcium carbonate [60].

The green algae *Ostreobium quekettii* in the skeleton of living corals grows outward from the coral skeletons, whereas green algae *Chlorophyta* bored inward from the surface of skeletons of dead coral colonies, followed by cyanobacteria [40]. Like chlorophytes, endolithic cyanobacteria are primary producers in dead corals [55]. However, in living corals, cyanobacteria provide their organic compounds to the coral tissue, highlighting the importance of cyanobacteria for survival of the coral host during coral bleaching [3].

Bacteria are most abundant and diverse among the coral-associated microbial consortia [61]. In that regard, Rohwer et al. obtained 430 distinct bacterial ribotypes from 14 coral specimens of 3 coral species [62]. Corals provide three microhabitats, the surface mucus, coral tissue, and coral skeleton for bacteria, and each habitat has a distinct bacterial composition [3]. Thus, with the exception of cyanobacteria, there are other bacteria that may possess other functions in coral skeletons [44].

In previous studies, methods for identification of endolithic algae, fungi, or cyanobacteria were mainly microscopic observation and culture-based methods that made it difficult to characterize endolithic bacterial diversity in coral skeletons. More recently, many culture-independent methods for 16S ribosomal DNA survey have been reported, including restriction fragment length polymorphism (RFLP), denaturing gradient gel electrophoresis (DGGE), clone library [62–65], 454 pyrosequencing [66, 67], and fluorescence in situ hybridization (FISH) [68]. Although bacterial diversity and potential function in coral skeleton has been investigated [4, 69, 70], most studies examined mixed bacterial communities collected from the entire coral samples, including coral tissues and mucus, and did not distinguish among bacterial groups in coral skeletons. Recently, next-generation sequencing, which is able to define minor microbial populations [71], characterized bacterial compositions in surface mucus, coral tissue, and skeleton (as distinct entities) and detected bacteria not previously reported from the skeleton of *Porites lutea* [72].

The variance of bacterial diversity in specific colored bands (layers) within coral skeleton indicates the microgeographical heterogeneity of this habit [47]. Differing bacterial composition in various colored layer highlights their potential function. In addition, due to the presence of bacteriochlorophyll signatures in coral skeletons, it has been suggested that there are anoxygenic phototrophic bacteria that can fix nitrogen in the coral skeleton [29, 48]. Recently, Yang et al. [47] reported that green sulfur bacteria (GSB) are the dominant bacterial group in the green layer within skeletons of *Isopora palifera*, although this was previously reported to contain algae and cyanobacteria, revealing more potential nitrogen fixers in coral skeletons and emphasizing the importance of endolithic bacteria to the coral holobiont. Regarding other colored layers in coral skeletons, Yuen et al. [73] detected sulfate-reducing bacteria in the black bands within the skeleton of coral *Goniastrea aspera*.

4.4 Conclusions and Future Research

In addition to endolithic algae and cyanobacteria, a recent study characterized endolithic bacteria in coral skeletons, including the prevalence of GSB within coral skeletons [47], highlighting endolithic bacteria and their role in nutrient regeneration and nutrient supply for the coral holobiont. Furthermore, it would be interesting to characterize endolithic bacteria within coral skeletons from locations large distances apart. In our view, future investigations should address three important questions. First, isolation and examination of specific endolithic bacterial strains is suggested to understand specific functions and evolutionary advantages of dominant bacteria in the green layer. Second, it is also suggested to use stable isotope labelling and nanoscale secondary ion mass spectrometry (NanoSIMS) to characterize, at the level of nutrients, relationships between coral tissues and endolithic microorganisms, which could expand knowledge of nutrient transfer within coral ecosystems. Third, bacterial communities associated with coral skeletons with large geographic distances, as well as from various depths, should be studied, which should help to characterize the diversity of bacteria associated with coral skeletons.

References

1. Azam F, Worden A. Microbes, molecules and marine ecosystems. Science. 2004;303:1622–4.
2. Reshef L, Koren O, Loya Y, Zilber-Rosenberg I, Rosenberg E. The coral probiotic hypothesis. Environ Microbiol. 2006;8:2068–73.
3. Rosenberg E, Koren O, Reshef L, Efrony R, Zilber-Rosenberg I. The role of microorganisms in coral health, disease and evolution. Nat Rev Microbiol. 2007;5:355–62.
4. Wegley L, Edwards R, Rodriguez-Brito B, Liu H, Rohwer F. Metagenomic analysis of the microbial community associated with the coral *Porites astreoides*. Environ Microbiol. 2007;9:2707–19.
5. Rohwer F, Kelly S. Culture independent analyses of coral associated microbes. In: Rosenberg E, Loya Y, editors. Coral health and disease. Berlin: Springer; 2004. p. 265–78.
6. Chazottes V, Le-Campion-Alsumard T, Peyrot-Clausade M. Bioerosion rates on coral reefs: interaction between macroborers, microborers and grazers (Moorea, French Polynesia). Palaeogeogr Palaeoclimatol Palaeoecol. 1995;113:189–98.
7. Radtke G, Le-Campion-Alsumard T, Golubic S. Microbial assemblages of the bioerosional "notch" along tropical limestone coasts. Algol Stud. 1996;83:469–582.
8. Radtke G, Le-Campion-Alsumard T, Golubic S. Microbial assemblages involved in tropical coastal bioerosion: an Atlantic- Pacific comparison. Proc 8th Int Coral Reef Symp. 1997;1:1825–30.
9. Ghirardelli LA. Endolithic microorganisms in live and dead thalli of coralline red algae (Corallinales, Rhodophyta) in the northern Adriatic Sea. Acta Geol Hisp. 2002;37:53–60.
10. De Los Rios A, Wierzchos J, Sancho LG, Green TGA, Ascaso C. Ecology of endolithic lichens colonizing granite in continental Antarctica. Lichenologist. 2005;37:383–95.
11. Garcia-Pichel F. Plausible mechanisms for the boring on carbonates by microbial phototrophs. Sediment Geol. 2006;185:205–13.
12. Tribollet A, Golubic S. Cross-shelf differences in the pattern and pace of bioerosion of experimental carbonate substrates exposed for 3 years on the northern great barrier reef, Australia. Coral Reefs. 2005;24:422–34.
13. Hutchings P. Biological destruction of coral reefs. Coral Reefs. 1986;4:239–52.
14. Bruggemann JH, van Oppen MJH, Breeman AN. Foraging by the stoplight parrotfish *Sparisoma viride*. I. Food selection in different, socially determined habitats. Mar Ecol Prog Ser. 1994;106:41–55.
15. Schlichter D, Kampmann H, Conrady S. Trophic potential and photoecology of endolithic algae living within coral skeletons. Mar Ecol. 1997;18:299–317.
16. Ferrer LM, Szmant AM. Nutrient regeneration by the endolithic community in coral skeletons. Proc 6th Int Coral Reef Symp. 1988;1:1–4.
17. Fine M, Loya Y. Endolithic algae: an alternative source of photoassimilates during coral bleaching. Proc Biol Sci. 2002;269:1205–10.
18. Shashar N, Cohen Y, Loya Y, Sar N. Nitrogen fixation (acetylene reduction) in stony corals: evidence for coral-bacteria interactions. Mar Ecol Prog Ser. 1994;111:259–64.
19. Wendt-Potthoff K, Koschorreck M, Diez-Ercilla M, Sanchez-Espana J. Microbial activity and biogeochemical cycling in a nutrient-rich meromictic acid pit lake. Limnologica. 2012;42:175–88.
20. van Gemerden H. Microbial mats: A joint venture. Mar Geol. 1993;113:3–25.
21. Paerl HW, Pinckney JL. A mini-review of microbial consortia: their roles in aquatic production and biogeochemical cycling. Microb Ecol. 1996;31:225–47.
22. Jørgensen BB, Kuenen JG, Cohen Y. Microbial transformations of sulfur compounds in a stratified Lake (Solar Lake, Sinai). Limnol Oceanogr. 1979;24:799–822.
23. Tonolla M, Peduzzi S, Demarta A, Peduzzi R, Hahn D. Phototropic sulfur and sulfate-reducing bacteria in the chemocline of meromictic Lake Cadagno, Switzerland. J Limnol. 2004;63:161–70.

24. Muscatine L, Porter JW. Reef corals: mutualistic symbioses adapted to nutrient-poor environments. Bioscience. 1977;27:454–9.
25. Thompson JR, Rivera HE, Closek CJ, Medina M. Microbes in the coral holobiont: partners through evolution, development, and ecological interactions. Front Cell Infect Microbiol. 2015;4:176.
26. Shashar N, Stabmler N. Endolithic algae within corals-life in an extreme environment. J Exp Mar Biol Ecol. 1992;163:277–86.
27. Roth MS. The engine of the reef: photobiology of the coral-algal symbiosis. Front Microbiol. 2014;5:422.
28. Muko S, Kawasaki K, Sakai K, Takasu F, Shigesada N. Morphological plasticity in the coral *Porites sillimaniani* and its adaptive significance. Bull Mar Sci. 2000;66:225–39.
29. Magnusson SH, Fine M, Kuhl M. Light microclimate of endolithic phototrophs in the scleractinian corals *Montipora monasteriata* and *Porites cylindrica*. Mar Ecol Prog Ser. 2007;332:119–28.
30. Kanwisher J, Wainwright SA. Oxygen balance in some reef corals. Biol Bull. 1967;133:378–90.
31. Halldal P. Photosynthetic capacities and photosynthetic action spectra of endozoic algae of the massive coral *Favia*. Biol Bull. 1968;134:411–24.
32. Fork DC, Larkum AWD. Light harvesting in the green alga *Ostreobium* sp., a coral symbiont adapted to extreme shade. Mar Biol. 1989;103:381–5.
33. Behrendt L, Larkum AW, Norman A, Qvortrup K, Chen M, Ralph P, et al. Endolithic chlorophyll *d*-containing phototrophs. ISME J. 2011;5:1072–6.
34. Bellamy N, Risk MJ. Coral gas: oxygen production in *Millepora* on the great barrier reef. Science. 1982;215:1618–9.
35. Gille JJ, Jonje H. Biological significance of oxygen toxicity: an introduction. In: Vigo-Pelirey C, editor. Membrane lipid oxidation, vol. 3. Boca Raton: CRC Press; 1991. p. 1–32.
36. Ingraham LL, Meyer DL. Biochemistry of dioxygen. New York: Plenum Press; 1985. p. 255.
37. Shashar N, Banaszak AT, Lesser MP, Amrami D. Coral endolithic algae: life in a protected environment. Pac Sci. 1997;51:167–73.
38. Schönberg CHL, Wisshak M. The perks of being endolithic. Aquat Biol. 2012;17:1–5.
39. Risk MJ, Muller HR. Pore water in coral heads: evidence for nutrient regeneration. Limnol Oceanogr. 1983;28:1004–8.
40. Le-Campion-Alsumard T, Golubic S, Hutchings P. Microbial endoliths in skeletons of live and dead corals: *Porites lobata* (Moorea, French Polynesia). Mar Ecol Prog Ser. 1995;117:149–57.
41. Tribollet A. The boring microflora in modern coral reef ecosystems: a review of its roles. In: Wisshak M, Tapanila L, editors. Current developments in bioerosion. Berlin: Springer-Verlag; 2008. p. 67–94.
42. Duerden JE. Endolithic algae as agents in the disintegration of corals. Bull Am Mus Nat Hist. 1902;16:323–32.
43. Schenider J, Le Campio-Alsumard T. Construction and destruction of carbonates by marine and freshwater cyanobacteria. Eur J Phycol. 1999;34:417–26.
44. DiSalvo LH. Isolation of bacteria from corallum of *Porites lobata* (Vaughn) and its possible significance. Am Zool. 1969;9:735–40.
45. Priess K, Le Campion-Alsumard T, Golubic S, Fadel F, Thomassin BA. Fungi in corals: black bands and density-banding of *Porites lutea* and *P. lobata* skeleton. Mar Biol. 2000;136:19–27.
46. Yamazaki SS, Nakamura T, Yamasaki H. Photoprotective role of endolithic algae colonized in coral skeleton for the host photosynthesis. In: Allen JF, Gantt E, Golbeck JH, Osmond B, editors. Photosynthesis. Energy from the sun: 14th international congress on photosynthesis. Dordrecht: Springer; 2008. p. 1391–5.
47. Yang SH, Lee STM, Huang CR, Tseng CH, Chiang PW, Chen CP, et al. Prevalence of potential nitrogen-fixing, green sulfur bacteria in the skeleton of reef-building coral *Isopora palifera*. Limnol Oceanogr. 2016;61:1078–86.
48. Ralph PJ, Larkum AWD, Kühl M. Photobiology of endolithic microorganisms in living coral skeletons: 1. Pigmentation, spectral reflectance and variable chlorophyll fluorescence analysis

of endoliths in the massive corals *Cyphastrea serailia*, *Porites lutea* and *Goniastrea australensis*. Mar Biol. 2007;152:395–404.
49. Lukas KJ. Two species of the chlorophyte genus *Ostreobium* from skeletons of Atlantic and Caribbean reef corals. J Phycol. 1974;10:331–5.
50. Highsmith RC. Lime-boring algae in hermatypic coral skeletons. J Exp Mar Biol Ecol. 1981;55:267–81.
51. Golubic S, Perkins RD, Lukas KJ. Boring microorganisms and microborings in carbonate substrates. In: Frey RW, editor. The study of trace fossils. New York: Springer; 1975. p. 229–59.
52. Al-Thukair A, Golubic S. Five new *Hyella* species from the Arabian Gulf. Algol Stud. 1991a;64:167–97.
53. Al-Thukair A, Golubic S. New endolithic cyanobacteria from the Arabian Gulf. I. *Hyella immanis* sp. nov. J Phycol. 1991b;27:167–97.
54. Golubic S, Radtke G, Le-Campion-Alsumard T. Endolithic fungi in marine ecosystems. Trends Microbiol. 2005;13:229–35.
55. Tribollet A, Langdon C, Golubic S, Atkinson MJ. Endolithic microflora are major primary producers in dead carbonate substrates of Hawaiian coral reefs. J Phycol. 2006;42:292–303.
56. Crossland CJ, Barnes DJ. Acetylene reduction by coral skeletons. Limnol Oceanogr. 1976;21:153–6.
57. Bak RPM, Laane RWPM. Annual black bands in skeletons of reef corals (Scleractinia). Mar Ecol Prog Ser. 1987;38:169–75.
58. Tribollet A, Payri C. Bioerosion of the crustose coralline alga *Hydrolithon onkodes* by microborers in the coral reefs of Moorea, French Polynesia. Oceanol Acta. 2001;24:329–42.
59. Kamennaya NA, Ajo-Franklin CM, Northen T, Jansson C. Cyanobacteria as biocatalysts for carbonate mineralization. Fortschr Mineral. 2012;2:338–64.
60. Arp G, Reimer A, Reitner J. Calcification in cyanobacterial biofilms of alkaline salt lakes. Eur J Phycol. 1999;34:393–403.
61. Mouchka ME, Hewson I, Harvell CD. Coral-associated bacterial assemblages: current knowledge and the potential for climate-driven impacts. Integr Comp Biol. 2010;50:662–74.
62. Rohwer F, Seguritan V, Azam F, Knowlton N. Diversity and distribution of coral-associated bacteria. Mar Ecol Prog Ser. 2002;243:1–10.
63. Hong MJ, Yu YT, Chen CA, Chiang PW, Tang SL. Influence of species specificity and other factors on bacteria associated with the coral Stylophora pistillata in Taiwan. Appl Environ Microbiol. 2009;75:7797–806.
64. Kvennefors EC, Sampayo E, Ridgway T, Barnes AC, Hoegh-Guldberg O. Bacterial communities of two ubiquitous great barrier reef corals reveals both site and species-specificity of common bacterial associates. PLoS One. 2010;5:e10401.
65. Kimes NE, Johnson WR, Torralba M, Nelson KE, Weil E, Morris PJ. The *Montastraea faveolata* microbiome: ecological and temporal influences on a Caribbean reef-building coral in decline. Environ Microbiol. 2013;15:2082–94.
66. Bayer T, Neave MJ, Alsheikh-Hussain A, Aranda M, Yum LK, Mincer T, et al. The microbiome of the Red Sea coral *Stylophora pistillata* is dominated by tissue-associated *Endozoicomonas* bacteria. Appl Environ Microbiol. 2013;79:4759–62.
67. Lema KA, Bourne DG, Willis BL. Onset and establishment of diazotrophs and other bacterial associates in the early life history stages of the coral *Acropora millepora*. Mol Ecol. 2014;23:4682–95.
68. Ainsworth TD, Fine M, Blackall LL, Hoegh-Guldberg O. Fluorescence in situ hybridization and spectral imaging of coral-associated bacterial communities. Appl Environ Microbiol. 2006;72:3016–20.
69. Lee OO, Yang J, Bougouffa S, Wang Y, Batang Z, Tian R, et al. Spatial and species variations in bacterial communities associated with corals from the Red Sea as revealed by pyrosequencing. Appl Environ Microbiol. 2012;78:7173–84.
70. Santos HF, Carmo FL, Duarte G, Dini-Andreote F, Castro CB, Rosado AS, et al. Climate change affects key nitrogen-fixing bacterial populations on coral reefs. ISME J. 2014;8:2272–9.

71. Sogin ML, Morrison HG, Huber JA, Welch DM, Huse SM, Neal PR, et al. Microbial diversity in the deep sea and the underexplored "rare biosphere". Proc Natl Acad Sci U S A. 2006;103:12115–20.
72. Li J, Chen Q, Long LJ, Dong JD, Yang J, Zhang S. Bacterial dynamics within the mucus, tissue and skeleton of the coral *Porites lutea* during different seasons. Sci Rep. 2014;4:7320.
73. Yuen YS, Yamazaki SS, Baird AH, Nakamura T, Yamasaki H. Sulfate-reducing bacteria in the skeleton of the massive coral *Goniastrea aspera* from the great barrier reef. Galaxea, J Coral Reef Stud. 2013;15:154–9.
74. Kendrick B, Risk MJ, Michaelides J, Bergman K. Amphibious microborers: bioeroding fungi isolated from live corals. Bull Mar Sci. 1982;32:862–7.
75. Bentis CJ, Kaufman L, Golubic S. Endolithic fungi in reef-building corals (order: Scleractinia) are common, cosmopolitan, and potentially pathogenic. Biol Bull. 2000;198:254–60.
76. Shibata K, Haxo FT. Light transmission and spectral distribution through epi- and endozoic algal layers in the brain coral *Favia*. Biol Bull. 1969;136:461–8.

Chapter 5
The Bacteria *Endozoicomonas*: Community Dynamics, Diversity, Genomes, and Potential Impacts on Corals

Jia-Ho Shiu and Sen-Lin Tang

Contents

Abstract *Endozoicomonas*, a recently identified bacterial genus, is commonly detected in corals, sponges, and other reef invertebrates. These bacteria have attracted attention, as there are many indications that they have important roles in coral health. This chapter includes introduction to *Endozoicomonas* and is divided into four main sections – (1) introducing history and type strain of *Endozoicomonas*; (2) ecological distribution; (3) abundance, diversity, and phylogeny; and (4) genomes – followed by a general discussion and future research directions in the end. In the first section, the taxonomy and classification of the genus *Endozoicomonas* are clarified and discussed. Important characteristics of all cultivable *Endozoicomonas* strains isolated from marine invertebrates are also described. The studies of *Endozoicomonas* in different hosts, locations, and time are described in the second section. The third section will discuss the variation of *Endozoicomonas* in abundance, phylogeny, and

J.-H. Shiu · S.-L. Tang (✉)
Biodiversity Research Center, Academia Sinica, Nan-gang, Taipei, Taiwan
e-mail: sltang@gate.sinica.edu.tw

Z. Li (ed.), *Symbiotic Microbiomes of Coral Reefs Sponges and Corals*,
https://doi.org/10.1007/978-94-024-1612-1_5

diversity. Finally, the genomes of *Endozoicomonas* and possible interactions between these bacteria and reef invertebrates, especially coral holobiont, are elucidated that paves the way for the future study of *Endozoicomonas*.

Keywords *Endozoicomonas* · Bacteria · Reef invertebrate · Coral

5.1 Introduction to *Endozoicomonas*

5.1.1 History of Endozoicomonas *and* Spongiobacter, *Family* Hahellaceae

The genus *Endozoicomonas*, under the family *Hahellaceae*, the order *Oceanospirillales*, and the class *Gammaproteobacteria*, was widely detected in various marine environments and abundant in various marine invertebrates. However, the taxonomic name of this genus is confusing, due to the use of alternate terminology in some reports, that is, *Spongiobacter* (genus) and *Endozoicimonaceae* or *Endozoicomonaceae* (family), transition errors, or provisional usage.

The genus name *Endozoicomonas* was first proposed by Kurahashi and Yokota in 2007 [1]. Prior to that, sequences similar to *Endozoicomonas* were annotated or referred to as *Spongiobacter*, the other "genus" name proposed by Nishijima (2005) in an unpublished conference paper, in which she introduced a nickel-tolerant bacterial isolate from a marine sponge (information unavailable). Unfortunately, the "genus" name *Spongiobacter* (should be *Candidatus Spongiobacter*) was never formally registered as a taxon, there was no description of characteristics, and no culture of this bacterium was deposited in an authorized collection center or institute, all of which are essential for publication of a new bacterial species. Furthermore, the genera *Endozoicomonas* and *Spongiobacter* are phylogenetically closely related; consequently, their 16S rRNA gene sequences are always entangled in phylogenetic analyses (Fig. 5.1) [2, 3], indicating that presumably they are the same genus (i.e., *Endozoicomonas*). Nonetheless, some sequences, with great similarity to *Endozoicomonas*, were recently annotated as *Spongiobacter* and deposited in public databases.

Regarding the family name, both *Endozoicomonaceae* and *Endozoicimonaceae* were used in recent reports [4, 5]. However, neither of those family names has been accepted as an official name of the taxonomic rank, resulting in confusion during data mining. For example, informal family names are included in the bacterial database of Greengenes (greengenes.lbl.gov), but they are absent in Silva (www.arb-silva.de) and RDP (rdp.cme.msu.edu). Furthermore, the same sequences assigned to *Endozoicomonaceae* in Greengenes were assigned to *Hahellaceae* in RDP [6]. Apparently, taxonomic issues of *Endozoicomonas* remain unsettled, but it should be clarified as soon as possible, to minimize further confusion. Due to a paucity of evidence to distinguish those informal taxonomic names, sequences designated *Spongiobacter*, *Endozoicomonaceae*, or *Endozoicimonaceae* are all referred as the genus *Endozoicomonas* in this chapter.

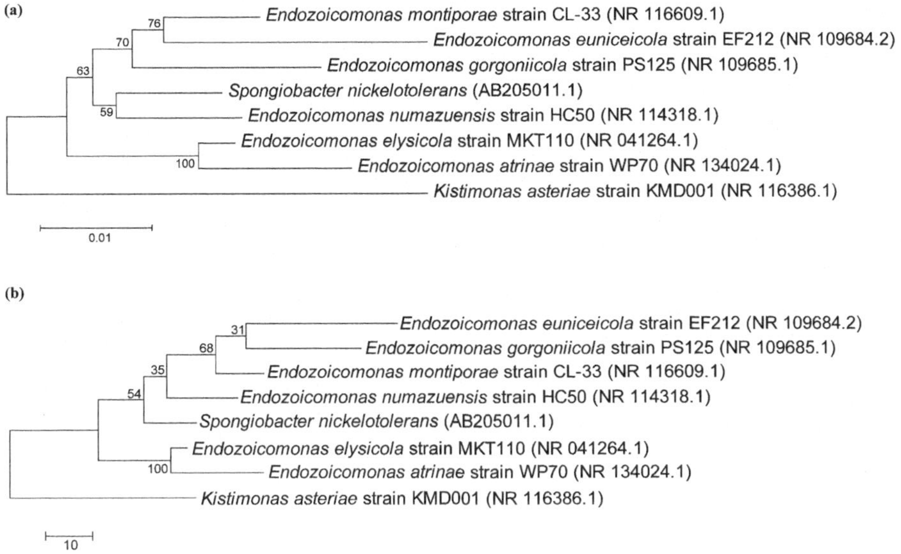

Fig. 5.1 Phylogenetic trees of partial 16S rRNA gene sequences (1397 sites) were constructed with 1000 bootstrap replicates using (**a**) neighbor-joining method based on Kimura 2P + G model and (**b**) maximum parsimony method. Bootstrap values are shown on nodes. *Kistimonas asteriae* was used as an outgroup

5.1.2 Cultivable Species of Endozoicomonas

Eight cultivable species in this genus have been isolated and published, including *E. elysicola* [1], *E. montiporae* [7], *E. numazuensis* [8], *E. euniceicola*, *E. gorgoniicola* [9], *E. atrinae* [10], *E. arenosclerae* [11], and *E. cretensis* [12]. With the exception of *E. cretensis* that was isolated from fish, all other species were originally isolated from marine invertebrates: three from corals, two from sponges, and one each from sea slugs and pen shells (Table 5.1). Some unique features of each species are described below.

E. elysicola MKT110^{T} was the first *Endozoicomonas* isolated from the sea slug *Elysia ornate* at depth of 15 m near the coast of Izu-Miyake Island, Japan. The bacterium was Gram-negative, rod-shaped, approximately 0.4–0.6 μm in diameter, and 1.8–2.2 μm long, with a single polar flagellum, pili, and releasable vesicles. The colony on marine agar (Difco Laboratories, Detroit, MI, USA) was 4–5 mm in diameter, circular, convex, and beige with intact edges. The optimal temperature range for aerobic growth conditions on agar plates was 25–30 °C in the presence of seawater-like salt [1].

E. montiporae CL-33^{T} was the first cultivable coral-associated *Endozoicomonas* from the reef-building coral *Montipora aequituberculata*, subclass Hexacorallia, collected at a depth of 10–15 m in tropical Taiwan [7]. Cultivation conditions were more fastidious and stringent than other species. For example, a sugar supply (e.g., glucose) is necessary for culture. Moreover, the bacterium grows relatively slower

Table 5.1 Differential phenotypic and genotypic characteristics of type strains of *Endozoicomonas* species

Characteristic	*E. elysicola*	*E. montiporae*	*E. gorgoniicola*	*E. euniceicola*	*E. numazuensis*	*E. arenosclerae*	*E. atrinae*
Strain	DSM22380	LMG24815	CECT8353	DSM26535	DSM25634	CBAS572	JCM19190
Isolation source	Sea slug	Hexacoral	Octocoral	Octocoral	Sponge	Sponge	Pen shell
Motility	+	+	+	+	–	+	–
Cell length (μm)	0.4–0.6	0.5–0.7	0.4–0.9	0.6–0.9	3.0–10.0	N.D.	1.2–3.6
Cell diameter (μm)	1.8–2.2	1.0–3.0	1.7–2.5	1.7–2.6	0.4–0.8	0.5–1.0	0.7–1.0
Growth temp. (°C)							
Range	4–37	15–35	15–30	15–30	15–37	12–35	15–37
Optimum	25–30	25	22–30	22–30	25	20–30	30
Growth NaCl (%, w/v)							
Range	>0	1–3	1–4	1–4	1–5	2–5	1–4
Optimum	>0	2–3	2–3	2–3	2	3	2
pH optimum	N.D.	8	8	8	7.5–8.0	N.D.	7
Aerobic/anaerobic	Aerobic	Aerobic	Both	Both	Both	Aerobic	Aerobic
Nitrate reduction	+	+	–	–	+	N.D.	+
Major fatty acids	$C_{16:\ 1}\omega 7c$	$C_{16:\ 1}\omega 6c$ and/or $C_{16:\ 1}\omega 7c$	$C_{16:\ 1}\omega 6c$ and/or $C_{16:\ 1}\omega 7c$	$C_{16:\ 1}\omega 6c$ and/or $C_{16:\ 1}\omega 7c$	$C_{18:\ 1}\omega 7c$	N.D.	$C_{16:\ 1}\omega 6c$ and/or $C_{16:\ 1}\omega 7c$
	$C_{16:\ 0}$	$C_{18:\ 1}\omega 6c$ and/or $C_{18:\ 1}\omega 7c$	$C_{18:\ 1}\omega 6c$ and/or $C_{18:\ 1}\omega 7c$	$C_{16:\ 0}$	$C_{16:\ 1}\omega 7c$	N.D.	$C_{18:\ 1}\omega 6c$ and/or $C_{18:\ 1}\omega 7c$
	$C_{14:\ 0}$	$C_{16:\ 0}$	$C_{16:\ 0}$	$C_{18:\ 1}\omega 6c$ and/or $C_{18:\ 1}\omega 7c$	$C_{16:\ 0}$	N.D.	$C_{16:\ 0}$
Quinones	Q-9	Q-9, Q-8	Q-9, Q-8	Q-9, Q-8	Q-9, Q-8, MK-9	N.D.	Q-9
G + C content	50.4	50.0	47.5	48.6	48.7	47.6	50.5
References	[1]	[7]	[9]	[9]	[8]	[11]	[10]

than other isolates (e.g., *E. elysicola* and *E. numazuensis*) [13]. Colonies growing on marine agar (Difco Laboratories) were 1–2 mm in diameter (smaller than *E. elysicola*). Optimal aerobic growth conditions on agar plates were 25 °C at pH 8.0 and 2–3% NaCl [7].

E. euniceicola and *E. gorgoniicola* [9] were isolated from the octocorals, *Eunicea fusca* and *Plexaura* sp. collected in Florida, USA, and Bimini, Bahamas, respectively. Unlike *E. montiporae* and *E. elysicola*, colonies on agar plates are white and creamy white for *E. euniceicola* and *E. gorgoniicola*, respectively. Moreover, they are facultative anaerobic but lack the ability to reduce nitrates to nitrites.

E. numazuensis [8] and *E. arenosclerae* [11] were both isolated from sponges, the purple sponge *Haliclona* sp. at Numazu in Japan, and the sponge *Arenosclera brasiliensis* in Rio de Janeiro, Brazil, respectively. *E. numazuensis* is longer (3–10 μm) and non-motile, facultative anaerobic, whereas *E. arenosclerae* is motile and aerobic. Although their phenotype profiles differ, they share >99% in sequence identity of the 16S rRNA gene (i.e., 1427 informative sites) [11].

E. atrinae [10] was isolated from the gut of the comb pen shell *Atrina pectinate* in the southern sea of Yeosu in Korea. Similar to *E. elysicola*, this is the other *Endozoicomonas* isolated from mollusks. This bacterium is non-motile, like *E. numazuensis*. Furthermore, it has the smallest colony size on agar plates, 0.6–1.1 μm, and the highest DNA GC content (50.5) of all seven *Endozoicomonas* species.

Although these seven *Endozoicomonas* species were all isolated from marine invertebrates, there were many differences in their distribution, phenotypic characteristics, and metabolic potentials. Therefore, we inferred that diversity is a characteristic of this genus.

5.2 Distribution of *Endozoicomonas*

5.2.1 Host Variation

Culture-independent bacterial community studies have provided valuable ecological information regarding distribution of *Endozoicomonas*. Using 16S rRNA gene sequencing, several surveys of *Endozoicomonas*-related bacteria have been reported from various marine invertebrates, including hexacorals [4, 14, 15], octocorals [16, 17], sea anemones [18], hydras [19], sponges [20–22], polychaetas [23], ascidians [5], sea slugs [1], oysters [3, 24], and bivalves [2]. These bacteria were associated with marine animals and were also detected in some environmental niches, including sediment [25] and seawater [26], although relative abundance of *Endozoicomonas* was much lower in environments than in marine animals [16].

5.2.2 *Spatial Discovery*

Regarding their geographic distribution, *Endozoicomonas*-related bacteria were widely detected in various regions, including South Africa (Mayotte); Asia (Japan and Taiwan); North, Middle, and South America (Florida, Caribbean, Belize, and Brazil); Europe (Mediterranean, Rockall Banks, and Norway); Red Sea; and Great Barrier Reef [1, 2, 7, 15–17, 19, 21, 22, 27–30]. Their geographic habitats were also variable, from intertidal zones [7, 31] to ocean locations at depths exceeding 700 m [6, 17, 26].

5.2.3 *Temporal Distribution*

Endozoicomonas species associated with corals have been discovered in various seasons and climate zones. For example, in the tropic zone, *Endozoicomonas* were regularly identified in a time-series survey of the coral *Isopora*, collected from the southern coast of Taiwan [32]. In the temperate zone, *Endozoicomonas* was also detected in summer and winter or summer and autumn in various gorgonian octocoral species, *Paramuricea clavata* and *Eunicella verrucosa*, respectively [27, 33]. In the Great Barrier Reef, *Endozoicomonas* were dominant in all seasons in the coral *Acropora muricata* of inshore reefs [34].

Regardless of spatial and temporal factors, *Endozoicomonas* were detected not only in various regions with wide longitudes and latitudes but also from intertidal areas to deep oceans and in various marine invertebrates and corals. Therefore, *Endozoicomonas* is a common resident bacterial group associated with marine invertebrates around the world, particularly invertebrates in coral reefs.

5.3 Abundance, Phylogeny, and Diversity of *Endozoicomonas* in Corals

5.3.1 *Abundance*

Although absolute abundance of *Endozoicomonas* in their hosts has not been well characterized, there is evidence of changes in relative abundance of the bacteria, often in association with specific factors. For example, relative abundance of *Endozoicomonas* was correlated with their habitats [35]. Similarly, bacterial abundance in fungid corals (*Ctenactis echinata*) was only dominant in sheltered sides of the offshore (i.e., open rocky substrates and clear water habitats) but less abundant in the nearshore (characterized by loose substrates and turbid water) [36], suggesting that these bacteria had environmental preferences that matched those of coral species in the central Red Sea. There was also habitat specificity of *Endozoicomonas* in a dominant reef-building coral (*Acropora millepora*) in the Great Barrier Reef [34], although the association was opposite to the other study [36], as there was higher abundance of *Endozoicomonas* on the midshore reef than the offshore reef [34]. There are many potential explanations for these apparently contradictory results, including differences in *Endozoicomonas* species, coral hosts, and environmental conditions. In addition, in some studies, *Endozoicomonas* in octocorals had stable abundance in different seasons [27, 33]. Although these studies have provided preliminary insights, much more work is needed to characterize changes in abundance of *Endozoicomonas* in corals.

There are some reports on the effects of stress and environmental factors in abundance of *Endozoicomonas* of corals. For example, relative abundance of *Endozoicomonas* was considerably decreased in response to abiotic stresses, e.g., temperature increases [37], ocean acidification [38], or anthropogenic impacts (viz., sedimentation and sewage) [39]. Furthermore, bleaching of *Acropora* corals in the Great Barrier Reef caused *Endozoicomonas* to dynamically disappear, although it largely recovered during the coral's resurgence over the summer of 2001 to 2002 [40]. Furthermore, loss of *Endozoicomonas* from the surface mucus layer was also a characteristic of lesions in *Pocillopora* in Belize [15]. Similarly, Vezzulli et al. (2013) reported that *Endozoicomonas* were a predominant group on healthy Mediterranean gorgonians but declined greatly when the host was compromised [41]. In addition, *Endozoicomonas* had higher relative abundance in new mucus of healthy coral *Porites astreoides* but was less abundant in aged mucus and disturbed coral [42]. These results strongly support that *Endozoicomonas* is highly associated with coral health (alternatively, host heath). Furthermore, we inferred that *Endozoicomonas* may have important roles in marine invertebrates or their holobionts, although it is noteworthy that these bacteria have also been implicated as a potential cause of disease in fish [12].

5.3.2 Phylogeny

Phylogenetic analysis of *Endozoicomonas* may provide clues regarding relationships between these bacteria and their hosts or habitats. For example, some specific *Endozoicomonas* species were present due to adaptation to the environment or host [17]. In that study, *Endozoicomonas* populations in the coral *Madrepora oculata* were grouped together in a phylogenetic tree of 16S rRNA gene that differed from *Endozoicomonas* detected in other species of octocorals and sponges from other places. Similarly, other studies provided evidence that *Endozoicomonas* was host specific [29, 30, 43], whereas it was even proposed that the relationship between the *Endozoicomonas* species and their gorgonian host *Eunicella cavolini* was an ancient evolutionary association, as two *Endozoicomonas* populations, both collected from gorgonians albeit from different locations in the Mediterranean and Caribbean, were closely related in a phylogenetic analysis [16]. The same team reported a similar pattern in another study with two hexacorals, *Porites damicornis* and *Acropora* spp., from the Red Sea and the Great Barrier Reef [31]. However, in the latter study, some *Endozoicomonas* sequences were mixed with the *Endozoicomonas* from different host species (e.g., *Stylophora pistillata* with *Goniastrea edwardsi* or *Pocillopora damicornis*) in a monophyletic branch of the phylogenetic tree with good bootstrapping values (93 and 78, respectively).

Endozoicomonas species in the hexacoral *Seriatopora hystrix* also clustered with *Endozoicomonas*, not only from other corals but also other marine invertebrates, e.g., sea slugs, sea cucumber, and sea anemones [35]. Besides, *Endozoicomonas* was only detected in three *Acropora* species at Magnetic Island, but not in the same coral species on Orpheus Island, <80 km away [28]. Both studies concluded that location or habitat effects were more important than species effects on bacterial community in host. Hence, host specificity of *Endozoicomonas* is still a complex and unresolved question.

5.3.3 Diversity

Detailed studies of diversity and absolute abundance of *Endozoicomonas* are lacking and cannot be easily done using current data sets. One reason is that the approaches used to conduct community surveys varied among studies. Therefore, there is an urgent need to use only a standardized method. For example, sequencing the same regions or the full length of 16S rRNA gene will facilitate comparative analysis. If possible, the development of universal or common primers to detect the *Endozoicomonas* community should be even more sensitive and helpful to characterize the diversity of these bacteria. Furthermore, measuring absolute abundance of *Endozoicomonas* should be based on the same method of normalization (e.g., equivalent numbers of host cells). Additionally, similar molecular or bioinformatic methods should be used to minimize bias among studies.

5.4 Genomes of *Endozoicomonas*

Several studies proposed or emphasized potential functions of *Endozoicomonas* or their interactive relationships with marine invertebrates, for example, an intimate relationship with the coral host, as *Endozoicomonas* cells were present in coral cells [2, 31]. In addition, *Endozoicomonas* may have a role in sulfur cycling [44], DMSP degradation [45], and production of antimicrobial compounds in its coral host [22]. However, there is no direct evidence that these actually occur.

Genomic approaches are useful to identify potential metabolic and other functions of *Endozoicomonas*. Three cultivable strains of *Endozoicomonas*, *E. elysicola*, *E. montiporae*, and *E. numazuensis*, were sequenced [46]. All three genomes were estimated to exceed 5 Mbp (Table 5.2), with potential capacity for the Embden-Meyerhof-Parnas (EMP) glycolytic pathway, tricarboxylic acid cycle, in addition to

Table 5.2 Genome information from *Endozoicomonas*-type strains

Characteristic	*E. elysicola*	*E. elysicola*	*E. numazuensis*	*E. montiporae*	*E. montiporae*
Strain	DSM 22380	DSM 22380	DSM 25634	LMG 24815	CL-33^T
Host	Sea slug, *Elysia ornata*	Sea slug, *Elysia ornata*	Sponge, cf. *Haliclona* spp.	Hexacoral, *Montipora aequituberculata*	Hexacoral, *Montipora aequituberculata*
Genome (Mbp)	5.55	5.61	6.34	5.6	5.43
Gene	4669	4693	5405	5113	5033
Protein	4515	4532	5129	4837	4761
16S rRNA	N/A	6	5	4	7
23S rRNA	N/A	6	2	4	7
5S rRNA	6	8	3	8	8
tRNA	78	85	90	104	109
GC content (%)	46.7	46.8	47.1	48.5	48.37
Pseudogene	72	55	175	155	198
References	[47]	[46]	[46]	[46]	[13]

genes for the conversion and assimilation of nitrate. Ding and co-workers [13] provided a high-quality, nearly completed genome (e.g., 99.8%) of *E. montiporae* and detailed characterizations, including comparative analysis of the three species. All of these had the capacity to synthesize all proteinogenic amino acids and most cofactors, prosthetic groups, and electron carriers required for growth, except vitamin B12 [13]. The researcher provided evidence-based inferences and speculation regarding how *E. montiporae* interacted with its host. Based on a physiological experiment, all three bacteria had the genes to degrade testosterone, implicating this male sex hormone as a potential "animal sign" for attracting *Endozoicomonas* [13]. Unique genes detected in *E. montiporae* for N-deglycosylation enzyme might be able to partially dissociate glycoproteins inside the coral mucus (without harming the host) and thereby enable bacteria to penetrate through the mucus layer to reach specific ephrin receptors on the coral cell membrane; an interaction of these receptors and ephrin ligands of *E. montiporae* enables the bacterium to enter the coral cell by endocytosis [13]. Moreover, a secreted protein of *E. montiporae* might modulate trafficking inside the host's cell and prevent attacks by the lysosome inside the host cells. More interestingly, various type III secretion effectors (e.g., T3SS: involved in survival inside hosts, regulating metabolism and increasing the fitness of the host) were detected in the *E. montiporae* genome; perhaps they are able to interact with hosts or provide certain responses when the coral host under stress (e.g., thermal-induced mitochondrial dysfunction). Finally, based on several striking features, including unusual high-repeat sequences, mobile elements, pseudogenes, and several eukaryotic genes detected in the genome, the authors inferred that *E. montiporae* was involved in genomic erosion and gene exchange and could be a facultative endosymbiont [13].

5.5 Future Directions

Although there are many reports of *Endozoicomonas*, their ecology and evolution remain unclear. Due to variations in regions of 16S rRNA gene that have been sequenced, and inconsistencies among reports, it is hard to clarify how abiotic or biotic factors affect distribution, composition, relative abundance, or phylogeny of *Endozoicomonas* communities. Nevertheless, some hypotheses can still be constructed. For example, occurrences of *Endozoicomonas* populations in marine invertebrates are not randomly distributed, and various marine invertebrates may provide distinct niches for certain *Endozoicomonas* spp. or strains. In addition, we suggest that some *Endozoicomonas* sp. have specific roles, including being endosymbiotic with marine invertebrates, pathogenic for fish, or free-living in seawater or sediment [12, 13, 23, 25, 31].

To demonstrate or comprehend biotic (e.g., host species) and abiotic effects on communities of *Endozoicomonas* and propose potential relationships of *Endozoicomonas* community with habitats or host species specificity and temporal variations, it is necessary to conduct large-scale surveys of *Endozoicomonas* community in various environments, habitats, or hosts and monitor the temporal transition of the *Endozoicomonas* community in long-term studies.

In addition, detailed molecular and physiological experiments, based on genomic information, should be conducted to clarify physiological functions or ecological roles of *Endozoicomonas*. These experiments should not only be done on cultivable-type strains in vitro but also be examined in vivo, i.e., inside host cells. These studies will provide insights into coral microbiology, as well as a big impetus to conduct additional studies of *Endozoicomonas* and interactions with their hosts.

5.6 Summary

(1) *Endozoicomonas* bacteria are highly diverse and widely dispersed across various hosts, geographies, and times.
(2) The relative abundance of *Endozoicomonas* is affected by habitat, environmental stress, or health of their host, suggesting *Endozoicomonas* spp. may have habitat or host specificity.
(3) Based on genomic results, *E. montiporae* may be a facultative symbiont.
(4) We hypothesize that occurrences of *Endozoicomonas* populations in marine invertebrates are not randomly distributed. Various marine invertebrates may serve as distinct niches for certain *Endozoicomonas* spp., whereas specific *Endozoicomonas* sp. has separate ecological roles.
(5) Standardization of the strategy and methods to characterize *Endozoicomonas* is essential for effective comparative analyses.

Acknowledgments We appreciate the financial support from Academia Sinica and Ministry of Science and Technology, Taiwan (Grant MOST 101-2628-B-001-001-MY3). We are also thankful for the helpful suggestions and comments of Dr. Jiun-Yan Ding.

References

1. Kurahashi M, Yokota A. *Endozoicomonas elysicola* gen. Nov., sp nov., a gammaproteobacterium isolated from the sea slug *Elysia ornata*. Syst Appl Microbiol. 2007;30:202–6.
2. Jensen S, Duperron S, Birkeland NK, Hovland M. Intracellular *Oceanospirillales* bacteria inhabit gills of *Acesta* bivalves. FEMS Microbiol Ecol. 2010;74:523–33.
3. Zurel D, Benayahu Y, Or A, Kovacs A, Gophna U. Composition and dynamics of the gill microbiota of an invasive Indo-Pacific oyster in the eastern Mediterranean Sea. Environ Microbiol. 2011;13:1467–76.
4. Jessen C, Lizcano JFV, Bayer T, Roder C, Aranda M, Wild C, et al. In-situ effects of eutrophication and overfishing on physiology and bacterial diversity of the Red Sea coral *Acropora hemprichii*. PLoS One. 2013;8:e62091.
5. Dishaw LJ, Flores-Torres J, Lax S, Gemayel K, Leigh B, Melillo D, et al. The gut of geographically disparate *Ciona intestinalis* harbors a core microbiota. PLoS One. 2014;9:e93386.
6. Lawler SN, Kellogg CA, France SC, Clostio RW, Brooke SD, Ross SW. Coral-associated bacterial diversity is conserved across two deep-sea *Anthothela* species. Front Microbiol. 2016;7:458.
7. Yang CS, Chen MH, Arun AB, Chen CA, Wang JT, Chen WM. *Endozoicomonas montiporae* sp. Nov., isolated from the encrusting pore coral *Montipora aequituberculata*. Int J Syst Evol Microbiol. 2010;60:1158–62.
8. Nishijima M, Adachi K, Katsuta A, Shizuri Y, Yamasato K. *Endozoicomonas numazuensis* sp. nov., a gammaproteobacterium isolated from marine sponges, and emended description of the genus *Endozoicomonas* kurahashi and yokota 2007. Int J Syst Evol Microbiol. 2013;63:709–14.
9. Pike RE, Haltli B, Kerr RG. Description of *Endozoicomonas euniceicola* sp. nov. and *Endozoicomonas gorgoniicola* sp. nov., bacteria isolated from the octocorals *Eunicea fusca* and *plexaura* sp., and an emended description of the genus *Endozoicomonas*. Int J Syst Evol Microbiol. 2013;63:4294–302.
10. Hyun DW, Shin NR, Kim MS, Oh SJ, Kim PS, Whon TW, et al. *Endozoicomonas atrinae* sp. nov., isolated from the intestine of a comb pen shell *Atrina pectinata*. Int J Syst Evol Microbiol. 2014;64:2312–8.
11. Appolinario LR, Tschoeke DA, Rua CPJ, Venas T, Campeao ME, Amaral GRS, et al. Description of *Endozoicomonas arenosclerae* sp. nov. using a genomic taxonomy approach. Anton Leeuw Int J G. 2016;109:431–8.
12. Katharios P, Seth-Smith HMB, Fehr A, Mateos JM, Qi WH, Richter D, et al. Environmental marine pathogen isolation using mesocosm culture of sharpsnout seabream: striking genomic and morphological features of novel *Endozoicomonas* sp. Sci Rep. 2015;5:17609.
13. Ding JY, Shiu JH, Chen WM, Chiang YR, Tang SL. Genomic insight into the host-endosymbiont relationship of *Endozoicomonas montiporae* CL-33(T) with its coral host. Front Microbiol. 2016;7:251.
14. Kvennefors ECE, Sampayo EM, Ridgway T, Barnes AC, Hoegh-Guldberg O. Bacterial communities of two ubiquitous Great Barrier Reef corals reveals both site- and species-specificity of common bacterial associates. PLoS One. 2010;5:e10401.
15. Meyer JL, Paul VJ, Teplitski M. Community shifts in the surface microbiomes of the coral *Porites astreoides* with unusual lesions. PLoS One. 2014;9:e100316.

16. Bayer T, Arif C, Ferrier-Pages C, Zoccola D, Aranda M, Voolstra CR. Bacteria of the genus *Endozoicomonas* dominate the microbiome of the Mediterranean gorgonian coral *Eunicella cavolini*. Mar Ecol Prog Ser. 2013;479:75–84.
17. Hansson L, Agis M, Maier C, Weinbauer MG. Community composition of bacteria associated with cold-water coral *Madrepora oculata*: within and between colony variability. Mar Ecol Prog Ser. 2009;397:89–102.
18. Du ZJ, Zhang WY, Xia HJ, Lu GQ, Chen GJ. Isolation and diversity analysis of heterotrophic bacteria associated with sea anemones. Acta Oceanol Sin. 2010;29:62–9.
19. Schuett C, Doepke H. Endobiotic bacteria and their pathogenic potential in cnidarian tentacles. Helgol Mar Res. 2010;64:205–12.
20. Fan L, Liu M, Simister R, Webster NS, Thomas T. Marine microbial symbiosis heats up: the phylogenetic and functional response of a sponge holobiont to thermal stress. ISME J. 2013;7:991–1002.
21. Montalvo NF, Davis J, Vicente J, Pittiglio R, Ravel J, Hill RT. Integration of culture-based and molecular analysis of a complex sponge-associated bacterial community. PLoS One. 2014;9:e90517.
22. Rua CP, Trindade-Silva AE, Appolinario LR, Venas TM, Garcia GD, Carvalho LS, et al. Diversity and antimicrobial potential of culturable heterotrophic bacteria associated with the endemic marine sponge *Arenosclera brasiliensis*. Peer J. 2014;2:e419.
23. Plante CJ, Coe KM, Plante RG. Isolation of surfactant-resistant bacteria from natural, surfactant-rich marine habitats. Appl Environ Microbiol. 2008;74:5093–9.
24. Roterman YR, Benayahu Y, Reshef L, Gophna U. The gill microbiota of invasive and indigenous spondylus oysters from the Mediterranean Sea and Northern Red Sea. Environ Microbiol Rep. 2015;7:860–7.
25. Chiellini C, Iannelli R, Verni F, Petroni G. Bacterial communities in polluted seabed sediments: a molecular biology assay in leghorn harbor. Sci World J. 2013;2013:165706.
26. van Bleijswijk JDL, Whalen C, Duineveld GCA, Lavaleye MSS, Witte HJ, Mienis F. Microbial assemblages on a cold-water coral mound at the se rockall bank (ne atlantic): interactions with hydrography and topography. Biogeosciences. 2015;12:4483–96.
27. La Riviére M, Roumagnac M, Garrabou J, Bally M. Transient shifts in bacterial communities associated with the temperate gorgonian *Paramuricea clavata* in the Northwestern Mediterranean Sea. PLoS One. 2013;8:e57385.
28. Littman RA, Willis BL, Pfeffer C, Bourne DG. Diversities of coral-associated bacteria differ with location, but not species, for three acroporid corals on the Great Barrier Reef. FEMS Microbiol Ecol. 2009;68:152–63.
29. Morrow KM, Moss AG, Chadwick NE, Liles MR. Bacterial associates of two Caribbean coral species reveal species-specific distribution and geographic variability. Appl Environ Microbiol. 2012;78:6438–49.
30. Séré MG, Tortosa P, Chabanet P, Turquet J, Quod JP, Schleyer MH. Bacterial communities associated with *Porites* white patch syndrome (PWPS) on three western Indian Ocean (WIO) coral reefs. PLoS One. 2013;8:e83746.
31. Bayer T, Neave MJ, Alsheikh-Hussain A, Aranda M, Yum LK, Mincer T, et al. The microbiome of the Red Sea coral *Stylophora pistillata* is dominated by tissue-associated *Endozoicomonas* bacteria. Appl Environ Microbiol. 2013;79:4759–62.
32. Chen CP, Tseng CH, Chen CA, Tang SL. The dynamics of microbial partnerships in the coral *Isopora palifera*. ISME J. 2011;5:728–40.
33. Ransome E, Rowley SJ, Thomas S, Tait K, Munn CB. Disturbance to conserved bacterial communities in the cold-water gorgonian coral *Eunicella verrucosa*. FEMS Microbiol Ecol. 2014;90:404–16.
34. Lema KA, Willis BL, Bourne DG. Amplicon pyrosequencing reveals spatial and temporal consistency in diazotroph assemblages of the *Acropora millepora* microbiome. Environ Microbiol. 2014;16:3345–59.

35. Pantos O, Bongaerts P, Dennis PG, Tyson GW, Hoegh-Guldberg O. Habitat-specific environmental conditions primarily control the microbiomes of the coral *Seriatopora hystrix*. ISME J. 2015;9:1916–27.
36. Roder C, Bayer T, Aranda M, Kruse M, Voolstra CR. Microbiome structure of the fungid coral *Ctenactis echinata* aligns with environmental differences. Mol Ecol. 2015;24:3501–11.
37. Tout J, Siboni N, Messer LF, Garren M, Stocker R, Webster NS, et al. Increased seawater temperature increases the abundance and alters the structure of natural *vibrio* populations associated with the coral *Pocillopora damicomis*. Front Microbiol. 2015;6:432.
38. Webster NS, Negri AP, Botte ES, Laffy PW, Flores F, Noonan S, et al. Host-associated coral reef microbes respond to the cumulative pressures of ocean warming and ocean acidification. Sci Rep. 2016;6:19324.
39. Ziegler M, Roik A, Porter A, Zubier K, Mudarris MS, Ormond R, et al. Coral microbial community dynamics in response to anthropogenic impacts near a major city in the central Red Sea. Mar Pollut Bull. 2016;105:629–40.
40. Bourne D, Iida Y, Uthicke S, Smith-Keune C. Changes in coral-associated microbial communities during a bleaching event. ISME J. 2008;2:350–63.
41. Vezzulli L, Pezzati E, Huete-Stauffer C, Pruzzo C, Cerrano C. 16S rDNA pyrosequencing of the Mediterranean gorgonian *Paramuricea clavata* reveals a link among alterations in bacterial holobiont members, anthropogenic influence and disease outbreaks. PLoS One. 2013;8:e67745.
42. Glasl B, Herndl GJ, Frade PR. The microbiome of coral surface mucus has a key role in mediating holobiont health and survival upon disturbance. ISME J. 2016;10:2280–92.
43. Lee OO, Yang JK, Bougouffa S, Wang Y, Batang Z, Tian RM, et al. Spatial and species variations in bacterial communities associated with corals from the Red Sea as revealed by pyrosequencing. Appl Environ Microbiol. 2012;78:7173–84.
44. Raina JB, Tapiolas D, Willis BL, Bourne DG. Coral-associated bacteria and their role in the biogeochemical cycling of sulfur. Appl Environ Microbiol. 2009;75:3492–501.
45. Bourne DG, Dennis PG, Uthicke S, Soo RM, Tyson GW, Webster N. Coral reef invertebrate microbiomes correlate with the presence of photosymbionts. ISME J. 2013;7:1452–8.
46. Neave MJ, Michell CT, Apprill A, Voolstra CR. Whole-genome sequences of three symbiotic *Endozoicomonas* strains. Genome Announc. 2014;2:4.
47. Kyrpides NC, Hugenholtz P, Eisen JA, Woyke T, Goker M, Parker CT, et al. Genomic encyclopedia of bacteria and archaea: sequencing a myriad of type strains. PLoS Biol. 2014;12:e1001920.

Chapter 6
Microbes in Gorgonian and Soft Corals

Xiao-Yong Zhang and Shu-Hua Qi

Contents

Abstract Corals provide a structurally and environmentally complex array of habitats, supporting a broad microbial diversity that influences both host physiology and ultimately ecosystem processes. Many studies have indicated that microbial communities occupy a range of niches on corals, from within the surface mucus layer to on and within the coral tissue layers. Furthermore, a variety of studies have suggested that coral-associated microorganisms may be saprophytic or pathogenic or may provide other important functions for corals. In the previous studies, most of them mainly focused on the structure and diversity of microbes from gorgonian corals; however, information about the microbial diversity in soft corals is relatively limited. So in this chapter, we mainly summarized the structure and diversity of microbes from gorgonian corals.

Keywords Microbiomes · Gorgonian coral · Soft corals · Symbionts

---

X.-Y. Zhang · S.-H. Qi (✉)
CAS Key Laboratory of Tropical Marine Bio–resources and Ecology, South China Sea Institute of Oceanology, Chinese Academy of Sciences, Guangzhou, China
e-mail: shuhuaqi@scsio.ac.cn

Z. Li (ed.), *Symbiotic Microbiomes of Coral Reefs Sponges and Corals*,
https://doi.org/10.1007/978-94-024-1612-1_6

6.1 Phylogeny of Prokaryotic Symbionts

Recently, an increasing number of studies on the diversity of bacteria have reported that the bacterial communities in gorgonian corals are very diverse and differ from the bacterial communities in the surrounding environment [1, 2]. The bacterial communities in healthy gorgonian corals were mainly composed of these phyla of *Actinobacteria*, *Bacteroidetes*, *Firmicutes*, *Fusobacteria*, *Lentisphaerae*, *Planctomycetes*, *Proteobacteria* (classes *Alpha-*, *Beta-*, *Delta-*, and *Gammaproteobacteria*), *Tenericutes*, and *Verrucomicrobia* [2–4] (Fig. 6.1). The bacterial class *Gammaproteobacteria* was consistently found in almost all of these gorgonian corals reported. Among this class, the genus *Endozoicomonas* was the most dominant bacteria [5]. The bacterial genus *Endozoicomonas* was frequently found in marine invertebrates, and the species *Endozoicomonas elysicola* was firstly recovered in the sea slug *Elysia ornata* off the coast of Izu-Miyake Island, Japan [6]. Interestingly, *Endozoicomonas* has been found to represent the dominant bacterial genus associated with gorgonian coral *Eunicella cavolini* from the Mediterranean [7]. Phyla *Actinobacteria* and *Firmicutes* are frequently found in gorgonian corals, in both shallow [8] and deep water [9], which also are a target for novel or bioactive natural product research [10]. In a recent study, Zhang et al. [8] investigated the diversity and antibacterial activity of culturable *Actinobacteria* from the South China Sea gorgonian corals (*Echinogorgia aurantiaca*, *Melitodes squamata*, *Muricella flexuosa*, *Subergorgia suberosa*, and *Verrucella umbraculum*). A total of 123 *Actinobacteria* isolates were recovered and assigned to 11 genera. Antibacterial activities of all the isolates against three marine pathogenic bacteria were tested using a double-layer technique. It was found that 42 isolates displayed antibacterial activity against at least 1 marine pathogenic bacterium. Peng et al. [11] reported

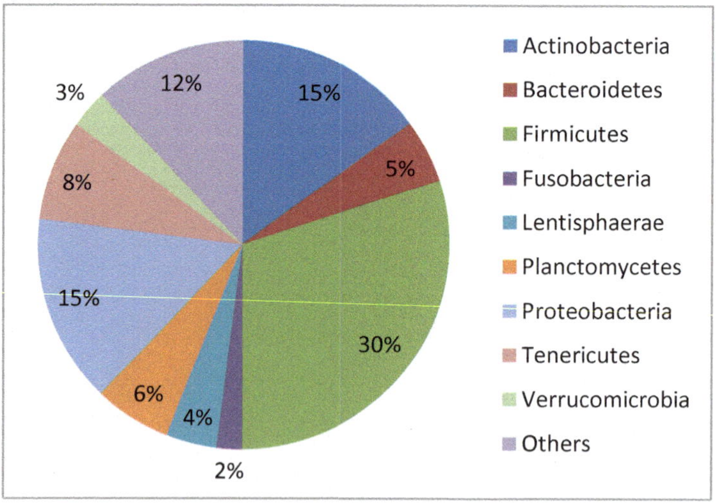

Fig. 6.1 Bacterial diversity of gorgonian corals

many culturable *Firmicutes* isolates from several gorgonian corals collected from the South China Sea that showed distinct antibacterial activity against two marine pathogenic bacteria. These studies suggested that many gorgonian-associated bacteria could produce some antimicrobial agents to protect their hosts against pathogens.

Bacterial communities in diseased gorgonian corals differ from that in healthy conditions [12]. Vezzulli et al. [1] investigated the bacterial communities in purple gorgonian coral *Paramuricea clavata* from different geographic locations and depth. Their results showed that the bacterial communities in healthy gorgonian coral *P. clavata* in pristine locations were dominated by *Endozoicomonas* within the order *Oceanospirillales* (belonged to the phylum of *Proteobacteria*). Gorgonian coral *P. clavata* collected in human-impacted areas and during disease events had more diverse bacterial communities and abundance of disease-related bacteria than those gorgonian coral samples collected in pristine locations while showing a reduced dominance of *Endozoicomonas* spp. In another study, Ransome et al. [2] found that the phyla *Cyanobacteria* and *Chloroflexi* that could be recovered in diseased gorgonian coral *Eunicella verrucosa* were not present in their healthy counterpart. Harvell et al. [13] found that the cyanobacterium *Scytonema* sp. had been associated with a disease gorgonian corals (*Briareum asbestinum*) from the Caribbean during a bleaching event. Cerrano et al. [14] reported extensive gorgonian coral mortality during unusually warm water conditions in 1999, which Martin et al. [15] linked to infection by *Vibrio* spp. bacteria at elevated temperatures. In another study, Hall-Spencer et al. [16] isolated 15 strains of *Vibrio splendidus* and 2 other closely matched *V. tasmaniensis* from the diseased gorgonian (*E. cavolini*) tissues. These *Vibrio* spp. found in diseased gorgonian corals produced proteolytic and cytolytic enzymes that damaged the tissues of gorgonian corals and may be responsible for the necrosis observed [16].

In contrast to gorgonian corals, information on the bacterial diversity in soft corals is relatively limited. To date, most of these studies on the bacterial communities in soft corals have employed culture-based techniques. A novel *Erythrobacter* species has been isolated from an unidentified soft coral [17], and a *Flavobacterium* species has been cultured from the soft coral *Paragorgia arborea* [18]. In a recent study, Chen et al. [19] isolated 1526 and 1138 heterotrophic bacteria from eight field-collected and five cultured soft coral species. Unfortunately, only 19 isolates with high antimicrobial activity were identified, which belonged to seven genera: *Marinobacterium*, *Pseudoalteromonas*, *Vibrio*, *Enterovibrio*, *Tateyamaria*, *Labrenzia*, and *Pseudovibrio*.

No comprehensive molecular surveys of microbial communities associated with a soft coral species currently exist [20]. The majority of bacterial communities in soft corals is the *Gammaproteobacteria*, although other phylogenetic groups including *Alpha-* and *Betaproteobacteria*, *Actinobacteria*, *Bacteroidetes*, *Firmicutes*, *Planctomycete*, and *Chlorobi* and bacteria from the functional group of sulfate-reducing bacteria are present [20].

6.2 Phylogeny of Eukaryotic Symbionts

Some fungi in gorgonian corals are now known to cause gorgonian diseases. Since Smith et al. [21] initially reported that *Aspergillus sydowii* was the causal agent of the aspergillosis of gorgonian corals (*G. ventalina* and *G. flabellum*) in the Caribbean, a few studies have found that widespread diseases and mortality in Caribbean and Western Indian gorgonian corals were also caused by the fungus *A. sydowii* [22, 23]. In addition, a recent disease outbreak of *A. versicolor*, affecting 250 fallen gorgonian corals in Southeast Asia, indicated that other fungal species were susceptible to gorgonian coral diseases [24]. However, some of the fungi that have been identified in diseased colonies and implicated in gorgonian diseases have also been found in healthy gorgonian corals, suggesting that they are components of healthy microbial communities [25, 26]. Koh et al. [25] recovered 51 fungal species from 10 different gorgonian coral species in Singapore by conventional culture-dependent methods. Toledo-Hernandez et al. [26] investigated the fungal communities from gorgonian coral *G. ventalina* from Puerto Rico and isolated 15 species of terrestrial fungi. In a further study, Toledo-Hernandez et al. [27] compared fungal communities in aspergillosis-affected and healthy gorgonian coral *G. ventalina* collected from 15 reefs around Puerto Rico by sequencing the fungal ITS (internal transcribed spacer) sequences. Thirty fungal species belonged to 15 genera were recovered from 203 *G. ventalina* colonies. Genera *Aspergillus* and *Penicillum* were the most common fungi recovered from diseased and healthy gorgonian corals. However, the fungal communities in healthy gorgonian corals were distinct and more diverse than that in diseased ones. In diseased gorgonian corals, the fungal communities in diseased tissues were different and more diverse than that in healthy tissues. The reduction of the fungal communities in diseased gorgonian coral colonies may occur prior to infection due to environmental changes affecting the host or after infection due to increase in dominance of the pathogen or because of host responses to infection [27]. In a recent study, Zhang et al. [28] described the diversity of culturable fungi isolated from six species of healthy gorgonian corals, collected in shallow water of the South China Sea. At least 19 fungal genera and 41 species isolated from the six gorgonians were identified. Of these, 12 genera and 30 species are new reports for marine gorgonian corals. These studies, including Koh et al. [25], Toledo-Hernandez et al. [26, 27], Zuluaga-Montero et al. [29], Wang et al. [30], and Zhang et al. [28], have shown that gorgonian corals have large and diverse fungal communities. To date, 43 fungal genera and 120 species had previously been identified from 16 gorgonian coral species all over the world using culture-dependent techniques (Fig. 6.2, Table 6.1) [25–32]. Of these 43 fungal genera, *Aspergillus* and *Penicillium* were the most diverse and common in gorgonians. These could be isolated from 12 of 16 healthy gorgonian species.

Despite the numerous fungal species successfully isolated from different species of gorgonians by culture-dependent techniques, greater fungal diversity is likely to be revealed by combining the culture-dependent approach and with a culture-independent method (such as direct amplification of DNA from gorgonians).

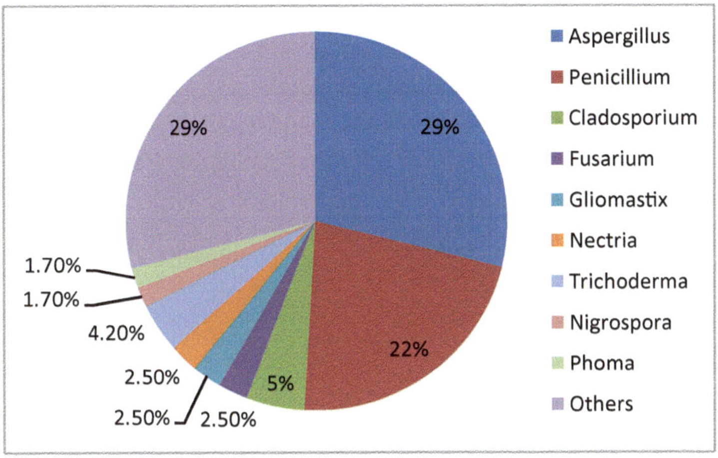

Fig. 6.2 Relative abundance of fungal genera from gorgonian corals

Recently, an increasing number of studies have indicated that comparisons between fungal community compositions obtained by culture-dependent and culture-independent methods highlight different fungi, emphasizing the need of complementary approaches to assess the fungal assemblage within unusual environments [33–36]. Furthermore, greater fungal diversity would probably be obtained by a rigorous sampling strategy. Toledo-Hernandez et al. [26] reported that when sampling the fungal communities in gorgonian corals, four points should be considered: sample size, fragment size, use of smaller tissue fragments that increases the number and frequency of fungi isolated and minimizes damage to the gorgonians, and the method of tissue processing and tissue collecting that influences the results.

6.3 Core Microbiome and Host-Specific Symbionts: Vertical Inheritance and Horizontal Transfer

The communities of microbes in gorgonian corals appear to be host species-specific and differ from those dominating the surrounding reef water. Bayer et al. [7] investigated the bacterial composition and relative abundances in gorgonian coral *Eunicella cavolini* from the French Mediterranean coast by pyrosequencing of 16S rDNA. Their results showed that the composition and abundance of the bacterial communities in gorgonian coral *E. cavolini* from the Mediterranean Sea differed from that in the surrounding water. The microbiome of gorgonian coral *E. cavolini* contained only a few shared species and that it was highly dominated by *Endozoicomonas* spp. [7]. In the Caribbean gorgonian *G. ventalina*, the most common bacteria were also *Endozoicomonas* spp. [37]. In a recent study, ordination by multidimensional scaling (MDS) confirmed that gorgonian coral *P. clavata*

Table 6.1 Fungal species isolated from different gorgonian corals

Fungal species	Host/gorgonian coral species[a]	References
Aspergillus aculeatus	SS, JSP, GV	[25, 27]
A. carneus	MS	[28]
A. cervinus	JG, CU	[25]
A. ficuum	SS, SM, ESP	[25]
A. flavipes	ER	[30]
A. flavus	DG, EP, GV	[25, 27, 28]
A. foetidus	JSP, CU	[25]
A. fumigatus	DG, EA, SS	[28]
A. furcatum	ESP, EP	[25]
A. gracilis	DG	[28]
A. insulicola	MS	[28]
A. kangawensis	SS	[25]
A. melleus	GV	[27]
A. niger	EA, GV	[26–28]
A. nomius	DG	[28]
A. nutans	ESP	[25]
A. ochraceopetaliformis	DG, MS	[28]
A. ochraceus	GV	[27]
A. ochraceus	CU	[25]
A. ornatus	JSP, CU, ESP, EP	[25]
A. penicillioides	DG, MF	[28]
A. pulverulentus	SS	[25]
A. sclertoiorum	DG, MF, VU	[28]
A. strictum	JSP, CU, EP	[25]
A. sydowii	DG, EA, MS, SS, VU, GV	[26–28]
A. tamarii	GV	[27]
A. terreus	EA, MS, MF, SS, GV	[27, 28]
A. terricola	CU	[25]
A. tubingensis	EA, MS, MF, SS	[28]
A. unguis	GV	[26, 27]
A. ustus	GV	[26, 27]
A. versicolor	MS, SS, VU, ER, GV	[27, 28, 30]
A. wentii	ESP	[25]
A. westerdijkiae	ER	[30]
Acremonium butryi	CU	[25]
Acremonium polychromum	VU	[28]
Alternaria alternata	ER	[30]
Ascochyta manawaorae	VU	[28]
Cladosporium cladosporioides	DG, ER, GV	[27, 28, 30]

(continued)

Table 6.1 (continued)

Fungal species	Host/gorgonian coral species[a]	References
C. cucumerinum	ER	[30]
C. musae	SS, ESP, EP	[25]
C. sphaerospermum	SS, SM, JSP, JG, CU, ESP, EP, GV, ME, ER	[25–28]
C. uredinicola	ER	[30]
C. uredinicola	VU	[28]
Cladosporium sp.	GV	[27]
Candida sp.	GV	[27]
Chaetophoma sp.	ESP	[25]
Chalaropsis sp.	GV	[27]
Cochliobolus lunatus	DG	[31]
Davidiella tassiana	GV	[27]
Debaryomyces subglobosus	DG, SS, VU	[28]
Fusarium chlamydosporum	MS	[28]
F. proliferatum	DG	[28]
Fusarium sp.	SS, EP	[25]
Gliomastix luzulae	ESP	[25]
G. cerealis	ESP	[25]
G. murorum	SS	[25]
Gloeotiniatemulenta	GV	[26, 27]
Helotiaceae	GV	[27]
Hymenula sp.	SS	[25]
Hypocrea lixii	GV	[27]
Isaria tenuipes	MF	[28]
Lecanicillium fusisporum	DG	[28]
Massarina corticola	MF, SS	[28]
Microascus triganosporus	EP	[25]
Microsphaeropsis arundinis	MS, MF, VU	[28]
Myrmecridium schulzeri	EA	[28]
Myrothecium inundatum	MS	[28]
Nectria gliocladioides	GV	[27]
N. haematococca	GV, ER	[27, 30]
Nectria sp.	GV	[27]
Nigrospora oryzae	VU	[28]
Nigrospora sp.	ER	[30]
Oidiodendron griseum	JSP, ESP	[25]
Penicillium brevicompactum	EP	[25]
P. camemberti	SM, ESP, EP	[25]
P. canescens	ESP, EP	[25]
P. chrysogenum	GV, ER	[27, 30]
P. citreonigum	GV	[26, 27]

(continued)

Table 6.1 (continued)

Fungal species	Host/gorgonian coral species[a]	References
P. citrinum	SS, JG, CU, ESP, EP, GV, DG, EA, MF, MS, SS, VU	[25–28]
P. coffeae	GV	[26]
P. commune	GV	[27]
P. crustosum	ER	[30]
P. decumbens	JG	[25]
P. frequentans	SS, ESP, EP	[25]
P. glabrum	ER	[30]
P. implicatum	SS, SM, CU, ESP, EP	[25]
P. janthinellum	SS, SM, JSP, JG, CU, ESP, EP	[25]
P. lanoso	CU	[25]
P. lilacinum	SS, ESP	[25]
P. minioluteum	GV	[27]
P. notatum	CU	[25]
P. oxalicum	DG, EA, MF, SS, VU	[28]
P. paxilli	SS	[28]
P. polonicum	ER	[30]
P. putaminum	VU	[28]
P. radicum	MS	[28]
P. steckii	GV, DG, ESP	[25, 26, 28]
P. verruculosum	MS	[28]
P. oxalicum	SS	[25]
Paecilomyces lilacinus	DG, SS	[28]
Paraconiothyrium brasiliense	DG, MF	[28]
Phoma	JSP	[25]
Phoma glomerata	DG	[28]
Pichia guilliermondii	GV	[27]
Ramichloridium apiculatum	DG, SS	[28]
Rhodotorula nymphaeae	GV	[26]
Rhodotorula nymphaeae	GV	[27]
Scolecobasidium humicola	SS, SM, JSP, JG, ESP, EP	[25]
Scopulariopsis sp.	CSP	[32]
Sporotrichum sp.	SS	[25]
Stachybotrys chartarum	GV	[26, 27]
Stachybotrys chlorohalonata	GV	[27]
Tilletiopsis albescens	MF, SS	[28]
Trichoderma harzianum	SS, ESP, EP, GV	[25, 27]
T. koningii	SS	[25]
T. longibrachiatum	SS, ESP, EP	[25]
T. pseduokoningii	SS, SM, ESP, EP	[25]
Tritirachium sp.	SS, ESP, GV	[25, 27]
Verticillium sp.	ESP, EP	[25]

(continued)

Table 6.1 (continued)

Fungal species	Host/gorgonian coral species[a]	References
Virgaria sp.	JG, EP	[25]
Xylaria hypoxylon	GV	[26, 27]
Black yeast	SM, JSP, JG, CU, ESP, EP	[25]
White yeast	SM, ESP, EP	[25]

[a]Gorgonian coral species: CSP *Carijoa* sp., CU *Ctenocella* cf. *umbraculum*, DG *Dichotellagemmacea*, EA *Echinogorgia aurantiaca*, EP *Euplexaura pinnata*, ESP *Echinogorgia* sp., GV *Gorgonia ventalina*, JG *Junceella gemmacea*, JSP *Junceella* sp., MF *Muricella flexuosa*, MS *Melitodes squamata*, SM *Subergorgia mollis*, SS *Subergorgia suberosa*, VU *Verrucella umbraculum*

collected from different areas of the Northwestern Mediterranean harbored closely related microbial communities [38]. Bayer et al. [7] compared the bacterial community composition of gorgonian corals from three different depths (24, 30, 41 m); their results showed that the gorgonian corals provided a very stable environment to the bacterial communities that did not change with water depth. However, there might be greater changes in bacterial communities as the depth increased beyond 41 m [7].

Riviere et al. [38] investigated the seasonality of the bacterial communities in gorgonian coral *P. clavata* sampled at three different study sites (in the Northwestern Mediterranean) in the winter and the summer of 2007 by analyzing the 16S rDNA sequences. Their results showed that the bacterial compositions differed greatly between the two seasons. The bacterial communities in gorgonian coral *P. clavata* sampled in the winter were dominated by the sequences of a unique bacterium affiliated with *Gammaproteobacteria*. In summer, a larger diversity of bacteria was detected in gorgonian coral *P. clavata*, and the bacterial communities were dominated by the members of the phylum *Firmicutes* [38]. Furthermore, the bacterial communities were no significant differences between the gorgonian coral samples from winter 2007 and winter 2008 [38]. Although the roles of associations between corals and bacteria are still unknown, understanding how the coral-associated bacteria change through time may represent a potential indicator of the coral health status. A recent study on microbial communities in tropical corals has demonstrated the shift of the microbial communities in affected coral colonies prior to visual signs of disease [12]. Thus, these changes in the normal microbial communities could serve as an early signal of stressful environmental conditions.

Acknowledgment This chapter was modified from the paper published by our group in Microbial Ecology (Zhang et al. 2012; 64:617–627) and *World Journal of Microbiology and Biotechnology* (Zhang et al. 2013; 29:1107-1116). The related contents are reused with the permission.

References

1. Vezzulli L, Pezzati E, Huete-Stauffer C, Cerrano C. 16S rDNA pyrosequencing of the Mediterranean gorgonian *Paramuricea clavata* reveals a link among alterations in bacterial holobiont members, anthropogenic influence and disease outbreaks. PLoS One. 2013;8:e67745.
2. Ransome E, Rowley SJ, Thomas S, Tait K, Munn BC. Disturbance to conserved bacterial communities in the cold-water gorgonian coral *Eunicella verrucosa*. FEMS Microbiol Ecol. 2014;90:404–16.
3. Correa H, Haltli B, Duque C, Kerr R. Bacterial communities of the gorgonian octocoral *Pseudopterogorgia elisabethae*. Microb Ecol. 2013;66:972–85.
4. Gray MA, Stone RP, Mclaughlin MR, Kellogg AC. Microbial consortia of gorgonian corals from the Aleutian Islands. FEMS Microbiol Ecol. 2011;76:109–20.
5. Raina J, Dinsdale EA, Willis BL, Bourne GD. Do the organic sulfur compounds DMSP and DMS drive coral microbial associations. Trends Microbiol. 2010;18:101–8.
6. Kurahashi M, Yokota A. *Endozoicomonas elysicola* gen. Nov., sp. nov., a c-proteobacterium isolated from the sea slug *Elysia ornate*. Syst Appl Microbiol. 2007;30:202–6.
7. Bayer T, Arif C, Ferrier-Pages C, Zoccola D, Aranda M, Voolstra RC. Bacteria of the genus Endozoicomonas dominate the microbiome of the Mediterranean gorgonian coral *Eunicella cavolini*. Mar Ecol Prog Ser. 2013;479:75–84.
8. Zhang XY, He F, Wang GH, Bao J, Xu XY, Qi SH. Diversity and antibacterial activity of culturable actinobacteria isolated from five species of the South China Sea gorgonian corals. World J Microbiol Biotechnol. 2013;29:1107–16.
9. Penn K, Wu D, Eisen JA, Ward N. Characterization of bacterial communities associated with deep-sea coral on Gulf of Alaska seamounts. Appl Environ Microbiol. 2006;72:1680–3.
10. Bull AT, Stach EMJ. Marine Actinobacteria: new opportunities for natural product search and discovery. Trends Microbiol. 2007;15:491–9.
11. Peng J, Zhang XY, Xu XY, He F, Qi SH. Diversity and chemical defense role of culturable non-actinobacterial bacteria isolated from the South China Sea gorgonians. J Microbiol Biotechnol. 2013;23:437–43.
12. Gil-Agudelo DL, Myers C, Smith GW, Kim K. Changes in the microbial communities associated with *Gorgonia ventalina* during aspergillosis infection. Dis Aquat Org. 2006;69:89–94.
13. Harvell CD, Kim K, Quirolo C, Weir J, Smith G. Coral bleaching and disease: contributors to 1998 mass mortality in *Briareum asbestinum* (*Octocorallia, Gorgonacea*). Hydrobiologia. 2001;460:97–104.
14. Cerrano C, Bavestrello G, Bianchi CN, Cattaneo-Vietti R, Bava S, Morganti C, et al. A catastrophic mass-mortality episode of gorgonians and other organisms in the Ligurian Sea (Northwestern Mediterranean), summer 1999. Ecol Lett. 2000;3:284–93.
15. Martin Y, Bonnefort JL, Chancerelle L. Gorgonians mass mortality during the 1999 late summer in French Mediterranean coastal waters: the bacterial hypothesis. Water Res. 2002;36:779–82.
16. Hall-Spencer J, Allain V, Fossa JH. Trawling damage to Northeast Atlantic ancient coral reefs. Proc R Soc Lond B. 2002;269:507–11.
17. Ivanova EP, Bowman JP, Lysenko AM, Zhukova NV, Gorshkova NM, Kuznetsova TA, et al. *Erythrobacter vulgaris* sp. nov., a novel organism isolated from the marine invertebrates. Syst Appl Microbiol. 2005;28:123–30.
18. Nedashkovskaya OI, Kim SB, Lysenko AM, Frolova GM, Mikhailov VV, Bae KS. *Bizionia paragorgiae* gen. Nov., sp. Nov., a novel member of the family Flavobacteriaceae isolated from the soft coral Paragorgia arborea. Int J Syst Evol Micr. 2005;55:375–8.
19. Chen YH, Kuo J, Sung PJ, Chang YC, Lu MC, Wong TY, et al. Isolation of marine bacteria with antimicrobial activities from cultured and field-collected soft corals. World J Microbiol Biotechnol. 2012;28:3269–79.
20. Webster NS, Bourne D. Bacterial community structure associated with the Antarctic soft coral, *Alcyonium antarcticum*. FEMS Microbiol Ecol. 2007;59:81–94.

21. Smith GW, Ives LD, Nagelkerken IA, Ritchie BK. Caribbean Sea fan mortalities. Nature. 1996;383:487.
22. Nagelkerken IK, Smith GW, Bonair K, Bush P, Garzón-Ferriera J, Botero L, et al. Widespread disease in Caribbean Sea fans: II. Patterns of infection and tissue loss. Mar Ecol Prog Ser. 1997;160:255–63.
23. Geiser DM, Taylor JW, Ritchie KB, Smith WG. Cause of sea fan death in the West Indies. Nature. 1998;394:137–8.
24. Sakayaroj J, Benzies C, Chuaypat J, Plathong S. Aspergillosis of the gorgonian sea fan *Annella* sp. after the 2004 tsunami at Mu Ko Similan National Park, Andaman Sea, Thailand. Coral Reefs. 2006;25:296.
25. Koh LL, Tan TK, Chou LM, Goh NKC Fungi associated with gorgonians in Singapore. In: Proceedings of the 9th International Coral Reef Symposium. 2000, vol. 1, pp. 521–526.
26. Toledo-Hernandez C, Bones-Gonzalez A, Oritz-Vazquez OE, Sabat AM, Bayman P. Fungi in the sea fan *Gorgonia ventalina*: diversity and sampling strategies. Coral Reefs. 2007;26:725–30.
27. Toledo-Hernandez C, Zuluaga-Montero A, Bones-González A, Sabat AM, Bayman P. Fungi in healthy and diseased sea fans (*Gorgonia ventalina)*: Is *Aspergillus sydowii* always the pathogen. Coral Reefs. 2008;27:707–14.
28. Zhang XY, Bao J, Wang GH, He F, Xu XY, Qi SH. Diversity and antimicrobial activity of culturable fungi isolated from six species of the South China Sea gorgonians. Microb Ecol. 2012;64:617–27.
29. Zuluaga-Montero A, Toledo-Hernandez C, Rodriguez JA, Sabat AM, Bayman P. Spatial variation in fungal communities isolated from healthy and diseased sea fans *Gorgonia ventalina* and seawater. Aquat Biol. 2010;8:151–60.
30. Wang YN, Shao CL, Zheng CJ, Chen YY, Wang CY. Diversity and antibacterial activities of fungi derived from the gorgonian *Echinogorgia rebekka* from the South China Sea. Mar Drug. 2011;9:1379–90.
31. Shao CL, Wu HX, Wang CY, Liu QA, Xu Y, Wei MY, et al. Potent antifouling resorcylic acid lactones from the gorgonian-derived fungus *Cochliobolus lunatus*. J Nat Prod. 2011;74:629–33.
32. Shao CL, Xu RF, Wei MY, She ZG, Wang CY. Structure and absolute configuration of fumiquinazoline L, an alkaloid from a gorgonian-derived *Scopulariopsis* sp. fungus. J Nat Prod. 2013;76:779–82.
33. Sette LD, Passarini MRZ, Rodrigues A, Leal RR, Simioni KCM, Nobre FS, et al. Fungal diversity associated with Brazilian energy transmission towers. Fungal Divers. 2010;44:53–63.
34. Liu WC, Li CQ, Zhu P, Yang JL, Cheng KD. Phylogenetic diversity of culturable fungi associated with two marine sponges: *Haliclona simulans* and *Gelliodes carnosa*, collected from Hainan Island coastal waters of the South China Sea, Fungal Divers. 2010;42:1–15.
35. Mouhamadou B, Molitor C, Baptist F, Sage L, Clement JC, Lavorel S, et al. Differences in fungal communities associated to *Festuca paniculata* roots in subalpine grasslands. Fungal Divers. 2011;47:55–63.
36. Tejesvi MV, Kajula M, Mattila S, Pirttila AM. Bioactivity and genetic of endophytic fungi in *Rhododendron tomentosum* Harmaja. Fungal Divers. 2011;47:97–107.
37. Sunagawa S, Woodley CM, Medina M. Threatened corals provide underexplored microbial habitats. PLoS One. 2010;5:e9554.
38. Riviere ML, Roumagnac M, Garrabou J, Bally M. Transient shifts in bacterial communities associated with the temperate gorgonian *Paramuricea clavata* in the Northwestern Mediterranean Sea. PLoS One. 2013;8:e57385.

Chapter 7
Marine Sponge Holobionts in Health and Disease

Beate M. Slaby, Andrea Franke, Laura Rix, Lucia Pita, Kristina Bayer, Martin T. Jahn, and Ute Hentschel

Contents

Abstract Sponges—like all multicellular organisms—are holobionts, complex ecosystems comprising the host and its microbiota. The symbiosis of sponges with their microbial communities is a highly complex system, requiring interaction mechanisms and adaptation on both sides. The microbiome seems to rely on

B. M. Slaby (✉) · A. Franke · L. Pita · K. Bayer
GEOMAR Helmholtz Centre for Ocean Research, Marine Symbioses, Kiel, Germany
e-mail: bslaby@geomar.de

L. Rix
The University of Queensland, School of Biological Sciences, Brisbane, Australia

M. T. Jahn
GEOMAR Helmholtz Centre for Ocean Research, Marine Symbioses, Kiel, Germany

U. Hentschel
GEOMAR Helmholtz Centre for Ocean Research, Marine Symbioses, Kiel, Germany

Christian-Albrechts-University of Kiel, Kiel, Germany

Z. Li (ed.), *Symbiotic Microbiomes of Coral Reefs Sponges and Corals*,
https://doi.org/10.1007/978-94-024-1612-1_7

eukaryotic-like protein domains, such as ankyrins, modifications of the lipopolysaccharide structure, CRISPR-Cas, toxin-antitoxin, and restriction-modification systems, as well as secondary metabolism to communicate with the host and within the microbial community, evade phagocytosis, and defend itself against foreign DNA. Secondary metabolites produced by certain symbionts may even defend the entire holobiont against predators. On the other hand, the immune system of the sponge itself has evolved to discriminate not only between self and nonself but also between its associated microbiota and foreign microbes, such as food bacteria. Sponge holobionts are inextricably dependent on the surrounding environmental conditions due to their sessile nature. Thus, we discuss the link between environmental stress and sponge disease and dysbiosis, with a particular focus on the holobiont's response to ongoing global change. While some species may be the "winners of climate change," other species are adversely affected, e.g., by metabolic and immune suppression, as well as microbiome shifts resulting in loss of symbiotic functions. Hence, a much better understanding of sponge holobionts and the underlying molecular mechanisms of host-microbe interaction is required before the fate of sponge holobionts in a changing ocean can finally be validated.

Keywords Sponge · Porifera · Holobiont · Symbiosis · Microbial consortia · Interaction mechanisms · Disease · Dysbiosis · Environmental change

7.1 Introduction

7.1.1 Marine Sponges as Ancient Animal Hosts

Marine sponges (phylum Porifera) are the oldest extant multicellular animals with a fossil record dating back to the Precambrian [1–6]. Throughout Earth's history, sponges played an important role as reef builders, and they survived a number of global mass extinction events [7, 8]. Thus, they are an evolutionary successful phylum able to cope with very different and also changing conditions. To this day, they are present in a wide range of marine ecosystems from shallow tropical reefs to the deep sea, and they still dominate the benthic community in specific regions, e.g., deep sea sponge grounds [9, 10]. Sponges are highly diverse, spanning an estimated number of 15,000 species [11]. They differ in size from a few millimeters to meters, they show a range of growth forms from bowl- or vase-shaped to encrusting and branching, and they can have a wide variety of colors. Taxonomically, sponges are divided into the four classes Demospongiae, Calcarea, Hexactinellida, and Homoscleromorpha that differ in the materials used to build the spicules that form the sponge skeleton, the presence or absence of spongin fibers, the cell type and the body form [11, 12]. The majority of extant sponges are demosponges.

Marine sponges are among the structurally simplest multicellular organisms. The sponge body (except that of Hexactinellida) possesses two types of barrier-forming cell layers, namely, pinacoderm and choanoderm, that consist of pinacocyte and choanocyte cells, respectively [13]. The pinacoderm forms the outer surface of the sponge body and lines the aquiferous canal system, while the choanocyte cells are located in choanocyte chambers [14]. The space between the external pinacoderm and the canal system is filled by the mesohyl matrix that is mainly composed of collagen, galectin, and glycoconjugates [14]. The archaeocytes, totipotent cells that are phagocytotically active and amoeboid, move freely through the mesohyl [15]. While sponges do not contain muscles, a gut, or a nervous system, they possess epithelia that are able to seal and control their internal milieu [16, 17]. Additionally, sequencing data has revealed a hidden genetic complexity that is still largely unexplored regarding its functions [18–20].

The vast majority of sponges are filter-feeders that pump thousands of liters of seawater per kg sponge per day through inhalant pores (ostia) in their outer pinacoderm layer (Fig. 7.1) and through a system of canals into choanocyte chambers [21]. Specialized flagellated choanocyte cells create the water current by beating their flagella and capture food particles out of the water [21, 22]. The food particles are moved from the aquiferous system into the mesohyl interior where they are phagocytosed by the archaeocyte cells [23]. The filtered water that is deprived of up to

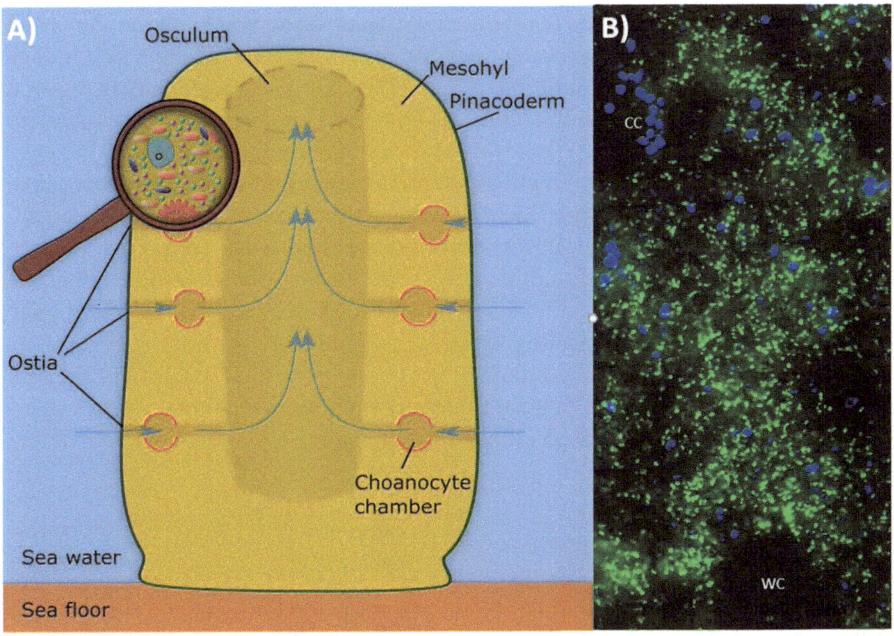

Fig. 7.1 (**a**) Schematic cross section through a demosponge. Blue arrows indicate water flow produced by choanocyte cells lining choanocyte chambers (red). The magnifying glass indicates a zoom-in on the mesohyl, where the totipotent amoeboid cells (turquoise) and the associated microbiota (various shapes and colors) are located. The array of flagellated cells (red) at the bottom of the magnification are choanocyte cells. (**b**) Fluorescent *in situ* hybridization on *Aplysina aerophoba* mesohyl cryo-section using universal eubacterial probe (EUB338 mix) in green and DAPI for blue nucleolus staining as described in Bayer *et al.* [25]; *CC* choanocyte chamber, *WC* water canal

89% of bacteria depending on the sponge species [24] is pumped into the central cavity and emerges through an exhalant opening (osculum). Sponges reproduce both sexually and asexually. In terms of asexual, clonal reproduction, sponges can fragment, bud, or produce gemmules [26, 27]. For sexual reproduction, a sponge individual can possess either both male and female gametes (hermaphroditic) or only one (gonochoristic) [26]. Demosponges can be ovoviviparous, oviparous, and even viviparous [14]. The reproductive strategies are polyphyletic, and even mainly oviparous orders like Astrophorida include ovoviviparous genera [14, 28]. The mode of reproduction influences the diversity and connectivity of sponge populations [29–31].

The first published sponge genome was that of the Great Barrier Reef sponge *Amphimedon queenslandica* [32]. Its publication represented a great step forward in exploring metazoan origin and early diversification. As of today, six sponge genomes of the classes Demospongiae, Homoscleromorpha, and Calcarea have been sequenced (Table 7.1), and a number of transcriptomes from sponges of all four classes have been studied (e.g., [33–35]). Sponge genome sizes range from 125 Mbp in *Tethya wilhelma* to 386 Mbp in *Stylissa carteri* [36, 37]. Sequencing of these nuclear and additional mitochondrial genomes has enabled phylogenomic analyses comparing sponges with other non-bilaterian organisms to shed light on the origin of all metazoan life [5, 38, 39]. Genomics has further revealed insights into sponge immunity. As microbes play multiple roles in the sponge context, namely, as food source, as pathogens, or as commensal or beneficial symbionts, sponges need to discriminate between (at least) these three types of microorganisms.

Sponges fulfill a variety of functions in marine ecosystems that affect ecosystem biodiversity and nutrient cycling. They provide habitat and structure for other organisms, such as crustaceans, molluscs, bryozoans, polychaetes, cnidarians, echinoderms, and fishes [40–43]. In coral reefs, sponges consolidate and stabilize the calcium carbonate substrate on which corals grow, thereby enhancing reef growth [44]. On the other hand, excavating sponges can hinder reef growth by boring into the reef structure and adversely affecting the corals' structural integrity [45]. They also affect biodiversity through spatial competition [46, 47] and facilitative interspecific interactions. For example, the coral reef sponge *Lissodendoryx colombien-*

Table 7.1 Sponge genome studies

Sponge species	References
Demospongiae	
Amphimedon queenslandica	[32]
Stylissa carteri	[36]
Tethya wilhelma	[37]
Xestospongia testudinaria	[36]
Homoscleromorpha	
Oscarella carmela/*O. pearsei*	[181, 183]
Calcarea	
Sycon ciliatum	[182]

sis was able to expand its habitat range to seagrass meadows by growing together with seagrass sponge species that deterred predation by large starfish [48]. Sponges create a trophic link between the water column and benthos by coupling carbon fluxes via their filtering of food particles [49]. Remineralization of these food particles further provides an important source of inorganic nitrogen in benthic habitats [50]. Additionally, sponges are able to take up dissolved organic matter (DOM) and—by shedding large amounts of cells due to their rapid cell turnover rates—render particulate organic matter (POM) available to benthic detritivores, such as crustaceans and polychaetes [51–53]. The process of recycling DOM derived from other members of the coral reef community, such as corals and macroalgae, and passing it on to detritivores as POM was termed the "sponge loop" [52, 54, 55] in analogy to the established "microbial loop" [56]. The sponge loop has been shown to play a role in oligotrophic tropical reef environments as well as the deep sea [52, 54, 55].

7.1.2 *Sponge-Microbe Symbiosis*

Sponges host highly diverse and distinct microbial communities that can constitute up to 40% of the sponge volume [27, 57–59]. Thus, sponges are considered "holobionts," a term that comprises the sponge and its associated microbiota [26, 60]. The sponge microbiota is located mainly extracellularly within the animal's mesohyl matrix. 16S rRNA gene amplicon studies have discovered high phylum-level diversity and stability of microbial associations in marine sponges comprising phototrophic as well as heterotrophic symbionts [58, 61–63]. The sponge microbiota spans as many as 52 microbial phyla and candidate phyla with the diversity and abundance varying between sponge species [27, 63, 64]. The most dominant symbiont groups belong to the phyla *Proteobacteria* (mainly *Gammaproteobacteria* and *Alphaproteobacteria*), *Actinobacteria*, *Chloroflexi*, *Nitrospirae*, *Cyanobacteria*, candidatus phylum *Poribacteria*, and *Thaumarchaea*. Most of these symbionts seem to be sponge species-specific and vertically transmitted to the next generation of sponges via their larvae [61, 65–68]. The microbial communities of sponges are hypothesized to play a crucial role in the sponges' success, e.g., by supplying supplemental nutrition to the host [59, 69].

Based on the abundance of microbes, two groups of sponges are observed: high microbial abundance (HMA) and low microbial abundance (LMA) sponges [15]. HMA sponges harbor microbial communities that are 2–4 orders of magnitude denser than in LMA sponges [15, 70]. The microbial communities in HMA sponges are also more diverse than those of LMA sponges, and specific taxa are enriched in either HMA sponges (e.g., *Poribacteria*, *Chloroflexi*, and *Acidobacteria*) or LMA sponges (*Proteobacteria* and *Cyanobacteria*) [71]. The HMA or LMA status has been predicted by machine learning based on microbial diversity and composition and appears to be moderately related to host phylogeny [71].

Virus like particles were described in sponges by electron microscopy early on in 1978 [72] but their diversity and function remained unclear. More recently, a first approach to sponge virome sequencing, termed HoloVir, was established [73] and applied to study viruses from four Great Barrier Reef sponges (*Amphimedon queenslandica, Xestospongia testudinaria, Ianthella basta* and *Rhopaloeides odorabile*) [74]. The comparison of the four viral community profiles indicated species specific viral signatures in sponges sharing low identity to known viral genomes. As expected for a bacterium dominated system tailed bacteriophages of the order Caudovirales (dsDNA) and Microviridae (ssDNA) were the most diverse taxa in all sponges [75]. However, there is also evidence for eukaryote infecting viruses such as members of Megavirales and Parvoviridae. Whether these infect sponge cells, sponge associated eukaryotes, or are external input from filtration activity remains to be investigated. A recent TEM based study assayed the morphology of viral-like particles from 15 sponge species. This revealed a diversity of 50 viral morphotypes, some of which were congruent with virome sequencing. Specifically, non-enveloped, non-tailed icosahedral particles were most abundant but also geminate, filamentous and brick-shaped viruses indicative for Poxviridae were detected.

7.2 The Molecular Basis for the Symbiosis: The Microbial Side

The great majority of sponge symbionts are as yet uncultivable [27], and therefore culture-independent approaches are used to gain genomic and functional information. General patterns have been revealed by analyzing metagenomes, metaproteomes, and metatranscriptomes (e.g., [36, 76–80]). A comparison of the metagenomic profiles of several sponge microbiomes with the profile of seawater microbiomes provided evidence of evolutionary convergence in the microbial consortia of different sponge species [77]. These commonly sponge symbiont-enriched functions or related genes include (i) eukaryotic-like proteins [27, 35, 76, 81], (ii) defense features against foreign DNA [27, 82, 83], (iii) secondary metabolism [84–87], and (iv) adaptations of the central metabolism, such as the nitrogen metabolism. In particular, denitrification [79, 88, 89], ammonia oxidation [89, 90], and ammonia assimilation via the glutamine synthase-glutamine oxoglutarate aminotransferase (GS-GOGAT) pathway seem to play an important role in sponge symbionts [76, 91]. The findings on these general patterns were confirmed and expanded by single-cell genomics and metagenomic binning of a number of symbiont genomes including even new bacterial clades and candidate phyla (e.g., [81, 83, 92–98]). In the following paragraphs, we discuss the role of eukaryotic-like protein domains, defense features, and secondary metabolism.

Table 7.2 Summary of bacterial and sponge features in the light of host-microbiota interaction

Feature	Putative role	References
Bacterial features		
Eukaryotic-like protein domains	Phagocytosis evasion, bacteria-host communication	[35, 76, 101, 102]
Altered lipopolysaccharide (Ca. *Synechococcus spongiarum*)	Phagocytosis evasion, symbiont recognition	[103]
CRISPR-Cas systems	Defense against foreign DNA	[76, 82, 83]
Toxin-antitoxin systems	Defense against foreign DNA	[27, 82, 83]
Restriction-modification systems	Defense against foreign DNA	[27, 82, 83]
Increased genome sizes	Genetic flexibility	[82]
Secondary metabolism (PKS I, NRPS)	Communication within the holobiont, defense of the holobiont, e.g., against predators	[23, 87, 108, 109]
Sponge features		
TLRs and poriferan TLR-like molecules	PRR, immune system	[36, 116–118]
NOD-like receptors/NLRs	PRR, immune system	[36, 113, 116]
AFs	Allorecognition, immune system	[122]
SRCRs	PRR, allorecognition, immune system	[36, 112, 114, 116]

7.2.1 *Eukaryotic-Like Proteins: A Possible Means for Phagocytosis Evasion*

One of the features frequently found enriched in sponge symbionts (Table 7.2) are eukaryotic-like proteins, particularly ankyrin and tetratricopeptide repeat (TPR) domains [27]. Since they were first found to be enriched in the metagenome of *Cymbastela concentrica* [76], their likely importance for sponge symbiosis has been fortified by their documentation in a number of additional studies (e.g., [35, 93, 95–100]). Ankyrin repeat domains are known to be involved in host-microbe interactions and, more specifically, have been proposed to function in phagocytosis evasion in the sponge host context [35, 76, 101]. TPR domains are generally involved in protein-protein interactions, and bacterial TPR domains have been hypothesized to be involved in prokaryote-eukaryote interactions [102]. Their role in the sponge-bacterial symbiosis needs to be further elucidated.

In addition to ankyrin repeat domains, a modified lipopolysaccharide (LPS) antigen has been proposed as a phagocytosis evasion strategy for the cyanobacterial sponge symbiont *Candidatus Synechococcus spongiarum* [103]. In contrast to closely related free-living *Synechococcus* spp., this symbiont seems to have lost genes involved in the production of dTDP-L-rhamnose, a residue of the O antigen of LPS ultimately leading to an altered Gram-negative bacterial cell wall [103]. In the context of the sponge host environment, this may represent a means of discrimination between free-living *Synechococcus*—a food source of the sponge—and the symbionts.

7.2.2 *Bacterial Defense Against Foreign DNA*

Sponge symbionts are likely exposed to high levels of free foreign DNA and high numbers of bacteriophages and other viruses due to the hosts' pumping and phagocytosis activity. The larger genome sizes estimated for sponge symbionts in comparison to seawater microbes may be an adaptation to the variable sponge microenvironment and likely a result of higher horizontal gene transfer (HGT) levels in comparison to the seawater environment [82]. A high number of mobile genetic elements and transposases are common to the genomes of sponge symbionts supporting this hypothesis [27, 77, 94, 99]. Bacteriophage-mediated HGT has been shown in the microbial community of the marine sponge *Hymeniacidon perlevis* [104]. Thus, genetic transfer may have been essential for the sponge microbiome to adapt to their symbiotic lifestyle [27, 82].

The genomic repertoire of sponge symbionts is comparatively rich in functions related to defense against foreign DNA, mainly the bacteriophages and other viruses that are reported from marine sponges [23, 82, 105], implying that they require these protective mechanisms due to their sponge host-related microenvironment. Such defense-related features are restriction-modification (RM) systems, toxin-antitoxin (TA) systems, and CRISPR-Cas systems [76, 79, 82, 83, 95]. Clustered regularly interspaced short palindromic repeats (CRISPRs) and CRISPR-associated proteins (Cas) represent a prokaryotic adaptive immune system against integration of extrachromosomal DNA. These genomic elements were found in high abundance and overrepresented in sponge metagenomes compared to seawater metagenomics data [76, 77, 82]. Also in a comparison between genomes of the cyanobacterial sponge symbiont *Candidatus Synechococcus spongiarum* and closely related seawater *Synechococcus* sp., CRISPR-Cas systems were found in the sponge symbionts, while they are known to be almost completely lacking from the remainder of the *Synechoc occus/Prochlorococcus* subclade [103]. Furthermore, TA and RM systems were found enriched in sponge symbiont genomes and may help to control the amounts and nature of foreign DNA implemented into symbiont genomes [27, 82, 83]. This stresses the potential influence of viruses on the microbial community of sponges. Yet, little is known about viruses in sponges and their influence on the holobiont.

7.2.3 *Secondary Metabolism*

Marine sponges are known to be rich sources of secondary metabolites with various activities including antimicrobial, antifungal, antiviral, cytotoxic, or anticancer effects [23, 106, 107]. Many of these compounds are in fact produced by the microbiome [84–87]. Especially polyketide synthases (PKS) and non-ribosomal peptide synthetases (NRPS) are often enriched in sponge symbionts, often with new structures and putatively novel activities [23, 108, 109]. An example of a sponge symbiont producing various and unusual bioactive compounds is *Candidatus Entotheonella factor* of the candidate phylum *Tectomicrobia*. This bacterium seems to be the source of nearly all polyketide and modified peptide families known from its host *Theonella swinhoei* [86, 87]. In the sponge holobiont context, bioactive compounds could exhibit a variety of functions. They may be means of communication between microbes and with the host. They are also a means of chemical defense against predation on the holobiont [110].

7.3 The Molecular Basis for the Symbiosis: The Host Side

The innate immune response of sponges is still largely uncharted territory. Important aspects of immunology, specifically the recognition of self and nonself, were originally discovered in sponges more than 100 years ago [111]. In spite of these early groundbreaking discoveries, our current knowledge of immunity in sponges is still very limited. Genomic studies have shown that sponges harbor immunological molecules conserved across the animal kingdom [91, 112, 113]. Sponges express a variety of pattern recognition receptors (PRRs) including Toll-like and NOD-like receptors (TLRs and NLRs) and the scavenger receptor cysteine-rich (SRCR) family [36, 113], whose role in the sponge's response to microbes is beginning to be elucidated (Table 7.2) [114–116]. Homologs of relevant molecular signaling components of the innate immune system, such as TLRs, caspases, IL-1 receptor-associated kinase-4-like protein (IRAK-4l), and the myeloid differentiation factor 88 (myD88) were identified in phylogenetically diverse sponge species [32, 33, 117–119]. While these molecules display the specific characteristics of immunological patterns of higher metazoans, various sponge-specific alterations and uncommon variations have been noticed which most likely relate to poriferan modifications of the "classic" innate immune pathways [118]. Moreover, it was shown that exposure of *Suberites domuncula* tissues to bacterial lipopolysaccharide (LPS) resulted in an activation and phosphorylation of two stress response kinases, p38 and JNK, within a few hours [120]. A comparison of field-collected *S. domuncula* specimens revealed higher levels of LPS-induced p38 kinase phosphorylation than in specimens collected from the protected coastal canal [121]. Gene expression patterns from specimens from the two sites showed that the allograft inflammatory factor-1 (AIF-1) homolog, involved in the activation of macrophages, was highly differentially expressed [121]. Also, aggregation factor (AF) genes are involved in allorecognition in demosponges [122]. They are hypothesized to have developed their role in self/nonself recognition by increasing specificity from their ancient role in cell-cell interaction [122].

Steindler and colleagues [114] employed suppression subtractive hybridization to investigate host gene expression in relation to the presence/absence of a specific clade of intracellular cyanobacterial symbionts. After a 5-month maintenance of the sponge *Petrosia ficiformis* in an illuminated natural habitat (symbiotic) and control specimens in a dark cave (aposymbiotic), seven genes were found to be upregulated in the symbiotic sponge including an SRCR gene [114]. SRCR genes were also upregulated during *S. domuncula* grafting experiments, suggesting they are important components of self/nonself recognition [112] and host-microbe interactions [114]. Indeed, potential immune system components, especially SRCR-like domains, have been shown to be expanded in marine sponges compared to other eukaryotes, probably as a result of host-microbe coevolution [36, 123]. For example, in *Amphimedon queenslandica*, the transcription factors *FoxO* and *NFκβ* were upregulated upon exposure to native but not to foreign bacteria [116]. These findings suggest that sponges actively recognize and discriminate symbionts vs. non-symbionts via immune signaling. However, unlike in cnidarians [124–126], the sponge (eco)immunology remains largely unexplored, and further research is crucially needed to unravel the mechanisms behind the coexistence between sponges and their microbial consortia. The recent establishment of a transgenic sponge cell line marks a breakthrough towards this goal [127].

7.4 Response of the Sponge Holobiont to Environmental Change

7.4.1 Infectious Disease and Dysbiosis

Surprisingly little is known about sponge diseases, especially considering their remarkable ecological relevance, for example with respect to benthic-pelagic coupling. Outbreaks of sponge epidemics resulting in severely decimated population sizes have been recorded for various marine areas such as the Mediterranean and Caribbean Seas (part of a name) [128, 129]. From the Great Barrier Reef, sponge mass mortalities have not yet been reported; however, a widely distributed disease-like syndrome has been recognized in the common Indo-Pacific sponge species *Ianthella basta* [130, 131]. Various microbial organisms including symbiotic cyanobacteria [132], *Bacillus* spp., and *Pseudomonas* spp. [133], as well as filamentous fungi [134, 135] have been suggested to cause disease outbreaks in sponges, while the role of viruses has hardly been explored until now. Isolating and identifying specific etiological agents has not been successful so far, the only exception being the pathogenic *Pseudoalteromonas agarivorans* strain NW4327 from unhealthy Great Barrier Reef sponges (*Rhopaloides odorabile*) [136, 137]. For instance, for the Caribbean sponge species *Xestospongia muta* and *Amphimedon compressa* suffering from orange band disease and white patch disease, respectively (Fig. 7.2), no evidence of a specific pathogen as the causative agent could be identified [138, 139].

Fig. 7.2 (**a**) *Amphimedon compressa* infected with white patch disease and (**b**) *Xestospongia muta* infected with orange band disease. Photos courtesy of Hilde Angermeier

Generally, it has been observed that, with increasing environmental pressure, opportunistic microbes (being part of the sponge's native microbiota and/or the surrounding water) can switch from a mutualistic to a pathogenic lifestyle [140] causing severe epidemics as reported for various marine organisms ranging from invertebrates (e.g., corals and oysters) to fish [141, 142]. Thus, whether sponge diseases are caused by true pathogens or rather by opportunistic or polymicrobial infections remains to be identified. A growing body of literature suggests that sponge diseases result from dysbiosis, which is defined as an imbalance of the resident microbiota that might give rise to opportunistic and/or polymicrobial infections [140, 143]. For example, in the deep sea sponge *Geodia barretti*, highly divergent microbial community profiles were found between different health states with distinct community shifts toward higher relative abundances of *Firmicutes*, *Deltaproteobacteria*, and *Bacteroidetes* in diseased specimens [144]. In the Red Sea sponge *Crella cyathophora*, a compositional shift in the microbiota of unhealthy sponges enriched with a novel clade affiliated with the phylum *Verrucomicrobia* was noticed [145]. In *Aplysina aerophoba* and *Ircinia fasciculata*, the overall bacterial diversity was considerably higher in diseased than in healthy sponges [146, 147]. In polymicrobial infections, consortia of microorganisms rather than single pathogens have been shown to cause diseases [148, 149]. Consequently, the term "pathobiome" specifically describing the dynamics of the microbiome in response to stress and the onset of disease was introduced and recently adopted by the holobiont research community [150, 151].

7.4.2 *Global Environmental Change*

Sponges may be adversely affected on different levels by global change stressors such as ocean warming, acidification, eutrophication, increasing sedimentation, and habitat destruction, which are often cumulative [152]. Observed adverse effects on the sponge holobiont are species-specific and include physiological stress causing metabolic and immune suppression as well as disadvantageous microbial community shifts, e.g., resulting in loss of symbiotic functions [128, 153, 154]. That the sponge's associated microbiota is not only driven by interactions with the host but also by environmental factors and is therefore sensitive to global change has been shown in several experimental and field studies [155–157]. Hence, the functioning of the associated microbiota is one of the keys for the survival and fitness of sponges in the Anthropocene. In the following, we discuss the environmental stressors that have been shown to adversely affect sponge holobiont composition and function.

7.4.2.1 Ocean Warming and Ocean Acidification

Calcifying sponges, like members of the classes Calcarea and Demospongiae, will likely be vulnerable to ocean acidification due to reduced calcification rates at lower seawater pH [158]. However, ocean warming resulting in thermal stress is likely to have the severest impact on sponge communities by causing diseases and mass mortalities [152]. In the Caribbean, for example, almost 30 sponge epidemics associated with elevated temperatures have been reported since the 1980s. In the Mediterranean Sea, up to 95% of the specimens of two important sponge species (*Ircinia fasciculata* and *Sarcotragus spinosulum)* died in the summers of 2008 and 2009 which were characterized by anomalous high temperature conditions [159]. Experiments investigating the effect of ocean warming in the reef sponges *Haliclona tubifera* revealed that the expression of stress and immune genes was altered [154]. The Great Barrier Reef sponge *Rhopaloeides odorabile* was also found to be highly sensitive to thermal stress [160]. While the heat shock protein 70 was induced, the expression of several genes involved in homeostasis was negatively correlated with temperature increase [153], and a shift in the bacterial community composition was detected [161]. The microbial diversity increased in an intermediate health state but then decreased in the subsequent necrotic state, as archetypal symbionts were outcompeted by new opportunistic microorganisms with a scavenging lifestyle and high growth rates [162]. Hence, Fan and colleagues [162] hypothesized that the interruption of symbiotic interactions is very likely a key factor in the loss of holobiont function causing disease and mortality in sponges.

The cumulative effect of environmental stressors, such as ocean warming and acidification, has recently been addressed in several studies. For example, the energy budget of the excavating Great Barrier Reef sponge *Cliona orientalis* was investigated under two possible future climate change scenarios predicted for the year 2100: a "reduced" and a "business-as-usual" CO_2 emission scenario [163]. While *C.*

orientalis maintained a positive energy budget under the "reduced" CO_2 emission scenario, the energy budget was negative under the "business-as-usual" scenario. These data indicate that the increased bioerosion capacity of *C. orientalis* previously described [164] might not hold true through a "business-as-usual" scenario since the sponge's metabolic maintenance and growth might be adversely affected [163]. Bennett and colleagues [165] found a species-specific response in four Great Barrier Reef species to future ocean scenarios depending on their nutritional mode, which is highly influenced by their symbionts. While heterotrophic species suffered from the additive effects of ocean warming and acidification, in phototrophic species, an increased pCO_2 mitigated thermal stress indicating that climate change may cause a shift in sponge community compositions toward phototrophic species [165]. In the Caribbean barrel sponge *Xestospongia muta*, the cumulative effect of ocean warming and acidification was experimentally tested and revealed significant effects on the sponge microbiota including symbiotic cyanobacteria [166].

7.4.2.2 Eutrophication and Sedimentation

Eutrophication (e.g., through river discharge of nutrients) and sedimentation (e.g., through dredging activities) are examples of local environmental stressors particularly affecting sponges due to their sessile, filter-feeding lifestyle [91]. The impact of eutrophication, such as increased nitrogen levels, on the microbial composition of sponges was analyzed in the species *Cymbastela stipitata* and *Rhopaloeides odorabile* [167, 168]. Both studies showed that the sponges' microbial communities remained stable upon short-term exposure to elevated nutrient concentrations. In the Atlantic pollution-tolerant sponge species *Paraleucilla magna* and *Hymeniacidon heliophila*, an alteration of the associated archaeal community composition was demonstrated which likely enhances the sponges' resistance toward eutrophic conditions [169]. In Australian *Carteriospongia foliascens*, a consistent increase in the proportion of *Cyanobacteria* over *Bacteroidetes* in turbid inshore waters (high-nutrient and low-light conditions) in contrast to oligotrophic offshore locations was demonstrated, suggesting that the specialist microbiota of *C. foliascens* is driven by environmental factors [157]. In a sedimentation experiment, five sponge species (including heterotrophic and phototrophic nutritional modes) were exposed to multiple sediment deposition events over a period of 30 days [170]. While the growth rate was negatively affected in the phototrophic sponge *Cliona orientalis*, the health status was not altered in any of the species [170].

Taken together, the results of environmental stress studies highlight that the effect on sponge-microbe interactions is highly species-specific. Accordingly, not all sponge species might be able to cope well with climate and global change [152]. In any case, long-term experiments are crucially needed to elucidate the response of the sponge holobiont to scenarios projected for the future.

7.4.2.3 Coping with Environmental Stress

The microbial contribution to the persistence of the sponge holobiont under environmental stress appears to be essential, since the observed microbial community dynamics upon environmental perturbation can drive acclimatization and adaptation [171, 172]. Microorganisms adapt much more rapidly than the host itself to stressful conditions, such as rising temperature or decreasing pH, and therefore facilitate holobiont evolution [173]. The generation time for most sponges is years to decades and thus is considered too long for genetic adaptation on the host level to cope with present and future rates of climate change [174]. In contrast, bacteria have very short generation times (minutes to hours), enabling symbiotic microorganisms to undergo adaptive evolution within weeks to months [175]. The holobiont's microbiota can either be altered by shifts in its taxonomic composition (acquisition of novel environmental microorganisms) or by genetic changes (mutation and/or horizontal gene transfer) resulting in the acquisition of novel functions [176, 177]. Newly acquired microbial functions and traits might considerably influence the holobiont's phenotype leading to microbiota-mediated acclimatization. Subsequent vertical transmission may facilitate transgenerational acclimatization, upon which selection can act potentially leading to host adaptation [172]. New microbes contribute to the functional stability of the sponge holobiont and hence its ability to respond to environmental change. This was demonstrated in the Mediterranean sponges *Dysidea avara*, *Agelas oroides*, and *Chondrosia reniformis* affected by ocean acidification [178]. The species-specific acquirement of novel microbes was found to be high in *D. avara*, moderate in *A. oroides*, and null in *C. reniformis* and was inversely correlated with the sponges' growth rates being not influenced in *D. avara*, decreased in *A. oroides*, and severely decreased in *C. reniformis*. At a CO_2 seep in Papua New Guinea, the two sponge species *Coelocarteria singaporensis* and *Cinachyra* sp. were approximately 40-fold more abundant and hosted a significantly higher relative number of symbiotic *Synechococcus* than specimens at control sites less than 500 m away [179]. However, in the species *Stylissa massa*, which was less abundant at the seep compared to the control site, the microbial community did not differ significantly between sites highlighting the interspecies variability in coping with abiotic conditions [179]. The importance of the dynamic relationship between the environment and symbiotic microbes for driving selection toward the most advantageous holobiont has also been noticed in corals [171]. In conclusion, to predict the effects of global change on the sponge holobiont, it is essential to unravel the underlying functional mechanisms of host-microbe interactions during health, dysbiosis, and disease and moreover to better understand the role of microbial restructuring in adaptation and survival.

7.5 Conclusions and Future Perspectives

Bacteria as well as the sponge host show a number of adaptations enabling their symbiosis. Bacterial symbionts appear to possess mechanisms to evade phagocytosis, defend themselves against foreign DNA, and produce a wide range of secondary metabolites. Likewise, sponges have a complex innate immune system that enables them to discriminate between food bacteria, pathogens, and their own microbiota. Sponge holobionts are multilevel systems comprising the host sponge as well as bacterial, archaeal, viral, and fungal symbionts, each of which are a node in a highly complex interactive network. Changes in environmental conditions (e.g., in temperature, pH, or light conditions) may affect all of these nodes to a different extent. For example, changes in light availability would affect only phototrophic symbionts and consequently the sponges that host them.

It was proposed that sponges may be "winners of climate change" [180]. However, certain species seem to be adversely affected on different levels including metabolic and immune suppression, as well as microbiome shifts resulting in loss of symbiotic functions. In fact, the reaction of sponges to climate change and short-term stressors, such as increased turbidity or heat waves, appears to be rather species-specific [128, 165]. Thus, a much better understanding of sponge holobionts and the underlying molecular mechanisms of host-microbe interaction is required before final conclusions on the response of sponges to a changing environment can be drawn.

Acknowledgments We gratefully acknowledge financial support from the DFG (CRC1182-TPB1) and from the European Union's Horizon 2020 research and innovation program under Grant Agreement No. 679849 ("SponGES"). AF was supported by the strategic research initiative "Ocean Health" of the Cluster of Excellence "The Future Ocean," and LP was awarded a postdoctoral fellowship from the Alexander von Humboldt Foundation, which was sponsored by "The Future Ocean" Cluster.

Parts of this chapter were previously published by BS in the framework of her PhD thesis (University of Würzburg, Germany): https://opus.bibliothek.uni-wuerzburg.de/frontdoor/index/index/docId/15186.

References

1. Brain CK, Prave AR, Hoffmann KH, Fallick AE, Botha A, Herd DA, et al. The first animals: ca. 760-million-year-old sponge-like fossils from Namibia. S Afr J Sci. 2012;108:83–90.
2. Antcliffe JB, Callow RHT, Brasier MD. Giving the early fossil record of sponges a squeeze. Biol Rev. 2014;89:972–1004.
3. Du W, Wang XL, Komiya T. Potential Ediacaran sponge gemmules from the Yangtze Gorges area in South China. Gondwana Res. 2015;28:1246–54.
4. Gold DA, Grabenstatter J, de Mendoza A, Riesgo A, Ruiz-Trillo I, Summons RE. Sterol and genomic analyses validate the sponge biomarker hypothesis. Proc Natl Acad Sci USA. 2016;113:2684–9.

5. Feuda R, Dohrmann M, Pett W, Philippe H, Rota-Stabelli O, Lartillot N, et al. Improved modeling of compositional heterogeneity supports sponges as sister to all other animals. Curr Biol. 2017;27:3864–70.
6. Zumberge JA, Love GD, Cárdenas P, Sperling EA, Gunasekera S, Rohrssen M, Grosjean E, Grotzinger JP, Summons RE. Demosponge steroid biomarker 26-methylstigmastane provides evidence for Neoproterozoic animals. Nature Ecol Evol. 2018;2:1709–14.
7. Heckel PH. Carbonate buildups in the geologic record: a review. Soc Econ Paleontol Mineral. 1974;18:90–154.
8. Muir LA, Botting JP, Beresi MS. Lessons from the past: sponges and the geological record. Climate change, ocean acidification and sponges. Cham: Springer; 2017.
9. Cárdenas P, Rapp HT, Klitgaard AB, Best M, Thollesson M, Tendal OS. Taxonomy, biogeography and DNA barcodes of *Geodia* species (Porifera, Demospongiae, Tetractinellida) in the Atlantic boreo-arctic region. Zool J Linnean Soc. 2013;169:251–311.
10. Maldonado M, Aguilar R, Bannister RJ, Bell JJ, Conway KW, Dayton PK, et al. Sponge grounds as key marine habitats: a synthetic review of types, structure, functional roles, and conservation concerns. Cham: Springer; 2016.
11. Hooper J, Van Soest RWM. Systema Porifera: a guide to the classification of sponges. New York: Kluwer Academic/Plenum Publishers; 2002.
12. Bergquist P, Porifera R. In: Anderson DT editor. Invertebrate Zoology. Oxford: Oxford University Press; 1998. p. 10–27.
13. Simpson TL. The cell biology of sponges. New York: Springer; 1984.
14. Ereskovsky AV. The comparative embryology of sponges. Dordrecht: Springer; 2010.
15. Hentschel U, Fieseler L, Wehrl M, Gernert C, Steinert M, Hacker J, et al. Microbial diversity of marine sponges. In: Müller W, editor. Sponges (Porifera). Berlin/Heidelberg/New York: Springer; 2003. p. 59–88.
16. Adams EDM, Goss GG, Leys SP. Freshwater sponges have functional, sealing epithelia with high transepithelial resistance and negative transepithelial potential. Plos One. 2010;5:e15040.
17. Leys SP, Riesgo A. Epithelia. An evolutionary novelty of metazoans. J Exp Zool Part B. 2012;318b:438–47.
18. Dunn CW, Leys SP, Haddock SHD. The hidden biology of sponges and ctenophores. Trends Ecol Evol. 2015;30:282–91.
19. Ludeman DA, Reidenbach MA, Leys SP. The energetic cost of filtration by demosponges and their behavioural response to ambient currents. J Exp Biol. 2017;220:995–1007.
20. Mah JL, Leys SP. Think like a sponge: the genetic signal of sensory cells in sponges. Dev Biol. 2017;431:93–100.
21. Vogel S. Current-induced flow through living sponges in nature. Proc Natl Acad Sci U S A. 1977;74:2069–71.
22. Wehrl M, Steinert M, Hentschel U. Bacterial uptake by the marine sponge *Aplysina aerophoba*. Microb Ecol. 2007;53:355–65.
23. Taylor MW, Radax R, Steger D, Wagner M. Sponge-associated microorganisms: evolution, ecology, and biotechnological potential. Microbiol Mol Biol R. 2007;71:295–347.
24. Morganti T, Coma R, Yahel G, Ribes M. Trophic niche separation that facilitates co-existence of high and low microbial abundance sponges is revealed by in situ study of carbon and nitrogen fluxes. Limnol Oceanogr. 2017;62:1963–83.
25. Bayer K, Kamke J, Hentschel U. Quantification of bacterial and archaeal symbionts in high and low microbial abundance sponges using real-time PCR. FEMS Microbiol Ecol. 2014;89:679–690.
26. Ayling AL. Patterns of sexuality, asexual reproduction and recruitment in some subtidal marine Demospongiae. Biol Bull. 1980;158:271–82.
27. Webster NS, Thomas T. The sponge hologenome. Mbio. 2016;7:e00135–16.

28. Vacelet J. Planktonic armoured propagules of the excavating sponge *Alectona* (Porifera: Demospongiae) are larvae: evidence from *Alectona wallichii* and *A. mesatlantica* sp. nov. Mem Qld Museum. 1999;44:627–42.
29. Calderon I, Ortega N, Duran S, Becerro M, Pascual M, Turon X. Finding the relevant scale: clonality and genetic structure in a marine invertebrate (*Crambe crambe*, Porifera). Mol Ecol. 2007;16:1799–810.
30. Chaves-Fonnegra A, Feldheim KA, Secord J, Lopez JV. Population structure and dispersal of the coral-excavating sponge *Cliona delitrix*. Mol Ecol. 2015;24:1447–66.
31. Riesgo A, Perez-Portela R, Pita L, Blasco G, Erwin PM, Lopez-Legentil S. Population structure and connectivity in the Mediterranean sponge *Ircinia fasciculata* are affected by mass mortalities and hybridization. Heredity. 2016;117:427–39.
32. Srivastava M, Simakov O, Chapman J, Fahey B, Gauthier MEA, Mitros T, et al. The *Amphimedon queenslandica* genome and the evolution of animal complexity. Nature. 2010;466:720–3.
33. Riesgo A, Farrar N, Windsor PJ, Giribet G, Leys SP. The analysis of eight transcriptomes from all poriferan classes reveals surprising genetic complexity in sponges. Mol Biol Evol. 2014;31:1102–20.
34. Guzman C, Conaco C. Comparative transcriptome analysis reveals insights into the streamlined genomes of haplosclerid demosponges. Sci Rep-UK. 2016;6:18774.
35. Diez-Vives C, Moitinho-Silva L, Nielsen S, Reynolds D, Thomas T. Expression of eukaryotic-like protein in the microbiome of sponges. Mol Ecol. 2017;26:1432–51.
36. Ryu T, Seridi L, Moitinho-Silva L, Oates M, Liew YJ, Mavromatis C, et al. Hologenome analysis of two marine sponges with different microbiomes. BMC Genomics. 2016;17:158.
37. Francis WR, Eitel M, Vargas S, Adamski M, Haddock SHD, Krebs S, et al. The genome of the contractile demosponge *Tethya wilhelma* and the evolution of metazoan neural signalling pathways. bioRxiv; 2017.
38. Ma JY, Yang Q. Early divergence dates of demosponges based on mitogenomics and evaluated fossil calibrations. Palaeoworld. 2016;25:292–302.
39. Simion P, Philippe H, Baurain D, Jager M, Richter DJ, Di Franco A, et al. A large and consistent phylogenomic dataset supports sponges as the sister group to all other animals. Curr Biol. 2017;27:958–67.
40. Pearse AS. Notes on the inhabitants of certain sponges at Bimini. Ecology. 1950;31:149.
41. Westinga E, Hoetjes PC. The intrasponge fauna of *Spheciospongia vesparia* (Porifera, Demospongiae) at Curacao and Bonaire. Mar Biol. 1981;62:139–50.
42. Duarte L, Nalesso R. The sponge *Zygomycale parishii* (Bowerbank) and its endobiotic fauna. Estuar Coast Shelf Sci. 1996;42:139–51.
43. Wulff JL. Ecological interactions of marine sponges. Can J Zool. 2006;84:146–66.
44. Wulff JL. Sponge-mediated coral reef growth and rejuvenation. Coral Reefs. 1984;3:157–63.
45. Diaz MC, Rutzler K. Sponges: an essential component of Caribbean coral reefs. B Mar Sci. 2001;69:535–46.
46. Rützler K. Spatial competition among porifera: solution by epizoism. Oecologia. 1970;5:85–95.
47. Aerts LAM. Dynamics behind standoff interactions in three reef sponge species and the coral *Montastraea cavernosa*. Mar Ecol. 2000;21:191–204.
48. Wulff JL. Collaboration among sponge species increases sponge diversity and abundance in a seagrass meadow. Mar Ecol-Evol Persp. 2008;29:193–204.
49. Gili JM, Coma R. Benthic suspension feeders: their paramount role in littoral marine food webs. Trends Ecol Evol. 1998;13:316–21.
50. Southwell MW, Weisz JB, Martens CS, Lindquist N. In situ fluxes of dissolved inorganic nitrogen from the sponge community on Conch Reef, Key Largo, Florida. Limnol Oceanogr. 2008;53:986–96.
51. Rix L, de Goeij JM, van Oevelen D, Struck U, Al-Horani FA, Wild C, et al. Reef sponges facilitate the transfer of coral-derived organic matter to their associated fauna via the sponge loop. Mar Ecol Prog Ser. 2018;589:85–96.

52. de Goeij JM, van Oevelen D, Vermeij MJA, Osinga R, Middelburg JJ, de Goeij AFPM, et al. Surviving in a marine desert: the sponge loop retains resources within coral reefs. Science. 2013;342:108–10.
53. Alexander BE, Liebrand K, Osinga R, van der Geest HG, Admiraal W, Cleutjens JPM, et al. Cell turnover and detritus production in marine sponges from tropical and temperate benthic ecosystems. PLoS One. 2014;9:e109486.
54. Rix L, de Goeij JM, Mueller CE, Struck U, Middelburg JJ, van Duyl FC, et al. Coral mucus fuels the sponge loop in warm- and cold-water coral reef ecosystems. Sci Rep-UK. 2016;6:18715.
55. Rix L, de Goeij JM, van Oevelen D, Struck U, Al-Horani FA, Wild C, et al. Differential recycling of coral and algal dissolved organic matter via the sponge loop. Funct Ecol. 2017;31:778–89.
56. Azam F, Fenchel T, Field JG, et al. The ecological role of water-column microbes in the sea. Mar Ecol Prog Ser. 1983;10:257–63.
57. Vacelet J. Etude en microscopie electronique de l'association entre bacteries et spongiaires du genre Verongia (Dictyoceratida). J Microsc Biol Cell. 1975;23:271–88.
58. Easson CG, Thacker RW. Phylogenetic signal in the community structure of host-specific microbiomes of tropical marine sponges. Front Microbiol. 2014;5:532.
59. Tian RM, Wang Y, Bougouffa S, Gao ZM, Cai L, Bajic V, et al. Genomic analysis reveals versatile heterotrophic capacity of a potentially symbiotic sulfur-oxidizing bacterium in sponge. Environ Microbiol. 2014;16:3548–61.
60. Pita L, Rix L, Slaby BM, Franke A, Hentschel U. The sponge holobiont in a changing ocean: from microbes to ecosystems. Microbiome. 2018;6:46.
61. Schmitt S, Tsai P, Bell J, Fromont J, Ilan M, Lindquist N, et al. Assessing the complex sponge microbiota: core, variable and species-specific bacterial communities in marine sponges. ISME J. 2012;6:564–76.
62. Erwin PM, Pita L, Lopez-Legentil S, Turon X. Stability of sponge-associated bacteria over large seasonal shifts in temperature and irradiance. Appl Environ Microb. 2012;78:7358–68.
63. Thomas T, Moitinho-Silva L, Lurgi M, Bjork JR, Easson C, Astudillo-Garcia C, et al. Diversity, structure and convergent evolution of the global sponge microbiome. Nat Commun. 2016;7:11870.
64. Moitinho-Silva L, Nielsen S, Amir A, Gonzalez A, Ackermann GL, Cerrano C, et al. The sponge microbiome project. Gigascience. 2017;6:1–7.
65. Usher KM, Kuo J, Fromont J, Sutton DC. Vertical transmission of cyanobacterial symbionts in the marine sponge *Chondrilla australiensis* (Demospongiae). Hydrobiologia. 2001;461:15–23.
66. Oren M, Steindler L, Ilan M. Transmission, plasticity and the molecular identification of cyanobacterial symbionts in the Red Sea sponge *Diacarnus erythraeus*. Mar Biol. 2005;148:35–41.
67. Schmitt S, Angermeier H, Schiller R, Lindquist N, Hentschel U. Molecular microbial diversity survey of sponge reproductive stages and mechanistic insights into vertical transmission of microbial symbionts. Appl Environ Microb. 2008;74:7694–708.
68. Webster NS, Taylor MW, Behnam F, Lucker S, Rattei T, Whalan S, et al. Deep sequencing reveals exceptional diversity and modes of transmission for bacterial sponge symbionts. Environ Microbiol. 2010;12:2070–82.
69. Erwin PM, Thacker RW. Phototrophic nutrition and symbiont diversity of two Caribbean sponge-cyanobacteria symbioses. Mar Ecol Prog Ser. 2008;362:139–47.
70. Gloeckner V, Wehrl M, Moitinho-Silva L, Gernert C, Schupp P, Pawlik JR, et al. The HMA-LMA dichotomy revisited: an electron microscopical survey of 56 sponge species. Biol Bull-US. 2014;227:78–88.
71. Moitinho-Silva L, Steinert G, Nielsen S, Hardoim CCP, Wu YC, McCormack GP, et al. Predicting the HMA-LMA status in marine sponges by machine learning. Front Microbiol. 2017;8:752.

72. Vacelet J, Gallissian M-F. Virus-like particles in cells of the sponge Verongia cavernicola (demospongiae, dictyoceratida) and accompanying tissues changes. J Invertebr Pathol. 1978;31:246–054 (1978).
73. Laffy PW, et al. HoloVir: a workflow for investigating the diversity and function of viruses in invertebrate holobionts. Front Microbiol. 2016;7:822.
74. Laffy PW, et al. Reef invertebrate viromics: diversity, host specificity and functional capacity. Environ Microbiol. 2018.
75. Pascelli C, Laffy PW, Kupresanin M, Ravasi T, Webster NS. Morphological characterization of virus-like particles in coral reef sponges. PeerJ. 2018;6:e5625.
76. Thomas T, Rusch D, DeMaere MZ, Yung PY, Lewis M, Halpern A, et al. Functional genomic signatures of sponge bacteria reveal unique and shared features of symbiosis. ISME J. 2010;4:1557–67.
77. Fan L, Reynolds D, Liu M, Stark M, Kjelleberg S, Webster NS, et al. Functional equivalence and evolutionary convergence in complex communities of microbial sponge symbionts. Proc Natl Acad Sci USA. 2012;109:E1878–87.
78. Radax R, Rattei T, Lanzen A, Bayer C, Rapp HT, Urich T, et al. Metatranscriptomics of the marine sponge *Geodia barretti*: tackling phylogeny and function of its microbial community. Environ Microbiol. 2012;14:1308–24.
79. Liu M, Fan L, Zhong L, Kjelleberg S, Thomas T. Metaproteogenomic analysis of a community of sponge symbionts. ISME J. 2012;6:1515–25.
80. Fiore CL, Labrie M, Jarett JK, Lesser MP. Transcriptional activity of the giant barrel sponge, *Xestospongia muta* Holobiont: molecular evidence for metabolic interchange. Front Microbiol. 2015;6:364.
81. Kamke J, Rinke C, Schwientek P, Mavromatis K, Ivanova N, Sczyrba A, et al. The candidate phylum Poribacteria by single-cell genomics: new insights into phylogeny, cell-compartmentation, eukaryote-like repeat proteins, and other genomic features. PLoS One. 2014;9:e87353.
82. Horn H, Slaby BM, Jahn MT, Bayer K, Moitinho-Silva L, Förster F, et al. An enrichment of CRISPR and other defense-related features in marine sponge-associated microbial metagenomes. Front Microbiol. 2016;7:1751.
83. Slaby BM, Hackl T, Horn H, Bayer K, Hentschel U. Metagenomic binning of a marine sponge microbiome reveals unity in defense but metabolic specialization. ISME J. 2017;11:2465–78.
84. Santos OCS, Pontes PVML, Santos JFM, Muricy G, Giambiagi-deMarval M, Laport MS. Isolation, characterization and phylogeny of sponge-associated bacteria with antimicrobial activities from Brazil. Res Microbiol. 2010;161:604–12.
85. Abdelmohsen UR, Yang C, Horn H, Hajjar D, Ravasi T, Hentschel U. Actinomycetes from Red Sea sponges: sources for chemical and phylogenetic diversity. Mar Drugs. 2014;12:2771–89.
86. Wilson MC, Mori T, Ruckert C, Uria AR, Helf MJ, Takada K, et al. An environmental bacterial taxon with a large and distinct metabolic repertoire. Nature. 2014;506:58–62.
87. Lackner G, Peters EE, Helfrich EJN, Piel J. Insights into the lifestyle of uncultured bacterial natural product factories associated with marine sponges. Proc Natl Acad Sci USA. 2017;114:E347–56.
88. Hoffmann F, Radax R, Woebken D, Holtappels M, Lavik G, Rapp HT, et al. Complex nitrogen cycling in the sponge *Geodia barretti*. Environ Microbiol. 2009;11:2228–43.
89. Fiore CL, Baker DM, Lesser MP. Nitrogen biogeochemistry in the Caribbean sponge, *Xestospongia muta*: a source or sink of dissolved inorganic nitrogen? PLoS One. 2013;8:e72961.
90. Radax R, Hoffmann F, Rapp HT, Leininger S, Schleper C. Ammonia-oxidizing archaea as main drivers of nitrification in cold-water sponges. Environ Microbiol. 2012;14:909–23.
91. Hentschel U, Piel J, Degnan SM, Taylor MW. Genomic insights into the marine sponge microbiome. Nat Rev Microbiol. 2012;10:641–75.
92. Liu MY, Kjelleberg S, Thomas T. Functional genomic analysis of an uncultured δ-proteobacterium in the sponge *Cymbastela concentrica*. ISME J. 2011;5:427–35.

93. Liu F, Li J, Feng G, Li Z. New genomic insights into "Entotheonella" symbionts in *Theonella swinhoei*: Mixotrophy, anaerobic adaptation, resilience, and interaction. Front Microbiol. 2016;7:1333.
94. Gao ZM, Wang Y, Tian RM, Wong YH, Batang ZB, Al-Suwailem AM, et al. Symbiotic adaptation drives genome streamlining of the cyanobacterial sponge symbiont "Candidatus Synechococcus spongiarum". MBio. 2014;5:e00079–14.
95. Astudillo-Garcia C, Slaby BM, Waite DW, Bayer K, Hentschel U, Taylor MW. Phylogeny and genomics of SAUL, an enigmatic bacterial lineage frequently associated with marine sponges. Environ Microbiol. 2018;20:561–76.
96. Moitinho-Silva L, Díez-Vives C, Batani G, Esteves AI, Jahn MT, Thomas T. Integrated metabolism in sponge-microbe symbiosis revealed by genome-centered metatranscriptomics. ISME J. 2017;11(7):1651–66.
97. Karimi E, Slaby BM, Soares AR, Blom J, Hentschel U, Costa R. Metagenomic binning reveals versatile nutrient cycling and distinct adaptive features in alphaproteobacterial symbionts of marine sponges. FEMS Microbiol Ecol. 2018;94(6):fiy074.
98. Jahn MT, Markert SM, Ryu T, Ravasi T, Stigloher C, Hentschel U, Moitinho-Silva L. Shedding light on cell compartmentation in the candidate phylum Poribacteria by high resolution visualisation and transcriptional profiling. Sci Rep. 2016;6:35860.
99. Gauthier MEA, Watson JR, Degnan SM. Draft genomes shed light on the dual bacterial symbiosis that dominates the microbiome of the coral reef sponge *Amphimedon queenslandica*. Front Mar Sci. 2016;3:196.
100. Gao ZM, Zhou GW, Huang H, Wang Y. The cyanobacteria-dominated sponge *Dactylospongia elegans* in the South China Sea: prokaryotic community and metagenomic insights. Front Microbiol. 2017;8:1387.
101. Nguyen MTHD, Liu M, Thomas T. Ankyrin-repeat proteins from sponge symbionts modulate amoebal phagocytosis. Mol Ecol. 2014;23:1635–45.
102. Mittl PRE, Schneider-Brachert W. Sel1-like repeat proteins in signal transduction. Cell Signal. 2007;19:20–31.
103. Burgsdorf I, Slaby BM, Handley KM, Haber M, Blom J, Marshall CW, et al. Lifestyle evolution in cyanobacterial symbionts of sponges. MBio. 2015;6:e00391–15.
104. Harrington C, Del Casale A, Kennedy J, Neve H, Picton BE, Mooij MJ, et al. Evidence of bacteriophage-mediated horizontal transfer of bacterial 16S rRNA genes in the viral metagenome of the marine sponge *Hymeniacidon perlevis*. Microbiology. 2012;158:2789–95.
105. Hadas E, Marie D, Shpigel M, Ilan M. Virus predation by sponges is a new nutrient-flow pathway in coral reef food webs. Limnol Oceanogr. 2006;51:1548–50.
106. Belarbi E, Gomez AC, Chisti Y, Camacho FG, Grima EM. Producing drugs from marine sponges. Biotechnol Adv. 2003;21:585–98.
107. Mehbub MF, Lei J, Franco C, Zhang W. Marine sponge derived natural products between 2001 and 2010: trends and opportunities for discovery of bioactives. Mar Drugs. 2014;12:4539–77.
108. Siegl A, Hentschel U. PKS and NRPS gene clusters from microbial symbiont cells of marine sponges by whole genome amplification. Env Microbiol Rep. 2010;2:507–13.
109. Pimentel-Elardo SM, Grozdanov L, Proksch S, Hentschel U. Diversity of nonribosomal peptide synthetase genes in the microbial metagenomes of marine sponges. Mar Drugs. 2012;10:1192–202.
110. Rohde S, Nietzer S, Schupp PJ. Prevalence and mechanisms of dynamic chemical defenses in tropical sponges. PLoS One. 2015;10:e0132236.
111. Wilson HV. On some phenomena of coalescence and regeneration in sponges. J Exp Zool. 1907;5:245–58.
112. Muller WEG, Blumbach B, Muller IM. Evolution of the innate and adaptive immune systems – relationships between potential immune molecules in the lowest metazoan phylum (Porifera) and those in vertebrates. Transplantation. 1999;68:1215–27.
113. Degnan SM. The surprisingly complex immune gene repertoire of a simple sponge, exemplified by the NLR genes: a capacity for specificity? Dev Comp Immunol. 2015;48:269–74.

114. Steindler L, Schuster S, Ilan M, Avni A, Cerrano C, Beer S. Differential gene expression in a marine sponge in relation to its symbiotic state. Mar Biotechnol. 2007;9:543–9.
115. Riesgo A, Peterson K, Richardson C, Heist T, Strehlow B, McCauley M, et al. Transcriptomic analysis of differential host gene expression upon uptake of symbionts: a case study with *Symbiodinium* and the major bioeroding sponge *Cliona varians*. BMC Genomics. 2014;15:376.
116. Yuen B. Deciphering the genomic toolkit underlying animal-bacteria interactions – insights through the demosponge *Amphimedon queenslandica*. The University of Queensland; 2016.
117. Wiens M, Korzhev M, Krasko A, Thakur NL, Perovic-Ottstadt S, Breter HJ, et al. Innate immune defense of the sponge *Suberites domuncula* against bacteria involves a MyD88-dependent signaling pathway – induction of a perforin-like molecule. J Biol Chem. 2005;280:27949–59.
118. Wiens M, Korzhev M, Perovic-Ottstadt S, Luthringer B, Brandt D, Klein S, et al. Toll-like receptors are part of the innate immune defense system of sponges (Demospongiae: Porifera). Mol Biol Evol. 2007;24:792–804.
119. Germer J, Cerveau N, Jackson DJ. The holo-transcriptome of a calcified early branching metazoan. Front Mar Sci. 2017;4:81.
120. Bohm M, Hentschel U, Friedrich AB, Fieseler L, Steffen R, Gamulin V, et al. Molecular response of the sponge *Suberites domuncula* to bacterial infection. Mar Biol. 2001;139:1037–45.
121. Schroder HC, Grebenjuk VA, Binder M, Skorokhod A, Batel R, Hassanein H, et al. Functional molecular biodiversity: assessing the immune status of two sponge populations (*Suberites domuncula*) on the molecular level. Mar Ecol. 2004;25:93–108.
122. Grice LF, Gauthier MEA, Roper KE, Fernandez-Busquets X, Degnan SM, Degnan BM. Origin and evolution of the sponge aggregation factor gene family. Mol Biol Evol. 2017;34:1083–99.
123. Buckley KM, Rast JP. Diversity of animal immune receptors and the origins of recognition complexity in the deuterostomes. Dev Comp Immunol. 2015;49:179–89.
124. Palmer CV, McGinty ES, Cummings DJ, Smith SM, Bartels E, Mydlarz LD. Patterns of coral ecological immunology: variation in the responses of Caribbean corals to elevated temperature and a pathogen elicitor. J Exp Biol. 2011;214:4240–9.
125. Burge CA, Mouchka ME, Harvell CD, Roberts S. Immune response of the Caribbean Sea fan, *Gorgonia ventalina*, exposed to an *Aplanochytrium* parasite as revealed by transcriptome sequencing. Front Physiol. 2013;4:180.
126. van de Water JAJM, Lamb JB, Heron SF, van Oppen MJH, Willis BL. Temporal patterns in innate immunity parameters in reef-building corals and linkages with local climatic conditions. Ecosphere. 2016;7:e01505.
127. Revilla-i-Domingo R, Schmidt C, Zifko C, Raible F. Establishment of transgenesis in the demosponge suberites domuncula. Genetics. 2018;210(2):435–43.
128. Webster NS. Sponge disease: a global threat? Environ Microbiol. 2007;9:1363–75.
129. Maldonado M, Sanchez-Tocino L, Navarro C. Recurrent disease outbreaks in corneous demosponges of the genus *Ircinia*: epidemic incidence and defense mechanisms. Mar Biol. 2010;157:1577–90.
130. Luter HM, Whalan S, Webster NS. Prevalence of tissue necrosis and brown spot lesions in a common marine sponge. Mar Freshw Res. 2010a;61:484–9.
131. Luter HM, Whalan S, Webster NS. Exploring the role of microorganisms in the disease-like syndrome affecting the sponge *Ianthella basta*. Appl Environ Microbiol. 2010b;76:5736–44.
132. Rützler K. Mangrove sponge disease induced by cyanobacterial symbionts: failure of a primitive immune system? Dis Aquat Org. 1988;5:143–9.
133. Cervino JM, Winiarski-Cervino K, Polson SW, Goreau T, Smith GW. Identification of bacteria associated with a disease affecting the marine sponge *Ianthella basta* in New Britain, Papua New Guinea. Mar Ecol Prog Ser. 2006;324:139–50.

134. Smith FGW. Sponge disease in British Honduras, and its transmission by water currents. Ecology. 1941;22:415–21.
135. Storr JF. Ecology of the Gulf of Mexico commercial sponges and its relation to the fishery. Washington, DC: US Department of the Interior, Bureau of Commercial Fisheries; 1964.
136. Webster NS, Negri AP, Webb RI, Hill RT. A spongin-boring α-proteobacterium is the etiological agent of disease in the Great Barrier Reef sponge *Rhopaloeides odorabile*. Mar Ecol Prog Ser. 2002;232:305–9.
137. Choudhury JD, Pramanik A, Webster NS, Llewellyn LE, Gachhui R, Mukherjee J. The pathogen of the Great Barrier Reef sponge *Rhopaloeides odorabile* is a new strain of *Pseudoalteromonas agarivorans* containing abundant and diverse virulence-related genes. Mar Biotechnol. 2015;17:463–78.
138. Angermeier H, Kamke J, Abdelmohsen UR, Krohne G, Pawlik JR, Lindquist NL, et al. The pathology of sponge orange band disease affecting the Caribbean barrel sponge *Xestospongia muta*. FEMS Microbiol Ecol. 2011;75:218–30.
139. Angermeier H, Glockner V, Pawlik JR, Lindquist NL, Hentschel U. Sponge white patch disease affecting the Caribbean sponge *Amphimedon compressa*. Dis Aquat Org. 2012;99:95–102.
140. Egan S, Gardiner M. Microbial dysbiosis: rethinking disease in marine ecosystems. Front Microbiol. 2016;7:991.
141. Vadstein O, Bergh O, Gatesoupe FJ, Galindo-Villegas J, Mulero V, Picchietti S, et al. Microbiology and immunology of fish larvae. Rev Aquacult. 2013;5:S1–S25.
142. Wendling CC, Batista FM, Wegner KM. Persistence, seasonal dynamics and pathogenic potential of vibrio communities from Pacific oyster hemolymph. PLoS One. 2014;9:e94256.
143. Lesser MP, Bythell JC, Gates RD, Johnstone RW, Hoegh-Guldberg O, et al. Are infectious diseases really killing corals? Alternative interpretations of the experimental and ecological data. J Exp Mar Bio Ecol. 2007;346:36–44.
144. Luter HM, Bannister RJ, Whalan S, Kutti T, Pineda MC, Webster NS. Microbiome analysis of a disease affecting the deep-sea sponge *Geodia barretti*. FEMS Microbiol Ecol 2017;93:fix074.
145. Gao ZM, Wang Y, Tian RM, Lee OO, Wong YH, Batang ZB, et al. Pyrosequencing revealed shifts of prokaryotic communities between healthy and disease-like tissues of the Red Sea sponge *Crella cyathophora*. PeerJ. 2015;3:e890.
146. Webster NS, Xavier JR, Freckelton M, Motti CA, Cobb R. Shifts in microbial and chemical patterns within the marine sponge *Aplysina aerophoba* during a disease outbreak. Environ Microbiol. 2008;10:3366–76.
147. Blanquer A, Uriz MJ, Cebrian E, Galand PE. Snapshot of a bacterial microbiome shift during the early symptoms of a massive sponge die-off in the Western Mediterranean. Front Microbiol. 2016;7:752.
148. Bakaletz LO. Developing animal models for polymicrobial diseases. Nat Rev Microbiol. 2004;2:552–68.
149. Sibley CD, Duan KM, Fischer C, Parkins MD, Storey DG, Rabin HR, et al. Discerning the complexity of community interactions using a *Drosophila* model of Polymicrobial infections. PLoS Pathog. 2008;4:e1000184.
150. Ryan ET. The intestinal pathobiome: its reality and consequences among infants and young children in resource-limited settings. J Infect Dis. 2013;208:1732–3.
151. Sweet MJ, Bulling MT. On the importance of the microbiome and pathobiome in coral health and disease. Front Mar Sci. 2017;4:9.
152. Carballo JL, Bell JJ. Climate change, ocean acidification and sponges. Cham: Springer; 2017.
153. Webster N, Pantile R, Botte E, Abdo D, Andreakis N, Whalan S. A complex life cycle in a warming planet: gene expression in thermally stressed sponges. Mol Ecol. 2013;22:1854–68.
154. Guzman C, Conaco C. Gene expression dynamics accompanying the sponge thermal stress response. PLoS One. 2016;11:e0165368.
155. Webster NS, Cobb RE, Negri AP. Temperature thresholds for bacterial symbiosis with a sponge. ISME J. 2008;2:830–42.

156. Freeman CJ, Baker DM, Easson CG, Thacker RW. Shifts in sponge-microbe mutualisms across an experimental irradiance gradient. Mar Ecol Prog Ser. 2015;526:41–53.
157. Luter HM, Widder S, Botté ES, Wahab MA, Whalan S, Moitinho-Silva L, et al. Biogeographic variation in the microbiome of the ecologically important sponge, *Carteriospongia foliascens*. PeerJ. 2015;3:e1435.
158. Smith AM, Berman J, Key MM, Winter DJ. Not all sponges will thrive in a high-CO_2 ocean: review of the mineralogy of calcifying sponges. Palaeogeogr Palaeoclimatol Palaeoecol. 2013;392:463–72.
159. Cebrian E, Uriz MJ, Garrabou J, Ballesteros E. Sponge mass mortalities in a warming Mediterranean Sea: are cyanobacteria-harboring species worse off? PLoS One. 2011;6:e20211.
160. Pantile R, Webster N. Strict thermal threshold identified by quantitative PCR in the sponge *Rhopaloeides odorabile*. Mar Ecol Prog Ser. 2011;431:97–105.
161. Simister R, Taylor MW, Tsai P, Fan L, Bruxner TJ, Crowe ML, et al. Thermal stress responses in the bacterial biosphere of the Great Barrier Reef sponge, *Rhopaloeides odorabile*. Environ Microbiol. 2012;14:3232–46.
162. Fan L, Liu M, Simister R, Webster NS, Thomas T. Marine microbial symbiosis heats up: the phylogenetic and functional response of a sponge holobiont to thermal stress. ISME J. 2013;7:991–1002.
163. Fang JKH, Schonberg CHL, Mello-Athayde MA, Hoegh-Guldberg O, Dove S. Effects of ocean warming and acidification on the energy budget of an excavating sponge. Glob Chang Biol. 2014;20:1043–54.
164. Fang JKH, Mello-Athayde MA, Schonberg CHL, Kline DI, Hoegh-Guldberg O, Dove S. Sponge biomass and bioerosion rates increase under ocean warming and acidification. Glob Chang Biol. 2013;19:3581–91.
165. Bennett HM, Altenrath C, Woods L, Davy SK, Webster NS, Bell JJ. Interactive effects of temperature and pCO_2 on sponges: from the cradle to the grave. Glob Chang Biol. 2017;23:2031–46.
166. Lesser MP, Fiore C, Slattery M, Zaneveld J. Climate change stressors destabilize the microbiome of the Caribbean barrel sponge, *Xestospongia muta*. J Exp Mar Biol Ecol. 2016;475:11–8.
167. Simister R, Taylor MW, Tsai P, Webster N. Sponge-microbe associations survive high nutrients and temperatures. PLoS One. 2012;7:e52220.
168. Luter HM, Gibb K, Webster NS. Eutrophication has no short-term effect on the *Cymbastela stipitata* holobiont. Front Microbiol. 2014;5:216.
169. Turque AS, Batista D, Silveira CB, Cardoso AM, Vieira RP, Moraes FC, et al. Environmental shaping of sponge associated archaeal communities. PLoS One. 2010;5:e15774.
170. Pineda MC, Strehlow B, Sternel M, Duckworth A, den Haan J, Jones R, et al. Effects of sediment smothering on the sponge holobiont with implications for dredging management. Sci Rep. 2017;7:5156.
171. Reshef L, Koren O, Loya Y, Zilber-Rosenberg I, Rosenberg E. The coral probiotic hypothesis. Environ Microbiol. 2006;8:2068–73.
172. Webster NS, Reusch TBH. Microbial contributions to the persistence of coral reefs. ISME J. 2017;11:2167–74.
173. McFall-Ngai M, Hadfield MG, Bosch TCG, Carey HV, Domazet-Loso T, Douglas AE, et al. Animals in a bacterial world, a new imperative for the life sciences. Proc Natl Acad Sci USA. 2013;110:3229–36.
174. Reusch TB. Climate change in the oceans: evolutionary versus phenotypically plastic responses of marine animals and plants. Evol Appl. 2014;7:104–22.
175. Elena SF, Lenski REM g. Evolution experiments with microorganisms: the dynamics and genetic bases of adaptation. Nat Rev Genet. 2003;4:457–69.

176. Rosenberg E, Koren O, Reshef L, Efrony R, Zilber-Rosenberg I. The role of microorganisms in coral health, disease and evolution. Nat Rev Microbiol. 2007;5:355–62.
177. Putnam HM, Barott KL, Ainsworth TD, Gates RD. The vulnerability and resilience of reef-building corals. Curr Biol. 2017;27:R528–40.
178. Ribes M, Calvo E, Movilla J, Logares R, Coma R, Pelejero C. Restructuring of the sponge microbiome favors tolerance to ocean acidification. Environ Microbiol Rep. 2016;8:536–44.
179. Morrow KM, Bourne DG, Humphrey C, Botte ES, Laffy P, Zaneveld J, et al. Natural volcanic CO_2 seeps reveal future trajectories for host-microbial associations in corals and sponges. ISME J. 2015;9:894–908.
180. Bell JJ, Davy SK, Jones T, Taylor MW, Webster NS. Could some coral reefs become sponge reefs as our climate changes? Glob Chang Biol. 2013;19:2613–24.
181. Ereskovsky AV, Richter DJ, Lavrov DV, Schippers KJ, Nichols SA. Transcriptome sequencing and delimitation of sympatric *Oscarella* species (*O. carmela* and *O. pearsei* sp nov) from California, USA. PLoS One. 2017;12:e0183002.
182. Fortunato SAV, Adamski M, Ramos OM, Leininger S, Liu J, Ferrier DEK, et al. Calcisponges have a ParaHox gene and dynamic expression of dispersed NK homeobox genes. Nature. 2014;514:620–3.
183. Nichols SA, Roberts BW, Richter DJ, Fairclough SR, King N. Origin of metazoan cadherin diversity and the antiquity of the classical cadherin/beta-catenin complex. Proc Natl Acad Sci U S A. 2012;109:13046–51.

Chapter 8
After the Taxonomic Identification Phase: Addressing the Functions of Symbiotic Communities Within Marine Invertebrates

Jose V. Lopez

> *"What am I living for and what am I dying for are the same question."*
> – Margaret Atwood, *The Year of the Flood*

> *A man said to the universe:*
> *"Sir, I exist!"*
> *"However," replied the universe,*
> *"The fact has not created in me*
> *A sense of obligation."*
> – Stephen Crane, *War Is Kind and Other Poems*

Contents

Abstract Characterizations of the identity and diversity of microbial symbiotic communities ("microbiomes") within different sponges have advanced considerably over the last two decades. Thousands of microbes, mostly unculturable,

J. V. Lopez (✉)
Halmos College of Natural Sciences and Oceanography, Guy Harvey Oceanographic Center, Nova Southeastern University, Dania Beach, FL, USA
e-mail: joslo@nova.edu

Z. Li (ed.), *Symbiotic Microbiomes of Coral Reefs Sponges and Corals*,
https://doi.org/10.1007/978-94-024-1612-1_8

operational taxonomic units (OTUs) have been identified through the advances of high-throughput DNA sequencing. However, in spite of compelling data pointing to bona fide symbioses between many microbes and the sponge host, determination of specific microbial symbiont functions remains difficult to pinpoint and equivocal in many cases. In this chapter, I highlight past and present approaches toward addressing the potential functions of microbial symbionts (mostly bacterial) found in marine sponges and invertebrates. In an interesting irony, one barrier to effective definition of some symbiont microbial functions stems from their obligate dependence on their host. Investigations suggest that microbes significantly contribute to fundamental processes such as elemental cycling, anabolism, and catabolism. An additional likely role for symbionts is the biosynthesis of unique secondary metabolites (SMs) and vitamins, exhibited in many sponge species. These can be used as defensive or communication factors increasing fitness and thus benefiting the holobiont, which appears more and more reminiscent of a vibrant community than the traditional notion of an individual sponge. One approach to circumvent the dearth of empirical evidence on specific symbiont functions is to apply modern *-omics* methods: for example, sequencing the entire sponge holobiont (host and microbiome) as a metagenome and metatranscriptome can reveal potential functional genetic information. Together with computational tools, one can infer function from biological sequence data, although rigorous experimentation is still needed for verifications. Newer combinations of older, sophisticated technologies such as fluorescence in situ hybridization-correlative light and electron microscopy (FISH-CLEM), stable isotope tracking, and nanoscale secondary ion mass spectrometry (NanoSIMS) now promise to reveal more potential symbiont functions. Metaproteomics will also help further advance the understanding of the relationships within the holobiont community, but its wide applicability still remains mostly on the horizon. Other pervasive questions on the origins, coevolution, and fitness of specific symbiont-host partners include relevant microbiome functions within their orbit.

Keywords Symbiotic microbes · Marine invertebrates · Function

8.1 Introduction

Besides being one of the premier evolutionary biologists of his time, Alfred Russel Wallace during his career also questioned on where the seat of consciousness lays – in single cells or only through the confederation of cells into complex (neural) tissue. In a parallel fashion, one could also rhetorically ask whether organismal identity stems mostly from being in some association, a coordinated biological system, or within the single-celled state which often has no allegiance or responsibility other than to its own reproductive fission event and singular lineage.

Species identities are often delineated by their observable characteristics described and measured in detail by expert taxonomists. Defining the function of species becomes central and ecological when they play pivotal keystone roles in their specific environment. Most species appear dispensable in any given environment, and their radiation occurred due to an abundance of either resources or ecological opportunity. From this, "what's in a name?" is the proverbial loaded question, yet highly relevant and applies poignantly to the realm of taxonomy and classification. Although unicellular evolution and radiation have marched on for eons, bacterial taxonomy still lags behind the advances and classification studies of their macroscopic descendants. The best example for this point is that only 13,000 bacteria have Latin binomial nomenclature ([1]; also see http://www.bacterio.net/).

As a host to many diverse microorganisms, phylum Porifera is the main subject of this volume and encompasses a wide range of over 8000 species, which can include important roles in benthic ecosystems [2]. The unique bauplan and lifestyle of sponges as sessile filter feeders makes them simultaneously distinct yet also poignantly integrated with its surroundings, the water column. Few organisms or particularly other animal species allow their external environment, known to be replete with many planktonic microbes, to flow intimately within and through their bodies without some transformation of the contents. In fact, if virtually any other multicellular organism slightly loosened its strict enforcement of homeostasis, it would probably be devoured by many of nature's opportunists that were kept at bay by only the of thinnest margins. From this basic question of defining the sponge's "identity" and its preservation, e.g., how it determines self from nonself, stems other fundamental conundrums: what ecological or *function role* does the sponge's own symbiotic microbial community have for its host, if any? And as one last extension – are identity and function two sides of the same coin and thus inseparable?

Dictionaries define the word *function* as "the special purpose or activity for which a thing exists or is used." This utility can appear selectively within animate biological entities or serendipitously and purposely in inanimate objects based on the user and the design. In the biological sense, many functions contribute to survival and diversity. Before delving deeper into the details of symbiont function in this chapter, I will provide further background and theoretical contexts for explaining function.

The main goal of the present discourse is to identify and describe the potential or probable function, raison d'etre, for microbial symbionts and their microbial communities (also known as "microbiomes") within various marine invertebrates, with particular focus on sponges ([3–5]). To reiterate, the idea of symbiosis means the living together of more of than one organism within a single body (holobiont) or space [6]. Secondly, the latest approaches to measuring and assessing possible functions will be discussed. This takes into consideration that more sponge biologists now have a better working knowledge of which bacterial taxa may compose a particular symbiotic community. For example, 16S small subunit rRNA-based approaches assist in microbial identification and partially reflect metagenomic information (discussed in more detail later); sometimes the applications can overinflate taxonomic representation, so the ubiquitous rRNA gene's limitations have been acknowledged [7]. In essence, the realization represents a kind of rite of passage, consequently invoking the inevitability that we are approaching, if not already in the midst of a "post"-16S rRNA taxonomy phase [8, 9], at least for sponge microbial symbiosis. As novel, often even larger datasets emerge (e.g., metagenomes, metabolomes), the community should prepare for changes, including likely new workflows, hypothetical frameworks, or paradigm shifts [13].

From the biological standpoint, function can also connote *adaptation*, which is "a characteristic that enhances the survival or reproduction of organisms that bear it, relative to alternative character states (especially the ancestral condition in the population in which the adaptation evolved)" [11]. Consequently, from an evolutionary viewpoint, biological functions arise mostly by necessity and as a result of interactions with the environment through the process of natural selection. Adaptive functions can then contribute to the reproductive success (fitness) of the organism. For example, ribosomes and all their associated rRNA molecules and proteins are required for protein translation; enzymes breakdown macromolecules to fuel metabolism; chromatophores, iridophores and leucophores help modulate vacuoles of dye which give octopi the ability to camouflage; the adipose tissue of a melon in the foreheads of toothed cetaceans help focus sound modulations during echolocations. These few examples portray the breadth of functions, from molecules to differentiated cell assemblages (tissues) and how adaptations with specific biological functions in nature abound and occur at all levels and biomes. Utility of the characteristic trait remains central, compared to vestigial parts whose function but not form was lost over time. Many more functions will be discussed throughout this chapter, though direct evidence of each adaptive connotation may not be fully proven. Nonetheless, evolutionary theory provides a key for interpretation, which can assist in the search for meaningful patterns and rationale.

There is little doubt that microbial processes provide important ecosystem services (functions), on regional or global scales, providing the foundation for sustainability and many biogeochemical cycles [10, 12, 13]. The awareness and recognition of beneficial microbial functions continues to grow, propelled notably by long-term research investments such as NSF Microbial Observatories in the 2000s and coordinated research efforts such as the "Microbial Census" [14] and the International Census of Marine Microbes (ICoMM). The enthusiasm and call for microbiome studies continue to be fueled by stories of human microbiomes in the scientific literature and popular press [15, 16], along with recent calls for microbiome-centered research communities and proposals for a US$121-million National Microbiome Initiative (NMI) [17, 18]. Some researchers even call for dropping the term "pathogen" altogether, which may be premature and probably ill-fated [19].

Microbial symbiotic activities may be viewed from the detrimental or the positive, as well as from the host or microbial, perspectives. The range of possible host-microbe partner interactions runs from beneficial (mutualistic) to neutral (commensal) to destructive (parasitic, pathogen), from each member's perspective, and sometimes a biological continuum of these states may exist. One of the best-characterized examples of such a continuum can be illustrated by gut microbes [20]. Enteric bacteria within *Bacteroidales* and *Enterobacteria* such as *E. coli* perform beneficial activities in mammalian digestive tracts such as pH regulation and vitamin production. These are important functions from the holobiont perspective. However, these beneficial roles fade when a strain serendipitously takes up and integrates pathogenic genetic loci (pathogenicity islands or PAIs) into their genomes via horizontal gene transfer [21, 22]. Parallel examples of transformations of other symbionts abound in nature and the literature. Reversal of the detrimental traits can be achieved when the offending coding sequences or gene products become attenuated or are deleted from the host genome. Symbiosis discussions should be viewed through this prism of dynamic complexity and a framework of exceptions (environmental factors) [6, 23, 24].

As introduced above, a strong consensus exists that symbiosis mostly serves as a vital function for survival in many instances. "Nothing exists in a vacuum," with "exists" as the operative term. The coexistence of both eukaryotic host and microbial symbiont actually defines most macroorganisms as holobionts (macroorganismal host + its microbial symbionts) [25–27]. The holobiont concept has evolved further to encompass seeing a composite genome that together can be seen as a possible unit of selection, albeit with caveats [28, 81]. The growing tandem focus on microbes and their community functions has elevated in recent years. That is, few axenic organisms are known to exist in nature, and in turn microbes occur in practically every habitat on the planet, whether commensal or beneficial [24]. Compared with arctic/Antarctic low temperatures and highly acidic or anoxic conditions in harsh biomes, living obligately and contributing to the fitness of a macroorganismal host probably represents the path of least resistance (and survival) for some microbial species.

Perhaps one could also ask the question, "Are sponge microbial communities required to have a defined function?" This proposition implies that essential functions for the holobiont cannot be provided by only one or two microbial species and rather that multiple species provide a larger repertoire of enhancements to the basic holobiont function. Yet free-living bacterial genomes typically hold only a few thousand genes that are mostly for basic housekeeping functions of the symbiont itself.

Biological functions for some marine microbial symbionts can appear clear and obvious, and a smattering appear in Table 8.1: providing nutrients [3, 29], catabolic/anabolic metabolism, bioluminescence in squid, and other marine animals [24, 30], nitrogen metabolism [31, 32], and defenses (via secondary metabolites or CRISPRs) [33, 34].

8.2 The View from a Benthic Resident: Marine Sponge Perspectives

What phylum Porifera lacks in charismatic movements and behavior is compensated by the group's collective biochemical and physical brilliance, ecological flexibility, evolutionary longevity, and hospitality for microbial partners. In selected topics that will be expanded further in this chapter, it will be shown that sponge natural history can elicit fundamental questions relevant to ecology and evolution. For example, a 1 kg sponge holobiont can filter up to 24,000 L of seawater per day

Table 8.1 Examples of symbiont functions across diverse marine organisms

Function or trait	Symbiont taxa	Host	References
Nutrient exchange and sequestration	Cyanobacteria Endosymbiont Endoriftia *Oceanospirillales*	*Aplysina cauliformis*, *Xestospongia muta* Tubeworm *Riftia pachyptila* *Osedax* worms	[3, 29, 41, 92, 279, 281]
Nitrogen metabolism, fixation	*Alpha*- and *Gammaproteobacteria* Cyanobacteria	Porifera (*Ircinia strobilina*)	[31, 32, 59, 214]
Bioluminescence for predation, camouflage	*Vibrio fischeri* and *Entovibrio* spp.	Deep-sea Anglerfish Bobtail squid *Euprymna scolopes*	[116, 217, 279, 280, 282]
Body structure, biomineralization	Endosymbiotic bacteria (calcibacteria)	Porifera, Cnidaria	[93, 96, 100, 281]
Biochemical defenses (CRISPRs, SMs, natural products)	Mostly uncultured bacteria	Various Bryozoa and Porifera	[34, 139, 195, 267, 269]

and provide a physical bridge between the pelagic realm and the benthos [35–37]. Sponge taxa such as demosponge *Cliona* have the capacity to reshape coral reefs, through their bioerosion of calcium carbonate ([38, 39]. Moreover, evidence continues to grow that sponges are intimately involved in nutrient cycling, as they facilitate transfer and retention of essential carbon and other elements on oligotrophic reefs [40, 41]. The wide biological diversity of sponges has been expertly reviewed [2] and should not be surprising for a phylum with over 500 million years of evolution [42, 43]. A lively debate has erupted regarding the primary ancestor of the Metazoa, which includes Ctenophora and Porifera, long considered to hold the most ancestral position beginning with Haeckel (1878) [44–49]. The discourse provides an opportunity to enhance phylogenetic algorithms and evolutionary models, while also upending current paradigms.

There has been much effort to characterize the composition and relative abundances of specific symbiont taxa in sponge microbiomes. With regard to hosting diverse microbial microcosms, sponges can be viewed as a microbial niche, incubator, and generator par excellence. Over the past two decades, the sponge research community has described and cataloged a large dataset of the microbial taxa that reside and appear to be symbiotic within this unique marine invertebrate. Since Wilkinson's pioneering papers describing the ultrastructure of sponge-associated microbes, many studies have since followed. Applying modern molecular tools and culture-independent methods based on 16S rRNA and metagenomic gene sequences to characterize diverse sponge microbial communities has ushered in a highly productive period of sponge "symbiology" research [10, 51–53, 56–68]. Several extensive reviews underscore the major research advances in sponge symbiosis research [3, 5, 22, 52].

Sponge symbiotic systems are clearly complex, based on the high number of actual bacterial symbionts in many demosponge species, compared to other marine organismal systems [22, 64]. Besides identifying microbial symbiont classification at higher resolution, symbiotic relationships need further clarification and have spawned wider interest even outside sponge research circles. So in spite of recent advances, much still needs to be learned regarding how community members specifically interact with each other, along with the ensuing ecological and evolutionary impacts, should a biological "function" actually arise.

For example, while nutrient exchange is expected between symbionts and hosts who have developed gut cavities, ample evidence also exists for nutrient transfer within Porifera, which lack true gut cavities [69]. Antibiotic defenses as part of competitive mechanisms are likely one form of interaction, but cooperative or commensal activities may also be present.

"Next-generation" high-throughput DNA sequencing (HT) advances, with less expensive and more rapid sequencing technologies, have helped usher in better characterizations of sponge microbial communities [167]. Webster et al. [71] represent one of the first studies to apply HT DNA sequencing methods (454 pyrosequencing, Illumina, Oxford Nanopore) to obtain comprehensive taxonomic profiles of sponge microbiomes. This was soon followed by other studies that provided wide profiles across diverse sponge taxa, including all three classes [22, 63, 72]. The sponge research community's collective effort to characterize sponge microbial communities culminated in a wide collaboration with the Earth Microbiome Project [64]. Twenty different laboratories around the world contributed samples for the microbiome analyses of more than 81 different sponge species derived from the Atlantic, Pacific, and Indian oceans as well as the Mediterranean and Red seas. General conclusions were that some sponge symbiont communities could be greatly influenced by both the environment and host phylogeny; microbial species richness could range from 50 to 3820 genetically distinct OTUs, but which to appear assemble independently.

Many sponge symbiotic habitats conform to the archetypal microbiome structure which is dominated by a relative few taxa, followed by a much longer "tail" of less common microbes that make up the "rare biosphere" first described by Sogin (181). Much of this tail can consist of singleton (or doubleton) reads, which are sequences that occur only once or twice, respectively, among the total large output (millions of reads). This renders them unique but also possibly due to machine, enzymatic, or random errors. In the same context, more precise quantitation of symbiont taxa can provide further circumstantial evidence of function. Overwhelming biomass and proportions cannot be ignored. When a specific microbial taxon or a few taxa show clear numerical domination in a habitat, functional importance may be inferred. In one specific sponge example, qPCR helped to verify distinct differences in the relative abundance and ability to quantitatively detect culturable vs unculturable microbial taxa in situ [73]. Signatures of previously cultured isolates from the deep-sea sponge, *Vetulina*, appeared in low numbers, with several isolates below the detectable range of <100 copies of template per reaction.

A more recent quantitative study showed that applying allometry, the method of measuring various physiological features – e.g., metabolic rate, organ size, and scaling with body size – can be illustrative. This novel approach compared the abundances and diversity of animal-associated microbes across various host species (holobionts) [74]. They found a surprising lack of pattern in the relationships between animal-associated microbial species diversity and animal mass. Interestingly, this finding also seems to counter the basic precepts of MacArthur and Wilson's "island biogeography theory," which posits that species richness correlates positively with habitat size [75]. Another salient point to consider is that similar essential functions may be provided by different microbial species and thus becomes a factor to lower the expectations or requirements for precise taxonomic identifications [76]. In this context, perhaps we should emphasize that function refers to an activity that serves the community, over any individual species or group identity. In other words, the fitness of an individual might be best enhanced through aiding the collective more so than competing within the collective [77]. This also highlights the value of a "gene-centric" approach to characterize function, via methods such as RNA-seq of community metatranscriptomes (see below), instead of by inferences from taxonomy. I can also introduce the idea that a complex organismal community could be viewed through a prism of an ecology of its collective molecules (e.g., an ecology and regulated equilibrium of genes, gene products, proteins, etc.). Basically all or most molecules within a cell appear for a specific reason and are subject to feedback inhibition, and thus a cellular "niche" is filled by their presence and function. Molecular ecology further comes into play when multiple molecules must interact with each other, which is a necessary occurrence for successful cellular metabolism; most cells are comprised of DNA-binding transcription factors, multimeric enzymes, RNA-protein-constituted ribosomes, etc. This view borrows partially from Dawkins' concept of the selfish gene archetype that organisms/cells will always be subservient vehicles to the master heredity code they transmit from generation to generation [78].

Nonetheless, rare symbionts have unequivocally been shown to reside within most sponges [22, 64, 79] and may be actively maintained by the holobiont in order to respond quickly to environmental perturbations. We then naturally should ask if there is an adaptive value to possess or maintain rare symbiont taxa. One possibility alluded to earlier is the potential for novel survival mechanisms if the dominant microbes succumbed to some density-dependent type of disturbance (toxin, temperature, etc.). Akin to a sudden "superbloom" of wildflowers that can occur after a torrential rain in a desert, perhaps the majority of sponge symbionts also lay dormant or in low numbers until conditions change and become more favorable.

8.3 Sponge Symbiosis and Evolution

If symbionts do indeed provide benefits or lead to adaptations for sponge survival, symbiont placement and function within sponge holobiont evolution should also be taken into account. Of course, because most microbes leave no fossil evidence, tracking past specific fluxes and compositional changes of sponge-specific microbial communities is even more difficult than inferring the precise sponge host evolution itself [46, 80].

On another evolutionary front, a growing body of literature points to the role of symbiosis in holobiont speciation [70, 81]. The subject necessarily evokes the question of how specific symbionts or inherent communities affect the inclusive fitness of each holobiont individual. This chapter has highlighted some of the advantages of having specific microbial symbionts carry out specific, beneficial activities. It has been proposed that microbial symbionts may exert multiple factors that can affect speciation, especially in sexually outcrossing species classified under the biological species concept: microbe-specific signals, microbe-assisted modifications of mating signals, and mate choice [70].

On a broader scale, there is little doubt that microorganisms, as tiny and inconspicuous as they are physically, can still affect and create whole ecosystems and biomes as an aggregate. The Great Oxidation Event was marked by a rapid rise in the partial pressure of atmospheric O2 (PO2) between 2.45 and 2.22 billion years ago (Gyr) [82], with larger increases in oxygen by 700 MYA and near modern levels at 543 MYA. This habitable atmosphere was driven by microbial photosynthesis [83]. Likewise the nitrogen cycle in the atmosphere and the soil has microbial activity and "fingerprints" at both ends through nitrogen fixation, nitrifying, and denitrification pathways. On reef environments, the well-known benefits and effects of primary production via photosynthesis are integral and iconic, since modern reef-building depends on the coral-dinoflagellate partnership [84, 85]. Reef building corals owe their existence to the partnership with photosynthetic, single-celled *Symbiodinium* dinoflagellates, another example of framework reliance on microbial function. The recognition and more precise characterization of microbial roles in molding reef ecosystems continue to be elaborated [86, 87]. Modeling will also likely acknowledge a shift toward more sponge (or non-coral)-dominated reefs, as rising ocean surface temperatures continue to rise and stress shallow coral reef ecosystems [39, 88].

In a similar context, the role of photosynthetic microbes or symbionts in sponges cannot be overlooked. Wilkinson [66] was one of the first to recognize the possibility of photosynthetic partners in the holobiont, and further support has now been garnered [89, 90]. Coevolution between sponge host and symbiont has been demonstrated for some groups such as cyanobacteria *Oscillatoria spongeliae* found in many *Dysidea* sponge species [89]. Interestingly, siliceous spicule production in sponges has no similar dependency for its formation, so sponge dominated reefs occur via heterotrophy. Nonetheless, several sponge species actively appear to have mutualistic associations with cyanobacteria [92].

In the same vein, it does not take a great leap of the imagination to posit that sponge-associated microbial communities represent separate, unique ecosystems unto themselves. Sponge symbiont composition is typically distinct from the surrounding seawater and sediment [22, 63, 64]. This finding cautions for the careful use of sponge microbiome data as environmental indicators or proxies for the surrounding ambient water.

8.4 Symbiont Contribution to Basic Poriferan Body Structure

A basic and obvious role of microbes for some sponges is helping to compose the poriferan organism's partial body structure, soft biomass or *mesohyl*, the matrix of collagen, siliceous spicules, and extracellular material that compose the basic body form [93]. Host genomes code for silicatein genes or carbonic anhydrases that produce the skeletal framework of spicules [94, 95]. However, some bacteria may actually contribute to sponge skeleton calcification, such as in the sponge species *Hemimycale* [96]. Because sponges lack true tissues, their symbionts in aggregate can provide structure/mass, biochemical flexibility, and ecological adaptability. Several early studies have shown that microbes contribute significantly to sponge body mass, sometimes surpassing 50% in some species [97]. This view was extended into a quasi-classification scheme of "low" and "high" microbial abundance sponges (LMA and HMA, respectively) [53, 98, 99]. In LMA sponge genera such as *Callyspongia* or *Niphates*, microbial contributions appear minimal to biomass and more closely reflect seawater content. Yet microbes represent important components of sponge mesohyl or affect structure in various ways. For example, Fig. 8.1 shows a bacteriocyte structure, with a specific membrane that contains bacterial cells, whose origin is unclear. Leys and Hill [100] detailed a compendium on poriferan bauplan and ultrastructure, which greatly extended earlier descriptions. Examination of TEM images from these many studies indicates the lack of flagella in the bacteria and encapsulation as protection from sponge archeocyte digestion [68], features which suggest a relatively comfortable residency within the host mesohyl. In social evolution, static individuals should not be over-exploitative of their environment since it can result in a tragedy of the commons they cannot escape from. Mobile individuals can escape the fate of their local exploitation. This is why altruism is more likely to evolve into a more viscous than freely mixing populations. This trait somewhat alludes to the lack of a highly developed immune system within sponges, which is partially supported by molecular studies that only detect basic innate immunity modes in genome sequences [101, 102]. Thus, sponges appear to

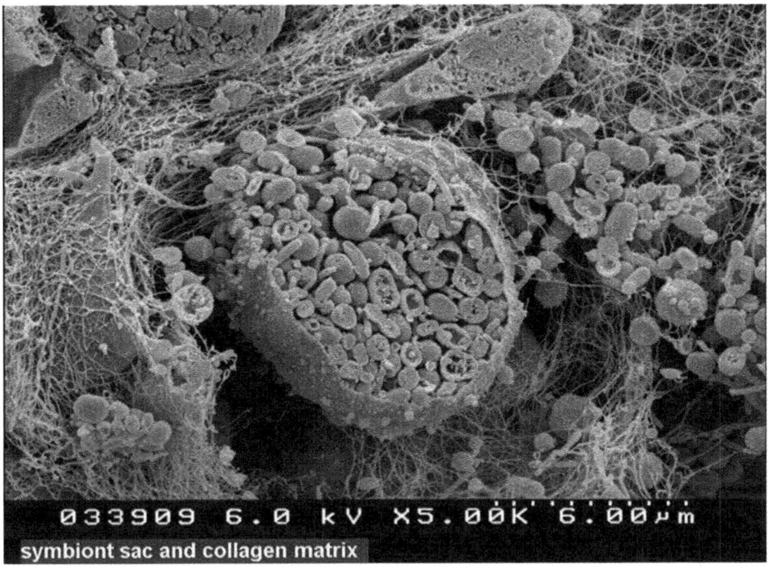

Fig. 8.1 Scanning electron micrograph of a likely bacteriocyte from *Spongosorites* sp., courtesy of Dr. Susan Sennett (HBOI@FAU). Note the membrane encasing multiple scores of bacterial cells (~1 μm)

have evolved into more of a tolerance to their microbial passengers. Even further, perhaps since microbes live within the sponge, it behooves the host to not have stringent defenses. The sponge does not need an immune system if some of the microbial symbionts themselves take on the role of defense. Just as metabolic functions are shared, defensive responsibilities may be deferred or shared among all established resident inhabitants.

Yet, an interesting question poses itself by pondering the origin of such a revolutionary system, the multicellular heterotrophic animal: how did the ancestral sponge develop? Perhaps, a more precise question for Porifera and the early functional role of its symbionts is "in its process of becoming multicellular, how and when did it obtain prokaryotes as an apparently willing partner?" This may be a crux of being a sponge and help partly answer the role of microbes in the symbiotic relationship.

Could symbiotic microbial communities have served as an original nucleation point for a sponge body, beginning as independent biofilms, before or possibly leading up to metazoan multicellularity itself [103]? I will return to this fundamental concept further along. For now, biofilm research has advanced extensively as a mature field unto itself [104–106], so the presently merging concepts regarding biofilms and sponge biology stems primarily as a thought exercise with only limited empirical data [107]. Nevertheless, although biofilms probably evolved without eukaryotic inputs per se, the phenomenon may have a bearing on the origins, evolution, and function of sponge microbial communities. Consider the parallels: bacterial biofilms are three-dimensional structures comprised of glycoproteins, glycolipids, and polysaccharides matrix (but without collagen) [108], formed primarily by bacteria to allow colonization, communication, and coping with hostile external environments. Bacterial biofilms have long been known to attract larval recruits of various marine invertebrates for settlement [109, 110]. Recently compounds such as tetrabromopyrrole have been shown to induce settlement of multiple coral species such as *Acropora palmata* and *Orbicella franksi* [111]; other biofilms can also promote horizontal gene transfer (HGT) ([113]; and see below), indicating a plurality of different potential functions based on the microbial community.

Biofilm formation could be self-serving for each microbial community member. For example, the structures could enhance HGT (via bacterial conjugation, transformation, or phage transduction). HGT can serve as an evolutionary mechanism but has only been sporadically studied or observed in marine biofilms [112–114, 116]. Ehrlich et al. [105] posit that multiple, closely related strains of bacteria in biofilms or infective swarms help to confuse and overwhelm host immune systems. The phenomenon encompasses an increase in microbial genetic diversity, which provides sources and platforms for gene transfer and ensuing genomic diversification. It is well known (e.g., via conversions of a beneficial or neutral symbiont to a pathogen) that microbes may rapidly manifest phenotypic plasticity and local adaptation via HGT. Moreover, transposases have been considered as one of the most common proteins in nature and corroborated by a study of the model Australian sponge *Cymbastela concentrica* [115]. They found about 50x more bacterial insertion elements and transposes in symbionts than planktonic bacteria in metagenomic comparisons. Similarly transposes appear common and relatively active in the gutless marine worm *Olavius algarvensis*, which likely plays a role in the transition of free-living bacteria to host restriction and obligate symbiosis [117].

It would be interesting to determine what could have constituted an initial attraction of those early colonizing bacteria and the enticement to lose mobility and become permanent residents.

Perhaps bacteria and their ability to form biofilms was the essential first step, i.e., where ancestral free-living eukaryotic cells (choanoflagellates) used the "glue" of an extracellular exudate or matrix (primordial mesohyl) of previously formed bacterial biofilms as nucleation and colonization centers in aquatic habitats. It has been shown that the choanoflagellate, *Salpingoeca rosetta*, initiates colony formation in response to bacterial signals [118]. Furthermore, perhaps the primordial sponge cell haphazardly coordinated with early biofilm bacteria, which included the pre-symbiotic ancestor of *Poribacteria*. The glue and adhesion of some sponge-microbe interactions may be partially facilitated by eukaryotic-like, ankyrin-repeat proteins (ARP), which have been found in several species of sponges including their microbiomes [119, 120].

One central point in the discussion of the poriferan bauplan is that the sponge extracellular matrix (ECM) consists primarily of the triple helical collagen, known chiefly as an animal product [121], not a regular component of a biofilm and rarely expressed in bacteria. Perhaps similar to ARPs, collagen genes could be horizontally transferred to bacteria. Possible collagen-like domains may have also evolved earlier in various bacteria, and gene sequences do appear sporadically, such as in the archaean *Nitrosopumilus maritimus* [122, 123]. Alternatively, a protozoan partner could have provided a collagen-like precursor, since the choanoflagellate genome codes for C-propeptide domain that could have been reshuffled to generate more derived fibril-forming collagens [124, 125].

The aggregation imperative can clearly have defensive purposes, and evidence exists for specific aggregations of microbes within modern sponges [62, 64]. Sponges fit snugly in the realm of chemical ecology and have long been a target of marine bioprospecting efforts focused on identifying and developing novel compounds with pharmaceutical potential (discussed later). Metabolism of intricate compounds matured through the evolutionary duration of bacterial genes and pathways, as certain lineages evolved potent bioactive natural products. Matz et al. [126] have shown that some marine biofilm communities generate compounds that ward off protistan predation. This type of scenario could be envisioned for the origin of sponges – i.e., those primordial free-living choanoflagellate-like cells settling on suitable surface but requiring protection from hostile predators. If biofilms were also present, it could have been advantageous to co-opt any toxic secondary metabolite (SM) production for the purposes of protecting a nascent colony of neo-sponge cells.

Consequently, the concept of how metazoan multicellularity originated can now be briefly revisited. This important physiological and evolutionary milestone has been discussed in depth elsewhere but can be distilled briefly into three primary hypotheses: (1) syncytial, (2) colonial, and (3) symbiotic [127–129, 132]. New hypotheses regarding the primordial ancestor have also emerged [130]. Bonner describes various examples of multicellular origins, including stalked colony formation from single-celled amoeboid forms of ciliates such as *Zoothamnium* and *Sorogena*; this behavior also resembles the well-known fruiting bodies of dictyostelid slime molds that aggregate when food sources become depleted. *Dictyostelium* does not appear to include bacteria as active cooperating partners. However, the slime mold has recently been shown to actively carry and culture their food supply to the next stage, prior to the aggregation event [131]. Eventually Margulis [132] formulated through extensive writings that multiple species cooperated to the point that group fitness exceeded the fitness of living separately. In the context that a larger, internally coordinated organism may better escape a predator or be able to control its own internal environment, the fitness arguments have plausibility. However, it must also take into account that the fitness of the entire holobiont, including the genetic content of both host and multiple symbionts, becomes affected under a single selective force. This does not conflict with standard natural selection as the traits with the highest overall fitness in the population are favored by natural selection; it is simply that when comparing fitness within and between holobionts, the highest fitness might be achieved through aiding the collective holobiont than going solo. This can be likened to the phenomenon of individual genes that benefit from forming collective chromosomes, individual cells benefiting from the formation of multicellular bodies, and individual social insects that benefit by aiding the colony [77]. These ideas have developed further in recent years toward a "hologenome" concept, including symbionts, which possibly exists for most organisms [24, 81, 133–135]. Indeed, Margulis referred to existence of multiple "supernumerary" genomes. This concept is complex, with many evolutionary ramifications, and thus still somewhat controversial and requires further empirical support and discussion [136]. Nonetheless, the realization that many extant species now have their own, sometimes coadapted microbial communities helps support the symbiotic hypothesis, which may eventually hold sway.

Interestingly, the development of multicellularity again evokes the establishment of some type of mechanism for identity, an immunological system, which at its essence is the recognition of "self vs. nonself." The latter, separate, nonself represents the environment. We do run into the danger of circularity at this point, since free-living microbes now and in the ancient past could be viewed as part of the external environment. If so, then incorporating microbes and their conversion to symbionts as an essential component for the generation of metazoan multicellularity gets very complicated, in both the figurative and literal sense. Can a sponge still be considered a sponge without its symbionts? Ryu et al. [135] did find an increase in immunological gene presence in the LMA sponge *Stylissa carteri*. However, the full topic of multicellularity origins goes the beyond the scope of this chapter, and the reader can be directed to other discourses [137, 138].

On the other hand, some studies show a lack of biofilm behavior (e.g., quorum sensing (QS) products) in sponge microbiomes and their metagenomes [139, 140]. Modern sponge symbiont microbiomes may have evolved past basic biofilm features and reflect the unique adaptations and characteristics of symbiont dependence. Host-specific factors, for example, could counter bacterial QS factors, as a way to control their specifically adapted community of symbionts.

8.5 Various Approaches to Determining Symbiont Function

Some of the discussion above remains within the realm of speculation. To determine specific functions in an experimental system, a specific assay is often employed to prove causality. However, the canonical dilemma of unculturability that has hampered traditional, terrestrial microbiology research for decades [9, 13, 141] also can stymie aspects of sponge and marine microbiology. The problem is compounded by the fact that some sponge species inhabit relatively inaccessible, sometimes very deep underwater habitats. Essential host-derived factors that maintain symbiotic partnerships cannot be separated from the novel microbes found within many sponge species if we wish to determine symbiont functions. Until biologists have better tools to study unculturable microbes in "real time," inference will continue to predominate. We cannot manipulate what we cannot yet grow.

For now, inference from DNA or protein sequence similarities (and consequently homology, though they are not the same) can effectively substitute for empirical data in some cases. However, inference will likely not match experimentation, especially if phylogenetic contexts and evolutionary time are too distant. If some sequence homologies were originally built upon in silico comparisons, further inferences without experimental ground truthing could result upon "hypotheses built upon hypotheses." At some point, concrete demonstration of the biology is required.

In this context, we can partition our methods of determining function into either direct (experimental) or indirect (in silico or informatics base inferences) approaches. Direct or experimental determination encompasses visualization (microscopy) or measurable physiological changes, effects in controlled laboratory conditions, or transferable metabolites (isotopes). Direct measurements have lately seen a revolution in new technologies, some of which are discussed next.

8.5.1 Direct Evidence of Specific Symbiont Activity

Fluorescence in situ hybridization (FISH) [112] has been a very useful tool to microbiologists for over two decades. This method combines fluorescent microscopy with molecular methods, whereby a specific oligonucleotide probe is hybridized to fixed tissues. DNA probes are typically designed from a database of microbial sequences, which allows the distinction at various taxonomic levels, as the probe must hybridize to a complementary sequence in the fixed sample based on stringency conditions set by the researcher. Specific FISH probe design and selection is facilitated by available databases of variable sequences, such as 16S rRNA probes, called proBase [142, 143]. Precisely localizing specific bacteria can provide evidence into their possible roles via their spatial localization and specific aggregations [144].

Moreover FISH techniques have also been modified to expand microbial diagnostics, as FISH now encompasses DOPE-FISH (double labeling of oligonucleotide probe-FISH) [145], CARD-FISH (catalyzed reporter deposition-FISH) [146], fluorescence in situ hybridization-correlative light and electron microscopy (FISH-CLEM) [147], CLASI-FISH (combinatorial labeling and spectral imaging-FISH) [148], and HCR-FISH (hybridization chain reaction-FISH) [149]. The latter HCR-FISH method was able to demonstrate spatiotemporal regulation of targeted gene expression between *E. scolopes* and *V. fischeri*. Because there is clear evidence that some microbial symbionts appear to form spatial aggregations or [62, 144] or thrive in bacterial microcompartments or BMCs [93, 150, 151] within various animal tissue including the sponge mesohyl, the newer methods could be applied to better mapping sponge mesohyl microenvironments and possibly reveal symbiont functions. The resulting chemical and genetic profiles would be overlaid on sponge ultrastructure to represent a type of geographic information system at the cellular level. Overall, these methods help to directly label specific microbes in their natural microcosms.

One example of this idea can be portrayed through the identification and characterization of a sponge-specific symbiont phylum, *Poribacteria*, which had several important milestones for characterizing sponge symbiosis [56, 152]. Through single amplified genome (SAG) sequencing, about 66% of a poribacterial genome from Mediterranean sponge *Aplysina aerophoba* was pyrosequenced to reveal putative microbial functions. The partial *Poribacteria* genomes yielded genes for sterol biosynthesis, carbon autotrophy through the Wood-Ljungdahl pathway, urea utilization, partial polyketide biosynthesis, and possible host interaction factors such as adhesion and laminin G domain proteins. The eukaryotic repeat structures were confirmed by sequencing other Poribacteria cells [153]. Moreover, the Poribacteria genome lacked chemotaxis and flagellar coding sequences which supported the microbial role as an obligate symbiont, incapable of a free-living lifestyle. This study provided valuable insights by focusing on the metabolic capacity of an important sponge symbiont. Indeed, the miniaturized "single-cell" approach has the potential to elucidate many other important functions in other host species, as the biotechnology improves [154, 155]. Conversely, because sponge microbiomes can comprise up to thousands of OTUs [64, 71], providing the same level of genome annotation across all member genomes will likely await further major advances in DNA sequencing methods and bioinformatics.

Poribacteria has been investigated further with fluorescence in situ hybridization-correlative light and electron microscopy (FISH-CLEM) mentioned earlier [147]. Jahn et al. [156] have characterized both metatranscriptomes and ultrastructure of *Poribacteria* in the model sponge *Aplysina aerophoba* and found evidence for gene expression related to bacterial microcompartments (BMC). Firstly, gene expression analyses showed high transcriptional levels of BMC-shell marker (1.3 $FPKM_{HK}$) and gas vesicle protein (GvP) (2.9 $FPKM_{HK}$). Then, adding three-dimensional FISH-CLEM methods with gene expression data, expression of the gas vesicle protein was further localized with other highly expressed *Poribacteria* gene products to specific locations in the sponge.

In a different type of direct assay, stable isotope ratios (e.g., isotopes of C and N ($\delta^{13}C$ and $\delta^{15}N$)) have now been routinely applied to better understand microbial niches [157]. For example, Freeman and Thacker [92] demonstrated in shading experiments that differential uptake of both carbon and nitrogen from photosymbionts to sponges occurred, depending on the host species: *Aplysina cauliformis* appeared to take up both carbon and nitrogen, *Neopetrosia subtriangularis* took up more carbon, while *A. fulva* appeared to preferentially accept more nitrogen. These were all photosymbiont-dependent sponge species, revealing a complex interaction with symbionts.

Single-cell analyses using high-resolution "nanoscale secondary ion mass spectrometry," also known as NanoSIMS, have risen in importance for microbial studies [158–161]. This technique can be applied to study stable isotopes and minerals in uncultivated microbes of specific habitats. For example, methods have been developed to sufficiently concentrate cell fractions from soil particles, which would allow the inference of carbon isotope composition from the signal intensities obtained from detection of $^{12}C^-$ and $^{13}C^-$secondary ions and nitrogen isotopes from $^{12}C^{14}N^-$ and $^{12}C^{15}N^-$ secondary ions [162]. In a similar application, dissolved sulfide in the form of $\delta^{34}S$ was shown to decrease due to bacterial sulfate reduction within hypersaline microbial mats from Baja California Sur.

8.5.2 Omics Approaches

The modern molecular genetics era began in 1953 with the unraveled structure of the universal heredity material, deoxyribonucleic acid [164]. Decoding the meaning of nucleotide sequences, gene structure, and consequently a slate of "first" whole genomes then followed this milestone elucidation. Comparable to the trajectory of human flight technology starting from the Wright brothers at Kitty Hawk to the Apollo moon landings (accomplished in only slightly more than 60 years), long-standing questions on the essential features of DNA, gene structure, and function were disposed and answered in a similar time frame. For those interested, the history of early molecular biology and genetics has been well documented by Judson [165] and more recently by Mukherjee [166]. Consequently, advances to reveal other parameters of genetic codes (such as epigenetics) ensued along with the quest to integrate genetics with other fields of biology. This intent can be fulfilled by the creation of "-omics" technologies, which represent the aim to grasp a comprehensive view of a given biological realm – e.g., whole complement of genes in a cell ("genomes") or exons ("exomes,") proteins (proteomes) and so on. The concomitant higher volume of DNA sequence data will be facilitated by increasingly lower sequencing costs and enhanced computer processing with increasing memory and storage capacities [167, 168].

However, generating larger volumes of genome sequence evokes the accompanying requirement of gene *annotation*, which is the deciphering of sequence homology, predicting and designating gene function and context for the cell and organism. Gene annotation represents another vast and emergent endeavor, with continual expansion and enhancement of databases, methods, and algorithms [169–172]. Yet gene prediction may be a bioinformatics task that may never be completely finished, for even to this day, about 50% of fully sequenced *E. coli* genes have no experimental support, and 20% have no annotation at all. As alluded to earlier, annotation can pertain directly to determining the function of symbiotic microbial genes, individually or in the context of the community genomics. Various innovative and modern -omics methods have been developed and brought more answers in recent years, yet the large cache of data is the not the final panacaea. Again, functional determination will ultimately require experimental and physical confirmation. Perhaps a political slogan "trust but verify" ("Доверяй, но проверяй" {translated Doveryai, no proveryai}) can be borrowed and applied scientifically, since more funding agencies themselves are requesting data beyond the sequences. Evidence of biological presence can take the form of PCR amplification of a gene's existence. Functional gene predictions from raw sequences will rely on comparisons and inferences from well-curated databases such as Gene Ontology (GO) orthologs and connections to KEGG (Kyoto Encyclopedia of Genes and Genomes) and proteomics resources [173–175]. Toward these goals, many efforts abound. Kanehisa et al. [176] describe automated servers, BlastKOALA and GhostKOALA, that can reconstruct metabolic pathways from gene sequences and amino acid best hit scores.

In this context, *metagenomics* (also known as "environmental genomics, molecular microbial ecology, or community genomics") is the study of (microbial) genomes or gene sequences recovered from environmental samples as opposed to isolated clonal cultures in the laboratory [141, 177–180]. These samples ostensibly represent an entire microbial community. Community genomics was preceded by the insightful pioneering work of Woese and others who sequenced ubiquitous rRNA molecules for the identification of unculturable microbes [8, 9, 181]. Since then, molecular determination of the taxonomic composition of microbial consortia has rapidly advanced, especially with increased sequencing technologies and capacity. The metagenomics field partly arose because up to 99% of microbes in many natural habitats cannot be cultured in the laboratory [182–184], although several innovative methods have tried to advance microbial culturing methods. Different sponge laboratories have also applied diverse media formulations to increase the diversity of isolates [57, 58, 185]. It may also be necessary to develop artificial microbial consortia in sponges that mimic complex and pluralistic natural communities in order to characterize the synergistic metabolic activities [186]. In this context, co-culturing was applied to >1 bacterial species (*Pseudomonas aeruginosa* and *Roseobacter denitrificans*) in order to demonstrate defensive pathway activation of metallo-beta-lactamases [91].

Bacterial genomes are highly efficient with respect to genome organization and relative length (averaging a few million base pairs and 2–3 orders of magnitude smaller than eukaryotic genomes) [187], being gene-rich and lacking introns and large tracts of repetitive DNA. Thus interpreting and annotating bacterial metagenomes is generally less bioinformatically intensive compared to eukaryotic genomes, and full metagenomic sequencing initiatives to fully understand symbiont functions will likely continue to be slated to expand future horizons.

After creating representative DNA libraries of a community, metagenome projects can theoretically target specific functional pathways for isolation and characterization: SM biosynthesis genes [188–193], ammonia oxidation in *Xestospongia muta* [194], cadmium accumulation from *Stylissa massa* [195], kasumigamide production by *Entotheonella* symbionts in *Discodermia* [196], and degradative capabilities or specific enzymes such as lipases from *Haliclona* [197, 198]. Horn et al. [34] sequenced the metagenomes of three Mediterranean sponge species, *Petrosia ficiformis*, *Sarcotragus foetidus*, and *Aplysina aerophoba*, to reveal the presence of a formidable array of defensive systems: clustered regularly interspaced short palindromic repeats (CRISPRs), restriction modification, DNA phosphorothioation, and phage growth limitation systems. In another study, "eukaryotic-like proteins" (ELPs) were identified in metagenomic libraries and tested with recombinant approaches to possibly affect amoeboid phagocytosis in the sponge *Cymbastela concentrica* [199]. Metagenomic studies will likely continue to reveal many interesting new genes related to the microbial integration and holobiont function.

Although ancient, the community milieu within sponge mesohyl does appear relatively stable or without a major flux of microbial dynamics, as revealed by recent temporal-based metagenome studies. For example, a fairly consistent microbiome in the Caribbean reef sponge, *Axinella corrugata*, was found across two different seasons in Florida [63]. Erwin et al. [200] further tested the questions of microbiome stability of six different species (three LMA and three HMA) and corroborated stability of the microbiome core across seasons in all of them. Metagenomics continues to be a busy field, and at this time we cannot cover all ongoing sponge-related studies.

8.5.3 Metatranscriptomes

Over the past several years, a great deal of effort has been expended in sequencing messenger RNA (mRNA), which is the genetic intermediate to the final protein product translated from a gene's open reading frame (ORF). High-throughput sequencing (e.g., pyrosequencing, Illumina, PacBio, Oxford Nanopore) of RNA (also known as RNA-seq) has further permitted reading virtually all transcribed (and non-protein coding) RNA molecules in a particular tissue or cell [163]. This allows the evaluation of the complete catalog of RNA transcripts, the "transcriptome" from an organismal or tissue sample [201].

In general, transcribed regions have long been viewed as only a small fraction of the genome, with mRNA from protein coding sequences perhaps comprising much less than 10% of the animal genome. However, the total volume of transcription and its significance in a cell has now entered a period of reinterpretation, created partially from evidence stemming from the large dataset of the human ENCODE project [202–204]. ENCODE initially claimed that 80% of the human genome may be transcribed and therefore "functional," though this conclusion has been challenged by evolutionary questions and criteria [205, 206]. Kellis et al. [207] wrote a review that helped clarify the debate by categorizing transcript functions into discrete biochemical, genetic, and evolutionary aspects. A recent analysis of nonhuman, nonmodel whole animal genomes appears to confirm some of ENCODE's results, by finding that up to 50% of these genomes may be transcribed [208]. Indeed a great deal of information has been gathered regarding the identity and function of various noncoding, non-mRNA species (e.g., small nuclear RNA, piRNA, siRNA, rasiRNA, etc.) over the past decade [209]. For example, microRNAs (miRNA) represent a small noncoding RNA (ncRNA) that has regulatory functions and forms secondary structure hairpin loops. miRNA also have limited phylogenetic utility based on their ubiquity across diverse metazoan groups [210]. Consistent with the evolutionary perspective, which indicates that selection and adaptive criteria should be applied to the new RNA data, echoes Theodosius Dobzhansky's famous quote: "Nothing biology makes sense except in the light of evolution." Moreover, it is judicious to consider that transcription in itself should not be equated with biological "function," while noting that there is often not a 1:1 correspondence between the amount of any one mRNA molecule and its corresponding protein product in a cell (discussed further under proteomics later; [211]). Turnover rates of the two different types of molecules can vary greatly. These disparities can further complicate interpretations of whether a particular RNA sequence or molecule's presence can be linked to function.

With regard to bacterial community analyses, environmental genomics would theoretically capture multiple transcriptomes (the *metatranscriptome*), which includes mRNA and most noncoding RNAs, of the uncultured microbial taxa. These metatranscriptome (cDNA) approaches are a logical extension of metagenomics and can provide more accurate, "instantaneous" profiles of host and microbial metabolic activities and the possible interplay between them [186, 212–214]. Sequencing metatranscriptomes should require less effort and overall coverage, since not all genes of a microbial community are expected to be expressed at any given time, or rare member activity will likely not be detected. Metatranscriptome analyses verify which genes are active [215, 209]. Transcriptome profiles represent an intermediate expression and bridge to the phenotype (of a microbial community) and thus can be viewed as valuable biological resource and lead to useful taxonomic markers. Moreover, these efforts can yield a large amount of putatively functional sequence data which can be subsequently applied in multiple biological contexts [215–217], while not requiring as much sequencing effort (coverage) compared to a full community metagenome. Because of the lower amount of "junk DNA," with no clear function, bacterial metagenomics and metatranscriptomics datasets may also find fewer pitfalls in relating transcriptional presence with function. Nevertheless, transcriptomes cannot represent the final goal of functional genomics, since mRNA remains only an "intermediate" to the phenotype. The final phenotype of any tissue or an organism is mostly produced by the enzymatic action of protein products and their derived metabolites.

For example, in the human gut microbiome, up to 1000 different bacterial species provide an expanded spectrum of metabolic activities and capacity, often beneficial for degrading compounds or xenobiotics that host enzymes do not recognize [218]. Thus in any discussion of symbiosis, there exist not only microbiomes and the interaction of taxa and single cells within microbial communities but also an ecology of genes, gene function, and proteins. *Faecalibacterium prausnitzii* produces butyrate and may have anti-inflammatory function in the ileal human intestines [219]. In contrast to the beneficial functions mentioned earlier, symbiotic microbiota can also affect drug metabolism in unexpected ways, such as the inhibition of a human anticancer drug 5-fluorouracil breakdown that led to accumulation, toxicity, and tragic fatalities [220]. Gilbert et al. [219] have summarized some of these multiple dimensional scales and interactions as "microbiome-wide association studies" (MWAS), analogous to the genome wide-association studies (GWAS) which has assisted tracking disease loci in humans ([221]; also see http://www.ebi.ac.uk/gwas). However, because microbial genes outnumber host genes even within a single sponge individual, new methodological and analytical approaches, such as enhanced bioinformatics data visualization and large sample curation via the Microbiome Quality Control Project and Earth Microbiome Projects, are suggested to better understand complex microbiomes [222].

In an ongoing research theme, the Palumbi laboratory has rigorously applied transcriptomics to understand coral holobiont responses to changing temperatures

[169, 223–225]. For example, sequencing the transcriptome of two symbiotic *Symbiodinium* dinoflagellates of the scleractinian coral *Acropora hyacinthus* showed differences in expression patterns that may be associated with thermotolerance [169]. The putatively thermotolerant *Symbiodinium* type D2 and the more susceptible type C3K *Symbiodinium* appeared to share hundreds of gene orthologs, which exhibited different patterns of expression after heat stress treatments. Many of the genes were characterized to act in functions related to thermotolerance, such as heat shock, chloroplast and thylakoid membrane thermostability, and maintenance of photosynthesis in the increased clade D tolerance. Similar types of experiments and transcriptome profiles could be applied to sponge microbial communities in order to decipher their function for individual sponge species in different environmental scenarios or experimental treatments.

Gene expression studies of host sponge transcriptomes, especially via RNA-seq methods, have advanced considerably over the past decade. For example, Guzman and Conaco [226] showed changes in heat shock protein, signal transduction, and antioxidant genes after short-term temperature changes on the sponge *Haliclona tubifera* (order Haplosclerida, class Demospongiae); different pathways appeared activated when temperatures were increased for longer periods. With a single-cell RNA-seq method, host genes likely to be interacting with the microbiome involved either innate immunity or had upregulated scavenger receptor cysteine-rich (SRCR) domains characterized during transitions from larval to postlarval stages in *Amphimedon queenslandica* [61]. SRCRs were also identified in hologenome analyses of *Stylissa carteri* [135]. This is consistent within the context of microbial community functions, and understanding the total gene expression landscape of both host and its unique microbiome within any particular sponge holobiont community requires similar study of gene expression changes of the whole metatranscriptome.

Furthermore, exome sequencing is a subset of metatranscriptomes, which focuses primarily on protein-coding genes [227]. This global view of gene expression intends to give clues as to which fraction of an organism's (or specific tissue) genome actually generated active protein products. Single-cell RNA profiling also appears to be on the verge of wider acceptability and usage [155, 228, 229].

Interpretation of burgeoning metagenomic, metatranscriptome, and metaproteome data continues to develop for the full elucidation of the roles that microbial symbionts can provide for marine sponges. As mentioned above, because the vast majority (~90%) of microbes and microbial symbionts cannot be cultured in the laboratory for direct experimentation and measurements, symbiologists must depend on inference. For this, bioinformatics and computational innovations can help fill the gaps and provide essential tools [13, 230].

There have been several highly informative transcriptome analyses of specific sponge species in recent years. These have elucidated interesting functional homologs from different sponge species, supporting the evolutionary connection of sponges with many different metazoan lineages, no matter how distant [102, 194]. With regard to better understanding sponge symbiotic bacterial communities, the transcriptome of not just one symbiont species but multiple species must be sequenced and then interpreted. And similar to the challenge of metagenomics, the community *metatranscriptome* is often constructed by de novo sequence assembly and without the availability of full reference genomes. At the present time, these are not trivial bioinformatics problems for any single laboratory or project, since metatranscriptomics aims to represent and encompass total community activity [154, 231].

Nonetheless, Fiore et al. [194] carried out an extensive study showing gene expression of key sponge enzymes involved in metabolism of thiamin (i.e., phosphorylation reactions) and riboflavin (vitamin B2) biosynthesis. This profile also included bacterial transcripts corresponding to the production of riboflavin and sponge transcripts corresponding to the conversion between riboflavin and the cofactors flavin mononucleotide (FMN) and flavin adenine dinucleotide (FAD). These results support the transfer of compounds between symbionts and host.

Ideally, many types of biologists (as well as biochemists) would welcome a full integration of transcriptome, proteomics, and even metabolomics data for a realistic view of the operational phenotype [232, 233]. Proteins represent one of the checkpoints for gene function, but do not always translate to the ideal phenotypic manifestation (typically a complex combination of multiple interactions of expressed genes and downstream products of active metabolism). Thus, integration is not trivial, but progress has been made through creative computational methods. For example, the community *metabolome* could be modeled with tools such as the Predicted Relative Metabolomic Turnover (PRMT) method [234, 235]. PRMT uses the relative abundance of enzyme functions across multiple sequences to generate a quantification of relative and encoded metabolic capacity of an organism or microbial community. The algorithm translates relative transcript abundances into the potential relative consumption of production of the metabolites mediated by the enzymes coded for by the different genes and transcripts (Fig. 8.2) A positive PRMT score predicts increased metabolic turnover and relatively greater consumption of a metabolite. A negative PRMT score predicts decreased turnover and relatively greater accumulation of a metabolite. It is important to note that PRMT scores do not predict net production or consumption of a metabolite or predict concentrations of specific metabolites.

In a preliminary study performed in our lab following the Deepwater Horizon oil spill (DWHOS), we tested the effects of oil, dispersant, and mixtures on local sponges. Host mRNA was isolated (oligo-dT selected) from four *Cinachyrella* spp. individuals exposed to the oil experiments described above – oil (o), Corexit dispersant (d), or a mixture of both (od) and untreated controls [236]. Approximately 50 million mRNA sequence reads were generated using the Illumina HiSeq platform from the sponge host (poly-A RNA-seq analysis). These reads were assembled into 39 million contigs (assembled fragments), which were annotated to identify likely genes for proteins with specific enzymatic functions. To translate these functional assignments into potential metabolic turnover scores, PRMT [235] was applied to these data. BLASTX (via MG-RAST) was used to search for homology of translated assembled nucleotides against a database of protein sequences for enzymes that catalyze specific reactions in the set of KEGG metabolic pathways. Total detected unique enzyme functions were used to construct a sponge metabolo-

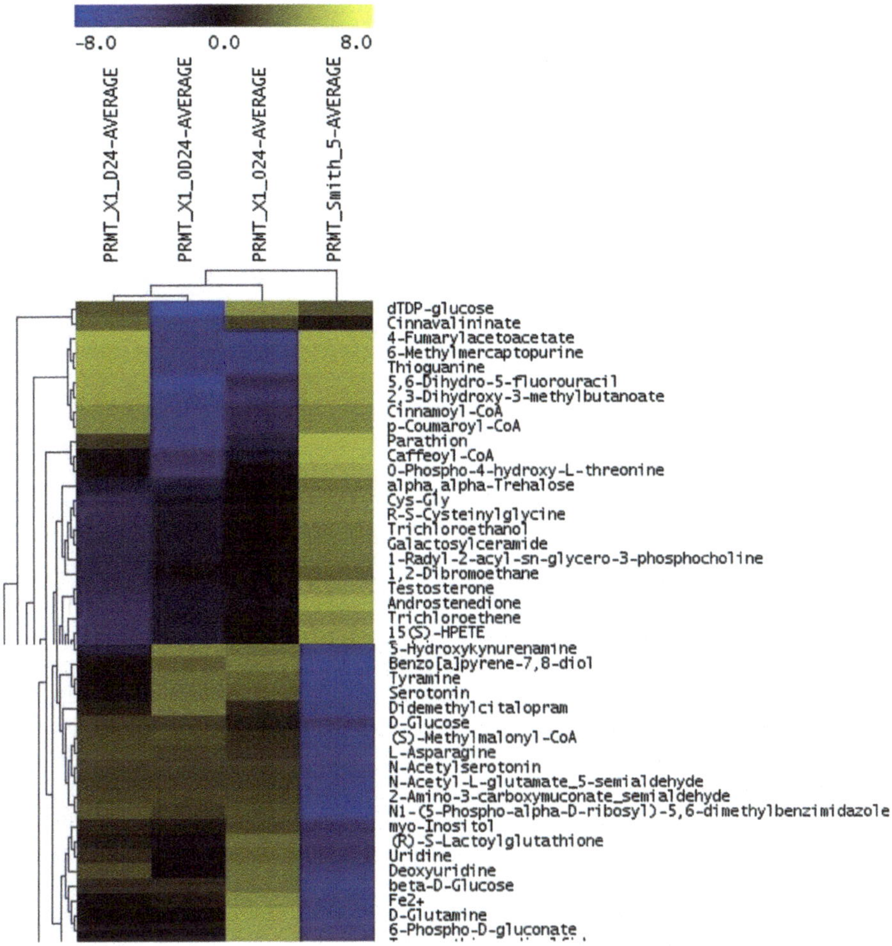

Fig. 8.2 Partial sample heat map showing PRMT scores for sponge metabolome analyses of RNA-seq data from 24 h of oil-dosing pilot studies. Blue = relative increase in synthesis, while yellow = relative increase in degradation. D24 = dispersant, OD24 = oil + dispersant, O24 = oil, Smith_5 = control. (Figure courtesy of Peter Larsen)

mic model. The resulting model contained 1627 predicted metabolites, connected by 2107 reactions mediated by 945 unique enzyme functions. Metabolome model and quantile-normalized, log-transformed enzyme function counts were used to calculate predicted relative metabolic turnover (PRMT) scores for all compounds in the metabolome model, relative to average abundance of enzyme function across all samples. Sets of strongly differentially metabolized compounds (top and bottom 5th percentile of calculated PRMT scores) were collected for each treatment (Table 8.2). In hierarchical clustering of differentially metabolized compound PRMT scores (Fig. 8.2), the control sample clustered alone. Dispersant-containing samples cluster together, suggesting that dispersant has a greater effect on sponge metabolome

Table 8.2 Differentially metabolized compounds

	Treatment[a]			
	Disp	Oil + Disp	Oil	Control
No. of differentially metabolized compounds	122	96	63	291

[a]Disp = dispersant, control = seawater only

than presence of oil. All treatments shared a differential ability to metabolize phenolic compounds and xenobiotics relative to control, but oil treatment had the fewest specific metabolites affected, suggesting that xenobiotic metabolism may be a general response to environmental stress and that non-dispersed oil has the least tendency to be directly metabolized by sponge. This also potentially identified a role for the microbiome which could result in a general phenotypic response of sponges to oil pollution. Classifications of biologically active compounds (BRITE annotation) and KEGG metabolic pathways enriched for strongly differentially metabolized compounds were identified. In dispersant and oil plus dispersant-treated samples, strongly differentially metabolized compounds were enriched for nucleotides and nucleosides, suggesting possible activation of DNA repair responses. Oligosaccharide metabolism distinguished control treatments from experimental treatments. Overall this exercise shows how transcriptome profiles can reveal potential activity in diverse holobiont systems.

For example, the PICRUSt algorithm ("phylogenetic investigation of communities by reconstruction of unobserved states") represents one of these tools and attempts to predict metabolic functions based on reference genome or taxonomic marker genes (such as 16S rRNA) data [237, 238]. This contrasts with the aforementioned idea that taxonomy may be less of a determinant for specific community function. For example, PICRUSt was applied to show distinct metabolic differences between eight different herbivorous and carnivorous fish species in the Yangtze River Basin [239]. After high-throughput 16S V4 sequencing, the total OTU table was input into PICRUSt. The results showed that herbivores possessed a gut microbiome enriched for carbohydrate-related metabolism such as starch and sucrose, fructose and mannose, and galactose and glycolysis/gluconeogenesis. This group included cellulose-degrading bacteria *Clostridium*, *Citrobacter*, and *Leptotrichia*. By contrast carnivorous fish had taxa that thrived in trypsin environments. In the marine area, Yilmaz et al. [238] applied PICRUSt to the International Census of Marine Microbes 16S rRNA dataset, to verify and place their previous bacterial clade classifications into ecological and biogeographical contexts.

Of course, caution should be exercised before relying exclusively on hypotheses built on a single genetic locus, which is the greatest drawback for PICRUSt. Its utility may likely decline as high-throughput sequencing costs continue to decline and accuracy increases. Alternatively, the growing number of sequences, when coupled with verified metabolic functions, can also hone PICRUSt's predictive capacity. Overall, the ability to read larger swaths of environmental metagenomes will reveal more potential functions.

8.5.4 Microbial Protein Function and Metaproteomics

The preceding sections have visited the first two nucleic acid phases of the canonical central dogma: DNA→ RNA → protein. Yet the flow of biological information in macromolecular sequences ends with the protein, originally encoded in discrete genes. Thus, -omics research has naturally extended into proteomics, and the global view of translated gene products within a cell or organism, which could be viewed colloquially as the "business end" of the dogma [240, 241]. Many proteins act as catalytic enzymes carrying out the actual functions we are now trying to describe. Although this is no trivial task, since the translation products of multiple taxa can be involved, a census and characterization of proteins in the symbiotic community cannot be disregarded – they are part of the previously mentioned "ecology of holobiont molecules." Moreover, genetic characterizations will not detect posttranslational modifications, such as hydroxylation, acetylation, methylation, citrullination, phosphorylation, methylthiolation, S-nitrosylation, and nitration, which can be characterized with multiple protease digestions, optimized high-resolution mass spectrometry, and high-performance computing [242, 243].

In spite of proteomics maturity in parallel with genomics and transcriptomics, throughput of raw sequence data from peptides and protein molecules cannot yet compete with nucleic acid methods at present. Given that novel liquid chromatography/mass spectrometry (LC/MS) has increased throughput, limitations remain. Technically, determining the total or global abundance of proteins in a symbiont community to represent the *metaproteome* does not present itself as an easy task from a biochemical viewpoint. As Wang et al. [244] lamented, obtaining sufficient protein biomass in samples can often limit progress, since many bacterial taxa in a community are rare, and proteins cannot be readily amplified in vitro like DNA in PCR. The samples must also be properly preserved for laboratory analyses, since protein functions stem mostly from their three-dimensional structures. Nonetheless, several studies have taken the plunge. For example, proteomics was applied to a microbial community from Oregon Coast summer surface waters and found a predominance of SAR11 bacteria and evidence of transporters for amino acids, taurine, and polyamines [245].

Due to the relatively breakneck pace of biotechnological progress, it may still be premature at this time to delve too deeply into the topic of metaproteomics with respect to invertebrate symbiosis [246]. As seen previously, community metaproteomics will likely follow and benefit from the advances in more highly funded, human biomedical sciences. Obvious differences will of course exist between different organismal communities, but mutual problems and challenges can also be recognized [244]. Proteins represent one of the checkpoints for gene function, but do not always translate to the ultimate phenotypic manifestation.

8.6 Secondary Metabolism by Symbionts

Invertebrates are rivaled perhaps only by microbes with respect to the diverse metabolic potential encoded within genomes. This includes the biosynthesis of unique, bioactive secondary metabolites (SM). For example, several gastropod *Conus* species produce more than 100 bioactive peptides (conopeptides or conotoxins) as an arsenal of potent venoms, with little overlap in sequence and function among species [247, 248].

Microorganisms have had millions of years practicing and honing biochemical syntheses via the evolution of peculiar metabolisms, and these compounds can provide clues to their function. A large portion of the biochemical diversity found in nature can often be traced to microbial organisms and communities, which are poised to provide an enormous boon to biomedicine [249–252]. It is well known that many secondary metabolites or "natural products" with therapeutic potential or bioactive effects derive from bacterial and fungal microorganisms [170, 253–256]. Providing a reliable microbial source and obtaining consistent production of secondary metabolites from microbes, however, continues to be a challenge for experimentation and pharmaceutical bioprospecting [257, 258]. In 2015 an apparent microbial culturing breakthrough appeared, which awaits further validation and application: the innovative method of the "iChip" helped isolate a previously uncultivable microbe, which produced the novel antibiotic, teixobactin [259]. The method was based on allowing resident microbes' natural habitat growth factors diffuse through an artificial semipermeable membrane, to nourish previously diluted and isolated candidate microbes [260]. In a recent application of mass spectrophotometry to the human microbiome, chemicals were spatially linked to the presence of specific microbes. Moreover, Versluis et al. [216] showed that antibiotic resistance gene expression could be detected via metatranscriptome libraries and in many different habitats.

As mentioned earlier in the context of microbial biofilms, a possible role for retaining original microbial colonizers in the holobiont is the production of novel secondary products that could be used to dissuade would-be predators from attacking unprotected nascent sponge ancestors. Several studies [261, 262] have documented examples of marine invertebrates hosting microbial symbionts that appear responsible for bioactive secondary metabolite biosynthesis. -Omics approaches have confirmed the presence of non-ribosomal peptide synthetase (NRPS) pathway genes for the tetrahydroisoquinoline ET-743, which is an approved anticancer agent (Yondelis®) originally isolated from the tunicate *Ecteinascidia turbinata* [263]. These metabolic genes appear to have the microbial source, *Candidatus Endoecteinascidia frumentensis*.

A fairly well-known case of a potent secondary metabolite produced by a confirmed symbiont occurs in the bryozoan, *Bugula neritina*. In the 1960s this invertebrate was found to hold bryostatins, a group of macrolide lactones with several modes of action. For example, bryostatin 1 can modulate protein kinase C (PKC) immune pathways and inhibit cell growth and angiogenesis [264]. 'Candidatus Endobugula sertula,' the uncultivated gammaproteobacterial symbiont of the marine bryozoan *Bugula neritina* [185], synthesizes bryostatins, complex polyketides that render *B. neritina* larvae unpalatable to predators. The bioactive bryostatins appear to sequester specifically in reproductive portions of the host colony. Gene expression data also points to coordination whereby the endosymbiont production of bryostatin protects host larvae [265]. The bryostatins exemplify a common problem in marine natural products – low quantities and unculturable microbial symbiont origins. This feature makes commercial development and thus widespread distribution of the potentially therapeutic compounds difficult.

Moreover, SM biosynthesis may be more dependent on complex environmental stimuli, such as nutrient deprivation, responses to stress or competition, and quorum sensing than is primary metabolism [107, 170, 266]. While hosting the wide diversity of microbial symbionts, marine sponges produce some of the widest diversity of bioactive secondary metabolites among marine invertebrate phyla which have perennially made them targets for potential therapeutic drug development [4, 267, 268]. Many of these compounds have complex biochemical structures, such as alkaloids, polyketides, or non-ribosomal peptides; the latter two categories point to at least a circumstantial connection with microbial metabolism when fully validated by gene sequencing [188, 189, 261].

For example, the tetractinellid sponge *Theonella swinhoei* has been a central focus of sponge natural products bioprospecting for over a decade, and biosynthetic genes have been identified [190, 269]. Piel et al. [190] have also identified polyketide biosynthetic genes from the sponge, *Theonella swinhoei*. They have continued characterizing symbiont activity by showing an ability of microbes to produce amino acids and rare cofactors like coenzyme F_{420} [270, 271].

In another example, Caribbean sponges of the lithistid genus *Discodermia* have been shown to produce a wide range of natural products including polyketides, peptides and non-ribosomal peptides, compounds of mixed biosynthesis, and simple alkaloids. The activities of the compounds are equally varied including induction of calcium release, cytotoxicity, antifungal, and antibacterial. The literature appears replete with examples showing sponge-associated, microbial antifeedant properties 272–274], but showing direct proof for microbial production remains difficult and more scarce [252]. The *Discodermia* sponge polyketide compound, discodermolide, showed early promise as a potential antitumor compound, due to its microtubule-binding mode of action, similar to Taxol [275, 276]. However, lack of reliable in vitro production and high toxicity in preclinical trials doomed its possibility as a marketable drug [277]. In a more recent paper and in contrast to the complexity of many of these SMs, Freeman et al. [278] have characterized a pathway of 7 pivotal enzymes that produce 50 site-specific modifications to create the membrane-spanning β-helical structure of polytheonamides, potent peptide cytotoxins from *Theonella swinhoei*. Again, in spite of the likely microbial origin and identification of possible synthetic genes, a producing microorganism has rarely been isolated in culture at the time of writing [140, 233]. Linking ecological parameters specifically to potentially producing symbiont partners has not been straightforward. If more novel microbial taxa could be cultured, then the pathway to therapeutic testing and functional assays could be widened experimentally.

8.7 Concluding Remarks

I have attempted to show various approaches to discerning the possible functions of microbial symbionts within sponges from various perspectives. Perhaps by now enough evidence from multiple sponge species has accumulated to show the primary cohesion between host and multiple microbial symbionts. The examples of integration emphasize that retention of the symbionts often provides tangible benefits for the eukaryotic host and supports the idea that the symbiont community composes the holobiont. Yet, this discussion also shows that many details still remain to be learned with regard to specific microbial taxa and their functions across the phylum. Symbiotic bacteria, fungi, and protozoans coexist and can be viewed as omnipresent with many sponge holobionts, but universal designation of microbiome functions still eludes a comprehensive or satisfactory explanation. For such a relatively simple body plan by metazoan standards, along with a primitive immunological repertoire, this conclusion remains incomplete and somewhat counterintuitive. Yet, considering a greater than 500 million year history of evolutionary radiation, the recalcitrance for a full revelation of all past and present microbial symbiont functions should make sense.

Acknowledgments I thank Dr. Susan Sennett, FAU at Harbor Branch Oceanographic Institute, for the archived SEM micrographs. I also thank Dr. Peter Larsen for his PRMT analyses of preliminary RNA-seq data. Part of the preliminary data was also partly supported by a Year 1 BP Gulf of Mexico Research Initiative grant to JVL by the Florida Institute of Oceanography. I thank the following colleagues for feedback on early drafts of the manuscript: Dr. Omar Eldekar, Dr. Cole Easson, and Dr. Hidetoshi Urekawa. I am also grateful to Amy Doyle for final proofreading of the manuscript.

References

1. Chun J, Rainey FA. Integrating genomics into the taxonomy and systematics of the bacteria and archaea. Int J Syst Evol Microbiol. 2014;64:316–24.
2. Van Soest RWM, Boury-Esnault N, Vacelet J, Dohrmann M, Erpenbeck D, et al. Global diversity of sponges (Porifera). PLoS One. 2012;7:e35105.
3. Taylor MW, Radax R, Steger D, Wagner M. Sponge-associated microorganisms: evolution, ecology, and biotechnological potential. Microbiol Mol Biol Rev. 2007;71:295–347.
4. Hentschel U, Fieseler L, Wehrl M, Gernert C, Steinert M, Hacker J, et al. Microbial diversity of marine sponges. In: Müller WEG, editor. Sponges (Porifera). Berlin/Heidelberg: Springer; 2003. p. 59–88.
5. Webster NS, Taylor MW. Marine sponges and their microbial symbionts: love and other relationships. Environ Microbiol. 2012;14:335–346.1.
6. Douglas AE. Symbiotic interactions. Oxon Great Britain: Oxford University Press; 1994.
7. Poretsky R, Rodriguez-R LM, Luo C, Tsementzi D, Konstantinidis KT. Strengths and limitations of 16S rRNA gene amplicon sequencing in revealing temporal microbial community dynamics. PLoS One. 2014;9:e93827.
8. Woese CR. Bacterial evolution. Microbiol Rev. 1987;51:221–71.
9. Pace NR. A molecular view of microbial diversity and the biosphere. Science. 1997;276:734–40.

10. Wilkinson CR, Nowak M, Austin B, Colwell RR. Specificity of bacterial symbionts in Mediterranean and Great Barrier reef sponges. Microb Ecol. 1981;7:13–21.
11. Futuyma D. Evolution. 2nd ed. Sunderland: Sinauer Assoc; 2009.
12. Colwell RR. Microbial biodiversity and biotechnology. In: Reaka-Kudla ML, Wilson DE, Wilson EO, editors. Biodiversity II. Washington, DC: Joseph Henry; 1997. p. 279–87.
13. Woese CR. A new biology for a new century. Microbiol Mol Biol R. 2004;68:173–86.
14. Schloss PD, Handelsman J. Status of the microbial census. Microbiol Mol Biol Rev. 2004;68:686–91.
15. HMP- The Human Microbiome Project Consortium. A framework for human microbiome research. Nature. 2012;486:215–21.
16. Smith PA. Can the bacteria in your gut explain your mood? New York Times, June 23 2015.
17. Dubilier N, McFall-Ngai M, Zhao L. Microbiology: create a global microbiome effort. Nature. 2015;526:631–4.
18. Reardon S. White house goes big on microbiome. Scientific American. http://www.scientificamerican.com/article/white-house-goes-big-on-microbiome-research/. March 2016.
19. Casadevell A, Pirofski LA. Ditch the term pathogen. Nature. 2014;516:163.
20. Qin J, Li R, Raes J, Arumugam M, Burgdorf KS, Manichanh C, et al. A human gut microbial gene catalogue established by metagenomic sequencing. Nature. 2010;464:59–65.
21. Coyne MJ, Zitomersky NL, McGuire AM, Earl AM, Comstock LE. Evidence of extensive DNA transfer between Bacteroidales species within the human Gut. MBio. 2014;5:e01305–14.
22. Schmitt S, Tsai P, Bell J, Fromont J, Ilan M, Lindquist N. Assessing the complex sponge microbiota: core, variable and species-specific bacterial communities in marine sponges. ISME J. 2012;6:564–76.
23. Martiny AC, Treseder K, Pusch G. Phylogenetic conservatism of functional traits in microorganisms. ISME J. 2013;7:830–8.
24. McFall-Ngai M, Hadfield MG, Bosch TC, Carey HV, Domazet-Lošo T, Douglas AE, et al. Animals in a bacterial world, a new imperative for the life sciences. Proc Natl Acad Sci U S A. 2013;110:3229–36.
25. Mindell DP. Phylogenetic consequences of symbioses: eukarya and eubacteria are not monophyletic taxa. Bio Systems. 1992;27:53–62.
26. Rohwer F, Seguritan V, Azam F, Knowlton N. Diversity and distribution of coral-associated bacteria. Mar Ecol Prog Ser. 2002;243:1–10.
27. Rosenberg E, Koren O, Reshef L, Efrony R, Zilber-Rosenberg I. The role of microorganisms in coral health, disease and evolution. Nat Rev Microbiol. 2007;5:355–62.
28. Bordenstein SR, Theis KR. Host biology in light of the microbiome: ten principles of holobionts and hologenomes. PLoS Biol. 2015;13:e1002226.
29. Freeman CJ, Easson CG, Baker DM. Metabolic diversity and niche structure in sponges from the Miskito Cays, Honduras. Hay M, ed. PeerJ. 2014;2:e695.
30. Nyholm SV, McFall-Ngai MJ. The winnowing: establishing the squid-vibrio symbiosis. Nat Rev Microbiol. 2004;2:632–42.
31. Fiore CL, Jarett JK, Olson ND, Lesser MP. Nitrogen fixation and nitrogen transformations in marine symbioses. Trends Microbiol. 2010;18:455–63.
32. Fan L, Reynolds D, Liu M, Stark M, Kjelleberg S, Webster NS, et al. Functional equivalence and evolutionary convergence in complex communities of microbial sponge symbionts. Proc Natl Acad Sci. 2012;109:E1878–87.
33. Schirmer A, Gadkari R, Reeves CD, Ibrahim F, DeLong EF, Hutchinson CR. Metagenomic analysis reveals diverse polyketide synthase gene clusters in microorganisms associated with the marine sponge *Discodermia dissoluta*. Appl Environ Microbiol. 2005;71:4840–9.
34. Horn H, Slaby BM, Jahn MT, Moitinhosilva L, Forster F, et al. An enrichment of CRISPR and other defense-related features in marine sponge-associated microbial metagenomes. Front Microbiol. 2016;7:1751.
35. Reiswig HM. In situ pumping activities of tropical Demospongiae. Mar Biol. 1971;9:38–50.
36. Reiswig HM. Particle feeding in natural populations of three marine demosponges. Biol Bull. 1971;141:568–91.

37. Pile AJ, Patterson MR, Witman JD. In situ grazing on plankton <10 um by the boreal sponge *Mycale lingua*. Mar Ecol Prog Ser. 1996;141:95–102.
38. Schönberg CHL. A history of sponge erosion: from past myths and hypotheses to recent approaches. In: Wisshak M, Tapanila L, editors. Current developments in bioerosion. Berlin/Heidelberg: Springer; 2008. p. 165–202.
39. Chaves-Fonnegra A, Riegl B, Zea Z, Lopez JV, Giliam DS. Bleaching events regulate shifts from corals to excavating sponges in algae dominated reefs. Glob Chang Biol. 2017;24:773–85. https://doi.org/10.1111/gcb.13962.
40. de Goeij JM, van Oevelen D, Vermeij MJ, Osinga R, Middelburg JJ, de Goeij AFPM, et al. Surviving in a marine desert: the sponge loop retains resources within coral reefs. Science. 2013;342:108–10.
41. Zhang F, Blasiak LC, Karolin JO, Powell RJ, Geddes CD, Hill RT. Phosphorus sequestration in the form of polyphosphate by microbial symbionts in marine sponges. Proc Nat Aca Sci USA. 2015;112:4381–6.
42. Erwin DH, Valentine JW. The Cambrian explosion. The construction of animal biodiversity. Greenwood Village: Roberts and Company Publishers; 2013. p. 416.
43. Antcliffe JB, Callow RH, Brasier MD. Giving the early fossil record of sponges a squeeze. Biol Rev Camb Philos Soc. 2014;89:972–1004.
44. Dunn CW, Hejnol A, Matus DQ, Pang K, Browne WE, Smith SA, et al. Broad phylogenomic sampling improves resolution of the animal tree of life. Nature. 2008;452:745–9.
45. Haeckel E. Das Protestenreich. Leipzig: Gunther; 1878.
46. Pisani D, Pett W, Dohrmann M, Feuda R, Rota-Stabelli O, Philippe H, et al. Genomic data do not support comb jellies as the sister group to all other animals. Proc Natl Acad Sci. 2015;112:15402–7.
47. Ryan JF, Pang K, Schnitzler CE, Nguyen A-D, Moreland RT, Simmons DK, et al. The genome of the ctenophore *Mnemiopsis leidyi* and its implications for cell type evolution. Science. 2013;342:1242592.
48. Simion P, Philippe H, Baurain D, Jager M, Richter DJ, Di Franco A, et al. A large and consistent phylogenomic dataset supports sponges as the sister group to all other animals. Curr Biol. 2017.; pii: S0960-9822 (17), 30199-9.
49. Whelan NV, Kocot KM, Moroz LL, Halanych KM. Error, signal, and the placement of Ctenophora sister to all other animals. Proc Nat Aca Sci USA. 2015;112:5773–8.
50. Moroz LL, Kocot KM, Citarella MR, Dosung S, Norekian TP, Povolotskaya IS, et al. The ctenophore genome and the evolutionary origins of neural systems. Nature. 2014;510:109–114
51. Lopez JV, McCarthy PJ, Janda KE, Willoughby R, Pomponi SA. Molecular techniques reveal wide phyletic diversity of heterotrophic microbes associated with the sponge genus *Discodermia* (Porifera: Demospongiae). Proceedings of the 5th International Sponge symposium. Mem Qld Mus. 1999;44:329–41.
52. Webster NS, Wilson KJ, Blackall LL, Hill RT. Phylogenetic diversity of bacteria associated with the marine sponge *Rhopaloeides odorabile*. Appl Environ Microbiol. 2001;67:434–44.
53. Hentschel U, Usher KM, Taylor MW. Marine sponges as microbial fermenters. FEMS Microbiol Ecol. 2006;55:167–77.
54. Hill RT. Microbes from marine sponges: a treasure trove of biodiversity for natural products discovery. In: Bull AT, editor. Microbial diversity and bioprospecting. Washington, DC: ASM Press; 2004.
55. Sfanos KAS, Harmody DK, McCarthy PJ, Dang P, Pomponi SA, Lopez JV. A molecular systematic survey of cultured microbial associates of deep water marine invertebrates. Syst Appl Microbiol. 2005;28:242–64.
56. Fieseler L, Horn M, Wagner M, Hentschel U. Discovery of the novel candidate phylum "Poribacteria" in marine sponges. Appl Environ Microbiol. 2004;70:3724–32.
57. Montalvo NF, Davis J, Vicente J, Pittiglio R, Ravel J, Hill RT. Integration of culture-based and molecular analysis of a complex sponge-associated bacterial community. PLoS One. 2014;9:e90517.
58. Olson JB, McCarthy PJ. Associated bacterial communities of two deep-water sponges. Aquat Microb Ecol. 2005;39:47–55.

59. Mohamed NM, Colman AS, Tal Y, Hill RT. Diversity and expression of nitrogen fixation genes in bacterial symbionts of marine sponges. Environ Microbiol. 2008a;10:2910–21.
60. Mohamed NM, Enticknap JJ, Lohr JE, McIntosh SM, Hill RT. Changes in bacterial communities of the marine sponge *Mycale laxissima* on transfer into aquaculture. Appl Environ Microbiol. 2008;74:1209–22.
61. Mohamed NM, Rao V, Hamann MT, Kelly M, Hill RT. Monitoring bacterial diversity of the marine sponge *Ircinia strobilina* upon transfer into aquaculture. Appl Environ Microbiol. 2008;74:4133–43.
62. Negandhi K, Blackwelder P, Ereskovsky AV, Lopez JV. Florida reef sponges harbor coral disease-associated bacteria. Symbiosis. 2010;51:117–29.
63. White J, Patel J, Ottesen A, Arce C, Blackwelder P, Lopez JV. Pyrosequencing of microbes within *Axinella corrugata* sponges: diversity and seasonal variability. PLoS One. 2012;7:e38204.
64. Thomas T, Moitinho-Silva L, Lurgi M, Easson CG, Björk J, Astudillo C, et al. The global sponge microbiome: symbiosis insights derived from a basal metazoan phylum. Nat Commun. 2016;7:1–12.
65. Wilkinson CR. Microbial associations in sponges. III. Ultrastructure of the in situ associations in coral reef sponges. Mar Biol. 1978;49:177–85.
66. Wilkinson CR. Net primary productivity in coral reef sponges. Science. 1983;219:410–2.
67. Wilkinson CR. Significance of microbial symbionts in sponge evolution and ecology. Symbiosis. 1987;4:135–46.
68. Wilkinson CR, Garrone R, Vacelet J. Marine sponges discriminate between food bacteria and bacterial symbionts: electron microscope radioautography and in situ evidence. Proc R Soc Lond B. 1984;220:519–28.
69. Fiore CL, Baker DM, Lesser MP. Nitrogen biogeochemistry in the Caribbean sponge, *Xestospongia muta*: a source or sink of dissolved inorganic nitrogen? PLoS One. 2013;8:e72961.
70. Shropshire JD, Bordenstein SR. Speciation by symbiosis: the microbiome and behavior. MBio. 2016;7:e01785.
71. Webster NS, Taylor MW, Behnam F, Lücker S, Rattei T, Whalan S, et al. Deep sequencing reveals exceptional diversity and modes of transmission for bacterial sponge symbionts. Environ Microbiol. 2010;12:2070–82.
72. Reveillaud J, Maignien L, Murat EA, Huber JA, Apprill A, Sogin ML, et al. Host-specificity among abundant and rare taxa in the sponge microbiome. ISME J. 2014;8:1198–209.
73. Cassler M, Winegar R, McCarthy PJ, Pomponi SA, Peterson CL, Wright AE, et al. Use of Real-Time PCR to quantitate the unculturable heterotrophic bacterial community in deep sea marine sponges of the family Lithistidae. Microb Ecol. 2008;55:384–94.
74. Kieft TL, Simmons KA. Allometry of animal-microbe interactions and global census of animal-associated microbes. Proc Biol Sci. 2015;282:20150702.
75. MacArthur RH, Wilson EO. The theory of island biogeography. Princeton: Princeton University Press; 1967. p. 20.
76. Oh S, Caro-Quintero A, Tsementzi D, DeLeon-Rodriguez N, Luo C, et al. Metagenomic insights into the evolution, function, and complexity of the planktonic microbial community of Lake Lanier, a temperate freshwater ecosystem. Appl Environ Microbiol. 2011;77:6000–11.
77. Wilson DS, Wilson EO. Rethinking the theoretical foundation of sociobiology. Q Rev of Biology. 2007;82:327–48.
78. Dawkins R. The selfish gene. London: Oxford University Press; 1976.
79. Easson CG, Thacker RW. Phylogenetic signal in the community structure of host-specific microbiomes of tropical marine sponges. Front Microbiol. 2014;5:532.
80. Dohrmann M, Wörheide G. Novel scenarios of early animal evolution–is it time to rewrite textbooks? Integr Comp Biol. 2013;53:503–11.
81. Webster NS, Thomas T. The sponge hologenome. MBio. 2016;7:e00135–16.
82. Bekker A, Holland HD, Wang PL, Rumble D, Stein HJ, Hannah JL, et al. Dating the rise of atmospheric oxygen. Nature. 2004;427:117–20.
83. De Marais DJ. When did photosynthesis emerge on Earth? Science. 2000;289:1703–5.

84. Goreau TF, Goreau NI. The physiology of skeletal formation in corals. II. Calcium deposition by hermatypic corals under various conditions in the reef. Biol Bull. 1959;117:239–50.
85. Stanley GD. Early history of scleractinian corals and its geological consequences. Geology. 1981;9:507–11.
86. Pawlik JR, Burkepile DE, Thurber RV. Vicious circle. Altered carbon and nutrient cycling may explain the low resilience of Caribbean coral reefs. Bioscience. 2016;66:470–6.
87. Zaneveld JR, Burkepile DE, Shantz AA, Pritchard, CE, Mcminds, R, Payet J. Overfishing and nutrient pollution interact with temperature to disrupt coral reefs down to microbial scales. Nat Commun. 2016;7:11833.
88. Hughes TP, Kerry JT, Álvarez-Noriega M, Álvarez-Romero JG, Anderson KD, et al. Global warming and recurrent mass bleaching of corals. Nature. 2017;543:373–7.
89. Thacker RW, Starnes S. Host specificity of the symbiotic cyanobacterium *Oscillatoria spongeliae* in marine sponges, *Dysidea* spp. Mar Biol. 2003;142:643–8.
90. Thacker RW. Impacts of shading on sponge-cyanobacteria symbioses: a comparison between host-specific and generalist associations. Integr Comp Biol. 2005;45:369–76.
91. Conway CA, Esiobu N, Lopez JV. Co-cultures of *Pseudomonas aeruginosa* and Roseobacter denitrificans reveal shifts in gene expression levels compared to solo cultures. Sci World J. 2012;2012:120108.
92. Freeman CJ, Thacker RW. Complex interactions between marine sponges and their symbiotic microbial communities. Limnol Oceanogr. 2011;56:1577–86.
93. Simpson TL. The cell biology of sponges. New York: Springer; 1984.
94. Mohri K, Nakatsukasa M, Masuda Y, Agata K, Funayama N. Toward understanding the morphogenesis of siliceous spicules in freshwater sponge: differential mRNA expression of spicule-type-specific silicatein genes in *Ephydatia fluviatilis*. Dev Dyn. 2008;237(10):3024–39.
95. Voigt O, Adamski M, Sluzek K, Adamska M. Calcareous sponge genomes reveal complex evolution of α-carbonic anhydrases and two key biomineralization enzymes. BMC Evol Biol. 2014;14:230.
96. Uriz MJ, Agell G, Blanquer A, Turon X, Casamayor EO. Endosymbiotic calcifying bacteria: a new cue to the origin of calcification in metazoa? Evolution. 2012;66:2993–9.
97. Santavy DL, Willenz P, Colwell RR. Phenotypic study of bacteria associated with the Caribbean sclerosponge, Ceratoporella nicholsoni. Appl Environ Microbiol. 1990;56:1750–62.
98. Gloeckner V, Wehrl M, Moitinho-Silva L, Gernert C, Schupp P, Pawlik JR, et al. The HMA-LMA dichotomy revisited: an electron microscopical survey of 56 sponge species. Biol Bull. 2014;227:78–88.
99. Moitinho-Silva L, Steinert G, Nielsen S, Hardoim CCP, Wu Y-C, McCormack GP, et al. Predicting the HMA-LMA status in marine sponges by machine learning. Front Microbiol. 2017;8:752.
100. Leys SP, Hill A. The physiology and molecular biology of sponge tissues. Adv Mar Biol. 2012;62:1.
101. Srivastava M, Simakov O, Chapman J, Fahey B, Gauthier MEA, Mitros T, et al. The *Amphimedon queenslandica* genome and the evolution of animal complexity. Nature. 2010;466:720–6.
102. Riesgo A, Peterson K, Richardson C, Heist T, Strehlow B, McCauley M, et al. Transcriptomic analysis of differential host gene expression upon uptake of symbionts: a case study with Symbiodinium and the major bioeroding sponge *Cliona varians*. BMC Genomics. 2014;16:376.
103. Watnick P, Kolter R. Biofilm, city of microbes. J Bacteriol. 2000;182:2675–9.
104. Hall-Stoodley L, Costerton JW, Stoodley P. Bacterial biofilms: from the natural environment to infectious diseases. Nat Rev Microbiol. 2004;2:95–10.
105. Ehrlich GD, Ahmed A, Earl J, Hiller NL, Costerton JW, Stoodley P, et al. The distributed genome hypothesis as a rubric for understanding evolution in situ during chronic bacterial biofilm infectious processes. FEMS Immunol Med Microbiol. 2010;59:269–79.
106. Estrela S, Kerr B, Morris JJ. Transitions in individuality through symbiosis. Curr Opin Microbiol. 2016;31:191–8.

107. Wagner-Döbler I. Biofilm transplantation in the deep sea. Mol Ecol. 2016;25:1905–7.
108. Frølund B, Palmgren R, Keiding K, Nielsen PH. Extraction of extracellular polymers from activated sludge using a cation exchange resin. Water Res. 1996;30:1749–58.
109. Crisp DJ, Ryland JS. Influence of filming and of surface texture on the settlement of marine organisms. Nature. 1960;185:119.
110. Whalan S, Webster NS. Sponge larval settlement cues: the role of microbial biofilms in a warming ocean. Sci Rep. 2014;4:4072.
111. Sneed JM, Sharp KH, Ritchie KB, Paul VJ. The chemical cue tetrabromopyrrole from a biofilm bacterium induces settlement of multiple Caribbean corals. Proc R Soc Biol. 2014;281:20133086.
112. Amann R, Fuchs BM. Single-cell identification in microbial communities by improved fluorescence in situ hybridization techniques. Nat Rev Microbiol. 2008;6:339–48.
113. Aminov RI. Horizontal gene exchange in environmental microbiota. Front Microbiol. 2011;26:158.
114. Angles ML,Marshall KC, Goodman AE. Plasmid transfer between marine bacteriain the aqueous phase and biofilms in reactormicrocosms. Appl Environ Microbiol. 1993; 59:843–50.
115. Thomas T, Rusch D, DeMaere MZ, Yung PY, Lewis M, Halpern A et al. Functional genomic signatures of sponge bacteria reveal unique and shared features of symbiosis. ISME J. 2010;4:1557–67.
116. Dunlap PV, Takami M, Wakatsuki S, Hendry TA, Sezaki K, Fukui A. Inception of bioluminescent symbiosis in early developmental stages of the deep-sea fish, *Coelorinchus kishinouyei* (Gadiformes: Macrouridae). Ichthyol Res. 2014;61:59–67.
117. Kleiner M, Young JC, Shah M, VerBerkmoes NC, Dubilier N. Metaproteomics reveals abundant transposase expression in mutualistic endosymbionts. MBio. 2013;4:e00223–13.
118. Alegado RA, Brown LW, Cao S, Dermenjian RK, Zuzow R, Fairclough SR, et al. A bacterial sulfonolipid triggers multicellular development in the closest living relatives of animals. elife. 2012;1:e00013.
119. Díez-Vives C, Moitinho-Silva L, Nielsen S, Reynolds D, Thomas T. Expression of eukaryotic-like protein in the microbiome of sponges. Mol Ecol. 2017;26:1432–51.
120. Nguyen MT, Liu M, Thomas T. Ankyrin-repeat proteins from sponge symbionts modulate amoebal phagocytosis. Mol Ecol. 2014;23:1635–45.
121. Exposito JY, Garrone R. Characterization of a fibrillar collagen gene in sponges reveals the early evolutionary appearance of two collagen gene families. Proc Natl Acad Sci U S A. 1990;87:6669–73.
122. Keeley FW, Mecham R. Evolution of extracellular matrix. New York: Springer; 2013.
123. Walker CB, de la Torre JR, Klotz MG, Urakawa H, Pinel N, Arp DJ, et al. *Nitrosopumilus maritimus* genome reveals unique mechanisms for nitrification and autotrophy in globally distributed marine crenarchaea. Proc Natl Acad Sci U S A. 2010;107:8818–23.
124. Exposito JY, Valcourt U, Cluzel C, Lethias C. The Fibrillar collagen family. Int J Mol Sci. 2010;11:407–26.
125. King N, Westbrook J, Young SL, Kuo A, Abedin M, Chapman J, et al. The genome of the choanoflagellate *Monosiga brevicollis* and the origin of metazoans. Nature. 2008;451:783–8.
126. Matz C, Webb JS, Schupp PJ, Phang SY, Penesyan A, Egan S, et al. Marine biofilm bacteria Evade eukaryotic predation by targeted chemical defense. PLoS One. 2008;3:e2744.
127. Bonner JT. The origins of multicellularity. Integr Biol. 1998;1:27–36.
128. Margulis L. Origin of eukaryotic cells. New Haven: Yale University Press; 1971.
129. King N. The unicellular ancestry review of animal development. Dev Cell. 2004;7:313–25.
130. Sebé-Pedrós A, Degnan BM, Ruiz-Trillo I. The origin of metazoa: a unicellular perspective. Nat Rev Genet. 2017;18:498–512.
131. Brock DA, Douglas TE, Queller DC, Strassmann JE. Primitive agriculture in a social amoeba. Nature. 2011;469:393–6.
132. Margulis L. Symbiosis in cell evolution. San Francisco: W.H. Freeman and Co; 1981.
133. Brucker RM, Bordenstein SR. The capacious hologenome. Zoology (Jena). 2013;116:260–1.
134. Zilber-Rosenberg I, Rosenberg E. Role of microorganisms in the evolution of animals and plants: the hologenome theory of evolution. FEMS Microbiol Rev. 2008;32:723–35.

135. Ryu T, Seridi L, Moitinho-Silva L, Oates M, Liew YJ, Mavromatis C, et al. Hologenome analysis of two marine sponges with different microbiomes. BMC Genomics. 2016;17:158.
136. Moran NA, Sloan DB. The hologenome concept: helpful or hollow? PLoS Biol. 2015;13:e1002311.
137. Mueller WEG. Molecular phylogeny of eumetazoa: genes in sponges (Porifera) give evidence for monophyly of animals. In: Mueller WEG, editor. Molecular evolution: evidence for monophyly of metazoa, Progress in molecular and subcellular biology, vol. 19. Berlin: Springer; 1998. p. 89–132.
138. Grosberg RK, Strathmann RR. The evolution of multicellularity: a minor major transition? Annu Rev Ecol Evol Syst. 2007;38:621–54.
139. Britstein M, Devescovi G, Handley KM, Malik A, Haber M, Saurav K, et al. A new N-acyl homoserine lactone synthase in an uncultured symbiont of the Red Sea sponge *Theonella swinhoei*. Appl Environ Microbiol. 2015;8:1274–85.
140. Hardoim CP, Costa R. Microbial communities and bioactive compounds in marine sponges of the family Irciniidae. Mar Drugs. 2014;12:5089–122.
141. Handelsman J. Metagenomics: application of genomics to uncultured microorganisms. Microbiol Mol Biol Rev. 2004;4:669–85.
142. Greuter D, Loy A, Horn M, Rattei T. ProbeBase—an online resource for rRNA-targeted oligonucleotide probes and primers: new features 2016. Nucleic Acids Res. 2016;44:D586–9.
143. Loy A, Maixner F, Wagner M, Horn M. probeBase – an online resource for rRNA-targeted oligonucleotide probes: new features 2007. Nucleic Acids Res. 2007;35:D800–4.
144. Manz W, Arp G, Schumann-Kindel G, Szewzyk U, Reitner J. Widefield, deconvolution epifluorescence microscopy combined with fluorescence in situ hybridization reveals the spatial arrangement of bacteria in sponge tissue. J Microbiol Methods. 2000;40:125–34.
145. Behnam F, Vilcinskas A, Wagner M, Stoecker KA. Straightforward DOPE (double labeling of oligonucleotide probes)-FISH (fluorescence in situ hybridization) method for simultaneous multicolor detection of six microbial populations. Appl Environ Microbiol. 2012;78:5138–42.
146. Pernthaler A, Pernthaler J, Amann R. Fluorescence in situ hybridization and catalyzed reporter deposition for the identification of marine bacteria. Appl Environ Microbiol. 2002;68:3094–101.
147. Laming SR, Duperron S. A correlative light-electron microscopy (CLEM) protocol for the identification of bacteria in animal tissue, exemplified by methanotrophic symbionts of deep-sea mussels. In: McGenity T, Timmis K, Nogales B, editors. Hydrocarbon and lipid microbiology protocols. Berlin/Heidelberg: Springer Protocols Handbooks Springer; 2015. p. 163–74.
148. Valm AM, Welch JLM, Rieken CW, Hasegawa Y, Sogin ML, Oldenbourg R, et al. Systems-level analysis of microbial community organization through combinatorial labeling and spectral imaging. Proc Natl Acad Sci U S A. 2011;108:4152–7.
149. Nikolakakis K, Lehnert E, McFall-Ngai MJ, Ruby EG. Use of hybridization chain reaction-fluorescent in situ hybridization to track gene expression by both partners during initiation of symbiosis. Appl Environ Microbiol. 2015;81:4728–35.
150. Chowdhury C, Sinha S, Chun S, Yeates TO, Bobik TA. Diverse bacterial microcompartment organelles. Microbiol Mol Biol Rev. 2014;78:438–68.
151. Kerfeld CA, Erbilgin O. Bacterial microcompartments and the modular construction of microbial metabolism. Trends Microbiol. 2015;23:22–34.
152. Siegl A, Kamke J, Hochmuth T, Piel J, Richter M, Liang C, et al. Single-cell genomics reveals the lifestyle of Poribacteria, a candidate phylum symbiotically associated with marine sponges. ISME J. 2010;5:61–70.
153. Kamke J, Rinke C, Schwientek P, Mavromatis K, Ivanova N, Sczyrba A, et al. The candidate phylum Poribacteria by single-cell genomics: new insights into phylogeny, cell-compartmentation, eukaryote-like repeat proteins, and other genomic features. PLoS One. 2014;9:e87353.

154. Mason OU, Hazen TC, Borglin S, Chain PS, Dubinsky EA, Fortney JL, et al. Metagenome, metatranscriptome and single-cell sequencing reveal microbial response to deepwater horizon oil spill. ISME J. 2012;6:1715–27.
155. Marr C, Zhou JX, Huang S. Single-cell gene expression profiling and cell state dynamics: collecting data, correlating data points and connecting the dots. Curr Opin Biotechnol. 2016;39:207–14.
156. Jahn MT, Markert SM, Ryu T, Ravasi T, Stigloher C, Hentschel U, et al. Shedding light on cell compartmentation in the candidate phylum Poribacteria by high resolution visualisation and transcriptional profiling. Sci Rep. 2016;6:35860.
157. Fry B. Stable isotope ecology. New York: Springer; 2006.
158. Herrmann AM, Ritz K, Nunan N, Clode PL, Pett-Ridge J, Kilburn MR, et al. Nano-scale secondary ion mass spectrometry: a new analytical tool in biogeochemistry and soil ecology: a review article. Soil Biol Biochem. 2007;39:1835–50.
159. Musat N, Halm H, Winterholler B, Hoppe P, Peduzzi S, Hillion F, et al. A single-cell view on the ecophysiology of anaerobic phototrophic bacteria. Proc Natl Acad Sci. 2008;105:17861–6.
160. Musat N, Musat F, Weber PK, Pett-Ridge J. Tracking microbial interactions with NanoSIMS. Curr Opin Biotechnol. 2016;41:114–21.
161. Polerecky L, Adam B, Milucka J, Musat N, Vagner T, Kuypers MMM. Look @ NanoSIMS – a tool for the analysis of nanoSIMS data in environmental microbiology. Environ Microbiol. 2012;14:1009–23.
162. Eichorst SA, Strasser F, Woyke T, Schintlmeister A, Wagner M, Woebken D. Advancements in the application of NanoSIMS and Raman microspectroscopy to investigate the activity of microbial cells in soils. FEMS Microbiol Ecol. 2015;91:fiv106.
163. da Fonseca RR, Albrechtsen A, Themudo GE, Ramos-Madrigal J, Sibbesen JA, Maretty L, et al. Next-generation biology: sequencing and data analysis approaches for non-model organisms. Mar Genomics. 2016;30:3–13.
164. Watson JD, Crick FHC. Molecular structure of nucleic acids: a structure for deoxyribose nucleic acid. Nature. 1953;171:737–8.
165. Judson HF. The eighth day of creation: makers of the revolution in biology. New York: Cold Spring Harbor Laboratory Press; 1996.
166. Mukherjee S. The gene: an intimate history. New York: Scribner; 2016.
167. Shendure J, Lieberman AE. The expanding scope of DNA sequencing. Nat Biotechnol. 2012;30:1084–94.
168. Stephens ZD, Lee SY, Faghri F, Campbell RH, Zhai C, Efron MJ, et al. Big data: astronomical or genomical? PLoS Biol. 2015;13:e1002195.
169. Barshis DJ, Ladner JT, Oliver TA, Palumbi SR. Lineage-specific transcriptional profiles of *Symbiodinium* spp. unaltered by heat stress in a coral host. Mol Biol Evol. 2014;31:1343–52.
170. Chadwick DJ, Whelan J. Secondary metabolites: their function and evolution. Chichester: Ciba Foundation/Wiley; 1992.
171. Niehaus TD, Thamm AM, de Crécy-Lagard V, Hanson AD. Proteins of unknown biochemical function: a persistent problem and a roadmap to help overcome it. Plant Physiol. 2015;169:1436–42.
172. The UniProt Consortium. UniProt: the universal protein knowledgebase. Nucleic Acids Res. 2017;45:D158–69.
173. The Gene Ontology Consortium, Ashburner M, Ball CA, Blake JA, Botstein D, Butler H, et al. Gene ontology: tool for the unification of biology. Nat Genet. 2000;25:25–9.
174. Chen IA, Markowitz VM, Chu K, Palaniappan K, Szeto E, Pillay M, et al. IMG/M: integrated genome and metagenome comparative data analysis system. Nucleic Acids Res. 2017;45:D507–16.
175. Gene Ontology Consortium. The gene ontology in 2010: extensions and refinements. Nucleic Acids Res. 2010;38:D331–5.
176. Kanehisa M, Sato Y, Morishima K. BlastKOALA and GhostKOALA: KEGG tools for functional characterization of genome and metagenome sequences. J Mol Biol. 2016;428:726–31.
177. Riesenfeld CS, Schloss PD, Handelsman J. Metagenomics: genomic analysis of microbial communities. Annu Rev Genet. 2004;38:525–52.

178. Tringe SG, Rubin EM. Metagenomics: DNA sequencing of environmental samples. Nat Rev Genet. 2005;6:805–14.
179. Tringe SG, von Mering C, Kobayashi A, Salamov AA, Chen K, Chang HW. Comparative metagenomics of microbial communities. Science. 2005;308:554–7.
180. Kennedy J, Flemer B, Jackson SA, Lejon DPH, Morrissey JP, O'Gara F, et al. Marine metagenomics: new tools for the study and exploitation of marine microbial metabolism. Mar Drugs. 2010;8:608–28.
181. Sogin ML, Morrison HG, Huber JA, Mark WD, Huse SM, Neal PR, et al. Microbial diversity in the deep sea and the underexplored rare biosphere. Proc Natl Acad Sci U S A. 2006;103(32):12115–20. Epub 2006 Jul 31.
182. Giovannoni SJ, Rappe MS. Evolution, diversity, and molecular ecology of marine prokaryotes. In: Kirchman DL, editor. Microbial ecology of the oceans. New York: Wiley-Liss Inc; 2000. p. 47–84.
183. Hiraoka S, Yang CC, Iwasaki W. Metagenomics and bioinformatics in microbial ecology: current status and beyond. Microbiol Environ. 2016;31:204–12.
184. Rappé MS, Giovannoni SJ. The uncultured microbial majority. Annu Rev Microbiol. 2003;57:369–94.
185. Sharp KH, Davidson SK, Haygood MG. Localization of 'Candidatus Endobugula sertula' and the bryostatins throughout the life cycle of the bryozoan *Bugula neritina*. ISME J. 2007;1:693–702.
186. Delong EF. The microbial ocean from genomes to biomes. Nature. 2009;459:200–6.
187. Brown TA. Genomes. 3rd ed. Oxford: Wiley-Liss; 2007.
188. Della Sala G, Hochmuth T, Costantino V, Teta R, Gerwick W, Gerwick L, et al. Polyketide genes in the marine sponge *Plakortis simplex*: a new group of mono-modular type I polyketide synthases from sponge symbionts. Environ Microbiol Rep. 2013;5:809–18.
189. Freeman MF, Gurgui C, Helf MJ, Morinaka BI, Uria AR, Oldham NJ, et al. Metagenome mining reveals polytheonamides as posttranslationally modified ribosomal peptides. Science. 2012;338:387–90.
190. Piel J, Hui D, Wen G, Butzke D, Platzer M, Fusetani N, et al. Antitumor polyketide biosynthesis by an uncultivated bacterial symbiont of the marine sponge *Theonella swinhoei*. Proc Natl Acad Sci U S A. 2004;101:16222–7.
191. Schirmer A, Gadkari R, Reeves CD, Ibrahim F, DeLong EF, Hutchinson CR. Metagenomic analysis reveals diverse polyketide synthase gene clusters in microorganisms associated with the marine sponge Discodermia dissoluta. Appl Environ Microbiol. 2005;71:4840–9.
192. Sudek S, Lopanik NB, Waggoner LE, Hildebrand M, Anderson C, Liu H, et al. Identification of the putative bryostatin polyketide synthase gene cluster from "Candidatus Endobugula sertula", the uncultivated microbial symbiont of the marine bryozoan *Bugula neritina*. J Nat Prod. 2007;70:67–74.
193. Wilson MC, Piel J. Metagenomic approaches for exploiting uncultivated bacteria as a resource for novel biosynthetic enzymology. Chem Biol. 2013;20:636–47.
194. Fiore CL, Labrie M, Jarett JK, Lesser MP. Transcriptional activity of the giant barrel sponge, *Xestospongia muta* holobiont: molecular evidence for metabolic interchange. Front Microbiol. 2015;6:364.
195. Mori T, Iwamoto K, Wakaoji S, Araie H, Kohara Y, Okamura Y, et al. Characterization of a novel gene involved in cadmium accumulation screened from sponge-associated bacterial metagenome. Gene. 2016;576:618–25.
196. Nakashima Y, Egami Y, Kimura M, Wakimoto T, Abe I. Metagenomic analysis of the sponge *Discodermia* reveals the production of the cyanobacterial natural product kasumigamide by 'Entotheonella'. PLoS One. 2016;11:e0164468.
197. Graf J. Lessons from digestive-tract symbioses between bacteria and invertebrates. Annu Rev Microbiol. 2016;70:375–93.
198. Selvin J, Kennedy J, Lejon DP, Kiran GS, Dobson AD. Isolation identification and biochemical characterization of a novel halo-tolerant lipase from the metagenome of the marine sponge *Haliclona simulans*. Microb Cell Factories. 2012;1:72.

199. Reynolds D, Thomas T. Evolution and function of eukaryotic-like proteins from sponge symbionts. Mol Ecol. 2016;25:5242–53.
200. Erwin PM, Coma R, López-Sendino P, Serrano E, Ribes M. Stable symbionts across the HMA-LMA dichotomy: low seasonal and interannual variation in sponge-associated bacteria from taxonomically diverse hosts. FEMS Microbiol Ecol. 2015;91:fiv115.
201. Hrdlickova R, Toloue M, Tian B. RNA-Seq methods for transcriptome analysis. Wiley Interdiscip Rev RNA. 2016;8:e1364.
202. Diehl AG, Boyle AP. Deciphering ENCODE. Trends Genet. 2016;32:238–49.
203. ENCODE Project Consortium An integrated encyclopedia of DNA elements in the human genome. Nature. 2012;489:57–74.
204. Qu H, Fang X. A brief review on the Human Encyclopedia of DNA Elements (ENCODE) project. Genomics Proteomics Bioinform. 2013;11:135–41.
205. Graur D, Zheng Y, Price N, Azevedo RB, Zufall RA, Elhaik E. On the immortality of television sets: "function" in the human genome according to the evolution-free gospel of ENCODE. Genome Biol Evol. 2013;5:578–90.
206. Graur D, Zheng Y, Azevedo RB. An evolutionary classification of genomic function. Genome Biol Evol. 2015;7:642–5.
207. Kellis M, Wold B, Snyder MP, Bernstein BE, Kundajea A, Marinov GK, et al. Defining functional DNA elements in the human genome. Proc Natl Acad Sci U S A. 2014;111:6131–8.
208. Francis WR, Woerheide G. Animals actively use at least half of the genome. BioRxiv. 2016:068627.
209. Tanzer A, Riester M, Hertel J, Bermudez-Santana, Gorodkin J, Hofacker, et al. Evolutionary genomics of microRNAs and their relatives. In: Caetano-Anoles G, editor. Evolutionary genomics and systems biology. Oxford: Wiley; 2010.
210. Tarver JE, Sperling EA, Nailor A, Heimberg AM, Robinson JM, King BL, et al. miRNAs: small genes with big potential in metazoan phylogenetics. Mol Biol Evol. 2013;30:2369–82.
211. Vogel C, Marcotte EM. Insights into the regulation of protein abundance from proteomic and transcriptomic analyses. Nat Rev Genet. 2012;13:227–32.
212. John DE, Zielinski BL, Paul JH. Creation of a pilot metatranscriptome library from eukaryotic plankton of a eutrophic bay (Tampa Bay, Florida). Limnol Oceanogr Methods. 2009;7:249–59.
213. McCarren J, Becker JW, Repeta DJ, Shi Y, Young CR, Malmstrom RR, et al. Microbial community transcriptomes reveal microbes and metabolic pathways associated with dissolved organic matter turnover in the sea. Proc Natl Acad Sci U S A. 2010;107:16420–7.
214. Moitinho-Silva L, Díez-Vives C, Batani G, Esteves AI, Jahn MT, Thomas T. Integrated metabolism in sponge-microbe symbiosis revealed by genome-centered metatranscriptomics. ISME J. 2017;11:1651–66.
215. Bashiardes S, Zilberman-Schapira G, Elinav E. Use of metatranscriptomics in microbiome research. Bioinform Biol Insights. 2016;10:19–25.
216. Versluis D, D'Andrea MM, Ramiro Garcia J, Leimena MM, Hugenholtz F, Zhang J. Mining microbial metatranscriptomes for expression of antibiotic resistance genes under natural conditions. Sci Rep. 2015;5:11981.
217. McFall-Ngai M. Divining the essence of symbiosis: insights from the squid-vibrio model. PLoS Biol. 2014;12:e1001783.
218. Gilbert JA, Quinn RA, Debelius JX, Z Z, Morton J, Garg N, et al. Microbiome-wide association studies link dynamic microbial consortia to disease. Nature. 2016;535:94–103.
219. Sokol H, Pigneur B, Watterlot L, Lakhdari O, Bermúdez-Humarán LG, Gratadoux J-J, et al. Faecalibacterium prausnitzii is an anti-inflammatory commensal bacterium identified by gut microbiota analysis of Crohn disease patients. Proc Natl Acad Sci U S A. 2008;105:16731–6.
220. Sousa T, Paterson R, Moore V, Carlsson A, Abrahamsson B, Basit AW. The gastrointestinal microbiota as a site for the biotransformation of drugs. Int J Pharm. 2008;363:1–25.
221. Johnson AD, O'Donnell CJ. An open access database of genome-wide association results. BMC Med Genet. 2009;10:6.

222. Gilbert JA, Meyer F, Antonopoulos D, Balaji P, Brown CT, Brown CT, et al. Meeting report: the terabase metagenomics workshop and the vision of an Earth microbiome project. Stand Genomic Sci. 2010;3:243–8.
223. Seneca FO, Palumbi SR. The role of transcriptome resilience in resistance of corals to bleaching. Mol Ecol. 2015;24:1467–84.
224. Ruiz-Jones LJ, Palumbi SR. Transcriptome-wide changes in coral gene expression at noon and midnight under field conditions. Biol Bull. 2015;228:227–41.
225. Bay RA, Palumbi SR. Rapid acclimation ability mediated by transcriptome changes in reef-building corals. Genome Biol Evol. 2015;7:1602–12.
226. Guzman C, Conaco C. Gene expression dynamics accompanying the sponge thermal stress response. PLoS One. 2016;11:e0165368.
227. Hintzsche JD, Robinson WA, Tan AC. A survey of computational tools to analyze and interpret whole exome sequencing data. Int J Genomics. 2016;2016:7983236.
228. Jaitin DA, Kenigsberg E, Keren-Shaul H, Elefant N, Paul F, Zaretsky I, et al. Massively parallel single-cell RNA-seq for marker-free decomposition of tissues into cell types. Science. 2014;343:776–9.
229. Macosko EZ, Basu A, Satija R, Nemesh J, Shekhar K, Goldman M, et al. Highly parallel genome-wide expression profiling of individual cells using nanoliter droplets. Cell. 2015;161:1202–14.
230. Markowetz F. All biology is computational biology. PLoS Biol. 2017;15:e2002050.
231. Har JY, Helbig T, Lim JH, Fernando SC, Reitzel AM, Penn K, et al. Microbial diversity and activity in the *Nematostella vectensis* holobiont: insights from 16S rRNA gene sequencing, isolate genomes, and a pilot-scale survey of gene expression. Front Microbiol. 2015;6:818.
232. Kumar D, Bansal G, Narang A, Basak T, Abbas T, Dash D. Integrating transcriptome and proteome profiling: strategies and applications. Proteomics. 2016;16:2533–44.
233. Rocha-Martin J, Harrington C, Dobson ADW, O'Gara F. Emerging strategies and integrated systems microbiology technologies for biodiscovery of marine bioactive compounds. Mar Drugs. 2014;12:3516–59.
234. Larsen PE, Dai Y. Metabolome of human gut microbiome is predictive of host dysbiosis. Gigascience. 2015;4:42.
235. Larsen P, Collart F, Field D, Meyer F, Keegan K, Henry C, et al. Predicted relative metabolomic turnover (PRMT): determining metabolic turnover from a coastal marine metagenomic dataset. Microb Inf Exp. 2011;1:4.
236. Smith E. De novo transcriptome analysis of the marine sponge *Cinachyrella* spp: a potential model organism for oil and dispersant ecotoxicology. Masters thesis. Nova Southeastern University Oceanographic Center. 2013.
237. Langille MG, Zaneveld J, Caporaso JG, McDonald D, Knights D, Reyes JA, et al. Predictive functional profiling of microbial communities using 16S rRNA marker gene sequences. Nat Biotechnol. 2013;31:814–21.
238. Yilmaz P, Yarza P, Rapp JZ, Glöckner FO. Expanding the world of marine bacterial and archaeal clades. Front Microbiol. 2016;6:1524.
239. Liu H, Guo X, Gooneratne R, Lai R, Zeng C, Zhan F, et al. The gut microbiome and degradation enzyme activity of wild freshwater fishes influenced by their trophic levels. Sci Rep. 2016;6:24340.
240. Alma'abadi AD, Gojobori T, Mineta K. Marine metagenome as a resource for novel enzymes. Genomics Proteomics Bioinform. 2015;13:290–5.
241. Kamisoglu K, Acevedo A, Almon RR, Coyle S, Corbett S, Dubois DC, et al. Understanding physiology in the continuum: integration of information from multiple -omics levels. Front Pharmacol. 2017;8:91.
242. Li Z, Wang Y, Yao Q, Justice NB, Ahn T-H, Xu D, et al. Diverse and divergent protein post-translational modifications in two growth stages of a natural microbial community. Nat Commun. 2014;5:4405.
243. Xiong W, Abraham PE, Li Z, Pan C, Hettich RL. Microbial metaproteomics for characterizing the range of metabolic functions and activities of human gut microbiota. Proteomics. 2015;15:3424–38.

244. Wang DZ, Kong LF, Li YY, Xie ZX. Environmental microbial community proteomics: status, challenges and perspectives. Int J Mol Sci. 2016;17:E1275.
245. Sowell SM, Abraham PE, Shah M, Verberkmoes NC, Smith DP, Barofsky DF, et al. Environmental proteomics of microbial plankton in a highly productive coastal upwelling system. ISME J. 2011;5:856–65.
246. Slattery M, Ankisetty S, Corrales J, Marsh-Hunkin KE, Gochfeld DJ, Willett KL, et al. Marine proteomics: a critical assessment of an emerging technology. J Nat Prod. 2012;75:1833–77.
247. Barghi N, Concepcion GP, Olivera BM, Lluisma AO. Comparison of the venom peptides and their expression in closely related conus species: insights into adaptive post-speciation evolution of conus exogenomes. Genome Biol Evol. 2015;7:1797–814.
248. GIGA Community of Scientists (COS). The global invertebrate genome alliance (GIGA): developing community resources to study diverse invertebrates. J Heredity. 2014;105:1–18.
249. Jensen PR, Mincer TJ, Williams PG, Fenical W. Marine actinomycete diversity and natural product discovery. Antonie Van Leeuwenhoek. 2005;87:43–8.
250. Engel S, Jensen PR, Fenical W. Chemical ecology of marine microbial defense. J Chem Ecol. 2002;28:1971–85.
251. Trindade M, van Zyl LJ, Navarro-Fernández J, Abd Elrazak A. Targeted metagenomics as a tool to tap into marine natural product diversity for the discovery and production of drug candidates. Front Microbiol. 2015;6:890.
252. Ziemert N, Lechner A, Wietz M, Millán-Aguiñaga N, Chavarria KL, Jensen PR. Diversity and evolution of secondary metabolism in the marine actinomycete genus Salinispora. Proc Natl Acad Sci U S A. 2014;111:E1130–9.
253. Berdy J. Bioactive microbial metabolites: a personal view. J Antibiot. 2005;58:1–26.
254. Blunt JW, Copp BR, Keyzers RA, Munro MH, Prinsep MR. Marine natural products. Nat Prod Rep. 2015;32:116–211.
255. Imhoff JF. Natural products from marine fungi--still an underrepresented resource. Mar Drugs. 2016;14:19.
256. Tyc O, Song C, Dickschat JS, Vos M, Garbeva P. The ecological role of volatile and soluble secondary metabolites produced by soil bacteria. Trends Microbiol. 2016;25:280–92.
257. Gomes NG, Dasari R, Chandra S, Kiss R, Kornienko A. Marine invertebrate metabolites with anticancer activities: solutions to the "supply problem". Mar Drugs. 2016;14:E98.
258. Hoppers A, Stoudenmire J, Wu S, Lopanik NB. Antibiotic activity and microbial community of the temperate sponge, *Haliclona* sp. J Appl Microbiol. 2015;118:419–30.
259. Ling LL, Schneider T, Peoples AJ, Spoering AL, Engels I, Conlon BP, et al. A new antibiotic kills pathogens without detectable resistance. Nature. 2015;517:455–9.
260. Kaeberlein T, Lewis K, Epstein SS. Isolating "uncultivable" microorganisms in pure culture in a simulated natural environment. Science. 2002;296:1127–9.
261. Donia MS, Fricke WF, Partensky F, Cox J, Elshahawi SI, White JR, et al. Complex microbiome underlying secondary and primary metabolism in the tunicate-Prochloron symbiosis. Proc Natl Acad Sci U S A. 2011;108:E1423–32.
262. Piel J. Bacterial symbionts: prospects for the sustainable production of invertebrate-derived pharmaceuticals. Curr Med Chem. 2006;13:39–50.
263. Rath CM, Janto B, Earl J, Ahmed A, Hu FZ, Hiller L, et al. Meta-omic characterization of the marine invertebrate microbial consortium that produces the chemotherapeutic natural product ET-743. ACS Chem Biol. 2011;6:1244–56.
264. Mackay HJ, Twelves CJ. Targeting the protein kinase C family: are we there yet? Nat Rev Cancer. 2007;7:554–62.
265. Mathew M, Lopanik NB. Host differentially expressed genes during association with its defensive endosymbiont. Biol Bull. 2014;226:152–63.
266. Stone MJ, Williams DH. On the evolution of functional secondary metabolites (natural products). Mol Microbiol. 1992;6:29–34.
267. Garson MJ. Biosynthesis of sponge secondary metabolites: why is it important? In: Van Soest RWM, Van Kempen TMG, Braekman J-C, editors. Sponges in time and space. Roterdam: AA Balkema; 1994. p. 427–40.

268. Pomponi SA. The bioprocess–technological potential of the sea. J Biotechnol. 1999;70:5–13.
269. Hamada T, Matsunaga S, Yano G, Fusetani NJ. Polytheonamides A and B, highly cytotoxic, linear polypeptides with unprecedented structural features, from the marine sponge, *Theonella swinhoei*. Am Chem Soc. 2005;127:110.
270. Helf M, Jud A, Piel J. Enzyme from an uncultivated sponge bacterium catalyzes S-methylation in a ribosomal peptide. Chem Bio Chem. 2017;18:444–50.
271. Lackner G, Peters EE, Helfrich EJ, Piel J. Insights into the lifestyle of uncultured bacterial natural product factories associated with marine sponges. Proc Natl Acad Sci U S A. 2017;114:E347–56.
272. Kubanek J, Fenical W, Pawlik JR. New antifeedant triterpene glycosides from the Caribbean sponge *Erylus formosus*. Nat Prod Lett. 2001;15:275–85.
273. Roué M, Quévrain E, Domart-Coulon I, Bourguet-Kondracki ML. Assessing calcareous sponges and their associated bacteria for the discovery of new bioactive natural products. Nat Prod Rep. 2012;29:739–51.
274. Loh TL, Pawlik JR. Chemical defenses and resource trade-offs structure sponge communities on Caribbean coral reefs. Proc Natl Acad Sci U S A. 2014;111:4151–6.
275. Gunasekera SP, Gunasekera M, Longley R, Schulte G. Discodermolide, a new bioactive polyhydroxylated lactone from the marine sponge, *Discodermia dissoluta*. J Organomet Chem. 1991;55:4912.
276. Longley RE, Caddigan D, Harmody D, Gunasekera M, Gunasekera SP. Discodermolide – a new, marine-derived immunosuppressive compound. II. In vivo studies. Transplantation. 1991;52:656–61.
277. Molinski TF, Dalisay DS, Lievens SL, Saludes JP. Drug development from marine natural products. Nat Rev Drug Discov. 2009;8:69.
278. Freeman MF, Helf MJ, Bhushan A, Morinaka BI, Piel J. Seven enzymes create extraordinary molecular complexity in an uncultivated bacterium. Nat Chem. 2017;9:387–95.
279. Goffredi SK, Yi H, Zhang Q, Klann JE, Struve IA, Vrijenhoek RC, et al. Genomic versatility and functional variation between two dominant heterotrophic symbionts of deep-sea Osedax worms. ISME J. 2013;8:908–24.
280. Hendry T, Freed L, Fader D, Fenolio D, Lopez J. Recently evolved host dependence in the luminous symbionts of deep sea anglerfish. mBio. 2018;9:e01033-18.
281. Klose J, Polz MF, Wagner M, Schimak MP, Gollner S, Bright M. Endosymbionts escape dead hydrothermal vent tubeworms to enrich the free-living population. Proc Natl Acad Sci U S A. 2015;112:11300–5.
282. Haddock SH, Moline MA, Case JF. Bioluminescence in the sea. Annu Rev Mar Sci. 2010;2:443–93.
283. Nagasawa H. The molecular mechanism of calcification in aquatic organisms. Biosci Biotechnol Biochem. 2013;77:1991–6.

Chapter 9
Carbon and Nitrogen Metabolism of Sponge Microbiome

Guofang Feng and Zhiyong Li

Contents

Abstract Sponges represent an evolutionarily divergent group of species with widespread physiological and ecological traits. Spongology has grown into a discipline attracting a progressively growing population of hundreds of scientists across the world. Sponges host complex communities of microbial symbionts and thus are ideal model to test functional equivalence and evolutionary convergence that exists in complex symbiont communities across phylogenetically divergent hosts. Many studies have demonstrated the tremendous advances in our understanding of the composition and phylogenetic diversity of sponge-associated microbes. As a comparison, the in situ activity and function of these microbes has become a major research focus. Already the rewards of this new emphasis are evident, with

G. Feng · Z. Li (✉)
Marine Biotechnology Laboratory, State Key Laboratory of Microbial Metabolism, School of Life Sciences and Biotechnology, Shanghai Jiao Tong University, Shanghai, People's Republic of China
e-mail: zyli@sjtu.edu.cn

Z. Li (ed.), *Symbiotic Microbiomes of Coral Reefs Sponges and Corals*, https://doi.org/10.1007/978-94-024-1612-1_9

cultivation-independent genomic and experimental approaches yielding novel insights into symbiont function. Herein, this review highlights the largest part of the available knowledge on recent developments about the sponge-meditated nutrient fluxes and their ecological implications of carbon and nitrogen. Gene, genome, transcriptome, and next-generation sequencing (NGS) analyses have provided extraordinary insights into the sponge microbial functions as well as the ecological roles. This review has covered the recent findings regarding dynamics of sponge microbiome, and several interesting research areas, that we believe are deserving of increased attention.

Keywords Sponge · Microbiome · Nitrogen metabolism · Carbon metabolism

9.1 Background of Sponge Microbiome

Sponges are among the most ancient of the multicellular organisms and possess relatively little in the way of differentiation and coordination of tissues [1]. More than 13,000 estimated species of sponges are distributed across a wide variety of marine and some freshwater ecosystems throughout tropical, temperate, as well as polar regions [2]. Sponges represent the ecologically significant constituents of marine benthic communities in terms of both biomass and their potential to influence benthic or pelagic processes [3, 4]. These sessile, filter-feeding animals can provide habitat for a wide range of infauna species, couple the benthic and pelagic zones through their filtration of enormous quantities of seawater, mediate biogeochemical fluxes, and facilitate the consumption and release of nutrients [4, 5]. Despite their simplistic body whose cells directly contact with the ambient seawater, sponges are known to host dense, diverse, and highly specific microbial assemblages which are constituted of archaea, bacteria, unicellular algae, and fungi [6–10]. So far, symbiotic microbes can span remarkable 52 different bacterial phyla (including candidate phyla) and 2 archaeal phyla [9]. The levels of richness and diversity of these symbiont communities vary significantly between different sponge species, ranging from a few distinct microbial species to thousands of genetically distinct symbionts per host taxon [11–13], most of which are considered metabolically active [14, 15]. In some sponges, these microbes are found at densities exceeding 10^9 microbial cells per cubic centimeter of sponge tissue, which are greater than the microbial density in the surrounding seawater by 3–4 orders of magnitude [16]. These microbial symbionts can comprise up to 60% of the total sponge biomass and make valuable contributions to many aspects of the sponge's ecology and physiology [17]. For this reason, sponges are described as holobionts, which are comprised of the sponge and the consortium of microbial symbionts. Compared to the considerable knowledge about the diversity and composition of microbes within sponges, our understanding on the functional roles of the symbionts that play within sponge holobiont is relatively understudied despite their multiple ecological derivations [18]. Nowadays, element metabolism in sponges has come under particularly intense scrutiny and has become a major focus of recent studies. Sponge symbionts are presumed to benefit from a stable supply of nutrients, partially of which are excreted from the host, while the host is presumed to benefit from supple nutrition or waste scavenging derived via a diverse range of metabolic processes provided by the symbiont community. Microbial symbionts are hypothesized to contribute to the health and nutrition of sponge hosts in different ways, including acquisition of limiting nutrients, and processing of metabolic waste [18, 19]. Therefore, it is urgent to assay and identify how the symbiotic microbes use and affect the availability and cycling of the chemical elements, such as carbon and nitrogen. Although the proposed symbiotic functions are extensive, examples of specific symbionts being unequivocally assigned functional roles are actually quite rare. Abundant sponge symbionts are so far recalcitrant to cultivation, and the lack of an experimental model of microbial strain-sponge host or axenic cell cultures for experimental manipulation is major constraint to identify the function of symbionts and their interaction with sponges. Despite these limitations, gene-centric, genomic, single-cell genomic, metagenomic, metatranscriptomic, and metaproteomic sequencing data, combined with physiological experiments, have provided insights into the importance of the symbiont-driven metabolic pathways within sponges, including carbon and nitrogen metabolism. Here, we review the available information with the objective of providing a better global understanding of the recent significant developments in sponge microbiome, providing a comprehensive knowledge of the ecological functions of sponge-associated microbes in carbon and nitrogen metabolisms, and highlight several fields that urgently require additional research attention.

9.2 Carbon Metabolism

With respect to carbon metabolism, heterotrophy is a common form of carbon metabolism for sponges, either via consumption of microbes filtered from seawater or via microbial uptake of dissolved organic carbon [20, 21]. Many studies have shown that the ability of sponges to release large quantities of organic and inorganic dissolved compounds may be related to the photoauto-, chemoauto-, hetero-, and mixotrophic processes mediated by sponge symbiotic microbial communities. These metabolic pathways include a variety of processes with alternative energy (photo- or chemotrophic) and carbon (auto- or heterotrophic) sources under different conditions [18, 22–25] (Fig. 9.1).

9.2.1 CO_2 Assimilation

Many sponges, particularly those in tropical regions, harbor substantial populations of cyanobacteria and zooxanthellae which have photosynthetic properties [26–34]. Sponges may benefit from their photosynthetic associates. For example, some

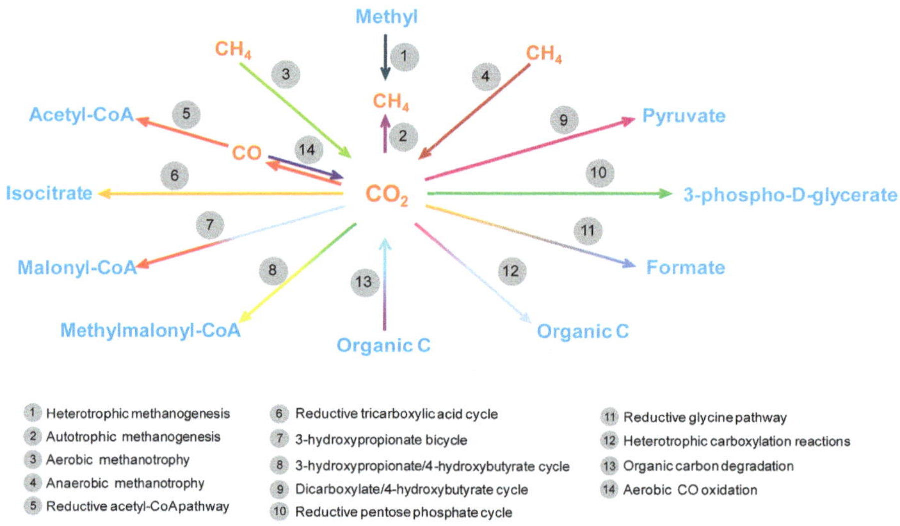

Fig. 9.1 Microbially mediated carbon transforms

sponges are apparently depending on their symbiotic phototrophs, with their low depth limits determined by the availability of light for photosynthesis [35]. Symbiotic dinoflagellates have been found in the Mediterranean sponge *Cliona viridis*, and the growth of *C. viridis* is greater for individuals under natural light conditions than those in constant darkness, reflecting the contribution of photosynthetic symbionts to host metabolism [36]. In addition, Great Barrier Reef sponges may derive much of their nutrition from photosynthetic symbionts in the depth of 15–30 m seawater, due to the clear seawater that can decrease light attenuation [35]. Phototrophic microbes have not only been detected in the outer tissue portion of the sponge hosts but also in deeper tissue regions. For example, the finding of phototrophic organisms seemingly assembled around spicules led to the hypothesis of a siliceous light transmission system in sponges [37]. In addition, translocation of photosynthates from the autotrophic symbionts to the host has now been demonstrated for several marine sponge species. For instance, in an earlier study, glucose produced by a chlorella-like green alga was confirmed to be passed to its freshwater sponge host, *Ephydatia fluviatilis* [38]. The specialized filamentous cyanobacterial symbiont *Oscillatoria spongiarum* benefits its sponge host *Lamellodysidea chlorea* with respect to growth [28]. Also, translocation of carbon nutrients from sponge-associated zooxanthellae (*Symbiodinium* spp.) to sponge cells has been confirmed for the sponges *Cliona* spp. [32]. In another study, nutrient translocation of photosynthates, such as glycerol and organic phosphate from bacterial and algal symbionts to their sponge hosts, can significantly contribute to host metabolism and growth [32, 39]. Certainly, carbon nutrition of some sponge species may depend heavily on photoautotroph-derived carbon sources. For example, contributions of phototrophic cyanobacteria to the carbon assimilation account for up to 80% of the carbon budget of the sponges [28, 40], allowing these species to thrive in the low-nutrient, high-light areas commonly found on tropical reefs. Moreover, microbial CO_2-assimilation functions may affect the colonization of sponges in changed environments, such as the ocean acidification due to the massive production of CO_2 [41]. For example, the sponges *Coelocarteria singaporensis* and *Cinachyra* sp. near the natural volcanic CO_2 seeps host a significantly higher abundance of *Synechococcus* than those far from the seeps, suggesting that symbiotic photoautotrophs may provide their sponge hosts with a nutritional benefit and enhance scope for growth under higher CO_2 concentrations [42]. In addition, there are sponges harboring chemoautotrophic prokaryotes with the capacity of fixing CO_2. For example, ammonia-oxidizing bacteria/archaea, nitrite-oxidizing bacteria, and sulfur-oxidizing bacteria as well as other chemoautotrophic clades have been detected from different sponge species from various geographic locations by using molecular sequencing approaches [23, 43–53]. These studies indicated the unique existence of chemoautotrophic microbes inhabiting sponges. CO_2 fixation capacity of chemoautotrophs within sponges has been also confirmed in dark habitats. For example, the carbon incorporation rate of CO_2 in the dark by symbiotic microbes within the sponges *Higginsia thielei* and *Rossella nodastrella* in cold-water coral reefs represented approximately 10% to the total net carbon production of the sponges [54].

Functional genes based surveys have also provided insights into the pathways for carbon metabolism in sponge holobionts. For example, genes encoding various enzymes involved in the major reactions in different CO_2 assimilation pathways have been identified in the genomes of sponge-associated microbes. Genomic information of cyanobacterial *Synechococcus spongiarum* and *Myxosarcina* sp. identified from sponges has demonstrated the photoautotrophic reductive pentose phosphate (rPP) pathway of these isolates [55, 56]. These cyanobacteria occur in many different sponge species from multiple geographic locations and express low-molecular-weight peptides of photosystem II, enabling low-light photosynthesis [55, 56], indicating genome streamlining as an adaptation to the sponge's intercellular microenvironment. Genomic construction of the archaeon *Cenarchaeum symbiosum* harvested from the sponge *Axinella mexicana* contains genes encoding the proteins involving autotrophic carbon fixation via the 3-hydroxypropionate/4-hydroxybutyrate (3-HP/4-HB) cycle [57]. This chemoautotrophic CO_2 fixation pathway has also been revealed from other *Thaumarchaeota* species from various niches, such as *Nitrosopumilus maritimus*, *Nitrosoarchaeum koreensis*, and *Nitrososphaera viennensis* [58–60]. In addition, partial genes involving the tricarboxylic acid (TCA) cycle have also been discovered from the genome, indicating that *C. symbiosum* may use the metabolic intermediates from the incomplete TCA pathway for the synthesis of some cofactors [57]. Besides, metagenomic reconstruction of the deep-sea sponge *Lophophysema eversa* has revealed partial genome information of a relative high-abundant chemoautotrophic archaeon *Nitrosopumilus* sp. [61]. CO_2 fixation pathway of this archaeon is also the 3-HP/4-HB cycle, which is identical to that of *C. symbiosum*, but different to the chemoautotrophic rPP pathway in ammonia-oxidizing bacteria (AOB), such as *Nitrosomonas europaea* and *Nitrosococcus oceani* [62, 63]. Single-cell sequencing of the sorted *Poribacteria* sp.

obtained from the sponge *Verongia aerophoba* by fluorescence-based cell sorting [64] showed that this bacterial clade is mixotrophic that can undertake autotrophic carbon fixation via the reductive acetyl-CoA (rACA) pathway and alternative heterotrophic metabolisms. Representative genes involved in the rACA pathway were identified in both the genome of *Poribacteria* sp. sorted from the sponge *Aplysina aerophoba* and the metagenome of the sponge *Cymbastela concentrica* holobiont [53, 64]. Genome revelation of the sulfur-oxidizing bacterium *Thioalkalivibrio nitratireducens* identified from the sponge *Haliclona cymaeformis* has illustrated the chemoautotrophic rPP pathway and versatile heterotrophic metabolisms of this bacterial clade in the sponge host [65]. Genome reconstruction of the unclassified *Nitrospira* from the sponge *Lophophysema eversa* has revealed the genes encoding the subunits of nitrate oxidoreductase, which share high peptide identity to the homologous in the genome of *Nitrospira gracilis* 3/211. In addition, gene-related functions in reductive TCA (rTCA) cycle may be employed for CO_2 fixation by this bacterial clade based on genomic analysis [61]. Genetic potentials for the chemoautotrophic rPP lifestyle and sulfur-oxidation/reduction have been identified from the metagenomic reconstruction of the symbiotic *Entotheonella* spp. within the sponge *Theonella swinhoei*, indicating the chemoautotrophic potential of this taxon in their sponge host; in addition, genomic potentials for highly efficient utilization of chitin and N-acetylglucosamine have also been identified from *Entotheonella*, indicating the mixotrophic lifestyle of this bacterial genus [66]. Additionally, NGS applications have given unprecedented insights into the functional potentials of sponge microbiota. For example, comparative metagenomic analysis of the shallow-water sponge *Theonella swinhoei* and the deep-sea sponge *Neamphius huxleyi* showed that a relatively higher ratio of CO_2 uptake carboxylase and respiration-related proteins in *T. swinhoei*, while a higher number of chemoautotrophic rPP cycle-related genes as well as a lower ratio of CO_2 uptake carboxylase were detected in *N. huxleyi* [67]. Transcribed functional genes related to cyanobacterial *Synechococcus* photosynthesis via the photoautotrophic rPP pathway have been explored in the metatranscriptome of the sponge *Stylissa carteri* [68]. GeoChip analysis of the sponges *Aplysina aerophoba*, *Dysidea avara*, *Xestospongia testudinaria*, and *Stylissa carteri* has comprehensively revealed the genes, such as the ones encoding ribulose-1, 5-bisphosphate carboxylase/oxygenase for carbon fixation in the rPP cycle [69].

9.2.2 CO Oxidation

For microbial CO metabolism, some microbial lineages in sponges can oxidize CO into CO_2 by the aerobic CO dehydrogenase (CODH)-mediated pathway. For example, bacterial CO oxidation potentials were suggested in the metagenome of the sponge *Neamphius huxleyi* by the identification of abundant *coxLMS* genes encoding the aerobic CODH [70], indicating that sponge symbiotic bacteria would perform CO consumption to reduce its concentration in the microenvironment of sponge. Transcripts matching with genes such as *coxL* genes involved in CO metabolism were recovered in the metatranscriptome of the sponge *Xestospongia muta* [14]. Metagenomic analysis of the sponge *Arenosclera brasiliensis* has revealed multiple kinds of genes related to carbon metabolism of symbiotic microbes. For instance, abundant genes encoding uroporphyrinogen methyltransferase and CODH have been revealed from the metagenome; the former is a key enzyme involved in the synthesis of a cofactor of the rACA pathway, and the latter is a key enzyme involved in CO metabolism in the facultative anaerobic *Proteobacteria* taxa involved in energy and CO_2 generation [71]. Such complexity reveals a diversity of niches and microbial metabolism pathways in the sponge. *A. brasiliensis* accordingly, metagenome of the sponge *Neamphius huxleyi* has revealed that the symbionts probably use two alternative pathways, the chemoautotrophic rPP and the rTCA cycle which does not depend directly on the presence of light, for CO_2 assimilation; in addition, abundant *cox* genes were also detected in this sponge metagenome, indicating the complex carbon metabolic pathways by symbiotic microflora in sponges [70]. Moreover, metatranscriptomic analysis of the sponge *Geodia barretti* showed that the representative potential of microbial functions in this sponge is consisted of key enzymes involved in diverse metabolic pathways, including aerobic CODH [51].

9.2.3 CH_4 Metabolism

Other chemoautotrophic microbial processes that have been observed in sponges may also contribute to sponge nutrition. Methanotrophic symbiosis was confirmed in the sponge *Cladorhiza methanophila* by the detection of abundant *Methylococcales* 16S rRNA (*rrs*) genes [72]. Methanotrophic *Methylophylales* and photosynthetic *Chromatiales* were also identified in the metagenome of the sponge *Arenosclera brasiliensis* [71]. Moreover, *pmoA* gene encoding the subunit of particulate methane monooxygenase has been detected from the sponge *Desmacidon* sp., which is closely related to the homologous sequence of *Methyloparacoccus murrellii* [73]. Like the symbiotic photoautotrophs, some methanotrophs are believed to translocate a significant portion of their nutrition to their sponge hosts. For example, stable carbon isotope analysis and methanol dehydrogenase activity assay combined with ultrastructural observation on the sponge *Cladorhiza* sp. have demonstrated that at least some of the sponge symbionts are methanotrophic which could directly transmit through generations in brooded embryos and provide a substantial portion of their carbon nutrition to their sponge host [22]. Similarly, carbon isotope analysis showed 38–100% methane-derived carbon from methanotrophic symbionts to the sponge *Pseudosuberites* sp. [74]. Additionally, transcribed functional genes encoding methanol dehydrogenase have been confirmed in the metatranscriptome of the sponge *Stylissa carteri* [68]. Metabolically diverse symbionts, including anaerobic bacteria *Methylophylales* using methane as electron acceptors, were detected from the sponge *Arenosclera brasiliensis*, revealing a diversity of microbial metabolism pathways in this sponge [71]. GeoChip analysis of the sponges *Aplysina aerophoba*,

Dysidea avara, *Xestospongia testudinaria*, and *Stylissa carteri* comprehensively revealed the *mcrA* genes encoding the subunit of methyl-coenzyme M reductase for methane production [69]. Functional transcripts encoding heterodisulfide reductase, methenyl tetrahydromethanopterin cyclohydrolase, and phosphosulfolactate synthase were identified for methane metabolism in the metatranscriptome of the sponge *Xestospongia muta* [14].

9.2.4 Organic Carbon Degradation and Assimilation

With regard to organic carbon degradation and assimilation, carbon degradation repertoire of *Poribacteria* would further be consistent with degradation of compounds in sponges [75]. In addition, a study on a random set of sponge species collected from mangrove roots and adjacent reefs suggested that sponges, which were commonly associated with mangrove roots, contained bacteria that were capable of degrading mangrove-derived organic carbon, while bacterial communities associated with sponges that were more typical to reef environments appeared less proficient in degrading mangrove-derived organic carbon nutrients. Thus, host specificity of bacterial endobionts capable of degrading mangrove-derived organic carbon and the presence of high concentrations of recalcitrant organic compounds may lead to ecological separation between mangrove and reef sponge communities [76]. Labeling of the sponge with diatom-derived ^{13}C-organic carbon demonstrated instantaneous incorporation of ^{13}C in bacterium-specific as well as sponge-related fatty acids of the sponge *Halisarca caerulea*, implying uptake of ambient organic carbon by both sponge cells and sponge-associated microbes [20]. Therefore, it is evident that sponge cells take up organic carbon directly from the ambient water, as do the sponge-associated prokaryotes. GeoChip analysis of the sponges *Aplysina aerophoba*, *Dysidea avara*, *Xestospongia testudinaria*, and *Stylissa carteri* has comprehensively revealed the genes for degradation of chitin (*chiA* encoding exochitinase) and hemicellulose (*xylA* encoding xylanase) [69]. The *chiA* gene pyrosequencing data showed the existence across habitats of core bacterial communities responsible for chitin-degrading microbial communities in the freshwater sponge *Ephydatia fluviatilis* [77]. Most of these *chiA* gene reads were from *Actinobacteria*, *Bacteroidetes*, *Chloroflexi*, *Deinococcus-Thermus*, *Dictyoglomi*, *Firmicutes*, and *Proteobacteria* and the sequences affiliated with the homologous of *Aeromonas veronii* B565 and *Lysobacter enzymogenes* covered the majority of reads [77]. Degradation of aromatic and benzoic compounds by sponge symbionts was also evidenced by the identification of the abundant enzymes from the metal-dependent hydrolase family, including 2-amino-3-carboxymuconate-6-semialdehyde decarboxylase, and the glyoxalase/bleomycin resistance protein/dioxygenase superfamily, including ring-cleavage extradiol dioxygenase, in the sponges *Cymbastela coralliophila* and *Rhopaloeides odorabile* [78].

9.3 Nitrogen Metabolism

Besides carbon, nitrogen is the most important nutrient for life, as it is required for the synthesis of amino acids. Essential and unique steps in the nitrogen cycle are performed by a variety of bacteria, archaea, and eukaryotes [79]. The nitrogen cycle controls the availability of nitrogenous nutrients, hence markedly influencing the biological productivity in marine systems. Nitrogen metabolism and cycling have been extensively studied in microbial communities of sponges. Moreover, different steps of the nitrogen cycle have been identified in the same sponge species, suggesting that nitrogen cycling in sponges goes through a complex network of metabolic pathways [23]. So far, several steps including nitrogen fixation, nitrification, denitrification, anammox, ammonia assimilation, ammonia mineralization, and assimilatory/dissimilatory nitrate to ammonium (ANRA/DNRA) processes have been identified in nitrogen cycling [80]. Many studies are based on the detection and description of microbial community structure that could mediate different nitrogen cycling steps [18] (Fig. 9.2). Herein, we review the microbially mediated nitrogen cycle processes noted in marine sponges.

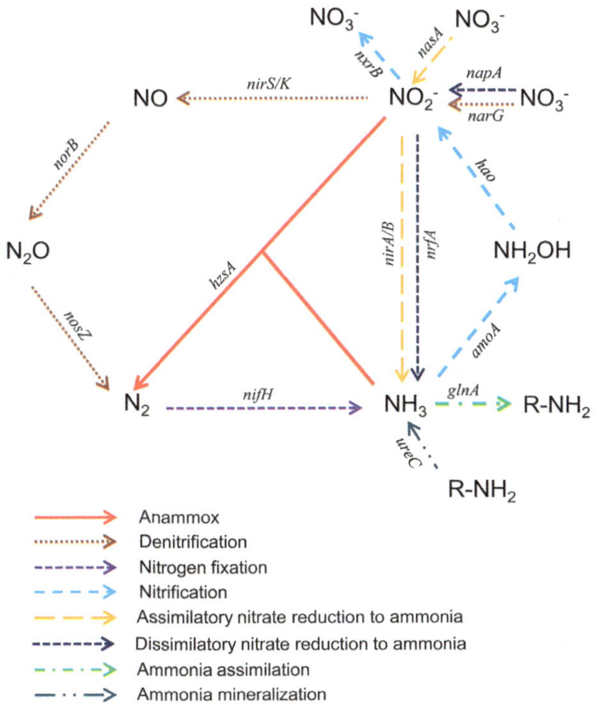

Fig. 9.2 Microbially mediated nitrogen cycle

9.3.1 Nitrogen Fixation

Nitrogen fixation is the key step of nitrogen element from inactive state to active state. In the ocean, prokaryote involved nitrogen fixation providing at least 50% of nitrogen needs for the oceanic primary production [81]. Nitrogen fixation by sponge-associated microbes may contribute to the nitrogen budget of sponges nitrogen needs in the oligotrophic environments with low nitrogen availability [18]. Sponge-mediated nitrogen fixation was first reported in the Red Sea sponges *Siphonochalina tabernacula* and *Theonella swinhoei* [82] and in a tropical species of *Halichondria* sp. [83]. However, in situ studies showed that nitrogen fixation rates in several sponges of Florida Bay were very low compared to the ambient water. Sponge-mediated nitrogen fixation in Caribbean reefs probably contributes neither to the sponge nutrition nor to nitrogen inputs to the reef. Nevertheless, the result may be different in other geographical regions with stronger nitrogen limitation. A subsequent study provided more concrete proof of nitrogen fixation in sponges by demonstrating incorporation of the stable isotope $^{15}N_2$ into various amino acids in the sponge *Callyspongia muricina* [84]. Moreover, using the *nifH* (encoding the nitrogenase subunit) gene as target, nitrogen fixation potentials of specific microbial lineages have been detected in sponges. For example, PCR targeting *nifH* genes and their transcribed counterparts have revealed the presence and activity of diazotrophs from the sponges *Ircinia strobilina* and *Mycale laxissima* offshore the Key Largo Island; these *nifH* genes belonged to the *Cyanobacteria* and *Proteobacteria* lineages, while their transcribed counterparts belonged to the *Cyanobacteria* lineages [85]. NGS analysis of *nifH* genes and their transcribed counterparts from the Caribbean Sea sponges *Ircinia strobilina* and *Mycale laxissima* showed that *nifH* genes and transcripts were mainly gathered into the *Cyanobacteria*, *Proteobacteria*, and *Verrucomicrobia* lineages, while seawater-derived *nifH* genes were mainly belonging to the *Proteobacteria* and *Verrucomicrobia* lineages [86]. *nifH* genes were detected in the metagenome of the deep-sea sponge *Neamphius huxleyi* [70]. Similarly, the *nifD*, *nifH*, and *nifK* genes encoding the subunits of nitrogenase have been uncovered from the metagenome of the sponge *Hymeniacidon heliophila* [87]. Metatranscriptomic analysis of the sponge *Xestospongia muta* has uncovered the transcribed *nifH* genes from the symbiotic prokaryotes, which fell into *Cyanobacteria* and *Proteobacteria* groups. Moreover, the cyanobacterial *nifH* sequences were similar to *Xenococcus* sp., *Myxosarcina* sp., and the cyanobacterium UCYN-A, while the proteobacterial *nifH* sequences were most similar to either *Alphaproteobacteria* such as *Bradyrhizobium japonicum* or to *Gammaproteobacteria* such as *Vibrio* spp. and *Azotobacter chroococcum* [14]. Comparative metagenomic analysis of the shallow-water sponge *Theonella swinhoei* and the deep-sea sponge *Neamphius huxleyi* showed that nitrogenase genes were detected in both sponge species yet showed varied diversity [67]. Moreover, genes related to nitrogen-fixing *Mesorhizobium* species have been identified by a combined metagenomic and metaproteomic approach in the sponge *Cymbastela concentrica* [88]. These findings suggest that nitrogen fixation by symbiotic microbes results in an important input of nitrogen into sponge symbionts and it is possible that any additional fixed nitrogen would be beneficial to sponges in habitats with low available nitrogen nutrients.

9.3.2 Nitrification

Like many marine invertebrates, sponges can produce ammonia as their metabolic wastes, which should make them a particularly attractive niche for microbes in nitrogen-poor oceanic waters. Sponge symbiotic nitrifiers would play an important role in ammonia transformation and consumption. The most evident nitrogen flux in sponges mediating nitrification is a net production of nitrite and nitrate, irrespective of habitat and geographic area. Ammonia and nitrite are suspected to accumulate within the sponge body during periods of pumping arrest. Conversion to nitrate would be a solution for eliminating these compounds which are toxic to the sponge if accumulated above a certain concentration threshold [89]. The study of ammonia oxidation in sponges provides an elegant example of the effective combination of molecular genomic-, transcriptomic-, and proteomic-derived data with physiological experiments and field studies. So far, sponge-mediated nitrification has been detected in many sponges from many geographic areas [5, 23, 50, 90–94]. Also, vertical transmission of ammonia-oxidizing archaea (AOA) has been described in the sponges *Luffariella variabilis* and *Rophaloeides odorabile* from the Great Barrier Reef [89]. In addition, some transcription-level revelation has reflected the metabolic activity potential of the functional community in sponges. For example, AOA and AOB *amoA* (encoding the subunit of ammonia monooxygenase) genes have been detected in the cold-water sponges *Phakellia ventilabrum*, *Geodia barrette*, *Antho dichotoma*, and *Tentorium semisuberites*; however, transcribed AOA *amoA* genes were several orders of magnitude higher than the transcribed AOB *amoA* genes in abundance [50], indicating that AOA may be the main ammonia-oxidizing functional population in these sponges. High rates of nitrification in sponges of Florida Keys reef was confirmed using incubation experiments, in agreement with previous work on a number of Caribbean species [5, 91]. Beside nitrite and nitrate production, nitrification can also result in ammonia uptake, being this combined nitrogen flux especially evident in sponges. For example, a seasonal variability existed in those fluxes, as reported in the Mediterranean sponge *Aplysina aerophoba* [93]. Metagenomic reconstruction of *Cenarchaeum symbiosum* revealed the genes encoding various key enzymes, including ammonia monooxygenase for nitrification [57]. Besides, metagenomic reconstruction of a relative high-abundant archaeon *Nitrosopumilus* sp. within the deep-sea sponge *Lophophysema eversa* revealed genes related to nitrification, which is most similar to the homologous *Nitrosopumilus maritimus* SCM1 [61]. Since the first confirmation of nitrite-oxidizing bacteria (NOB) within sponge [95], NOB have been recovered by molecular surveys of sponges from various areas. Most of the detected NOB lineage in sponges belong to the *Nitrospira* lineage [17, 18, 95]. Therefore, *Nitrospira* lineage might be the dominant NOB group in sponges. The *Nitrospira* lineage has been unraveled from various niches, such as sewage treatment plant [96, 97], biofilter of aquaculture [98], freshwater systems [99], and soils as well as hot springs [100, 101], as the dominant NOB group therein. As yet, only one strain of *Nitrospira* has been enriched from the sponge *Aplysina aerophoba*, namely, *Nitrospira* sp. A01 [102]. Physiological analysis of this strain demonstrated that this strain showed limited concentration tolerance to nitrite, with their suitable concentration as 0.5–0.75 mM, while higher than 1.5 mM of nitrite would inhabit the growth of this species. As a comparison, *Nitrospira marina* isolated from seawater could tolerate as high as 6 mM nitrite [103], while the *Nitrospira defluvii* from sewage treatment plant could tolerate the nitrite up to 25 mM [104]. This feature would relate to the adaptability of this NOB type to the low nitrite concentration in sponges. Surprisingly, low oxygen concentration would be suitable for the growth of this strain. Since no sponge-derived NOB strain has been sequenced so far, a lot of unsuspected aspects about such functional group need further research.

Customarily, most of the AOA, AOB, or NOB within sponges cannot be isolated. Therefore, their community structure and diversity as well as activity would be detected by *rrs* gene or functional gene assays. For example, previous works on *rrs* gene sequencing illustrated that several types of AOB in the genera *Nitrosospira* have been identified from the sponges *Aplysina aerophoba*, *Ircinia strobilina*, and *Mycale laxissima* [48, 93], while phylotypes belonging to the *Nitrosomonas eutropha/europaea*-affiliated ammonia oxidizers have been recovered from various mangrove sponges [105]. In addition, *amoA* gene from AOA or AOB and *nxrA* or *nxrB* gene (encoding the subunit of nitrite oxidoreductase) from NOB have been commonly used as phylogenetic marker to reveal the nitrifying lineages in the sponge microbiomes. For example, *amoA* gene amplification has discovered diversified AOA taxa in the sponge *Phakellia fusca* [46] and *Nitrosospira* AOB groups from the sponges *Ircinia strobilina* and *Mycale laxissima* [48]. Similarly, *rrs* and *amoA* gene assays have revealed the ammonia-oxidizing group consisting of *Nitrosospira* AOB and *Thaumarchaeota* AOA from the sponge *Aplysina aerophoba* [93]. AOA which could vertically transmit from sponge parental generation to their offspring has been certified in different geographic sponges from the South Pacific, Caribbean Sea, and Mediterranean Sea by *amoA* assays [89], indicating this transmission style of AOA in sponges would be an ubiquitous way in sponges and the compacted relationship between sponges and their AOA symbionts. The emerging NGS techniques performed in the sponge microbiology have provided new insights for the revelation of the multiple ecological roles, like the nitrification function, of sponge symbionts. For instance, *rrs* gene pyrosequencing revealed the ammonia-oxidizing *Nitrosomonadales* associated with the endemic sponge *Arenosclera brasiliensis* [71]. Similarly, pyrosequencing of *rrs* gene has found the *Nitrosopumilus* group from sponge microbiomes [106]. Metagenomic sequencing has revealed the functional community including *Nitrosopumilus* AOA in the deep-sea sponge *Neamphius huxleyi* [70]. Besides that, GeoChip analysis of the sponges *Aplysina aerophoba*, *Dysidea avara*, *Xestospongia testudinaria*, and *Stylissa carteri* has comprehensively revealed diverse nitrifying genes such as *amoA* and *hao* (encoding hydroxylamine oxidoreductase) genes [69]. The *hao* gene has been also uncovered from the metagenome of the sponge *Hymeniacidon heliophila* [87]. The *amoA* genes falling into the genera *Nitrosopumilus* and *Cenarchaeum* were detected in the metagenome of the deep-sea sponge *Neamphius huxleyi* [70]. Besides, metagenomic analysis has revealed the complex nitrifying population consisting of *Nitrosopumilus*, *Nitrosococcus*, *Nitrosomonas*, and *Nitrosospira* in sponges [107]. Metatranscriptomic analysis showed that the representative potential nitrifying organisms in the sponge *Geodia barretti* are consisted of ammonia-oxidizing *Thaumarchaeota* and the nitrite-oxidizing *Nitrospirae*, which have been confirmed by functional profiles including archaeal *amo* transcripts and bacterial *nxr* transcripts [51]. Similarly, transcribed functional genes related to active archaeal ammonia oxidation have been indicated by the high expression of archaeal *amo* genes in the metatranscriptome of the sponge *Stylissa carteri* [68]. Metatranscriptomic analysis of the sponge *Xestospongia muta* uncovered the transcripts from the symbiotic prokaryotes, including *nxr* genes which showed high amino acid similarity to the homologue of *Nitrospira fluvii*, and the archaeal *amo* genes which had significantly higher expression [14]. Comparative metagenomic analysis of the shallow-water sponge *Theonella swinhoei* and the deep-sea sponge *Neamphius huxleyi* demonstrated that nitrogen cycle-related genes encoding *amo*, *hao*, and *nxr* were detected in both sponge species yet showed varied composition and diversity [67]. Metagenomic and metaproteomic analysis recovered the expression of the a*moABC* and *amtB* (encoding the subunit of ammonia transporter) genes in the microbial community of the sponge *Cymbastela concentrica*, which were most closely related to those of the archaeon *Nitrosopumilus maritimus*, showing that aerobic nitrification is being carried out by a *Nitrosopumilus*-like AOA in this sponge species [88]. These studies demonstrated that the nitrifying function of microbes may play an important role in sponge-mediated nitrogen cycling in the ocean.

Consistently, a single microbe which completes the nitrification process would be more advantageous and more effective than two distinct microbes, i.e., the AOA/AOB and NOB, especially when the ammonia concentration is considerably low. In a recent research, two bacterial strains, namely, Ca. *Nitrospira nitrosa* and Ca. *Nitrospira nitrificans*, were found with the potential to complete the whole nitrification process in one cell [108]. Genome sequencing analysis showed that these two bacteria contained a complete set of the nitrification-related genes, including *amoABC*, *hao*, and *nxrABC* genes. Their *amoABC* genes are significantly different from the reported AOA or AOB *amoABC* genes; rather, they were distantly related to the *pmoA* gene of *Crenothrix polyspora*, with a nucleotide sequence similarity as 97–98%. Further alignment analysis hitting the NCBI nucleotide database showed that such *amoA* homogenous has been detected from many metagenomes and metatranscriptomes of soils, sediments, and sewage treatment plant deposits. There is no report of the similar nitrifying *Nitrospira* identified in sponges, which need further attention.

9.3.3 Denitrification

Denitrification is the conversion of nitrate to net N_2 production via several steps by bacteria or archaea (Fig. 9.2). Denitrification is catalyzed by a phylogenetically diverse range of microbes from at least 50 genera, interspersing among the lineages of the phyla *Proteobacteria*, *Actinobacteria*, *Firmicutes*, *Bacteroidetes*, *Aquificae*, and *Deinococcus-Thermus* clades [109]. Limited oxygen diffusion (e.g., caused by ceased pumping) can rapidly create anoxic conditions in sponge tissue, thus facilitating anaerobic denitrification. Denitrification evidence in sponges was firstly discovered in the sponge *Geodia barretti* [23] and then in the Mediterranean species *Dysidea avara* and *Chondrosia reniformis* [90]. This procedure could provide the symbiotic microbes using sponge metabolic waste products, which is a way to clean the sponge tissue from noxious compounds [23]. The *rrs* gene analysis showed that a common alphaproteobacterial lineage within marine sponges is very closely related to the marine denitrifier *Pseudovibrio denitrificans* [110], and at least some of the sponge-derived bacterial strains have also been tested to be positive for denitrification [111]. The *narG* genes encoding the subunit of respiratory nitrate reductase have been detected from the metagenomes of six sponge species [78]. Comparative metagenomic analysis of the shallow-water sponge *Theonella swinhoei* and the deep-sea sponge *Neamphius huxleyi* showed that *narG* genes were detected in both sponge species yet showed varied composition and diversity [67]. In the cold-water sponges *Astrosclera willeyana*, *Geodia barretti*, *Lamellomorpha* sp., and *Spirastrellidae diplastrella*, *nirS* and *nirK* genes (encoding the subunit of nitrite reductase) have been detected falling into various *Betaproteobacteria* and *Gammaproteobacteria* lineages [23, 45, 112]. Analysis of denitrifying community in five South China Sea sponge species showed that *nirK* genes were only detected in *Spheciospongia vesparium*, while *nosZ* (encoding the subunit of nitrous oxide reductase) genes were discovered from *Amphimedon queenslandica*, *S. vesparium*, *Xestospongia testudinaria*, *Cinachyrella* sp., and *Cinachyrella australiensis*; these denitrifying genes interspersed among the lineages of *Proteobacteria*, and their diversity and richness were varied among these sponges [113]. Metagenomic reconstruction of the sponge-associated *Cenarchaeum symbiosum* revealed the genes encoding various key enzymes, including a nitrite reductase and nitric-oxide reductase for denitrification [57]. Several genes encoding the denitrifying enzymes have been identified by a single-cell genomics analysis of *Poribacteria* sp. [64], and proteins related to both nitrification and denitrification have been identified through a combined metagenomic and metaproteomic analysis of the microbial consortium in the sponge *Cymbastela concentrica* [88]. In addition, the *nirB*, *nirD*, *narG*, *narH*, *narJ*, and *narI* genes encoding nitrite reductase, *nirK* gene encoding nitrite reductase, *nosZ* gene encoding nitrous oxide reductase, and *norB*, *norC*, and *norF* genes encoding nitric-oxide reductase in denitrification process have been uncovered from the metagenome of the sponge *Hymeniacidon heliophila* [87]. Metatranscriptomic analysis of the sponges *Stylissa massa* and *Xestospongia testudinaria* has revealed relative high-abundant nitrogen-fixing and denitrifying genes, including *nifH*, *nirK*, *nosZ*, and *norB* transcripts in these sponges [114]. Moreover, *nirK* gene has been identified from the *Thaumarchaeota*, which are commonly present in sponges [115], indicating the significant functions of these archaeal lineages in nitrification and denitrification within sponges. Besides that, GeoChip analysis of the sponges *Aplysina aerophoba*, *Dysidea avara*, *Xestospongia testudinaria*, and *Stylissa carteri* has comprehensively revealed *narG*, *nirS*, and *nosZ* genes for denitrification [69]. In addition, a nitrate reductase gene cluster including *narG*, *narH*, *narI*, and *narY* genes was present in the partial genome of the *Phyllobacteriaceae* phylotype highlighting its potential for nitrate reduction in the sponge *Cymbastela concentrica* [88]. Similarly, transcribed functional gene encoding copper-containing nitrite reductase has been unraveled in the metatranscriptome of the sponge *Stylissa carteri* [68]. Metatranscriptomic analysis of the sponge *Xestospongia muta* has uncovered the denitrifying transcripts from the symbiotic prokaryotes including the *nirK* genes related to *Thaumarchaeota*, *Oleispira antarctica*, and *Mesorhizobium amorphae*, *nirS* genes similar to *Pseudomonas aeruginosa* and *Sinorhizobium* sp., and *norB* genes related to *Alcaligenes faecalis* [14]. Denitrifying genes including *nirS*, *nirK*, *norB*, and *nosZ* have been identified from the sponge-associated microbial communities from the metagenomes of six sponge species [78], as well as the shallow-water sponge *Theonella swinhoei* and the deep-sea sponge *Neamphius huxleyi*, which showed varied diversity between these sponge species [67]. Interestingly, *norQ* and *norD* genes encoding putative nitric-oxide reductase subunits are also encoded in the metagenome of the sponge *Cymbastela concentrica* [88]. These findings demonstrated the important ecological function of denitrifiers in sponges.

9.3.4 Anammox

Anammox reaction is the conversion of ammonia and nitrite to dinitrogen in the anaerobic conditions. Microbes that carry out this reaction are bacteria special from the phylum *Planctomycetales*. The first evidence of anammox activity in sponges was described in the sponge *Geodia barretti* whose symbiotic anammox bacteria were closely related to the *Planctomycetes Scalindua sorokinii* and *Scalindua brodae* [23]. Anammox bacteria may uniquely exist in sponges. For example, *rrs* gene amplification has represented the existence of anammox bacteria in the sponges *Discodermia dissoluta* [116], *Antho chartacea* [18], *Tethya aurantium* [117], *Ircinia strobilina* [118], *Mycale laxissima* [119], *Ircinia strobilina*, *Mycale laxissima* [48], *Xestospongia testudinaria*, *Geodia barretti* [120], *Crambe crambe* [121], and *Hymeniacidon heliophila* [122] from intertidal and subtidal habitats by *rrs* pyrosequencing analysis. In addition, *Planctomycetes* strains have been isolated from sponges. For instance, three *Planctomycetes* strains were isolated firstly from the sponges *Aplysina* spp., which were closely related to the genus *Pirellula* [123], while 29 species have been isolated from the sponge *Niphates* sp., which are related to the *Planctomycetes* species *Blastopirellula marina*, *Rhodopirellula baltica*, and *Planctomyces brasiliensis* [124].

9.3.5 ANRA and DNRA

ANRA is the process that nitrate is reduced to ammonia via nitrite, which is carried out by bacteria to incorporate ammonia for cell growth. This process has been detected from the lineages of *Proteobacteria*, *Firmicutes*, *Bacteroidetes*, and *Planctomycetes* [125]. The *nasA* gene encoding nitrate reductase, and *nirA*/*nirB* genes encoding ferredoxin-nitrite reductase in ANRA process have been uncovered from the metagenome of the sponge *Hymeniacidon heliophila* [87]. The DNRA process employs identical chemistry to ANRA, but it is facilitated by evolutionarily unrelated nitrite reductases. DNRA can generate a cellular protonmotive force and thus conserve energy to support cellular growth, although this feature depends on which enzymes are coupled. In the sponge metatranscriptomes of *Stylissa massa* and *Xestospongia testudinaria*, *nrfA* gene encoding the subunit of membrane-binding cytochrome c nitrite reductase which catalyzes the reduction of nitrite to ammonia has been identified with relative high abundance [114], indicating that microbial DNRA action might play an important role in sponge nitrogen cycle. Besides that, GeoChip analysis of the sponges *Aplysina aerophoba*, *Dysidea avara*, *Xestospongia testudinaria*, and *Stylissa carteri* has comprehensively revealed the *nrfA* genes for DNRA in these species [69]. Metagenomic analysis showed periplasmic dissimilatory nitrate reductase gene *napA* in six sponge species [78] as well as the shallow-water sponge *Theonella swinhoei* and the deep-sea sponge *Neamphius huxleyi* [67, 70]. The *napA*, *napB*, and *napC* genes encoding nitrate reductase and *nrfA* and *nrfD* genes encoding nitrate reductase in DNRA processes have been uncovered from the metagenome of the sponge *Hymeniacidon heliophila* [87]. Metatranscriptomic analysis of the sponge *Xestospongia muta* has uncovered the DNRA-related transcripts from the symbiotic prokaryotes including the *nrfA* genes associated with *Shewanella* spp. [14]. Comparative metagenomic analysis of the shallow-water sponge *Theonella swinhoei* and the deep-sea sponge *Neamphius huxleyi* showed *nrfA* genes in both sponge species yet showed varied diversity [67].

9.3.6 Ammonia Assimilation

Ammonia assimilation represents the incorporation of ammonia into organic nitrogen compounds. This process is performed via the glutamine synthase-glutamine oxoglutarate aminotransferase (GS-GOGAT) pathway in organisms [126]. Abundant actively transcribed genes encoding glutamine synthetase from the microbial assemblage of marine microplankton indicated that microbial ammonia assimilation may be an important pathway for nitrogen uptake [127]. The efficient recycling of sponge-excreted nitrogenous wastes is one obvious benefit for symbionts in return for an existence in sponge tissues. In the metagenome of the sponge *Cymbastela concentrica*, abundant bacterial *gln* (glutamine synthetase) genes were detected, demonstrating that sponge symbionts were with the great potential for organic nitrogen metabolism via the GS-GOGAT pathway for ammonia assimilation [53]. Similarly, transcribed *gln* genes have been revealed by functional transcript-based analysis from the sponges *Theonella swinhoei*, *Plakortis simplex*, and *Phakellia fusca* [128] and in the metatranscriptome of the sponge *Stylissa carteri* [68]. In addition, metatranscriptomic analysis of the sponge *Xestospongia muta* has uncovered the transcripts from the symbiotic prokaryotes involving in nitrogen assimilation pathways that incorporate ammonia into amino acids (e.g., glutamate synthase, glutamate dehydrogenase, glutamine synthetase) [14]. The *glnA* gene has been uncovered from the metagenome of the sponge *Hymeniacidon heliophila* [87]. Comparative metagenomic analysis of the shallow-water sponge *Theonella swinhoei* and the deep-sea sponge *Neamphius huxleyi* showed that nitrogen cycle-related genes encoding glutamate dehydrogenase and glutamate synthase were detected in both sponge species yet showed varied composition and diversity [67]. Highly regulated ammonia assimilation in the bacterial community of the sponge *Cymbastela concentrica* is indicated by the abundance of nitrogen regulatory protein PII, which controls the transcription of the glutamine synthetase gene. This suggests that specific pathways and regulation of assimilation are preferred in the sponge-associated community and that assimilation processes rather than oxidation might be important for the ammonium utilization by the bacterial community in *C. concentrica* [53]. In addition, genes involved in assimilation which were closely related to the homologous of *Agrobacterium* and *Sphingomonadaceae* were detected in the metagenome indicating a complex nitrogen cycle in the sponge *Neamphius huxleyi* [70]. Conclusively, ammonia assimilation pathway is one of the important ammonia-scavenging procedures in sponges.

9.3.7 Ammonia Mineralization

The process of organic ammonia mineralization involves the conversion of the dissolved organic nitrogen to inorganic ammonia by symbiotic microbes in sponges. One representative process is the urea hydrolysis to ammonia. Urea is one of the dominant organic nitrogenous compounds in oligotrophic oceans and likely serves as an alternative nitrogen source to ammonia, nitrate, and nitrite within the sponge holobionts [129, 130]. Microbes contain urease which can degrade urea to produce ammonia. The *ureC* gene encoding the subunit of urease was identified from various ecosystems, such as surface seawaters, ocean, and estuary planktonic microbes, and showed unexpected diversity [130]. Using the isolating cultivation method, a bacterium *Marinobacter litoralis* has been isolated from the sponge *Xestospongia testudinaria* with urea hydrolysis activity; in addition, PCR amplification has revealed diverse *ureC* genes and transcripts from the metagenome and metatranscriptome of this sponge [131]; subsequent phylogenetic analysis showed that most of the *ureC* genes and transcripts are from the lineages of *Proteobacteria* and the remaining ones, which take a minor proportion, are from the taxa of *Magnetococcus*, *Cyanobacteria*, and *Actinobacteria* [131]. Besides that, GeoChip analysis of the microbiomes associated with the sponges *Aplysina aerophoba*, *Dysidea avara*, *Xestospongia testudinaria*, and *Stylissa carteri* has comprehensively revealed the *ureC* genes for ammonia mineralization [69]. Metagenomic reconstruction of *Cenarchaeum symbiosum* revealed the genes encoding various key enzymes, including urease for organic nitrogen degradation [57]. Genes encoding the proteins related to urease accessory proteins, urea channels, and urea transporters were found in the genomes of *Entotheonella* spp. identified from the sponge *Theonella swinhoei* [66]. Comparative metagenomic analysis of the shallow-water sponge *Theonella swinhoei* and the deep-sea sponge *Neamphius huxleyi* showed that nitrogen cycle-related genes encoding urea transporter in both sponge species yet showed varied composition and diversity [67]. Therefore, ammonia mineralization as well as other organic nitrogen degradation processes would contribute to the organic nitrogen scavenging and the nitrogen needs for sponges and their microbial symbionts.

9.4 Conclusions and Perspectives

The role of sponges is much related to the activity of their microbial associates. Here, we summarize the roles of sponge microbiomes in element cycling. Carbon and nitrogen metabolisms derive from a complex combination of metabolic processes and are performed by large microbial populations within the sponge body. Sponge symbiotic microbes are capable for inorganic carbon metabolisms, including CO_2 assimilation, CO oxidation, and CH_4 metabolism, composing the carbon metabolic networks in sponge tissues. Regarding nitrogen, the situation is more complicated, as most studies are still attempting to unravel the basic metabolic pathways. Microbes with the potential to mediate in each of the steps known in nitrogen cycling have been molecularly identified in marine sponges. Preliminary assessments of these processes suggest that they contribute significantly either to sponge and symbionts nutrition or to induce relevant changes in concentration of nitrogen compounds in the sponge tissues. The emerging researches have greatly improved our knowledge about sponge-microbe symbioses and interactions. We should note that sequencing advances are facilitating rapid growth in the field of sponge symbiotic functions. However, analyses are still somewhat constrained by the often distinct phylogenetic positioning of sponge symbionts, making assessment of symbionts' functions difficult. Furthermore, the physiology of most sponge-associated microbes remains unclear, as do many fundamental aspects of sponge symbiont ecology. Though a lot of data from molecular studies have been provided to give a clear picture of microbial diversity and functions in their sponge hosts, further investigations are still needed to enhance our insights into the comprehensive understanding of the ecological functions of sponge symbionts as well as the sponge symbiont interplay.

Acknowledgments We gratefully acknowledge financial supports from the Natural Science Foundation of China (NSFC) (31861143020, 41776138, 41742002, U1301131, 41176127, 41076077), the High-Tech Research and Development Program of China (2013AA092901, 2011AA090702, 2007AA09Z447, 2004AA628060, 2002AA608080), and the National Major Scientific Research Program of China (2013CB956103).

References

1. Cavalier-Smith T. Origin of animal multicellularity: precursors, causes, consequences-the choanoflagellate/sponge transition, neurogenesis and the Cambrian explosion. Phil Trans R Soc B. 2017;372:20150476.
2. Leys SP, Rohksar DS, Degnan BM. Sponges. Curr Biol. 2005;15:R114–5.
3. Bell JJ. Functional roles of marine sponges. Estuar Coast Shelf Sci. 2008;79:341–53.
4. Maldonado M, Ribes M, van Duyl FC. Nutrient fluxes through sponges: biology, budgets, and ecological implications. Adv Mar Biol. 2012;62:113–82.
5. Southwell MW, Weisz JB, Martens CS, Lindquist N. In situ fluxes of dissolved inorganic nitrogen from the sponge community on Conch Reef, Key Largo, Florida. Limnol Oceanogr. 2008;53:986.
6. Jin L, Liu F, Sun W, Zhang F, Karuppiah V, Li Z. Pezizomycotina dominates the fungal communities of South China Sea sponges *Theonella swinhoei* and *Xestospongia testudinaria*. FEMS Microbiol Ecol. 2014;90:935–45.
7. He L, Liu F, Karuppiah V, Ren Y, Li Z. Comparisons of the fungal and protistan communities among different marine sponge holobionts by pyrosequencing. Microb Ecol. 2014;67:951–61.
8. Rodriguez-Marconi S, De la Iglesia R, Diez B, Fonseca CA, Hajdu E, Trefault N. Characterization of bacterial, archaeal and eukaryote symbionts from antarctic sponges reveals a high diversity at a three-domain level and a particular signature for this ecosystem. PLoS One. 2015;10:e0138837.
9. Thomas T, Moitinho-Silva L, Lurgi M, Bjork JR, Easson C, Astudillo-Garcia C, et al. Diversity, structure and convergent evolution of the global sponge microbiome. Nat Commun. 2016;7:11870.
10. Polonia AR, Cleary DF, Freitas R, Coelho FJ, de Voogd NJ, Gomes NC. Comparison of archaeal and bacterial communities in two sponge species and seawater from an Indonesian coral reef environment. Mar Genomics. 2016;29:69–80.
11. Webster NS, Taylor MW, Behnam F, Lucker S, Rattei T, Whalan S, et al. Deep sequencing reveals exceptional diversity and modes of transmission for bacterial sponge symbionts. Environ Microbiol. 2010;12:2070–82.
12. Reveillaud J, Maignien L, Eren AM, Huber JA, Apprill A, Sogin ML, et al. Host-specificity among abundant and rare taxa in the sponge microbiome. ISME J. 2014;8:1198–209.
13. Lee OO, Wang Y, Yang J, Lafi FF, Al-Suwailem A, Qian PY. Pyrosequencing reveals highly diverse and species-specific microbial communities in sponges from the Red Sea. ISME J. 2011;5:650–64.
14. Fiore CL, Labrie M, Jarett JK, Lesser MP. Transcriptional activity of the giant barrel sponge, *Xestospongia muta* Holobiont: molecular evidence for metabolic interchange. Front Microbiol. 2015;6:364.
15. Kamke J, Taylor MW, Schmitt S. Activity profiles for marine sponge-associated bacteria obtained by 16S rRNA vs 16S rRNA gene comparisons. ISME J. 2010;4:498–508.

16. Webster NS, Taylor MW. Marine sponges and their microbial symbionts: love and other relationships. Environ Microbiol. 2012;14:335–46.
17. Schmitt S, Weisz JB, Lindquist N, Hentschel U. Vertical transmission of a phylogenetically complex microbial consortium in the viviparous sponge *Ircinia felix*. Appl Environ Microbiol. 2007;73:2067–78.
18. Taylor MW, Radax R, Steger D, Wagner M. Sponge-associated microorganisms: evolution, ecology, and biotechnological potential. Microbiol Mol Biol Rev. 2007;71:295–347.
19. Hentschel U, Piel J, Degnan SM, Taylor MW. Genomic insights into the marine sponge microbiome. Nat Rev Microbiol. 2012;10:641–54.
20. de Goeij JM, Moodley L, Houtekamer M, Carballeira NM, van Duyl FC. Tracing ^{13}C-enriched dissolved and particulate organic carbon in the bacteria-containing coral reef sponge *Halisarca caerulea*: evidence for DOM-feeding. Limnol Oceanogr. 2008;53:1376–86.
21. de Goeij JM, van Oevelen D, Vermeij MJ, Osinga R, Middelburg JJ, de Goeij AF, et al. Surviving in a marine desert: the sponge loop retains resources within coral reefs. Science. 2013;342:108–10.
22. Vacelet J, Fiala-Medioni A, Fisher CR, Boury-Esnault N. Symbiosis between methane-oxidizing bacteria and a deep-sea carnivorous cladorhizid sponge. Mar Ecol Prog Ser. 1996;145:77–85.
23. Hoffmann F, Radax R, Woebken D, Holtappels M, Lavik G, Rapp HT, et al. Complex nitrogen cycling in the sponge *Geodia barretti*. Environ Microbiol. 2009;11:2228–43.
24. Hentschel U, Usher KM, Taylor MW. Marine sponges as microbial fermenters. FEMS Microbiol Ecol. 2006;55:167–77.
25. Figueroa IA, Barnum TP, Somasekhar PY, Carlström CI, Engelbrektson AL, Coates JD. Metagenomics-guided analysis of microbial chemolithoautotrophic phosphite oxidation yields evidence of a seventh natural CO_2 fixation pathway. Proc Natl Acad Sci U S A. 2018;115:E92–E101.
26. Burgsdorf I, Erwin PM, Lopez-Legentil S, Cerrano C, Haber M, Frenk S, et al. Biogeography rather than association with cyanobacteria structures symbiotic microbial communities in the marine sponge Petrosia ficiformis. Front Microbiol. 2014;5:529.
27. Cebrian E, Uriz MJ, Garrabou J, Ballesteros E. Sponge mass mortalities in a warming Mediterranean Sea: are cyanobacteria-harboring species worse off? PLoS One. 2011;6:e20211.
28. Thacker RW. Impacts of shading on sponge-cyanobacteria symbioses: a comparison between host-specific and generalist associations. Integr Comp Biol. 2005;45:369–76.
29. Webb VL, Maas EW. Sequence analysis of 16S rRNA gene of cyanobacteria associated with the marine sponge *Mycale* (*Carmia*) *hentscheli*. FEMS Microbiol Lett. 2002;207:43–7.
30. Granados C, Camargo C, Zea S, Sanchez JA. Phylogenetic relationships among zooxanthellae (*Symbiodinium*) associated to excavating sponges (*Cliona* spp.) reveal an unexpected lineage in the Caribbean. Mol Phylogenet Evol. 2008;49:554–60.
31. Hill M, Allenby A, Ramsby B, Schonberg C, Hill A. Symbiodinium diversity among host clionaid sponges from Caribbean and Pacific reefs: evidence of heteroplasmy and putative host-specific symbiont lineages. Mol Phylogenet Evol. 2011;59:81–8.
32. Weisz JB, Massaro AJ, Ramsby BD, Hill MS. Zooxanthellar symbionts shape host sponge trophic status through translocation of carbon. Biol Bull. 2010;219:189–97.
33. Zea S, Lopez-Victoria M. *Cliona acephala* (Porifera: Demospongiae: Clionaida), a new encrusting excavating reef sponge from the Colombian Caribbean belonging to the *Cliona viridis* species complex. Zootaxa. 2016;4178:583–92.
34. Zundelevich A, Lazar B, Ilan M. Chemical versus mechanical bioerosion of coral reefs by boring sponges-lessons from *Pione* cf. *vastifica*. J Exp Biol. 2007;210:91–6.
35. Cheshire AC, Wilkinson CR. Modelling the photosynthetic production by sponges on Davies Reef, Great Barrier Reef. Mar Biol. 1991;109:13–8.

36. Rosell D, Uriz MJ. Do associated zooxanthellae and the nature of the substratum affect survival, attachment and growth of *Cliona viridis* (Porifera: Hadromerida)? Mar Biol. 1992;114:503–7.
37. Brümmer F, Pfannkuchen M, Baltz A, Hauser T, Thiel V. Light inside sponges. J Exp Mar Biol Ecol. 2008;367:61–4.
38. Wilkinson CR. Nutrient translocation from green algal symbionts to the freshwater sponge *Ephydatia fluviatilis*. Hydrobiologia. 1980;75:241–50.
39. Koopmans M, van Rijswijk P, Martens D, Egorova-Zachernyuk TA, Middelburg JJ, Wijffels RH. Carbon conversion and metabolic rate in two marine sponges. Mar Biol. 2011;158:9–20.
40. Wilkinson CR. Net primary productivity in coral reef sponges. Science. 1983;219:410–2.
41. Murray F, Widdicombe S, McNeill CL, Solan M. Consequences of a simulated rapid ocean acidification event for benthic ecosystem processes and functions. Mar Pollut Bull. 2013;73:435–42.
42. Morrow KM, Bourne DG, Humphrey C, Botte ES, Laffy P, Zaneveld J, et al. Natural volcanic CO_2 seeps reveal future trajectories for host-microbial associations in corals and sponges. ISME J. 2015;9:894–908.
43. Jensen S, Fortunato SA, Hoffmann F, Hoem S, Rapp HT, Ovreas L, et al. The relative abundance and transcriptional activity of marine sponge-associated microorganisms emphasizing groups involved in sulfur cycle. Microb Ecol. 2017;73:668–76.
44. Feng G, Sun W, Zhang F, Karthik L, Li Z. Inhabitancy of active *Nitrosopumilus*-like ammonia-oxidizing archaea and *Nitrospira* nitrite-oxidizing bacteria in the sponge *Theonella swinhoei*. Sci Rep. 2016;6:24966.
45. Han M, Li Z, Zhang F. The ammonia oxidizing and denitrifying prokaryotes associated with sponges from different sea areas. Microb Ecol. 2013;66:427–36.
46. Han M, Liu F, Zhang F, Li Z, Lin H. Bacterial and archaeal symbionts in the South China Sea sponge *Phakellia fusca*: community structure, relative abundance, and ammonia-oxidizing populations. Mar Biotechnol. 2012;14:701–13.
47. Lopez-Legentil S, Erwin PM, Pawlik JR, Song B. Effects of sponge bleaching on ammonia-oxidizing Archaea: distribution and relative expression of ammonia monooxygenase genes associated with the barrel sponge *Xestospongia muta*. Microb Ecol. 2010;60:561–71.
48. Mohamed NM, Saito K, Tal Y, Hill RT. Diversity of aerobic and anaerobic ammonia-oxidizing bacteria in marine sponges. ISME J. 2010;4:38–48.
49. Nishijima M, Lindsay DJ, Hata J, Nakamura A, Kasai H, Ise Y, et al. Association of thioautotrophic bacteria with deep-sea sponges. Mar Biotechnol. 2010;12:253–60.
50. Radax R, Hoffmann F, Rapp HT, Leininger S, Schleper C. Ammonia-oxidizing archaea as main drivers of nitrification in cold-water sponges. Environ Microbiol. 2012;14:909–23.
51. Radax R, Rattei T, Lanzen A, Bayer C, Rapp HT, Urich T, et al. Metatranscriptomics of the marine sponge *Geodia barretti*: tackling phylogeny and function of its microbial community. Environ Microbiol. 2012;14:1308–24.
52. Ribes M, Jimenez E, Yahel G, Lopez-Sendino P, Diez B, Massana R, et al. Functional convergence of microbes associated with temperate marine sponges. Environ Microbiol. 2012;14:1224–39.
53. Thomas T, Rusch D, DeMaere MZ, Yung PY, Lewis M, Halpern A, et al. Functional genomic signatures of sponge bacteria reveal unique and shared features of symbiosis. ISME J. 2010;4:1557–67.
54. van Duyl FC, Hegeman J, Hoogstraten A, Maier C. Dissolved carbon fixation by sponge-microbe consortia of deep water coral mounds in the northeastern Atlantic Ocean. Mar Ecol Prog Ser. 2008;358:137–50.
55. Burgsdorf I, Slaby BM, Handley KM, Haber M, Blom J, Marshall CW, et al. Lifestyle evolution in cyanobacterial symbionts of sponges. MBio. 2015;6:e00391–15.
56. Yu CH, Lu CK, Su HM, Chiang TY, Hwang CC, Liu T, et al. Draft genome of *Myxosarcina* sp. strain GI1, a baeocytous cyanobacterium associated with the marine sponge *Terpios hoshinota*. Stand Genomic Sci. 2015;10:28.

57. Hallam SJ, Konstantinidis KT, Putnam N, Schleper C, Watanabe Y, Sugahara J, et al. Genomic analysis of the uncultivated marine crenarchaeote *Cenarchaeum symbiosum*. Proc Natl Acad Sci U S A. 2006;103:18296–301.
58. Walker CB, de la Torre JR, Klotz MG, Urakawa H, Pinel N, Arp DJ, et al. *Nitrosopumilus maritimus* genome reveals unique mechanisms for nitrification and autotrophy in globally distributed marine crenarchaea. Proc Natl Acad Sci U S A. 2010;107:8818–23.
59. Kim BK, Jung MY, Yu DS, Park SJ, Oh TK, Rhee SK, et al. Genome sequence of an ammonia-oxidizing soil archaeon, "Candidatus *Nitrosoarchaeum koreensis*" MY1. J Bacteriol. 2011;193:5539–40.
60. Tourna M, Stieglmeier M, Spang A, Konneke M, Schintlmeister A, Urich T, et al. *Nitrososphaera viennensis*, an ammonia oxidizing archaeon from soil. Proc Natl Acad Sci U S A. 2011;108:8420–5.
61. Tian RM, Sun J, Cai L, Zhang WP, Zhou GW, Qiu JW, et al. The deep-sea glass sponge *Lophophysema eversa* harbors potential symbionts responsible for the nutrient conversions of carbon, nitrogen and sulfur. Environ Microbiol. 2016;18:2481–94.
62. Chain P, Lamerdin J, Larimer F, Regala W, Lao V, Land M, et al. Complete genome sequence of the ammonia-oxidizing bacterium and obligate chemolithoautotroph *Nitrosomonas europaea*. J Bacteriol. 2003;185:2759–73.
63. Klotz MG, Arp DJ, Chain PS, El-Sheikh AF, Hauser LJ, Hommes NG, et al. Complete genome sequence of the marine, chemolithoautotrophic, ammonia-oxidizing bacterium *Nitrosococcus oceani* ATCC 19707. Appl Environ Microbiol. 2006;72:6299–315.
64. Siegl A, Kamke J, Hochmuth T, Piel J, Richter M, Liang C, et al. Single-cell genomics reveals the lifestyle of Poribacteria, a candidate phylum symbiotically associated with marine sponges. ISME J. 2011;5:61–70.
65. Tian RM, Wang Y, Bougouffa S, Gao ZM, Cai L, Bajic V, et al. Genomic analysis reveals versatile heterotrophic capacity of a potentially symbiotic sulfur-oxidizing bacterium in sponge. Environ Microbiol. 2014;16:3548–61.
66. Liu F, Li J, Feng G, Li Z. New genomic insights into "*Entotheonella*" symbionts in *Theonella swinhoei*: mixotrophy, anaerobic adaptation, resilience, and interaction. Front Microbiol. 2016;7:1333.
67. Li Z, Wang Y, Li J, Liu F, He L, He Y, et al. Metagenomic analysis of genes encoding nutrient cycling pathways in the microbiota of deep-sea and shallow-water sponges. Mar Biotechnol. 2016;18:659–71.
68. Moitinho-Silva L, Seridi L, Ryu T, Voolstra CR, Ravasi T, Hentschel U. Revealing microbial functional activities in the Red Sea sponge *Stylissa carteri* by metatranscriptomics. Environ Microbiol. 2014;16:3683–98.
69. Bayer K, Moitinho-Silva L, Brummer F, Cannistraci CV, Ravasi T, Hentschel U. GeoChip-based insights into the microbial functional gene repertoire of marine sponges (high microbial abundance, low microbial abundance) and seawater. FEMS Microbiol Ecol. 2014;90:832–43.
70. Li ZY, Wang YZ, He LM, Zheng HJ. Metabolic profiles of prokaryotic and eukaryotic communities in deep-sea sponge *Neamphius huxleyi* indicated by metagenomics. Sci Rep. 2014;4:3895.
71. Trindade-Silva AE, Rua C, Silva GG, Dutilh BE, Moreira AP, Edwards RA, et al. Taxonomic and functional microbial signatures of the endemic marine sponge *Arenosclera brasiliensis*. PLoS One. 2012;7:e39905.
72. Hestetun JT, Dahle H, Jorgensen SL, Olsen BR, Rapp HT. The microbiome and occurrence of methanotrophy in carnivorous sponges. Front Microbiol. 2016;7:1781.
73. Jensen S, Neufeld JD, Birkeland N-K, Hovland M, Murrell JC. Insight into the microbial community structure of a Norwegian deep-water coral reef environment. Deep-Sea Res Pt I. 2008;55:1554–63.
74. Thurber AR, Kröger K, Neira C, Wiklund H, Levin LA. Stable isotope signatures and methane use by New Zealand cold seep benthos. Mar Geol. 2010;272:260–9.

75. Kamke J, Sczyrba A, Ivanova N, Schwientek P, Rinke C, Mavromatis K, et al. Single-cell genomics reveals complex carbohydrate degradation patterns in poribacterial symbionts of marine sponges. ISME J. 2013;7:2287–300.
76. Hunting ER, de Goeij JM, Asselman M, van Soest RW, van der Geest HG. Degradation of mangrove-derived organic matter in mangrove associated sponges. B Mar Sci. 2010;86:871–7.
77. Cretoiu MS, Kielak AM, Al-Soud WA, Sørensen SJ, van Elsas JD. Mining of unexplored habitats for novel chitinases—*chiA* as a helper gene proxy in metagenomics. Appl Microbiol Biotechnol. 2012;94:1347–58.
78. Fan L, Reynolds D, Liu M, Stark M, Kjelleberg S, Webster NS, et al. Functional equivalence and evolutionary convergence in complex communities of microbial sponge symbionts. Proc Natl Acad Sci U S A. 2012;109:E1878–87.
79. Ward BB, Capone DG, Zehr JP. What's new in the nitrogen cycle? Oceanography. 2007;20:101–9.
80. Stein LY, Klotz MG. The nitrogen cycle. Curr Biol. 2016;26:R94–8.
81. Gruber N, Galloway JN. An Earth-system perspective of the global nitrogen cycle. Nature. 2008;451:293–6.
82. Wilkinson CR, Fay P. Nitrogen fixation in coral reef sponges with symbiotic cyanobacteria. Nature. 1979;279:527–9.
83. Shieh WY, Lin YM. Association of heterotrophic nitrogen-fixing bacteria with a marine sponge of *Halichondria* sp. Bull Mar Sci. 1994;54:557–64.
84. Wilkinson CR, Summons RE, Evans E. Nitrogen fixation in symbiotic marine sponges: ecological significance and difficulties in detection. Mem Qld Mus. 1999;44:667–73.
85. Mohamed NM, Colman AS, Tal Y, Hill RT. Diversity and expression of nitrogen fixation genes in bacterial symbionts of marine sponges. Environ Microbiol. 2008;10:2910–21.
86. Zhang F, Vicente J, Hill RT. Temporal changes in the diazotrophic bacterial communities associated with Caribbean sponges *Ircinia stroblina* and *Mycale laxissima*. Front Microbiol. 2014;5:561.
87. Weigel BL, Erwin PM. Effects of reciprocal transplantation on the microbiome and putative nitrogen cycling functions of the intertidal sponge, *Hymeniacidon heliophila*. Sci Rep. 2017;7:43247.
88. Liu M, Fan L, Zhong L, Kjelleberg S, Thomas T. Metaproteogenomic analysis of a community of sponge symbionts. ISME J. 2012;6:1515–25.
89. Steger D, Ettinger-Epstein P, Whalan S, Hentschel U, de Nys R, Wagner M, et al. Diversity and mode of transmission of ammonia-oxidizing archaea in marine sponges. Environ Microbiol. 2008;10:1087–94.
90. Schlappy ML, Schottner SI, Lavik G, Kuypers MM, de Beer D, Hoffmann F. Evidence of nitrification and denitrification in high and low microbial abundance sponges. Mar Biol. 2010;157:593–602.
91. Diaz MC, Ward BB. Sponge-mediated nitrification in tropical benthic communities. Mar Ecol Prog Ser. 1997;156:97–107.
92. Jiménez E, Ribes M. Sponges as a source of dissolved inorganic nitrogen: nitrification mediated by temperate sponges. Limnol Oceanogr. 2007;52:948–58.
93. Bayer K, Schmitt S, Hentschel U. Physiology, phylogeny and in situ evidence for bacterial and archaeal nitrifiers in the marine sponge *Aplysina aerophoba*. Environ Microbiol. 2008;10:2942–55.
94. Fiore CL, Baker DM, Lesser MP. Nitrogen biogeochemistry in the Caribbean sponge, *Xestospongia muta*: a source or sink of dissolved inorganic nitrogen? PLoS One. 2013;8:e72961.
95. Hentschel U, Hopke J, Horn M, Friedrich AB, Wagner M, Hacker J, et al. Molecular evidence for a uniform microbial community in sponges from different oceans. Appl Environ Microbiol. 2002;68:4431–40.
96. Juretschko S, Timmermann G, Schmid M, Schleifer KH, Pommerening-Roser A, Koops HP, et al. Combined molecular and conventional analyses of nitrifying bacterium diversity in

activated sludge: *Nitrosococcus mobilis* and *Nitrospira*-like bacteria as dominant populations. Appl Environ Microbiol. 1998;64:3042–51.

97. Spieck E, Hartwig C, McCormack I, Maixner F, Wagner M, Lipski A, et al. Selective enrichment and molecular characterization of a previously uncultured *Nitrospira*-like bacterium from activated sludge. Environ Microbiol. 2006;8:405–15.
98. Foesel BU, Gieseke A, Schwermer C, Stief P, Koch L, Cytryn E, et al. *Nitrosomonas* Nm143-like ammonia oxidizers and *Nitrospira marina*-like nitrite oxidizers dominate the nitrifier community in a marine aquaculture biofilm. FEMS Microbiol Ecol. 2008;63:192–204.
99. Hovanec TA, Taylor LT, Blakis A, Delong EF. *Nitrospira*-like bacteria associated with nitrite oxidation in freshwater aquaria. Appl Environ Microbiol. 1998;64:258–64.
100. Bartosch S, Hartwig C, Spieck E, Bock E. Immunological detection of *Nitrospira*-like bacteria in various soils. Microb Ecol. 2002;43:26–33.
101. Lebedeva EV, Alawi M, Fiencke C, Namsaraev B, Bock E, Spieck E. Moderately thermophilic nitrifying bacteria from a hot spring of the Baikal rift zone. FEMS Microbiol Ecol. 2005;54:297–306.
102. Off S, Alawi M, Spieck E. Enrichment and physiological characterization of a novel *Nitrospira*-like bacterium obtained from a marine sponge. Appl Environ Microbiol. 2010;76:4640–6.
103. Keuter S, Kruse M, Lipski A, Spieck E. Relevance of *Nitrospira* for nitrite oxidation in a marine recirculation aquaculture system and physiological features of a *Nitrospira marina*-like isolate. Environ Microbiol. 2011;13:2536–47.
104. Maixner F, Wagner M, Lucker S, Pelletier E, Schmitz-Esser S, Hace K, et al. Environmental genomics reveals a functional chlorite dismutase in the nitrite-oxidizing bacterium 'Candidatus *Nitrospira defluvii*'. Environ Microbiol. 2008;10:3043–56.
105. Diaz MC, Akob D, Cary CS. Denaturing gradient gel electrophoresis of nitrifying microbes associated with tropical sponges. Boll Mus Ist Biol Univ Genova. 2004;68:279–89.
106. Polonia AR, Cleary DF, Freitas R, de Voogd NJ, Gomes NC. The putative functional ecology and distribution of archaeal communities in sponges, sediment and seawater in a coral reef environment. Mol Ecol. 2015;24:409–23.
107. Rua CP, Gregoracci GB, Santos EO, Soares AC, Francini-Filho RB, Thompson F. Potential metabolic strategies of widely distributed holobionts in the oceanic archipelago of St Peter and St Paul (Brazil). FEMS Microbiol Ecol. 2015;91:fiv043.
108. van Kessel MA, Speth DR, Albertsen M, Nielsen PH, Op den Camp HJ, Kartal B, et al. Complete nitrification by a single microorganism. Nature. 2015;528:555–9.
109. Zumft WG. Cell biology and molecular basis of denitrification. Microbiol Mol Biol Rev. 1997;61:533–616.
110. Shieh WY, Lin YT, Jean WD. *Pseudovibrio denitrificans* gen. nov., sp. nov., a marine, facultatively anaerobic, fermentative bacterium capable of denitrification. Int J Syst Evol Microbiol. 2004;54:2307–12.
111. Enticknap JJ, Kelly M, Peraud O, Hill RT. Characterization of a culturable alphaproteobacterial symbiont common to many marine sponges and evidence for vertical transmission via sponge larvae. Appl Environ Microbiol. 2006;72:3724–32.
112. Yang Z, Li Z. Spatial distribution of prokaryotic symbionts and ammoxidation, denitrifier bacteria in marine sponge *Astrosclera willeyana*. Sci Rep. 2012;2:528.
113. Zhang X, He L, Zhang F, Sun W, Li Z. The different potential of sponge bacterial symbionts in N_2 release indicated by the phylogenetic diversity and abundance analyses of denitrification genes, *nirK* and *nosZ*. PLoS One. 2013;8:e65142.
114. de Voogd NJ, Cleary DF, Polonia AR, Gomes NC. Bacterial community composition and predicted functional ecology of sponges, sediment and seawater from the thousand islands reef complex, West Java, Indonesia. FEMS Microbiol Ecol. 2015;91:fiv019.
115. Lund MB, Smith JM, Francis CA. Diversity, abundance and expression of nitrite reductase (*nirK*)-like genes in marine thaumarchaea. ISME J. 2012;6:1966–77.

116. Schirmer A, Gadkari R, Reeves CD, Ibrahim F, DeLong EF, Hutchinson CR. Metagenomic analysis reveals diverse polyketide synthase gene clusters in microorganisms associated with the marine sponge *Discodermia dissoluta*. Appl Environ Microbiol. 2005;71:4840–9.
117. Thiel V, Neulinger SC, Staufenberger T, Schmaljohann R, Imhoff JF. Spatial distribution of sponge-associated bacteria in the Mediterranean sponge *Tethya aurantium*. FEMS Microbiol Ecol. 2007;59:47–63.
118. Mohamed NM, Rao V, Hamann MT, Kelly M, Hill RT. Monitoring bacterial diversity of the marine sponge *Ircinia strobilina* upon transfer into aquaculture. Appl Environ Microbiol. 2008;74:4133–43.
119. Mohamed NM, Enticknap JJ, Lohr JE, McIntosh SM, Hill RT. Changes in bacterial communities of the marine sponge *Mycale laxissima* on transfer into aquaculture. Appl Environ Microbiol. 2008;74:1209–22.
120. Montalvo NF, Hill RT. Sponge-associated bacteria are strictly maintained in two closely related but geographically distant sponge hosts. Appl Environ Microbiol. 2011;77:7207–16.
121. Croue J, West NJ, Escande ML, Intertaglia L, Lebaron P, Suzuki MT. A single betaproteobacterium dominates the microbial community of the crambescidine-containing sponge *Crambe crambe*. Sci Rep. 2013;3:2583.
122. Weigel BL, Erwin PM. Intraspecific variation in microbial symbiont communities of the sun sponge, *Hymeniacidon heliophila*, from intertidal and subtidal habitats. Appl Environ Microbiol. 2015;82:650–8.
123. Pimentel-Elardo S, Wehrl M, Friedrich AB, Jensen PR, Hentschel U. Isolation of planctomycetes from *Aplysina* sponges. Aquat Microb Eco. 2003;33:239–45.
124. Izumi H, Sagulenko E, Webb RI, Fuerst JA. Isolation and diversity of planctomycetes from the sponge *Niphates* sp., seawater, and sediment of Moreton Bay, Australia. Antonie Van Leeuwenhoek. 2013;104:533–46.
125. Mohan SB, Schmid M, Jetten M, Cole J. Detection and widespread distribution of the *nrfA* gene encoding nitrite reduction to ammonia, a short circuit in the biological nitrogen cycle that competes with denitrification. FEMS Microbiol Ecol. 2004;49:433–43.
126. Herrero A, Muro-Pastor AM, Flores E. Nitrogen control in cyanobacteria. J Bacteriol. 2001;183:411–25.
127. Gibson AH, Jenkins BD, Wilkerson FP, Short SM, Zehr JP. Characterization of cyanobacterial *glnA* gene diversity and gene expression in marine environments. FEMS Microbiol Ecol. 2006;55:391–402.
128. Feng G, Sun W, Zhang F, Orlic S, Li Z. Functional transcripts indicate phylogenetically diverse active ammonia-scavenging microbiota in sympatric sponges. Mar Biotechnol. 2018;20:131–43.
129. Glibert PM, Azanza R, Burford M, Furuya K, Abal E, Al-Azri A, et al. Ocean urea fertilization for carbon credits poses high ecological risks. Mar Pollut Bull. 2008;56:1049–56.
130. Collier JL, Baker KM, Bell SL. Diversity of urea-degrading microorganisms in open-ocean and estuarine planktonic communities. Environ Microbiol. 2009;11:3118–31.
131. Su J, Jin L, Jiang Q, Sun W, Zhang F, Li Z. Phylogenetically diverse *ureC* genes and their expression suggest the urea utilization by bacterial symbionts in marine sponge *Xestospongia testudinaria*. PLoS One. 2013;8:e64848.

Chapter 10
Integrative Omics Approach for the Community Function Evaluation of Sponge and Coral Microbiomes

Fang Liu and Zhiyong Li

Contents

Abstract Sponges and corals are often found to be associated with symbionts that are phylogenetically diverse and ecologically important. Assessing the functions of symbionts in the context of host-microbe interaction and microbial community dynamics has emerged as a new frontier of sponge microbiology. Culture-independent molecular methods, such as 16S rRNA gene sequencing and metagenome sequencing, have been the main tools to tackle the diversity and function of sponge-/coral-associated microbiota. Nonetheless, using one or two methods might not be able to generate comprehensive insights into the cross talk between host and microbes, as well as the dynamics of symbiotic community. Here we present two hypothesized cases focusing on the integration of various high-throughput techniques to demonstrate the methodological framework and potential outcome of integrative omics approach. We chosed two existing researches as basis for design, raised new scientific questions, and developed integrative research plans using state-of-the-art techniques such as single-cell sequencing, metagenome binning, GeoChip, RNA-Seq, etc.

F. Liu · Z. Li (✉)
Marine Biotechnology Laboratory, State Key Laboratory of Microbial Metabolism, School of Life Sciences and Biotechnology, Shanghai Jiao Tong University, Shanghai, People's Republic of China
e-mail: zyli@sjtu.edu.cn

Z. Li (ed.), *Symbiotic Microbiomes of Coral Reefs Sponges and Corals*,
https://doi.org/10.1007/978-94-024-1612-1_10

Keywords Symbionts · High-throughput sequencing · Functional DNA array · Host-microbe interaction

10.1 Introduction

In the era of high-throughput omics study, not only the phylogenetic diversity but also the genomic features and ecological roles of sponge/coral microbiomes have been illustrated in an unprecedented way. In previous chapters, we have reviewed various aspects of the community function of sponge/coral microbiomes derived from omics studies, especially metagenomic studies, e.g., biogeochemical cycling, chemical defense, response to environmental stress, and host-symbiont recognition. Due to historical and technical reasons, most insights into sponge/coral microbiomes were derived from studies based on 16S rRNA gene amplicon sequencing. Nevertheless, the uncultured status of most microbial symbionts, the complexity of microbiomes, and the diverse evolutionary trajectories of microbes often resulted in gaps between the rRNA-based phylogenetic diversity and the functional annotations of metagenome/metatranscriptome assemblies [1–4]. Besides the challenges presented by natural microbiomes, scientists often have to face the constraints from funds, facilities, and personnel. As pointed out in a recent review, "Only by applying a diverse range of methodological approaches to a broad suite of model and non-model systems studied by a well-networked community of interdisciplinary researchers will we truly be able to reveal the extraordinary extent of symbiotic interactions that exist in nature" [5].

It is important to choose the omics tools wisely when they are available. In general, target gene sequencing (e.g., 16S rRNA gene, *amoA*, *nifH*) is most suitable for the in-depth exploration of microbial diversity or a specific function; shotgun metagenome sequencing serves the need for unveiling genetic diversity of unknown microbial communities and discovering novel genes; metatranscriptome sequencing and metaproteomics analysis are feasible for surveys of functional activities of unknown microbial communities and validation of metagenomic data; metabolomics can be used to describe the metabolic profiles of microbial communities and identify biomarkers; functional DNA array (e.g., GeoChip) can be applied to compare the functional diversity and activity of microbial communities across many samples [6, 7]. With the aim to enlighten readers on designing applicable and cost-effective integrative omics approaches, we herein present the possible strategies for integrating different approaches to explore the functional traits of sponge/coral microbiomes by constructing hypothesized cases based on previous studies. The omics approaches discussed below are not strange to most microbial ecologists, but we rarely see these approaches being integrated to study the sponge/coral microbiomes. The reason we choose certain studies [8, 9] as the basis to construct our hypothesized cases is because we believe these studies provide important materials and experiment foundation for further exploration of the sponge/coral microbiomes.

10.2 Case 1: Functional Analysis of Vertically Transmitted *Proteobacteria* throughout the Life Cycle of the Demosponge *Amphimedon queenslandica*

Amphimedon queenslandica is a low-microbial-abundance (LMA) sponge with a fully sequenced genome [10, 11]. As the first sponge species with its genome sequenced, *A. queenslandica* represents a good model for metazoan evolution study and innate immune research [12]. The LMA feature also makes *A. queenslandica* a good model for the study of host-microbe interaction. According to the research from Bayes [8], five putative vertically transmitted proteobacterial OTUs (hereafter *AqVTPs*) are present throughout the sponge life cycle. Notably, three putative vertically transmitted OTUs have low (< 90%) sequence similarities to their closest relatives and likely represent new orders of the *Proteobacteria*. Furthermore, a complete community shift from the *Proteobacteria*-dominated community present in sponges collected from their natural habitat to a community dominated by representatives of the *Chlamydiae* in the sponges maintained in aquaria was observed.

To further our understanding toward the function of vertically transmitted symbionts and the cross talk between sponges and symbionts, it is necessary to obtain the genomes of *AqVTPs* and analyze their activities along the sponge life cycle. We herein present a design of an integrative omics approach to study the subject topic.

In the beginning, we could obtain the genomes of those *AqVTPs* via either single-cell genome sequencing [13] or metagenome binning [14]. Then, the single-cell genomes or metagenome bins will serve as references for metatranscriptome analysis along the sponge life cycle. The genome-centered metatranscriptome analysis can reduce the requirement for computer resources and shorten the bioinformatics process. Since *A. queenslandica* genome and transcriptome are annotated extensively, it is possible to develop a functional DNA array to monitor the response of *A. queenslandica* during the establishment of symbiotic communities and the metabolism characteristics at different life stages. Using functional DNA array rather than transcriptome sequencing to monitor the sponge activities can reduce the research cost and accelerate the data analysis. In the past, functional DNA array data have already provided some insights into how corals and their dinoflagellate symbionts may communicate [15]. Thus, by incorporating single-cell genomics, metatranscriptomics, and microarray analysis, we can provide insights into (1) the genomic features and metabolism repertoire of *AqVTPs*; (2) the phylogenetic affiliation of *AqVTPs* based on multigene phylogeny; (3) the activity and roles of *AqVTPs* along the sponge life cycle; and (4) the mechanisms of *AqVTPs* – *A. queenslandica* metabolic interaction and immune recognition. In the next paragraphs, we will present a phase-wise description of this integrative omics approach. The key steps are illustrated in Fig. 10.1.

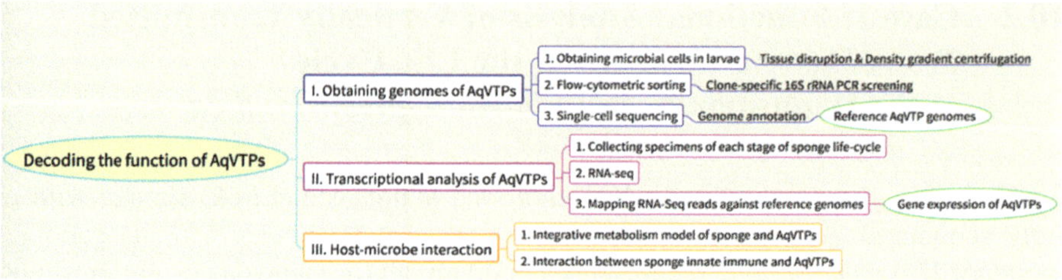

Fig. 10.1 The workflow of decoding the functions of vertically transmitted *Proteobacteria* (*AqVTPs*) throughout the life cycle of the demosponge *Amphimedon queenslandica*

Phase 1 Obtaining the *AqVTP* Genomes
A. queenslandica larvae are collected and disrupted in calcium-magnesium free artificial seawater followed by the density-gradient centrifugation to separate bacterial cells and sponge cells [16]. The fraction of bacterial cells is then submitted to flow cytometric sorting and whole-genome amplification [17]. According to the results from Bayes [8], specific primers, which will be used for screening *AqVTP* cells, can be designed based on the representative sequences of OTUs. After phylogenetic screening and verification, amplified genomes of *AqVTPs* are submitted to sequencing platform. Alternatively, if flow cytometric sorting is not available, we can sequence the metagenome of *A. queenslandica* and then extract the genomes of interest using metagenome binning approach [14, 18, 19]. Nonetheless, this method requires deep sequencing depth and considerable computer resources. There is a chance that not all the *AqVTP* genomes could be recovered from metagenomic datasets. Sometimes, the 16S rRNA operons might be missing in the binned genomes, which make it challenging to screen the *AqVTPs*.

Once the *AqVTP* genomes are available, we can perform the genome annotation and multigene phylogeny analysis. There are many automated platforms and software to make genome annotation and phylogenetic analysis easier for biologists, e.g., IMG 4 [20] and CheckM [21].

Phase 2 Transcriptional Activities of *AqVTPs*
A. queenslandica embryos, larvae, postlarvae, and adults are collected in biological replicates and preserved in RNA later. Unlike the transcriptome sequencing of sponges, sample preparation for prokaryotic symbiont mRNA remains to be complex and challenging [16, 22]. The yield of prokaryotic mRNA is generally lower than the optimal amount due to the loss during the removal of rRNA and eukaryotic mRNA. Given the LMA status of *A. queenslandica*, it is crucial to collect enough specimens for multiple rounds of RNA extraction. Once the RNA-Seq is finished, we can use the metatranscriptomic reads to estimate the gene expression of *AqVTPs* without de novo assembling of metatranscriptome. Then we can analyze the differential gene expression pattern of *AqVTPs* in the context of sponge life cycle.

Phase 3 Interaction Between *AqVTPs* and *A. queenslandica*
Based on the genomic and transcriptomic information of *AqVTP*s, it is possible to develop an integrative model for sponge-microbe symbiosis which also incorporates the expression of sponge genes (indicated by functional DNA array data). Furthermore, it is also possible to analyze the relation between sponge innate immune gene expression (NLRs, SRCRs, etc.) [12, 16] and the activities of *AqVTP*s along the sponge life cycle.

10.3 Case 2: Response of Coral Symbionts to the Cumulative Pressures of Climate Change

How coral-associated microbial communities respond to the global climate change has always been a focus of marine microbiology. Yet our understandings of how coral-associated microbes respond to elevated sea surface temperature (SST) and ocean acidification (OA) are limited, and we have barely any knowledge of how they respond to the interactive effects of these climate stressors. A recent study has conducted an 8-week experimental exposure to near-future climate change conditions and analyzed the bacterial community response of the corals *Acropora millepora* and *Seriatopora hystrix* using 16S rRNA gene amplicon pyrosequencing [9]. This study showed no visible signs of compromised host health, and coral-associated bacterial communities remained stable under the cumulative pressures. Nonetheless, it is still unknown how symbiotic communities respond to the cumulative pressures at functional gene level. Here we present an integrative omics approach that employs open (RNA-Seq) and closed format (GeoChip) high-throughput technologies [7] to extend the research and provide insights into the questions above.

The experimental design is based on the research from Webster et al. [9]. The workflow is showed in Fig. 10.2. In brief, *Acropora millepora* are deployed into triplicate aquaria for different pH/temperature treatments (8.1/28 °C, 8.1/31 °C, 7.9/28 °C, and 7.9/31 °C) and incubated for 8 weeks. We will focus on the gene expression of *Symbiodinium* symbionts (RNA-Seq) and the functional gene repertoire and expression of symbiotic bacteria, archaea, and fungi (GeoChip). The reason we use *Acropora millepora* as the material is because the genome of *Acropora digitifera* is available [23], and the *Acropora millepora genome* will be available soon in McDonnell Genome Institute. Therefore, we can extract the genetic information of dinoflagellate symbionts from the metatranscriptome datasets using *Acropora digitifera* genome and *Symbiodinium kawagutii* genome [24] as references. Considering the total RNA of *Acropora millepora* will probably be dominated by coral and dinoflagellate RNA, we will use GeoChip [25], which is insensitive to nontargeted nucleic acids, to detect the gene expression of other symbionts.

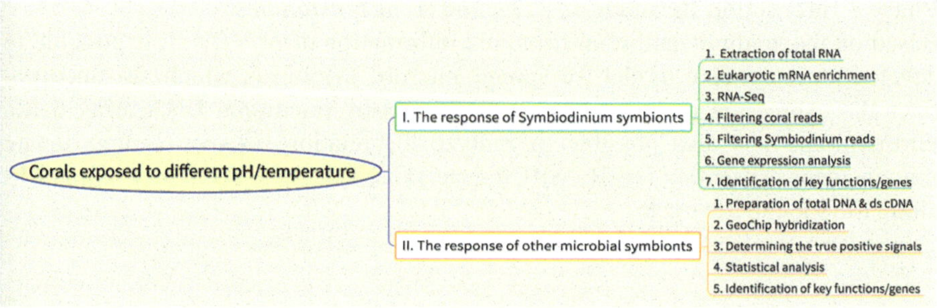

Fig. 10.2 The workflow of revealing the response of coral symbionts to the cumulative pressures of climate change

Part 1: The Response of *Symbiodinium* Symbionts
Although Webster et al. [9] did not observe any signs of compromised health in any coral hosts, investigating the molecular response of *Symbiodinium* is still beneficial for our understanding of the mechanisms related to the homeostasis of coral-associated dinoflagellate symbionts. Firstly, we will perform eukaryotic metatranscriptome sequencing. The oligo-d(T)-based eukaryotic mRNA enrichment is a mature approach available in many sequencing facilities. Thus, sequencing the eukaryotic metatranscriptome is less challenging than sequencing the prokaryotic metatranscriptome. Secondly, we map the reads against *Acropora digitifera* genome (or *Acropora millepora* genome when it is available) to filter out the host reads, then using *Symbiodinium kawagutii* genome as the reference to screen and quantify the *Symbiodinium* reads. The read count provides a measure of gene expression. Using DESeq [26], we can identify the differentially expressed genes (DEGs) across all the treatments. Thirdly, the enriched biological functions and pathways indicated by DEGs can be illustrated in several bioinformatics platforms, e.g., DAVID [27], PANTHER [28], REVIGO [29], etc.

Part 2: The Response of Other Microbial Symbionts
GeoChip, the most comprehensive functional gene array available, provides a rapid and cost-effective way to investigate the functional gene repertoire of microbiomes. Since its invention, GeoChip has been constantly updated and has been widely used in both terrestrial and ocean environments [7]. The latest version, GeoChip 5.0, contains 167,044 distinct probes, which are designed based on 395,894 coding sequences (CDS) from ~1500 functional gene families involved in microbial geochemical cycling, energy metabolism, metal homeostasis, organic remediation, secondary metabolism, stress responses, viruses, and virulence (http://glomics.com/gch-tech.html). Here we use GeoChip to assess the functional gene repertoire and transcriptional activities of *A. millepora* microbiomes. Firstly, the total DNA and RNA of *A. millepora* specimens are extracted simultaneously using Qiagen AllPrep DNA/RNA Kit or something similar. This operation can ensure the cohesiveness of the genetic information recovered between DNA and RNA samples. Then the total RNA are converted into ds cDNA for GeoChip hybridization. Secondly, the DNA

and ds cDNA samples are labeled and then loaded onto GeoChip for hybridization. This step is usually performed in Glomics Inc. along with data normalization as customer service. In the end, we will receive a data sheet recording the signal intensity of positive probes and the metadata of the positive probes (e.g., GenBank ID, gene name, functional category, phylogenetic affiliation). The data sheet can be directly imported into R environment (http://www.R-project.org/) or similar software for statistical analysis and visualization.

Determining positive signals is crucial for GeoChip analysis. In our case, a minimum of two positive spots out of three biological replicates is required for each probe/gene to be considered for data analysis. Only spots for which both DNA and RNA signals are detected are considered positive. Combining DNA and RNA data is a more conservative approach than analyzing DNA/RNA separately but will gain confidence that detected signals originate from sizable and active populations [30].

Analysis of GeoChip data usually can be divided into three levels: whole dataset (whole community structure), gene categories/functional genes, and probe/sequence levels. The statistical analysis should be carried out from high level (whole community) to low level (probe). Different statistical methods might be good for different levels. In our case, we can use clustering and ordination algorithms (e.g., UPGAMA, nMDS, PCA) to check if there is any grouping pattern in accordance with different treatments. Then using pairwise test (e.g., t-test, ANOVA), we can find the genes whose relative abundance or expression levels vary significantly between different treatments. Last but not least, we can analyze the genes of interest to investigate subcommunities that carry out special ecological functions (e.g., carbon degradation, stress response, secondary metabolism).

Acknowledgments We gratefully acknowledge the financial supports from the National Natural Science Foundation of China (NSFC) (31861143020, 41776138, 41742002, U1301131, 41176127, 41076077).

References

1. Burke C, Steinberg P, Rusch D, Kjelleberg S, Thomas T. Bacterial community assembly based on functional genes rather than species. Proc Natl Acad Sci U S A. 2011;108:14288–93.
2. Overballe-Petersen S, Willerslev E. Horizontal transfer of short and degraded DNA has evolutionary implications for microbes and eukaryotic sexual reproduction. BioEssays. 2014;36:1005–10.
3. Shapiro BJ, Polz MF. Microbial speciation. Cold Spring Harb Perspect Biol. 2015;7:a018143.
4. Bendall ML, Stevens SL, Chan LK, Malfatti S, Schwientek P, Tremblay J, et al. Genome-wide selective sweeps and gene-specific sweeps in natural bacterial populations. ISME J. 2016;10:1589–601.
5. Webster NS. Cooperation, communication, and co-evolution: grand challenges in microbial symbiosis research. Front Microbiol. 2014;5:16.
6. Ottman N, Smidt H, de Vos W, Belzer C. The function of our microbiota: who is out there and what do they do? Front Cell Infect Microbiol. 2012;2:104.

7. Zhou J, He Z, Yang Y, Deng Y, Tringe SG, Alvarez-Cohen L. High-throughput metagenomic technologies for complex microbial community analysis: open and closed formats. MBio. 2015;6:e02288–14.
8. Bayes JM. The characterisation of the microbiome of the demosponge Amphimedon queenslandica reveals host immune responses to changes in the symbiotic bacterial community. MPhil. The University of Queensland. 2013.
9. Webster NS, Negri AP, Botte ES, Laffy PW, Flores F, Noonan S, et al. Host-associated coral reef microbes respond to the cumulative pressures of ocean warming and ocean acidification. Sci Rep. 2016;6:19324.
10. Srivastava M, Simakov O, Chapman J, Fahey B, Gauthier MEA, Mitros T, et al. The Amphimedon queenslandica genome and the evolution of animal complexity. Nature. 2010;466:720–6.
11. Fernandez-Valverde SL, Calcino AD, Degnan BM. Deep developmental transcriptome sequencing uncovers numerous new genes and enhances gene annotation in the sponge Amphimedon queenslandica. BMC Genomics. 2015;16:387.
12. Degnan SM. The surprisingly complex immune gene repertoire of a simple sponge, exemplified by the NLR genes: a capacity for specificity? Dev Comp Immunol. 2015;48:269–74.
13. Rinke C, Lee J, Nath N, Goudeau D, Thompson B, Poulton N, et al. Obtaining genomes from uncultivated environmental microorganisms using FACS-based single-cell genomics. Nat Protoc. 2014;9:1038–48.
14. Albertsen M, Hugenholtz P, Skarshewski A, Nielsen KL, Tyson GW, Nielsen PH. Genome sequences of rare, uncultured bacteria obtained by differential coverage binning of multiple metagenomes. Nat Biotechnol. 2013;31:533–8.
15. Voolstra CR, Schwarz JA, Schnetzer J, Sunagawa S, Desalvo MK, Szmant AM, et al. The host transcriptome remains unaltered during the establishment of coral–algal symbioses. Mol Ecol. 2009;18:1823–33.
16. Ryu T, Seridi L, Moitinho-Silva L, Oates M, Liew YJ, Mavromatis C, et al. Hologenome analysis of two marine sponges with different microbiomes. BMC Genomics. 2016;17:1–11.
17. Siegl A, Kamke J, Hochmuth T, Piel J, Richter M, Liang C, et al. Single-cell genomics reveals the lifestyle of Poribacteria, a candidate phylum symbiotically associated with marine sponges. ISME J. 2011;5:61–70.
18. Burgsdorf I, Slaby BM, Handley KM, Haber M, Blom J, Marshall CW, et al. Lifestyle evolution in cyanobacterial symbionts of sponges. MBio. 2015;6:e00391–15.
19. Kang DD, Froula J, Egan R, Wang Z. MetaBAT, an efficient tool for accurately reconstructing single genomes from complex microbial communities. PeerJ. 2015;3:e1165.
20. Markowitz VM, Chen IM, Palaniappan K, Chu K, Szeto E, Pillay M, et al. IMG 4 version of the integrated microbial genomes comparative analysis system. Nucleic Acids Res. 2014;42:D560–7.
21. Parks DH, Imelfort M, Skennerton CT, Hugenholtz P, Tyson GWCM. Assessing the quality of microbial genomes recovered from isolates, single cells, and metagenomes. Genome Res. 2015;25:1043–55.
22. Moitinho-Silva L, Seridi L, Ryu T, Voolstra CR, Ravasi T, Hentschel U. Revealing microbial functional activities in the Red Sea sponge *Stylissa carteri* by metatranscriptomics. Environ Microbiol. 2014;16:3683–98.
23. Shinzato C, Shoguchi E, Kawashima T, Hamada M, Hisata K, Tanaka M, et al. Using the Acropora digitifera genome to understand coral responses to environmental change. Nature. 2011;476:320–32.
24. Lin S, Cheng S, Song B, Zhong X, Lin X, Li W, et al. The Symbiodinium kawagutii genome illuminates dinoflagellate gene expression and coral symbiosis. Science. 2015;350:691.
25. Gao Y, Wang S, Xu D, Yu H, Wu L, Lin Q, et al. GeoChip as a metagenomics tool to analyze the microbial gene diversity along an elevation gradient. Genomics Data. 2014;2:132–4.
26. Love MI, Huber W, Anders S. Moderated estimation of fold change and dispersion for RNA-seq data with DESeq2. Genome Biol. 2014;15:1–21.

27. Huang DW, Sherman BT, Lempicki RA. Systematic and integrative analysis of large gene lists using DAVID bioinformatics resources. Nat Protoc. 2008;4:44–57.
28. Mi H, Huang X, Muruganujan A, Tang H, Mills C, Kang D, et al. PANTHER version 11: expanded annotation data from Gene Ontology and Reactome pathways, and data analysis tool enhancements. Nucleic Acids Res. 2017;45:D183–9.
29. Supek F, Bošnjak M, Škunca N, Šmuc T. REVIGO summarizes and visualizes long lists of Gene Ontology terms. PLoS One. 2011;6:e21800.
30. Wawrik B, Boling WB, Van Nostrand JD, Xie J, Zhou J, Bronk DA. Assimilatory nitrate utilization by bacteria on the West Florida Shelf as determined by stable isotope probing and functional microarray analysis. FEMS Microbiol Ecol. 2012;79:400–11.

Chapter 11
Response of Sponge Microbiomes to Environmental Variations

Qi Yang, Wei Zhang, and Christopher M. M. Franco

Contents

Q. Yang · W. Zhang (✉)
Center for Marine Drugs, State Key Laboratory of Oncogene and Related Genes, Department of Pharmacy, Renji Hospital, School of Medicine, Shanghai Jiao Tong University, Shanghai, China

Centre for Marine Bioproducts Development, Flinders University, Adelaide, South Australia, Australia

Medical Biotechnology, College of Medicine and Public Health, Flinders University, Adelaide, South Australia, Australia
e-mail: wei.zhang@flinders.edu.au

C. M. M. Franco
Centre for Marine Bioproducts Development, Flinders University, Adelaide, South Australia, Australia

Medical Biotechnology, College of Medicine and Public Health, Flinders University, Adelaide, South Australia, Australia

Z. Li (ed.), *Symbiotic Microbiomes of Coral Reefs Sponges and Corals*,
https://doi.org/10.1007/978-94-024-1612-1_11

Abstract Sponges (phylum Porifera), sessile invertebrates, are the oldest multicellular animals that play an important role in evolutionary study. Thanks to their efficient filter-feeding capabilities, sponges have important ecological and biotechnological functions in nutrient cycles within marine ecosystems. Sponges permanently host remarkable microbial taxa with high diversity and complex structure. The associated microbes have been proved to highly contribute to the host growth and metabolite production, chemical defence, and susceptibility to biotic and abiotic stressors. This chapter will provide a systematic review on the variations of sponge microbiomes in relation to environmental stressors, including physical, chemical, and biological factors; as well as on how the changes in microbial composition cause the host sponge to suffer diseases and the consequent variations of the associated microbial community during a disease outbreak.

Keywords Sponge microbiomes · Physical factors · Chemical factors · Biological factors · Microbe-related diseases

Sponges (Porifera), sessile invertebrates, are the oldest multicellular animals that play an important role in evolutionary study [1]. To date (November 2018), 9083 valid species have been reported by World Porifera Database [2], and among them 7300 species belong to the class Demospongiae. Thanks to their efficient filter-feeding capabilities, sponges have important ecological and biotechnological functions in nutrient cycles within marine ecosystem [3, 4] and bioactive secondary metabolite production. Sponges are the most prolific producers of marine natural products with the contribution of almost 30% of all new marine natural products discovered by 2010 [5]. Importantly, on average, more than 290 new compounds continue to be derived from sponge each year since 2011: 296 (2011), 355 (2012), 243 (2013), 283 (2014), and 291 (2015) [6–10].

Sponges permanently host abundant microbial taxa with high diversity and complicated structure [11, 12]. The sponge-associated microbes include diverse phylogenetic lineages of Archaea and bacteria, fungi, and viruses [13, 14]. Thousands of taxa have been reported to belong to bacteria. An extensive sponge microbiome survey reported 41 microbial phyla from 81 sponge species collected globally [15]. Apart from the dominant microbial taxa, more unique and specific types have been detected and studied, e.g. *Cyanobacteria* [11, 16]. With the development of molecular techniques, functional gene pathways in these communities have been focused on, e.g. ammonia oxidation [17]. A most recent study characterized the genomic repertoire of *Aplysina aerophoba* symbionts at an unprecedented resolution, and it provides greater insights into the molecular mechanisms underlying sponge-microbe symbiosis [18].

Sponge-symbiotic microbes have been proved to be highly relevant to the host metabolism and growth (e.g. Freeman and Thacker [19]), chemical defence (e.g. Flatt et al. [20]; Marty et al. [21]), and susceptibility to biotic (e.g. disease in the study of Webster et al. [22]) and abiotic stressors (e.g. temperature stress in studies of Webster et al. [23]; Guzman and Conaco [24]).

This chapter will offer a systematic review on the variations of sponge microbiomes in relation to environmental stressors, including physical, chemical, and biological factors, as well as how the changed microbial composition causes the host sponge to suffer diseases and the consequent variations of the associated microbial community during a disease outbreak.

11.1 Relationship of Sponge Microbiomes with Environment Factors

The knowledge of how the composition and structure of the sponge microbial communities change over time is the basis to get a deep understanding of this biologically complex community. Generally, the main scenario is that sponge-microbe associations are largely stable over temporal scales [25], including epibionts [26], cultivatable symbionts [27], and entire bacterial communities [28]. Different extent variations linking with seasonal changes have been reported when repeatedly examining the same individuals in a period of time [29, 30]. Driven by the interests of the bioactive secondary metabolites, sponge cultivation was carried out in the artificially designed aquarium to produce the compounds from sponge-associated microbes. The high stability of the microbial community in short-term scales (11 days to 12 weeks) was observed [31, 32]; however, the community composition performed substantial variations during a longer-term aquaculture (6 months to 2 years) [33, 34].

This section offers an insight into how different environmental factors impact sponge microbiomes, including physical factors such as geographic and seasonal variations, temperature fluctuation, irradiance and depth variations, and cultivation method, chemical factors including heavy metal and nutrients/chemicals, and biological factors such as diseases and predators. In particular, there are three main scenarios about the sponge microbiome shift with geographic and seasonal variations: highly host-specific, habitat-relevant, and host- and habitat-related variations. For the microbial community response to temperature fluctuations, three widely studied model sponge species are described in this section.

11.1.1 Physical Factors

11.1.1.1 Sponge Microbiomes Impacted by Geographic and Seasonal Variations

The biologically complex sponge-microbe association is generally specific to the hosts and is impacted by various environmental conditions [35]. An integrated review is presented here to illustrate how these associations respond to the geographic and seasonal variations. To date, the most comprehensive biogeographical study investigated 32 sponge species located in different regions with geographic differences between each other. The results demonstrated that their microbial communities shared core bacterial taxa with small percentage and hosted a large percentage of species-specific taxa [36], which also was confirmed by the study of Schmitt et al. [37] for other sponge species. In contrast, a representative study analysed two sponges with distinct colours belonging to *Petrosia ficiformis* and demonstrated that biogeographic factor could be more responsible for the shifts observed in the microbial community than the host phylogeny [38]. Overall, the reported studies revealing the sponge microbiomes' response to different environmental conditions come down to three main scenarios: sponges maintain highly stable and specific microbial communities; sponges possess highly variable microbial communities in terms of their diversity and structure driven by the environmental factors or stresses.

1. **Scenario I: Host-specific, minor impact by the geographic and seasonal variations**

A comprehensive survey of sponges *Aplysina aerophoba* and *Theonella swinhoei* from nonoverlapping geographic regions demonstrated a sponge host-specific microbial community and first introduced the concept of "uniform microbial community" [39]. It was defined as "a community with distinct difference from that existing in marine plankton or marine sediments". Sponge *Cymbastela concentrica* was demonstrated to contain a highly stable bacterial community for most of the individuals living within the temperate range in Australia using 16S rDNA denaturing gradient gel electrophoresis (DGGE) [40]. Cárdenas et al. [41] used pyrosequencing to investigate the microbial communities of sponges *Tethya bergquistae* and *Ecionemia alata*. The bacterial communities of two species showed substantial differences, and the communities could remain conserved after moving into the aquarium system.

The microbial communities of a certain sponge species have demonstrated a stable host-specific profile, regardless of different living environmental conditions. Meanwhile, the question arises as to whether the associated microbes can also be maintained for the sponges that are phylogenetically close to each other but in distinct locations. The bacterial communities associated with sponges *Xestospongia muta* (living in tropical reefs of Atlantic Ocean) and *X. testudinaria* (living in tropical reefs of Pacific) showed a remarkable similarity [42]. It suggested the phenomenon of vertical transmission of the symbiotic microbes within sponges.

The stability of sponge-microbe associations and their response to different environmental conditions is important for the ecological and economic role of the symbionts in nature and society. Studies further proved that the sponge-associated bacterial community is stable not only between different biogeographic locations but also after a temporal change, such as seasonal shift. For example, Olson et al. Olson, Thacker [43] assessed the microbial communities associated with healthy and *Aplysina* red band syndrome (ARBS)-affected sponges *Aplysina cauliformis* from two locations over 2 years. The sponges were observed to maintain a general stability in microbial communities across time and space; however, a significant shift due to the disease was recorded. Furthermore, over 3 years of investigation, the microbial communities in the sponges belonging to six orders that have been identified as high or low microbial abundance sponges were illustrated to show a high degree of host specificity and low seasonal shift using terminal restriction fragment length polymorphism (TRFLP) analysis [44].

So far, it is confirmed that sponge-associated microbial community shows high diversity and complexity with host specificity. It is not only for the sponges belonging to the same species living in distant locations but also for the species with phylogenetically close relationships. However, how the abiotic factors impact the intraspecific variations in sponge microbial community structure is still not demonstrated. It is valuable for a deeper understanding on the host specificity of sponge microbial community. The microbial communities of seven different individuals belonging to sponge *Axinella corrugata* in two different Florida reef locations were revealed by 16S rRNA amplicon pyrosequencing [28]. The conserved microbial profiles indicated the stability of interspecific sponge-associated symbionts across seasonal variations.

- **Host species specificity supported by evolutionary mechanism**

Driven by the stable host specificity of the sponge microbiome, the question arises as to which evolutionary mechanism results in the sponge species-specific microbial communities. From the evolutionary aspect, sponges are the oldest multicellular animals; sponge-microbe association therefore could principally result from a coevolution event [45].

Generally, it is found that a given sponge species hosts a specific microbial community but lacks dominant taxon/taxa shared within the sponge phylum. One proposed explanation is the particular reproduction strategies of sponges, referring to their sexual and asexual reproductions, make the symbiotic microbes show host specificity [46]. For sexual reproduction, bacteria have been detected during the oocyte stage [47, 48] as it will allow bacteria to be transmitted vertically from generation to generation. In terms of asexual reproduction, it is conducted via the formation of gemmules, buds, or branches that can develop into viable adults when sponges are in the stringent conditions. Accordingly, the bacteria need to perform an accompany transmission via the germ lines. Therefore, a concept of convergent evolution was proposed, referring to "the development of similar structures in phylogenetically unrelated organisms as a result of adapting to the same environment" [39], which was also hypothesized to play a role in shaping the microbial community of sponges.

Moreover, the marine environment may provide the selection parameters to build up sponge-specific microbial community, which is also considered to be an explanation to their evolutionary mechanism. The effective water pumping system makes every single kg of sponge biomass filter 24,000 litres of seawater per day [49]. Based on the special characters, the monophyletic 16S rRNA gene sequence clusters of sponge-specific bacterial types have been revealed from seawater samples [50], freshwater lake samples [51], and marine sediment samples [52] with very low abundances.

Again, as filter feeders, sponges could positively process a selective filtration [39, 53]. In fact, the most abundant bacteria detected as sponge-specific taxa typically contain thickened cell walls, multiple membranes, and slime capsules, which are revealed by electron microscopy. These morphological characters probably help themselves to prevent phagocytosis by sponge archaeocytes [54, 55]. It further explains the reason why sponge and associated microbes can live in such a stable relationship to shape the host-specific microbial community.

- **A deep insight to the structure of sponge microbial community**

It has been confirmed that the sponge-associated microbial communities enable the maintenance of a stable and specific profile, which is highly relevant with the phylogeny of the host species regardless of the geographical and temporal variations. In addition, a deeper inspection to understand whether the so-called uniform community is always dominant or is a minority for the particular sponge species is essential. The bacterial community-associated sponges *Hymeniacidon heliophila* and *Polymastia janeirensis* were revealed by 16S rRNA gene sequencing and transmission electron microscopy (TEM) [56]. The sponge species-specific bacteria are shared among many sponge specimens [39, 57]. A consistent conclusion was also found in other studies focusing on different sponge species, e.g. Sharp et al. [58], Zhu et al. [59], and Alex et al. [60].

The concepts of "specialists" and "contaminants" were first proposed by Meyer and Kuever [61] when revealing the spatial distribution of the microorganisms within the sponge body. In the sponge *Polymastia* cf. *corticata*, papillae, outer and inner cortices, and choanosome tissue sections showed distinct bacterial populations. Among the identified archaeal and bacterial taxa, most are sponge-specific microbial taxa, also referring to uniform community, which are shared with other sponge species. A small part composes a group called "specialists" that are unique for the sponge *P.* cf. *corticata*; a smallest percentage of the microbial taxa are defined as "contaminants" that are enriched during the water filtering process and mainly composed of different types of *Proteobacteria*.

2. **Scenario II: Habitat-relevant, impacted by the geographic and seasonal variations**

Other recent studies reported that some particular environmental factors, including locations and living conditions, are responsible for driving sponge microbial community shifts. For example, an ecologically important sponge species *Carteriospongia foliascens* was investigated to understand how its microbial community responses to the variations between inshore and offshore locations [62]. The *Cyanobacteria* biomass consistently increased over *Bacteroidetes* in the examined locations, which suggested that the *C. foliascens* is induced by the specific environmental factors.

The sponges *Hymeniacidon heliophila*, *Paraleucilla magna*, and *Petromica citrina* in two environments with different degrees of pollution were studied [63]. It compared the diversity and composition of archaea associated with seawater and sponges. As a result, the complexity of the archaeal community for the sponges in the inner bay seawater was higher than that of coastal Cagarras Archipelago by principal component analysis (PCA) plots and rarefaction analyses. Crenarchaeota showed higher diversity in the sponges living in polluted area, which could be considered to indicate their strategy to adapt to the impacted environments by rearranging the structure of their associated microbial communities.

The microbial communities of sponge *Hymeniacidon heliophila* were compared between intertidal and subtidal individuals by amplicon-based sequencing and clone library analyses of full-length 16S rRNA gene [64]. The diversity, structure, and complexity of the microbial communities performed significant differences between individual sponges in subtidal and intertidal regions. It was the first report for intraspecific differences. The key differences were represented by changes of the relative abundance of a few dominant microbial taxa, as well as the presence or absence of numerous rare microbial taxa, which were recently discovered to display host specificity [65]. The results indicated that the extreme abiotic fluctuation (periodic air exposure in intertidal habitats in this case) can induce sponge microbial community shifts between different individuals within the species.

The bacterial communities associated with different individuals of the sponge *Mycale hentscheli* showed a spatial and temporal complexity using TRFLP and DGGE analyses [30]. Over a 21-month period, multiple individuals were collected from one location. Both TRFLP profiles and DGGE bands for each sponge individual showed substantial differences, though some microbial taxa appeared to be spatially conserved through all *M. hentscheli* populations.

3. **Scenario III: Habitat- and host-related variations**

Different from the previous two scenarios, some sponge microbiomes showed habitat- and host-related variations, which implies the importance of both host phylogeny and living conditions in shaping the composition and structure of the microbial community. For example, two sponge species (*Suberites diversicolor* and *Cinachyrella australiensis*) inhabiting both marine lakes and adjacent open coastal systems were selected to characterize their microbial composition (Cleary, Becking [35]). Between the species, a significant difference in microbial diversity was revealed. Within *S. diversicolor*, their bacterial communities showed a slight difference between the samples inside and outside the lakes. In contrast, within *C. australiensis*, the variation of the habitats led to a remarkable shift of their associated microbial communities (only 9.4% of OTUs shared). Therefore, the sponges living in the same environmental condition showed a microbial community specific to the host species instead of the shared living condition; different sponge species located at different habitats performed various degrees of responses to the environmental factors.

This observation was also supported by Noyer et al. Noyer, Casamayor [66] who reported the core microbial taxa and substantial differences between the associated microbial profiles of mesohyl, embryos, and larvae of sponge *Spongia lamella* by both DGGE and quantitative polymerase chain reaction (qPCR) analyses. Similarly, both ecological and biological sponge features were indicated to impact the shaping of the sponge microbial composition.

The mechanism of the variability observed in this study is unclear, but some proposed assumptions can be considered: a combination of environmental (horizontal) and parental (vertical) transmissions [25, 36] could both impact and determine the variation. Firstly, the idea of "sponge positively selective enrichment" has been explained previously in *host species specificity supported by evolutionary mechanism*. Moreover, competition between each of the associated microbial taxon is also believed to shape the microbial community [67]. "Genetic variability of sponge populations" [68] could also contribute to the observed variability in microbial communities. A combination of "filtration concentration" and "hereditary colonization" may be another possible mechanism.

11.1.1.2 Various Tolerances to Temperature Fluctuation

In addition to the widespread biogeographic stability of tropical and temperate sponge-microbe associations, correlations of stability with temperature fluctuation have also been explored to obtain a better understanding of the specific symbiotic associations. The shifts of symbiotic microbial communities driven by climate change or environmental stress will affect sponge health, growth rates, or their capacity for defence from predation, fouling, and disease. Meanwhile, the adjustment reaction of the microbial communities may also help sponges to adapt to changing environmental conditions [69, 70]. However, there is limited information of the adaptive capacity of sponge-microbe association.

Sea surface temperature (SST) has been predicted to increase up to 4 °C in this century [71], which is a significant environmental threat to coral reef populations and marine sponge populations [72]. Because the elevated SST could induce the shifts of sponge-associated microbial community in terms of its diversity, function, and structure [73]; it could directly/indirectly lead to a damage of sponge body [17, 74, 75]. Typically, sponge health will link with the specificity and stability of the symbionts in the sponge over temporal, geographic, and environmental gradients, as well as during their vertical transmission from generation to generation [76, 77]. For example, the Crenarchaeota community associated with giant barrel sponge *Xestospongia muta* could keep stable during sponge mortality, and its composition was found to be similar to that of the ambient seawater and sediments. This change indicated the relation between the reduced diversity of microbial community and the declined sponge health condition [17].

Our knowledge of sponge-microbe association and its response to environmental stresses remains rudimentary. Based on the reported studies so far, there are three main model species/genera investigated for evaluating tolerances of the sponges and their associated microbial communities to temperature fluctuation.

1. **Model species I: *Rhopaloeides odorabile***

Sponge *Rhopaloeides odorabile*, a common and ecologically important sponge in Great Barrier Reef, has been selected as a model for thermal stress research due to its diverse and stable microbial community [27, 76, 78–80].

R. odorabile was exposed to temperatures in the range of 27–33 °C [23]. No differences in microbial community composition or sponge health were detected in treatments between 27 and 31 °C. Once the temperature increased up to 33 °C, sponges underwent a complete decline of the symbionts which led to cellular necrosis after 3 days. Between 31 and 33 °C, a dramatic shift of the composition and the structure of microbial community was revealed by DGGE, clone sequence analysis, and pure cultures 16S rRNA sequencing. Most of the microbes from sponges treated under 27–31 °C showed a high similarity to known sponge-derived bacterial taxa. However, many of the microbes from sponges exposed to 33 °C were similar to the ones retrieved from diseased and bleached corals. Moreover, Pantile and Webster Pantile and Webster [81] confirmed this sensitive thermal threshold of 32 °C for *R. odorabile* by qPCR assay. In addition, 16S rRNA gene-based amplicon pyrosequencing was employed and again demonstrated that microbial communities in sponge *R. odorabile* are highly stable in a temperature range of 27–31 °C, with bacterial community composition only shifting when the necrotic syndrome happened in the tissue [82]. Therefore, sponge *R. odorabile* was considered to be highly vulnerable to the influence of global climate change as a 1 °C temperature increase could cause a rapid decline of host health so as to make symbiont loss.

In contrast to the strict thermal threshold of 32 °C identified in adult *R. odorabile* [23], larvae exhibited a higher thermal tolerance, without any adverse health effects detected at temperatures of only less than 36 °C [79]. Therefore, sponge larvae enable to maintain a highly stable symbiosis in facing the changes of seawater temperatures up to those that are predicted under current climate conditions.

- **Decline in symbiotic interactions is the key factor for temperature-induced sponge diseases**

The mathematical model predicted that "a temperature-induced decrease in antibiotic production in corals would cause a general shift in microbial communities and the invasion of pathogens that effectively compete for nutrients" [83]. The investigation of the sponge *R. odorabile* [84] supported some aspects of this model; however, it was also indicated that a decline in symbiotic interactions is likely to be a key factor for temperature-induced sponge diseases. In fact, the elevated temperature could lead to the changes of gene expression in both the host and the symbiotic community. Therefore, the variation of the water temperature has a great influence on sponge health, represented by the shifts of the symbiotic microbes.

From an ecological aspect, a general conclusion has been made that the stability of a microbial community requires a certain degree of taxa diversity and functional redundancy [84–88]. In sponge-microbe association, each microbial taxon is highly interdependent and potentially very specific to the host sponge species [89–91]. If an essential symbiotic microbial taxon was missing, the sponge microbial community would have a very small chance to be recovered and balanced, even a functionally equivalent species existed in this community[83, 86]. However, if the communities have many functionally redundant members, they would probably still maintain their function under stressful conditions, even with a low total species richness. Therefore, sponge microbiomes are likely to be very sensitive to environmental stresses [23].

Considering both genetic mechanism and ecological explanation, we found that for the sponge-microbe community with similar highly interdependent microbial members, environmental change (e.g. elevated temperature) may irreversibly disrupt the symbiosis with significant implications for host health.

2. **Model species II: *Ircinia* spp. (*Ircinia fasciculata*, *I. variabilis*, and *I. oros*)**

For the sponge species in the genus *Ircinia*, the sponge-associated microbes exhibited host-specific structure and remarkable stability over large seasonal shifts in temperature and irradiance, rather than a sensitive and strict temperature threshold (e.g. adult *R. odorabile*).

Sponges *I. fasciculata*, *I. variabilis*, and *I. oros* were investigated to perform a stable microbial community with host-specific structure during large fluctuations in temperature and irradiance within 1.5 years [92]. The variation in the sponge microbiota was only restricted to rare symbionts and occurred most prominently in warmer seasons. Similarly, the sponges *I. fasciculata* and *I. oros* were reported to harbour highly stable microbial communities under thermal and food shortage stresses [93]. The findings are consistent with the trends observed in the sponges across spatial and temporal scales [28, 94, 95]. In these studies, the three sympatric host species belonging to the genus *Ircinia* were characterized to understand their microbial communities using electron microscopy and replicated 16S rRNA gene sequence clone libraries. As a result, a main part called "specific mix of generalists" was found to be common between the microbial communities of *Ircinia* hosts, and low abundant "distinct symbiont mix" was observed to be specific to each host species.

In addition, *Ircinia* spp. have also been reported to suffer temperature-induced diseases and mass mortality. Unlike the small impact of direct fluctuations in temperature and irradiance on the sponge microbiome, a bigger affect could be caused as a certain length of high-temperature exposure will make the system exceeding the threshold to get sick. Some microbial taxa were observed to be replaced by pathogenic microbes, and the elevated seawater temperature was hypothesized to trigger such episodic mortality events [96, 97]. In addition to tissue necrosis, affected sponges also exhibit characteristic changes in their associated microbiota, including the loss of stable symbionts [98].

3. **Model species III: *Halichondria bowerbanki***

The sponge *Halichondria bowerbanki*, a common species in the Chesapeake Bay (Virginia, USA), was selected as a model to study global climate by investigating its microbial community shifts responding to the changing environmental factors [75]. The text was designed to mimic the predicted SST increases over the next 50 and 100 years. Under the potentially stressful thermal environments, the microbial communities of the studied sponge *H. bowerbanki* were found to be stable. There were no significant shifts in terms of the community composition and structure revealed by TEM. However, DGGE analysis demonstrated consistent changes induced by different thermally stressful treatments. A loss or significant reduction in some particular microbial taxa was detected by observing that some DGGE bands disappeared when sponges were exposed to high temperatures. The unique bands present only in the samples under highest temperature may indicate that the extreme temperature could increase the relative frequency of rare members within the microbial community.

- **Mode of transmission for sponge symbionts**

The presence of putative ammonia-oxidizing archaea (AOA) has been revealed to exist in sponge larvae by 16S rRNA, *amoA* PCR assays, and fluorescence in situ hybridization (FISH) [77]. It indicated that sponge-derived archaeal taxa are capable of vertical transmission. This behaviour could be considered to explain why host sponges stably maintain diverse and abundant microbial taxa [58, 77, 99]. In addition, high-throughput sequencing by next-generation sequencing (NGS) technique further demonstrated the role of the rare seawater flora and fauna in environmental transmission of sponge symbionts. For example, Webster et al. [76] showed that microbial diversity was unparalleled in an invertebrate host, with more than 250,000 sponge-derived sequence tags. For each sponge species, there are 2996 operational taxonomic units (OTUs) (95% sequence similarity) identified, which belong to 23 bacterial phyla. Against the previously described "sponge-specific" gene clusters, 48% of the sequences reported in this study were found exclusively in adults and larvae.

11.1.1.3 Highly Stable Response to Irradiance and Depth Variations

Do the different levels of light exposure affect the microbial profile? The sponges *Ircinia* spp. were selected to investigate their microbial communities' response to changing irradiance [92]. Results indicated the microbial communities maintain species-specific stability over the monitoring period with significant fluctuations of irradiance. Moreover, the variation in light intensity was tested to understand how the microbial community of sponge *Aplysina aerophoba* responded [100]. It was found that the microbial profiles were stable with a limited effect of light. In addition, the abundance of most secondary metabolites increased regardless of the illumination regime.

Does the depth affect the microbial community composition? A recent study analysed the microbial communities of different individuals of three sponge species collected in the Caribbean regions across a depth range from 10 to 100 m [101]. They perform significantly different morphological characters. The results indicated that the stability and specificity of the associated microbial communities varied with the host phylogeny, but each species supported a distinct community, which suggested that different sponge species exhibit different degrees of stability in their associated microbial communities with depth. Notably, each sponge species shared "core" microbial taxa across a depth range with the composition of the remaining community potentially influenced by both biotic and abiotic factors.

11.1.1.4 Sponge Cultivation Method Impacted Microbial Community

The stability of microbial communities during sponge cultivation varies in different host species. For sponges *Aplysina cavernicola* and *A. aerophoba*, their microbial communities are not affected by cultivation with different conditions [31, 102, 103]. There were three types of setting for the cultivation system: (1) untreated blank control condition with recirculating seawater only, (2) starvation condition by filtering seawater using 0.45 μm filter, and (3) antibiotic exposure by adding antibiotics into 0.45 μm filtered seawater. The sponges were cultivated in these three designed aquarium systems [31], and a highly stable microbial community was revealed for 11-day observation. In addition, sponge *A. aerophoba* was analysed over a period of 6 months of maintenance under different artificial cultivation conditions [103]. Variations of the cultivation conditions are (1) water temperature 20 ± 2 °C, (2) water temperature 25 ± 2, (3) illumination, and (4) feeding of the sponges. As a result, the treatments did not induce a significant shift of microbial community. Even though the sponges suffered tissue degradation during the cultivation period, the overall microbial taxa were still highly stable.

In contrast, for sponges *Mycale laxissima*, *Ircinia strobilina*, and *Clathria prolifera*, a distinct shift was observed once transferred into artificially controlled aquarium systems [33, 34, 104]. DGGE and clone libraries of bacterial 16S rRNA analysis revealed a significantly increased diversity of the microbial community associated with sponge *M. laxissima* after 1–3 years of aquaculture [34]. The diversity of microbial community associated with sponge *I. strobilina* showed a quick increase once they were transferred into the aquarium system. However, after 9 months of maintenance, the microbial community recovered to the condition the same as the wild sponges [33]. For sponge *C. prolifera*, substantial shifts of the microbial communities were revealed for each of the sponge individuals over 6 months of investigation [104].

In terms of sponge *Rhopaloeides odorabile*, a complex response was performed by their microbial communities when they were cultivated under different nutrient concentrations and seawater chemistries [32]. The conditions are (1) natural environmental condition (*in situ*), (2) small flow-through aquarium system (*ex situ*), and (3) large aquarium system (*ex situ*). A stable microbial community was observed for sponges cultured in situ and ex situ in small aquarium system (12 weeks). In contrast, a shift in the microbial community was detected in sponges cultivated *ex situ* in a large aquarium system (12 months). For the sponges maintained in the large aquarium system, their microbial community varied as a result of different nutrient concentrations and seawater chemistries.

11.1.2 Chemical Factors

Sponges are described as "dynamic multicellular systems" [105] based on their adaptation capacity to changing environmental conditions. Moreover, there is evidence that sponge-symbiotic microbes play an important role in nutrition and secondary metabolite biosynthesis of sponge-microbe association [25]. However, it is still unclear about how the interactions between host sponges and their associated microbes induce the production of secondary metabolites. Further complexity is added by sponges conducting chemical defences simultaneously against various threats on various sizes of scale (e.g. against predators, competitors, biofouling, pathogens) and using metabolites in multiple ways. The resulting interactions may obscure linkages between single effect due to single cause and complicate discerning clear defence concepts [106]. So far, there are no confirmed results to show sponge microbiome variations are induced by sponges' chemical repertoire. It is still not clear whether the sponge-associated microbial community is involved in the sponge chemical defence.

11.1.2.1 Bioindicator of Heavy Metals

Marine sponges have been proposed as sentinels of heavy metal pollution [107–109]. Some studies have demonstrated that they can accumulate high amounts of metals depending on the contamination in their environment [110, 111].

Sponge *Rhopaloeides odorabile* showed a decline in density and diversity of its total microbial community under the treatment of cupric ion ($Cu2^{+}$), in particular for the bacteria with high tolerance to copper [112]. Under the Cu^{2+} treatment ($\geq$1.7 μg/L) after a 14-day exposure, the abundance of sponge-associated microbes was significantly reduced. Only three types did not reduce regarding their numbers. Archaea could not be detected any more after a treatment (Cu^{2+} concentration to 19.4 μg/L) for 14 days. Many sponge *R. odorabile*-associated microorganisms were indicated to be sensitive to free copper ions. Sponges could accumulate copper (142 ± 18 mg kg^{-1}) when exposed to a condition with 223 μg/L Cu^{2+}, though they started becoming highly necrosed after 48-h treatment. When Cu^{2+} concentration is reduced to 19.4 μg/L, the sponge could grow well without apoptosis or mortality after 14-day treatment. Notably, they accumulated 306 ± 15 mg kg^{-1} copper during the treatment. Therefore, sponges could be applied as monitors for increased concentrations of heavy metals. In fact, their symbiotic microbes, in particular for certain taxa, performed sensitively for the changed heavy metal concentrations and have been proposed to be used as a novel bioindicator to assess metal pollution of seawater.

Sponge *Haliclona cymaeformis* displayed a selective response to copper treatment on its associated microbial community [113]. Two dominant associated microbial taxa (sulphur-oxidizing *Ectothiorhodospiraceae* and photosynthetic *Cyanobacteria*) decreased substantially after treatment by a high copper concentration. The shift of the sponge-associated microbial community driven by copper was revealed in terms of restructuring its composition and functional diversity. Certain microbial taxa were enriched to rearrange the community for surviving copper stress. The mechanism of this quick responding strategy could be indicated by metagenomic analysis, which demonstrated a varied profile of functional genes, such as virulence-associated genes, and cell signalling and regulation genes, as well as enriched functions linking with bacterial motility, extracellular polysaccharide, and capsule synthesis. Moreover, microscopic observation and comparison revealed dynamic bacterial aggregation during the copper treatment. Apart from copper sensitivity, a recent study reported that the bacteria from the Brazilian sponge *Polymastia janeirensis* had a high resistance to heavy metals $HgCl_2$ and MeHg [114].

The culturable bacterial community associated with sponges was also reported to be tolerant to heavy metals [115–117]. The culturable bacteria associated with sponge *Spongia officinalis* were isolated using metal-enriched microbiological media (high concentrations of zinc, nickel, lead, and copper) and showed high tolerances to the selected heavy metals [115]. Eight strains were isolated from sponge *Fasciospongia cavernosa* and performed to be sensitively resistance to the heavy metals Cd and Hg [118]. Microbial community associated with sponge *Hemigellius pilosus* also showed response to heavy metals (mercury between 10 and 500 ppm), antibiotics, and enzymatic profile [116]. Metal tolerance is often considered to be relevant to antibiotic resistance and can also affect biochemical activities within the microbial community. In addition, the comparison of the heavy metal and antibiotic resistance patterns at phylogenetic level of the associated microbes revealed some different characters. Interestingly, different strains vary on the growth patterns to perform distinct tolerance of the heavy metal. Therefore, the heavy metal tolerance patterns for the sponge are more likely to be specific to the symbiotic microbial community.

11.1.2.2 Large Threshold of Nutrients/Chemicals

In addition to the high tolerance to environmental heavy metals, sponge-microbe associations also have a high threshold of concentrations of nutrients, such as the microbial communities of sponges *Aplysina cauliformis* [119], *Rhopaloeides odorabile* [120], *Ircinia fasciculata*, and *I. oros* [93], as well as *Cymbastela stipitata* [121].

The Caribbean sponge *A. cauliformis* was assessed to demonstrate the independent and interacting effects of nutrient enrichment and the disease *Aplysina* red band syndrome (ARBS) on its microbial community (Gochfeld, Easson) [119]. In the study, nutrient enrichment had no effects on sponge or symbiont microbial community. In contrast, disease is a much greater stressor than eutrophication on the host sponge growth and the microbial composition. The symbiotic microbial community of the Great Barrier Reef sponge *R. odorabile*, which was living in the environment with the elevated nutrient levels of 10 mmol/L total nitrogen at 31 °C, showed high stability at both phylum and operational taxonomic unit (OTU) (97% sequence similarity) levels [120]. Apart from the bacterial community, the eukaryotic taxa were also found to be stable across all nutrient and temperature treatments by DGGE. The findings indicated that the microbial community of sponge *R. odorabile* is able to survive a short-term exposure to elevated nutrient concentrations. The Mediterranean sponges *Ircinia fasciculata* and *I. oros* were studied on how anomalous environmental conditions (temperature and food shortage) impact the composition and structure of the associated microbial communities [93]. The sponges were treated with (1) 13 °C with unfiltered seawater (control condition), (2) 13 °C with 0.1 μm filtered seawater (low food condition), (3) 25 °C with unfiltered seawater, and (4) 25 °C with 0.1 μm filtered seawater. As a result, there are no significant differences in microbial TRFLP profiles among treatments for both sponge species, suggesting that high temperatures and food shortage had no effect on symbiont community. Only the communities in *I. fasciculata* were found to be significantly affected by host mass mortalities due to increased temperature and food shortage. Coastal area sponge *Cymbastela stipitata* was investigated to examine the potential effect of excess nitrogen (up to 240 μM) on its microbial symbiosis [121]. The sponges performed unaffected at all treatment concentrations despite the nitrogen exposure increasing up to 124-fold above ambient. The conserved microbial community in all sponge individuals regardless of the various nitrogen treatments indicated the stability of the sponge-microbe relationship and demonstrated that the association is capable to survive a short period of high nitrogen concentration (eutrophication).

Higher partial pressure of CO_2 (pCO_2) in seawater caused by increase of atmospheric carbon dioxide (CO_2) will lead to the decline of seawater pH, called ocean acidification (OA) [122, 123]. Tropical coral reef communities existing within natural volcanic seeps (within Papua New Guinea) provide a unique opportunity to study organisms in an unmanipulated natural pCO_2/pH experiment, as 99% pure CO_2 was continuously expelled into the sediments and water column in this area [123]. Microbial communities of two sponge species *Coelocarteria singaporensis* and *Cinachyra* sp. in tropical coral reef ecosystem were reported to respond to elevated pCO_2 with host species specificity [124]. It was found the stability and flexibility of the associated microbial community may have an important role in helping the host adapt to new environmental condition. In addition, the analysis of one pCO_2-sensitive sponge species *Stylissa massa* also demonstrated that the functional stability of the sponge microbial community facing OA was species-specific as the adaptation capacity is considered to be highly relevant to the horizontal transmission of associated microbiome [125].

Polybrominated diphenyl ethers (PBDEs) are used as flame retardants, which are cytotoxic and genotoxic to organisms living in the ocean. The sponge *Haliclona cymaeformis* was investigated on how the community structure and functional gene composition of the microbial community respond to PBDEs [126]. As a result, the composition of the microbial community shifted from an autotrophic bacteria-dominated community to a heterotrophic bacteria-dominated community. A substantial decline of the symbiont abundance and an enrichment of heterotrophic bacteria were observed when the sponges were exposed in a high PBDE condition (1 μg/L). A selective effect of the high concentration treatment on the functional gene composition of the enriched heterotrophic bacteria was revealed by metagenomic data.

11.1.3 Biological Factors

11.1.3.1 Microbial Community Changes for Sponges Suffering Diseases

Sponges not only enable to protect themselves from harmful microbes but also to select harmless or even beneficial microbes by producing bioactive compounds [10]. Interestingly, the symbiotic microbes may also contribute to the host defence [127, 128]. Moreover, the innate immune systems of sponges are also believed to play a role in the prevention of microbial invasion [25]. Many studies have focused on the shifts in microbial and chemical patterns during a sponge disease outbreak. We offer a brief introduction here, specific to the representative cases about microbial community variations when sponges suffer the diseases; a more detailed review is documented in Sect. 11.3 in this chapter.

The microbial community composition of sponge *Aplysina aerophoba* was assessed to ascertain the role of microbes in the disease process by analysing affected and unaffected portions of diseased sponges [22]. The microbial diversity was found to be higher in diseased sponges than in healthy ones. *Deltaproteobacteria*, *Epsilonproteobacteria*, and *Firmicutes* were only found in affected tissues of diseased sponges. *Bacteroidetes* and the prokaryote community had significant shifts and performed higher abundance in sponges affected by the disease. In contrast, healthy sponges were dominated by *Chloroflexi* and *Gammaproteobacteria*. They contained cyanobacteria and "OP1" taxa that were only detected from healthy tissues. The study of sponge *A. cauliformis* also demonstrated that the disease had a significant detrimental impact on sponge-associated microbial community [119]. The microbial community associated with *A. cauliformis* was further studied on its response to disease accompanied with temporal and spatial variations [43]. The data demonstrated that the disease-affected sponges hosted a significant microbial profile compared to healthy ones but shared some common microbial taxa with the healthy sponges.

Another prominent sponge species on Caribbean reefs, *Xestospongia muta*, was reported to undergo two types of bleaching: (1) cyclic, a recoverable condition, and (2) fatal, a condition associated with the disease-like sponge orange band (SOB) syndrome and sponge death [17]. The bleached tissue contained Crenarchaeota taxa that are similar to those found in sediment and sand but are different with those in non-bleached and type I bleached tissues. [17].

The sponge *Ianthella basta*, common to the Great Barrier Reef and Torres Strait, is often affected by a disease-like syndrome (brown spot lesions). A comprehensive analysis of the microbial communities associated with affected and healthy sponges was determined [129]. The microbial communities associated with healthy and diseased sponge all showed low diversity.

The sponge *Lubomirskia baicalensis*, an endemic species in Baikalian, suffered the syndrome of the bleached patches of tissue [130, 131]. *Cyanobacteria* was found to be dominant in the microbial community. The microbiota showing a high similarity with the ones inhabiting substrates rich in organic matter were also found.

The bleaching processes of Baikal sponges were proven to affect the composition and the structure of the major taxonomic groups of sponge-associated bacteria.

Red Sea sponge *Crella cyathophora* with disease-like characteristics was analysed for its prokaryotic communities' shift [132]. As a result, a significant community shift was observed, and a novel clade belonging to phylum *Verrucomicrobia* was detected only in the disease-like sponge individuals.

Die-offs of Western Mediterranean sponge *Ircinia fasciculata* have been frequently reported. The microbial community shift happened in early diseased tissues [133]. Compared with healthy tissues, the abundance of *Gammaproteobacteria* and *Acidobacteria* increased and that of *Deltaproteobacteria* decreased in diseased tissues. Microbial diversity increased substantially in diseased tissues, in particular including more low-abundance OTUs.

The first disease affecting the deep-water sponge *Geodia barretti* was described by Luter et al. [134]. Highly divergent community profiles were found between the different health states, with distinct community shifts involving higher relative abundances of *Bacteroidetes*, *Firmicutes*, and *Deltaproteobacteria* in diseased individuals. In addition, three OTUs were absent in diseased individuals but were shared between the disease lesions and the apparently healthy tissue of diseased individuals, which could be considered as a non-localized infection.

11.1.3.2 Defensive Function of Symbiotic Microbe Against Predators

In benthic ecosystem, sponges share their habitat with a group of potential predators [135–139]. As sessile, filter feeders, sponges are required to develop defence mechanisms to survive from the competition with other benthic organisms (e.g. algae, corals, ascidians, and bryozoans) for living space. The defence mechanisms include structuring external and internal skeletons, as well as producing spicules [140]. In fact, production of chemical compounds [140] and cryptic growth habits [141] are the most effective strategies. Most of the secondary metabolites produced by sponges have been proved to perform antimitotic, cytotoxic, antibacterial, or antitumor activities [10] that could inhibit the settlement of foreign larvae [142] or deter predation [143].

In addition to the complexity of defence mechanisms, sponges host highly diverse microbial communities, which could form stable symbiotic associations and help host sponge maintain a healthy growth [11, 25]. Importantly, most of the bioactive secondary metabolites protecting sponges from predation have been reported to be actually produced by their symbiotic microbes [144].

Mediterranean sponge *Hemimycale columella* was found to accumulate thousands of calcite spherules at the body periphery to form a kind of skin as a protection from predators [145]. These structural materials have been proposed to be produced by sponge-symbiotic bacteria. One hypothesis is that these symbiotic bacteria can be transmitted from generation to generation within the *H. columella* community. The relative contributions of the calcite spherules produced by endosymbiotic calcibacteria and the bioactive compounds derived from *H. columella* to sponge defence were further assessed to test whether the mixture of the secondary metabolites and the calcibacterial calcareous cover contributes to a cooperative function in protecting the potential predators from feeding on the sponge [146]. The results revealed that the sponge conducts multiple defence mechanisms to protect themselves from a large array of potential predators, for example, sea urchin *Paracentrotus lividus* and some fish, including *Chromis chromis*, *Oblada melanura*, and *Diplodus vulgaris*.

11.2 Response of Sponge Microbiomes to Environmental Stresses

Sponge community, inhabiting the ocean bottom, is the second biggest ecosystem compared to coral reefs and plays an important ecological role in the marine ecosystem. The environmental stresses, for example, long-term impact of sea surface temperature (SST) increase and ocean acidification (OA), have been widely reported to affect the structure of coral reef community. However, there are very limited studies about marine sponge community responses to climate change. In addition, external microbes (bacteria, viruses, and protozoans) coming from human and animal waste could be pathogens for the flora and fauna living in the coastal marine environments [147]. As filter feeders, sponges organize a very effective water pumping system [49, 148, 149] to absorb and collect a wide range of particles with different sizes, such as organic materials, bacteria, and phytoplankton used as food sources [150–152]. Meanwhile, the free-living pathogenic microbes and microbial population blooming could be reduced and cleaned by sponges as food or nutrition [153]. In this section, we describe representative case studies with a longer-term perspective of sponge microbial community shifts driven by environmental stresses to help us get a deeper understanding about relationship of sponge microbiomes with environmental factors, which are documented in Sect. 11.1 in this chapter.

11.2.1 Sponge Microbiomes in a Changing Global Climate

To date, only a few studies have demonstrated the variations and stability of sponge microbial communities responding to long-term environmental factors, e.g. Friedrich et al. [31]; Tomas et al. [102]; Vicente [154]; Taylor et al. [155]. We have even less understanding on the mechanism of sponge-symbiotic microbial community adapting to the environmental stresses.

The Earth's average temperature had been reported to increase to about 0.8 °C since the start of the twentieth century and may exceed an increase up to 2 °C by 2050 [156, 157]. The prediction of SST for the next 100 years shows stressful conditions will cause deterioration of many marine habitats [158, 159]. A relatively small increase of 0.1 °C in SST could correlate strongly with increase of the bleached coral reefs, if considering that the mass bleaching events were only due to a 0.2 °C SST change [160]. In 1999, gorgonian and sponge populations suffered temperature-induced mass mortality in the Ligurian Sea of the Mediterranean [161]. Therefore, the global climate changes are causing serious consequences on the diversity and biomass of the marine ecosystem.

The temperate sponge *Halichondria bowerbanki* was assessed to study its microbial community shifts responding to potentially stressful thermal condition based on the prediction projection for SST increases in the next 50 and 100 years [75]. As a result, consistent changes were revealed among the three temperature treatments by DGGE analysis. When sponges were treated under thermally stressful temperatures, some microbial taxa performed a substantial decline in their abundances. Unique microbial taxa were detected only when the sponges were exposed under the highest temperature among the three treatments.

A long-lived and dominant Caribbean sponge *Xestospongia muta* [162] has been shown to harbour a diverse symbiotic microbial community, e.g. Fiore et al. [163] and Fiore et al. [164]. An SST increase and OA combined environmental event occurred in the Caribbean Sea and consequently affected the sponge *X. muta* living in this region [165]. The study demonstrated that OA condition alone, and its interaction with elevated SST, had significant effects on the sponge microbial community. The productivity potential of the symbiotic cyanobacteria substantially declined driven comprehensively by the impacts of both thermal stress and ocean acidification. The nutrient transfer (e.g. carbohydrate) between host and symbiont microbes was consequently impacted.

11.2.2 Sponge Microbiome Response to Marine Pollutions

The pathogens brought from human and animal sewage would result in two major consequences on marine environments [147]: (1) free pathogenic microbes in the seawater acting as a health hazard to cause recurrent occurrences of diseases for other aquatic organisms and (2) the seafood products contaminated by pathogenic microbes acting as biohazard to human health via human consumption [166]. Sponge could maintain microbes in two different ways: (1) absorbing and collecting free-living microorganisms, which could stay in the sponge body temporarily and could be utilized as food digested by sponge cells [55, 152, 167], and (2) selectively enriching certain groups of microbes, which could live within host sponges as extracellular and intercellular symbiont microbes [168–170].

Some of the microorganisms temporarily living in the sponge tissue have been identified as the contaminations from the environment. Therefore, the sponge microbial community shifts can be applied as contamination indicators [171, 172]. For example, for the commercial sponge *Spongia officinalis* living in the areas with various degrees of contaminations, its bacterial abundance and diversity were greater than that in ambient seawater [171]. The bacteria only detected from sponge tissues are the ones affiliated to genera *Escherichia*, *Morganella*, *Proteus*, *Pasteurella*, *Aeromonas*, *Pseudomonas*, and *Acinetobacter*, which were proposed to be applied as indicators of contamination in marine ecosystems.

Sponges *Petrosia testudinaria*, *Cinachyra cavernosa*, *Haliclona* sp., *Callyspongia fibrosa*, *Heteronema erecta*, *Fasciospongia cavernosa*, and *Callyspongia reticutis* var. *solomonensis* and two unidentified ones collected from Mandapam region in India (sewage disposal set) were analysed to retrieve microbial groups specific to the polluted condition [172]. Symbionts belonging to the genera *Loktanella* sp., *Pontibacillus* sp., *Planococcus* sp., and *Enterococcus* sp. were retrieved for the first time from sponges of Mandapam. Isolates belonging to genera *Enterococcus*, *Corynebacterium*, *Enterobacter*, and *Pseudomonas* are pathogenic bacteria to humans.

Moreover, selective enrichment and specific digestion for the associated microbes allow sponges to be utilized as a tool for bioremediation of polluted seawater. For example, sponges *Chondrilla nucula* and *Spongia officinalis* var. *adriatica* have been reported to successfully remove bacteria for marine environmental bioremediation [173, 174]. In addition, a common intertidal demosponge *Hymeniacidon perlevis* was found often in semi-enclosed or enclosed basins and in polluted seawaters [175]. The study demonstrated its capacity to survive for a relatively long time when exposed to air and suggested that *H. perlevis* may be applied as a useful bio-indicator and bio-remediator species for its performances on accumulation, remediation, and metabolizing microbes.

11.3 Microbe-Related Diseases of Sponges

Microorganisms have been shown to make up to 60% of the total biomass in healthy sponges [176]. The majority of these microorganisms are thought to be able to evade digestion and immune responses of the sponges. They normally grow and reside in the micro-environments of the mesophyll. The majority of these microorganisms are symbionts; however, some may contain pathogenic agents. Under certain conditions, for example, the stress of temperature, these agents may be activated and cause disease to their host. Recently in the Mediterranean and Caribbean Sea, sponge population has reduced significantly with reports of increasing sponge diseases. Worrisomely, reports of sponge disease were increasingly prevalent in areas of the Great Barrier Reefs, Papua New Guinea, and Cozumel. The sponge diseases imposed a huge impact not only on sponge itself but also on the ecology of the coral reefs and their associated marine organisms.

Our knowledge of sponge disease remains rudimentary, though the commercial and ecological values of sponges have been commonly recognised. Disease-induced mortalities of marine sponges have been correlated with environmental factors such as increased sea surface temperature (SST) and agriculture runoff; however, the causative agents of the diseases largely remain a mystery. A study in 2002 demonstrated that the primary sponge pathogen is a novel α-proteobacteria strain, which was the first to fulfil Henle-Koch's postulates. Viruses, cyanobacteria, and bacterial strains belonging to *Bacillus* and *Pseudomonas* genera, as well as some fungi, are indicated to be potential disease agents. In 2015, in sponge *Callyspongia* (*Euplacella*) aff *biru*, a novel sponge disease – sponge necrosis syndrome – was proved to be caused by a polymicrobial consortium. These confirmed causative agents fulfilled Henle-Koch's postulates. In this section, a summary is given on the global occurrence of sponge diseases from 1884 to 2015 and the suspected causative factors. The observed sponge disease-like syndromes and their potential pathogens are described, as well as the sponge recovery process after disease occurrence. The aim is to provide some potential factors to consider in order to develop a strategy for better management of sponge disease outbreaks.

11.3.1 Identified Diseases and Their Pathogenic Agents

Marine sponges have been studied intensively; most of the studies focused on the healthy sponges. However, the studies on the sponges suffered from diseases are also essential to gain a comprehensive understanding of the marine benthic ecology and the diversity of sponges for a sustainable development of the oceanic ecosystem. Additionally, a healthy and stable growth of the commercial sponge species is the key for the sponge industry.

Sponge diseases have gained considerable attention recently [177]. They have been increasingly reported with associated global population declines of ecologically and commercially important sponge species. The disease epidemics pose serious long-term threats to sponge populations, in particular for the long-lived, slow-growing species [178]. Several factors, such as ocean warming, acidification, overfishing, and dominant currents, have been considered to be relevant with the initial outbreak and spread of sponge disease [169, 179–183]. In the Mediterranean, Caribbean, Papua New Guinea (PNG), and the Great Barrier Reef (GBR) [43, 96, 97, 177, 184], the phenomenon of sponge mortality was recorded. Although the majority of the studies pointed out that microbiomes are likely the cause of sponge disease, a few of them isolated the potential pathogens [177]. Some investigators aimed to identify the putative pathogens causing sponge diseases by designing infection experiments. However, microbes may be detrimental to sponges in a direct (i.e. pathogenesis or parasitism) or indirect (e.g. microbial films promoting surface fouling) way. Other studies even suggested that microbes may not be a primary factor in the process of sponge pathogenesis. For example, the results in the study of Negri et al. (2009) Negri, Soo [185] indicated that insecticides containing endotoxin-producing *Bacillus* spp. are unlikely to be acutely pathogenic to sponges. Researchers used infection assay and compared bacteria, viruses, fungi, and micro-eukaryotes in healthy sponge *Ianthella basta* with those in diseased one. They found that microbes were not responsible for the brown spot lesions in this species [186]. Another study found that underwater pathogens were inadequate to infect healthy individuals, despite a shift of cyanobacteria between healthy and diseased *Xestospongia muta* [187].

Driven by the need for a comprehensive and systemic review on the diseases affecting sponges, we summarized the global occurrence of sponge diseases for the years 1884–2015 and the suspected causative factors in Table 11.1.

The sponge disease outbreaks have only been reported in saltwater, covering wide geographic locations. In 1938, a severe epidemic in the Caribbean was reported to wipe out 70–95% of sponge specimens [188]. In 1986, sponge disease caused catastrophic sponge mortalities in the Eastern Mediterranean, by reducing commercially important sponge fishery output from about 100 tonnes per year to a mere few tonnes [180]. In 1987, sponge disease affected over 60% of commercial sponge specimens in the Ligurian Sea [178]. The diseased sponges in Cozumel, Mexico, increased significantly from 2004 to 2005. For example, the affected *X. muta* went up from 50% in 2004 to 100% in 2005. Similarly, the affected *Geodia neptuni* increased from 25% to 75% and the damaged *Verongula gigantea* was from 5% to 50% [189]. In 2008, up to 66% of sponge *Ircinia fasciculata* population was affected by a syndrome characterized by brown spot lesions and necrotic tissue [186, 190].

Despite the implication of a large number of diverse pathogens in taxonomical level, our knowledge remains largely inadequate with regard to the pathogenesis of sponge – the identification of causative agents at taxonomical level, the transmission mechanism, the and virulence of sponge pathogen. It is worth noting that a spongin-boring α-proteobacterium strain NW4327, the first agent to satisfy Henle-Koch's postulates, was confirmed as the primary pathogen of the GBR sponge *Rhopaloeides odorabile* [191]. To establish a pathogen as the primary causative agent leading to disease, it is necessary to fulfil Henle-Koch's postulates, as stated below: "(1) the microorganism must always be associated with the disease; (2) the microorganism must be isolated from an infected sample and grown in pure culture; (3) the pure culture must induce the disease when inoculated into a healthy animal and (4) the microorganism must be re-isolated from the newly diseased animal and identified as the same microorganism as the original pathogen" [192]. When it comes to environmental samples, however, it can be challenging to culture disease-causing microorganisms and ascertain the pathogen doses at an ecologically relevant level. Mukherjee et al. (2009) Mukherjee, Webster [193] purified an enzyme that lyses collagen from a pathogen strain named NW4327, which promotes the pathogenic process of this particular bacterial strain against sponge. It was found in the study that the collagenase, partially purified, can either be a nonselective protease or, simply, tainted with nonspecific ones. The enzyme activity was highest at an internal pH of 5 of *R. odorabiles* and the mean seawater temperature of the GBR temperature at 30 °C [193]. More recently, Choudhury et al. [194] reported a draft genome of this confirmed sponge pathogen *Pseudoalteromonas* sp. strain NW 4327 (MTCC 11073, DSM 25418). A year later, Choudhury et al. [195] presented a complete polyphasic taxonomic identification of strain NW4327 as a novel *Pseudoalteromonas agarivorans* and described the virulence-related genes to further study the only primary sponge pathogen known so far. Comparisons with non-pathogenic strains were performed to highlight the differences in genotype and

Table 11.1 Global occurrence of sponge diseases from 1884 to 2015 and the suspected causative factors

Year	Diseased sponge species	Description of the affected sponge population/ individual	Suspected causative factors	Location	References
1884	*Ircinia* spp.	Destroyed sponge body leaving a crust of spongin fibres	Unclear. Fungal filaments were observed	Indian Ocean	[207]
1895	Commercial spp.	Loss of commercial sponges, internal tissue completely eroded	Unclear.	Florida	[208]
1906	*Hippospongia equina*	Dermal cortex covered in white, grey, or green liquid	Unclear. Attributed to shallow water and low water exchange	Tunisia	[209, 210]
1938	*Hippospongia* spp.	Heavy mortality	Unclear. Fungi always present	Caribbean	[188, 211]
1938–1939	Commercial spp.	70–95% mortality	Unclear. Unidentified fungal filaments in diseased tissue	Bahamas	[212, 213]
1939	Commercial spp.	Extensive mortality	Unclear. Unidentified fungal filaments in diseased tissue	Florida Keys	[213]
1939	Commercial spp.	70–95% mortality	Unclear. Unidentified fungal filaments in diseased tissue	Cuba	[212, 213]
1939	Commercial spp.	Loss of commercial sponges	Unclear. Unseasonably high temperature and salinity; unbranched fungal filaments in live, dead tissue	British Honduras	[213, 214]
1983	Unidentified sponges	Covered by whitish grey threads of mucus-like matter	Unclear. Key factors: (1) a strong thermocline development, (2) hydrology, (3) extremely high pelagic productivity, (4) bottom currents, (5) pollution	Gulf of Trieste	[215]

(continued)

Table 11.1 (continued)

Year	Diseased sponge species	Description of the affected sponge population/ individual	Suspected causative factors	Location	References
1984–1995	*Amphimedon compressa* *Aplysina fulva* *Callyspongia vaginalis* *Iotrochota* spp. *Ircinia* spp. *Niphates* spp. *Xestospongia* spp. *Verongula rigida*	Spreading lesions. *I. birotulata*: dulling of surface tissue, tissue loss, and white discolouration. *A. fulva* and *A. compressa*: glossing of surface and advance of a narrow brown band, exposure of skeletal fibres	Unclear. Disease symptoms evident over multiple census periods; cross-species contact infections unsuccessful; excision of diseased tissue often a successful treatment	Panama	[216, 217]
1985	*Geodia papyracea*	Extensive tissue decay, mutualism changed in high water temperature	Unclear. Potentially caused by proliferation of cyanobacterial symbiont	Belize	[197]
1986–1989	*Anchinoe paupertas* *Ircinia variabilis* *Petrosia ficiformis* *Spongia officinalis*	(1) Portofino Promontory: 60% of *S. officinalis* affected in 1987; 5% in 1988, and 20% in 1989; 5% mortality; *P. ficiformis*, *I. variabilis* occasionally affected. (2) Marsala Lagoon: *A. paupertas* and *I. variabilis* affected; white patches on the surface of the specimens	Unclear. Bacteria presenting fibres: Endobiotic cyanobacteria invading sponge tissue	Portofino Promontory and Marsala Lagoon (Italian coasts)	[178]

(continued)

Table 11.1 (continued)

Year	Diseased sponge species	Description of the affected sponge population/ individual	Suspected causative factors	Location	References
1986–1990	*Hippospongia* spp. *Spongia* spp.	Devastated commercial sponge populations and rapidly spreading white spots, brittle skeletons, putrefied tissue, white veil	Unclear. Unidentified bacteria tunnel through skeleton but more prevalent in shallow, warm water	Mediterranean	[170, 180, 218]
1986–1995	*Spongia officinalis* *Hippospongia* spp.	Loss of commercial sponges Recolonization of *S. officinalis* by 1995	Unclear. Ovoid bacteria filled the canaliculi of exposed skeletal fibres	Sicily and Ligurian coasts	[219]
1987	Commercial spp.	Loss of commercial sponges	Unclear. Pollution, environmental conditions, and neglect of fishery	Libya	[220]
1987	*Spongia officinalis*	The fibre skeleton was exposed to the environment. In some cases the whole sponge body was reduced to a mesh of brittle fibres, or only a small portion of living tissue was retained; 60% of the specimens were affected	Unclear. Could not be verified, but fibre-invading bacteria appear to be the most probable agents for this event	Coast of Portofino, Italy (Eastern Ligurian Sea)	[169]
1988–1989	*Haliclona oculata* *Halichondria panicea*	Brown decaying patches in *H. oculata* and decayed patches in *H. panacea* covered in white bacterial film	Unclear. A number of individuals detected; possibly algal overgrowth but no disease confirmed	North Wales	[177]

(continued)

Table 11.1 (continued)

Year	Diseased sponge species	Description of the affected sponge population/ individual	Suspected causative factors	Location	References
1994–1995	*Crambe crambe*	Loss in abundance: 69% sponges damaged in well-illuminated habitat, 38% in the shaded wall damaged	Unclear	Blanes, northeast Spain	[221]
1994–1996	*Ircinia spinosula* *Ircinia* sp.	Massive mortality in 1994 Decay of spongin fibres	Unclear	Mediterranean	[222]
1996	*Xestospongia muta*	Bleaching from the base up until the whole sponge is completely white, and then it just crumbles apart in about 2 weeks; a large bare rock with no growth where the sponge was previously attached to the substrate is left behind	Unclear	Belize Barrier Reef	[223]
1996–2000	*Ianthella basta* *Jaspis* sp. *Xestospongia* spp.	Decline in health and abundance of *I. basta*; brown lesions with rotted tissue, large holes, and brown biofilm	Five bacterial strains within the *Bacillus* and *Pseudomonas* genera were strongly correlated with disease	Papua New Guinea	[224]

(continued)

Table 11.1 (continued)

Year	Diseased sponge species	Description of the affected sponge population/ individual	Suspected causative factors	Location	References
1997	*Xestospongia muta*	Inside brown spot; the tissue at this spot turned white, brittle, and disintegrated; the appearance of a funnel-shaped hole; tissue around the hole started to regenerate	Unclear. Not caused by high seawater temperature and the blooms of cyanobacteria; potentially caused by frequent touching by scuba divers	San Juan Reef, Curacao	[225]
1998	*Petrosia ficiformis*	White superficial patches dispersed in the pigmented surface, which later spread across most of the sponge	Unclear. Not the primary consequence of bacterial and fungal infection as no sign of a massive invasion by these organisms was detected	Island of Gallinara (Ligurian Sea, Western Riviera)	[226]
1998	*Rhopaloeides odorabile*	Soft and fragile tissue with portions of pinacoderm eroded away to reveal the skeletal fibres; bacteria observed burrowing through the spongin fibres	Primary sponge pathogen (strain NW4327) is a novel α-proteobacteria; reisolated from sponges at all infection doses, fulfilling *Henle-Koch's postulates*	GBR, Australia	[191]

(continued)

Table 11.1 (continued)

Year	Diseased sponge species	Description of the affected sponge population/ individual	Suspected causative factors	Location	References
1998–1999	*Amphimedon compressa* *Aplysina fulva* *Iotrochota birotulata*	(1) *I. birotulata*: dulling of the surface, followed by loss of tissue from the purple-stained skeleton, which soon turned white. (2) *A. compressa*: glossing of the surface, and the subsequent advance of a narrow band of brown, followed by exposure of the completely bared skeleton. (3) *Aplysina fulva*: surfaces became glossy, and progression of a narrow band of brown was followed by exposure of the golden skeletal fibres	Unclear. Potentially a species-specific pathogen	Caribbean coral reef (shallow) in San Blas, Panama	[227, 228]
1999	*Spongia* sp. *Cacospongia* sp. *Hippospongia* sp.	A skeleton of spongin fibres	N/A	Gallinara Island, Ligurian Sea (Northwestern Mediterranean)	[229]
2001	*Suberites domuncula*	Not presented. Manually bacterial infection	Potential pathogenic *Vibrio* sp. was isolated	Adriatic Sea near Rovinj (Croatia)	[230]
2001	*Aplysina aerophoba*	Massive necrosis associated	Proposed a bacterial pathogen but more relevant with temperature increase	Greece	[231]

(continued)

Table 11.1 (continued)

Year	Diseased sponge species	Description of the affected sponge population/ individual	Suspected causative factors	Location	References
2003	*Spongia officinalis* *Ircinia dendroides* *Ircinia variabilis* *Ircinia oros* *Agelas oroides* *Crambe crambe* *Cacospongia* spp. *Hippospongia communis* *Petrosia ficiformis*	Partial mortality white parts or brown parts	The exceptional warming was observed in the depth ranges most affected by the mortality; it seems likely that the 2003 anomalous temperature played a key role in the observed mortality event	Rocky coasts, Northwestern Mediterranean	[232]
2004	*Aplysina cauliformis*	*Aplysina* red band syndrome. Rust coloured leading edge, necrosis.	Cyanobacterium was responsible for colouration but aetiological agent was uncertain.	Bahamas	[200]
2004–2005	*Callyspongia plicifera* *Geodia* spp. *Ircinia* spp. *Verongula gigantea* *Xestospongia muta*	100% of *X. muta* populations were affected by disease; white with brittle dead tissue around perimeter of lesions	Not known whether diseases are caused by several pathogens or just one pathogen or anything about the nature of the pathogen	Mexico	[189]
2005	*Xestospongia muta*	Orange band between healthy and bleached tissues	Putative agent unknown; no environmental trigger identified	Florida Keys, USA	[203]
2005–2006	51 sponge species and 2532 individuals	*Aplysina* red band syndrome (ARBS); reduction of the sponge communities varies between different species in different locations	Environmental factor impacted but require more detailed microbiological and microscopic study to confirm	Reefs in Bahía Almirante along the south shore of Isla Colon in Bocas del Toro, Panama	[201]

(continued)

Table 11.1 (continued)

Year	Diseased sponge species	Description of the affected sponge population/ individual	Suspected causative factors	Location	References
2006 (since 1991)	23 sponge species	Widespread mortality. Recovery of the biomass in the Florida Keys was slow, taking 10–15 years	Blooms of the picoplanktonic cyanobacterium *Synechococcus* sp.	Marathon and Long Key, Florida, USA	[204]
2006	*Xestospongia muta*	Fatal bleaching: a condition associated with the disease-like sponge orange band (SOB) syndrome and sponge death. (1) Early stage: with typical barrel shape and more than 50% of living tissue (2) Late stage: with less than 10% living tissue, disintegrating, losing its barrel shape. Cyclic bleached sponge: a recoverable condition	N/A	Conch Reef and Conch Wall, Key Largo, Florida	[17]
2007	*Aplysina aerophoba*	Black circular patches on the ectoderm at the initial stage of disease, developing into increasingly large areas of white-coloured necrotic tissue that led to exposure of the sponge skeletal fibres	Potential pathogens identified from sponges include a δ-*Proteobacteria*, a ε-*Proteobacteria*, and a *Cytophaga* strain	Pacug, Slovenia	[22]

(continued)

Table 11.1 (continued)

Year	Diseased sponge species	Description of the affected sponge population/ individual	Suspected causative factors	Location	References
2007–2010	*Ircinia fasciculata* *Sarcotragus spinosulum*	Total mortality ranging from 80 to 95% of specimens, the sponge disease led to a severe decrease in the abundance of the surveyed populations. It represents one of the most dramatic mass mortality events to date in the Mediterranean Sea	A significant positive correlation between the percentage of injured *I. fasciculata* specimens and exposure time to elevated temperature conditions in all populations, suggesting a key role of temperature in triggering mortality events	Western Mediterranean Sea: Cabrera National Park (Spain); Reserve Naturelle de Scandola (France)	[98]
2008	*Ircinia fasciculata*	Up to 66% of the *I. basta* population was afflicted by a syndrome characterized by brown spot lesions and necrotic tissue	Microbes are not responsible for the formation of brown spot lesions and necrosis; thermal and sedimentation stresses are not responsible for the brown spot lesions; true pathogen is unclear	Reefs of the Palm Islands, central GBR and Masig Island, Torres Strait, north-eastern Australia	[186, 190]
2008–2009	*Ircinia fasciculata* *Ircinia variabilis*	Small pustules on the sponge surface, which subsequently coalesced forming larger, extensive lesions, 27% mortality	An external bacterium: slightly twisted rod; a distinct central nucleoplasmic area; the riboplasm was peripheral, limited by a cytoplasmic membrane enclosed by a thin wall with typical Gram-positive organization with no obvious glycocalyx; no internal polarization and a flagellum	European (coast of Granada) and the African (Chafarinas Islands) side of the Alboran Sea (Western Mediterranean)	[96]

(continued)

Table 11.1 (continued)

Year	Diseased sponge species	Description of the affected sponge population/ individual	Suspected causative factors	Location	References
2008–2009	*Aplysina cauliformis*	*Aplysina* red band syndrome is characterized by a rust-coloured leading margin	Potentially, the strain of *Leptolyngbya* spp. is the pathogenic agent, but not confirmed	Patch reefs, Lee Stocking Island, Exuma Cays, Bahamas, and Carrie Bow Cay, Belize	[43]
2009	*Rhopaloeides odorabile*	Degradation of the spongin tissue of the infected sponges	A collagenolytic enzyme from a pathogen bacterium strain NW4327	N/A	[193]
2009	*Ianthella basta*	N/A	Bacteria closely related to *B. thuringiensis* associated with diseased sponges are unlikely to be acutely pathogenic to sponges	Davies Reef, Magnetic Island	[185]
2009	*Aplysina cauliformis*	*Aplysina* red band syndrome (ARBS); reduced sponge growth and survival	N/A	North Norman's reef, Bahamas	[119]
2009	*Ircinia variabilis*	Wide necrotic areas on the surface or disruption of the body in several portions; necrotic areas were whitish and covered with a thin mucous coat formed by bacteria; partial or complete loss, with the final exposure of spongin fibres to the environment	A hypothesis: under stressful physiological conditions, induces sponge pathogens *V. rotiferanius* to become virulent	Southern Adriatic and Ionian seas (Apulian coast)	[97]

(continued)

Table 11.1 (continued)

Year	Diseased sponge species	Description of the affected sponge population/ individual	Suspected causative factors	Location	References
2009	*Ircinia variabilis* *Sarcotragus spinosulus* *Spongia officinalis*	Necrotic areas and portions with bare skeleton; many specimens were covered with a white mat of cyanobacteria; about 22% suffered from this disease	High temperatures throughout summer created unfavourable environmental conditions	Conero Promontory, North Adriatic Sea	[206]
2009 (since 2007)	*Xestospongia muta*	Brown to a bleached white colour, accompanied by an orange band, massive tissue destruction and erosion; sponge orange band (SOB) disease	The aetiologic agent needs further identification	Florida Keys National Marine Sanctuary, USA	[187]
2010 (since 2007)	*Amphimedon compressa*	Sponge white patch (SWP) symptoms. Nearly 20% of the population with this symptom; about 21% of these biomass were bleached with degradation	A spongin-boring bacterial morphotype was revealed. A novel α-proteobacterium was isolated; but true pathogen needs confirmation	Dry Rocks Reef, Florida, USA	[184]
2011	*Hippospongia communis* *Spongia agaricina* *S. officinalis adriatica* *S. officinalis mollissima* *S. zimocca*	External corruption, deterioration	N/A	Coastal of the Central and Southeastern Aegean Sea	[233]

(continued)

Table 11.1 (continued)

Year	Diseased sponge species	Description of the affected sponge population/ individual	Suspected causative factors	Location	References
2011	*Chondrosia reniformis* *Tedania anhelans*	White necrotic areas, damaged surface up to 98% for *C. reniformis*, the abundance reduced from 43% to 6.5%; for *T. anhelans* slight damage on the surface, abundance reduced from 4.9% to 2.9%	The temperature conditions in summer 2011 were not so extreme to trigger the mass mortality, while other factors, probably of anthropic origin and the study area feature (a land-based pollution hot spot), were the potential causes	Conero Promontory, North Adriatic Sea	[234]
2012 (since 2008)	*Aplysina cauliformis*	*Aplysina* red band syndrome (ARBS) caused increased loss of healthy sponge tissue over time and a higher likelihood of individual mortality	Unconfirmed, but it is potentially triggered by a hurricane	Shallow reefs near Perry Institute for Marine Science, Lee Stocking Island, Exuma Cays, Bahamas	[202]
2014	*Rhopaloeides odorabile*	N/A	Draft genome sequence of the 1st marine sponge pathogen strain NW4327 (*Pseudoalteromonas* sp.) is released	Great Barrier Reef, Australia	[194]
2015	*Rhopaloeides odorabile*	N/A	The pathogen NW4327 is further identified as a new strain *Pseudoalteromonas agarivoran*, containing abundant and diverse virulence-related genes	Davies Reef, GBR, Australia	[195]

(continued)

Table 11.1 (continued)

Year	Diseased sponge species	Description of the affected sponge population/ individual	Suspected causative factors	Location	References
2015	*Callyspongia* (*Euplacella*) aff *biru*	A novel disease (sponge necrosis syndrome). The prevalence of the disease at the two sites surveyed ranged from 36% at Vavvaru Island to 30% at Komandoo Island	Six potential primary causal agents: *Rhodobacteraceae* sp., *Oscillatoria spongeliae*, *Ascomycota* sp., *Pleosporales* sp., *Rhabdocline* sp., *Cladosporium* sp. Only a combination of one bacterium and one fungus could replicate the disease, fulfilling *Henle-Koch's postulates*	Shallow water reefs at Vavvaru and Komandoo Island in Lhaviyani Atoll in Maldives	[196]

phenotype so as to distinguish the virulence of the sponge pathogen. The process of identifying the pathogenic agents also assisted the diagnosis of sponge diseases. Therefore, sponge necrosis syndrome was confirmed as a novel disease that affects the sponge *Callyspongia* (*Euplacella*) aff *biru*. By assessing fungal ITS and bacterial 16S rRNA gene diversities in healthy and diseased sponges, researchers highlighted six potential primary disease causative strains, including two bacteria and four fungi. Further histological analysis showed fungal hyphae, rather than bacteria, play the dominant role throughout the disease lesion. Only a combination of one fungus and one bacterium can replicate the disease, as evident on inoculation trails. It shows that this sponge disease is a result of a polymicrobial consortium, as it fulfils Henle-Koch's postulates [196].

Table 11.2 summarizes the observed sponge disease-like syndromes and the identification of their potential pathogens. The healthy and diseased tissues of affected sponge species were presented by the underwater pictures in Figs. 11.1, 11.2, 11.3, 11.4, 11.5, 11.6, 11.7, and 11.8.

Most of the studies reported that sponge disease first starts from white spots located at the upper surface. The sponges then display their brittle and fragile skeletons that decompose under water [180]. It remains unclear whether bacteria play a primary or secondary role in the pathogenic process. The sponge skeleton (collagenous fibres) of the experimentally infected GBR sponges [191] was observed to be digested by bacteria [169, 178, 180], which suggests the existence of a collagenous enzyme secreted by causative agent [193]. *Cyanobacteria*, the *Aphanocapsa feldmanni*-like symbionts on sponges, can multiply at a pace that far exceeds the sponge archaeocytes' ability to eliminate excess cells, leading to an extensive histolysis of their host tissue [197, 198]. There are reports that putrefied sponge tissue

Table 11.2 Observed sponge disease-like syndromes and their potential pathogens

Identified disease	Morphology of diseased sponge	Diseased sponge species	Healthy and diseased tissue pictures	Potential pathogen	Pathogenic agent isolation/gene analysis	References
Mangrove sponge disease (MSD)	"Yellow decay" under dark brown patches of cortex	*Geodia papyracea*	Fig. 11.1	*Oscillatoria* sp. (Cyanobacteria)	Observed	[197]
Spongin-boring necrosis (SBN)	(1) Soft and fragile with large portions of pinacoderm; (2) external tissue eroded away to reveal the collagenous skeletal fibres	*Hippospongia communis*, *Ircinia variabilis*, *Sarcotragus spinosula*, *Spongia officinalis*	N/A	Unknown	Observed	[170]
		Rhopaloeides odorabile	Fig. 11.2	A novel strain *Pseudoalteromonas agarivorans*	Isolated, confirmed, fulfilling Henle-Koch's postulates/whole genome available	[191, 194, 195]
Aplysina red band syndrome (ARBS)	One or more rust-coloured leading edges, with or without a trailing area of necrotic tissue, such that the lesion forms a contiguous band around part or all branches	*Aplysina cauliformis*	Fig. 11.3	A cyanobacterium may be the aetiologic agent	Observed	[200]
Sponge orange band (SOB)	A distinct orange band developed along the zone of healthy and dead tissue	*Xestospongia muta*	Fig. 11.4	Unknown, it resembles an infection	Unknown	[17, 187, 203]

Brown lesion necrosis (BLN)	Mottled with brown lesions, rotted tissue, and large holes	*Ianthella basta*	Fig. 11.5	2 *Bacillus* spp. and 3 *Pseudomonas* spp.	Isolated	[224]
Aplysina black patch syndrome (ABPS)	Black patches, white necrotic tissue, or exposed skeletal fibres	*Aplysina aerophoba*	N/A	*Deltaproteobacteria* sp.; *Cytophaga* sp.; *Epsilonproteobacteria* sp.	DGGE detected	[22]
Brown spot lesion (BSL)	Discoloured, necrotic spots leading to tissue degradation, exposure of the skeletal fibres, and disruption of the choanocyte chambers	*Ianthella basta*	N/A	Unknown, but microbes are not responsible for the formation of BSL	Unknown	[186, 190]
Pustule syndrome (PS)	Small pustules on the sponge surface, which subsequently coalesced forming larger, extensive lesions	*Ircinia fasciculata* *Ircinia variabilis*	Fig. 11.6	A twisted rod of external bacterium, but triggering factors not clear	TEM detected	[96]
Sponge white patch (SWP)	Distinctive white patches of variable size	*Amphimedon compressa*	Fig. 11.7	Unknown, but detected spongin-boring bacteria and disease-causing α -proteobacterium	Unknown	[184]
Sponge necrosis syndrome (SNS)	Necrotic edge, damaged fibres	*Callyspongia* (*Euplacella*) aff *biru*	Fig. 11.8	2 bacteria: *Rhodobacteraceae* sp., *Oscillatoria spongeliae* 4 fungi: *Ascomycota* sp. *Pleosporales* sp. *Rhabdocline* sp. *Cladosporium* sp.	Isolated, only a combination of one bacterium and one fungus could replicate the disease, fulfilling Henle-Koch's postulates	[196]

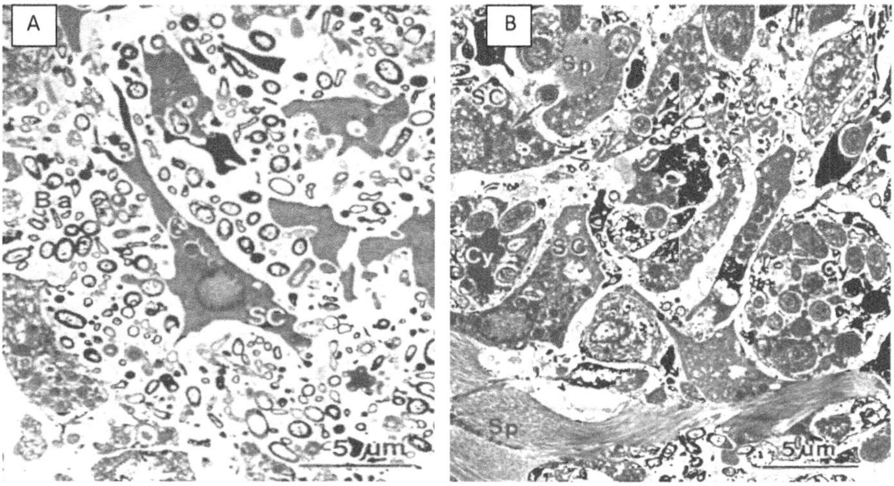

Fig. 11.1 Healthy and diseased tissue of sponge *Geodia papyracea*. (**a**) Histology of healthy sponge with bacteria (TEM). (**b**) Sponge-forming "pseudogemmule" to expel cyanobacteria trapped inside archaeocytes in diseased tissue zones under transmission electron microscopy (TEM). SC, sponge cell (arrow highlights the digestion of cyanobacterium); Ba, bacterium; Sp, sponging; Cy, cyanobacterium. (Figures are adapted based on the publication by Rützler [197])

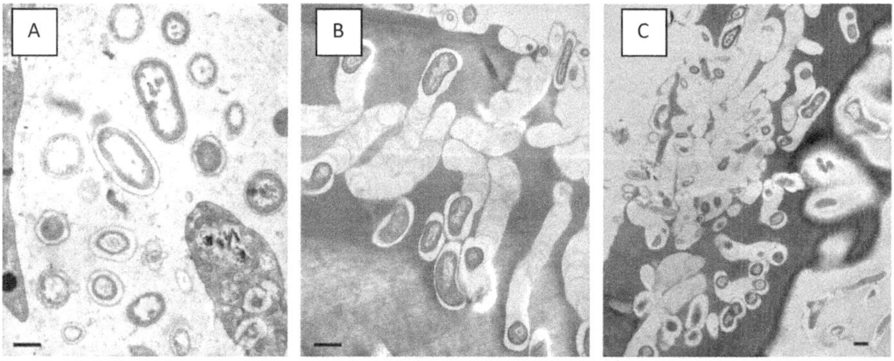

Fig. 11.2 Healthy and diseased tissue of sponge *Rhopaloeides odorabile*. (**a**) The diversity of bacterial morphotypes associated with healthy sponges under transmission electron microscopy. (**b**) Sponges infected with 10^6 CFU ml^{-1} of strain NW4327. (**c**) The diseased sponge with massive spongin-boring necrosis (SBN). The scale bar on all micrographs is 500 nm. (Figures are adapted based on the publications by Webster et al. [191], Choudhury et al. [194], and [195])

was covered by white veils of *Oscillatoria* sp., and it is considered as a secondary phenomenon and *Oscillatoria* sp. is not the primary pathogen [180]. There were also sporadic reports suggesting on viral and fungal cause of sponge disease [199]; however, these studies, based on histopathology, failed to identify the causative agents [188]. Other syndromes appear physiologically distinct from previous disease reports; however, no aetiological agents have yet been identified. There are also

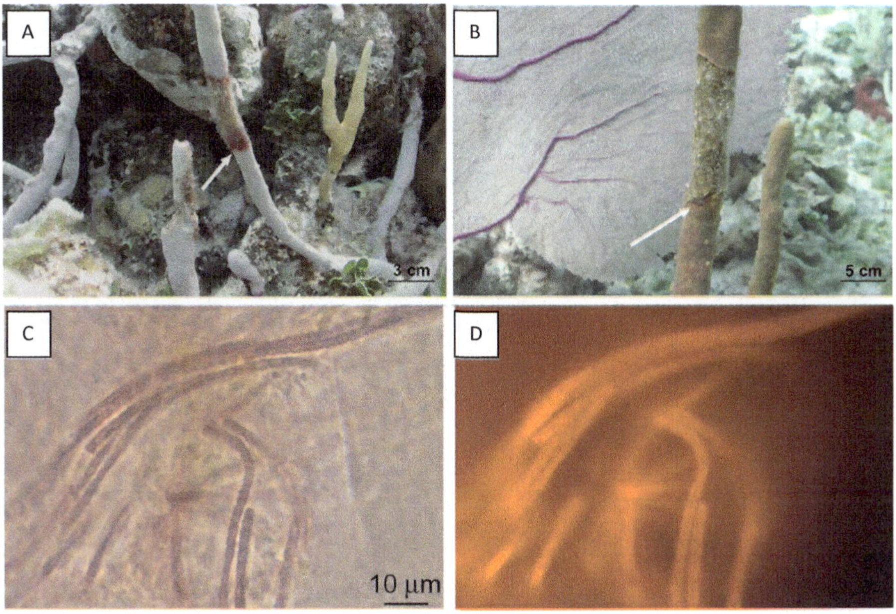

Fig. 11.3 Healthy and diseased tissue of sponge *Aplysina cauliformis*. (**a**, **b**) *Aplysina* red band syndrome (ARBS). Arrows highlight the leading edge of affected tissue; (**c**) cyanobacterium associated with margin of affected tissue; (**d**) same images under UV excitation, demonstrating the autofluorescence of photopigments. (Photos are adapted based on the publication by Olson, Gochfeld [200])

Fig. 11.4 Diseased tissue of sponge *Xestospongia muta* with sponge orange band (SOB). (**a**) An SOB advancing across a large sponge. Right edge of slate = 14 cm. (**b**) SOB advancing from below. (Photos are adapted based on the publications by Angermeier et al. [187], Cowart et al. [203], and López-Legentil et al. [17]

reports that describe syndromes distinct physiologically from the prior studies; however, they again failed to identify the pathogens. Descriptive names have been used to refer to sponge disease in a specific species, for example, jelly base rot disease, brown rot disease, white spot disease, and brown fringe yellow leather disease [189].

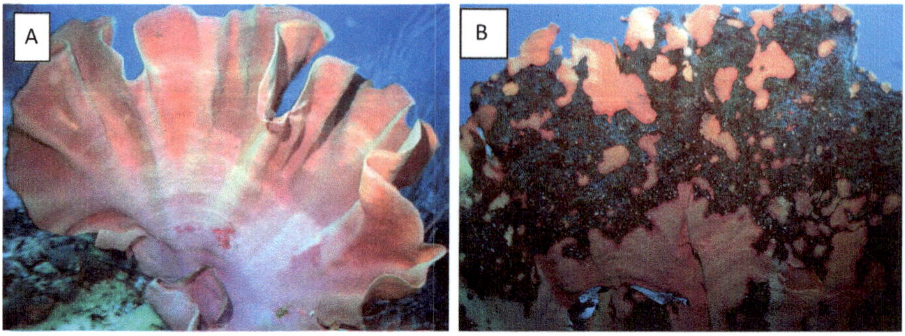

Fig. 11.5 Healthy and diseased tissue of sponge *Ianthella basta* with brown lesion necrosis (BLN). (**a**) Healthy specimen in year 1996. (**b**) Affected tissue in year 2000. (Photos are adapted based on the publication by Cervino et al. [224])

The monitoring programme did not utilize microbiological analysis; however, it is strongly suggested that more than a single pathogen exists in causing a variable presentation of disease symptoms.

Since 2006, a much better understanding of the role of microbes in the sponge-microbial association has been obtained, primarily due to the application of advanced technology. It allowed for a deeper identification of the sponge diseases and their pathogenic agents.

The disease termed *Aplysina* red band syndrome (ARBS) was firstly described in *Aplysina cauliformis* [200]. ARBS infection can lead to both visible and microscopic findings. Visible signs include a necrotic lesion that is rust-coloured and an adjacent band surrounding a part or the entire branch of sponge. Microscopic findings near the diseased ARBS margin suggest the discolouration involves a cyanobacterium, the role of which in the disease causation process is yet to be ascertained [200]. Analysis on 51 sponge species including 2532 individuals indicated the ARBS has sensitive reaction to the differences in environmental conditions on these reefs [201]. This is a reflection that sponges are regarded to be useful as bioindicators that react to changing environmental conditions. In a microbial and chemical pattern study for the ARBS-affected sponge *A. aerophoba*, a δ-proteobacteria sequence, one that has high homology compared to a coral black band disease strain, was only found in all sponge regions with lesion, whereas the proteobacteria was absent in unaffected regions of the same diseased sponge or healthy sponges. This pathogen was identified through a process called denaturing gradient gel electrophoresis (DGGE). Using DGGE, researchers also identified other potential pathogens including an environmental *Cytophaga* strain and a ε-proteobacteria strain, a novel strain that no close relatives have been identified so far [22]. The worldwide sponge mortality has been commonly recorded, but their dynamics are not known. In the study of the impact of disease ARBS on sponge *A. cauliformis* at individual to population level, the infection led to a decrease of healthy sponge tissue over time and an increase in the odds of individual mortality [202]. Researchers examined the fates of healthy and diseased sponges over a period of 1 year (2008–2009).

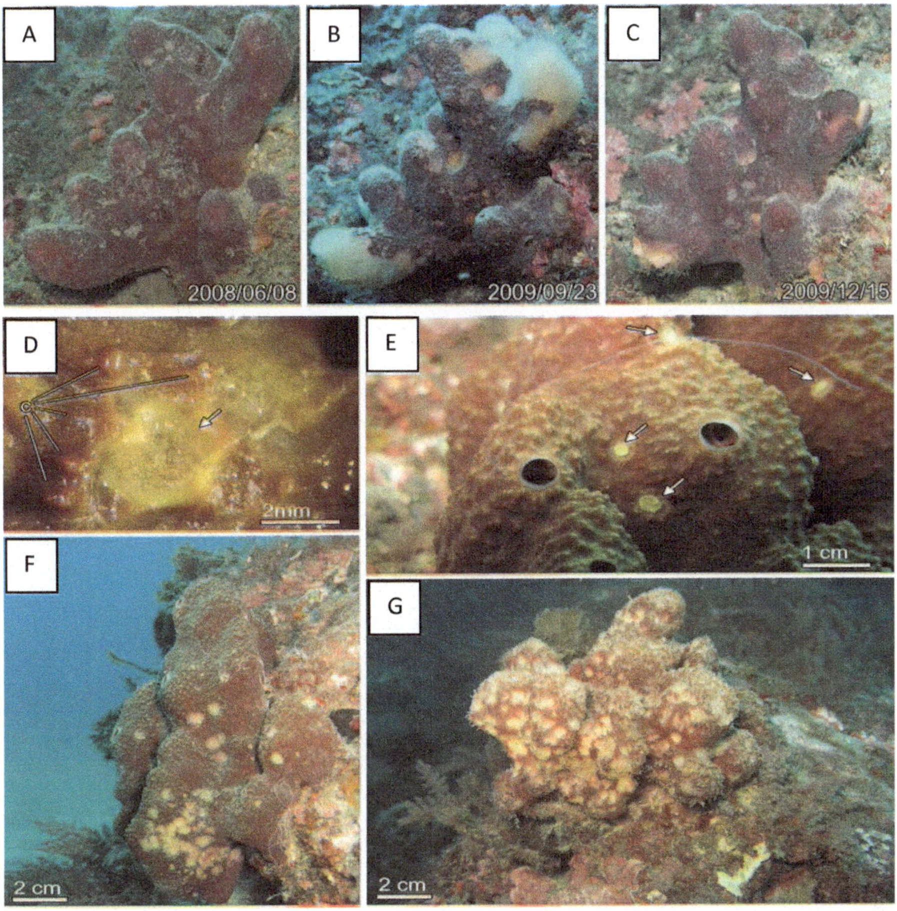

Fig. 11.6 Healthy and diseased *Ircinia* spp. with pustule syndrome (PS). (**a–c**) Healthy specimens; (**d**) detail of a pustule (arrow) surrounded by conules (c); (**e**) Disease level #1: initial symptoms characterized by few, small, subcircular pustules (arrows); (**f**) Disease level #2: large, coalescing pustules restricted to some sponge areas; (**g**) Disease level #3: large, coalescent pustules extended over large sponge areas. (Photos are adapted based on the publication by Maldonado et al. [96])

Population-level impacts and transmission mechanisms of ARBS were investigated by monitoring two populations of *A. cauliformis* over a 3-year period (2010–2012). The results indicated the biomass was greatly reduced in both healthy and ARBS-infected group after hurricane Irene. However, the biomass of the diseased sponge group reduced at higher level, likely as a result of direct contact transmission between a diseased individual and a healthy one within that group. Additionally, the application of terminal restriction fragment length polymorphism (TRFLP) on the healthy and ARBS-affected sponge *A. cauliformis* demonstrated that healthy and diseased sponges differ significantly in bacterial communities, which were maintained

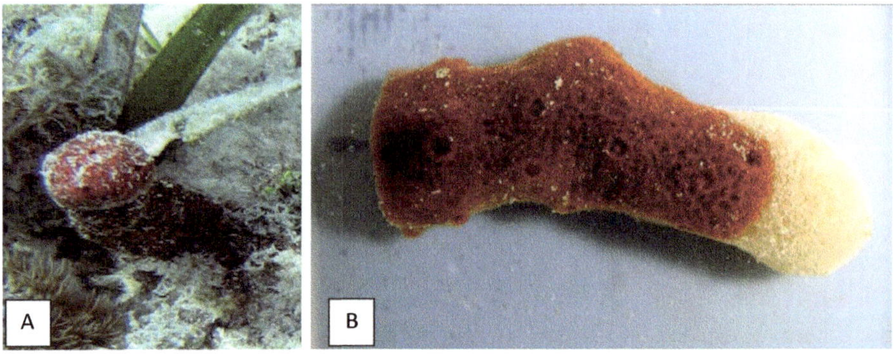

Fig. 11.7 Healthy and diseased *Amphimedon compressa*. (**a**) Healthy specimen; (**b**) Affected specimen with sponge white patch (SWP). (Photos are adapted based on publication by Angermeier et al. [184])

Fig. 11.8 Healthy and diseased *Callyspongia* (*Euplacella*) aff *biru*. (**a**) Healthy specimen; (**b**) Affected specimen. Scale bars 1 cm. (Photos are adapted based on the publication by Sweet et al. [196])

across space (two locations), time (over 2 years: 2008–2009), and health status [43]. Researchers found ten terminal restriction fragments markedly vary depending on sponge health conditions. Six of the ten increased in abundance in the diseased sponge and the other four decreased. ARBS-infected sponge has lower prevalence of the photo-symbiont *Synechococcus spongiarum* – a result of the functional consequences of disease process. *Leptolyngbya* spp., red-pigmented strains cultivated from ARBS lesion, increased in abundance in the diseased sponges. Nevertheless, researchers failed to identify it as a pathogen of ARBS, since no signs of infection were observed 24 days after healthy sponges were contacted with the cultured strain. In 2015, the pathogen of ARBS-infected sponge *R. odorabile* at the Great Barrier Reef was finally confirmed to be a new strain of *Pseudoalteromonas agarivorans*, which was found to contain diverse and abundant genes that are virulence-related [195].

The sponge orange band (SOB) disease was firstly described in sponge *Xestospongia muta* in Florida Keys, USA [203]. The SOB terminology describes a sequel rather than an actual disease process. It is hypothesized that the orange colouration is in fact an artefact of dying sponge. The disease-leading process occurs before the orange band appears. Widespread mortality of sponge was observed in the Florida Keys following a set of changing environmental conditions. As a result, a detailed and long-term (1991–2006) study on sponge population dynamics was conducted [204]. They studied the relationship between the mortality events and sponge community biomass and closely monitored the population response of 23 sponge species. It is found that there is a correlation between the mortality events and the consecutive peak of the picoplanktonic cyanobacterium *Synechococcus* sp.; however, the leading cause of sponge death is yet to be found. In contrast, another study of SOB-affected sponge *X. muta* in Florida Keys provided no evidence for the involvement of a specific microbial pathogen as an aetiologic agent of disease [187]. In this study, SOB infections showed obvious destruction of the pinacoderm observed by scanning and transmission electron microscopy (SEM and TEM). Chlorophyll *a* content in bleached sponges reduced significantly compared to healthy tissues. For the symbiotic microbial community, a distinct shift from *Synechococcus*/*Prochlorococcus* to several unspecific cyanobacterial taxa (e.g. *Leptolyngbya* sp.) was observed.

In a Florida reef, a novel sponge disease called sponge white patch (SWP) was first reported to affect sponge *Amphimedon compressa* [184]. SWP starts from the sign of distinctive white patches with variable size on the branches of diseased sponges. A spongin-boring bacterial morphotype was detected by TEM on the same tissues previously been implicated in sponge disease [191]. The disease-causing α-proteobacterium was also isolated from bleached tissues. So far, the cause of SWP in *A. compressa* still needs to be further confirmed as the tissue transplantation experiments could not demonstrate infectivity from diseased to healthy sponges.

Tissue necrosis and brown spot lesions (BSL) were firstly reported in a common marine sponge *Ianthella basta* [186]. Sponge *Ianthella basta* affected by BSL performed up to 66% of lesions and symptoms of necrosis at all studied sites. It is further demonstrated that microbial community of *I. basta* could maintain stably and the thermal and sedimentation stresses do not directly link with the BSL [190]. The experiment for 6 necrotic and 12 healthy *I. basta* specimens was conducted in the aquarium facilities. The tissue regression caused by collection and transportation process was observed, but the unhealthy sponges can rapidly recover [205].

Pustule disease (PD) was firstly described in sponge *Ircinia* spp. with the symptoms of small pustules on the sponge surface, which consequently coalesce forming larger, extensive lesions. The symptoms start from the outside of sponges and then move to kill the cells below the ectosome and penetrate deeper into the body [96]. A twisted rod is suggested to be the aetiological agent by TEM. Furthermore, the study on the affected sponge *I. variabilis* indicated that the stressful environmental conditions could induce sponge pathogens to become virulent, which makes sponges unable to control their proliferation [97]. In this case, *Vibrio rotiferanius* was indicated as the pathogen. Another mass mortality episode involving *I. variabilis* was observed in Conero Promontory, North Adriatic Sea, in the late summer of 2009 [206]. Evident necrotic areas and portions with bare skeleton were detected from affected specimens. The results indicated that the high temperatures in summer built unfavourable environmental conditions, which is the main reason to induce the disease outbreak.

11.3.2 Sponges' Recovery After the Disease-Caused Mortality

Studies have illustrated the dramatic effect of diseases on both populations and individuals, though there is no adequate information about disease causes. Diseases have been reported in some cases to lead certain affected species to almost extinction [235]. However, so far there are only ten sponge diseases with assigned names (Table 11.2) as observation in the natural environment is one of the many difficulties. Among of these diseases, only five have had likely aetiological agents identified, which highlights the urgent requirement for deeper understanding on sponge diseases. Loss of biomass to disease is a cause for concern, not only for the conservation of the sponge diversity but also for their important functional roles, which include enrichment of bacteria from the water, binding live corals to the reef frame, and facilitating reef regeneration [227, 236, 237].

Prevalence of the diseases in the sponge population can be triggered by multiple factors (Table 11.1), such as the increased water temperature due to the seasonal change, currents, and light condition [206, 233], as well as the changes of the associated microbial community and their secondary metabolites [96]. Normally, the sponges need a long time to recover, and they usually are not able to recover to the original condition in most cases. It is essential to investigate sponge recovery from the mortalities to help build up a better resource management strategy regarding the protection of sponge community biomass. If the dominate and long-lived sponge species enable to recover and achieve their former levels of abundance and community size, the ecological function of the sponge community will have the chance to be restored.

In Table 11.3, we summarize the representative cases of sponge recovery in natural environment from disease-caused mortality, the length of recovery period, and their recovery strategies. Species responded in different ways at different locations, and population change was sometimes dramatic and sometimes unpredictable. There are four patterns of change in abundance based on the studies so far: (1) rapid decline in the mortality event, followed by gradual recovery, (2) fluctuating abundance, (3) gradual decline, and (4) low abundance, no change. More importantly, a consensus has been concluded that the characters of different sponge species and the sponge population density in the natural environment play a critical role in determining their recovery strategy and speed, which were evidently demonstrated by Stevely et al. (2011) Stevely, Sweat [204], Castritsi-Catharios et al. (2011) Castritsi-Catharios, Miliou [233], and Di Camillo et al. (2013) Di Camillo, Bartolucci [206].

In many Caribbean reef areas, sponges are the most abundant sessile invertebrates in terms of biomass [236]. Disease outbreak has been reported to happen in unusually large sponges disproportionately frequently, for example, sponge *Spheciospongia vesparium* normally called as "loggerhead" and *Xestospongia muta* generally called "barrel" sponges [198, 238–240]. Meanwhile, the sponges with large sizes had more chance to suffer the disease as it normally takes longer period for a pathogen to progress through all of their tissues. It took a couple of months for a bathtub-sized *X. muta* to entirely succumb to the disease [227]. For a medium-sized sponge *Mycale laxissima*, it took 2 months to complete the degeneration [241]. Instead, for the smaller sponges, such as the species in genera *Spongia* and *Hippospongia*, it took only 1 week from initially internal lesions to the detectable degradation at the tissue surfaces [213]. Moreover, for the encrusting species with very small volumes, their tissue could be fully degraded within few days [227, 242].

On the other hand, the sponges with larger body size will also have a better chance to recover from disease infection because they have more time to deal with a pathogen and stop it before it spreads to the entire body. Widespread sponge mortalities occurred in Florida Keys, USA, in the early 1990s. During the observation period, hurricane Wilma hitting the study area in 2005 significantly impacted sponge abundance; however, the massive and long-lived sponges tended to be more resistant. Particularly, after a 2-year mortality event (1991–1993), sponge abundance and biomass at Marathon and Long Key had declined significantly. Considering the 1991–1993 blooms, a conclusion is obtained that the bloom could be the critical cause of the sponge mortality during that time. In fact, the sponge population at Marathon was highly impacted by the environmental event. However, there was only a little effect at Long Key. Overall, the affected 23 sponge species at Marathon and Long Key were observed for up to 15 years (1991–2006) [204]. The data showed it was not a rapid process for all species to recover. At least 10–15 years of the recovery process was required for the sponge population biomass in the studied area.

Another severe mass mortality was observed in the bay of Trieste in Northern Adriatic Sea in 1974. The sponge *Reniera* spp. accompanied with the brittle star *Ophiothrix quinquemaculata* and the ascidians *Microcosmus* spp. contributed the main population [243]. Several physical factors, including high water temperature, low water circulation, and oxygen depletion, were considered to be the main causes

Table 11.3 Sponge response to the disease-caused mortality in natural environment based on the recovery pattern

Affected sponge	Description of recovery	Recovery period	Recovery strategy	Suggested solution or consideration factors	References
Pattern I: Rapid decline and gradual recovery					
Spheciospongia vesparium	Slow: (1) Site Marathon: recruitment was recorded 7 years later the mortality, in another 3 years, population abundance recovered to original level, but the biomass did not. (2) Site Long Key: 13 years later, the abundance not recovered to original level	10–13 years	Asexual regeneration of remaining specimens or sexual reproduction and larval dispersion	It is a relatively long-lived and slow-growing species and had greater resistance to the algal blooms than most species	[204, 250]
Cliona varians	Rapid: stable increase in an 8-year period after the mortality and then a significant increase in another 4 years	12 years	Asexual regeneration of remaining specimens, species bore basally into calcareous substrate as refugia	N/A	[204, 251]
Ircinia variabilis *Ircinia fasciculata*	*I. variabilis*: individual mortality reduced from 4% to 1.5%, but not recovered to the original 0.5%; *I. fasciculata*: recovered to the original 5% from 15%	1–2 years	Sponge immune response stopped the disease in many instances, allowing regeneration of necrotic mesohyl areas and epithelia	(1) A recurrent epidemic, reappearing each year, typically after the hottest months. (2) The appearance of the diseased sponge appears to be related to proliferation of an external bacterium within the peripheral mesohyl.	[96]

(continued)

Table 11.3 (continued)

Affected sponge	Description of recovery	Recovery period	Recovery strategy	Suggested solution or consideration factors	References
Ircina variabilis Sarcotragus spinosulus Spongia officinalis	The damaged tissues of large specimens were able to recover.	Slow	N/A	Non-pathogenic bacteria usually associated with sponges may become harmful in response to temperature increase.	[206]
Chondrosia reniformis Tedania anhelans	The damaged tissues of large specimens were able to recover, but the abundance was much less than the original level	Slow	N/A	The abundance was impacted by other organisms in the studied area	[234]
Ircinia campana	Slow: (1) Site Marathon: recruitment was recorded 9 years later the mortality, in another 2 years, abundance increased significantly. (2) Site Long Key: the biomass increased significantly after 10 years and took 1 year to recover. But the abundance and biomass at both sites remained far below the original levels	11 years	Sexual reproduction and larval dispersion	N/A	[204]
Ircinia felix	Quickly and significantly, original levels on abundance and biomass were recovered	Within 1 year	Asexual regeneration of remaining specimens	N/A	[204]

(continued)

Table 11.3 (continued)

Affected sponge	Description of recovery	Recovery period	Recovery strategy	Suggested solution or consideration factors	References
Ircinia strobilina	Rapid and pronounced in Site Marathon, abundance exceeded the original level 7 times; limited in Site Long Key, 13 years later still less than the original level	Within 1 year or up to 13 years	Asexual regeneration of remaining specimens or sexual reproduction and larval dispersion	N/A	[204]
Spongia officinalis	60% mortality reduced to 5%	1–2 years	The ability to shed the damaged parts and to regenerate	N/A	[178]
Spongia officinalis	Settled after the epidemic resulting in a complete restoration of the population	5–8 years	Recolonization	Sponge farming	[219, 252]
Spongia cfr *zimocca* *Hippospongia communis*	Absence and then grow up slowly	Up to 10 years	Repopulation by adapting to the higher light condition	Prohibition of destructive fishing methods in the deeper waters	[181, 233, 253]
Spongia agaricina *S. zimocca* *S. officinalis adriatica* *S. officinalis mollissima* *Hippospongia communis*	Vary, depending on the sponge types, substratum type, depths (shallow, <40 m; or deep, 50–100 m), upwelling, temperature, light condition, predators, and the impact of fishing gears	N/A	Heal the damaged tissues and grow to a larger dimension	(1) Cultivation of sponges is a promising solution to relieve the impact of harvesting. (2) Selection of the food organisms to adapt to their natural diet and the appropriate environmental conditions	[148, 233, 252, 254]
Spongia barbara	Site Long Key: recovery started from 4 years after the mortality till reaching the original level	2–9 years	Asexual regeneration of remaining specimens	N/A	[204]

(continued)

Table 11.3 (continued)

Affected sponge	Description of recovery	Recovery period	Recovery strategy	Suggested solution or consideration factors	References
Spongia Barbara dura	Rapid recruitment in abundance at both sites (Marathon and Long Key)	2 years	N/A	No records for per mortality but showed significant increase	[204]
Spongia graminea	Recovery started from 2 years later the mortality; in another 9 years, the population gradually recovered, but abundance was still lower than the original level	Up to 11 years	Sexual reproduction and larval dispersion	N/A	[204]
Hippospongia lachne	Slow and limited recovery, no clear recovery record in 15-year observation	N/A	N/A	N/A	[204]
Aplysina cauliformis	Slight recovery	1-year observation	Recover and grow when no major storms hit	N/A	[200]
Aplysina fulva	Recovery started from 13 years later, from uncommon to the 5th most abundant species	13–15 years	N/A	N/A	[204]
Aplysina fulva Iotrochota birotulata	Individual diseased from 4% reduced to 1.5% for *I. birotulata*; a reduction from 15% to 5% for *A. fulva*	1–2 year in a 14-year observation period	Population density impacted the individual recovery	N/A	[228]

(continued)

Table 11.3 (continued)

Affected sponge	Description of recovery	Recovery period	Recovery strategy	Suggested solution or consideration factors	References
Pattern II: Fluctuating abundance					
Cinachyrella alloclada	Significant fluctuations in abundance; reduced in a 5-year period following the mortality, significantly increased in another 3 years, then declined in 1 year, then recovered in 1 year, then declined again	9–13 years	It is considered to be an "opportunistic" species as its reproduction can be the extrusion of numerous tiny sponge buds to lead to rapid increase in numbers of individuals	It was the most resistant sponge during the plankton bloom event	[255, 256]
Hyrtios proteus	Significant fluctuations in abundance in the Site Marathon with "W" trend	7 years	N/A	Complexity of sponge community dynamics in space and time. A variety of factors can influence their life histories	[204]
Chalinula molitba Niphates erecta	Significant fluctuations in abundance, rapid decreases and then increases	11 years	N/A	N/A	[204]
Halichondria melanodocia	(1) Site Marathon: recovery started from 2 years later the mortality. (2) Site Long Key: stable and recovered 8 years later to twofold abundance	2–8 years	Unstable recovery: (1) Site Marathon – declined in the next 5 years after the recovery (2) Site Long Key: declined after the recovery	N/A	[204]
Adocia implexiformis	Significant fluctuations in abundance in the Site Marathon with "M" trend	7–13 years	N/A	Complexity of sponge community dynamics in space and time. A variety of factors can influence their life histories	[204]

(continued)

Table 11.3 (continued)

Affected sponge	Description of recovery	Recovery period	Recovery strategy	Suggested solution or consideration factors	References
Lissodendoryx isodictyalis Amphimedon viridis	Significant fluctuations in abundance, a period recovery (significantly increase), followed by a rapid decline	8–11 years	N/A	N/A	[204]
Petrosia cf. *pellasarca*	Absent for many years, then started a new and common recruit 12 years later; abundance significantly decreased following the hurricanes	N/A	N/A	N/A	[204]
Pattern III: Gradual decline					
Tedania ignis	Site Long Key: it underwent a long-term gradual decline in abundance	N/A	N/A	One explanation for this general demise is that, with the increasing recruitment of other species into Long Key, opportunistic species *T. ignis* could no longer compete for food and space; it can reclaim bared space, is a fast grower, and can overgrow other species; but it is a poor recruiter to bare substrate and is vulnerable to predation, which could contribute to a long-term decrease in abundance	[204, 227, 257]

(continued)

Table 11.3 (continued)

Affected sponge	Description of recovery	Recovery period	Recovery strategy	Suggested solution or consideration factors	References
Adocia implexiformis	Site Long Key: gradual decline in abundance	N/A	N/A	Complexity of sponge community dynamics in space and time. A variety of factors can influence their life histories	[204]
Hyrtios proteus	Site Long Key: gradual decline in abundance	N/A	N/A	Complexity of sponge community dynamics in space and time. A variety of factors can influence their life histories	[204]
Pattern IV: Low abundance, no change					
Tectitethya crypta	Site Marathon: no significant change in its abundance was observed during the 15-year study	N/A	N/A	Resistant to bloom conditions; can cease pumping for considerable periods of time to provide protection against bloom; has stable local populations and several life history adaptations; however, unable to rapidly colonize newly available substrate	[167, 204, 256]

(continued)

Table 11.3 (continued)

Affected sponge	Description of recovery	Recovery period	Recovery strategy	Suggested solution or consideration factors	References
Callyspongia vaginalis	Site Marathon: no significant change in its abundance was observed during the 15-year study	N/A	A slow grower and it relies on the production of many larvae for colonization; it has no chemical defences against at least some predators	This poor defence system and dispersal ability may limit its abundance in areas where competition for substrate is strong	[204, 258–261]
Tedania ignis	Site Marathon: no significant trends	N/A	N/A	N/A	[204]
Hyrtios proteus	Site Long Key: from a rapid decline in the first 4 years after the mortality to the point that they were uncommon at Long Key	N/A	N/A	Complexity of sponge community dynamics in space and time. A variety of factors can influence their life histories	[204]

of mortality. The recovery of the benthic community was slow, and the biomass of the sponge population remained at least 50% below the original levels even after 3 years [244].

In Conero Promontory, North Adriatic Sea, three marine sponges, *Ircinia variabilis*, *Sarcotragus spinosulus*, and *Spongia officinalis*, were recorded in the late summer of 2009 to be affected by serious mortality [206]. The unfavourable environmental conditions (high temperatures and calm sea) in summer led to the outbreak of the disease. The symptom of necrotic areas and portions with bare skeleton was firstly detected in the affected *S. spinosulus* specimens, and many of them (about 22%) suffered from this disease showing a white mat of cyanobacteria on the surface. More than 900 damaged sponges were found along the 1-km-long beach. When the water temperature dropped below 20 °C, some large specimens started showing the evidence of recovery, however, the smaller ones became even more sensitive. Unfortunately, the disease outbreak came back to Conero Promontory again 3 months later and lasted about a month; a larger number of species were affected. Sponges *Ircinia variabilis*, *Sarcotragus spinosulus*, *Spongia officinalis*, and *Aplysina aerophoba* were entirely or partially bleached [234]. In contrast, some sponges like *Dysidea pallescens*, *Haliclona* sp., and *Tedania anhelans* survived from the disease. The most affected sponge was *Chondrosia reniformis* with 100% of the specimens damaged and only large bare areas left on the rock. In a 3 years' time after the mortality, the survived tissues of *C. reniformis* still could not recover significantly in size. The fast-growing species, such as stoloniferans, hydrozoans, mussels, algae, serpulids, and bryozoans, covered the bare areas.

In conclusion, sponge mortality has been identified to be mainly induced by water-warming anomalies that could trigger a complex cascading effect, which includes physiological response [245, 246], phenology [247], pathogens [248], and possible adaptation in situ [249]. The physical damage is tropically much easier to recover than the disease-caused damage, which takes years to increase to the original abundance and biomass by asexual regeneration of the remaining specimens or sexual reproduction and larval dispersion. Understanding the response patterns of different sponge species/populations after the mortality is essential to structure a better management strategy to prevent sponge loss.

Future research direction could focus on developing suitable approaches to assess sponge health status, determine the prevalence sponge diseases, isolate and identify pathogenic agents, as well as evaluate sponge immune system to uncover sponge cellular defence mechanism. Utilization of advanced technologies, such as single-cell technique, metagenomic sequencing, and transcriptome analysis, could explore the understanding on the mechanism of sponge response to the environmental anomalies at the cellular and molecular levels. Moreover, it is also valuable to investigate how the environmental factors, such as temperature, nutrients, pH, water flow, and bio-disturbance, impact on the occurrence of sponge disease and the outbreak of pathogenic microbes. Thanks to the development of sponge cultivation technology, it will allow us to monitor the impact of high-density aquaculture conditions on disease occurrence and spread. Fundamental research of sponge disease should focus more on developing a guidance on potential treatments and containment policies for sponge diseases. The risk minimization strategy would be essential for both sponge aquaculture and reef protection.

References

1. van Soest RWM, Boury-Esnault N, Vacelet J, Dohrmann M, Erpenbeck D, de Voogd NJ, et al. Global diversity of sponges (porifera). PLoS One. 2012;7:e35105.
2. van Soest RWM, Boury-Esnault N, Hooper JNA, Rützler K, de Voogd NJ, Alvarez B, et al. World porifera database. 2018. Accessed at http://www.marinespecies.org/porifera
3. Hentschel U, Piel J, Degnan SM, Taylor MW. Genomic insights into the marine sponge microbiome. Nat Rev Microbiol. 2012;10:641–54.
4. Hoffmann F, Radax R, Woebken D, Holtappels M, Lavik G, Rapp HT, et al. Complex nitrogen cycling in the sponge geodia barretti. Environ Microbiol. 2009;11:2228–43.
5. Mehbub MF, Lei J, Franco CMM, Zhang W. Marine sponge derived natural products between 2001 and 2010: trends and opportunities for discovery of bioactives. Mar Drugs. 2014;12:4539–77.
6. Blunt JW, Copp BR, Keyzers RA, Munro MHG, Prinsep MR. Marine natural products. Nat Prod Rep. 2013;30:237–323.
7. Blunt JW, Copp BR, Keyzers RA, Munro MHG, Prinsep MR. Marine natural products. Nat Prod Rep. 2014;31:160–258.

8. Blunt JW, Copp BR, Keyzers RA, Munro MHG, Prinsep MR. Marine natural products. Nat Prod Rep. 2015;32:116–211.
9. Blunt JW, Copp BR, Keyzers RA, Munro MHG, Prinsep MR. Marine natural products. Nat Prod Rep. 2016;33:382–431.
10. Blunt JW, Copp BR, Keyzers RA, Munro MHG, Prinsepd MR. Marine natural products. Nat Prod Rep. 2017;34:235–94.
11. Thacker RW, Freeman CJ. Sponge–microbe symbioses: recent advances and new directions. Adv Mar Biol. 2012;62:57–111.
12. Webster NS, Taylor MW. Marine sponges and their microbial symbionts: love and other relationships. Environ Microbiol. 2012;14:335–46.
13. Simister RL, Deines P, Botté ES, Webster NS, Taylor MW. Sponge-specific clusters revisited: a comprehensive phylogeny of sponge-associated microorganism. Environ Microbiol. 2012;14:517–24.
14. Moitinho-Silva L, Nielsen S, Amir A, Gonzalez A, Ackermann GL, Cerrano C, et al. The sponge microbiome project. GigaScience. 2017;6:1–7.
15. Thomas T, Moitinho-Silva L, Lurgi M, Björk JR, Easson C, Astudillo-García C, et al. Diversity, structure and convergent evolution of the global sponge microbiome. Nat Commun. 2016;7:11870.
16. Gao Z-M, Zhou G-W, Huang H, Wang Y. The cyanobacteria-dominated sponge dactylospongia elegans in the south china sea: Prokaryotic community and metagenomic insights. Front Microbiol. 2017;8:art1387.
17. López-Legentil S, Erwin PM, Pawlik JR, Song B. Effects of sponge bleaching on ammonia-oxidizing archaea: distribution and relative expression of ammonia monooxygenase genes associated with the barrel sponge *Xestospongia muta*. Microb Ecol. 2010;60:561–71.
18. Slaby BM, Hackl T, Horn H, Bayer K, Hentschel U. Metagenomic binning of a marine sponge microbiome reveals unity in defense but metabolic specialization. ISME J. 2017;11:2465–78.
19. Freeman CJ, Thacker RW. Complex interactions between marine sponges and their symbiotic microbial communities. Limnol Oceanogr. 2011;56:1577–86.
20. Flatt PM, Gautschi JT, Thacker RW, Musafija-Girt M, Crews P, Gerwick WH. Identification of the cellular site of polychlorinated peptide biosynthesis in the marine sponge *Dysidea (lamellodysidea) herbacea* and symbiotic cyanobacterium *Oscillatoria spongeliae* by card-fish analysis. Mar Biol. 2005;147:761–74.
21. Marty MJ, Vicente J, Oyler BL, Place A, Hill RT. Sponge symbioses between *Xestospongia deweerdtae* and *Plakortis* spp. Are not motivated by shared chemical defense against predators. PLoS One. 2017;12:e0174816.
22. Webster NS, Xavier JR, Freckelton M, Motti CA, Cobb R. Shifts in microbial and chemical patterns within the marine sponge *Aplysina aerophoba* during a disease outbreak. Environ Microbiol. 2008;10:3366–76.
23. Webster NS, Cobb RE, Negri AP. Temperature thresholds for bacterial symbiosis with a sponge. ISME J. 2008;2:830–42.
24. Guzman C. Conaco, C Gene expression dynamics accompanying the sponge thermal stress response. PLoS One. 2016;11:e0165368.
25. Taylor MW, Radax R, Steger D, Wagner M. Sponge-associated microorganisms: evolution, ecology, and biotechnological potential. Microbiol Mol Biol Rev. 2007;71:295–347.
26. Lee YK, Jung HJ, Lee HK. Marine bacteria associated with the korean brown alga, *Undaria pinnatifida*. J Microbiol. 2006;44:694–8.
27. Webster NS, Hill RT. The culturable microbial community of the great barrier reef sponge *Rhopaloeides odorabile* is dominated by an α-proteobacterium. Mar Biol. 2001;138:843–51.
28. White JR, Patel J, Ottesen A, Arce G, Blackwelder P, Lopez JV. Pyrosequencing of bacterial symbionts within *Axinella corrugata* sponges: diversity and seasonal variability. PLoS One. 2012;7:e38204.
29. Wichels A, Würtz S, Döpke H, Schütt C, Gerdts G. Bacterial diversity in the breadcrumb sponge *Halichondria panicea*. FEMS Microb Ecol. 2006;56:102–18.

30. Anderson SA, Northcote PT, Page MJ. Spatial and temporal variability of the bacterial community in different chemotypes of the New Zealand marine sponge *Mycale hentscheli*. FEMS Microbiol Ecol. 2010;72:328–42.
31. Friedrich AB, Fischer I, Proksch P, Hacker J, Hentschel U. Temporal variation of the microbial community associated with the mediterranean sponge *Aplysina aerophoba*. FEMS Microbiol Ecol. 2001;38:105–13.
32. Webster NS, Cobb RE, Soo R, Anthony SL, Battershill CN, Whalan S, et al. Bacterial community dynamics in the marine sponge *Rhopaloeides odorabile* under *in situ* and *ex situ* cultivation. Mar Biotechnol. 2011;13:296–304.
33. Mohamed NM, Rao V, Hamann MT, Kelly M, Hill RT. Monitoring bacterial diversity of the marine sponge *Ircinia strobilina* upon transfer into aquaculture. Appl Environ Microbiol. 2008;74:4133–43.
34. Mohamed NM, Enticknap JJ, Lohr JE, McIntosh SM, Hill RT. Changes in bacterial communities of the marine sponge *Mycale laxissima* on transfer into aquaculture. Appl Environ Microbiol. 2008;74:1209–22.
35. Cleary DFR, Becking LE, de Voogd NJ, Pires ACC, Polónia ARM, Egas C, et al. Habitat- and host-related variation in sponge bacterial symbiont communities in indonesian waters. FEMS Microbiol Ecol. 2013;85:465–82.
36. Schmitt S, Tsai P, Bell J, Fromont J, Ilan M, Lindquist N, et al. Assessing the complex sponge microbiota: core, variable and species-specific bacterial communities in marine sponges. ISME J. 2012;6:564–76.
37. Schmitt S, Hentschel U, Taylor M. Deep sequencing reveals diversity and community structure of complex microbiota in five mediterranean sponges. In: Maldonado M, Turon X, Becerro MA, Uriz MJ, editors. Ancient animals, new challenges: developments in sponge research. Dordrecht: Springer; 2012. p. 341–51.
38. Burgsdorf I, Erwin PM, López-Legentil S, Cerrano C, Haber M, Frenk S, et al. Biogeography rather than association with cyanobacteria structures symbiotic microbial communities in the marine sponge *Petrosia ficiformis*. Front Microbiol. 2014;10:1–11.
39. Hentschel U, Hopke J, Horn M, Friedrich AB, Wagner M, Hacker J, et al. Molecular evidence for a uniform microbial community in sponges from different oceans. Appl Environ Microbiol. 2002;68:4431–40.
40. Taylor MW, Schupp PJ, de Nys R, Kjelleberg S, Steinberg PD. Biogeography of bacteria associated with the marine sponge *Cymbastela concentrica*. Environ Microbiol. 2005;7:419–33.
41. Cárdenas CA, Bell JJ, Davy SK, Hoggard M, Taylor MW. Influence of environmental variation on symbiotic bacterial communities of two temperate sponges. FEMS Microbiol Ecol. 2014;88:516–27.
42. Montalvo NF, Hill RT. Sponge-associated bacteria are strictly maintained in two closely related but geographically distant sponge hosts. Appl Enrivon Micorbiol. 2011;77:7207–16.
43. Olson JB, Thacker RW, Gochfeld DJ. Molecular community profiling reveals impacts of time, space, and disease status on the bacterial community associated with the caribbean sponge aplysina cauliformis. FEMS Microbiol Ecol. 2013;87:268–79.
44. Erwin PM, Coma R, López-Sendino P, Serrano E, Ribes M. Stable symbionts across the hma-lma dichotomy: low seasonal and interannual variation in sponge-associated bacteria from taxonomically diverse hosts. FEMS Microbiol Ecol. 2015;91:1–11.
45. Peek AS, Feldman RA, Lutz R, Vrijenhoek RC. Cospeciation of bacteria and deep sea clams. Proc Natl Acad Sci U S A. 1998;95:9962–6.
46. Brusca RC, Brusca GJ. Phylum Porifera: the sponges. In: Sinauer AD, editor. Invertebrates. Cambridge, MA: Sinauer Press; 1990. p. 181–210.
47. Gallissian MF, Vacelet J. Ultrastructure de quelques stades de l'ovogénèse de spongiaires du genre verongia (dictyoceratida). Ann Sci Nat Zool. 1976;18:381–404.
48. Vacelet J, Boury-Esnault N, Fiala-Medioni A, Fisher CR. A methanotrophic carnivorous sponge. Nature. 1995;377:296.

49. Vogel S. Current-induced flow through living sponges in nature. Proc Natl Acad Sci U S A. 1977;74:2069–71.
50. Rappe MS, Vergin K, Giovannoni SJ. Phylogenetic comparisons of a coastal bacterioplankton community with its counterparts in open ocean and freshwater systems. FEMS Microbiol Ecol. 2000;33:219–32.
51. Glöckner FO, Zaichikov E, Belkova N, Denissova L, Pernthaler J, Pernthaler A, et al. Comparative 16s rRNA analysis of lake bacterioplankton reveals globally distributed phylogenetic clusters including an abundant group of actinobacteria. Appl Enrivon Micorbiol. 2000;66:5053–6.
52. Ravenschlag K, Sahm K, Pernthaler J, Amann R. High bacterial diversity in permanently cold marine sediments. Appl Enrivon Micorbiol. 1999;65:3982–9.
53. Turon X, Galera J, Uriz MJ. Clearance rates and aquiferous systems in two sponges with contrasting life-history strategies. J Exp Zool. 1997;278:22–36.
54. Friedrich AB, Merkert H, Fendert T, Hacker J, Proksch P, Hentschel U. Microbial diversity in the marine sponge *Aplysina cavernicola* (formerly *Verongia cavernicola*) analyzed by fluorescence in situ hybridization (FISH). Mar Biol. 1999;134:461–70.
55. Wilkinson CR, Garrone R, Vacelet J. Marine sponges discriminate between food bacteria and bacterial symbionts: electron microscope radioautography and *in situ* evidence. Proc R Soc Lond B. 1984;220:519–28.
56. Turque AS, Cardoso AM, Silveira CB, Vieira RP, Freitas FAD, Albano RM, et al. Bacterial communities of the marine sponges *Hymeniacidon heliophila* and *Polymastia janeirensis* and their environment in Rio de janeiro, Brazil. Mar Biol. 2008;155:135–46.
57. Lafi FF, Garson MJ, Fuerst JA. Culturable bacterial symbionts isolated from two distinct sponge species (*Pseudoceratina clavata* and *Rhabdastrella globostellata*) from the Great Barrier Reef display similar phylogenetic diversity. Microb Ecol. 2005;50:213–20.
58. Sharp KH, Eam B, Faulkner DJ, Haygood MG. Vertical transmission of diverse microbes in the tropical sponge *Corticium* sp. Appl Enrivon Micorbiol. 2007;73:622–9.
59. Zhu P, Li Q, Wang G. Unique microbial signatures of the alien hawaiian marine sponge *Suberites zeteki*. Micorbial Ecol. 2007;55:406–14.
60. Alex A, Silva V, Vasconcelos V, Antunes A. Evidence of unique and generalist microbes in distantly related sympatric intertidal marine sponges (Porifera: Demospongiae). PLoS One. 2013;8:e80653.
61. Meyer B, Kuever J. Phylogenetic diversity and spatial distribution of the microbial community associated with the caribbean deep-water sponge *Polymastia* cf. *Corticata* by 16S rRNA, apra, and amoa gene analysis. Micorbial Ecol. 2008;56:306–21.
62. Luter HM, Widder S, Botté ES, Wahab MA, Whalan S, Moitinho-Silva L, et al. Biogeographic variation in the microbiome of the ecologically important sponge, *Carteriospongia foliascens*. PeerJ. 2015;3:e1435.
63. Turque AS, Batista D, Silveira CB, Cardoso AM, Vieira RP, Moraes FC, et al. Environmental shaping of sponge associated archaeal communities. PLoS One. 2010;5:e15774.
64. Weigel BL, Erwin PM. Intraspecific variation in microbial symbiont communities of the sun sponge, *Hymeniacidon heliophila*, from intertidal and subtidal habitats. Appl Enrivon Micorbiol. 2016;82:650–8.
65. Reveillaud J, Maignien L, Eren AM, Huber JA, Apprill A, Sogin ML, et al. Host-specificity among abundant and rare taxa in the sponge microbiome. ISME J. 2014;8:1198–209.
66. Noyer C, Casamayor EO, Becerro MA. Environmental heterogeneity and microbial inheritance influence sponge-associated bacterial composition of *Spongia lamella*. Microb Ecol. 2014;68:611–20.
67. Dethlefsen L, McFall-Ngai MJ, Relman DA. An ecological and evolutionary perspective on human-microbe mutualism and disease. Nature. 2007;449:811–8.
68. Noyer C, Becerro MA. Relationship between genetic, chemical, and bacterial diversity in the atlanto-mediterranean bath sponge *Spongia lamella*. Hydrobiologia. 2012;687:85–99.

69. Rohwer F, Seguritan V, Azam F, Knowlton N. Diversity and distribution of coral-associated bacteria. Mar Ecol Prog Ser. 2002;243:1–10.
70. Reshef L, Koren O, Loya Y, Zilber-Rosenberg I, Rosenberg E. The coral probiotic hypothesis. Environ Microbiol. 2006;8:2068–73.
71. Climate change 2007: the physical science basis. In: Solomon S, Qin D, Manning M, Chen Z, Marquis M, Averyt KB, Tignor M, Miller HL, editors. Contribution of Working Group I to the fourth assessment report of the Intergovernmental Panel on Climate Change. Cambridge/New York: Cambridge University Press; 2007. p. 1–996.
72. Hoegh-Guldberg O, Mumby PJ, Hooten AJ, Steneck RS, Greenfield P, Gomez E, et al. Coral reefs under rapid climate change and ocean acidification. Science. 2007;318:1737–42.
73. Webster NS, Blackall LL. What do we really know about sponge-microbial symbioses? ISME J. 2009;3:1–3.
74. López-Legentil S, Song B, Mcmurray SE, Pawlik JR. Bleaching and stress in coral reef ecosystems: *Hsp70* expression by the giant barrel sponge *Xestospongia muta*. Mol Ecol. 2008;17:1840–9.
75. Lemoine N, Buell N, Hill A, Hill M. Assessing the utility of sponge microbial symbiont communities as models to study global climate change: a case study with *Halichondria bowerbanki*. In: Custódio MR, Lôbo-Hajdu G, Hajdu E, Muricy G, editors. Porifera research: biodiversity, innovation and sustainability. Rio de Janeiro: Série Livros/Museu Nacional; 2007. p. 419–25.
76. Webster NS, Taylor MW, Behnam F, Lücker S, Rattei T, Whalan S, et al. Deep sequencing reveals exceptional diversity and modes of transmission for bacterial sponge symbionts. Environ Microbiol. 2010;12:2070–82.
77. Steger D, Ettinger-Epstein P, Whalan S, Hentschel U, de Nys R, Wagner M, et al. Diversity and mode of transmission of ammonia-oxidizing archaea in marine sponges. Environ Microbiol. 2008;10:1087–94.
78. Webster NS, Wilson KJ, Blackall LL, Hill RT. Phylogenetic diversity of bacteria associated with the marine sponge *Rhopaloeides odorabile*. Appl Environ Microbiol. 2001;67:434–44.
79. Webster NS, Botté ES, Soo RM, Whalan S. The larval sponge holobiont exhibits high thermal tolerance. Environ Microbiol Rep. 2011;3:756–62.
80. Webster NS, Watts JEM, Hill RT. Detection and phylogenetic analysis of novel crenarchaeote and euryarchaeote 16S ribosomal RNA gene sequences from a Great Barrier Reef sponge. Mar Biotechnol. 2001;3:600–8.
81. Pantile R, Webster N. Strict thermal threshold identified by quantitative PCR in the sponge *Rhopaloeides odorabile*. Mar Ecol Prog Ser. 2011;431:97–105.
82. Simister R, Taylor MW, Tsai P, Fan L, Bruxner TJ, Crowe ML, et al. Thermal stress responses in the bacterial biosphere of the Great Barrier Reef sponge, *Rhopaloeides odorabile*. Environ Microbiol. 2012;14:3232–46.
83. Mao-Jones J, Ritchie KB, Jones LE, Ellner SP. How microbial community composition regulates coral disease development. PLoS Biol. 2010;8:e1000345.
84. Fan L, Liu M, Simister R, Webster NS, Thomas T. Marine microbial symbiosis heats up: the phylogenetic and functional response of a sponge holobiont to thermal stress. ISME J. 2013;7:991–1002.
85. Girvan MS, Campbell CD, Killham K, Prosser JI, Glover LA. Bacterial diversity promotes community stability and functional resilience after perturbation. Environ Microbiol. 2005;7:301–13.
86. Allison SD, Martiny JB. Colloquium paper: resistance, resilience, and redundancy in microbial communities. Proc Natl Acad Sci U S A. 2008;105:11512–9.
87. Jones SE, Newton RJ, McMahon KD. Evidence for structuring of bacterial community composition by organic carbon source in temperate lakes. Environ Microbiol. 2009;11:2463–72.
88. Robinson CJ, Bohannan BJ, Young VB. From structure to function: the ecology of host-associated microbial communities. Microbiol Mol Biol Rev. 2010;74:453–76.

89. Thomas T, Rusch D, DeMaere MZ, Yung PY, Lewis M, Halpern A, et al. Functional genomic signatures of sponge bacteria reveal unique and shared features of symbiosis. ISME J. 2010;4:1557–67.
90. Fan L, Reynolds D, Liu M, Stark M, Kjelleberg S, Webster NS, et al. Functional equivalence and evolutionary convergence in complex communities of microbial sponge symbionts. Proc Natl Acad Sci U S A. 2012;109:E1878–87.
91. Liu M, Fan L, Zhong L, Kjelleberg S, Thomas T. Metaproteogenomic analysis of a community of sponge symbionts. ISME J. 2012;6:1515–25.
92. Erwin PM, Pita L, López-Legentil S, Turona X. Stability of sponge-associated bacteria over large seasonal shifts in temperature and irradiance. Appl Environ Microbiol. 2012;78:7358–68.
93. Pita L, Erwin PM, Turon X, López-Legentil S. Till death do us part: stable sponge-bacteria associations under thermal and food shortage stresses. PLoS One. 2013;8:e80307.
94. Pita L, Turon X, López-Legentil S, Erwin PM. Host rules: spatial stability of bacterial communities associated with marine sponges (*Ircinia* spp.) in the western mediterranean sea. FEMS Microbiol Ecol. 2013;86:268–76.
95. Erwin PM, López-Legentil S, González-Pech R, Turon X. A specific mix of generalists: bacterial symbionts in mediterranean *Ircinia* spp. FEMS Microbiol Ecolol. 2012;79:619–37.
96. Maldonado M, Sánchez-Tocino L, Navarro C. Recurrent disease outbreaks in corneous demosponges of the genus *Ircinia*: epidemic incidence and defense mechanisms. Mar Biol. 2010;157:1577–90.
97. Stabili L, Cardone F, Alifano P, Tredici SM, Piraino S, Corriero G, et al. Epidemic mortality of the sponge *Ircinia variabilis* (schmidt, 1862) associated to proliferation of a *vibrio* bacterium. Microb Ecol. 2012;64:802–13.
98. Cebrian E, Uriz MJ, Garrabou J, Ballesteros E. Sponge mass mortalities in a warming mediterranean sea: are cyanobacteria-harboring species worse off? PLoS One. 2011;6:e20211.
99. Schmitt S, Wehrl M, Lindquist N, Weisz JB, Hentschel U. Morphological and molecular analyses of microorganisms in caribbean reef adult sponges and in corresponding reproductive material. In: Custódio MR, Lôbo-Hajdu G, Hajdu E, Muricy G, editors. Porifera research: biodiversity, innovation and sustainability. Rio de Janeiro: Série Livros, Museu Nacional; 2007. p. 561–8.
100. Sacristán-Soriano O, Banaigs B, Becerro MA. Can light intensity cause shifts in natural product and bacterial profiles of the sponge *Aplysina aerophoba*? Mar Ecol. 2016;37:88–105.
101. Olson JB, Gao X. Characterizing the bacterial associates of three caribbean sponges along a gradient from shallow to mesophotic depths. FEMS Microbiol Ecol. 2013;85:74–84.
102. Thoms C, Horn M, Wagner M, Hentschel U, Proksch P. Monitoring microbial diversity and natural product profiles of the sponge *Aplysina cavernicola* following transplantation. Mar Biol. 2003;142:685–92.
103. Gerce B, Schwartz T, Voigt M, Rühle S, Kirchen S, Putz A, et al. Morphological, bacterial, and secondary metabolite changes of aplysina aerophoba upon long-term maintenance under artificial conditions. Microb Ecol. 2009;58:865–78.
104. To Isaacs L, Kan J, Nguyen L, Videau P, Anderson MA, Wright TL, et al. Comparison of the bacterial communities of wild and captive sponge clathria prolifera from the chesapeake bay. Mar Biotechnol. 2009;11:758–70.
105. Gaino E, Magnino G. Dissociated cells of the calcareous sponge clathrina: a model for investigating cell adhesion and cell motility in vitro. Microsc Res Tech. 1999;15:279–92.
106. Thoms C, Schupp PJ. Chemical defense strategies in sponges: a review. In: Custódio MR, Lôbo-Hajdu G, Hajdu E, Muricy G, editors. Porifera research: biodiversity, innovation and sustainability. Rio de Janeiro: Série Livros/Museu Nacional; 2007. p. 627–37.
107. Pérez T, Longet D, Schembri T, Rebouillon P, Vacelet J. Effect of 12 years'operation of a sewage treatment plant on trace metal occurrence within a mediterranean commercial sponge (*Spongia officinalis*, Demospongiae). Marine Poll Bull. 2005;50:301–9.

108. Patel B, Balani MC, Patel S. Sponge 'sentinel' of heavy metals. Sci Total Environ. 1985;41:143–52.
109. de Mestre C, Maher W, Roberts D, Broad A, Krikowa F, Davis AR. Sponges as sentinels: patterns of spatial and intra-individual variation in trace metal concentration. Mar Pollut Bull. 2012;64:80–9.
110. Cebrian E, Uriz MJ, Turon X. Sponges as biomonitors of heavy metals ins patial and temporal surveys in northwestern mediterranean: multispecies comparison. Environ Toxicol Chem. 2007;26:2430–9.
111. Hansen IV, Weeks JM, Depledge MH. Accumulation of copper, zinc, cadmium and chromium by the marine sponge *Halichondria panacea* pallas and the implications for biomonitoring. Marine Poll Bull. 1995;31:133–8.
112. Webster NS, Webb RI, Ridd MJ, Hill RT, Negri AP. The effects of copper on the microbial community of a coral reef sponge. Environ Microbiol. 2001;3:19–31.
113. Tian R-M, Wang Y, Bougouffa S, Gao Z-M, Cai L, Zhang W-P, et al. Effect of copper treatment on the composition and function of the bacterial community in the sponge *Haliclona cymaeformis*. MBio. 2014;5:e01980–14.
114. Santos-Gandelman JF, Cruz K, Crane S, Muricy G, Giambiagi-deMarval M, Barkay T, et al. Potential application in mercury bioremediation of a marine sponge-isolated *bacilluscereus* strain pj1. Curr Microbiol. 2014;69:374–80.
115. Bauvais C, Zirah S, Piette L, Chaspoul F, Domart-Coulon I, Chapon V, et al. Sponging up metals: bacteria associated with the marine sponge *Spongia officinalis*. Mar Environ Res. 2015;104:20–30.
116. Mangano S, Michaud L, Caruso C, Giudice AL. Metal and antibiotic resistance in psychrotrophic bacteria associated with the antarctic sponge *Hemigellius pilosus* (kirkpatrick, 1907). Polar Biol. 2014;37:227–35.
117. Wanick RC, de Sousa Barbosa H, Frazão LR, Santelli RE, Arruda MAZ, Coutinho CC. Evaluation of differential protein expression in *Haliclona aquarius* and sponge-associated microorganisms under cadmium stress. Anal Bioanal Chem. 2013;405:7661–70.
118. Selvin J, Priya SS, Kiran GS, Thangavelu T, SapnaBai N. Sponge-associated marine bacteria as indicators of heavy metal pollution. Microbiol Res. 2009;164:352–63.
119. Gochfeld DJ, Easson CG, Freeman CJ, Thacker RW, Olson JB. Disease and nutrient enrichment as potential stressors on the caribbean sponge *Aplysina cauliformis* and its bacterial symbionts. Mar Ecol Prog Ser. 2012;456:101–11.
120. Simister R, Taylor MW, Tsai P, Webster N. Sponge-microbe associations survive high nutrients and temperatures. PLoS One. 2012;7:e52220.
121. Luter HM, Gibb K, Webster NS. Eutrophication has no short-term effect on the *Cymbastela stipitata* holobiont. Font Microbiol. 2014;5:art216.
122. Hall-Spencer JM, Rodolfo-Metalpa R, Martin S, Ransome E, Fine M, Turner SM, et al. Volcanic carbon dioxide vents show ecosystem effects of ocean acidification. Nature. 2008;454:96–9.
123. Fabricius KE, Langdon C, Uthicke S, Humphrey C, Noonan S, De'ath G, et al. Losers and winners in coral reefs acclimatized to elevated carbon dioxide concentrations. Nat Clim Chang. 2011;1:165–9.
124. Morrow KM, Bourne DG, Humphrey C, Botté ES, Laffy P, Zaneveld J, et al. Natural volcanic CO_2 seeps reveal future trajectories for host-microbial associations in corals and sponges. ISME J. 2015;9:894–908.
125. Ribes M, Calvo E, Movilla J, Logares R, Coma R, Pelejero C. Restructuring of the sponge microbiome favors tolerance to ocean acidification. Environ Microbiol Rep. 2016;8:536–44.
126. Tian R-M, Lee OO, Wang Y, Cai L, Bougouffa S, Chiu JMY, et al. Effect of polybrominated diphenyl ether (PBDE) treatment on the composition and function of the bacterial community in the sponge *Haliclona cymaeformis*. Front Microbiol. 2015;5:1–8.

127. Chelossi E, Milaneseb M, Milanoc A, Pronzatob P, Riccardi G. Characterisation and antimicrobial activity of epibiotic bacteria from petrosia ficiformis (Porifera, Demospongiae). J Exp Mar Biol Ecol. 2004;309:21–33.
128. Lee OO, Qian P-Y. Chemical control of bacterial epibiosis and larval settlement of *Hydroides elegans* in the red sponge *Mycale adherens*. Biofouling. 2003;19:171–80.
129. Luter HM. The effects of disease and stress on the microbial community of the sponge *Ianthella basta*. Townsville: School of Marine and Tropical Biology, James Cook University; 2011.
130. Kaluzhnaya OV, Itskovich VB. Bleaching of baikalian sponge affects the taxonomic composition of symbiotic microorganisms. Russ J Genet. 2015;51:1153–7.
131. Denikina NN, Dzyuba EV, Bel'kova NL, Khanaev IV, Feranchuk SI, Makarov MM, et al. The first case of disease of the sponge lubomirskia baicalensis: investigation of its microbiome. Biol Bull. 2016;43:263–70.
132. Gao Z-M, Wang Y, Tian R-M, Lee OO, Wong YM, Batang ZB, et al. Pyrosequencing revealed shifts of prokaryotic communities between healthy and disease-like tissues of the red sea sponge crella cyathophora. PeeJ. 2015;3:e890.
133. Blanquer A, Uriz MJ, Cebrian E, Galand PE. Snapshot of a bacterial microbiome shift during the early symptoms of a massive sponge die-off in the western mediterranean. Front Microbiol. 2016;7:752.
134. Luter HM, Bannister RJ, Whalan S, Kutti T, Pineda M-C, Webster NS. Microbiome analysis of a disease affecting the deep-sea sponge geodia barretti. FEMS Microbiol Ecol. 2017;93:1–6.
135. McClintock JB, Baker BJ, Slattery M, Hamann M, Kopitzke R, Heine J. Chemotactic tube-foot responses of a spongivorous sea star perknaster fuscus to organic extracts from antarctic sponges. J Chem Ecol. 1994;20:859–70.
136. Wuff JL. Sponge predators may determine differences in sponge fauna between two sets of mangrove cays, Belize Barrier Reef. Atoll Res Bull. 2000;477:251–63.
137. Santos CP, Coutinho AB, Hajdu E. Spongivory by eucidaris tribuloides from Salvador, Bahia (echinodermata: Echinoidea). J Mar Biol Assoc UK. 2002;82:295–7.
138. León YL, Bjorndal KA. Selective feeding in the hawksbill turtle, an important predator in coral reef ecosystems. Mar Ecol Prog Ser. 2002;245:249–58.
139. Knowlton A, Highsmith RC. Nudibranch-sponge feeding dynamics: benefits of symbiont-containing sponge to *Archidoris montereyensis* (Cooper, 1862) and recovery of nudibranch feeding scars by *Halichondria panicea* (Pallas, 1766). J Exp Mar Biol Ecol. 2005;327:36–46.
140. Jones AC, Blum JE, Pawlik JR. Testing for defensive synergy in caribbean sponges: bad taste or glass spicules? J Exp Mar Biol Ecol. 2005;322:67–81.
141. Bertolino M, Cerrano C, Bavestrello G, Carella M, Pansini M, Calcinai B. Diversity of porifera in the mediterranean coralligenous accretions, with description of a new species. ZooKeys. 2013;336:1–37.
142. De Caralt S, Bry D, Bontemps N, Turon X, Uriz M-J, Banaigs B. Sources of secondary metabolite variation in dysidea avara (Porifera: Demospongiae): the importance of having good neighbors. Mar Drugs. 2013;11:489–503.
143. Arias J, Santos-Acevedo M, Newmark F. Evaluation of the feeding deterrent potential of crude organic extracts from fifteen marine sponges. Bol Invest Mar Cost. 2011;40:293–308.
144. Esteves AIS, Hardoim CCP, Xavier JR, Gonçalves JMS, Costa R. Molecular richness and biotechnological potential of bacteria cultured from Irciniidae sponges in the Northeast Atlantic. FEMS Microbiol Ecol. 2013;85:519–36.
145. Uriz M-J, Agell G, Blanquer A, Turon X, Casamayor EO. Endosymbiotic calcifying bacteria: a new cue to the origin of calcification in metazoa? Evolution. 2012;66:2993–9.
146. Garate L, Blanquer A, Uriz M-J. Calcareous spherules produced by intracellular symbiotic bacteria protect the sponge *Hemimycale columella* from predation better than secondary metabolites. Mar Ecol Prog Ser. 2015;523:81–92.

147. Gifford S, Dunstan RH, O'Connor W, Roberts T, Toia R. Pearl aquaculture-profitable environmental remediation? Sci Total Environ. 2004;319:27–37.
148. Osinga R, Tramper J, Wijffels RH. Cultivation of marine sponges. Mar Biotechnol. 1999;1:509–32.
149. Reiswig HM. *In situ* feeding in two shallow water Hexactinellid sponges. In: Rützler K, editor. New perspectives in sponge biology. Washington, DC: Smithsonian institute; 1990. p. 204–510.
150. Larsen PS, Riisgård HU. The sponge pump. J Theor Biol. 1994;168:3–63.
151. Riisgård HU, Larsen PS. Filter-feeding in marine macro-invertebrates: pump characteristics, modelling and energy cost. Biol Rev Camb Philos Soc. 1995;70:67–106.
152. Simpson TL. The cell biology of sponges. New York: Springer-Verlag; 1984.
153. Gifford S, Dunstan RH, O'Connor W, Koller CE, MacFarlane GR. Aquatic zooremediation: deploying animals to remediate contaminated aquatic environments. Trends Biotechnol. 2006;25:60–5.
154. Vicente VP. Response of sponges with autotrophic endosymbionts during the coral-bleaching episode in Puerto Rico. Coral Reefs. 1990;8:199–202.
155. Taylor MW, Schupp PJ, Dahllöf I, Kjelleberg S, Steinberg PD. Host specificity in marine sponge-associated bacteria, and potential implications for marine microbial diversity. Environ Microbiol. 2004;6:121–30.
156. Climate change 2001: impacts, adaptation and vulnerability. In: McCarthy, JJ, Canziani OF, Leary NA, Dokken DJ, White KS editors. Contribution of Working Group II to the third assessment report of the Intergovernmental Panel on Climate Change. Cambridge/New York: Cambridge University Press; 2001. p. 1–1033.
157. Thomas CD, Cameron A, Green RE, Bakkenes M, Beaumont LJ, Collingham YC, et al. Extinction risk from climate change. Nature. 2004;427:145–8.
158. Bellwood DR, Hughes TP, Folke C, Nystrom M. Confronting the coral reef crisis. Nature. 2004;429:827–33.
159. Sheppard CRC, Rioja-Nieto R. Sea surface temperature 1871–2099 in 38 cells in the caribbean region. Mar Environ Res. 2005;60:389–96.
160. McWilliams JP, Côté IM, Gill JA, Sutherland WJ, Watkinson AR. Accelerating impacts of temperature-induced coral bleaching in the caribbean. Ecology. 2005;86:2055–60.
161. Cerrano C, Magnino G, Sara A, Bavestrello G, Gaino E. Necrosis in a population of petrosia ficiformis (Porifera, demospongiae) in relations with environmental stress. Ital J Zool. 2001;68:131–6.
162. McMurray SE, Henkel TP, Pawlik JR. Demographics of increasing populations of the giant barrel sponge *Xestospongia muta* in the Florida keys. Ecology. 2010;91:560–70.
163. Fiore CL, Baker DM, Lesser MP. Nitrogen biogeochemistry in the caribbean sponge, *Xestospongia muta*: a source or sink of dissolved inorganic nitrogen? PLoS One. 2013;8:e72961.
164. Fiore CL, Jarett JK, Lesser MP. Symbiotic prokaryotic communities from different populations of the giant barrel sponge, *Xestospongia muta*. Microbiology Open. 2013;2:938–52.
165. Lesser MP, Fiore C, Slattery M, Zaneveld J. Climate change stressors destabilize the microbiome of the caribbean barrel sponge, *Xestospongia muta*. J Exp Mar Biol Ecol. 2016;475:11–8.
166. Reilly A, Kaferstein F. Food safety hazards and the application of the principles of the hazard analysis and critical control point (HACCP) system for their control in aquaculture production. Aquac Res. 1997;28:735–52.
167. Reiswig HM. *In situ* pumping activities of tropical demospongiae. Mar Biol. 1971;9:38–50.
168. Vacelet J, Donadey C. Electron microscope study of the association between some sponges and bacteria. J Exp Mar Biol Ecol. 1977;30:301–14.
169. Gaino E, Pronzato R. Ultrastructural evidence of bacterial damage to *Spongia officinalis* (Porifera, Demospongiae) fibres. Dis Aquat Org. 1989;6:67–74.

170. Vacelet J, Vacelet E, Gaino E, Gallissian M-F. Bacterial attack of spongin skeleton during the 1986–1990 mediterranean sponge disease. In: van Soest RWM, van Kempen TMG, Braekman JC, editors. Sponges in time and space. Rotterdam: Balkema AA; 1994. p. 355–62.
171. Kefalas E, Castritsi-Catharios J, Miliou H. Bacteria associated with the sponge *Spongia officinalis* as indicators of contamination. Ecol Indic. 2003;2:339–43.
172. Velho-Pereira S, Furtado I. Retrieval of euryhaline eubacterial and haloarchaeal bionts from nine different benthic sponges: reflection of the bacteriological health of waters of mandapam, India. Indian J Mar Sci. 2014;43:773–83.
173. Milanese M, Chelossi E, Manconi R, Sarà A, Sidri M, Pronzato R. The marine sponge chondrilla nucula schmidt, 1862 as an elective candidate for bioremediation in integrated aquaculture. Biomol Eng. 2003;20:363–8.
174. Stabilia L, Liccianoa M, Giangrandea A, Longoc C, Mercurioc M, Marzanoc CN, et al. Filtering activity of *Spongia officinalis* var. *adriatica* (schmidt) (Porifera, Demospongiae) on bacterioplankton: implications for bioremediation of polluted seawater. Water Res. 2006;40:3083–90.
175. Xue L, Zhang X, Zhang W. Larval release and settlement of the marine sponge *Hymeniacidon perlevis* (Porifera, Demospongiae) under controlled laboratory conditions. Aquaculture. 2009;290:132–9.
176. de Voogd NJ, Cleary DFR, Polónia ARM, Gomes NCM. Bacterial community composition and predicted functional ecology of sponges, sediment and seawater from the thousand islands reef complex, west java, Indonesia. FEMS Microbiol Ecol. 2015;91:1–12.
177. Webster NS. Sponge disease: a global threst? Environ Microbiol. 2007;9:1363–75.
178. Gaino E, Pronzato R, Corriero G, Buffa P. Mortality of commercial sponges: incidence in two mediterranean areas. Ital J Zool. 1992;59:79–85.
179. Gaino E, Pronzato R. Epidemie e pesca intensive minacciano la sopravvirenza delle spugne commeciali del bacino mediterraneo. Bollettino dei Musei e degli Istituti Biologici dell'Universita di Genova. 1992;56:209–24.
180. Vacelet J. The struggle against the epidemic which is decimating mediterranean sponges. Technical report. Rome; 1994, FI:TCP/RAB/8853.
181. Castritsi-Catharios J, Kefalas E. Spreading of the sponge disease in the eastern mediterranean. In: 2nd workshop on sponge disease 'Mass mortality of mediterranean sponges'. University of Genova, Italy; 1999.
182. Fang JKH, Mello-Athayde MA, Schönberg CHL, Kline DI, Hoegh-Guldberg O, Dove S. Sponge biomass and bioerosion rates increase under ocean warming and acidification. Glob Chang Biol. 2013;19:3581–91.
183. Webster N, Pantile R, Botté E, Abdo D, Andreakis N, Whalan S. A complex life cycle in a warming planet: gene expression in thermally stressed sponges. Mol Ecol. 2013;22:1854–68.
184. Angermeier H, Glöckner V, Pawlik JR, Lindquist NL, Hentschel U. Sponge white patch disease affecting the caribbean sponge *Amphimedon compressa*. Dis Aquat Org. 2012;99:95–102.
185. Negri AP, Soo RM, Flores F, Webster NS. *Bacillus* insecticides are not acutely harmful to corals and sponges. Mar Ecol Prog Ser. 2009;381:157–65.
186. Luter HM, Whalan S, Webster NS. Prevalence of tissue necrosis and brown spot lesions in a common marine sponge. Mar Freshw Res. 2010;61:484–9.
187. Angermeier H, Kamke J, Abdelmohsen UR, Krohne G, Pawlik JR, Lindquist NL, et al. The pathology of sponge orange band disease affecting the caribbean barrel sponge *Xestospongia muta*. FEMS Microbiol Ecol. 2011;75:218–30.
188. Galstoff PS. Wasting disease causing mortality of sponges in the west indies and gulf of Mexico. Proc VIII Am Sci Congr. 1942;3:411–21.
189. Gammill ER, Fenner D. Disease threatens caribbean sponges: report and identification guide. ReefBase: A Global Information System for Coral Reefs; 2005. Available at http://www.reefbase.org/spongedisease/

190. Luter HM, Whalan S, Webster NS. Thermal and sedimentation stress are unlikely causes of brown spot syndrome in the coral reef sponge, *Ianthella basta*. PLoS One. 2012;7:e39779.
191. Webster NS, Negri AP, Webb RI, Hill RT. A spongin-boring α-proteobacterium is the etiological agent of disease in the great barrier reef sponge *Rhopaloeides odorabile*. Mar Ecol Prog Ser. 2002;232:305–9.
192. Brock TD. Milestones in microbiology. Englewood Cliffs: Prentice Hall; 1961.
193. Mukherjee J, Webster N, Llewellyn LE. Purification and characterization of a collagenolytic enzyme from a pathogen of the great barrier reef sponge, *Rhopaloeides odorabile*. PLoS One. 2009;4:e7177.
194. Choudhury JD, Pramanik A, Webster NS, Llewellyn LE, Gachhui R, Mukherjeea J. Draft genome sequence of *Pseudoalteromonas* sp. Strain nw 4327 (MTCC 11073, DSM 25418), a pathogen of the great barrier reef sponge *Rhopaloeides odorabile*. Genome Announc. 2014;2:e00001–14.
195. Choudhury JD, Pramanik A, Webster NS, Llewellyn LE, Gachhui R, Mukherjee J. The pathogen of the great barrier reef sponge *Rhopaloeides odorabile* is a new strain of *Pseudoalteromonas agarivorans* containing abundant and diverse virulence-related genes. Mar Biotechnol. 2015;17:463–78.
196. Sweet M, Bulling M, Cerrano C. A novel sponge disease caused by a consortium of micro-organisms. Coral Reefs. 2015;34:871–83.
197. Rützler K. Mangrove sponge disease induced by cyanobacterial symbionts: failure of a primitive immune system? Dis Aquat Org. 1988;5:143–9.
198. Butler MJ, Hunt JH, Herrnkind WF, Childress MJ, Bertelsen R, Sharp W. Cascading disturbances in florida bay, USA: cyanobacteria blooms, sponge mortality, and implications for juvenile spiny lobsters panulirus argus. Mar Ecol Prog Ser. 1995;129:119–25.
199. Vacelet J, Gallissian M-F. Virus-like particles in cells of the sponge *Verongia cavernicola* (Demospongiae, Dictyoceratida) and accompanying tissue changes. J Invertebr Pathol. 1978;31:246–54.
200. Olson JB, Gochfeld DJ, Slattery M. *Aplysina* red band syndrome: a new threat to caribbean sponges. Dis Aquat Org. 2006;72:163–8.
201. Gochfeld DJ, Schlöder C, Thacker RW. Sponge community structure and disease prevalence on coral reefs in Bocas del Toro, Panama. In: Custódio MR, Lôbo-Hajdu G, Hajdu E, Muricy G, editors. Porifera research: biodiversity, innovation and sustainability. Rio de Janeiro: Série Livros, Museu Nacional; 2007. p. 335–44.
202. Easson CG, Slattery M, Momm HG, Olson JB, Thacker RW, Gochfeld DJ. Exploring individual- to population-level impacts of disease on coral reef sponges: using spatial analysis to assess the fate, dynamics, and transmission of *Aplysina* red band syndrome (ARBS). PLoS One. 2013;8:e79976.
203. Cowart JD, Henkel TP, McMurray SE, Pawlik JR. Sponge orange band (SOB): a pathogenic-like condition of the giant barrel sponge *Xestospongia muta*. Coral Reefs. 2006;25:513.
204. Stevely JM, Sweat DE, Bert TM, Sim-Smith C, Kelly M. Sponge mortality at Marathon and Long key, Florida: patterns of species response and population recovery. Proceeding of the 63rd Gulf and Caribbean Fisheries Institute. San Juan, Puerto Rico; 2011. p. 384–400.
205. Luter HM, Whalan S, Webster NS. The marine sponge *Ianthella basta* can recover from stress-induced tissue regression. Hydrobiologia. 2012;687:227–35.
206. Di Camillo CG, Bartolucci I, Cerrano C, Bavestrello G. Sponge disease in the adriatic sea. Mar Ecol. 2013;34:62–7.
207. Carter HJ. Parasites of the spongida. Ann Mag Nat Hist. 1878;2:157–72.
208. Brice JJ. The fish and fisheries of the coastal waters of Florida. US Bur Fish Rept Comm Fish. 1896;22:263–342.
209. Allemand-Martin A. Étude de physiologie appliquÉe sur la spongiculture sur le côtes de tunisie. In: Lyon; 1906.
210. Allemand-Martin A. Contribution à l'Étude de la culture des Éponges. Cr Ass Advmt Sci Tunis. 1914;42:375–7.

211. Storr JF. Ecology of the gulf of Mexico commercial sponges and its relation to the fishery. US Fish Wildl Serv Spec Scient Rep. 1964;466:1–73.
212. Galstoff PS, Brown HH, Smith CL, Walton Smith FG. Sponge mortality in the Bahamas. Nature. 1939;143:807–8.
213. Smith FGW. Sponge disease in British Honduras, and its transmission by water currents. Ecology. 1941;22:415–21.
214. Smith FGW. Sponge mortality at British Honduras. Nature. 1939;143:785.
215. Stachowitsch M. Mass mortality in gulf of Trieste: the course of community destruction. Mar Ecol. 1984;5:243–64.
216. Wulff JL. Sponge systematics by starfish: predators distinguish cryptic sympatric species of caribbean fire sponges, *Tedania ignis* and *Tedania klausi* n. sp. (Demospongiae, Poecilosclerida). Biol Bull. 2006;211:83–94.
217. Wulff JL. Ecological interactions of marine sponges. Can J Zool. 2006;84:146–66.
218. Economou E, Konteatis D. Information on the sponge disease of 1986 in the waters of cyprus. Report of department of fisheries, ministry of agriculture and natural resources. Republic of Cyprus, Ministry of Agriculture and Natural Resources, Cyprus; 1988.
219. Rizzello R, Corriero G, Scalera-Liaci L, Pronzato R. Extinction and recolonization of spongia officinalis in the Marsala lagoon. Biol Mar Mediterr. 1997;4:443–4.
220. Gashout SF, Haddud DA, El-Zintani AA, Elbare RMA. Evidence for infection of libyan sponge grounds. In: International seminar on the combat of pollution and the conservation of marine wealth in the Mediterranean sea. Gulf of Sirte: Marine Biological Resources Centre; 1989. p. 100–13.
221. Turon X, Tarjuelo I, Uriz MJ. Growth dynamics and mortality of the encrusting sponge *Crambe crambe* (Poecilosclerida) in contrasting habitats: correlation with population structure and investment in defence. Funct Ecol. 1998;12:631–9.
222. Corriero G, Scalera-Liaci L, Rizzello R. Osservazioni sulla mortalita *Diircinia spinosula* (schmidt) and *Ircinia* sp. (Porifera, Demospongiae) nell′ insenatura della strea di porto cesareo. Thalassia Salent. 1996;22:51–62.
223. Paz M. New killer disease attacks giant barrel sponge; 1997. Available at: http://sanpedrosun.net/old/sponge.html
224. Cervino JM, Winiarski-Cervino K, Polson SW, Goreau T, Smith GW. Identification of bacteria associated with a disease affecting the marine sponge *Ianthella basta* in New Britain, Papua New Guinea. Mar Ecol Prog Ser. 2006;324:139–50.
225. Nagelkerken I, Aerts L, Pros LPJJ. Barrel sponge bows out. Reef Encounter. 2000;28:14–5.
226. Cerrano C, Magnino G, Sarà A, Bavestrello G, Gaino E. Necrosis in a population of *Petrosia ficiformis* (Porifera, Demospongiae) in relation with environmental stress. Ital J Zool. 2001;68:131–6.
227. Wulff JL. A simple model of growth form-dependent recovery from disease in coral reef sponges, and implications for monitoring. Coral Reefs. 2006;25:419–26.
228. Wulff JL. Disease prevalence and population density over time in three common caribbean coral reef sponge species. J Mar Biol Assoc UK. 2007;87:1715–20.
229. Cerrano C, Bavestrello G, Bianchi CK, Cattaneo-vietti R, Bava S, Morganti C, et al. A catastrophic mass-mortality episode of gorgonians and other organisms in the ligurian sea (northwestern mediterranean), summer 1999. Ecol Lett. 2000;3:284–93.
230. Böhm M, Hentschel U, Friedrich AB, Fiesler L, Steffen R, Gamulin V, et al. Molecular response of the sponge suberites domuncula to bacterial infection. Mar Biol. 2001;139:1037–45.
231. Skoufas G. Massive necrosis of sedentary benthic animal organisms in the north aegean sea. 7th Hellenic symposium on oceanography and fisheries. Chersonissos: National Centre for Marine Research; 2003.
232. Garrabou J, Coma R, Bensoussan N, Bally M, Chevaldonné P, Cigliano M, et al. Mass mortality in northwestern mediterranean rocky benthic communities: effects of the 2003 heat wave. Glob Chang Biol. 2009;15:1090–103.

233. Castritsi-Catharios J, Miliou H, Kapiris K, Kefalas E. Recovery of the commercial sponges in the central and southeastern aegean sea (NE Mediterranean) after an outbreak of sponge disease. Mediterr Mar Sci. 2011;12:5–20.
234. Di Camillo CG, Cerrano C. Mass mortality events in the NW Adriatic Sea: phase shift from slow- to fast-growing organisms. PLoS One. 2015;10:e0126689.
235. Cerrano C, Bavestrello G. Mass mortalities and extinctions. In: Wahl M, editor. Marine hard bottom communities ecological studies 206. Berlin/Heidelberg: Springer; 2009. p. 295–307.
236. Diaz MC, Ruetzler K. Sponges: an essential component of caribbean coral reefs. Bull Mar Sci. 2001;69:535–46.
237. Wulff J. Assessing and monitoring coral reef sponges: why and how? Bull Mar Sci. 2001;69:831–46.
238. Hayes RL, Goreau NI. The significance of emerging diseases in the tropical coral reef ecosystem. Rev Biol Trop. 1998;5:173–85.
239. Goreau TJ, Cervino J, Goreau M, Hayes R, Hayes M, Richardson L. Rapid spread of diseases in caribbean coral reefs. Rev Biol Trop. 1998;46:157–71.
240. Williams JEH, Bunkley-Williams L. Marine major ecological disturbances of the Caribbean. J Infect Dis Rev. 2000;2:110–27.
241. Reiswig HM. Population dynamics of three jamaican demospongiae. Bull Mar Sci. 1973;23:191–226.
242. Hummel H, Sepers ABJ, de Wolf L, Melissen FW. Bacterial growth on the marine sponge *Halichondria panicea* induced by reduced waterflow rate. Mar Ecol Prog Ser. 1988;42:195–8.
243. Fedra K, Ölscher EM, Scherübel C, Stachowitsch M, Wurzian RS. On the ecology of a north adriatic benthic community: distribution, standing crop and composition of the macrobenthos. Mar Biol. 1976;38:129–45.
244. Stachowitsch M. Anoxia in the northern adriatic sea: rapid death, slow recovery. In: Tyson RV, Pearson TH, editors. Modern and ancient continental shelf anoxia. London: Geological Society, Special Publications; 1991. p. 119–29.
245. Zocchi E, Basile G, Cerrano C, Bavestrello G, Giovine M, Bruzzone S, et al. Aba- and cadpr-mediated effects on respiration and filtration downstream of the temperature-signaling cascade in sponges. J Cell Sci. 2003;116:629–36.
246. Previati M, Scinto A, Cerrano C, Osinga R. Oxygen consumption in mediterranean octocorals under different temperatures. J Exp Mar Biol Ecol. 2010;390:39–48.
247. Bavestrello G, Puce S, Cerrano C, Zocchi E, Boero N. The problem of seasonality of benthic hydroids in temperate waters. Chem Ecol. 2006;22:197–205.
248. Vezzulli L, Previati M, Pruzzo C, Marchese A, Bourne DG, Cerrano C, et al. *Vibrio* infections triggering mass mortality events in a warming mediterranean sea. Environ Microbiol. 2010;12:2007–9.
249. Torrents O, Tambutté E, Caminiti N, Garrabou J. Upper thermal thresholds of shallow vs. deep populations of the precious mediterranean red coral *Corallium rubrum* (l.): assessing the potential effects of warming in the nw mediterranean. J Exp Mar Biol Ecol. 2008;357:7–19.
250. Kelly-Borges M, Bergquist PR. Sponges from motupore island, Papua New Guinea. Indo-Malayan Zool. 1988;5:121–59.
251. Vincente VP. An ecological evaluation of the west indian demosponge *Anthosigmella varians* (Hadromerida, Spriastrellidae). Bull Mar Sci. 1978;28:771–7.
252. Pronzato R. Sponge-fishing, disease and farming in the mediterranean sea. Aquat Conserv Mar Freshwat Ecosyst. 1999;9:485–93.
253. Castritsi-catharios J, Miliou H, Pantelis J. Experimental sponge fishery in Egypt during recovery from sponge disease. Aquat Conserv Mar Freshwat Ecosyst. 2005;15:109–16.
254. Corriero G, Longo C, Mercurio M, Marzano CN, Lembo G, Spedicato MT. Rearing performance of *Spongia officinalis* on suspended ropes off the southern italian coast (Central Mediterranean Sea). Aquaculture. 2004;238:195–205.

255. Rützler K, Smith KP. Guide to the Western Atlantic species of *Cinachyrella* (Porifera: Tetillidae). In: Proceedings of the biological Society of Washington; 1992.
256. Butler MJI. Algae bloom impacts on hard bottom communities. In: Donahue S, editor. Algae bloom workshop: re-evaluation of management needs in Florida Bay. Key West: Florida Keys National Marine Sanctuary Program; 2008. p. 10–1.
257. Wulff J. Regeneration of sponges in ecological context: is regeneration an integral part of life history and morphological strategies? Integr Comp Biol. 2010;50:494–505.
258. Waddell B, Pawlik JR. Defenses of caribbean sponges against invertebrate predators. I. Assays with hermit crabs. Mar Ecol Prog Ser. 2000;195:125–32.
259. Walters KD, Pawlik JR. Is there a trade-off between wound-healing and chemical defenses among caribbean reef sponges? Integr Comp Biol. 2005;45:352–8.
260. Leong W, Pawlik JR. Fragments or propagules? Reproductive tradeoffs among *Callyspongia* spp. from Florida coral reefs. Oikos. 2009;119:1417–22.
261. DeBiasse MB, Richards VP, Shivji MS. Genetic assessment of connectivity in the common reef sponge, *Callyspongia vaginalis* (Demospongiae: Haplosclerida) reveals high population structure along the Florida reef tract. Coral Reefs. 2010;29:47–55.

Chapter 12
Biosynthesis of Antibiotics from Microbial Symbionts of Sponges and Corals

Loganathan Karthik and Zhiyong Li

Contents

Abstract Sponges and corals are significant sources for marine natural products. They have a pool of novel microorganisms. Due to low cost of gene sequencing, in recent years, several reports are available for novel compounds from sponge- and coral-associated microorganisms. Still, most of the biosynthesis mechanisms are not revealed. There are only few reports on the biosynthesis mechanism of antibiotics from sponge- and coral-associated microorganisms. The scanty amount of antibiotic was produced by most of the strains; hence it is important to explore the biosynthesis of antibiotics to improve the production. In this chapter, we cover the important reports of biosynthesis of antibiotics from microbial symbionts especially sponges and corals.

Keywords Sponges · Corals · Symbionts · Biosynthesis · Antibiotics

L. Karthik · Z. Li (✉)
Marine Biotechnology Laboratory, State Key Laboratory of Microbial Metabolism, School of Life Sciences and Biotechnology, Shanghai Jiao Tong University, Shanghai, People's Republic of China
e-mail: zyli@sjtu.edu.cn

Z. Li (ed.), *Symbiotic Microbiomes of Coral Reefs Sponges and Corals*,
https://doi.org/10.1007/978-94-024-1612-1_12

12.1 Marine Invertebrates: A Treasure of Antibiotics

Due to a wide chemical diversity and potential activity, the natural products are best-selling drugs in clinical use [1]. The best example is Taxol (1, paclitaxel); it is isolated from the bark of the Pacific yew tree, *Taxus brevifolia*. David Yuon said "Since 2006, the annual total sales of the natural raw materials paclitaxel injection and semi-synthetic paclitaxel injection paclitaxel reached $ 3.7 billion in international market" (http://www.articlesfactory.com/articles/marketing/paclitaxel-ranks-first-among-worlds-anti-cancer-drugs.html) [2].

The earth's surface covers ca. 70% of seawater with a wide biodiversity potential. So far around 100,000 species was reported in the world's ocean and day-by-day lot of new species is reporting by researchers. The phyla *Bryozoa*, *Coelenterata*, *Porifera*, and *Echinodermata* are exist only in aquatic region. These invertebrates don't have any physical protection (shells or spines), but still they are fighting against predators using biologically active secondary metabolites [3]. Previously, it was taught that the marine invertebrates only produce secondary metabolites, but now it's reported that the invertebrate-associated microbiome is a key producer of active secondary metabolites. Several new lead molecules have been discovered from marine invertebrates such as diterpene glycoside; eleutherobin, from the Australian soft coral *Eleutherobia* sp. [4]; discodermolide from the Caribbean sponge *Discodermia dissoluta* [5].

12.2 Important Enzymes Involved in the Biosynthesis of Secondary Metabolites

12.2.1 Polyketide Synthases (PKSs)

It is a multi-domain enzyme responsible for synthesis of polyketide compounds. Polyketide biosynthesis is similar to the fatty acid biosynthesis [6, 7]. It is classified into three types:

A. PKS-I
A.1. Iterative PKS s
A.1.1. Nonreducing PKSs (NR-PKSs)
A.1.2. Partially reducing PKSs (PR-PKSs)
A.1.3. Fully reducing PKSs (FR-PKSs)

A.2. Modular PKSs
B. PKS-II
C. PKS-III

PKS-I has several domains with known functions. It has three important modules such as starting module (AT-ACP), elongation modules (KS-AT-[DH-ER-KR]-ACP), and termination module (TE). The domains are acyltransferase (AT), acyl carrier protein (ACP), ketosynthase (KS), ketoreductase (KR), dehydratase (DH), enoylreductase (ER), methyltransferase (MT), sulfhydrolase (SH), and thioesterase (TE).

Polyketides are group of secondary metabolites which have a unique structure and function. Their biological activities includes antimicrobial, antiparasitic, antitumor, etc. Examples for known polyketides include erythromycin A, avermectin, rifamycin, and lovastatin [8].

12.2.2 Nonribosomal Peptide Synthetases (NRPSs)

It is a multimodular enzyme, capable of synthesizing the nonribosomal peptide molecule, independent of ribosomal machinery. It is a class of secondary metabolites with a wide range of properties, like toxins, siderophores, pigments, antibiotics, etc. This enzyme is located at operon; hence its transcriptional or posttranscriptional regulation can be positive or negative [9, 10].

A peptide acts as a backbone in the amino acids inserted in a systemic manner by NRPS enzyme. Further, the module converted as domains leads to the nonribosomal peptide synthesis. One module contains three domains [11]:

1. Adenylation (A) domain
2. Peptidyl carrier protein (PCP) or thiolation (T) domain
3. Condensation (C) domain

The reaction is N- to C-terminal direction. The final peptide product size is 3–15 amino acids length; it is linear, cyclic, or branched cyclic form [12]. Figure 12.1 explains the biosynthesis of surfactin by NRPS enzyme [13].

12.2.3 Ribosomally Synthesized and Posttranslationally Modified Peptides (RiPPs)

It is recently identified as the major class of secondary metabolites with a wide variety of structural diversity due to extensive posttranslational modifications (PTMs) [14, 15]. The PTMs will play a major role in RiPP synthesis via expanded chemical functionalities, improved target recognition, and increased metabolic and chemical stability [14]. Figure 12.2 explains the general biosynthetic pathway of RiPPs, and Fig. 12.3 explains the biosynthesis of nisin A. Recently Tietz et al. [17] developed a new software called as RODEO (Rapid ORF Description and Evaluation Online) to identify the RiPP precursor peptides.

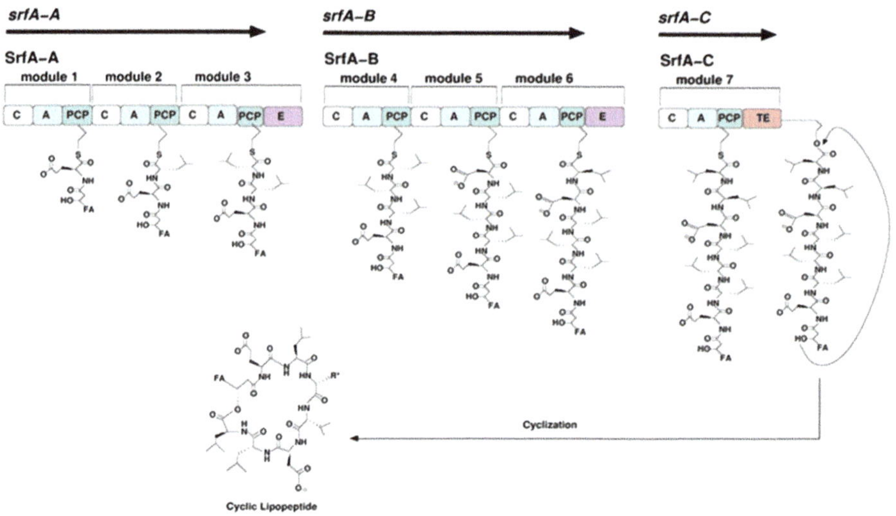

Fig. 12.1 Biosynthesis of surfactin [13]

Fig. 12.2 General biosynthetic pathway of RiPPs

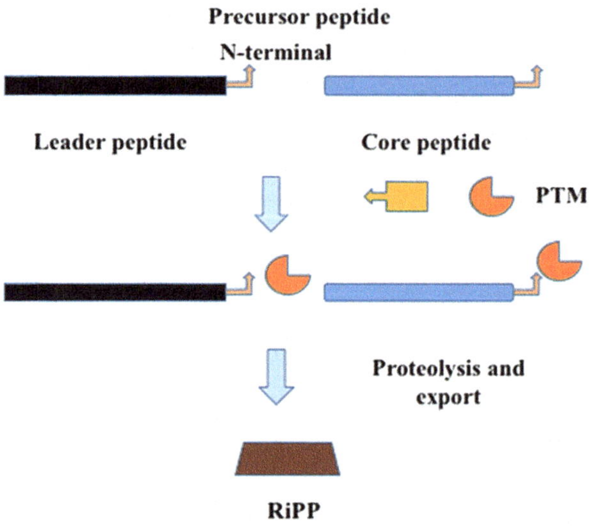

12.3 Revolution of Methods to Study the Biosynthesis of Natural Products

The classical method to study the biosynthesis of natural products is chemical degradation. In this method, compounds must be fully synthesized to assign the structure of compound. This method is vanished today because new technique arrivals.

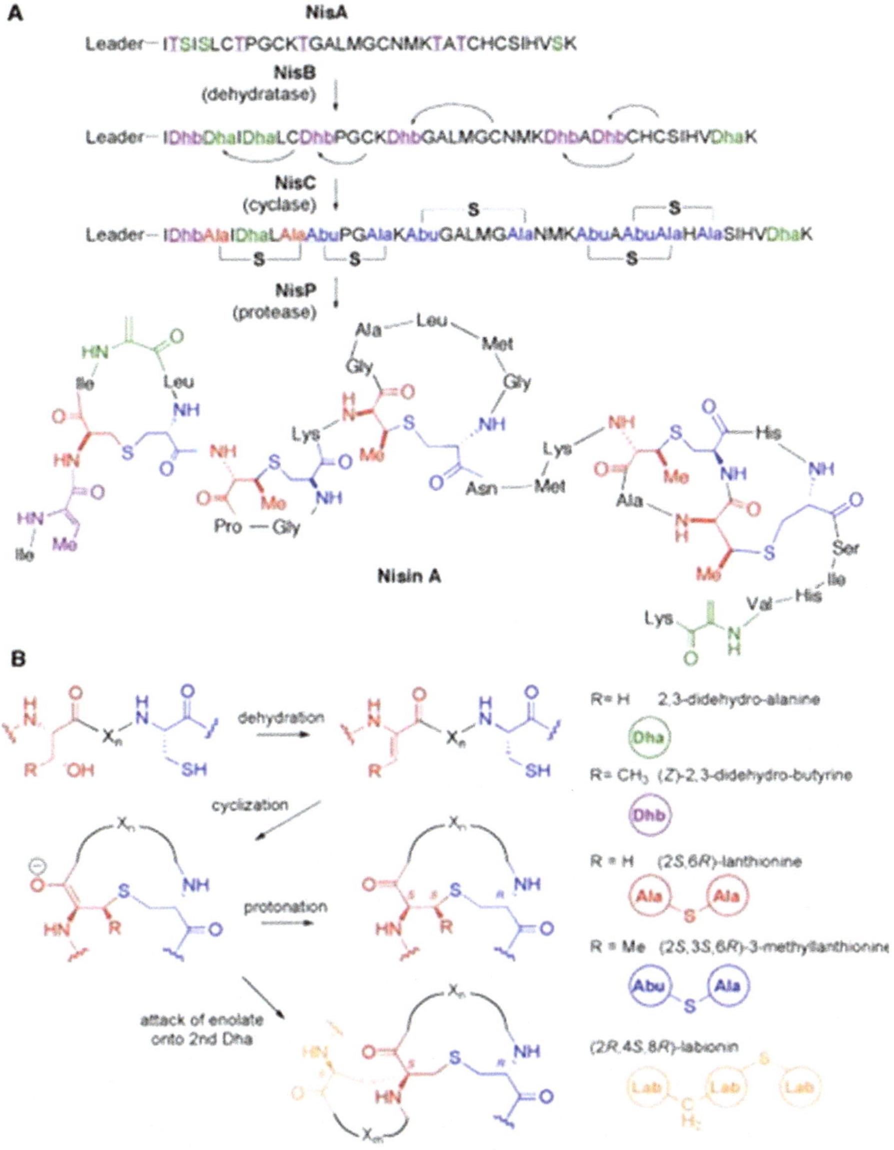

Fig. 12.3 (**a**) Biosynthesis of nisin A. (**b**) Generation of (Me)Lan and labionin motifs [16]

After the discovery of isotopes, the dimension of biosynthesis research was changed [18], and it led to the discovery of cholesterol biosynthesis [19].

The first metabolite investigated by isotope was polyketide compound [20]. The isotopes (1,2-^{13}C2) acetate and (1-^{13}C) or (2-^{13}C) acetate are sources for acetate units, chain direction, and modifications of PKS-derived natural products [21]. Bode et al. (2012) developed a method combination of isotope labeled with the

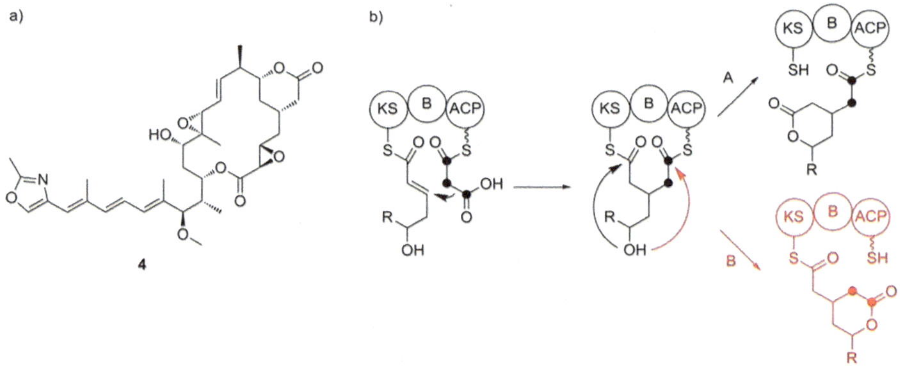

Fig. 12.4 Biosynthesis of rhizoxin using ^{13}C-labeled carbons [23]

bacterial strains and its transaminase mutants followed by MS analysis. Using this strategy, GameXPeptides, novel cyclopeptide structure was predicted in crude extract itself [22]. Figure 12.4 is an example for ^{13}C-labeled carbon used to study the biosynthesis of rhizoxin [23].

These techniques were used for terrestrial microbes and plants. But after the revolution of molecular biology techniques, the biosynthesis study went next level. The advantage and cost of DNA sequencing, lots of database with protein and gene information, and other advances in biological field expand our understanding of the biosynthesis of marine molecules [24].

The molecular techniques revealed the marine natural product "Dogma" from gene to products. It also differentiates the key steps and biosynthetic pathway that leads to the diverse structural diversity of marine natural products [25–27]. The molecular techniques only gave a clear picture on natural products from microbial origin, not from macroorganisms [28]. At first, genetic-level marine natural product biosynthesis was explained in actinomycetes and cyanobacteria. In 2000, Piel et al. explained the first marine actinomycete natural product biosynthesis (enterocins and wailupemycins). These compounds were isolated from *Streptomyces maritimus* (marine sediment) (Fig. 12.5) [29].

12.4 Biosynthetic Potential of Sponge-Associated Microbes

Most marine natural products are isolated from the marine sponges when compared to the other marine invertebrates [30]. So far, the natural products were reported from the class Demospongiae and particularly the 3 orders *Halichondrida*, *Poecilosclerida*, and *Dictyoceratida*. Many studies have proved that sponge-associated microbes are responsible for most of the natural product synthesis instead of sponges. In these, *Actinobacteria* and fungal division *Ascomycota* were potential producers of drugs. Sponges harbor large amount of gene diversity due to the

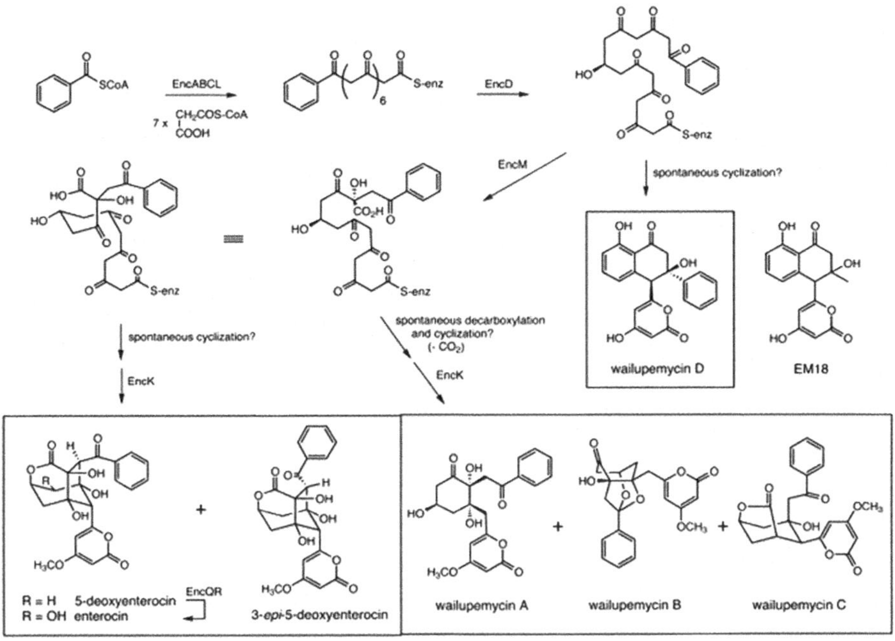

Fig. 12.5 Biosynthetic pathway of enterocins and wailupemycins [29]

localization of specific biosynthetic gene sequences [31]. Table 12.1 summarizes the list of marine natural product biosynthesis identified and characterized.

In microbial biodiversity, ca. 99% of microorganisms are unculturable. The novel chemical entity will be discovered followed by an identification of novel bacterial species. In the future, using in situ cultivation methods and growth factor for unculturable microbes, we can discover new secondary metabolites [62].

So far only five biosynthetic gene clusters were identified from uncultured microorganisms associated with marine organisms such as psymberin from uncultivated prokaryotic symbiont of *Psammocinia* aff. *bulbosa* (sponge), bryostatin from uncultivated prokaryotic symbiont of *Bugula neritina* (bryozoan), patellamide from *Prochloron didemni*, uncultivated cyanobacterial symbiont of *Lissoclinum patella* (ascidian), and onnamide/theopedrin from uncultivated prokaryotic symbiont of *Theonella swinhoei* (sponge). Out of five, three biosynthetic gene clusters were reported from sponge symbionts.

In 2004, Piel et al. reported the first genetic evidence of natural products from uncultured sponge-associated microbes. The PKS-NRPS hybrid gene was responsible for biosynthesis of onnamide/theopedrin which was isolated from *Theonella swinhoei* (sponge). The compound structure is similar with pederin which is originally isolated from *Paederus fuscipes*. Piel's group identified the PKS gene responsible for the onnamides and theopederins from a complex metagenome. These two compounds' gene clusters have unique property compared to pederin gene cluster. The pederin gene-encoding type I PKS megasynthases don't have a sequence of acyltransferase (AT) domains, but it is present in other two gene clusters (Fig. 12.6) [57].

Table 12.1 List of marine natural products' gene identified and characterized (2000–2016) [31]

Year published	Molecule	Organism	Molecule type class	References
2016	Thalassospiramide lipopeptides	*Rhodospirillaceae* strains	PKS-NRPS hybrid	Zhang et al. 2016 [32]
2016	Ammosamides A–C, pyrroloquinoline alkaloids	*Streptomyces* sp. CNR-698	NRPS	Jordan and Moore 2016 [33]
2016	Tetrabromopyrrole	*Pseudoalteromonas* sp.	Halogenase	Gamel et al. 2016 [34]
2015	Unusual thiotetronic acid	*Salinispora*	PKS-NRPS hybrid	Tang et al. 2015 [35]
2014	Polybrominated diphenyl ethers	*Pseudoalteromonas* spp.	Halogenase	Agarwal et al. 2014 [36]
2013	Novel cyanosporasides C–F	*Salinispora pacifica* CNS-143 and *Streptomyces* sp. CNT-179	PKS	Lane et al. 2013 [37]
2013	Thalassospiramide C Thalassospiramide F	Marine α-proteobacterium *Thalassospira* sp. CNJ-328	PKS-NRPS hybrid	Ross et al. 2013 [38]
2012	Didemnin	Marine α-proteobacteria *Tistrella mobilis*	PKS-NRPS hybrid	Xu et al. 2012 [39]
2011	Ansalactam A	*Streptomyces* sp.	PKS	Wilson et al. 2011 [40]
2010	ML-449	*Streptomyces* sp.	PKS	[41]
2010	Rifamycin/saliniketal	*Salinispora arenicola*	PKS	[42]
2010	Tirandamycin	*Streptomyces* sp.	PKS/NRPS	[43]
2010	TP-1161	*Streptomyces* sp.	Ribosomal peptide	[44]
2009	BE-14106	*Streptomyces* sp.	PKS	[45]
2009	Psymberin	Uncultivated prokaryotic symbiont of *Psammocinia* aff. *bulbosa* (sponge)	PKS	[46]
2008	Cyclomarin/ cyclomarazine	*Salinispora arenicola*	NRPS	[47]
2008	Napyradiomycin	*Streptomyces aculeolatus* NRRL 18422 and CNQ525	Polyketide/ terpenoid	[48]
2007	Bryostatin	Uncultivated prokaryotic symbiont of *Bugula neritina* (bryozoan)	PKS	[49]
2007	Hectochlorin	*Lyngbya majuscula*	PKS/NRPS	[50]
2007	Salinosporamide	*Salinispora tropica*	PKS/NRPS	[51]
2007	Sporolide	*Salinispora tropica*	Polyketide (enediyne)	[51]
2005	Patellamide	*Prochloron didemni*, uncultivated cyanobacterial symbiont of *Lissoclinum patella* (ascidian)	Ribosomal peptide	[52]

(continued)

Table 12.1 (continued)

Year published	Molecule	Organism	Molecule type class	References
2004	Curacin	*Lyngbya majuscula*	PKS/NRPS	[53]
2004	Jamaicamide	*Lyngbya majuscula*	PKS/NRPS	[54]
2004	Lyngbyatoxin	*Lyngbya majuscula*	Nonribosomal peptide/ terpenoid	[55]
2004	Nodularin	*Nodularia spumigena*	PKS/NRPS	[56]
2004	Onnamide/theopedrin	Uncultivated prokaryotic symbiont of *Theonella swinhoei* (sponge)	PKS/NRPS	[57]
2003	Barbamide	*Lyngbya majuscula*	PKS/NRPS	[58]
2002	Eicosapentaenoic acid	*Photobacterium profundum*	PKS	[59]
2002	Griseorhodin	*Streptomyces* sp.	PKS	[60]
2000	Docosahexaenoic acid	*Moritella marina*	PKS	[61]
2000	Enterocin/ wailupemycin	*Streptomyces maritimus*	PKS	[29]

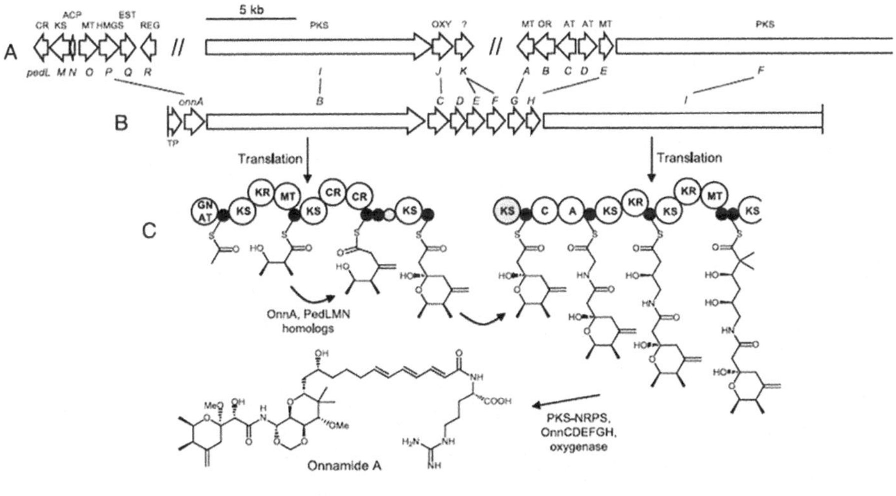

Fig. 12.6 Biosynthesis of onnamide A [57]

Until the microbes grow in laboratory condition, the natural products from microbial origin is remain a mystery, but the molecular evidence support the microbial biosynthesis. Recently, psymberin metabolites were reported from uncultured microbial symbiont of the sponge *Psammocinia* aff. *bulbosa*. Its structure analogs and genes are significantly similar to the theopederins, onnamides, and pederins [46, 63]. These findings show the importance of polyketide biosynthesis of prokaryotic symbionts.

12.5 Future Perspectives

Sponge- and coral-associated symbionts can produce diverse group of secondary metabolites, but its biosynthesis process was not explored fully. In recent decade, due to the revolution of genome mining and other molecular techniques, the marine natural product biosynthesis was revealed. In the future, direct cloning and heterologous expression of large biosynthetic pathways will lead to the next level of biosynthesis study. Recently, several studies showed the heterologous expression of biosynthetic gene cluster of actinomycetes [64, 65]. The discovery of biosynthetic potential of marine microbes will be the new era in pharmaceutical industry. Only less reports are available on biosynthetic potential of marine microbes, in the case of sponge- and coral-associated microbes. Hence, the researchers should turn their attention toward the biosynthesis of antibiotics from sponge- and coral-associated microbes.

Acknowledgments We gratefully acknowledge the financial supports from the National Natural Science Foundation of China (NSFC) (31861143020, 41776138), High-Tech Research and Development Program of China (2013AA092901, 2011AA090702, 2007AA09Z447, 2004AA628060, 2002AA608080) and Chinese Post-Doctoral Funding (No: 15005188).

References

1. Newmann DJ, Cragg GM, Snader KM. Natural products as sources of new drugs over the period 1981–2002. J Nat Prod. 2013;66:1022–37.
2. http://www.articlesfactory.com/articles/marketing/paclitaxel-ranks-first-among-worlds-anti-cancer-drugs.html
3. Andersen RJ, Williams DE. Pharmaceuticals from the sea. In: Harrison RM, Hester RE, editors. Environmental science and technology. Cambridge: The Royal society of Chemistry; 2000.
4. Lindel T, Jensenm PR, Fenical W, Long BH, Casazza AM, Carboni J, et al. Eleutherobin, a new cytotoxin that mimics paclitaxel (Taxol) by stabilizing microtubules. J Am Chem Soc. 1997;119:8744–5.
5. Ter Haar E, Kowalski RJ, Hamel E, Lin CM, Longley RE, Gunasekera SP, et al. Discodermolide, a cytotoxic marine agent that stabilizes microtubules more potently than taxol. Biochemist. 1996;35:243.
6. Khosla C, Gokhale RS, Jacobsen JR, Cane DE. Tolerance and specificity of polyketide synthases. Annu Rev Biochem. 1999;68:219–53.
7. Jenke-Kodama H, Sandmann A, Müller R, Dittmann E. Evolutionary implications of bacterial polyketide synthases. Mol Biol Evol. 2005;22:2027–39.
8. Cox RJ. Biosynthesis. Annu Rep Prog Chem Sect B. 2002;96:231–58.
9. Wang H, Fewer DP, Holm L, Rouhiainen L, Sivonen K. Atlas of nonribosomal peptide and polyketide biosynthetic pathways reveals common occurrence of nonmodular enzymes. Proc Natl Acad Sci. 2014;111:9259–64.
10. Bushley KE, Turgeon BG. Phylogenomics reveals subfamilies of fungal nonribosomal peptide synthetases and their evolutionary relationships. BMC Evol Biol. 2010;10:26.

11. Drake EJ, Miller BR, Shi C, Tarrasch JT, Sundlov JA, Leigh Allen C, Skiniotis G, et al. Structures of two distinct conformations of holo-non-ribosomal peptide synthetases. Nature. 2016;529:235–8.
12. Mootz HD, Schwarzer D, Marahiel MA. Ways of assembling complex natural products on modular nonribosomal peptide synthetases. Chem Bio Chem. 2002;3:490–504.
13. Martínez-Núñez MA, Lópezy López VE. Nonribosomal peptides synthetases and their applications in industry. Sus Chemi Proc. 2016;4:13.
14. Arnison PG, Bibb MJ, Bierbaum G, Bowers AA, Bugni TS, Bulaj G, Camarero JA, et al. Ribosomally synthesized and post-translationally modified peptide natural products: overview and recommendations for a universal nomenclature. Nat Prod Rep. 2013;30:108–60.
15. McIntosh JA, Donia MS, Schmidt EW. Ribosomal peptide natural products: bridging the ribosomal and nonribosomal worlds. Nat Prod Rep. 2009;26:537–59.
16. Yang X, van der Donk WA. Ribosomally synthesized and post-translationally modified peptide natural products: new insights into the role of leader and core peptides during biosynthesis. Chemistry. 2013;19:7662–77.
17. Tietz JI, Schwalen CJ, Patel PS, Maxson T, Blair PM, Tai H-C, Zakai UI, et al. A new genome-mining tool redefines the lasso peptide biosynthetic landscape. Nat Chem Biol. 2017;13:70–478.
18. Kennedy EP. Hitler's gift and the era of biosynthesis. J Biol Chem. 2001;276:42619–31.
19. Liscum L. Cholesterol biosynthesis. In: Vance JE, Vance DE, editors. Biochemistry of lipids, lipoproteins and membranes. Amsterdam: Elsevier; 2008. p. 399–421.
20. Bentley R. Secondary metabolite biosynthesis: the first century. Crit Rev Biotechnol. 1999;19:1–40.
21. Bretschneider T, Heim JB, Heine D, Winkler R, Busch B, Kusebauch B, et al. Vinylogous chain branching catalysed by a dedicated polyketide synthase module. Nature. 2013;502:124–8.
22. Bode HB, Daniela Reimer D, Fuchs SW, Kirchner F, Dauth C, Kegler C, et al. Determination of the absolute configuration of peptide natural products by using stable isotope labeling and mass spectrometry. Chem Eur J. 2012;18:2342–8.
23. Rinkel J, Dickschat JS. Recent highlights in biosynthesis research using stable isotopes. Beilstein J Org Chem. 2015;11:2493–508.
24. Bateman A, Coin L, Durbin R, Finn RD, Hollich V, Griffiths-Jones S, et al. The Pfam protein families database. Nucleic Acids Res. 2002;30:276–80.
25. Schmidt EW, Donia MS. Complex enzymes in microbial natural product biosynthesis, part A: overview articles and peptides. Methods Enzymol. 2009;458:575–95.
26. Kalaitzis JA, Lauro FM, Neilan BA. Mining cyanobacterial genomes for genes encoding complex biosynthetic pathways. Nat Prod Rep. 2009;26:1447–65.
27. Gross H. Genomic mining–a concept for the discovery of new bioactive natural products. Curr Opin Drug Discov Devel. 2009;12:207–19.
28. Piel J. Metabolites from symbiotic bacteria. Nat Prod Rep. 2009;26:338–62.
29. Piel J, Hertweck C, Shipley PR, Hunt DM, Newman MS, Moore BS. Cloning, sequencing and analysis of the enterocin biosynthesis gene cluster from the marine isolate '*Streptomyces maritimus*': evidence for the derailment of an aromatic polyketide synthase. Chem Biol. 2000;7:943–55.
30. Fusetani N, Matsunaga S. Bioactive sponge peptides. Chem Rev. 1993;93:1793–806.
31. Lane AL, Moore BS. A sea of biosynthesis: marine natural products meet the molecular age. Nat Prod Rep. 2011;28:411–28.
32. Zhang W, Lu L, Lai Q, Zhu B, Li Z, Ying Xu Y, et al. Family-wide structural characterization and genomic comparisons decode the diversity-oriented biosynthesis of thalassospiramides by marine proteobacteria. J Biol Chem. 2016;291:27228–38.
33. Jordan PA, Moore BS. Biosynthetic pathway connects cryptic ribosomally synthesized post-translationally modified peptide genes with Pyrroloquinoline alkaloids. Cell Chem Biol. 2016;23:1504–14.

34. El Gamal A, Agarwal V, Rahman I, Moore BS. Enzymatic reductive dehalogenation controls the biosynthesis of marine bacterial pyrroles. J Am Chem Soc. 2016;138:13167–70.
35. Tang X, Li J, Millán-Aguiñaga N, Zhang JJ, O'Neill EC, Ugalde JA, et al. Identification of thiotetronic acid antibiotic biosynthetic pathways by target-directed genome mining. ACS Chem Biol. 2015;10:2841–9.
36. Agarwal V, El Gamal AA, Yamanaka K, Poth D, Kersten RD, Schorn M, et al. Biosynthesis of polybrominated aromatic organic compounds by marine bacteria. Nat Chem Biol. 2014;10:640–7.
37. Lane AL, Nam S-J, Fukuda T, Yamanaka K, Kauffman CA, Jensen PR, et al. Structures and comparative characterization of biosynthetic gene clusters for cyanosporasides, enediyne-derived natural products from marine actinomycetes. J Am Chem Soc. 2013;135:4171–4.
38. Ross AC, Xu Y, Lu L, Kersten RD, Shao Z, Al-Suwailem AM, et al. Biosynthetic multitasking facilitates thalassospiramide structural diversity in marine bacteria. J Am Chem Soc. 2013;135:1155–62.
39. Xu Y, Kersten RD, Nam S-J, Lu L, Al-Suwailem AM, Zheng H, et al. Bacterial biosynthesis and maturation of the didemnin anti-cancer agents. J Am Chem Soc. 2012;134:8625–32.
40. Wilson MC, Nam S-J, Gulder TAM, Kauffman CA, Jensen PR, William Fenical W, et al. Structure and biosynthesis of the marine streptomycete ansamycin ansalactam A and its distinctive branched chain polyketide extender unit. J Am Chem Soc. 2011;133:1971–7.
41. Jørgensen H, Degnes KF, Dikiy A, Fjaervik E, Klinkenberg G, Zotchev SB. Insights into the evolution of macrolactam biosynthesis through cloning and comparative analysis of the biosynthetic gene cluster for a novel macrocyclic lactam, ML-449. Appl Environ Microbiol. 2010;76:283–93.
42. Wilson MC, Gulder TA, Mahmud T, Moore BS. Shared biosynthesis of the saliniketals and rifamycins in *Salinispora arenicola* is controlled by the sare1259-encoded cytochrome P450. J Am Chem Soc. 2010;132:12757–65.
43. Carlson JC, Fortman JL, Anzai Y, Li S, Burr DA, Sherman DH. Identification of the tirandamycin biosynthetic gene cluster from Streptomyces sp. 307-9. Chem Bio Chem. 2010;11:564–72.
44. Engelhardt K, Degnes KF, Zotchev SB. Isolation and characterization of the gene cluster for biosynthesis of the thiopeptide antibiotic TP-1161. Appl Environ Microbiol. 2010;76:7093–101.
45. Jørgensen H, et al. Biosynthesis of macrolactam BE-14106 involves two distinct PKS systems and amino acid processing enzymes for generation of the aminoacyl starter unit. Chem Biol. 2009;16:1109–21.
46. Fisch KM, Gurgui C, Heycke N, van der Sar SA, Anderson SA, Webb VL, et al. Polyketide assembly lines of uncultivated sponge symbionts from structure-based gene targeting. Nat Chem Biol. 2009;5:494–501.
47. Schultz AW, Oh D-C, Carney JR, Williamson RT, Udwary DW, Jensen PR, et al. Biosynthesis and structures of cyclomarins and cyclomarazines, prenylated cyclic peptides of marine actinobacterial origin. J Am Chem Soc. 2008;130:4507–16.
48. Winter JM, Moffitt MC, Zazopoulos E, McAlpine JB, Dorrestein PC, Moore BS. Molecular basis for chloronium-mediated meroterpene cyclization: cloning, sequencing, and heterologous expression of the napyradiomycin biosynthetic gene cluster. J Biol Chem. 2007;282:16362–8.
49. Sudek S, Lopanik NB, Waggoner LE, Hildebrand M, Anderson C, Haibin Liu H, et al. Identification of the putative bryostatin polyketide synthase gene cluster from "*Candidatus Endobugula* sertula", the uncultivated microbial symbiont of the marine bryozoan *Bugula neritina*. J Nat Prod. 2007;70:67–74.
50. Ramaswamy AV, Sorrels CM, Gerwick WH. Cloning and biochemical characterization of the hectochlorin biosynthetic gene cluster from the marine cyanobacterium *Lyngbya majuscula*. J Nat Prod. 2007;70:1977–86.
51. Udwary DW, Zeigler L, Asolkar RN, Singan V, Lapidus A, Fenical W, et al. Genome sequencing reveals complex secondary metabolome in the marine actinomycete *Salinispora tropica*. Proc Natl Acad Sci U S A. 2007;104:10376–81.

52. Schmidt EW, Nelson JT, Rasko DA, Sudek S, Eisen JA, Haygood MG, et al. Patellamide A and C biosynthesis by a microcin-like pathway in *Prochloron didemni*, the cyanobacterial symbiont of *Lissoclinum patella*. Proc Natl Acad Sci U S A. 2005;102:7315–20.
53. Chang Z. Biosynthetic pathway and gene cluster analysis of curacin A, an antitubulin natural product from the tropical marine cyanobacterium *Lyngbya majuscula*. J Nat Prod. 2004;67:1356–67.
54. Edwards DJ, Marquez BL, Nogle LM, McPhail K, Goeger DE, Ann Roberts M, et al. Structure and biosynthesis of the jamaicamides, new mixed polyketide-peptide neurotoxins from the marine cyanobacterium *Lyngbya majuscula*. Chem Biol. 2004;11:817–33.
55. Edwards DJ, Gerwick WH. Lyngbyatoxin biosynthesis: sequence of biosynthetic gene cluster and identification of a novel aromatic prenyltransferase. J Am Chem Soc. 2004;126:11432–3.
56. Moffitt MC, Neilan BA. Characterization of the nodularin synthetase gene cluster and proposed theory of the evolution of cyanobacterial hepatotoxins. Appl Environ Microbiol. 2004;70:6353–62.
57. Piel J, Hui D, Wen G, Butzke D, Platzer M, Fusetani N, Matsunaga S. Antitumor polyketide biosynthesis by an uncultivated bacterial symbiont of the marine sponge *Theonella swinhoei*. Proc Natl Acad Sci U S A. 2004;101:6222–7.
58. Chang Z, Flatt P, Gerwick WH, Nguyen VA, Willis CL, Sherman DH. The barbamide biosynthetic gene cluster: a novel marine cyanobacterial system of mixed polyketide synthase (PKS)-non-ribosomal peptide synthetase (NRPS) origin involving an unusual trichloroleucyl starter unit. Gene. 2002;296:235–47.
59. Allen EE, Bartlett DH. Structure and regulation of the omega-3 polyunsaturated fatty acid synthase genes from the deep-sea bacterium *Photobacterium profundum* strain SS9. Microbiol. 2002;148:1903–13.
60. Li A, Piel J. A gene cluster from a marine Streptomyces encoding the biosynthesis of the aromatic spiroketal polyketide griseorhodin A. Chem Biol. 2002;9:1017–26.
61. Morita N, Tanaka M, Okuyama H. Biosynthesis of fatty acids in the docosahexaenoic acid-producing bacterium *Moritella marina* strain MP-1. Biochem Soc Trans. 2000;28:943–5.
62. Lewis K, Epstein S, D'Onofrio A, Ling LL. Uncultured microorganisms as a source of secondary metabolites. J Antibiot (Tokyo). 2010;63:468–76.
63. Cichewicz RH, Valeriote FA, Crews P. Psymberin, a potent sponge-derived cytotoxin from *Psammocinia* distantly related to the pederin family. Org Lett. 2004;6:1951–4.
64. Komatsu M, Uchiyama T, Omura S, Cane DE, Ikeda H. Genome-minimized streptomyces host for the heterologous expression of secondary metabolism. Proc Natl Acad Sci U S A. 2010;107:2646–51.
65. Tan GY, Deng K, Liu X, Tao H, Chang Y, Chen J, Chen K, et al. Heterologous biosynthesis of spinosad: an omics-guided large polyketide synthase gene cluster reconstitution in *Streptomyces*. ACS Synth Biol. 2017;6:995–1005.

Chapter 13
Marine Natural Products from Marine Sponge Microorganisms

Cong Wang, Xiangui Mei, Dongyang Wang, and Weiming Zhu

Contents

Abstract Marine sponge microorganisms were a kind of important source for marine natural products (MNPs) due to their chemodiversity and good biological activities. They account for 19% MNPs of marine fungal origins from 1951 to 2014. This paper reviewed the sources, structures, and bioactivities of 519 sponge microbial MNPs reported in 1988 to the end of 2014. These new MNPs have a variety of chemical structures including nitrogen compounds, steroids and terpenoids,

C. Wang
Key Laboratory of Marine Drugs, Ministry of Education of China, School of Medicine and Pharmacy, Ocean University of China, Qingdao, China

Guangxi Key Laboratory of Chemistry and Engineering of Forest Products, School of Chemistry and Chemical Engineering, Guangxi University for Nationalities, Nanning, China

X. Mei · D. Wang
Key Laboratory of Marine Drugs, Ministry of Education of China, School of Medicine and Pharmacy, Ocean University of China, Qingdao, China

W. Zhu (✉)
Key Laboratory of Marine Drugs, Ministry of Education of China, School of Medicine and Pharmacy, Ocean University of China, Qingdao, China

Laboratory for Marine Drugs and Bioproducts, Qingdao National Laboratory for Marine Science and Technology, Qingdao, China
e-mail: weimingzhu@ouc.edu.cn

Z. Li (ed.), *Symbiotic Microbiomes of Coral Reefs Sponges and Corals*,
https://doi.org/10.1007/978-94-024-1612-1_13

polyketides, and so on. The half, of MNPs from marine sponge microorganisms, which are biological activities and the major bioactivities are, antimicrobial activity and cytotoxicity.

Keywords Marine sponge microorganisms · Natural products · Chemical structures · Bioactivities

13.1 Introduction

Since the last few years, marine natural products (MNPs) from marine sponge microorganisms were regarded as an important source for drug discovery which have attracted tremendous attention. Sponges (Porifera) are one of the most ancient multicellular Porifera animals, which contain dense and various microbial communities, including actinomycetes, bacteria, and fungi, having potential for bioactive natural products. Our previous reviews revealed that marine sponges are the main source of microorganisms for producing new and bioactive MNPs which account for 19% (= 392/2025) MNPs of marine fungal origins from 1951 to 2014 [1–3] and 12% (= 106/895) MNPs of marine microbial origins from 2010 to 2013 [4]. The sources, structures, and bioactivities of 519 compounds from sponge-associated bacteria, actinomycetes, and fungi reported up to August 2014 are reviewed.

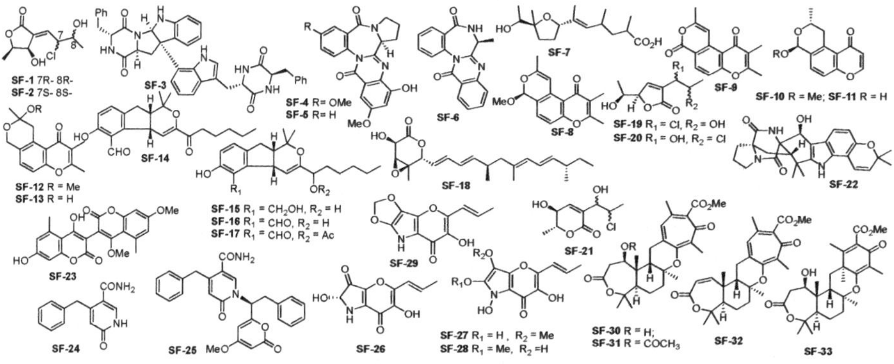

13.2 MNPs from Marine Sponge-Derived Fungi

13.2.1 MNPs from Marine Sponge-Derived Aspergillus *sp.*

Two novel chloropolyketides chlorocarolides A (**SF-1**) and B (**SF-2**) were obtained from *Aspergillus cf. ochraceus* 941026 (*Jaspis* cf. *coriacea*, Indo-Pacific Ocean) [5]. *Aspergillus niger* (*Hyrtios* sp., Dry Tortugas National Park, Florida) produced a dimeric diketopiperazine alkaloid, asperazine (**SF-3**), which is selectively cytotoxic to leukemia [6]. Circumdatins D–F (**SF-4**–**SF-6**) were separated from *A. ostianus* IBT 12704 (unidentified sponge, Pohnpei, Micronesia) [7]. The new compound, aspartic acid (**SF-7**), was isolated from *Aspergillus niger* 94-1212 (*Hyrtios proteus*, Dry Tortugas National Park, Florida) [8]. *Aspergillus versicolor* (*Xestospongia exigua*, Bali, Indonesia) yielded six new chromone derivatives, aspergiones A–F (**SF-8**–**SF-13**) [9] and aspergillone (**SF-14**), aspergillodiol (**SF-15**), aspergillol (**SF-16**), and 12-acetylaspergillol (**SF-17**) [10]. Nafuredin (**SF-18**) was obtained from *Aspergillus niger* FT-0554 (unknown sponge, Palau) which showed inhibition of NFRD of *Ascaris suum* (pig roundworm) at nM concentrations [11]. *Aspergillus ostianus* TUF 01F313 (unknown sponge, Pohnpei) yielded three new chlorine-containing compounds **SF-19**–**SF-21**. Compound **SF-19** inhibited the growth of *R. atlantica* at 5 μg/disk with inhibition zone diameter (IZD) of 12.7 mm, while compounds **SF-20** and **SF-21** were active at 25 μg/disk with IZD of 10.1 mm and 10.5 mm, respectively [12]. 21-Hydroxystephacidin A (**SF-22**) was obtained from *Aspergillus ostianus* 01F313 (unknown sponge, Pohnpei) [13]. Seven new compounds, bicoumanigrin (**SF-23**), aspernigrins A and B (**SF-24** and **SF-25**), and pyranonigrins A–D (**SF-26**–**SF-29**), were isolated from the cultures of *A. niger* (*Axinella damicornis*, Elba/Italy). Among them, compound **SF-23** showed cytotoxicity against human cancer cell lines at 1–20 μg/mL, while **SF-25** exhibited a strong neuroprotective effect against glutamic acid-induced neuronal cell death [14]. Four meroterpenoids, tropolactones A–D (**SF-30**–**SF-33**), originated from *Aspergillus* sp. CNK-371 (unknown sponge, Manele Bay, Hawaii). **SF-30**–**SF-32** showed cytotoxicity against HCT-116 with the IC_{50} values of 13.2, 10.9, and 13.9 μg/mL, respectively [15]. This fungal strain also produced three new pentaketides, aspinotriols A (**SF-34**) and B (**SF-35**) and aspinonediol (**SF-36**) [16], and three new 14-membered macrolides, aspergillides A–C (**SF-37**–**SF-39**). **SF-36**–**SF-38** exhibited cytotoxicity against mouse lymphocytic leukemia cells (L1210) with the LD_{50} values 2.1, 71.0, and 2.0 μg/mL, respectively [17]. *A. aculeatus* CRI323-04 (*Xestospongia testudinaria*, Phi Phi Islands, Thailand) produced a new tyrosine-derived metabolite, aspergillusol A (**SF-40**) [18], and a new sesquiterpenoid, asperaculin A (**SF-41**) [19]. Aspergillusol A (**SF-40**) inhibited α-glucosidase from the yeast *Saccharomyces cerevisiae* and *Bacillus stearothermophilus* with the IC_{50} values of 465 and 1060 μM, respectively. Seven new drimane sesquiterpenoids (**SF-42**–**SF-48**) were isolated from *A. ustus* 8009 (*Suberites domuncula*, Adriatic Sea). Compounds **SF-45** and **SF-46** showed cytotoxic activity against a panel of tumor cell lines, especially for compound **SF-46** with an EC_{50} value 0.6 μg/mL against L5178Y [20]. Five new ophiobolin-type sesterterpenoids **SF-49**–**SF-53** and two new pyrrolidine alkaloids **SF-54** and **SF-55** were obtained from the same strain [21]. *A. insuetus* (*Petrosia ficiformis*, Mediterranean Sea, Spain) yielded two new meroterpenoids, terretonins E (**SF-56**) and F (**SF-57**). Both showed activity against the mammalian mitochondrial respiratory chain with the IC_{50} values of 3.90 and 2.97 μM, respectively [22]. *A. sclerotiorum* Huber SP080903f04 (*Mycale* sp., Ishigaki Island, Japan) produced a new aspochracin derivative, JBIR-15 (**SF-58**) [23]. *A. versicolor* (*Petrosia* sp., Jeju Island, Korea) produced a new lipopetide, fellutamide C (**SF-59**), cytotoxic toward SK-MEL-2, XF498, and HCT15 with the IC_{50} values of 5.1, 3.9, and 3.1 μM, respectively [24]. JBIR-74 (**SF-60**) and JBIR-75 (**SF-61**) were isolated from *Aspergillus* sp. fs14 (unidentified sponge, Ishigaki Island, Okinawa, Japan) [25].

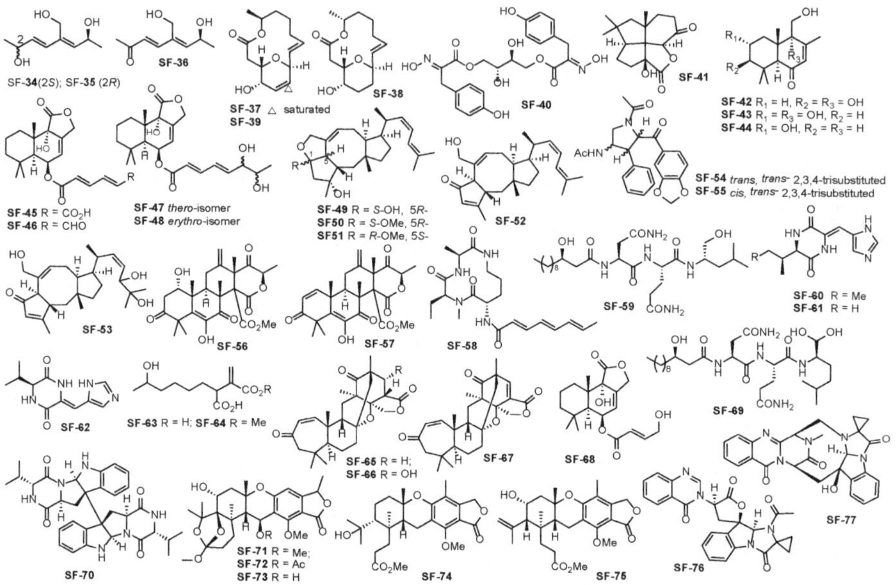

Pre-aurantiamine (**SF-62**) along with the compounds **SF-63** and **SF-64** was isolated from *A. aculeatus* CRI322-03 (*Stylissa flabelliformis*, Phi Phi Islands, Thailand) [26]. *A. insuetus* OY-207 (*Psammocinia* sp., Israel) yielded three novel meroterpenoids, insuetolides A–C (**SF-65**–**SF-67**), and a new sesquiterpene **SF-68**. **SF-65** showed antifungal activity toward *Neurospora crassa* with MIC value of 140 μM, while **SF-67** and **SF-68** showed 51% and 55% inhibitions on MOLT-4 cell line, respectively, at 50 mg/mL [27]. Fellutamide F (**SF-69**), a new lipopetide cytotoxic toward multiple human solid tumor cell lines with the EC_{50} values of 0.13–1.81 μg/mL, was obtained from the fermentation broth of *A. versicolor* PF10M (*Petrosia* sp., Jeju Island, Korea) [28]. Eurocristatine (**SF-70**), a new dimic diketopiperazine, was produced by *Eurotium cristatum* KUFC 7356 (*Mycale* sp., Bang Saen Beach, Thailand) [29]. Fermentation of *Aspergillus* sp. (*Tethya aurantium*, Mediterranean Sea, Italy) produced five new meroterpenoids, austalides M–Q (**SF-71**–**SF-75**) [30], as well as eight new alkaloids, tryptoquivaline K (**SF-76**) and fumiquinazolines K–P (**SF-77**–**SF-82**) [31], **SF-83,** and austalide R (**SF-84**) [32]. **SF-83** showed selectively antibacterial activity against *Vibrio* species, while **SF-84** exhibited a broad spectrum of antibacterial activity both with the MIC values from 0.01 to 0.1 μg/mL. *Aspergillus* sp. (*Xestospongia testudinaria*, South China Sea) produced four new bisabolane-type sesquiterpenoids, aspergiterpenoid A (**SF-85**), (-)-sydonol (**SF-86**), (-)-sydonic acid (**SF-87**), and **SF-88**, which selectively exhibited eight bacterial strains with the MIC values from 1.25 and 20.0 μM [33]. *Aspergillus* sp. (*Xestospongia testudinaria*, South China Sea) yielded three new phenolic bisabolane sesquiterpenoid dimers, disydonols A–C (**SF-89–SF-91**). Compounds **SF-89** and **SF-91** were cytotoxic toward HepG2 and Cashi with the IC_{50} values of 9.31/12.40 μg/mL and 2.91/10.20 μg/mL, respectively [34]. Three new depsidones **SF-92–SF-94**, a new diaryl ether **SF-95**, and a new pyrone **SF-96** were originated from *A. unguis* CRI282-03 (unidentified sponge CRI282, Thailand). Among them, **SF-92–SF-94** could inhibit aromatase with the IC_{50} values of 2.2, 4.1, and 0.7 μM, respectively. Compounds **SF-92** and **SF-93** exhibited XXO radical scavenging activity with the IC_{50} values of 16.0 μM and <15.6 μM, respectively [35]. *A. versicolor* MF359 (*Hymeniacidon perlevis*, Bohai Sea, China) produced hemiacetals, sterigmatocystin (**SF-97**), acyl sterigmatocystin (**SF-98**), and 5-methoxydihydrosterigmatocystin (**SF-99**), among which **SF-99** showed antibacterial activity against *S. aureus* and *B. subtilis* with the MIC values of 12.5 and 3.125 μg/mL, respectively [36].

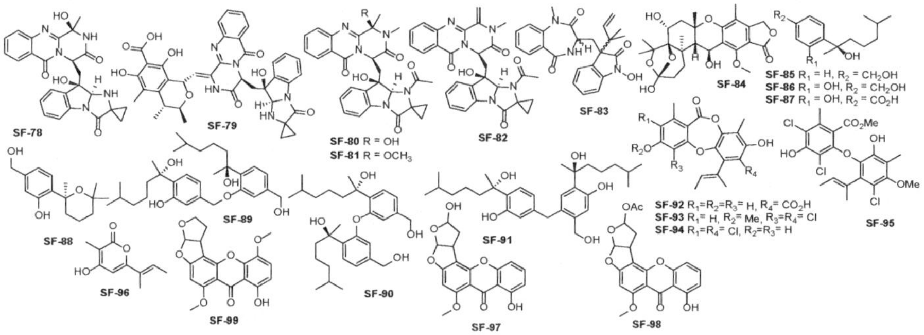

13.2.2 *MNPs from Marine Sponge-Derived* Penicillium *sp.*

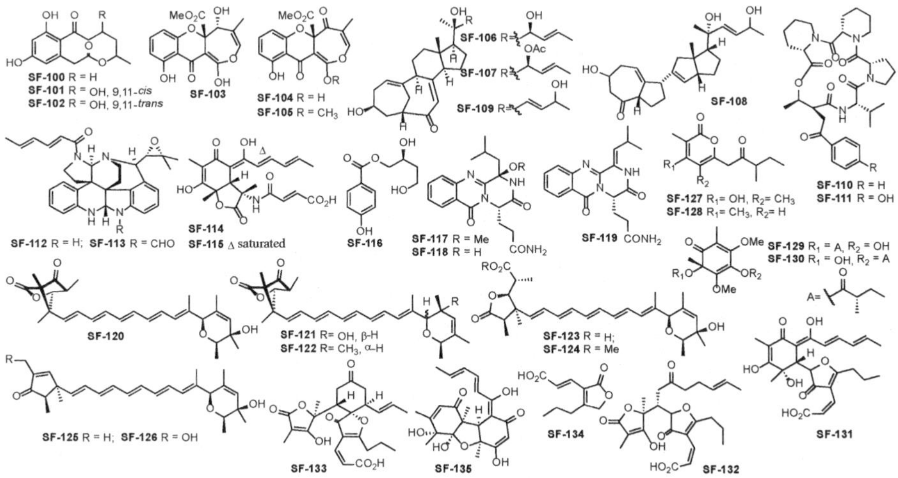

Penicillium cf. montanense (*Xestospongia exigua*, Bali Sea) produced three new decalactones with a 1,3-dihydroxybenzene fragment, xestodecalactones A–C (**SF-100–SF-102**). **SF-101** showed antibacterial activity against *Candida albicans* with 7 mm IZD at 20 μM [37]. Three novel polyketides, brocaenols A–C (**SF-103–SF-105**) [38], were identified from *P. brocae* (*Zyzzya* sp., Fijian, Japan). *P. citrinum* (*Axinella* sp., Papua New Guinea) [39] produced two new steroids isocyclocitrinol A (**SF-106**) and 22-acetylisocyclocitrinol A (**SF-107**), which displayed weak antibacterial activity against *Enterococcus durans* and *Staphylococcus epidermidis*. The structure of cyclocitrinol was revised as **SF-109** from the initial **SF-108** [40]. *P. brevicompactum* (*P. ficiformis*, Elba Island, Italy) [41] produced two new cyclodepsipeptides, petrosifungins A (**SF-110**) and B (**SF-111**). *Penicillium* sp. (*Axinella verrucosa*, Mediterranean Sea, Elba Island, Italy) [42] produced communesins C (**SF-112**) and D (**SF-113**). **SF-112** and **SF-113** showed growth inhibition against MOLT-3 and SUP-B15 tumor cell lines with the ED_{50} values of 8.6 and 9.0 μg/mL, respectively. Both compounds also showed activity against *Artemia salina* with LD_{50} values of 0.57 and 1.96 μg/mL, respectively. *P. chrysogenum* (*Ircinia fasciculata*, Fetovaia Bight, Italy) produced sorbicillactones A (**SF-114**) and B (**SF-115**), which showed strong anti-HIV activity at the concentrations ranging from 0.3 to 3.0 μg/mL [43]. (*S*)-2,4-Dihydroxy-1-butyl-4-hydroxybenzoate (**SF-116**) from *P. aurantiogriseum* (*Mycale plumose*, Qingdao, China) showed tsFT210 cell inhibition at 8.0 μg/mL [44]. This fungus also produced aurantiomides A–C (**SF-117–SF-119**), among which **SF-117** and **SF-118** showed cytotoxicities against P388, HL-60, and BEL-7402 tumor cell lines [45]. Prugosenes A1–A3 (**SF-120–SF-122**), B1 (**SF-123**), B2 (**SF-124**), C1 (**SF-125**), and C2 (**SF-126**) [46] were identified from *P. rugulosum* (*Chondrosia reniformis*, Elba, Italy). *P. lilacinus* (*Petrosia* sp., Jeju Island, Korea) [47] produced α-pyrones (**SF-127** and **SF-128**) and cyclohexenones (**SF-129** and **SF-130**). *P. chrysogenum* R03–8/4 (*Ircinia fasciculate*, Elba, Italy) produced sorbifuranones **SF-131–SF-133** [48]. *P. chrysogenum* LF066 (*Tethya aurantium*, Adriatic Sea, Croatia) [49] produced cillifuranone (**SF-134**). JBIR-59 (**SF-135**) was identified from *P. citrinum* SpI080624G1f01 (Ishigaki Island, Okinawa, Japan), which protected neuronal hybridoma N18-RE-105 cells with an EC_{50} value of 31 μM [50].

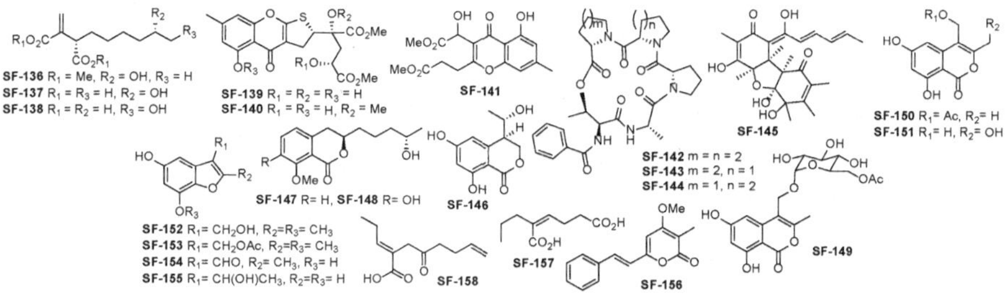

SF-136–SF-138 were isolated from *Penicillium* sp. J05B-3-F-1 (*Stelletta* sp., Jeju Island, Korea), among which **SF-136** showed weak activity against IL-1β expression in LPS-stimulated RAW264.7 at 200 μM [51]. *P. oxalicum* SCSGAF 0023 (unidentified sponge, South China Sea) [52] metabolized oxalicumones (**SF-139–SF-141**), among which **SF-139** showed cytotoxic activity on SW-620 and A375 cells with the IC_{50} values of 22.6 and 11.7 μM, respectively. Three new depsipeptides, JBIR-113–JBIR-115 (**SF-142–SF-144**) were from *Penicillium* sp. fs36 [53]. JBIR-124 (**SF-145**) from *P. citrinum* (Ishigaki Island, Okinawa, Japan) showed DPPH radical scavenging activity with an IC_{50} value of 30 μM [54]. Unidentified sponge-derived strain (South China Sea), *Penicillium* sp. MWZ14-4, produced penicimarins A–F (**SF-146–SF-151**) and penicifurans A–D (**SF-152–SF-155**). **SF-152** exhibited antibacterial activity against *Staphylococcus albus* with a MIC value of 3.13 μM [55]. *Penicillium* sp. JF-55 (unidentified sponge, Jeju Island, Korea) produced penstyrylpyrone (**SF-156**) which showed PTP1B inhibition [56]. (*Z*)-2-Ethylhex-2-enedioic acid (**SF-157**) and (*E*)-4-oxo-2-propylideneoct-7-enoic acid (**SF-158**) [57] were isolated from the co-cultivation of *Trichoderma* sp. Gc (M2)1 (*Geodia corticostylifera*, Sao Paulo, Brazil) with *Penicillium* sp. Ma (M3)V (*Mycale angulosa*, Sao Paulo, Brazil).

13.2.3 MNPs from Other Marine Sponge-Derived Fungi

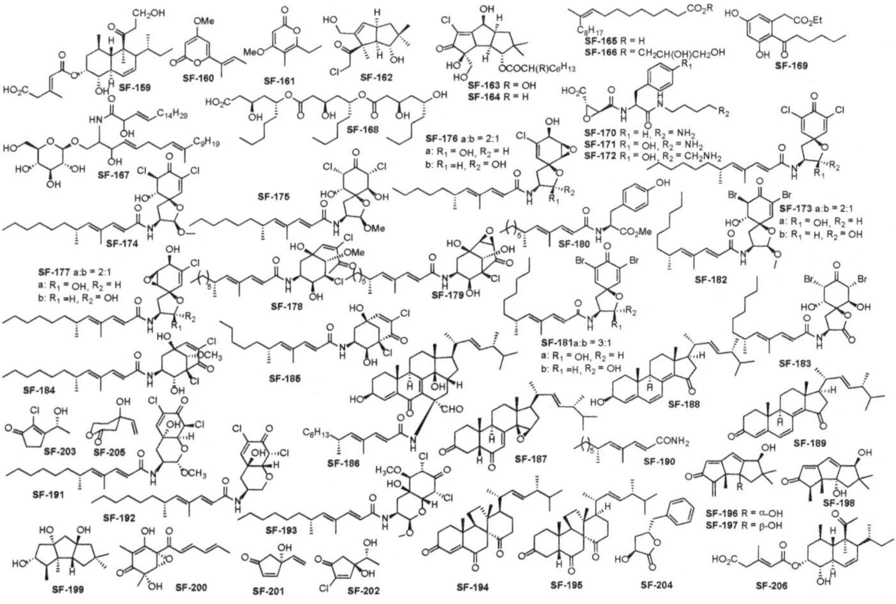

Trichoharzin (**SF-159**) was from *Trichoderma harzianum* (*Micale cecilia*, Japan) [58]. Demethylnectriapyrone (**SF-160**) and nectriapyrone (**SF-161**) sourced from an unidentified fungal strain (*Styloteflu* sp., Fiji) [59]. Chloriolins A–C (**SF-162–SF-164**) were from an unidentified fungal strain (*Jaspis* aff. *Johnstoni*, Pacific), among which **SF-162** was cytotoxic to T-47D and SNB-75 cells with the IC_{50} values of 0.7 and 0.5 μM, respectively [60]. *Microsphaeropsis olivacea* F010 (*Agelus* sp., Florida) produced **SF-165–SF-167** [61, 62]. *Exophiala pisciphila* NI10102 (*Mycale adhaerens*, Japan) yielded exophilin A (**SF-168**) that was active against *Enterococcus faecalis* IF 12964 and *E. faecalis* IF 12367 with the MIC values of 25 and 12.5 μg/mL, respectively [63]. Secocurvularin (**SF-169**) was metabolized by an unidentified fungal strain (#951014) derived from the sponge *Spirastrella vagabunda* (Indonesia) [64]. Cathestatins A–C (**SF-170–SF-172**) from *Microascus longirostris* SF-73 (a New Zealand sponge) was a protease inhibitor to papain and cathepsins L and B with the IC_{50} values of 1.4–177.6 nM [65]. *Gymnascella dankaliensis* (*Halichondria japonica*, Japan) produced gymnastatins A–K (**SF-173–SF-183**) [66–69], gymnastatins Q and R (**SF-184** and **SF-185**) [70], gymnasterones A–D (**SF-186–SF-189**) [71, 72], gymnamide (**SF-190**) [68], dankasterones A and B (**SF-191** and **SF-192**) [72, 73], and dankastatins A–C (**SF-193–SF-195**) [70, 74]. **SF-173–SF-175**, **SF-178**, **SF-179**, **SF-181–SF-183**, **SF-192**, and **SF-193** showed cytotoxic activity against P388 with the ED_{50} of 0.018–0.21 mg/ml, while **SF-177**, **SF-184–SF-189**, and **SF-194** were also cytotoxic to the same cells with the ED_{50} values of 0.9–10.8 μg/mL. **SF-194** showed appreciable growth inhibition against human cancer cell lines with the mean value (MG-MID) of log GI_{50} –5.41 [72, 73]. **SF-184** was cytotoxic to 39 human cancer cell lines (HCC panel) with MG-MID of log GI_{50} –4.81 and BSY-1 (breast) and MKN7 (stomach) cell lines with log GI_{50} –5.47 and –5.17, respectively [70]. MG-MID of log GI_{50} over the 39 human cancer cell lines of **SF-181** and **SF-182** were –5.77 and –5.71, respectively [69]. An unidentified fungal strain (no. 95–1005C) (*Haliclona* sp., Pacific) yielded hirsutanols A–C (**SF-196–SF-198**) and ent-gloeosteretriol (**SF-199**) [75]. *Trichoderma longibrachiatum* from the sponge *Haliclona* sp. (Italy) yielded epoxysorbicillinol (**SF-200**) [76], which was synthesized by 13 steps beginning from diethyl methylmalonate [77].

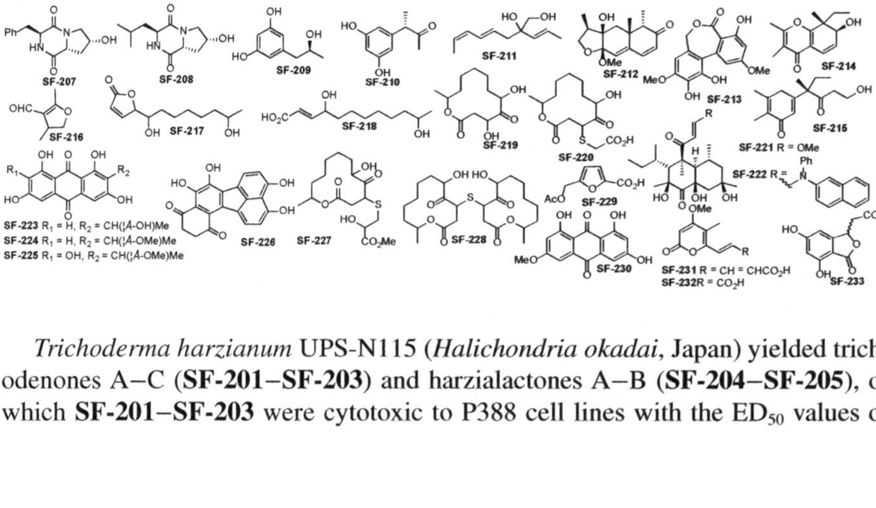

Trichoderma harzianum UPS-N115 (*Halichondria okadai*, Japan) yielded trichodenones A–C (**SF-201–SF-203**) and harzialactones A–B (**SF-204–SF-205**), of which **SF-201–SF-203** were cytotoxic to P388 cell lines with the ED_{50} values of

0.21, 1.21, and 1.45 μg/mL [78], respectively. Moreover, **SF-201** naturally occurs as a racemate [79]. Deoxynortrichoharzin (**SF-206**) was from *Paecilomyces* cf. *javanica* 961331 (*Jaspis coriacea*, Fiji) [80]. A sponge-derived *Aureobasidium pullulans* (Japan) yielded **SF-207–SF-208** and orcinotriol (**SF-209**) [81]. **SF-210–SF-211** were from *Coniothyrium* sp. 193H77 (*Ectyplasia perox*, Caribbean Sea) [82], and microsphaeropsisin (**SF-212**) was from *Microsphaeropsis* sp. H5-50 (*Myxilla incrustans*, German) [82]. *Ulocladium botrytis* 193A4 from sponge *Callyspongia vaginalis* (Caribbean Sea) yielded ulocladol (**SF-213**) [83]. A culture of *Drechslera hawaiiensis* P22-96 derived from *Callyspongia aerizusa* (Indonesia) led to the identification of **SF-214–SF-215** [84]. *Asteromyces cruciatus* H5-81 from *Myxilla incrustans* (German) yielded **SF-216** [85]. An unidentified sponge fungal strain I96S215 (Indonesia) produced isocladospolide (**SF-217**), seco-patulolide (**SF-218**), pandangolide 1 (**SF-219**), and pandangolide 2 (**SF-220**) [86]. **SF-221–SF-225** were from *Microsphaeropsis* sp. (*Aplysina aerophoba*, Mediterranean Sea), all of which except **SF-222** have inhibitory effects on protein-tyrosine kinases, PKC-ε, CDK4, and EGF, with the IC_{50} values of 8.5–54.0 μM [87]. Hortein (**SF-226**) was from *Hortaea werneckii* (*Aplysina aerophoba*, Mediterranean Sea) [88]. *Cladosporium herbarum* (*Callyspongia aerizusa*, Bali) yielded pandangolides 3 and 4 (**SF-227** and **SF-228**) and acetyl Sumiki's acid (**SF-229**) [89], among which **SF-229** showed antibacterial activity against *Staphylococcus aureus* and *Bacillus subtilis* with the IZD of 7 mm at 5 μg/disk. *Curvularia lunata* (*Niphates olemda*, Indonesia) yielded lunatin (**SF-230**) that was active against *S. aureus*, *E. coli*, *E. coli* HBI-101, and *B. subtilis* at the 5 μg/disk with IZDs of 8.5, 9, 8.0, and 7.5 mm, respectively [90]. *Cladosporium herbarum* (*Aplysina aerophoba*, Mediterranean Sea) yielded herbarins A and B (**SF-231** and **SF-232**) that showed inhibitions of *Artemia salina* with 75% and 65% mortality rates at 50 μg dose, respectively [90]. *Cladosporium herbarum* (*Callyspongia aerizusa*, Indonesia) yielded herbaric acid (**SF-233**).

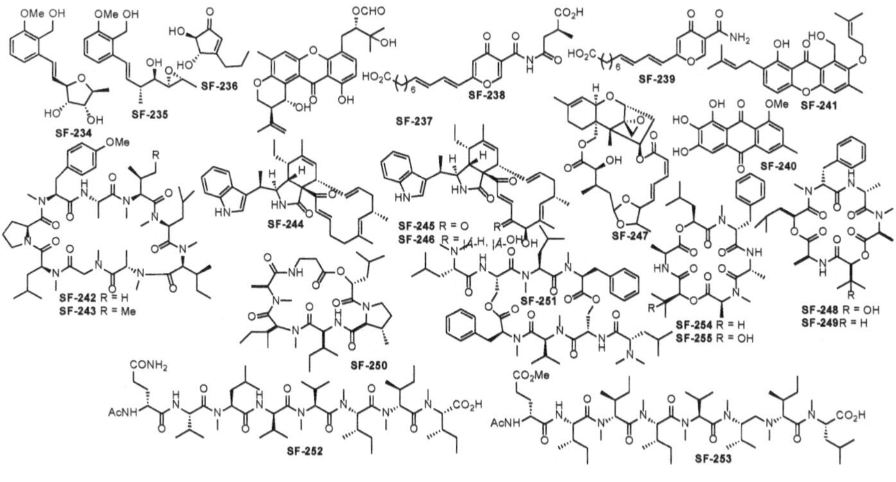

Emericella variecolor M75-2 (Caribbean waters of the Mochima Bay) yielded varitriol (**SF-234**), varioxirane (**SF-235**), dihydroterrein (**SF-236**), and varixanthone (**SF-237**). **SF-234** displayed a selective cytotoxicity to renal, CNS, and breast cancer cell lines among NCI-60 cell panel with the GI_{50} values of 0.163–95.9 nM, while **SF-237** showed antimicrobial activity with the MIC value of 12.5 μg/mL for *E. coli*, *Proteus* sp., *B. subtilis*, and *S. aureus* and 50 μg/mL for *E. faecalis* [91]. Microsphaerones A and B (**SF-238** and **SF-239**) were from *Microsphaeropsis* sp. (*Aplysina aerophoba*, France) [92]. *Emericella variecolor* (*Haliclona valliculata*, Elba/Italy) yielded evariquinone (**SF-240**) and isoemericellin (**SF-241**), between which **SF-240** showed 60% and 69% inhibitions on KB and NCI-H460 cells at 3.16 μg/mL, respectively [93]. Clonostachysins A and B (**SF-242** and **SF-243**) were isolated from *Clonostachys rogersoniana* HJK9 (*Halicondria japonica*, Numazu, Japan), both of which exhibited a selective inhibition on the dinoflagellate *Prorocentrum micans* at 30 μM [94]. *Phomopsis asparagi* 031113a (*Rhaphidophlus juniperina*, the US Virgin Islands) yielded chaetoglobosins 510, 540, and 542 (**SF-244–SF-246**), among which **SF-246** displayed antimicrofilament activity and cytotoxicity against L1210, C38, and CFU-GM cells [95]. Roridin R (**SF-247**) was from *Myrothecium* sp. TUF 02F6 (Manado, Indonesia) and active against the murine leukemia cell line L1210 with an IC_{50} of 0.45 μM [96]. The fungus 001314c derived from the sponge *Ianthella* sp. (Verongidae order, Ianthellidae family) (Papua New Guinea) yielded guangomides A and B (**SF-248** and **SF-249**) and homodestcardin (**SF-250**), among which **SF-248** and **SF-249** showed weak antibacterial activity against *Enterococcus durans* and *Staphylococcus epidermidis* with the MIC values of 100 μg/mL [97]. *Clonostachys* sp. ESNA-A009 (Japan) yielded IB-01212 (**SF-251**) that was active against LN-caP, SK-BR3, HT29, and HeLa cell lines with the GI_{50} value of 10 nM [98]. RHM1 (**SF-252**) and RHM2 (**SF-253**) were from *Acremonium* sp. (*Teichaxinella* sp., Papua New Guinea), which exhibited mild cytotoxicity against murine L1210 cells, and **SF-252** was active against *S. epidermidis* with a MIC value of 25–50 μg/mL [99]. Spicellamides A and B (**SF-254** and **SF-255**) were isolated from *Spicellum roseum* (*Ectyplasia perox*, Dominica), which showed cytotoxicity against neuroblastoma cells with the IC_{50} values of 30 and 6.2 μg/mL, respectively [100].

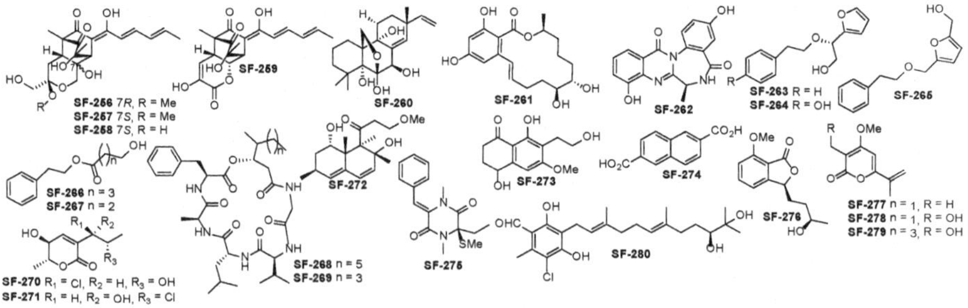

Trichoderma sp. (*Agelas dispar*, Dominica) yielded trichodermanones A–D (**SF-256**–**SF-259**) [101]. 11-Deoxydiaporthein A (**SF-260**) was from *Cryptosphaeria eunomia* var. *eunomia* (Pohnpei Island) [102]. 5′-Hydroxyzearalenol (**SF-261**) was isolated from the fermentation broth of *Fusarium* sp. 5ABR26 (Japan) [103]. *Exophiala* sp. (*Halichondria panicea*, Korea) yielded circumdatin I (**SF-262**) that showed an UVA protection with an ED_{50} value of 98 μM, while the ED_{50} value of oxybenzone (positive control, a currently used sunscreen agent) was 350 μM [104]. Pichiafurans A–C (**SF-263**–**SF-265**) and pichiacins A and B (**SF-266** and **SF-267**) were identified from the cultures of *Pichia membranifaciens* (J04J-1) F-9E (*Petrosia* sp., Jeju Island, Korea) [105, 106]. Two novel cyclodepsipeptides, scopularides A (**SF-268**) and B (**SF-269**) from the metabolites of *Scopulariopsis brevicaulis* (*Tethya aurantium*, Limski Fjord, Croatia), were weakly active against *Bacillus subtilis* and *Staphylococcus lentus* and cytotoxic against the Colo357, Panc89, and HT29 tumor cell lines [107]. Chlorohydroaspyrones A and B (**SF-270** and **SF-271**) were from *Exophiala* sp. (*Halichondria panicea*, Bogil Island, Korea), both of which showed antibacterial activity against *S. aureus* (SA), methicillin-resistant *S. aureus* (MRSA), and multidrug-resistant *S. aureus* (MDRSA), with the MIC values of 62.5, 125, and 125 μg/mL for **SF-270** and 62.5, 62.5, and 125 μg/mL for **SF-271**, respectively [108]. **SF-272**–**SF-274** were from *Mycelia sterilia* 97F49 [109]. **SF-275** and **SF-276** were from *Pleosporales* sp. CRIF2 (Surin Island, Phang Nga Province, Thailand), in which **SF-275** exhibited weak cytotoxic activity [110]. New infectopyrone derivatives **SF-277**–**SF-279** were from *Petriella* sp. TUBS 7961 (*Suberites domuncula*, Croatia), among which **SF-277** exhibited cytotoxic activity against the L5178Y cell line with an ED_{50} value of 0.2 mg/mL [111]. *Acremonium* sp. J05B-1-F-3 (*Stelletta* sp., Jeju Island, Korea) metabolized a chlorinated merosesquiterpenoid (**SF-280**), acremofuranones A and B (**SF-281** and **SF-282**), and dihydroxybergamotene (**SF-283**) [112].

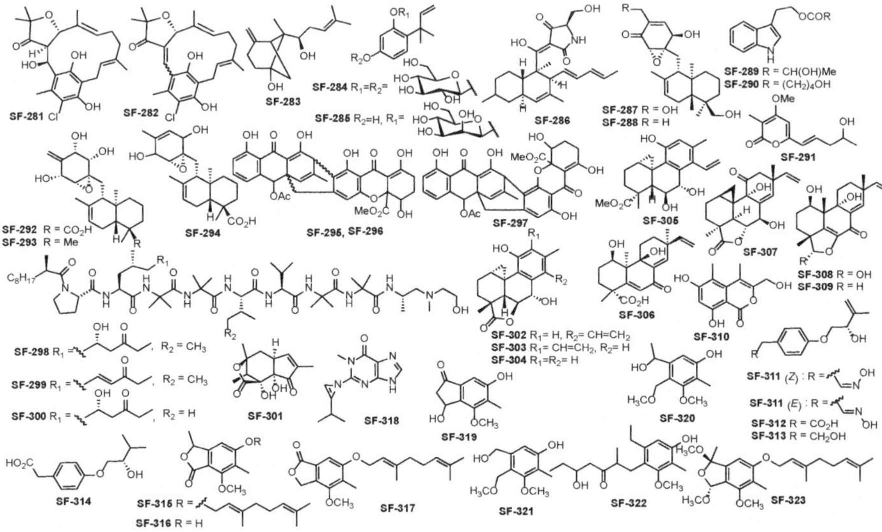

JBIR-37 (**SF-284**) and JBIR-38 (**SF-285**) were isolated from a culture broth of *Acremonium* sp. SpF080624G1f01 (*Demospongiae*, Ishigaki Island, Japan) [113]. *Beauveria bassiana* (Hale Granville Island, Germany) yielded a new equisetin-like tetramic acid derivative, beauversetin (**SF-286**) [114]. Culture of *Phoma* sp. (*Ectyplasia perox*, Caribbean Sea, Dominica) led to the discovery of epoxyphomalins A and B (**SF-287** and **SF-288**) that were cytotoxic against the 36 tumor cell lines with the mean IC_{50} values of 0.11 and 1.25 μg/mL, respectively [115]. Indole derivatives **SF-289** and **SF-290** that showed DPPH free radical scavenging activity were isolated from the cultures of *Pichia membranifaciens* USF-H25 (*Halichondria okadai*, Izu Peninsula, Japan) [116]. *Trichoderma viride* (*Agelas dispar*, Dominica) produced trichopyrone (**SF-291**) [117]. Fermentation of *Paraconiothyrium* sp. 193H12 derived from the sponge *Ectyplasia perox* (Dominica) led to identification of epoxyphomalins C–E (**SF-292–SF-294**), among which **SF-293** was cytotoxic toward prostate PC3M and bladder BXF 1218L cancer cell lines with the IC_{50} values of 0.72 and 1.43 μM, respectively [118]. JBIR-97 (**SF-295**), JBIR-98 (**SF-296**), and JBIR-99 (**SF-297**) produced by *Trichoderma* sp. SpB081112Mef2 (*Pseudoceratina purpurea*, Ishigaki Island) exhibited cytotoxic effects with the IC_{50} values of 11, 17, and 17 μM against HeLa cells and 31, 63, and 59 μM against ACC-MESO-1 cells, respectively [119]. Three new aminolipopeptides, trichoderins A (**SF-298**), A1 (**SF-299**), and B (**SF-300**), were isolated from a culture of *Trichoderma* sp. 05FI48, among which **SF-299** and **SF-300** showed potent anti-mycobacterial activity against *Mycobacterium smegmatis*, *M. bovis* BCG, and *M. tuberculosis* H37Rv with the MIC values of 0.02–2.0 μg/mL [120]. *Acremonium strictum* MB05005 (Choristida, Korea) produced acremostricin (**SF-301**) that was antibacterial to *Proteusbacillus vulgaris* ATCC 3851 with a MIC value of 12.5 μg/mL [121]. Arthrinins A–D (**SF-302–SF-305**) and myrocin D (**SF-306**) were isolated from a culture of *Arthrinium* sp. 9287 (*Geodia cydonium*, Italy) [122]. **SF-306** exhibited cytotoxic effects on L5178Y, K562, A2780, and A2780CisR cells with the IC_{50} values of 2.05, 50.3, 41.3, and 66.0 μM, respectively. **SF-306** also inhibited VEGF-A-dependent endothelial cell sprouting with an IC_{50} value of 2.6 μM. Myrocin D (**SF-307**), libertellenones E and F (**SF-308** and **SF-309**), and decarboxyhydroxycitrinone (**SF-310**) were isolated from a culture of *Arthrinium sacchari* (Japan) [123]. Although the structures **SF-306** and **SF-307** are different, both of them were the same name in the two references.

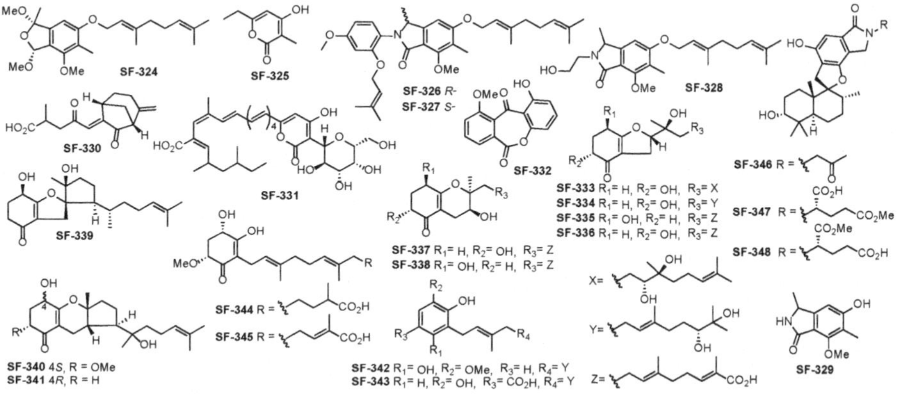

Stachylidium sp. 220 (*Callyspongia* cf. *C. flammea*, Sydney) yielded stachylines A–D (**SF-311–SF-314**) where **SF-311** is an equal *E*/*Z*-mixture [124] and marilones A–C (**SF-315–SF-317**) [125]. **SF-315** has activity against *Plasmodium berghei* liver stages with an IC_{50} value of 12.1 μM, while **SF-316** showed selective antagonistic activity toward receptor 5-HT2B with the K_i value of 7.7 μM. **SF-316** and **SF-317** were cytotoxic to NCI-H460, MCF-7, and SF268 cells with the mean GI_{50} values of 36.7 and 26.6 μM, respectively. Acremolin (**SF-318**) cytotoxic to A549 cell lines (IC_{50} 45.9 μg/mL) was from the marine fungus *Acremonium strictumsame* (*Choristida*, Korean) [126, 127]. The same strain *Stachylidium* sp. 220 (*Callyspongia* cf. *C. flammea*, Sydney) also produced cyclomarinone (**SF-319**), maristachones A–E (**SF-320–SF-324**), and marilactone (**SF-325**) [128], as well as marilines A_1 (**SF-326**), A_2 (**SF-327**), B (**SF-328**), and C (**SF-329**) [129]. **SF-326** and **SF-327** inhibited the serine protease HLE with the same IC_{50} value of 0.86 μM and the cholesterol esterase activity with the IC_{50} values of 0.18 and 0.63 μM, respectively. **SF-326** was cytotoxic to five cancer cell lines with a mean GI_{50} value of 24.4 μM, while the GI_{50} value for **SF-327** against 19 cancer cell lines was 11.02 μM. **SF-326–SF-328** displayed antiplasmodial activity with the IC_{50} values of 6.68, 11.61, and 13.84 μM, respectively. **SF-328** and **SF-329** were antagonistic on the cannabinoid receptor CB2, while **SF-327** was antagonistic on the histamine receptor H2, the dopamine receptor DAT, and the adrenergic receptor Beta3 with the K_i values of 5.97, 5.94, 5.92, 5.63, and 5.63 μM, respectively. **SF-326** showed antiparasitic activity against *Trypanosoma brucei* with an IC_{50} value of 17.7 μM. *Emericellopsis minima* derived from the marine sponge (*Hyrtios erecta*, Similan islands, Thailand) produced **SF-330** [130]. A new pyronepolyene C-glucoside, iso-D8646-2-6 (**SF-331**), sourced from *Epicoccum* sp. JJY40 (*Callyspongia* sp., Sanya, China) showed weak NF-κB inhibition and anti-H1N1 activities with the IC_{50} values of 40.0 and 91.5 μM, respectively [131]. *Beauveria bassiana* TPU942 (Iriomote Island in Okinawa, Japan) produced **SF-332** [132].

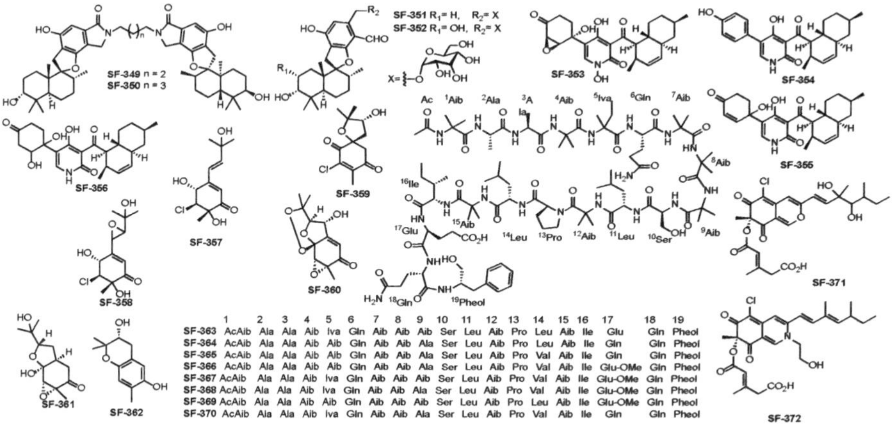

	1	2	3	4	5	6	7	8	9	10	11	12	13	14	15	16	17	18	19
SF-363	AcAib	Ala	Ala	Aib	Iva	Gln	Aib	Aib	Aib	Ser	Leu	Aib	Pro	Leu	Aib	Ile	Glu	Gln	Pheol
SF-364	AcAib	Ala	Ala	Aib	Aib	Gln	Aib	Aib	Ala	Ser	Leu	Aib	Pro	Leu	Aib	Ile	Gln	Gln	Pheol
SF-365	AcAib	Ala	Ala	Aib	Aib	Gln	Aib	Aib	Ala	Ser	Leu	Aib	Pro	Val	Aib	Ile	Gln	Gln	Pheol
SF-366	AcAib	Ala	Ala	Aib	Aib	Gln	Aib	Aib	Ala	Ser	Leu	Aib	Pro	Val	Aib	Ile	Glu-OMe	Gln	Pheol
SF-367	AcAib	Ala	Ala	Aib	Iva	Gln	Aib	Aib	Aib	Ser	Leu	Aib	Pro	Val	Aib	Ile	Glu-OMe	Gln	Pheol
SF-368	AcAib	Ala	Ala	Aib	Iva	Gln	Aib	Aib	Ala	Ser	Leu	Aib	Pro	Val	Aib	Ile	Glu-OMe	Gln	Pheol
SF-369	AcAib	Ala	Ala	Aib	Aib	Gln	Aib	Aib	Aib	Ser	Leu	Aib	Pro	Leu	Aib	Ile	Glu-OMe	Gln	Pheol
SF-370	AcAib	Ala	Ala	Aib	Iva	Gln	Aib	Aib	Ala	Ser	Leu	Aib	Pro	Val	Aib	Ile	Gln	Gln	Pheol

Bicycloalternarenes A–F (**SF-333**–**SF-338**), tricycloalternarenes A–C (**SF-339**–**SF-341**), and monocycloalternarenes A–D (**SF-342**–**SF-345**) were from *Alternaria* sp. JJY-32 (*Callyspongia* sp., Hainan Island, China). Among them, **SF-333**–**SF-336** and **SF-339**–**SF-345** showed anti-inflammatory activities against RAW264.7 cells with the IC_{50} values of 39–85 μM [133]. Stachybotrins D–F (**SF-346**–**SF-348**), stachybocins E (**SF-349**) and F (**SF-350**), and stachybosides A (**SF-351**) and B (**SF-352**) were from *Stachybotrys chartarum* MXH-X73 (*Xestospongia testudinaria*, Xisha Island, China). **SF-346** showed an inhibitory effect on HIV-1 replication with an EC_{50} value of 50 μM. **SF-346** could also block NNRTI-resistant strains, HIV-$1_{RT\text{-}K103N}$, HIV-$1_{RT\text{-}L100I,\ K103N}$, HIV-$1_{RT\text{-}K103N,\ V108I}$, HIV-$1_{RT\text{-}K103N,\ G190A}$, and HIV-$1_{RT\text{-}K103N,\ P225H}$, as well as wild HIV-$1_{wt}$ with the EC_{50} values of 7.0, 23.8, 13.3, 14.2, 6.2, and 8.4 μM, respectively [134]. Four new 4-hydroxy-2-pyridone alkaloids, didymellamides A–D (**SF-353**–**SF-356**), were isolated from the metabolites of *Stagonosporopsis cucurbitacearum* (Japan). **SF-353** inhibited the growth of azole-resistant and azole-sensitive *C. albicans*, *C. glabrata*, and *Cryptococcus neoformans* with the MIC of 3.1, 3.1, and 1.6 μg/mL, respectively. **SF-354** only inhibited *C. neoformans* with a MIC of 6.3 μg/mL [135]. Six new acremine compounds, 5-chloroacremines A (**SF-357**) and H (**SF-358**), as well as acremines O–R (**SF-359**–**SF-362**) were identified from the metabolites of *Acremonium persicinum* (*Anomoianthella rubra*, Mooloolaba, Australia) [136]. Trichorzianines 1938, 1909, 1895, 1896, 1924, 1910, 1924A, and 1909A (**SF-363**–**SF-370**) were isolated from the cultures of *Trichoderma atroviride* NF16 (*Axinella* sp., Israel) [137]. Except for **SF-367**, **SF-363**–**SF-370** exhibited antibacterial activity against *Sporosarcina* sp. NB90, *Bacillus* sp. NB36, *Microbacterium* sp. PII.14, *Rhodobacteraceae* (PI.03), and *Shewanella* sp. PIII.07 with the MIC values of 12.5–200 μg/mL. Some of the compounds were active against *S. albus* and *B. subtilis* with MIC values of 50–200 μg/mL. Helicusin E (**SF-371**) and isochromophilones X (**SF-372**) and XI (**SF-373**) along with bartanolide (**SF-374**) were isolated from *Bartalinia robillardoides* LF550 (*Tethya aurantium*, Mediterranean) [138]. **SF-373** showed weak antibacterial activities against *B. subtilis*, *S. lentus*, and

Trichophyton rubrum with the IC_{50} values of 55.6, 78.4, and 41.5 μM. **SF-372** and **SF-373** showed inhibition on phosphodiesterase 4 (PDE4) with the IC_{50} values of 11.7 and 8.30 μM, respectively.

Chrysoarticulins A–C (**SF-375–SF-377**) were isolated from the cultures of *Chrysosporium articulatum* (Korea) [139]. **SF-377** moderately inhibited sortase A and isocitrate lyase (ICL) with the IC_{50} values of 95.1 and 236.4 μM, respectively. Hypocreaterpenes A (**SF-378**) and B (**SF-379**) were from *Hypocreales* sp. HLS-104 (*Gelliodes carnosa*, South China Sea) [140]. *Dichotomomyces cejpii* (*Callyspongia* sp. cf. *C. flammea*, Australia) produced **SF-380** that was a CB2 antagonist and **SF-381** that was a selective GPR18 antagonist [141]. Two new oxaphenalenone dimers, talaromycesones A (**SF-382**) and B (**SF-383**), and a new isopentenyl xanthenone, talaroxanthenone (**SF-384**), were isolated from *Talaromyces* sp. LF458 (*Axinella verrucosa*, Italy) [142]. **SF-382** exhibited antibacterial activities against *S. epidermidis* and MRSA with the IC_{50} values of 3.7 and 5.48 μM, while the corresponding IC_{50} values for **SF-383** were 17.36 and 19.50 μM, respectively. In addition, the acetylcholinesterase (AchE) inhibition for **SF-382** and **SF-384** and the phosphodiesterase (PDE-4B2) inhibition for **SF-384** were observed with the IC_{50} values of 7.49, 1.61, and 7.25 μM, respectively. Xylarianaphthol-1 (**SF-385**) was isolated from *Xylariales* 05FI52 (Indonesia), which activated the p21 promoter in MG63^{luc+} cells with a dose-dependent manner, and the luciferase expression increased approximately 2.5-fold over the untreated cells at the 0.3 μM. When treated with **SF-385**, p21 protein expression in the wild-type MG63 cells also increased [143]. Six new caryophyllene-based sesquiterpenoids named punctaporonins H–M (**SF-386–SF-391**) were isolated from the metabolites of *Hansfordia sinuosae* (*Niphates* sp., Southern China Sea). Punctaporonin K (**SF-389**) could reduce the triglycerides and the total cholesterol in the intracellular levels [144]. A new merodi-terpene, sartorypyrone C (**SF-392**), was isolated from the culture of *Neosartorya paulistensis* KUFC 7897 (*Chondrilla australiensi*, Thailand) [145].

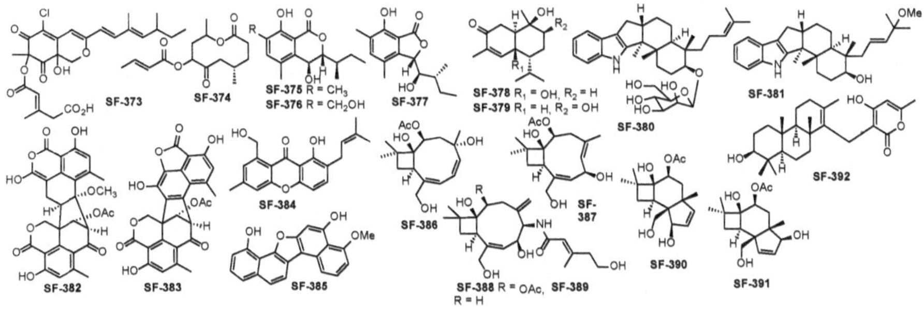

13.3 MNPs from Marine Sponge-Derived Actinobacteria

Diketopiperazines **SA-1**–**SA-3** were obtained from a culture of *Micrococcus* strain (*Tedania ignis*, Florida Keys) [146]. Culture of *Streptomyces* sp. Ni-80 (unidentified sponge, Iriomote, Japan) yielded urauchimycins A (**SA-4**) and B (**SA-5**), both of which inhibited the morphological differentiation of *C. albicans* at the concentration of 10 μg/mL [147]. Acyl-1-(acyl-6′-mannobiosyl)-3-glycerol (**SA-6**) was from *Micrococcus luteus* (*Xestospongia* sp., New Caledonia) [148]. Fatty acids **SA-7**–**SA-13** were from a *Streptomyces* sp. KM86-9B (Keomun Island, Korea), which showed the inhibition on topoisomerase I at the concentration of 100 μg/ml except **SA-8** [149]. Macrolide **SA-14** from *Micromonospora* sp. L-25-ES25-008 (unidentified sponge, Indian Ocean, Mozambique) showed cytotoxicities against A549, HT-29, MEL-28, and P388 cell lines with the IC_{50} values of 1, 1, 1, and 0.0001 μg/mL, respectively [150]. *Microbacterium* sp. (*Halichondria panicea*, Adriatic coast, Rovinj, Croatia) yielded glycoglycerolipids **SA-15**–**SA-18** and diphosphatidylglycerol **SA-19** [151]. Staurosporine derivatives **SA-20** and **SA-21** from *Micromonospora* sp. L-31-CLCO-02 (*Clathrina coriacea*, Fuerteventura Island, Canary Islands Archipelago) [152] showed strong cytotoxic effects on A549, HT-29, SK-MEL-28, and P388D1 cell lines with the IC_{50} values of 2, 4, 4, and 40 nM for **SA-20** and 2, 4, 2, and 20 nM for **SA-21**, respectively. *Streptomyces* sp. (unidentified sponge, Jeju Island, Korea) yielded dehydroxynocardamine (**SA-22**) and desmethylenylnocardamine (**SA-23**) that showed weak inhibition on the recombinant enzyme sortase B [153]. *Streptomyces* sp. HB202 (*Halichondria panicea*, Baltic Sea, Germany) produced antibacterial streptophenazines A–H (**SA-24**–**SA-31**). The MIC values of **SA-24**, **SA-26**–**SA-28**, and **SA-31** against *B. subtilis* and **SA-24**–**SA-26**, **SA-29**, and **SA-30** against *S. lentus* were 46.9, 15.6, 62.5, 62.5, 15.6, 62.5, 62.5, 46.9, 62.5, and 62.5 μg/mL, respectively [154]. Fermentation of *Saccharopolyspora cebuensis* SPE 10-1 (*Haliclona* sp., Cebu, Philippines) led to the identification of cebulactams A1 (**SA-32**) and A2 (**SA-33**) [155].

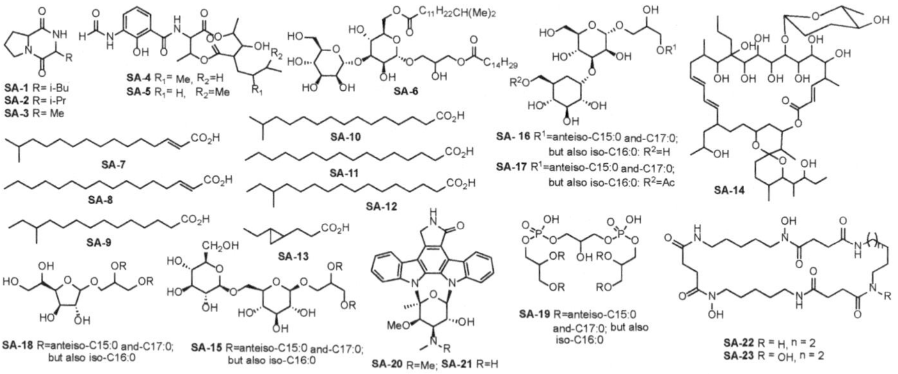

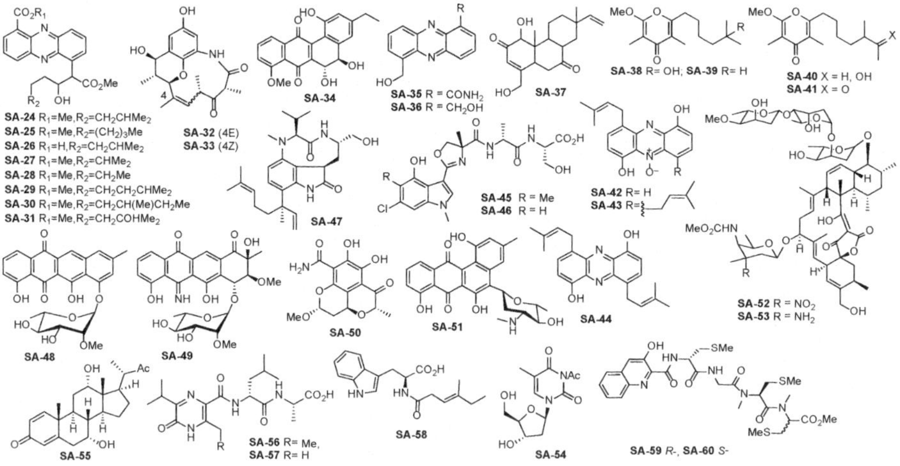

Saccharopolyspora taberi PEM-06-F23-019B (unidentified sponge, coast of Tanzania) produced angucyclinone **SA-34** [156]. Culture of *Brevibacterium* sp. KMD 003 (*Callyspongia* sp., Kyung Po Beach, Korea) yielded 6-hydroxymethyl-1-phenazine-carboxamide (**SA-35**) and 1,6-phenazinedimethanol (**SA-36**) that showed antimicrobial activity against *B. subtilis*, *E. hirae*, and *M. luteus* with the MIC_{50} values of 5–20 μM [157]. Diterpenene **SA-37** was from *Actinomadura* sp. SpB081030SC-15 (unidentified sponge, Ishigaki Island, Okinawa), which showed modest radical scavenging activity with an EC_{50} value of 6.3 μM [158]. *γ*-Pyrones **SA-38–SA-41** were from *Nocardiopsis* sp. HB383 (*Halichondria panicea*, Baltic Sea, Germany) [159]. Isoprenoids **SA-42–SA-44** were from *Streptomyces* sp. SpC080624SC-11 (*Cinachyra* sp., Nagura Bay, Ishigaki, Okinawa) [160]. *Streptomyces* sp. Sp080513GE-23 (*Haliclona* sp., Tateyama City, Chiba Prefecture, Japan) produced two chlorinated indolyltetrapeptides, **SA-45** and **SA-46**, which exhibited weak DPPH radical scavenging with the IC_{50} values of 1.0 and 2.5 mM, respectively [161]. *Streptomyces* sp. NBRC 105896 (*Haliclona* sp., Tateyama, Chiba Prefecture, Japan) yielded a teleocidin analog (**SA-47**)[162], while *Streptomyces* sp. Sp080513GE- 26 (*Haliclona* sp., Tateyama, Chiba Prefecture, Japan) produced anthracyclines **SA-48** and **SA-49** [163]. A salicylamide derivative, **SA-50**, was from *Streptomyces* sp. SpC080624SC-11 (*Demospongiae*, Ishigaki City, Okinawa), whose IC_{50} value against HeLa cell was 28 μM [164]. A benz[a] anthracene derivative, **SA-51**, was from *Streptomyces* sp. HB202 (*Halichondria panicea*, Baltic Sea), which showed cytotoxicity against eight human cancer cell lines with the IC_{50} values of 0.13–0.33 μM [165]. Culture of *Streptomyces carnosus* AZS17 (*Hymeniacidon* sp., coastal waters of the East China Sea) yielded lobophorins C and D (**SA-52** and **SA-53**), whose IC_{50} values against 7402 hepatoma cells for **SA-52** and MDA-MB 435 cells for **SA-53** were 0.6 μg/mL and 7.5 μM, respectively [166]. Nucleoside derivative **SA-54** was from *Streptomyces microflavus* (*Hymeniacidon perlevis*, coast of Dalian, China), and some of its analogs had previously been obtained from the Swedish sponge *Geodia barretti*. The isolation of

SA-54 hints that this class of compounds may be actually produced by sponge-associated microorganisms [167]. Guided by a novel high-content screen for NF-κB and glucocorticoid receptor (GR) activity, bendigole E (**SA-55**) was isolated from the cultures of *Actinomadura* sp. SBMs009 (*Suberites japonicus*, source unspecified). [168]. Culture of a new species of *Streptomyces* SpD081030SC-03 (*Demospongiae*, Ishigaki City, Okinawa, Japan) yielded pyrazinones **SA-56** and **SA-57** [169]. Culture of *Streptomyces* sp. DA22 (*Craniella australiensis*, South China Sea) yielded streptomycindole (**SA-58**) [170].

Thiocoraline analogs **SA-59–SA-63** were from *Verrucosispora* sp. strain WMMA107 (*Chondrilla caribensis f. caribensis*, Florida Keys, USA) [171], of which **SA-61–SA-63** demonstrated significant cytotoxicity against A549 cell line with the EC_{50} values of 0.13, 2.86, and 1.26 μM, respectively. Tetromycins **SA-64–SA-67** were isolated from the cultures of *Streptomyces axinellae* Pol001^{T} (*Axinella polypoides*, Banyuls-sur-Mer, France), which showed antiparasitic activities against *T. brucei* and time-dependent inhibition of cathepsin L-like proteases with the micromolar *K*i values [172]. Trichostatin analogs **SA-68–SA-70** from *Streptomyces* sp. RM72 (Takarajima Island, Kagoshima Prefecture, Japan) showed human HDAC1 inhibition with the IC_{50} values of 48, 74, and 57 μM, respectively [173]. Cyclic lipopeptides **SA-71–SA-74** were from *Streptomyces* sp. RV15 (*Dysidea tupha*, Rovinj, Croatia) [174]. The C-glycosylated benz[a]anthraquinone derivatives (**SA-75–SA-77**) were from a *Streptomyces* sp. BCC45596 (*Xestospongia* sp. Sichang Island, Chonburi, Thailand), which were antimalarial with the IC_{50} values of 0.053, 0.142, and 2.93 μg/mL and antitubercular with the MIC values of 3.13, 12.50, and 6.25 μg/mL, respectively [175]. In addition, **SA-75–SA-77** showed cytotoxicity against KB, MCF-7, NCl-H187, and Vero cells with the IC_{50} values of 0.179, 0.196, 0.092, and 1.71 μg/mL for **SA-75**; 0.324, 0.45, 0.242, and 3.05 μg/mL for **SA-76**; and 6.96, 3.41, 3.97, and 10.07 μg/mL for **SA-77**, respectively. By combination of genome mining on PKS-NRPs and antibiotic activity screening, a thiazolyl peptide (kocurin **SA-78**) was obtained from *Kocuria palustris* (unidentified sponge, Florida Keys, USA). **SA-78** displayed strong inhibition on the growth of MRSA MB5393 with a MIC value of 0.25 μg/mL. In addition, this compound also displayed antibacterial activity against *B. subtilis* and *E. faecium* [176]. Fermentation of *Streptomyces tateyamensis* NBRC 105047 (unidentified sponge) produced JBIR-107 (**SA-79**) [177]. Actinosporins **SA-80** and **SA-81** were isolated from the cultures of *Actinokineospora* sp. EG49 (*Spheciospongia vagabunda*, Rovinj), between which **SA-80** showed activity against *Trypanosoma brucei* with an IC_{50} value of 15 μM [178]. **SA-82** and **SA-83** were obtained by co-culture of two sponge-associated bacterial strains, *Actinokineospora* sp. EG49 (*Spheciospongia vagabunda*, Rovinj) and *Nocardiopsis* sp. RV163 (*Dysidea avara*) [179]. Compounds **SA-84–SA-86** were from *Amycolatopsis* sp. (unidentified sponge, Micronesia), which exhibited cytotoxic effects on A546, K562, and SK-HEP1 cells with the IC_{50} values of 13.7, 9.6, and 8.3 μM, respectively. And **SA-86** also displayed significant cytotoxicity against the gastric cancer cell line SNU638 and the colon cancer cell line HCT116 with the IC_{50} values of 0.8 and 2.0 μM, respectively [180].

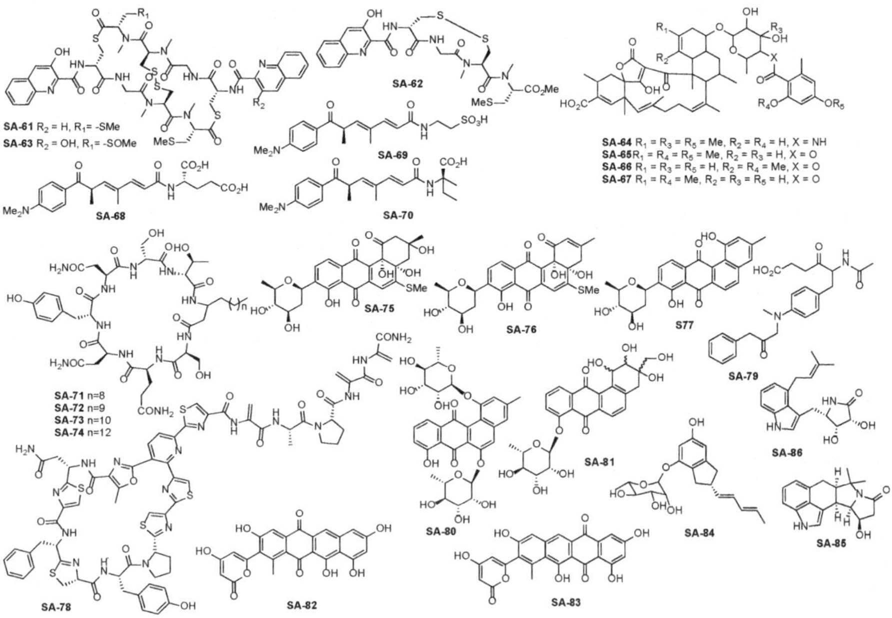

13.4 MNPs from Marine Sponge-Derived Bacteria

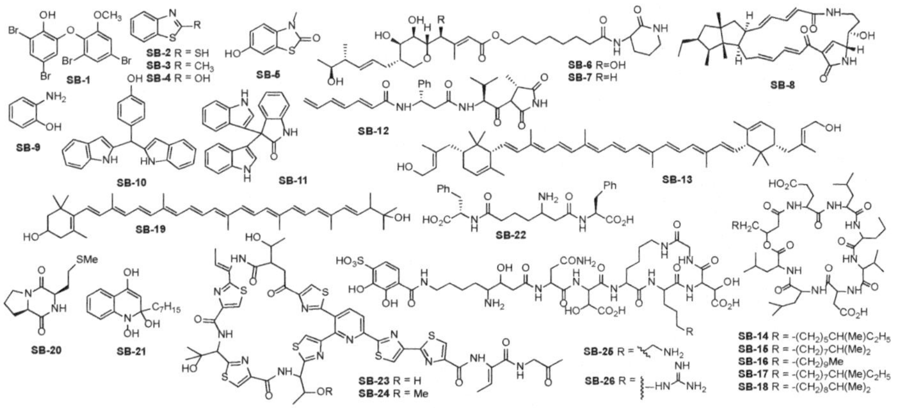

Vibrio sp. (*Dysidea* sp., Tutuila and Ofu islands, Eastern Samoa) produced a brominated diphenyl ether, 3,5-dibromo-2-(3′,5′-dibromo-2′-methoxyphenoxy)phenol (**SB-1**) [181]. Four benzothiazoles (**SB-2–SB-5**) were isolated from the fermentation broth of *Micrococcus* sp. SD.3 (*Tedania ignis*), and these compounds were first isolated from the marine biosphere [182]. Two antimicrobial pseudomonic acid derivatives (**SB-6** and **SB-7**) against *S. aureus* were identified from a culture of

Alteromonas sp. (*Darwinella rosacea*, Harrington Sound, Bermuda) [183]. Alteramide A (**SB-8**) from *Alteromonas* sp. (*Halichondria okadai*, Nagai, Kanagawa) showed cytotoxicity toward the murine leukemia P388 cells, the murine lymphoma L1210 cells, and the human epidermoid carcinoma KB cells with the IC_{50} values of 0.1, 1.7, and 5.0 pg/mL, respectively [184]. A purple bacterium M16-2 (*Adocia* sp., Nichinan-Oshima Island, Miyazaki, Japan) produced an antimicrobial agent *o*-aminophenol (**SB-9**) [185]. *Vibrio* sp. (*Hyatella* sp.) produced 4-bis(3-indolyl) methylphenol (**SB-10**) that showed inhibition of *B. subtilis* with a MIC value of 70 μg/mL [186]. Trisindoline (**SB-11**) was isolated from a culture of *Vibrio* sp. (*Hyrtios altum*, Aragusuku Island, Okinawa Prefecture) [187]. *Vibrio* sp. M22-I (*Hyatella* sp.) metabolized andrimid (**SB-12**) [188]. *Pseudomonas* sp. KK10206C (*Halichondria okadai*, Numazu area of Suruga Bay, Shizuoka Prefecture) produced an unusual CSO-carotenoid okadaxanthin (**SB-13**) [189]. *Bacillus pumilus* (*Ircinia* sponge) produced five surfactin-like cyclic depsipeptides, bacircines (**SB-14**–**SB-18**) [190]. Fermentation of *Havobacterium* sp. (*Homaxinella* sp., Palau) produced the carotenoid myxol **SB-19** [191]. Diketopiperazine **SB-20** was from *Pseudomonas aeruginosa* (*Isodictya setifera*, Hut Point and Danger Slopes on Ross Island, Antarctica) [192]. 2-Heptyl-1,2,4-trihydroxyquinoline (**SB-21**) was from *Pseudomonas* (*Suberea creba*, the eastern coast of New Caledonia) [193]. Tripeptide **SB-22** was from *Alteromonas* sp. DF-1 (*Dysidea fragilis*, the Bulgarian Black Sea) [194]. Thiopeptides **SB-23** and **SB-24** were isolated from the culture of *Bacillus cereus* (*Halichondria japonica*, Iriomote Island, Japan), both of which exhibited potent antibacterial activities against *Staphylococci* and *Enterococci* sp. with the MIC of 0.025 and 0.1 μg/mL for **SB-23** and 0.025 and 0.025 μg/mL for **SB-24**, respectively [195]. Pseudoalterobactins A (**SB-25**) and B (**SB-26**) were from *Pseudoalteromonas* sp. (*Cinachyrella australiensis*, Palau), both of which displayed strong binding affinity for the ferric ion with the ED_{50} value of 20 μM compared to enterobactin (ED_{50} 60 μM) and desferrioxamine B (ED_{50} 500 μM) [196]. *Ruegeria* species (*Suberites domuncula*, Gulf of Naples, Italy) produced the cyclic peptides **SB-27** and **SB-28** [197].

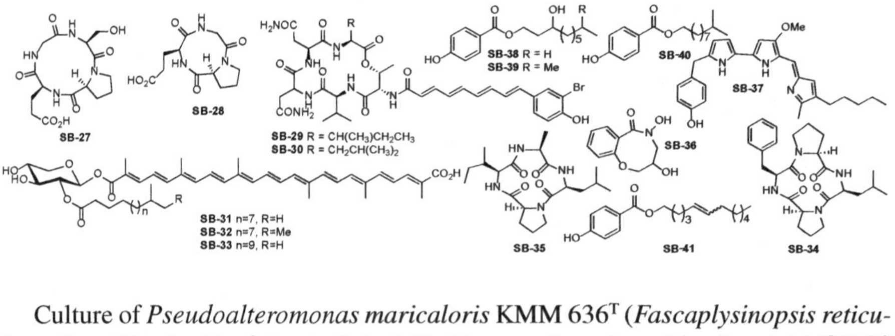

Culture of *Pseudoalteromonas maricaloris* KMM 636[T] (*Fascaplysinopsis reticulata*, Great Barrier Reef, Australia) yielded bromoalterochromides A and A′ (**SB-29**

and **SB-30**) with the ratio of 3:1, which displayed moderate toxicity to eggs of the sea urchin *Strongylocentrotus intermedius* [198]. *Rubritalea squalenifaciens* (*Halichondria okadai*, Miura Peninsula, Kanagawa, Japan) produced xylosyl diapolycopenediates A–C (**SB-31**–**SB-33**), among which **SB-31** displayed $^1O_2$ suppression activity with an IC_{50} of 5.1 μM [199]. Tetrapeptides **SB-34** and **SB-35** were from *Pseudoalteromonas* sp. S-9 (*Halisarca ectofibrosa*, Burapha, Thailand) [200]. Siderophore **SB-36** was from a bacterial strain (unidentified sponge, Japan and Indonesia) [201]. *Pseudoalteromonas rubra* (*Mycale armata*, Kaneohe Bay, Hawaii) produced 2-(4-hydroxybenzyl)prodigiosin (**SB-37**) that was cytotoxic to SK-OV-3 cells with an IC_{50} value of 1.3 μM [202]. Parabens **SB-38**–**SB-41** from *Microbulbifer* sp. L4-n2 (*Leuconia nivea*; Concarneau, France) exhibited bacteriocidal or bacteriostatic properties against *S. aureus* with the MIC values of 20.5–170 μM. This study first revealed that a paraben metabolite is persistently produced by a *Microbulbifer* bacterial strain within its sponge host [203].

13.5 Conclusions

According to the statistics (Tables 13.1, 13.2, 13.3, and 13.4, Fig. 13.1), the study on the sponge microbial MNPs could be traced back to 1988 when cyclo-(Pro-Leu), cyclo-(Pro-Val), and cyclo-(Pro-Ala) were identified from the metabolites of *Micrococcus* sp. associated with *Tedania ignis* [146]. Up to August 2014, 519 new MNPs sourced from the marine sponge microorganisms in the literatures were reported.

These new MNPs have a variety of chemical structures including nitrogen compounds, polyketides, steroids, terpenoids, and so on (Fig. 13.2), 50% of which showed bioactivities (Fig. 13.3). The amounts of the halogenated MNPs from marine sponge microorganisms are very small, whose major producers are *Aspergillus* sp. and other fungi (Fig. 13.4). Nitrogen compounds and polyketides are the two major structural types of MNPs from marine sponge microorganisms (Fig. 13.2). For each kind of microbes, 83% of new MNPs from *Penicillium* sp. are polyketides, while the ratios of *Aspergillus* sp., other fungi, actinobacteria, and bacteria producing new polyketides are 67%, 56%, 62%, and 41%, respectively (Fig. 13.5). And the ratios for nitrogen-producing compounds of *Aspergillus* sp., *Penicillium* sp., other fungi, actinobacteria, and bacteria are 29%, 20%, 33%, 60%, and 76%, respectively (Fig. 13.4). *Aspergillus* sp. and other fungi are the two main

Table 13.1 MNPs from marine sponge fungi (1996–2014)

Compound	Producing strain	Environment source	Bioactivity	Ref.
SF-1–SF-2	*A. cf. ochraceus* 941026	*Jaspis cf. coriacea*, Indian-Pacific Ocean	/	[5]
SF-3	*A. niger*	*Hyrtios* sp., Florida, USA	Cytotoxicity	[6]
SF-4–SF-6	*A. ostianus* IBT 12704	Unidentified sponge, Pohnpei, Micronesia	/	[7]
SF-7	*A. niger* 94-1212	*Hyrtios proteus*, Florida, USA	/	[8]
SF-8–SF-17	*A. versicolor* (Vuill) Triab	*Xestospongia exigua*, Bali Island, Indonesia	/	[9, 10]
SF-18	*A. niger* FT-0554	Unidentified sponge, Palau	Inhibited NFRD of *Ascaris suum*	[11]
SF-19–SF-22	*A. ostianus* TUF 01F313	Unidentified sponge, Pohnpei, Micronesia	**SF-19–SF-21**: antibacterial activity	[12, 13]
SF-23–SF-29	*A. niger*	*Axinella damicornis*, Elba, Italy	**SF-23**: cytotoxicity **SF-25**: neuroprotection	[14]
SF-30–SF-39	*Aspergillus* sp. CNK-371	Unidentified sponge, Hawaii State	**SF-36–SF-38**: cytotoxicity	[15–17]
SF-40 and **SF-41**	*A. aculeatus* CRI323-04	*Xestospongia testudinaria*, Phi Phi Island, Thailand	**SF-40**: α-glucosidase inhibition	[18, 19]
SF-42–SF-55	*A. ustus* 8009	*Suberites domuncula*, the Adriatic Sea	**SF-45** and **SF-46**: cytotoxicity	[20, 21]
SF-56 and **SF-57**	*A. insuetus*	*Petrosia ficiformis*, Santa Ana Alhambra Nestorius, Spain	Inhibitor of the mammalian mitochondrial respiratory chain	[22]
SF-58	*A. sclerotiorum* Huber SP080903f04	*Mycale* sp., Okinawa Island, Japan	/	[23]
SF-59	*A. versicolor*	*Petrosia* sp., Jeju Island, Korea	Cytotoxicity	[24]
SF-60 and **SF-61**	*Aspergillus* sp. fs14	Unidentified sponge, Okinawa Island, Japan	/	[25]
SF-62–SF-64	*A. aculeatus* CRI322-03	Unidentified sponge, Phi Phi Island, Thailand	/	[26]

(continued)

Table 13.1 (continued)

Compound	Producing strain	Environment source	Bioactivity	Ref.
SF-65–SF-68	*A. insuetus* OY-207	*Psammocinia* sp., Israel	**SF-65**: antibacterial activity; **SF-67** and **SF-68**: cytotoxicity	[27]
SF-69	*A. versicolor* PF10M	*Petrosia* sp., Jeju Island, Korea	Cytotoxicity	[28]
SF-70	*Eurotium cristatum* KUFC 7356	*Mycale* sp., State Beach, Thailand	/	[29]
SF-71–SF-84	*Aspergillus* sp.	*Tethya aurantium*, Mediterranean, Italy	**SF-83 and SF-84**: antibacterial activity	[30–32]
SF-85–SF-91	*Aspergillus* sp.	*Xestospongia testudinaria*, South China Sea, China	**SF-85-88**: antibacterial activity **SF-89**, **SF-91**: cytotoxicity	[33, 34]
SF-92–SF-96	*A. unguis* CRI282-03	Unidentified sponge CRI282, Thailand	**SF-92–SF-94**: aromatase inhibition **SF-92–SF-93**: XXO scavenging activity	[35]
SF-97–SF-99	*A. versicolor* MF359	*Hymeniacidon perlevis*, Bohai, China	**SF-99**: antibacterial activity	[36]
SF-100–SF-102	*P. montanense*	*Xestospongia exigua*, Bali Sea, Indonesia	**SF-101**: antibacterial activity	[37]
SF-103–SF 105	*P. brocae*	*Zyzzya* sp., Fijian, Japan	/	[38]
SF-106–SF-109	*P. citrinum*	*Axinella* sp., Papua New Guinea	**SF-106** and **SF-107**: antibacterial activity	[39, 40]
SF-110–SF-111	*P. brevicompactum*	*P. ficiformis*, Elba Island, Italy	/	[41]
SF-112 and **SF-113**	*Penicillium* sp.	*Axinella verrucosa*, Mediterranean Sea, Elba Island, Italy	Cytotoxicity; brine shrimp lethality	[42]
SF-114 and **SF-115**	*P. chrysogenum*	*Ircinia fasciculate*, Fetovaia Bight, Italy	Anti-HIV activity	[43]
SF-116–SF-119	*P. aurantiogriseum*	*Mycale plumose*, Jiaozhou Bay, China	**SF-116–SF-118**: cytotoxicity	[44, 45]
SF-120–SF-126	*P. rugulosum*	*Chondrosia reniformis*, Elba, Italy	/	[46]
SF-127–SF-130	*P. lilacinus*	*Petrosia* sp., Jeju Island, Korea	/	[47]

(continued)

Table 13.1 (continued)

Compound	Producing strain	Environment source	Bioactivity	Ref.
SF-131–SF-133	*P. chrysogenum* R03-8/4	*Ircinia fasciculate*, Elba, Italy	/	[48]
SF-134	*P. chrysogenum* LF066	*Tethya aurantium*, Adriatic Sea, Croatia	/	[49]
SF-135	*P. citrinum* SpI080624G1f01	Ishigaki Island, Okinawa, Japan	Neuroprotection	[50]
SF-136–SF-138	*Penicillium* sp. J05B-3-F-1	*Stelletta* sp., Jeju Island, Korea	**SF-136**: IL-1β inhibition	[51]
SF-139–SF-141	*P. oxalicum* SCSGAF 0023	Unidentified sponge, South China Sea	**SF-139**: cytotoxicity	[52]
SF-142–SF-144	*Penicillium* sp. fs36	Sponge, Takarajima Island, Japan	/	[53]
SF-145	*P. citrinum*	Ishigaki Island, Okinawa, Japan	DPPH radical scavenging activity	[54]
SF-146–SF-155	*Penicillium* sp. MWZ14-4	Sponge, South China Sea, China	**SF-152:** antibacterial activity	[55]
SF-156	*Penicillium* sp. JF-55	Unidentified sponge, Jeju Island, Korea	PTP1B inhibition; anti-inflammation	[56]
SF-157 and **SF-158**	*Penicillium* sp. Ma (M3)V and *Trichoderma* sp. Gc (M2)1	*Mycale angulosa*, and *Geodia corticostylifera*, Sao Paulo, Brazil	/	[57]
SF-159	*Trichoderma harzianum*	*Micale cecilia*, Japan	/	[58]
SF-160 and **SF-161**	Unidentified fungal strain	*Styloteflu* sp., Fiji	/	[59]
SF-162–SF-164	Unidentified fungal strain	*Jaspis aff. Johnstoni*, Pacific	**SF-162**: cytotoxicity	[60]
SF-165–SF-167	*Microsphaeropsis olivacea* F010	*Agelus* sp., Florida	/	[61, 62]
SF-168	*Exophiala pisciphila* NI10102	*Mycale adhaerens*, Japan	**SF-168**: antibacterial activity	[63]
SF-169	Unidentified fungal strain #951014	*Spirastrella vagabunda*, Indonesia	/	[64]
SF-170–SF-172	*Microascus longirostris* SF-73	Unidentified sponge, New Zealand	Inhibition of papain and cathepsins L and B	[65]
SF-173–SF-195	*Gymnascella dankaliensis*	*Halichondria japonica*, Japan	**SF-173–SF-175**, **SF-177–SF-179**, **SF-181–SF-189**, **and SF-192–SF-194**: cytotoxicity	[66–74]

(continued)

Table 13.1 (continued)

Compound	Producing strain	Environment source	Bioactivity	Ref.
SF-196–SF-199	Unidentified fungal strain no.95-1005C	*Haliclona* sp., Pacific	/	[75]
SF-200	*Trichoderma longibrachiatum*	*Haliclona* sp.	/	[76, 77]
SF-201–SF-205	*Trichoderma harzianum* UPS-N115	*Halichondria okadai*, Japan	**SF-201–SF-203**: cytotoxicity	[78, 79]
SF-206	*Paecilomyces* cf. *javanica* 961331	*Jaspis coriacea*, Fiji	/	[80]
SF-207–SF-209	*Aureobasidium pullulans*	The sponge, Japan	/	[81]
SF-210 and **SF-211**	*Coniothyrium* sp. 193H77	*E. perox*, Caribbean Sea	/	[82]
SF-212	*Microsphaeropsis* sp. H5-50	*M. incrustans*, German	/	[82]
SF-213	*Ulocladium botrytis* 193A4	*Callyspongia vaginalis*, Caribbean Sea	/	[83]
SF-214 and **SF-215**	*Drechslera hawaiiensis* P22-96	*Callyspongia aerizusa*, Indonesia	/	[84]
SF-216	*Asteromyces cruciatus* H5-81	*Myxilla incrustans*, German	/	[85]
SF-217–SF-220	Unidentified fungal strain I96S21	Unidentified sponge, Indonesia	/	[86]
SF-221–SF-225	*Microsphaeropsis* sp.	*Aplysina aerophoba*, Mediterranean Sea	**SF-221, SF-223–SF-225**: inhibition of protein-tyrosine kinases	[87]
SF-226	*Hortaea werneckii*	*Aplysina aerophoba*, Mediterranean Sea	/	[88]
SF-227–SF-229	*Cladosporium herbarum*	*Callyspongia aerizusa*, Bali	**SF-229**: antibacterial activity	[89]
SF-230	*Curvularia lunata*	*Niphates olemda*, Indonesia	**SF-230**: antibacterial activity	[90]
SF-231 and **SF-232**	*Cladosporium herbarum*	*Aplysina aerophoba*, Mediterranean Sea	Anti-*Artemia salina*	[90]
SF-233	*Cladosporium herbarum*	*Callyspongia aerizusa*, Indonesia	/	[90]
SF-234–SF-237	*Emericella variecolor* M75-2	Unidentified sponge, Caribbean waters of the Mochima Bay	**SF-234**: cytotoxicity **SF-237**: antibacterial activity	[91]
SF-238 and **SF-239**	*Microsphaeropsis* sp.	*Aplysina aerophoba*, France	/	[92]

(continued)

Table 13.1 (continued)

Compound	Producing strain	Environment source	Bioactivity	Ref.
SF-240 and **SF-241**	*Emericella variecolor*	*Haliclona valliculata*, Elba/Italy	**SF-240**: cytotoxicity	[93]
SF-242 and **SF-243**	*Clonostachys rogersoniana* HJK9	*Halicondria japonica*, Numazu, Japan	Selective inhibition of dinoflagellate, *Prorocentrum micans*	[94]
SF-244–**SF-246**	*Phomopsis asparagi* No. 031113a	*Rhaphidophlus juniperina*, US Virgin Islands	**SF-246**: antibacterial activity; cytotoxicity	[95]
SF-247	*Myrothecium* sp. TUF 02F6	Unidentified sponge, Manado, Indonesia	Cytotoxicity	[96]
SF-248–**SF-250**	The fungus 001314c	*Ianthella* sp., Papua New Guinea	**SF-248–SF-249**: antibacterial activity	[97]
SF-251	*Clonostachys* sp. ESNA-A009	Unidentified sponge, Japan	Cytotoxicity	[98]
SF-252 and **SF-253**	*Acremonium* sp.	*Eichaxinella* sp., Papua New Guinea	**SF-252**: antibacterial activity **SF-252–SF-253**: cytotoxicity	[99]
SF-254 and **SF-255**	*Spicellum roseum*	*Ectyplasia perox*, Dominica	Cytotoxicity	[100]
SF-256–**SF-260**	*Trichoderma* sp.	*Agelas dispar*, Dominica	/	[101, 102]
SF-261	*Cryptosphaeria eunomia* var. *eunomia*	Unidentified sponge, Pohnpei Island	/	[103]
SF-262	*Exophiala* sp.	*Halichondria panicea*, Korea	UVA protecting activity	[104]
SF-263–**SF-267**	*Pichia membranifaciens* (J04J-1)F-9E	*Petrosia* sp., Jeju Island Korea	/	[105; 106]
SF-268 and **SF-269**	*Scopulariopsis brevicaulis*	*Tethya aurantium*, Limski Fjord, Croatia	Antibacterial activity; cytotoxicity	[107]
SF-270 and **SF-271**	*Exophiala* sp.	*Halichondria panicea*, Bogil Island, Jeonnam Province, Korea	Antibacterial activity	[108]
SF-272–**SF-274**	*Mycelia sterilia* 97F49	Unidentified sponge, unknown place	/	[109]

(continued)

Table 13.1 (continued)

Compound	Producing strain	Environment source	Bioactivity	Ref.
SF-275 and **SF-276**	Unidentified fungus	Unidentified sponge, Surin Island, Phang Nga Province, Thailand	/	[110]
SF-277–SF-279	*Petriella* sp. TUBS 7961	*Suberites domuncula*, Croatia	**SF-277**: cytotoxicity	[111]
SF-280–SF-283	*Acremonium* sp. J05B-1-F-3	*Stelletta* sp., Jeju Island, Korea	/	[112]
SF-284 and **SF-285**	*Acremonium* sp. SpF080624G1f01	*Demospongiae*, Ishigaki Island, Japan	/	[113]
SF-286	*Beauveria bassiana*	Unidentified sponge, Hale Granville Island, Germany	/	[114]
SF-287 and **SF-288**	*Phoma* sp.	*Ectyplasia perox*, Caribbean Sea, Dominica	Cytotoxicity	[115]
SF-289 and **SF-290**	*Pichia membranifaciens* USF-H25	*Halichondria okadai*, Izu Peninsula in Shizuoka Prefecture, Japan	DPPH radical scavenging activity	[116]
SF-291	*Trichoderma viride*	*Agelas dispar*, Dominica	/	[117]
SF-292–SF-294	*Paraconiothyrium* sp. 193H12	*Ectyplasia perox*, Dominica	**SF-293**: cytotoxicity	[118]
SF-295–SF-297	*Trichoderma* sp. SpB081112Mef2	*Pseudoceratina purpurea*, Ishigaki Island	Cytotoxicity	[119]
SF-298–SF-300	*Trichoderma* sp. 05FI48	Unidentified sponge, unknown place	**SF-299–SF-300**: antibacterial activity, cytotoxicity	[120]
SF-301	*Acremonium strictum* MB05005	*Choristida*, Korea	Antibacterial activity	[121]
SF-302–SF-306	*Arthrinium* sp. 9287	*Geodia cydonium*, Italy	**SF-306**: cytotoxicity	[122]
SF-307–SF-310	*Arthrinium sacchari*	Unidentified sponge, Japan	/	[123]

(continued)

Table 13.1 (continued)

Compound	Producing strain	Environment source	Bioactivity	Ref.
SF-311–SF-329	*Stachylidium* sp. 220	*Callyspongia* cf. *C. flammea*, Sydney	**SF-315**: antiplasmodial activity	[124–129]
			SF-316: antagonistic activity, cytotoxicity	
			SF-317–SF-318: cytotoxicity	
			SF-326: antibacterial activity, cytotoxicity, cholesterol esterase activity, antiplasmodial activity	
			SF-327: cytotoxicity, cholesterol esterase activity, antagonistic activity	
			SF-328: antiplasmodial activity, antagonistic activity; **SF-329**: antagonistic activity	
SF-330	*Emericellopsis minima*	*Hyrtios erecta*, Similan Islands	/	[130]
SF-331	*Epicoccum* sp. JJY40	*Callyspongia* sp., Sanya, China	/	[131]
SF-332	*Beauveria bassiana* TPU942	Unidentified sponge, Iriomote Island in Okinawa	Cytotoxicity	[132]
SF-333–SF-345	*Alternaria* sp. JJY-32	*Callyspongia* sp., Hainan Island, China	**SF-339–SF345**: cytotoxicity	[133]
SF-346–SF-352	*Stachybotrys chartarum* MXH-X73	*Xestospongia testudinaria*, Xisha Island, China	**SF-346**: HIV activity	[134]
SF-353–SF-356	*Stagonosporopsis cucurbitacearum*	Unidentified sponge, Japan	**SF-354**: antibacterial activity	[135]
SF-357–SF-362	*Acremonium persicinum*	*Anomoianthella rubra*, Mooloolaba, Australia	/	[136]
SF-363–SF-370	*Trichoderma atroviride* NF16	*Axinella* sp., Israel	Antibacterial activity	[137]
SF-371–SF-374	*Bartalinia robillardoides* LF550	*Tethya aurantium*, Mediterranean	**SF-372**: phosphodiesterase 4 inhibition	[138]
			SF-373: antibacterial activity	
SF-375–SF-377	*Chrysosporium articulatum*	Unidentified sponge, Korea	**SF-377**: inhibited sortase A and isocitrate lyase	[139]

(continued)

Table 13.1 (continued)

Compound	Producing strain	Environment source	Bioactivity	Ref.
SF-378 and **SF-379**	*Hypocreales* sp. HLS-104	*Gelliodes carnosa*, South China Sea	/	[140]
SF-380 and **SF-381**	*Dichotomomyces cejpii*	*Callyspongia* sp. cf. *C. flammea*, Australia	**SF-380**: CB2 antagonist	[141]
SF-382–SF-384	*Talaromyces* sp. LF458	*Axinella verrucosa*, Italy	**SF-382** and **SF383**: antibacterial activity; **SF-382** and **SF-384**: acetylcholinesterase inhibition	[142]
SF-385	*Xylariales* 05FI52	Unidentified sponge, Indonesia	Activation of p21 promoter	[143]
SF-386–SF-391	*Hansfordia sinuosae*	*Niphates* sp., Southern China Sea	**SF-389**: reduce the triglycerides and total cholesterol	[144]
SF-392	*Neosartorya paulistensis* KUFC 7897	*Chondrilla australiensi*, Thailand	/	[145]

/ no bioactivity was reported

Table 13.2 MNPs from marine sponge actinobacteria (1988–2014)

Compound	Producing strain	Environment source	Bioactivity	Ref.
SA-1–SA-3	*Micrococcus*	*Tedania ignis*, Florida Keys	/	[146]
SA-4 and **SA-5**	*Streptomyces* sp. Ni-80	Unidentified sponge, Urauchicove, Japan	Antibacterial activity	[147]
SA-6	*Micrococcus luteus*	*Xestospongia* sp., New Caledonia	/	[148]
SA-7–SA-13	*Streptomyces* sp. KM86-9B	Unidentified sponge, Keomun Island, Korea	Topo I inhibition	[149]
SA-14	*Micromonospora* sp. L-25-ES25-008	Unidentified sponge, Indian Ocean, Mozambique	Cytotoxicity	[150]
SA-15– SA-19	*Microbacterium* sp.	*Halichondria panacea*, Adriatic coast, Rovinj, Croatia	/	[151]
SA-20 and **SA-21**	*Micromonospora* sp. L-31-CLCO-02	*Clathrina coriacea*, Fuerteventura Island, Canary Islands Archipelago	Cytotoxicity	[152]
SA-22 and **SA-23**	*Streptomyces* species	Unidentified sponge, Jaeju Island, Korea	Inhibition of recombinant enzyme sortase B	[153]

(continued)

Table 13.2 (continued)

Compound	Producing strain	Environment source	Bioactivity	Ref.
SA-24– SA-31	*Streptomyces* sp. HB202	*Halichondria panicea*, Baltic Sea, Germany	Antibacterial activity	[154]
SA-32 and **SA-33**	*Saccharopolyspora cebuensis* SPE 10-1	*Haliclona* sp., Cebu, Philippines	/	[155]
SA-34	*Saccharopolyspora taberi* PEM-06-F23-019B	*Clathrina coriacea*, coast of Tanzania	/	[156]
SA-35 and **SA-36**	*Brevibacterium* sp. KMD 003	*Callyspongia* sp., Kyung-po, S. Korea	Antibacterial activity	[157]
SA-37	*Actinomadura* sp. SpB081030SC-15	Unidentified sponge, Ishigaki Island., Okinawa	DPPH activity	[158]
SA-38– SA-41	*Nocardiopsis* sp. HB383	*Halichondria panicea*, Baltic Sea, Germany	/	[159]
SA-42– SA-44	*Streptomyces* sp. SpC080624SC-11	*Cinachyra* sp., Nagura Bay, Ishigaki, Okinawa	/	[160]
SA-45 and **SA-46**	*Streptomyces* sp. Sp080513GE-23	*Haliclona* sp., Chiba Prefecture, Japan.	DPPH activity	[161]
SA-47	*Streptomyces* sp. NBRC 105896	*Haliclona* sp., Tateyama, Chiba Prefecture, Japan	/	[162]
SA-48 and **SA-49**	*Streptomyces* sp. Sp080513GE- 26	*Haliclona* sp., Tateyama, Chiba Prefecture, Japan	/	[163]
SA-50	*Streptomyces* sp. SpC080624SC-11	*Demospongiae,* Ishigaki City, Okinawa	Cytotoxicity	[164]
SA-51	*Streptomyces* sp. strain HB202	*Halichondria panicea*, Baltic Sea	Cytotoxicity	[165]
SA-52 and **SA-53**	*Streptomyces carnosus* strain AZS17	*Hymeniacidon* sp., coastal waters of the East China Sea	Cytotoxicity	[166]
SA-54	*Streptomyces microflavus*	*Hymeniacidon perlevis*, coast of Dalian, China	/	[167]
SA-55	*Actinomadura* sp. SBMs009	*Suberites japonicus*, source unspecified	Inhibitor of NF-κB and glucocorticoid receptors	[168]
SA-56 and **SA-57**	*Streptomyces* sp. SpD081030SC-03	*Demospongiae* sponge, Ishigaki City, Okinawa, Japan	/	[169]
SA-58	*Streptomyces* sp. DA22	*Craniella australiensis*, South China Sea	/	[170]
SA-59– SA-63	*Verrucosispora* sp. Strain WMMA107	*Chondrilla caribensis f. caribensis*, Florida Keys, USA	**SA-61**–**SA-63**: cytotoxicity	[171]

(continued)

Table 13.2 (continued)

Compound	Producing strain	Environment source	Bioactivity	Ref.
SA-64– SA-67	*Streptomyces axinellae* Pol001^T	*Axinella polypoides*, Banyuls-sur-Mer, France	Antibacterial activity	[172]
SA-68– SA-70	*Streptomyces* sp. RM72	Unidentified sponge, Takarajima Island, Kagoshima Prefecture, Japan	Human HDAC1 inhibitory activity	[173]
SA-71– SA-74	*Streptomyces* sp. RV15	*Dysidea tupha*, Rovinj, Croatia	/	[174]
SA-75– SA-77	*Streptomyces* sp. BCC45596	*Xestospongia* sp., Sichang Island, Chonburi, Thailand	Cytotoxicity, antimalarialactivity, antitubercular activity	[175]
SA-78	*Kocuria palustris*	Unidentified sponge, Florida Keys, USA	Antibacterial activity	[176]
SA-79	*Streptomyces tateyamensis* NBRC 105047	Unidentified sponge, unknown place	/	[177]
SA-80 and **SA-81**	*Actinokineospora* sp. EG49	*Spheciospongia vagabunda*, Rovinj	**SA-80**: insecticidal activity	[178]
SA-82 and **SA-83**	*Actinokineospora* sp. EG49 and *Nocardiopsis* sp. RV163	*Spheciospongia vagabunda*, Rovinj, and *Dysidea avara*, Mediterranean	/	[179]
SA-84– SA-86	*Amycolatopsis* sp.	Unidentified sponge, Micronesia	Cytotoxicity	[180]

/ no bioactivity was reported

Table 13.3 MNPs from marine sponge bacteria (1990–2014)

Compound	Producing strain	Environment source	Bioactivity	Ref.
SB-1	*Vibrio* sp.	*Dysidea* sp., Tutuila and Ofu Islands, Eastern Samoa	/	[181]
SB-2–SB-5	*Micrococcus* sp. SD.3	*Tedania ignis* sponge, unknown place	/	[182]
SB-6 and **SB-7**	*Alteromonas* sp.	*Darwinella rosacea*, Harrington Sound, Bermuda	Antibacterial activity	[183]
SB-8	*Alteromonas* sp.	*Halichondria okadai*, Nagai, Kanagawa	Cytotoxicity	[184]
SB-9	Unidentified bacterium M16-2	*Adocia* sp., Nichinan-Ooshima Island, Miyazaki, Japan	Antibacterial activity	[185]
SB-10	*Vibrio sp.*	*Hyatella* sp., unknown place	Antibacterial activity	[186]
SB-11	*Vibrio* sp.	*Hyrtios altum*, Aragusuku Island, Okinawa Prefecture	/	[187]

(continued)

Table 13.3 (continued)

Compound	Producing strain	Environment source	Bioactivity	Ref.
SB-12	*Vibrio* sp. M22-I	*Hyatella* sp., unknown place	/	[188]
SB-13	*Pseudomonas* sp. KK10206C	*Halichondria okadai*, Numazu area of Suruga Bay, Shizuoka Prefecture	/	[189]
SB-14–SB-18	*Bacillus pumilus*	Genus *Ircinia* sponge, unknown place	/	[190]
SB-19	*Havobacterium* sp.	*Homaxinella* sp., Palau	/	[191]
SB-20	*Pseudomonas aeruginosa*	*Isodictya setifera*, Hut Point and Danger Slopes on Ross Island, Antarctica	/	[192]
SB-21	*Pseudomonas*	*Suberea creba*, the eastern coast of New Caledonia	/	[193]
SB-22	*Alteromonas* sp. DF-1	*Dysidea fragilis*, the southern part of the Bulgarian Black Sea	/	[194]
SB-23 and **SB-24**	*Bacillus cereus*	*Halichondria japonica*, Hoshisuna Beach, Iriomote Island, Okinawa Prefecture, Japan	Antibacterial activity	[195]
SB-25 and **SB-26**	*Pseudoalteromonas* sp.	*Cinachyrella australiensis*, Palau	Binding affinity for the ferricion activity	[196]
SB-27 and **SB-28**	*Ruegeria* specie	*Suberites domuncula*, Gulf of Naples, Italy	/	[197]
SB-29 and **SB-30**	*Pseudoalteromonas maricaloris* KMM 636^T	*Fascaplysinopsis reticulata*, Great Barrier Reef, Australia	/	[198]
SB-31–SB-33	*Rubritalea squalenifacien*	*Halichondria okadai*, Miura peninsula, Kanagawa, Japan	**SB-31:** ^{1}O2 suppression activity	[199]
SB-34 and **SB-35**	*Pseudoalteromonas* sp. S-9	*Halisarca ectofibrosa*, Burapha, Gulf of Thailand	/	[200]
SB-36	Unidentified bacterial strain	Unidentified sponge, Japan and Indonesia	/	[201]
SB-37	*Pseudoalteromonas rubra*	*Mycale armata*, Kaneohe Bay, Oahu, Hawaii	Cytotoxicity	[202]
SB-38–SB-41	*Microbulbifer* sp. L4-n2	*Leuconia nivea*, Concarneau, France	Antibacterial activity	[203]

/ no bioactivity was reported

Table 13.4 The initial MNP research from marine sponge microbes

First producing strain	Environment source	MNPs	Time
Aspergillus ochraceus 941026	*Jaspis coriacea*, Indian-Pacific Ocean	Chlorocarolides A and B (**SF-1** and **SF-2**)	1996
Penicillium montanense	*Xestospongia exigua*, Bali Sea, Indonesia	Xestodecalactones A–C (**SF-100–SF-102**)	2002
Other fungi (*Trichoderma harzianum*)	*Micale cecilia*, Japan	Trichoharzin (**SF-159**)	1993
Actinobacteria (*Micrococcus*)	*Tedania ignis*, Florida Keys	Diketopiperazines (**SA-1–SA-3**)	1988
Bacteria (*Vibrio* sp.)	*Dysidea* sp., Tutuila and Ofu islands, Eastern Samoa	3,5-Dibromo-2-(3′,5′-dibromo-2′-methoxyphenoxy)phenol (**SB-1**)	1990

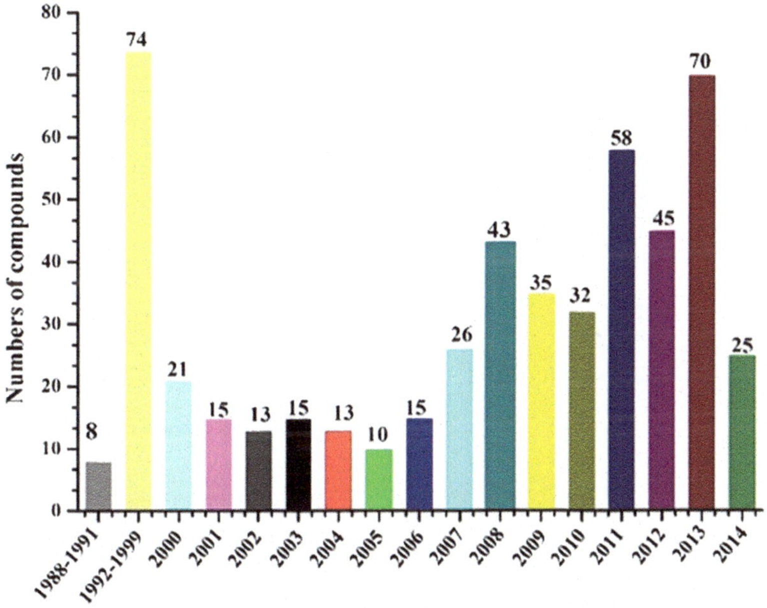

Fig. 13.1 Annual number of MNPs from marine sponge microbes since 1988

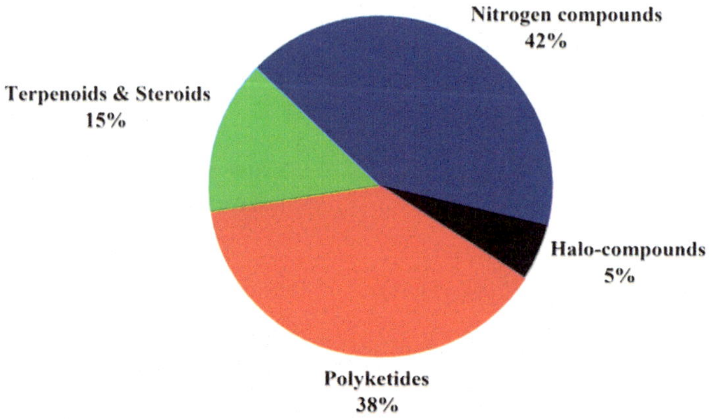

Fig. 13.2 Structural types of MNPs from marine sponge microbes

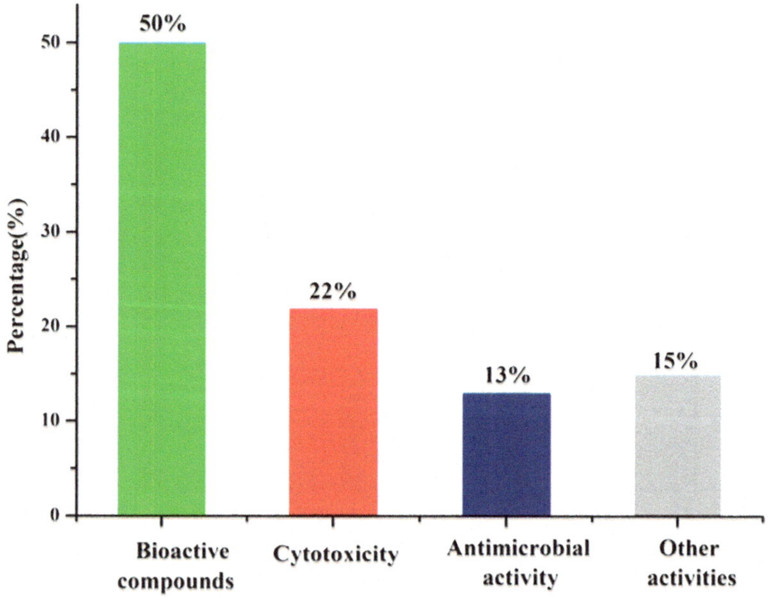

Fig. 13.3 Bioactive categories of MNPs from marine sponge microbes

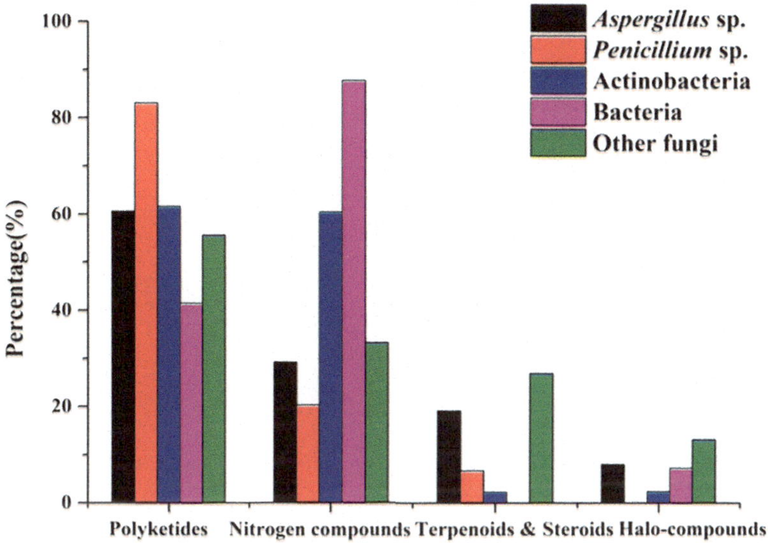

Fig. 13.4 Structural categories of MNPs from each kind of marine sponge microbes

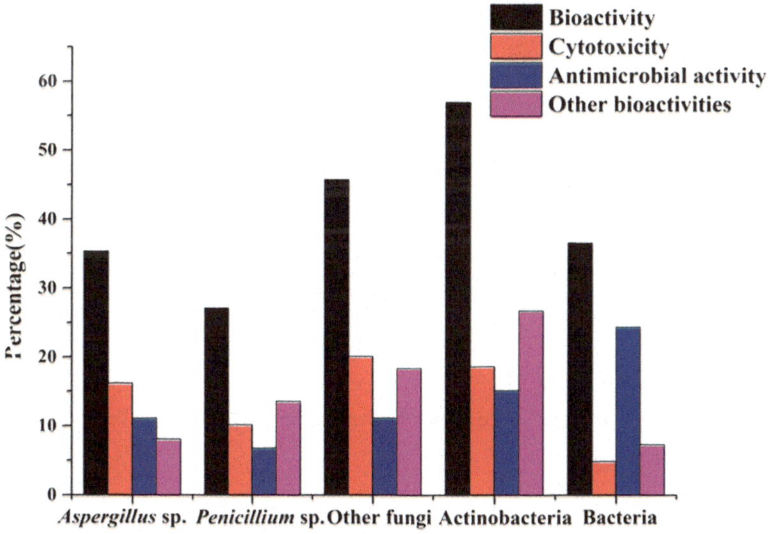

Fig. 13.5 Bioactivities of MNPs from each kind of marine sponge microbes

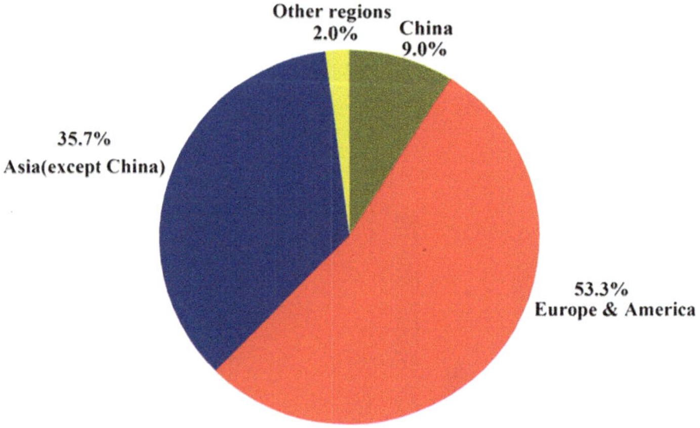

Fig. 13.6 Country categories of papers for publishing MNPs from marine sponge microbes

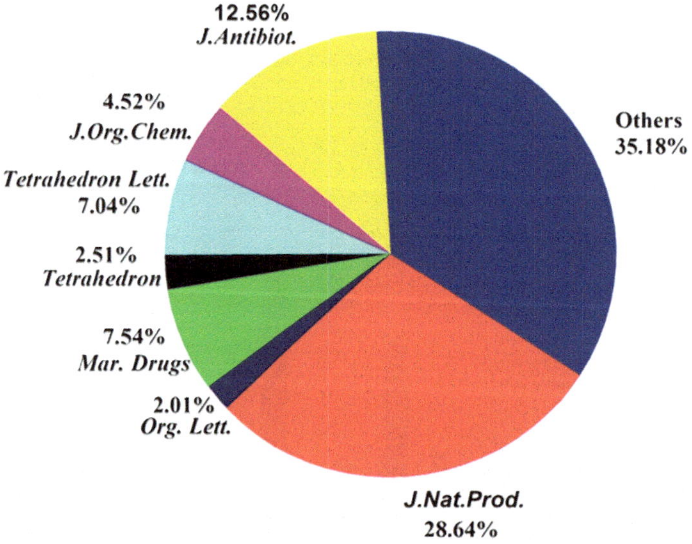

Fig. 13.7 The journal categories of the papers for publishing MNPs from marine sponge microbes

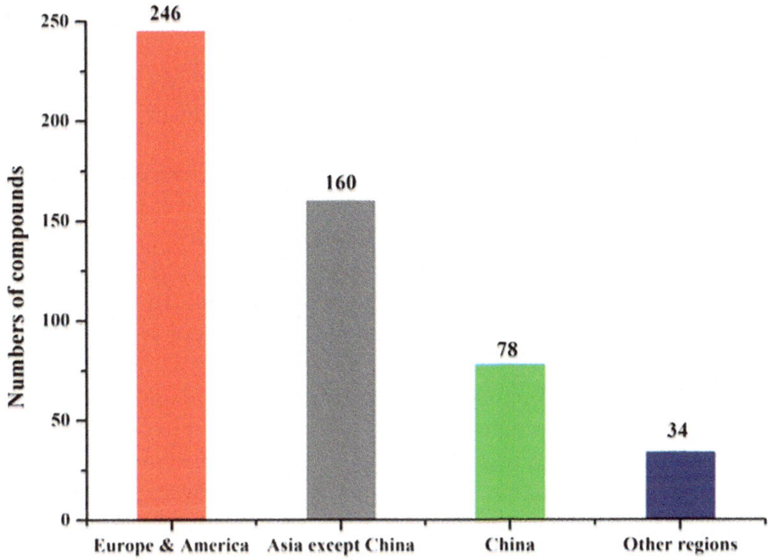

Fig. 13.8 The numbers of MNPs from marine sponge microbes by different countries

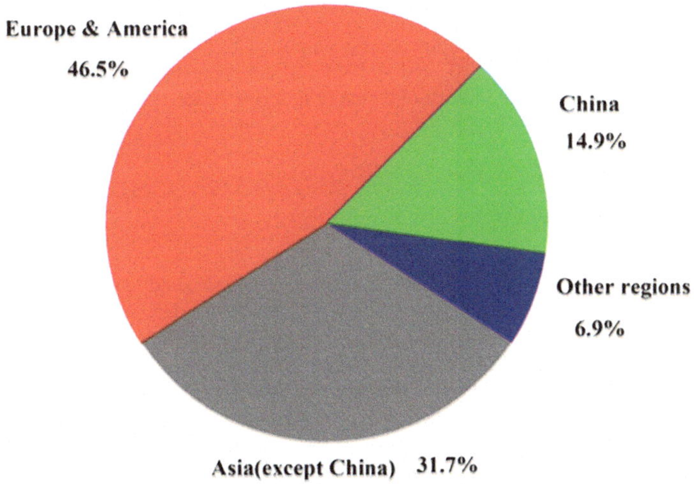

Fig. 13.9 The ratios for publishing MNPs from marine sponge microbes by different countries

producing microbes for steroids and terpenoids, 19% and 27% MNPs of which are steroids and terpenoids, respectively (Fig. 13.4).

Fifty percent of MNPs from marine sponge *microorganisms* are bioactive, and the major bioactivities are cytotoxic activity and antimicrobial activity with the proportion of 22% and 15%, respectively (Fig. 13.3). For each kind of microbes, 57% of MNPs from actinobacteria are bioactive, while the corresponding ratios for *Aspergillus* sp., *Penicillium* sp., other fungi, and bacteria are 35%, 27%, 46%, and 37%, respectively (Fig. 13.5). Apart from those from marine sponge bacteria, the major bioactivity of bioactive MNPs from actinobacteria, *Aspergillus* sp., *Penicillium* sp., and other fungi is cytotoxicity with the ratios of 19%, 16%, 10%, and 20% (Fig. 13.5), respectively. Antimicrobial activities are the major bioactivity of bacterial MNPs with the ratio of 24% (Fig. 13.5).

Europe, America, and Asia except China are the main regions for publishing MNPs from marine sponge microorganisms with the percentages of 53.3%, 35.7%, and 9% for published papers, respectively (Fig. 13.6). The lead journals for publishing MNPs sourced from marine sponge microorganisms were *J. Nat. Prod.* and *J. Antibiot.* with the ratios of 28.6% and 12.6%, respectively (Fig. 13.7). The scholars from Europe, America, and Asia except China published 246, 160, and 78 new MNPs from marine sponge microbes (Fig. 13.8) with the ratios of 47%, 31%, and 15% (Fig. 13.9), respectively. Sponge microorganisms will proceed to be a very potential resource of structurally novel bioactive compounds for drug discovery.

Acknowledgment This chapter was financially supported by the grants from the NSFC (Nos. 81561148012, U1501221, 81741150, and 41876172) and from the Special Fund for Marine Scientific Research in the Public Interest of China (No. 201405038).

References

1. Ma HG, Liu Q, Zhu GL, Liu HS, Zhu WM. Marine natural products sourced from marine-derived *Penicillium* fungi. J Asian Nat Prod Res. 2016;18:92–115.
2. Zhao C, Liu H, Zhu W. New natural products from the marine-derived *Aspergillus* fungi-a review. Acta Microbiol Sin. 2016;56:331–62.
3. Zhu TH, Ma YN, Wang WL, Chen ZB, Qin SD, Du YQ, et al. New marine natural products from the marine-derived fungi other than *Penicillium* sp. and *Aspergillus* sp. (1951–2014). Chin J Mar Drugs. 2015;34:56–108.
4. Zhao C, Zhu T, Zhu W. New marine natural products of microbial origin from 2010 to 2013. Chin J Org Chem. 2013;33:1195–234.
5. Abrell LM, Borgeson B, Crews P. Chloro polyketides from the cultured fungus (*Aspergillus*) separated from a marine sponge. Tetrahedron Lett. 1996;37:2331–4.
6. Varoglu M, Corbett TH, Valeriote FA, Crews P. Asperazine, a selective cytotoxic alkaloid from a sponge-derived culture of *Aspergillus niger*. J Org Chem. 1997;62:7078–9.
7. Rahbaek L, Breinholt J. Circumdatins. D, E, and F: further fungal benzodiazepine analogues from *Aspergillus ochraceus*. J Nat Prod. 1999;62:904–5.
8. Varoglu M, Crews P. Biosynthetically diverse compounds from a saltwater culture of sponge-derived *Aspergillus niger*. J Nat Prod. 2000;63:41–3.

9. Lin WH, Fu HZ, Li J, Proksch P. Novel chromone derivatives from marine fungus *Aspergillus versicolor* isolated from the sponge *Xestospongia exigua*. Chin Chem Lett. 2001;12:235–8.
10. Lin WH, Li J, Fu HZ, Proksch P. Four novel hydropyranoindeno-derivatives from marine fungus *Aspergillus versicolor*. Chin Chem Lett. 2001;12:435–8.
11. Ui H, Shiomi K, Yamaguchi Y, Masuma R, Nagamitsu T, Takano D, et al. Nafuredin, a novel inhibitor of NADH-fumarate reductase, produced by *Aspergillus niger* FT-0554. J Antibiot. 2001;54:234–8.
12. Namikoshi M, Negishi R, Nagai H, Dmitrenok A, Kobayashi H. Three new chlorine containing antibiotics from a marine-derived fungus *Aspergillus ostianus* collected in Pohnpei. J Antibiot. 2003;56:755–61.
13. Kito K, Ookura R, Kusumi T, Namikoshi M, Ooi T. X-ray structures of two stephacidins, heptacyclic alkaloids from the marine-derived fungus *Aspergillus ostianus*. Heterocycles. 2009;78:2101–6.
14. Hiort J, Maksimenka K, Reichert M, Perovic-Ottstadt S, Lin WH, Wray V, et al. New natural products from the sponge-derived fungus *Aspergillus niger*. J Nat Prod. 2004;67:1532–43.
15. Cueto M, MacMillan JB, Jensen PR, Fenical W. Tropolactones A–D, four meroterpenoids from a marine-derived fungus of the genus *Aspergillus*. Phytochemistry. 2006;67:1826–31.
16. Kito K, Ookura R, Yoshida S, Namikoshi M, Ooi T, Kusumi T. Pentaketides relating to aspinonene and dihydroaspyrone from a marine-derived fungus, *Aspergillus ostianus*. J Nat Prod. 2007;70:2022–5.
17. Kito K, Ookura R, Yoshida S, Namikoshi M, Ooi T, Kusumi T. New cytotoxic 14-membered macrolides from marine-derived fungus *Aspergillus ostianus*. Org Lett. 2008;10:225–8.
18. Ingavat N, Dobereiner J, Wiyakrutta S, Mahidol C, Ruchirawat S, Kittakoop P. Aspergillusol A, an α-glucosidase inhibitor from the marine-derived fungus *Aspergillus aculeatus*. J Nat Prod. 2009;72:2049–52.
19. Ingavat N, Mahidol C, Ruchirawat S, Kittakoop P. Asperaculin A, a sesquiterpenoid from a marine-derived fungus, *Aspergillus aculeatus*. J Nat Prod. 2011;74:1650–2.
20. Liu HB, Edrada-Ebel R, Ebel R, Wang Y, Schulz B, Draeger S, et al. Drimane sesquiterpenoids from the fungus *Aspergillus ustus* isolated from the marine sponge *Suberites domuncula*. J Nat Prod. 2009;72:1585–8.
21. Liu HB, Edrada-Ebel RA, Ebel R, Wang Y, Schulz B, Draeger S, et al. Ophiobolin sesterterpenoids and pyrrolidine alkaloids from the sponge-derived fungus *Aspergillus ustus*. Helv Chim Acta. 2011;94:623–31.
22. López-Gresa MP, Cabedo N, González-Mas MC, Ciavatta ML, Avila C, Primo J. Terretonins E and F, inhibitors of the mitochondrial respiratory chain from the marine-derived fungus *Aspergillus insuetus*. J Nat Prod. 2009;72:1348–51.
23. Motohashi K, Inaba S, Takagi M, Shin-Ya K. JBIR-15, a new aspochracin derivative, isolated from a sponge-derived fungus, *Aspergillus sclerotiorum* Huber Sp080903f04. Biosci Biotechnol Biochem. 2009;73:1898–900.
24. Lee Y, Dang HT, Hong J, Lee CO, Bae K, Kim D, et al. A cytotoxic lipopeptide from the sponge-derived fungus *Aspergillus versicolor*. Bull Kor Chem Soc. 2010;31:205–8.
25. Takagi M, Motohashi K, Shin-ya K. Isolation of 2 new metabolites, JBIR-74 and JBIR-75, from the sponge-derived *Aspergillus* sp. fS14. J Antibiot. 2010;63:393–5.
26. Antia BS, Aree T, Kasettrathat C, Wiyakrutta S, Ekpa OD, Ekpe UJ, et al. Itaconic acid derivatives and diketopiperazine from the marine-derived fungus *Aspergillus aculeatus* CRI322-03. Phytochemistry. 2011;72:816–20.
27. Cohen E, Koch L, Thu KM, Rahamim Y, Aluma Y, Ilan M, et al. Novel terpenoids of the fungus *Aspergillus insuetus* isolated from the Mediterranean sponge *Psammocinia* sp. collected along the coast of Israel. Bioorg Med Chem. 2011;19:6587–93.
28. Lee Y, Dang HT, Li J, Zhang P, Hong J, Lee CO, et al. A cytotoxic fellutamide analogue from the sponge-derived fungus *Aspergillus versicolor*. Bull Kor Chem Soc. 2011;32:3817–20.

29. Gomes NM, Dethoup T, Singburaudom N, Gales L, Silva A, Kijjoa A. Eurocristatine, a new diketopiperazine dimer from the marine sponge-associated fungus *Eurotium cristatum*. Phytochem Lett. 2012;5:717–20.
30. Zhou Y, Mándi A, Debbab A, Wray V, Schulz B, Müller WEG, et al. New Austalides from the sponge-associated fungus *Aspergillus* sp. Eur J Org Chem. 2011;2011:6009–19.
31. Zhou Y, Debbab A, Mándi A, Wray V, Schulz B, Müller WEG, et al. Alkaloids from the sponge-associated fungus *Aspergillus* sp. Eur J Org Chem. 2013;2013:894–906.
32. Zhou Y, Debbab A, Wray V, Lin W, Schulz B, Trepos R, et al. Marine bacterial inhibitors from the sponge-derived fungus *Aspergillus* sp. Tetrahedron Lett. 2014;55:2789–92.
33. Li D, Xu Y, Shao CL, Yang RY, Zheng CJ, Chen YY, et al. Antibacterial bisabolane-type sesquiterpenoids from the sponge-derived fungus *Aspergillus* sp. Mar Drugs. 2012;10:234–41.
34. Sun LL, Shao CL, Chen JF, Guo ZY, Fu XM, Chen M, et al. New bisabolane sesquiterpenoids from a marine-derived fungus *Aspergillus* sp. isolated from the sponge *Xestospongia testudinaria*. Bioorg Med Chem Lett. 2012;22:1326–9.
35. Sureram S, Wiyakrutta S, Ngamrojanavanich N, Mahidol C, Ruchirawat S, Kittakoop P. Depsidones, aromatase inhibitors and radical scavenging agents from the marine-derived fungus *Aspergillus unguis* CRI282-03. Planta Med. 2012;78:582–8.
36. Song F, Ren B, Chen C, Yu K, Liu X, Zhang Y, et al. Three new sterigmatocystin analogues from marine-derived fungus *Aspergillus versicolor* MF359. Appl Microbiol Biotechnol. 2014;98:3753–8.
37. Edrada RA, Heubes M, Brauers G, Wray V, Berg A, Gräfe U, et al. Online analysis of xestodecalactones A-C, novel bioactive metabolites from the fungus *Penicillium cf. montanense* and their subsequent isolation from the sponge *Xestospongia exigua*. J Nat Prod. 2002;65:1598–604.
38. Bugni TS, Bernan VS, Greenstein M, Janso JE, Maiese WM, Mayne CL, et al. Brocaenols A-C: novel polyketides from a marine-derived *Penicillium brocae*. J Org Chem. 2003;68:2014–7.
39. Kozlovsky AG, Zhelifonova VP, Ozerskaya SM, Vinokurova NG, Adanin VM, Grafe U. Cyclocitrinol, a new fungal metabolite from *Penicillium citrinum*. Pharmazie. 2000;55:470–1.
40. Amagata T, Amagata A, Tenney K, Valeriote FA, Lobkovsky E, Clardy J, et al. Unusual C25 steroids produced by a sponge-derived *Penicillium citrinum*. Org Lett. 2003;5:4393–6.
41. Bringmann G, Lang G, Steffens S, Schaumann K. Petrosifungins A and B, novel cyclodepsipeptides from a sponge-derived strain of *Penicillium brevicompactum*. J Nat Prod. 2004;67:311–5.
42. Jadulco R, Edrada RA, Ebel R, Berg A, Schaumann K, Wray V, et al. New communesin derivatives from the fungus *Penicillium* sp. derived from the mediterranean sponge *Axinella verrucosa*. J Nat Prod. 2004;67:78–81.
43. Bringmann G, Lang G, Gulder TAM, Tsuruta H, Mühlbacher J, Maksimenka K, et al. The first sorbicillinoid alkaloids, the antileukemic sorbicillactones A and B, from a sponge-derived *Penicillium chrysogenum* strain. Tetrahedron. 2005;61:7252–65.
44. Xin ZH, Zhu WM, Gu QQ, Fang YC, Duan L, Cui CB. A new cytotoxic compound from *Penicillium auratiogriseum*, symbiotic or epiphytic fungus of sponge *Mycale plumose*. Chin Chem Lett. 2005;16:1227–9.
45. Xin ZH, Fang Y, Du L, Zhu T, Duan L, Chen J, et al. Aurantiomides A-C, quinazoline alkaloids from the sponge-derived fungus *Penicillium aurantiogriseum* SP0-19. J Nat Prod. 2007;70:853–5.
46. Lang G, Wiese J, Schmaljohann R, Imhoff JF. New pentaenes from the sponge-derived marine fungus *Penicillium rugulosum*: structure determination and biosynthetic studies. Tetrahedron. 2007;63:11844–9.
47. Elbandy M, Shinde PB, Hong J, Bae KS, Kim MA, Lee SM, et al. α-Pyrones and yellow pigments from the sponge-derived fungus *Paecilomyces lilacinus*. Bull Kor Chem Soc. 2009;30:188–92.
48. Bringmann G, Lang G, Bruhn T, Schäffler K, Steffens S, Schmaljohann R, et al. Sorbifuranones A–C, sorbicillinoid metabolites from *Penicillium* strains isolated from Mediterranean sponges. Tetrahedron. 2010;66:9894–901.

49. Wiese J, Ohlendorf B, Blumel M, Schmaljohann R, Imhoff JF. Phylogenetic identification of fungi isolated from the marine sponge *Tethya aurantium* and identification of their secondary metabolites. Mar Drugs. 2011;9:561–85.
50. Ueda JY, Hashimoto J, Inaba S, Takagi M, Shin-ya K. JBIR-59, a new sorbicillinoid, from a marine-derived fungus *Penicillium citrinum* SpI080624G1f01. J Antibiot. 2010;63:203–5.
51. Li JL, Zhang P, Lee YM, Hong J, Yoo ES, Bae KS, Jung JH. Oxygenated hexylitaconates from a marine sponge-derived fungus *Penicillium* sp. Chem Pharm Bull. 2011;59:120–3.
52. Sun YL, He F, Liu KS, Zhang XY, Bao J, Wang YF, et al. Cytotoxic dihydrothiophene-condensed chromones from marine-derived fungus *Penicillium oxalicum*. Planta Med. 2012;78:1957–61.
53. Kawahara T, Takagi M, Shin-Ya K. Three new depsipeptides, JBIR-113, JBIR-114 and JBIR-115, isolated from a marine sponge-derived *Penicillium* sp. fS36. J Antibiot. 2012;65:147–50.
54. Kawahara T, Takagi M, Shin-ya K. JBIR-124: a novel antioxidative agent from a marine sponge-derived fungus *Penicillium citrinum* SpI080624G1f01. J Antibiot. 2012;65:45–7.
55. Qi J, Shao CL, Li ZY, Gan LS, Fu XM, Bian WT, et al. Isocoumarin derivatives and benzofurans from a sponge-derived *Penicillium* sp. fungus. J Nat Prod. 2013;76:571–9.
56. Lee D, Jang J, Ko W, Kim K, Sohn JH, Kang M, et al. PTP1B inhibitory and anti-inflammatory effects of secondary metabolites isolated from the marine-derived fungus *Penicillium* sp. JF-55. Mar Drugs. 2013;11:1409–26.
57. Kossuga MH, Ferreira AG, Sette LD, Berlinck RGS. Two polyketides from a co-culture of two marine-derived fungal strains. Nat Prod Commun. 2013;8:721–4.
58. Kobayashi M, Uehara H, Matsunami K, Aoki S, Kitagawa I. Trichoharzin, a new polyketide produced by the imperfect fungus *Trichoderma harzianum* separated from the marine sponge *Micale cecilia*. Tetrahedron Lett. 1993;34:7925–8.
59. Abrell LM, Cheng X, Crews P. New nectriapyrones by salt water culture of a fungus separated from an Indo-Pacific sponge. Tetrahedron Lett. 1994;35:9159–60.
60. Cheng X, Varoglu M, Abrell L, Crews P, Lobkovsky E, Clardy J. Chloriolins A-C, chlorinated sesquiterpenes produced by fungal cultures separated from a Jaspis marine sponge. J Org Chem. 1994;59:6344–8.
61. Keusgen M, Yu C, Curtis JM, Brewer D, Ayer SW. A cerebroside from the marine fungus *Microsphaeropsis olivacea* (Bonord) Höhn. Biochem Syst Ecol. 1996;24:465–8.
62. Yu CM, Curtis JM, Walter JA, Wright JG, Ayer SW, Kaleta J, et al. Potent inhibitors of cysteine proteases from the marine fungus *Microascus longirostris*. J Antibiot. 1996;49:395–7.
63. Doshida J, Hasegawa H, Onuki H, Shimidzu N. Exophilin A, a new antibiotic from a marine microorganism *Exophiala pisciphila*. J Antibiot. 1996;49:1105–9.
64. Abrell LM, Borgeson B, Crews P. A new polyketide, secocurvularin, from the salt water culture of a sponge derived fungus. Tetrahedron Lett. 1996;37:8983–4.
65. Yu C, Fathi-Afshar ZR, Curtis JM, Wright JLC, Ayer SW. An unusual fatty acid and its glyceride from the marine fungus *Microsphaeropis olivacea*. Can J Chem. 1996;74:730–5.
66. Numata A, Amagata T, Minoura K, Ito T. Gymnastatins, novel cytotoxic metabolites produced by a fungal strain from a sponge. Tetrahedron Lett. 1997;38:5675–8.
67. Amagata T, Doi M, Ohta T, Minoura K, Numata A. Absolute stereostructures of novel cytotoxic metabolites, gymnastatins A–E, from a *Gymnascella* species separated from a *Halichondria* sponge. J Chem Soc Perkin Trans. 1998;1:3585–600.
68. Amagata T, Minoura K, Numata A. Gymnastatins F-H, cytostatic metabolites from the sponge-derived fungus *Gymnascella dankaliensis*. J Nat Prod. 2006;69:1384–8.
69. Numata A, Amagata T, Takigawa K, Minoura K. Gymnastatins I-K, cancer cell growth inhibitors from a sponge-derived *Gymnascella dankaliensis*. Heterocycles. 2010;81:897–907.
70. Amagata T, Tanaka M, Yamada T, Minoura K. Gymnastatins and dankastatins, growth inhibitory metabolites of a gymnascella species from a Halichondria sponge. J Nat Prod. 2008;71:340–5.
71. Amagata T, Minoura K, Numata A. Structures for cytotoxic metabolites of a fungus separated from a sponge and total synthesis of gymnastatin I. Tennen Yuki Kagobutsu Toronkai Koen Yoshishu. 1998;40:115–20.

72. Amagata T, Tanaka M, Yamada T, Doi M, Minoura K, Ohishi H, et al. Variation in cytostatic constituents of a sponge-derived *Gymnascella dankaliensis* by manipulating the carbon source. J Nat Prod. 2007;70:1731–40.
73. Amagata T, Doi M, Tohgo M, Minoura K, Numata A. Dankasterone, a new class of cytotoxic steroid produced by a *Gymnascella* species from a marine sponge. Chem Commun. 1999;14:1321–2.
74. Amagata T, Tanaka M, Yamada T, Chen Y, Minoura K, Numata A. Additional cytotoxic substances isolated from the sponge-derived *Gymnascella dankaliensis*. Tetrahedron Lett. 2013;54:5960–2.
75. Wang G, Abrell LM, Avelar A, Borgeson BM, Crews P. New hirsutane based sesquiterpenes from salt water cultures of a marine sponge-derived fungus and the terrestrial fungus *Coriolus consors*. Tetrahedron. 1998;54:7335–42.
76. Sperry S, Samuels GJ, Crews P. Vertinoid polyketides from the saltwater culture of the fungus *Trichoderma longibrachiatum* separated from a *Haliclona* marine sponge. J Org Chem. 1998;63:10011–4.
77. Wood JL, Thompson BD, Yusuff N, Pflum DA, Matthus MSP. Total synthesis of (±)-epoxysorbicillinol. J Am Chem Soc. 2001;123:2097–8.
78. Amagata T, Usami Y, Minoura K, Ito T, Numata A. Cytotoxic substances produced by a fungal strain from a sponge: physico-chemical properties and structures. J Antibiot. 1998;51:33–40.
79. Usami Y, Ikura T, Amagata T, Numata A. First total syntheses and configurational assignments of cytotoxic trichodenones A-C. Tetrahedron Asymmetry. 2000;11:3711–25.
80. Rahbæk L, Sperry S, Piper JE, Crews P. Deoxynortrichoharzin, a new polyketide from the saltwater culture of a sponge-derived *Paecilomyces* fungus. J Nat Prod. 1998;61:1571–3.
81. Shigemori H, Tenma M, Shimazaki K, Kobayashi J. Three new metabolites from the marine yeast *Aureobasidium pullulans*. J Nat Prod. 1998;61:696–8.
82. Höller U, König GM, Wright AD. Three new metabolites from marine-derived fungi of the genera *Coniothyrium* and *Microsphaeropsis*. J Nat Prod. 1999;62:114–8.
83. Höller U, König GM, Wright AD. A new tyrosine kinase inhibitor from a marine isolate of *Ulocladium botrytis* and new metabolites from the marine fungi *Asteromyces cruciatus* and *Varicosporina ramulosa*. Eur J Org Chem. 1999;1999:2949–55.
84. Edrada RA, Wray V, Berg A, Gräfe U, Brauers G, Proksch P. Novel spiciferone derivatives from the fungus *Drechslera hawaiiensis* isolated from the marine sponge *Callyspongia aerizusa*. Z Naturforsch C Biosci. 1999;55:218–21.
85. Höller U, Wright AD, Matthee GF, Konig GM, Draeger S, Aust H, et al. Fungi from marine sponges: diversity, biological activity and secondary metabolites. Mycol Res. 2000;11:1354–65.
86. Smith CJ, Abbanat D, Bernan VS, Maiese WM, Greenstein M, Jompa J, et al. Novel polyketide metabolites from a species of marine fungi. J Nat Prod. 2000;63:142–5.
87. Brauers G, Edrada RA, Ebel R, Proksch P, Wray V, Berg A, et al. Anthraquinones and betaenone derivatives from the sponge-associated fungus *Microsphaeropsis* species: novel inhibitors of protein kinases. J Nat Prod. 2000;63:739–45.
88. Brauers G, Ebel R, Edrada R, Wray V, Berg A, Gräfe U, et al. Hortein, a new natural product from the fungus *Hortaea werneckii* associated with the sponge *Aplysina aerophoba*. J Nat Prod. 2001;64:651–2.
89. Jadulco R, Proksch P, Wray V, Sudarsono, Berg A, Gräfe U. New macrolides and furan carboxylic acid derivative from the sponge-derived fungus *Cladosporium herbarum*. J Nat Prod. 2001;64:527–30.
90. Jadulco R, Brauers G, Edrada RA, Ebel R, Wray V, Sudarsono S, et al. New metabolites from sponge-derived fungi *Curvularia lunata* and *Cladosporium herbarum*. J Nat Prod. 2002;65:730–3.
91. Malmstrøm J, Christophersen C, Barrero AF, Oltra JE, Justicia J, Rosales A. Bioactive metabolites from a marine-derived strain of the fungus *Emericella variecolor*. J Nat Prod. 2002;65:364–7.
92. Wang C, Wang B, Brauers G, Guan H, Proksch P, Ebel R. Microsphaerones A and B, two novel γ-pyrone derivatives from the sponge-derived fungus *Microsphaeropsis* sp. J Nat Prod. 2002;65:772–5.

93. Bringmann G, Lang G, Steffens S, Günther E, Schaumann K. Evariquinone, isoemericellin, and stromemycin from a sponge derived strain of the fungus *Emericella variecolor*. Phytochemistry. 2003;63:437–43.
94. Adachi K, Kanoh K, Wisespongp P, Nishijima M, Shizuri Y. Clonostachysins A and B, new anti-dinoflagellate cyclic peptides from a marine-derived fungus. J Antibiot. 2005;58:145–50.
95. Christian OE, Compton J, Christian KR, Mooberry SL, Valeriote FA, Crews P. Using jasplakinolide to turn on pathways that enable the isolation of new chaetoglobosins from *Phomospis asparagi*. J Nat Prod. 2005;68:1592–7.
96. Xu J, Takasaki A, Kobayashi H, Oda T, Yamada J, Mangindaan RE, et al. Four new macrocyclic trichothecenes from two strains of marine-derived fungi of the genus *Myrothecium*. J Antibiot. 2006;59:451–5.
97. Amagata T, Morinaka BI, Amagata A, Tenney K, Valeriote FA, Lobkovsky E, et al. A chemical study of cyclic depsipeptides produced by a sponge-derived fungus. J Nat Prod. 2006;69:1560–5.
98. Cruz LJ, Insua MM, Baz JP, Trujillo M, Rodriguez-Mias RA, Oliveira E, et al. PIB-01212, a new cytotoxic cyclodepsipeptide isolated from the marine fungus *Clonostachys* sp. ESNA-A009. J Org Chem. 2006;71:3335–8.
99. Boot CM, Tenney K, Valeriote FA, Crews P. Highly *N*-methylated linear peptides produced by an atypical sponge-derived *Acremonium* sp. J Nat Prod. 2006;69:83–92.
100. Kralj A, Kehraus S, Krick A, van Echten-Deckert G, König GM. Two new depsipeptides from the marine fungus *Spicellum roseum*. Planta Med. 2007;73:366–71.
101. Neumann K, Abdel-Lateff, Wright AD, Kehraus S, Krick A, König GM. Novel sorbicillin derivatives with an unprecedented carbon skeleton from the sponge-derived fungus *Trichoderma* species. Eur J Org Chem. 2007;2007:2268–75.
102. Yoshida S, Kito K, Ooi T, Kanoh K, Shizuri Y, Kusumi T. Four pimarane diterpenes from marine fungus: chloroform incorporated in crystal lattice for absolute configuration analysis by X-ray. Chem Lett. 2007;36:1386–7.
103. Zhao LL, Gai Y, Kobayashi H, Hu CQ, Zhang HP. 5′-Hydroxyzearalenol, a new β-resorcylic macrolide from *Fusarium* sp. 05ABR26. Chin Chem Lett. 2008;19:1089–92.
104. Zhang D, Yang X, Kang JS, Choi HD, Son BW. Circumdatin I, a new ultraviolet-A protecting benzodiazepine alkaloid from a marine isolate of the fungus *Exophiala*. J Antibiot. 2008;61:40–2.
105. Tricand de la Gôutte J, Khan JA, Vulfson EN. Identification of novel polyphenol oxidase inhibitors by enzymatic one-pot synthesis and deconvolution of combinatorial libraries. Biotechnol Bioeng. 2001;75:93–9.
106. Elbandy M, Shinde PB, Dang HT, Hong J, Bae KS, Jung JH. Furan metabolites from the sponge-derived yeast *Pichia membranifaciens*. J Nat Prod. 2008;71:869–72.
107. Yu Z, Lang G, Kajahn I, Schmaljohann R, Imhoff JF. Scopularides A and B, cyclodepsipeptides from a marine sponge-derived fungus, *Scopulariopsis brevicaulis*. J Nat Prod. 2008;71:1052–4.
108. Zhang D, Yang X, Kang JS, Choi HD, Son BW. Chlorohydroaspyrones A and B, antibacterial aspyrone derivatives from the marine-derived fungus *Exophiala* sp. J Nat Prod. 2008;71:1458–60.
109. Gao H, Zhang QH, Jiang MM, Tang JS, Du MC, Hong K, et al. Polyketides from a marine sponge-derived fungus *Mycelia sterilia* and proton–proton long-range coupling. Magn Reson Chem. 2008;46:1148–52.
110. Prachyawarakorn V, Mahidol C, Sureram S, Sangpetsiripan S, Wiyakrutta S, Ruchirawat S, et al. Diketopiperazines and phthalides from a marine derived fungus of the order Pleosporales. Planta Med. 2008;74:69–72.
111. Proksch P, Ebel R, Edrada R, Riebe F, Liu HB, Diesel A, et al. Sponge-associated fungi and their bioactive compounds: the Suberites case. Bot Mar. 2008;51:209–18.
112. Zhang P, Bao B, Dang HT, Hong J, Lee HJ, Yoo ES, et al. Anti-inflammatory sesquiterpenoids from a sponge-derived Fungus *Acremonium* sp. J Nat Prod. 2009;72:270–5.

113. Izumikawa M, Khan ST, Komaki H, Nagai A, Inaba S, Takagi M, et al. JBIR-37 and-38, novel glycosyl benzenediols, isolated from the sponge-derived fungus, *Acremonium* sp. SpF080624G1f01. Biosci Biotechnol Biochem. 2009;73:2138–40.
114. Neumann K, Kehraus S, Gütschow M, König GM. Cytotoxic and HLE-inhibitory tetramic acid derivatives from marine-derived fungi. Nat Prod Commun. 2009;4:347–54.
115. Mohamed IE, Gross H, Pontius A, Kehraus S, Krick A, Kelter G, et al. Epoxyphomalin A and B, prenylated polyketides with potent cytotoxicity from the marine-derived fungus *Phoma* sp. Org Lett. 2009;11:5014–7.
116. Sugiyama Y, Ito Y, Suzuki M, Hirota A. Indole derivatives from a marine sponge-derived yeast as DPPH radical scavengers. J Nat Prod. 2009;72:2069–71.
117. Abdel-Lateffa A, Fischa K, Wrightd AD. Trichopyrone and other constituents from the marine sponge-derived fungus *Trichoderma* sp. Z Naturforsch C. 2009;64:186–92.
118. Mohamed IE, Kehraus S, Krick A, Konig GM, Kelter G, Maier A, et al. Mode of action of epoxyphomalins A and B and characterization of related metabolites from the marine-derived fungus *Paraconiothyrium* sp. J Nat Prod. 2010;73:2053–6.
119. Ueda JY, Takagi M, Shin-ya K. New xanthoquinodin-like compounds, JBIR-97, -98 and -99, obtained from marine sponge-derived fungus *Tritirachium* sp. SpB081112MEf2. J Antibiot. 2010;63:615–8.
120. Pruksakorn P, Arai M, Kotoku N, Vilcheze C, Baughn AD, Moodley P, et al. Trichoderins, novel aminolipopeptides from a marine sponge-derived *Trichoderma* sp., are active against dormant mycobacteria. Bioorg Med Chem Lett. 2010;20:3658–63.
121. Julianti E, Oh H, Jang KH, Lee JK, Lee SK, Oh DC, et al. Acremostrictin, a highly oxygenated metabolite from the marine fungus *Acremonium strictum*. J Nat Prod. 2011;74:2592–4.
122. Ebada SS, Schulz B, Wray V, Totzke F, Kubbutat MH, Muller WE, et al. Arthrinins A-D: novel diterpenoids and further constituents from the sponge derived fungus *Arthrinium* sp. Bioorg Med Chem. 2011;19:4644–51.
123. Tsukada M, Fukai M, Miki K, Shiraishi T, Suzuki T, Nishio K, et al. Chemical constituents of a marine fungus, *Arthrinium sacchari*. J Nat Prod. 2011;74:1645–9.
124. Almeida C, Part N, Bouhired S, Kehraus S, König GM. Stachylines A-D from the sponge-derived fungus *Stachylidium* sp. J Nat Prod. 2011;74:21–5.
125. Almeida C, Kehraus S, Prudencio M, König GM. Marilones A-C, phthalides from the sponge-derived fungus *Stachylidium* sp. Beilstein J Org Chem. 2011;7:1636–42.
126. Julianti E, Oh H, Lee H, Oh D, Oh K, Shin J. Acremolin, a new 1*H*-azirine metabolite from the marine-derived fungus *Acremonium strictum*. Tetrahedron Lett. 2012;53:2885–6.
127. Januar LA, Molinski TF. Acremolin from *Acremonium* strictum is N^2,3-etheno-2′-isopropyl-1-methylguanine, not a 1*H*-azirine. Synthesis and structural revision. Org Lett. 2013;15:2370–3.
128. Almeida C, Eguereva E, Kehraus S, König GM. Unprecedented polyketides from a marine sponge-associated *Stachylidium* sp. J Nat Prod. 2012;76:322–6.
129. Almeida C, Hemberger Y, Schmitt SM, Bouhired S, Natesan L, Kehraus S, et al. Marilines A–C: novel phthalimidines from the sponge-derived fungus *Stachylidium* sp. Chem Eur J. 2012;18:8827–34.
130. Pinheiro A, Dethoup T, Bessa J, Silva A, Kijjoa A. A new bicyclic sesquiterpene from the marine sponge associated fungus *Emericellopsis minima*. Phytochem Lett. 2012;5:68–70.
131. Peng JX, Jiao JY, Li J, Wang W, Gu QQ, Zhu TJ, et al. Pyronepolyene C-glucosides with NF-κB inhibitory and anti-influenza A viral (H1N1) activities from the sponge-associated fungus *Epicoccum* sp. JJY40. Bioorg Med Chem Lett. 2012;22:3188–90.
132. Yamazaki H, Rotinsulu H, Kaneko T, Murakami K, Fujiwara H, Ukai K, et al. A new dibenz[b,e]oxepine derivative, 1-Hydroxy-10-methoxy-dibenz[b,e]oxepin-6,11-dione, from a marine-derived fungus, *Beauveria bassiana* TPU942. Mar Drugs. 2012;10:2691–7.
133. Zhang GJ, Wu GW, Zhu TJ, Kurtán T, Mándi A, Jiao JY, et al. Meroterpenoids with diverse ring systems from the sponge-associated fungus *Alternaria* sp. JJY-32. J Nat Prod. 2013;76:1946–57.

134. Ma XH, Li LT, Zhu TJ, Ba MY, Li GQ, Gu QQ, et al. Phenylspirodrimanes with anti-HIV activity from the sponge-derived fungus *Stachybotrys chartarum* MXH-X73. J Nat Prod. 2013;76:2298–306.
135. Haga A, Tamoto H, Ishino M, Kimura E, Sugita T, Kinoshita K, et al. Pyridone alkaloids from a marine-derived fungus, *Stagonosporopsis cucurbitacearum*, and their activities against azole-resistant *Candida albicans*. J Nat Prod. 2013;76:750–4.
136. Fraser JA, Lambert LK, Pierens GK, Bernhardt PV, Garson MJ. Secondary metabolites of the sponge-derived fungus *Acremonium persicinum*. J Nat Prod. 2013;76:1432–40.
137. Panizel I, Yarden O, Ilan M, Carmeli S. Eight new peptaibols from sponge-associated *Trichoderma atroviride*. Mar Drugs. 2013;11:4937–60.
138. Jansen N, Ohlendorf B, Erhard A, Bruhn T, Bringmann G, Imhoff JF. Helicusin E, isochromophilone X and isochromophilone XI: new chloroazaphilones produced by the fungus *Bartalinia robillardoides* strain LF550. Mar Drugs. 2013;11:800–16.
139. Jeon J, Julianti E, Oh H, Park W, Oh D, Oh K, et al. Stereochemistry of hydroxy-bearing benzolactones: isolation and structural determination of chrysoarticulins A–C from a marine-derived fungus *Chrysosporium articulatum*. Tetrahedron Lett. 2013;54:3111–5.
140. Zhu H, Hua XX, Gong T, Pang J, Hou Q, Zhu P. Hypocreaterpenes A and B, cadinane-type sesquiterpenes from a marine-derived fungus, *Hypocreales* sp. Phytochem Lett. 2013;6:392–6.
141. Harms H, Rempel V, Kehraus S, Kaiser M, Hufendiek P, Müller CE, et al. Indoloditerpenes from a marine-derived fungal strain of *Dichotomomyces cejpii* with antagonistic activity at GPR18 and cannabinoid receptors. J Nat Prod. 2014;77:673–7.
142. Wu B, Ohlendorf B, Oesker V, Wiese J, Malien S, Schmaljohann R, et al. Acetylcholinesterase inhibitors from a marine fungus *Talaromyces* sp. strain LF458. Mar Biotechnol. 2014;17:1–10.
143. Kotoku N, Higashimoto K, Kurioka M, Arai M, Fukuda A, Sumii Y, et al. Xylarianaphthol-1, a novel dinaphthofuran derivative, activates p21 promoter in a p53-independent manner. Bioorg Med Chem Lett. 2014;24:3389–91.
144. Wu ZH, Liu D, Proksch P, Guo P, Lin WH. Punctaporonins H–M: caryophyllene-type sesquiterpenoids from the sponge-associated fungus *Hansfordia sinuosae*. Mar Drugs. 2014;12:3904–16.
145. Gomes NM, Bessa LJ, Buttachon S, Costa PM, Buaruang J, Dethoup T, et al. Antibacterial and antibiofilm activities of tryptoquivalines and meroditerpenes isolated from the marine-derived fungi *Neosartorya paulistensis*, *N. laciniosa*, *N. tsunodae*, and the soil fungi *N. fischeri* and *N. siamensis*. Mar Drugs. 2014;12:822–39.
146. Stierle AC, Cardellina JH, Singleton FL. A marine micrococcus produces metabolites ascribed to the sponge *Tedania ignis*. Experientia. 1988;44:1021.
147. Imamura N, Nishijima M, Adachi K, Sano H. Novel antimycin antibiotics, urauchimycins A and B, produced by marine actinomycete. J Antibiot. 1993;46:241–6.
148. Bultel-Poncé VV, Debitus C, Berge JP, Cerceau C, Guyot M. Lutoside: an acyl-l-(acyl-6′-mannobiosyi)-3-giycerol isolated from the sponge-associated bacterium *Micrococcus luteus*. Tetrahedron Lett. 1997;38:5805–8.
149. Lee DS, Lim J, Kim JS, Im KS, Jung JH. Topoisomerase I inhibitors from the *Streptomyces* sp. strain KM86-9B isolated from a marine sponge. Arch Pharm Res. 1998;21:729–33.
150. Fernandez-Chimeno RI, Cañedo L, Espliego F, Gravalos D, Calle FDA, Fernandez-Puentes JL. IB-96212, a novel cytotoxic macrolide produced by a marine *Micromonospora*. I. Taxonomy, fermentation, isolation and biological activities. J Antibiot. 2000;53:479–83.
151. Wicke C, Hüners M, Wray V, Nimtz M, Bilitewski U, Lang S. Production and structure elucidation of glycoglycerolipids from a marine sponge-associated *Microbacterium* species. J Nat Prod. 2000;63:621–6.
152. Hernández LM, Blanco JA, Baz JP, Puentes JL, Millán FR, Vázquez FE, et al. 4′-*N*-Methyl-5′-hydroxystaurosporine and 5′-hydroxystaurosporine, new indolocarbazole alkaloids from a marine *Micromonospora* sp. strain. J Antibiot. 2000;53:895–902.
153. Lee HS, Shin HJ, Jang KH, Kim TS, Oh KB, Shin J. Cyclic peptides of the nocardamine class from a marine-derived bacterium of the genus *Streptomyces*. J Nat Prod. 2005;68:623–5.

154. Mitova MI, Lang G, Wiese J, Imhoff JF. Subinhibitory concentrations of antibiotics induce phenazine production in a marine *Streptomyces* sp. J Nat Prod. 2008;71:824–7.
155. Pimentel-Elardo SM, Gulder TAM, Hentschel U, Bringmann G. Cebulactams A1 and A2, new macrolactams isolated from *Saccharopolyspora cebuensis*, the first obligate marine strain of the genus *Saccharopolyspora*. Tetrahedron Lett. 2008;49:6889–92.
156. Perez M, Schleissner C, Rodriguez P, Zuniga P, Benedit G, Sanchez-Sancho F, et al. PM070747, a new cytotoxic angucyclinone from the marine-derived *Saccharopolyspora taberi* PEM-06-F23-019B. J Antibiot. 2009;62:167–9.
157. Choi EJ, Kwon HC, Ham J, Yang HO. 6-Hydroxymethyl-1-phenazine-carboxamide and 1,6-phenazinedimethanol from a marine bacterium, *Brevibacterium* sp. KMD 003, associated with marine purple vase sponge. J Antibiot. 2009;62:621–4.
158. Takagi M, Motohashi K, Khan ST, Hashimoto J, Shin-Ya K. JBIR-65, a new diterpene, isolated from a sponge-derived *Actinomadura* sp. SpB081030SC-15. J Antibiot. 2010;63:401–3.
159. Schneemann I, Ohlendorf B, Zinecker H, Nagel K, Wiese J, Imhoff JF. Nocapyrones A-D, γ-pyrones from a *Nocardiopsis* strain isolated from the marine sponge *Halichondria panicea*. J Nat Prod. 2010;73:1444–7.
160. Izumikawa M, Khan ST, Takagi M, Shin-ya K. Sponge-derived *Streptomyces* producing isoprenoids via the mevalonate pathway. J Nat Prod. 2010;73:208–12.
161. Motohashi K, Takagi M, Shin-Ya K. Tetrapeptides possessing a unique skeleton, JBIR-34 and JBIR-35, isolated from a sponge-derived actinomycete, *Streptomyces* sp. Sp080513GE-23. J Nat Prod. 2010;73:226–8.
162. Izumikawa M, Khan ST, Komaki H, Takagi M, Shin-Ya K. JBIR-31, a new teleocidin analog, produced by salt-requiring *Streptomyces* sp. NBRC 105896 isolated from a marine sponge. J Antibiot. 2010;63:33–6.
163. Motohashi K, Takagi M, Shin-Ya K. Tetracenoquinocin and 5-iminoaranciamycin from a sponge-derived *Streptomyces* sp. Sp080513GE-26. J Nat Prod. 2010;73:755–8.
164. Ueda JY, Khan ST, Takagi M, Shin-ya K. JBIR-58, a new salicylamide derivative, isolated from a marine sponge-derived *Streptomyces* sp. SpD081030ME-02. J Antibiot. 2010;63:267–9.
165. Schneemann I, Kajahn I, Ohlendorf B, Zinecker H, Erhard A, Nagel K, et al. Mayamycin, a cytotoxic polyketide from a *Streptomyces* strain isolated from the marine sponge *Halichondria panicea*. J Nat Prod. 2010;73:1309–12.
166. Wei RB, Xi T, Li J, Wang P, Li FC, Lin YC, et al. Lobophorin C and D, new kijanimicin derivatives from a marine sponge-associated actinomycetal strain AZS17. Mar Drugs. 2011;9:359–68.
167. Li K, Li QL, Ji NY, Liu B, Zhang W, Cao XP. Deoxyuridines from the marine sponge associated actinomycete *Streptomyces microflavus*. Mar Drugs. 2011;9:690–5.
168. Simmons L, Kaufmann K, Garcia R, Schwar G, Huch V, Muller R. Bendigoles D-F, bioactive sterols from the marine sponge-derived *Actinomadura* sp. SBMs009. Bioorg Med Chem. 2011;19:6570–5.
169. Motohashi K, Inaba K, Fuse S, Doi T, Izumikawa M, Khan ST, et al. JBIR-56 and JBIR-57, 2(1 H)-Pyrazinones from a marine sponge-derived *Streptomyces* sp. SpD081030SC-03. J Nat Prod. 2011;74:1630–5.
170. Huang XL, Gao Y, Xue DQ, Liu HL, Peng CS, Zhang FL, et al. Streptomycindole, an indole alkaloid from a marine *Streptomyces* sp. DA22 associated with South China Sea sponge *Craniella australiensis*. Helv Chim Acta. 2011;94:1838–42.
171. Wyche TP, Hou Y, Braun D, Cohen HC, Xiong MP, Bugni TS. First natural analogs of the cytotoxic thiodepsipeptide thiocoraline A from a marine *Verrucosispora* sp. J Org Chem. 2011;76:6542–7.
172. Pimentel-Elardo SM, Buback V, Gulder TA, Bugni TS, Reppart J, Bringmann G, et al. New tetromycin derivatives with anti-trypanosomal and protease inhibitory activities. Mar Drugs. 2011;9:1682–97.

173. Hosoya T, Hirokawa T, Takagi M, Shin-ya K. Trichostatin analogues JBIR-109, JBIR-110, and JBIR-111 from the marine sponge-derived *Streptomyces* sp. RM72. J Nat Prod. 2012;75:285–9.
174. Abdelmohsen UR, Zhang G, Philippe A, Schmitz W, Pimentel-Elardo SM, Hertlein-Amslinger B, et al. Cyclodysidins A–D, cyclic lipopeptides from the marine sponge-derived *Streptomyces* strain RV15. Tetrahedron Lett. 2012;53:23–9.
175. Supong K, Thawai C, Suwanborirux K, Choowong W, Supothina S, Pittayakhajonwut P. Antimalarial and antitubercular C-glycosylated benz[a]anthraquinones from the marine-derived *Streptomyces* sp. BCC45596. Phytochem Lett. 2012;5:651–6.
176. Martin J, da SST, Crespo G, Palomo S, Gonzalez I, Tormo JR, et al. Kocurin, the true structure of PM181104, an anti-methicillin-resistant *Staphylococcus aureus* (MRSA) thiazolyl peptide from the marine-derived bacterium *Kocuria palustris*. Mar Drugs. 2013;11:387–98.
177. Izumikawa M, Kawahara T, Hwang JH, Takagi M, Shin-ya K. JBIR-107, a new metabolite from the marine-sponge-derived actinomycete, *Streptomyces tateyamensis* NBRC 105047. Biosci Biotechnol Biochem. 2013;77:663–5.
178. Abdelmohsen UR, Cheng C, Viegelmann C, Zhang T, Grkovic T, Ahmed S, et al. Dereplication strategies for targeted isolation of new antitrypanosomal actinosporins A and B from a marine sponge associated *Actinokineospora* sp. EG49. Mar Drugs. 2014;12:1220–44.
179. Dashti Y, Grkovic T, Abdelmohsen UR, Hentschel U, Quinn RJ. Production of Induced secondary metabolites by a co-culture of sponge-associated actinomycetes, *Actinokineospora* sp. EG49 and *Nocardiopsis* sp. RV163. Mar Drugs. 2014;12:3046–59.
180. Kwon Y, Kim SH, Shin Y, Bae M, Kim BY, Lee SK, et al. A new benzofuran glycoside and indole alkaloids from a sponge-associated rare actinomycete, *Amycolatopsis* sp. Mar Drugs. 2014;12:2326–40.
181. Elyakov GB, Kuznetsova T, Mikhailov VV, Maltsev II, Voinov VG, Fedoreyev SA. Brominated diphenyl ethers from a marine bacterium associated with the sponge *Dysidea* sp. Experientia. 1990;47:632–3.
182. Stierle AA, Cardellina JH, Singleton FL. Benzothiazoles from a putatitve bacterial symbiont of the marine sponge *Tedania ignis*. Tetrahedron Lett. 1991;32:4847–8.
183. Stierle DB, Stierle AA. Pseudomonic acid derivatives from a marine bacterium. Experientia. 1992;48:1165–9.
184. Shigemori H, Bae MA, Yazawa K, Sasaki T, Kobayashi J. Alteramide A, a new tetracyclic alkaloid from a bacterium *Alteromonas* sp. associated with the marine sponge *Halichondria okadai*. J Org Chem. 1992;57:4320–3.
185. Oclarit JM, Ohta S, Kamimura K, Yarnaoka Y, Ikegami S. Production of an antibacterial agent, o-aminophenol, by a bacterium isolated from the marine sponge. Fish Sci. 1994;60:559–62.
186. Oclarit JM, Ohta S, Kamimura K, Yamaoka Y, Shimizu T, Ikegami S. A novel antimicrobial substance from a strain of the bacterium, *Vibrio* sp. Nat Prod Lett. 1994;4:309–12.
187. Kobayashi M, Aoki S, Gato K, Matsunami K, Kurosu M, Kitagawa I. Marine natural products. XXXIV. Trisindoline, a new antibiotic indole trimer, produced by a bacterium of *Vibrio* sp. separated from the marine sponge *Hyrtios altum*. Chem Pharm Bull. 1994;42:2449–51.
188. Oclarit JM, Okada H, Ohta S, Kaminura K, Yamaoka Y, Iizuka T, et al. Anti-bacillus substance in the marine sponge, *Hyatella* species, produced by an associated *Vibrio* species bacterium. Microbios. 1994;78:7–16.
189. Miki W, Otaki N, Yokoyama A, Izumida H, Shimidzu N. Okadaxanthin, a novel C50-carotenoid from a bacterium, *Pseudomonas* sp. KK10206C associated with marine sponge, *Halichondria okadai*. Experientia. 1994;50:684–6.
190. Kalinovskaya NI, Kuznetsova TA, Rashkes YV, Mil'grom YM, Mil'grom EG, Willis RH, et al. Surfactin-like structures of five cyclic depsipeptides from the marine isolate of *Bacillus pumUus*. Russ Chem Bull. 1995;44:951–5.
191. Yokoyama A, Miki W. Isolation of myxol from a marine bacterium *Flavobacterium* sp. associated with a marine sponge. Fish Sci. 1995;61:684–6.

192. Jayatilake GS, Thornton MP, Leonard AC, Grimwade JE, Baker BJ. Metabolites from an Antarctic sponge-associated bacterium, *Pseudomonas aeruginosa*. J Nat Prod. 1996;59:293–6.
193. Debitus C, Guella G, Mancini I, Waikedre J, Guemas J, Nicolas JL, et al. Quinolones from a bacterium and tyrosine metabolites from its host sponge, *Suberea creba* from the Coral Sea. Mar Biotechnol. 1998;6:136–41.
194. De Rosa S, De Giulio A, Tommonaro G, Popov S, Kujumgiev A. A β-amino acid containing tripeptide from a *Pseudomonas alteromonas* bacterium associated with a Black Sea sponge. J Nat Prod. 2000;64:1454–5.
195. Suzumura K, Yokoi T, Funatsu M, Nagai K, Tanaka K, Zhang H, et al. YM-266183 and YM-266184, novel thiopeptide antibiotics produced by *Bacillus cereus* isolated from a marine sponge II. Structure elucidation. J Antibiot. 2003;56:129–34.
196. Kanoh K, Kamino K, Leleo G, Adachi K, Shizuri Y. Pseudoalterobactin A and B, new siderophores excreted by marine bacterium *Pseudoalteromonas* sp. KP20-4. J Antibiot. 2003;56:871–5.
197. Mitova M, Tommonaro G, Hentschel U, Muller WE, De Rosa S. Exocellular cyclic dipeptides from a Ruegeria strain associated with cell cultures of *Suberites domuncula*. Mar Biotechnol. 2004;6:95–103.
198. Speitling M, Smetanina OF, Kuznetsova TA, Laatsch H. Bromoalterochromides A and A′, unprecedented chromopeptides from a marine *Pseudoalteromonas maricaloris* strain KMM 636T. J Antibiot. 2007;60:36–42.
199. Shindo K, Asagi E, Sano A, Hotta E, Minemura N, Mikami K, et al. Diapolycopenedioic acid xylosyl esters A, B, and C, novel antioxidative glyco-C30-carotenoic acids produced by a new marine bacterium *Rubritalea squalenifaciens*. J Antibiot. 2008;61:185–91.
200. Rungprom W, Siwu ERO, Lambert LK, Dechsakulwatana C, Barden MC, Kokpol U, et al. Cyclic tetrapeptides from marine bacteria associated with the seaweed *Diginea* sp. and the sponge *Halisarca ectofibrosa*. Mar Biotechnol. 2008;6:95–103.
201. You MX, Zhang HP, Hu CQ. Isolation and characterization of three siderophores from marine bacteria. Chin J Chem. 2008;26:1332–4.
202. Fehér D, Barlow RS, Lorenzo PS, Hemscheidt TK. A 2-substituted prodiginine, 2-(p-hydroxybenzyl)prodigiosin, from *Pseudoalteromonas rubra*. J Nat Prod. 2008;71:1970–2.
203. Quevrain E, Domart-Coulon I, Pernice M, Bourguet-Kondracki ML. Novel natural parabens produced by a *Microbulbifer* bacterium in its calcareous sponge host *Leuconia nivea*. Environ Microbiol. 2009;11:1527–39.

Chapter 14
Marine Natural Products from Marine Coral-Derived Microorganisms

Xuan Ma and Shu-Hua Qi

Contents

Abstract Marine coral-associated microorganisms played a critical role in chemical defense for their hosts, even considered as the true producers of some marine bioactive natural products. From the past 10 years, the studies on the secondary metabolites of these microbes (especially fungi) have been developed rapidly, which led to the obtainment of over 170 new compounds from microbes associated with gorgonians and soft corals. Some of these compounds showed strong bioactivities, such as cytotoxic, antimalarial, antibacterial, antifungal, and antifouling activity,

X. Ma · S.-H. Qi (✉)
Key Laboratory of Tropical Marine Bio-resources and Ecology, Guangdong Key Laboratory of Marine Materia Medica, RNAM Center for Marine Microbiology, South China Sea Institute of Oceanology, Chinese Academy of Sciences, Guangzhou, China
e-mail: shuhuaqi@scsio.ac.cn

Z. Li (ed.), *Symbiotic Microbiomes of Coral Reefs Sponges and Corals*,
https://doi.org/10.1007/978-94-024-1612-1_14

which indicate that coral-associated microbes are important resources for finding new drugs.

Keywords Coral-associated microbe · Secondary metabolite · Bioactivity

14.1 Introduction

Coral belonging to marine invertebrates is most abundant in the tropical ocean. On the basis of digestive cavity branch number, coral can be divided into Octocorallia and Hexactinellid coral subclasses. Among Octocorallia, soft coral and gorgonian are the research points of marine natural products. Microorganisms of the surface or inner tissue of coral are abundant. Some of the compounds isolated from the marine invertebrate-associated microorganisms were closely resembling those compounds identified from their host invertebrates. Some of these marine microbes were deemed to take part in the biosynthesis of some marine bioactive natural products and played a critical role in chemical defense for their hosts, even considered as the true producers of some marine bioactive natural products [1, 2]. The mutualistic symbiosis between symbiotic microorganisms and hosts makes corals and other invertebrates create a special chemical, ecological mechanism, and metabolic mechanism, which make them produce novel and diverse active secondary metabolites. As the limitation of coral resources and the microbial culturability, the study on the secondary metabolites of microbes associated with coral was valuable for screening active compound with chemical ecology for practical value. For a long time, the studies of secondary metabolites of microbes associated with coral were rare [3, 4]. Until the recent 10 years, it has been developed rapidly and has attracted the attentions of researchers toward these microbes, which have the ability to produce different types of known and novel secondary metabolites with different biological activities. In this book chapter, we reviewed the discovery of secondary metabolites of microbes associated with gorgonians and soft corals.

14.2 Marine Natural Products (MNPs) from Gorgonian-Associated Microorganisms

14.2.1 MNPs from Gorgonian-Derived Fungi

14.2.1.1 MNPs from Marine Gorgonian-Derived *Aspergillus* sp.

Three phenolic bisabolane-type sesquiterpenoids (+)-methyl sydowate (**1**), 7-deoxy-7, 14-didehydrosydonic acid (**2**), and 7-deoxy-7,8-didehydrosydonic acid (**3**) were obtained from a marine-derived fungus *Aspergillus* sp. isolated from the gorgonian *Dichotella gemmacea* [5]. Two novel benzylazaphilone derivatives with an unprecedented carbon skeleton, aspergilone A (**4**), and its symmetrical dimer with a unique methylene bridge, aspergilone B (**5**), were also obtained from a marine-derived fungus *Aspergillus* sp. isolated from *D. gemmacea* [6]. Aspergilone A (**4**) exhibited selective cytotoxicity toward human promyelocytic leukemia HL-60, breast adenocarcinoma MCF-7, and lung carcinoma A-549 cell lines. Aspergilone A also exhibited potential antifouling activity against the larval settlement of barnacle *B. amphitrite* [6]. Aspergillusenes A–B (**6–7**), (+)-(7*S*)-7-*O*-methylsydonic acid (**8**), and two hydrogenated xanthone derivatives, aspergillusone A (**9**) and aspergillusone B (**10**), were obtained from the fungus *A. sydowii* PSU-F154 isolated from a sea fan *Annella* sp. [7].

Three cyclic tetrapeptides, aspergillipeptides A–C (**11–13**), and asteltoxin B (**14**) were obtained from the fungal strain *Aspergillus* sp. SCSGAF 0076 that was isolated from the gorgonian *Melitodes squamata* [8]. Aspergillipeptide C had strong antifouling activity. The fungus *Aspergillus* sp. SCSGAF0093 yielded four mycotoxins aluminiumneoaspergillin (**15**), zirconiumneoaspergillin (**16**), aspergilliamide (**17**), and ochratoxin A n-butyl ester (**18**) [9]. Two prenylated indole alkaloids, 17-epinotoamides Q and M (**19** and **20**), and two phenyl etherderivatives, cordyols D and E (**21** and **22**), were obtained from a marine-derived *Aspergillus* sp. isolated from *D. gemmacea* [10]. Four lumazine peptides, penilumamides B–D (**23–25**) and penilumamide (**26**), together with cyclic pentapeptide asperpeptide A (**27**) were isolated from the gorgonian-derived fungus *Aspergillus* sp. XS-20090B15 [11]. One anthraquinone derivative, 8-*O*-methylnidurufin (**28**), was isolated from a gorgonian-derived fungus *Aspergillus* sp. associated with *D. gemmacea* [12]. Six steroid derivatives (**29–34**) and five butyrolactone derivatives (**35–39**) were isolated from a gorgonian-derived *Aspergillus* sp. [13].

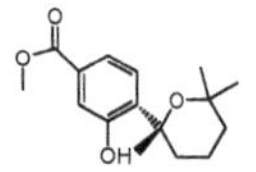
1
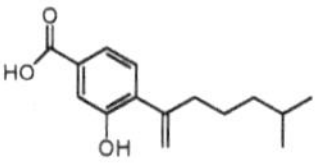
2
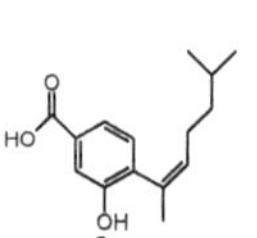
3
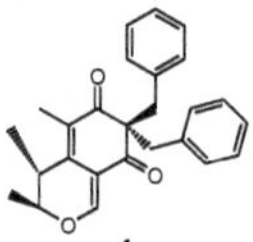
4
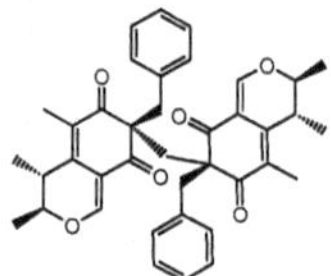
5
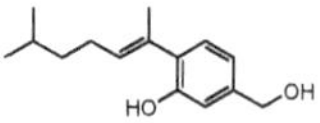
6
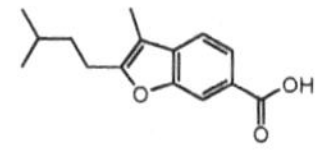
7
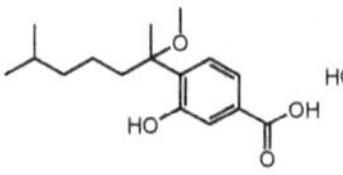
8
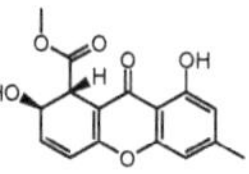
9
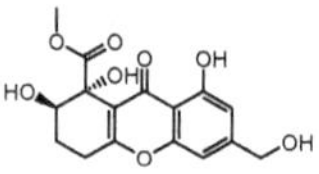
10
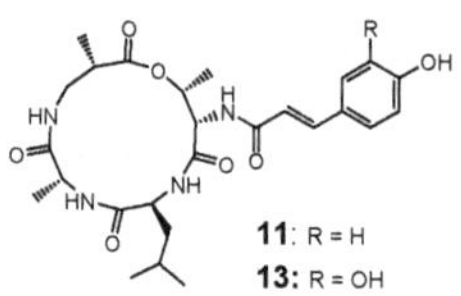
11: R = H
13: R = OH
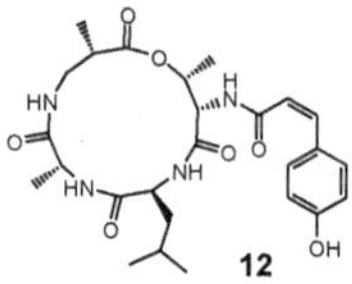
12
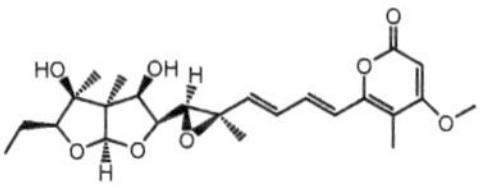
14
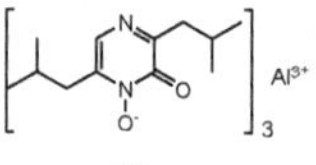
15
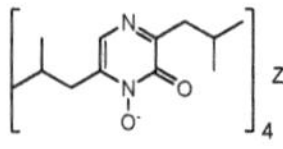
16
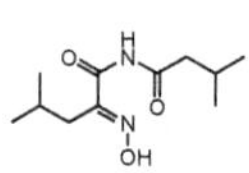
17
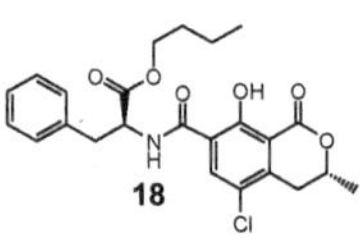
18
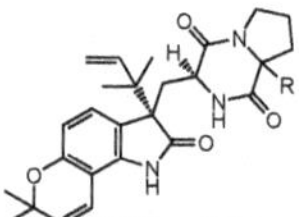
19: R = $-OCH_3$
20: R = -OH
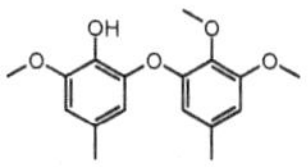
21
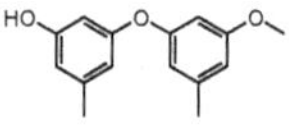
22
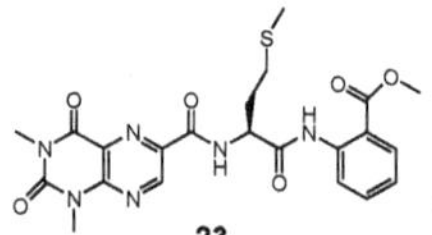
23
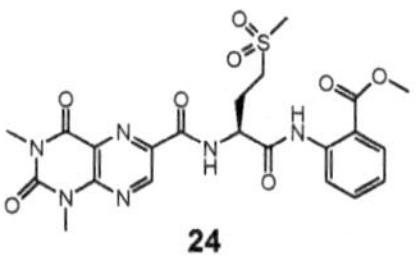
24
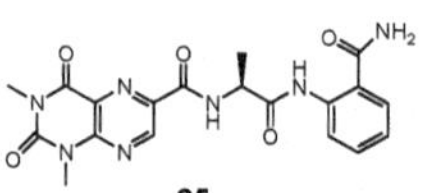
25
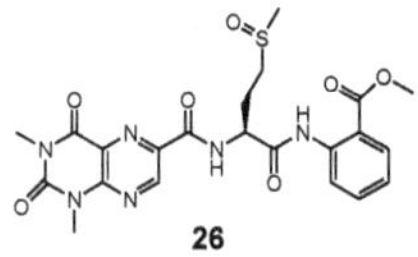
26
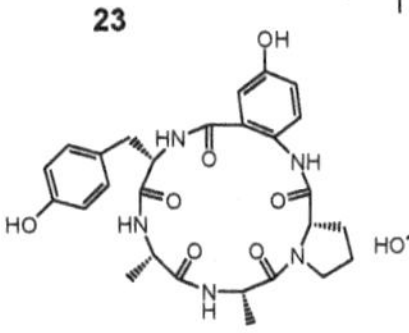
27
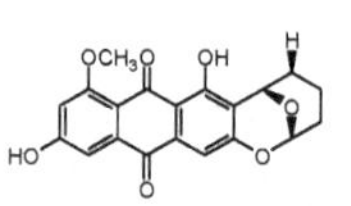
28
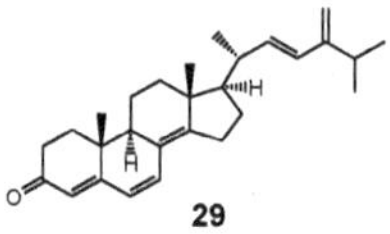
29
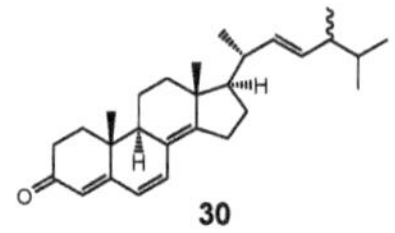
30
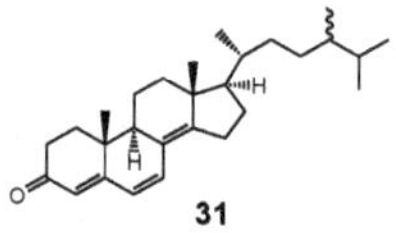
31
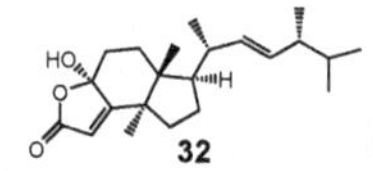
32
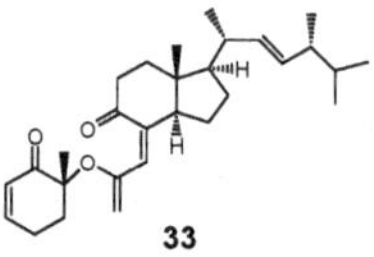
33
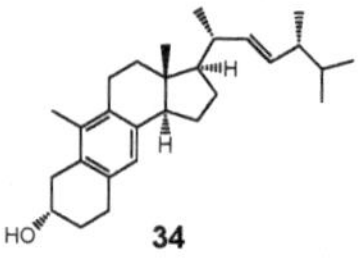
34
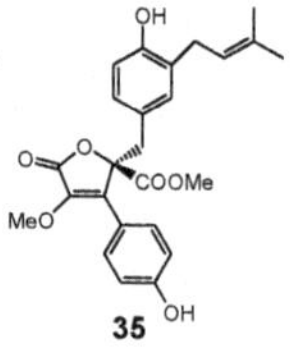
35
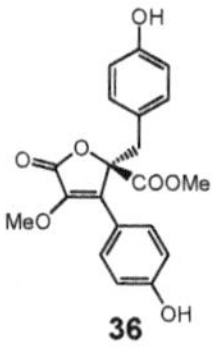
36
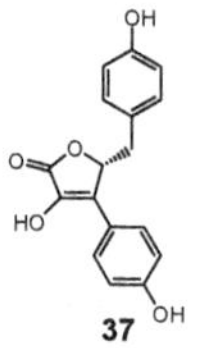
37
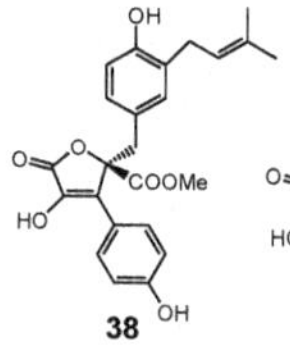
38
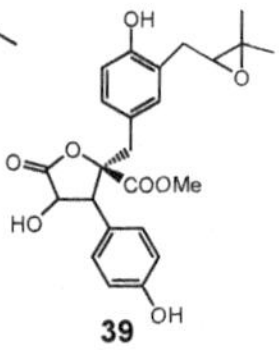
39

14.2.1.2 MNPs from Marine Gorgonian-Derived *Penicillium* sp.

Penicipyrone (**40**) and penicilactone (**41**) were obtained from the fungal strain *Penicillium* sp. PSU-F44 isolated from a sea fan *Annella* sp. [14]. Nine compounds including penicisochromans A–E (**42–46**), penicipyrone (**47**), penicipyranone (**48**), peniciphenol (**49**), and penicisoquinoline (**50**) were isolated from the fungus *Penicillium* sp. PSU-F40 isolated from a gorgonian *Annella* sp. [15]. Two benzopyranones, benzopyranones C and D (**51–52**); one isochroman, (3*R*,4*S*)-6,8-dihydroxy-1,1-dimethyl-3,4,5-trimethylisochroman (**53**); and two anthraquinone-citrinin derivatives, penicillanthranins A–B (**54–55**), were isolated from the fungus *P. citrinum* PSU-F51 isolated from a sea fan *Annella* sp. [16]. Penicillanthranins A displayed moderate antibacterial activity against *Staphylococcus aureus* ATCC25923 and *S. aureus* and indicated mild cytotoxicity toward KB cell line with IC_{50} values of 30 and 24 μM and weak cytotoxicity toward MCF7 and Vero cell lines [16].

Seven aromatic polyketides, communols A–G (**56–62**), were separated from *P. commune* 518 isolated from gorgonian *Muricella abnormalis* [17]. Communols A, F, and G showed weak antibacterial activities against *E. coli* and *E. aerogenes* [17]. Dihydrothiophene-condensed chromones, oxalicumones A–B (**63–64**), a natural chromone oxalicumone C (**65**), and oxalicumones D and E (**66–67**), were isolated from the marine-derived fungus *P. oxalicum* SCSGAF 0023 obtained from the gorgonian *D. gemmacea* [18]. Oxalicumone A and oxalicumone B displayed moderate and mild cytotoxicity against A375 and SW-620 cell lines, respectively. Oxalicumone E exhibited cytotoxicity against several cell lines with $IC_{50} \leq 10$ μM [19]. Three azaphilone derivatives, pinophilins D–F (**68–70**), and one diphenyl ether derivative, hydroxypenicillide (**71**), were isolated from the gorgonian-derived fungus *P. pinophilum* XS-20090E18 [20].

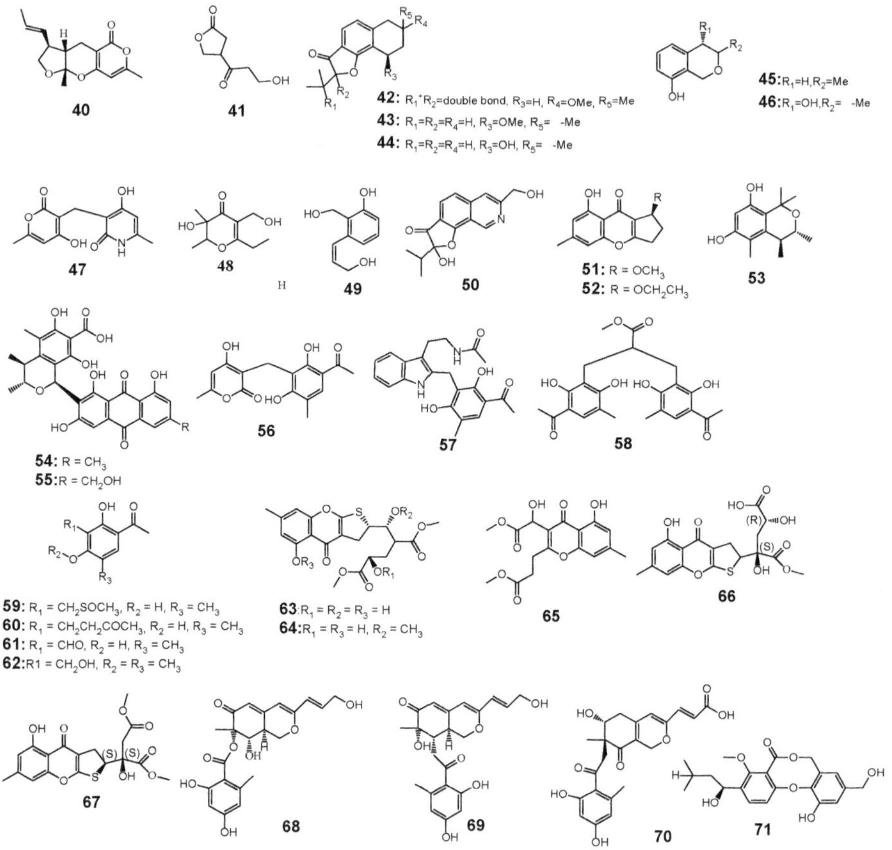

14.2.1.3 MNPs from Other Marine Gorgonian-Derived Fungi

Nigrospoxydons A–C (**72**–**74**) and nigrosporapyrone (**75**) were obtained from the marine-derived fungus *Nigrospora* sp. PSU-F5 isolated from a sea fan *Annella* sp. [21]. Nigrospoxydon A (**72**) showed antibacterial activity toward *S. aureus* with the MIC values of 64 and 128 μM, respectively. Four pyrones, nigrosporapyrones A–D (**76**–**79**), were isolated from the fungus *Nigrospora* sp. PSU-F18 isolated from a sea fan *Annella* sp. Nigrosporapyrone A (**76**) exhibited antibacterial activity against *S. aureus* with MIC value of 128 μM [22]. A cytochalasin derivative, xylarisin (**80**), was obtained from the fungus *Xylaria* sp. PSU-F100 isolated from the gorgonian *Annella* sp. [23]. Fusaranthraquinone (**81**), fusarnaphthoquinones A–C (**82–84**), and fusarone (**85**) were isolated from the fungi *Fusarium* spp. PSU-F14 and PSU-F135 isolated from a gorgonian *Annella* sp. [24]. Two cylohexene derivatives, nigrosporanenes A and B (**86–87**), were isolated from the fungi *Nigrospora* sp. PSU-F11 and PSU-F12 from a gorgonian *Annella* sp. Nigrosporanene A exhibited cytotoxicity against MCF-7 and Vero cell lines with IC_{50} values of 9.37 and 5.42 μM, respectively. Nigrosporanenes A–B showed weak radical scavenging activity with IC_{50} values of 0.34 and 0.24 μg/ml, respectively [25]. Three 14-membered resorcylic acid lactones cochliomycins A–C (**88–90**) were isolated from *Cochliobolus lunatus* obtained from the gorgonian *D. gemmacea.* Cochliomycin A showed potent antifouling activity against the larval settlement of barnacle *B. amphitrite* and moderate antibacterial activity against *S. aureus* [26].

Curvulapyrone (**91**), curvulalide (**92**), and curvulalic acid (**93**) were isolated from the fungus *Curvularia* sp. PSUF22 isolated from a gorgonian *Annella* sp. [27]. Trichodermaquinone (**94**) and trichodermaxanthone (**95**) were obtained from a broth extract of the fungus *Trichoderma aureoviride* PSU-F95 isolated from a gorgonian *Annella* sp. [28] Fumiquinazoline L (**96**) and sterigmatocystin were isolated from a fungus *Scopulariopsis* sp. isolated from a gorgonian *Carijoa* sp. [29]. Fumiquinazoline L had weak antibacterial activity against *E. coli*, *S. albus*, and *Vibrio parahaemolyticus*.

Four nucleoside derivatives, kipukasins D–E (**97–98**) and kipukasins H–I (**99–100**), were isolated from the fungus *A. versicolor* isolated from the gorgonian *D. gemmacea* [30]. Five pairs of dihydroisocoumarin enantiomers, (±)-eurotiumides A–E (**101–105**), and two related racemates, (±)-eurotiumides F and G, were isolated from the gorgonian-derived fungus *Eurotium* sp. XS-200900E6 [31].

A new phenolic glucoside, acremonoside (**106**), was isolated from the fungus *Acremonium polychromum* PSU-F125 from a sea fan *Annella* sp. [32]. Nigrospin (**107**), nigrospin A (**108**), and nigrospins B and C (**109–110**) were isolated from the fungal strain *Nigrospora oryzae* SCSGAF 0111 isolated from the gorgonian *Verrucella umbraculum* [33]. One bicyclic lactam, cladosporilactam A (**111**), was isolated from a gorgonian-derived fungus *Cladosporium* sp. [34]. Three diphenyl ether derivatives, talaromycins A–C (**112–114**), were isolated from a gorgonian-derived fungus *Talaromyces* sp. [35].

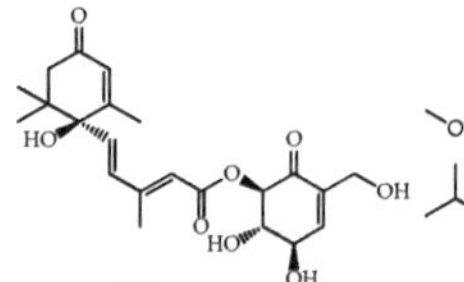
72

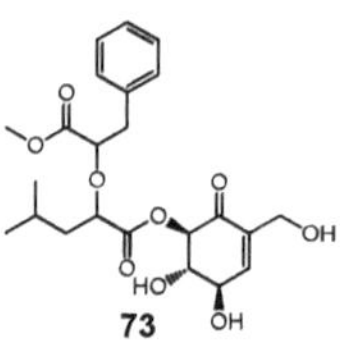
73

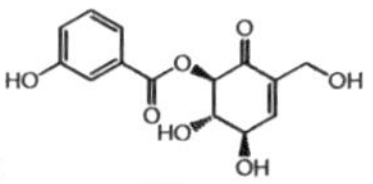
74

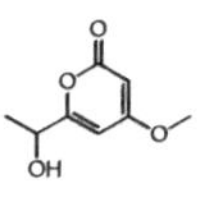
75

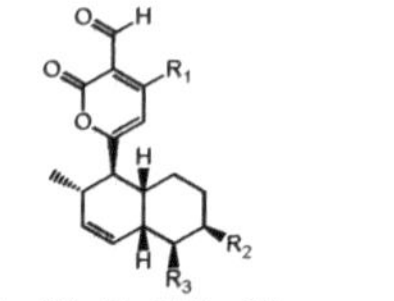
76: R_1 = OMe, R_2 = H, R_3 = OH
77: R_1 = $NHCH_2CH_2OH$, R_2 = H, R_3 = OH
78: R_1 = $NHCH_2CH_2OH$, R_2 = OH, R_3 = H

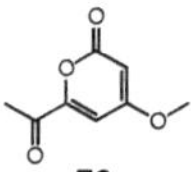
79

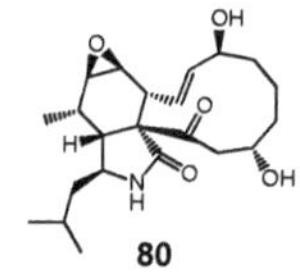
80

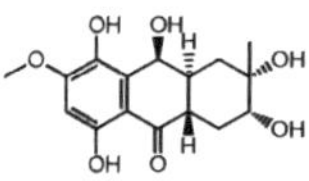
81

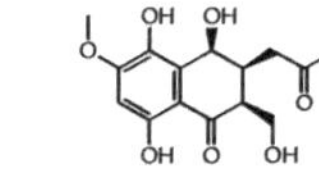
82

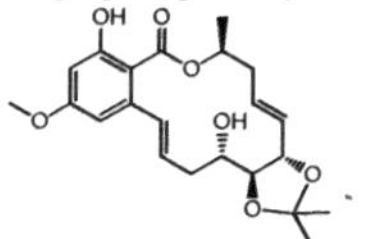
88

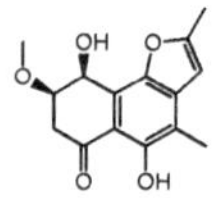
83

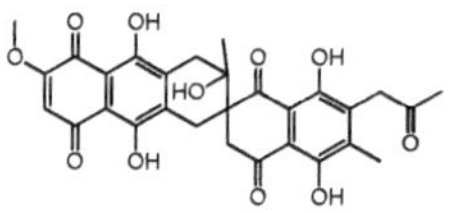
84

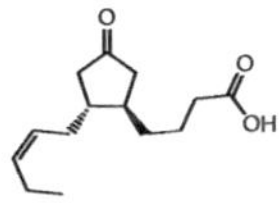
85

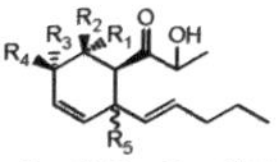
86: R_1 = R_3 = H, R_2 = R_4 = OH, R_5 = H
87: R_1 = R_4 = OH, R_2 = R_3 = H, R_5 = H

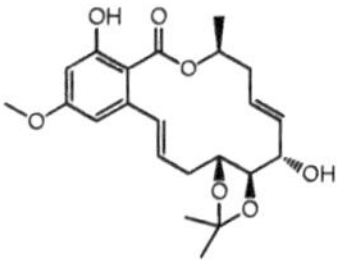
89

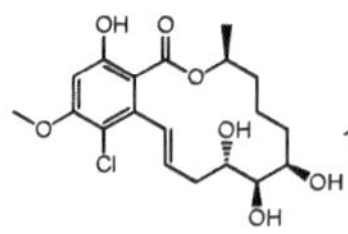
90

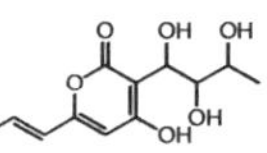
91

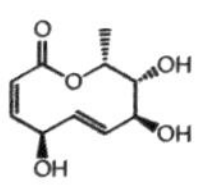
92

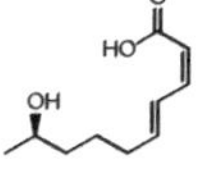
93

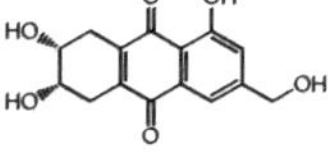
94

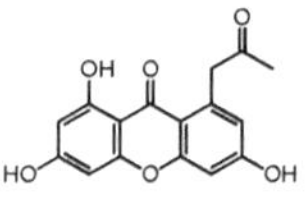
95

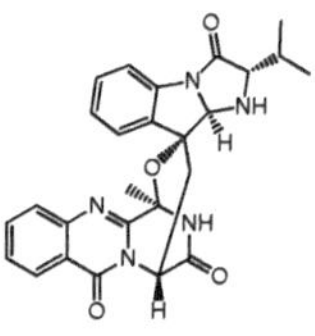
96

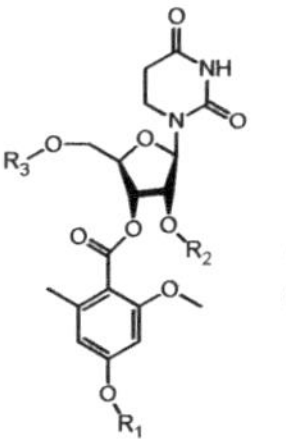
97: R_1=R_2=R_3=H
98: R_1=CH_3,R_2=R_3=H

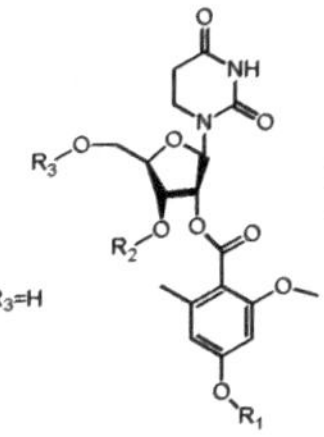
99: R_1=R_2=R_3=H
100: R_1=CH_3,R_2=R_3=H

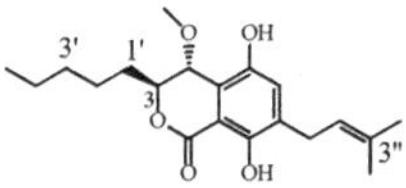
(+)-**101**
(+)-**103** △3'

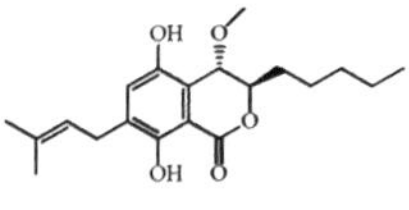
(-)-101
(-)-103 △3'

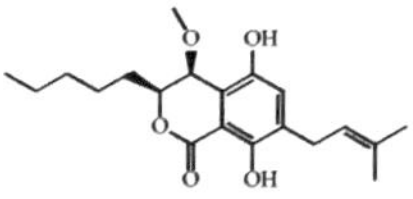
(+)-100
(+)-104 △3'
(+)-105 △1'

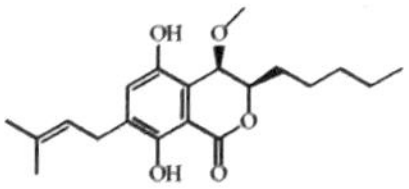
(-)-100
(-)-104 △3'
(-)-105 △1'

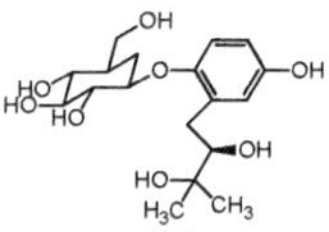
106

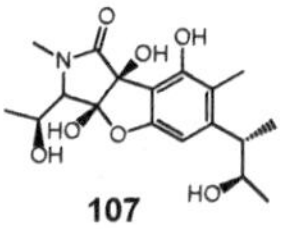
107

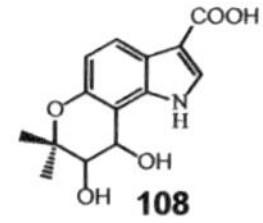
108

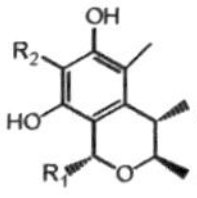
109: R_1= 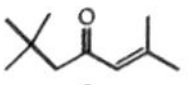 R2=COOH
110: R_1= 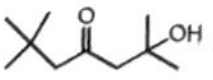 R2=H

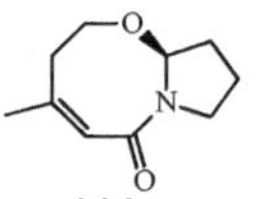
111

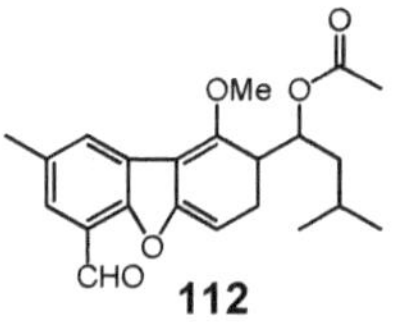
112

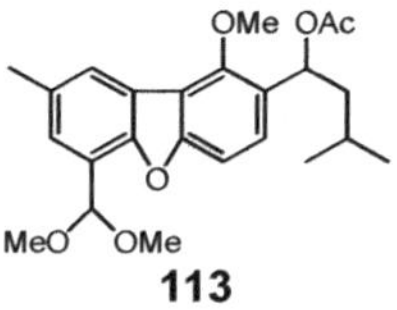
113

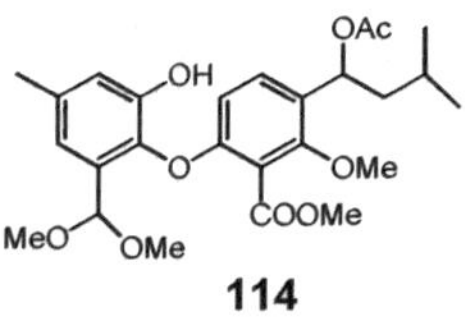
114

14.2.2 MNPs from Gorgonian-Derived Bacteria and Actinobacteria

Lactones macrolactin V (**115**) and macrolactin S were obtained from the bacterium *Bacillus amyloliquefaciens* SCSIO 00856 isolated from the gorgonian *Junceella juncea*. Macrolactin V had strong antibacterial activity against *E. coli*, *B. subtilis*, and *S. aureus* with an MIC value of 0.1 μM, but weak activity against *B. subtilis*, which indicated that the configuration of OH-7 could affect the antibacterial activity of macrolactins V and S [36]. Two octatomic ring lactones, octalactins A–B (**116–117**), were isolated from an actinomycete *Streptomyces* sp. isolated from a gorgonian *Pacifigorgia* sp. Octalactin A displayed strong cytotoxic activity toward human melanoma B-16-F10 and colon carcinoma HCT-116 cell lines with IC_{50} values of 7.2×10^{-3} and 0.5 μM, respectively [3].

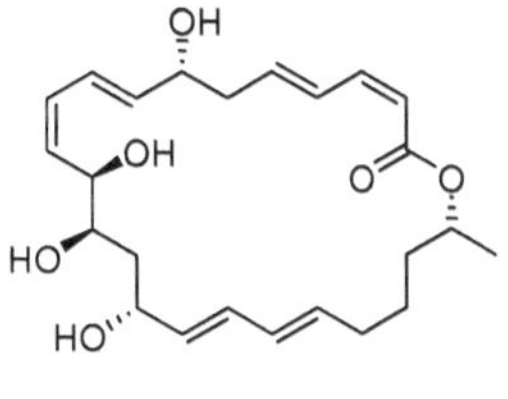

115

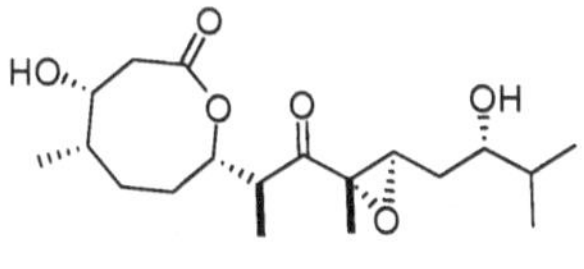

116

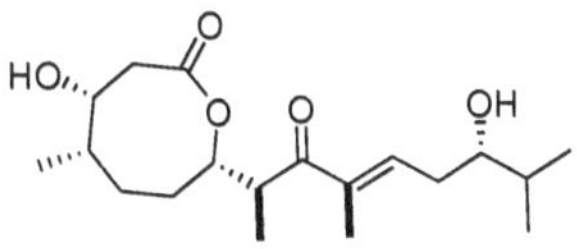

117

14.3 MNPs from Soft Coral-Derived Microorganisms

14.3.1 MNPs from Soft Coral-Derived Fungi

14.3.1.1 MNPs from Soft Coral-Derived *Aspergillus* sp.

Two aromatic butenolides, aspernolides A and B (**118–119**), were isolated from the fungus *A. terreus* isolated from the soft coral *Sinularia kavarattiensis*. Aspernolide A showed weak cytotoxicity against H460, ACHN, Calu, Panc1, and HCT116 cell lines [37]. 3,6-Diisobutyl-2(1H)-pyrazinone (**120**) was isolated from a fungus *Aspergillus* sp. isolated from the soft coral *Sarcophyton tortuosum* [38]. Three alkaloids, cottoquinazolines B–D (**121–123**), were isolated from the fungus *Aspergillus versicolor* LCJ-5-4 isolated from a soft coral *Cladiella* sp. Cottoquinazoline D exhibited moderate antifungal activity against *Candida albicans* [39]. Three cyclopentapeptides, versicoloritides A–C (**124–126**); a orcinol tetramer, tetraorcinol A (**127**); and two lactones, versicolactones A and B (**128, 129**), were isolated from the coral-associated fungus *A. versicolor* LCJ-5-4. Tetraorcinol A showed weak antioxidative activity against DPPH radical with an IC_{50} value of 67 μM [40]. An alkaloid fumiquinazoline K (**130**) and a nordammarane triterpenoid (**131**) were isolated from the marine fungal strain *A. fumigatus* KMM 4631 associated with a soft coral *Sinularia* sp. [41].

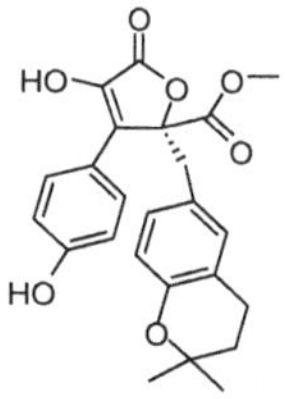

118

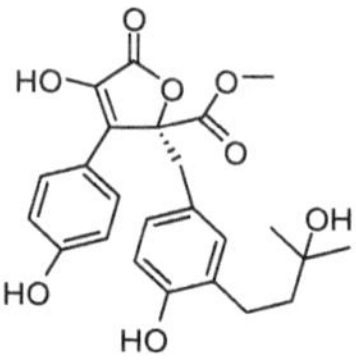

119

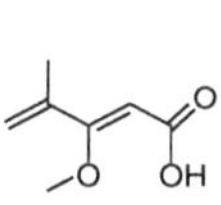

120

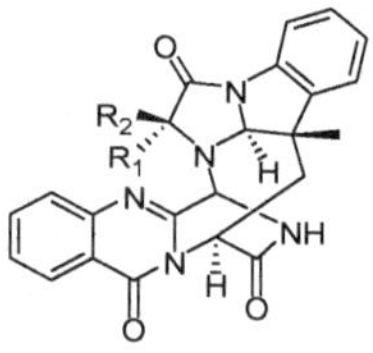

121: R_1=CH_3,R_2=H
122: R_1=H,R_2=
123: R_1+R_2=CH_2CH_2

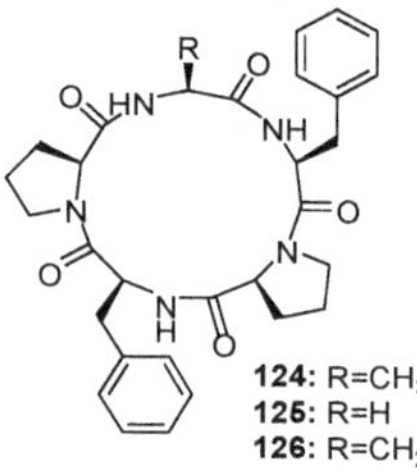

124: R=CH_3
125: R=H
126: R=CH_2OH

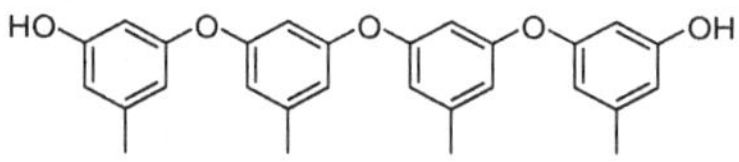

127

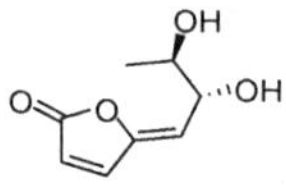

128

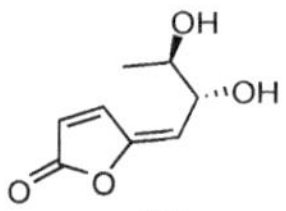

129

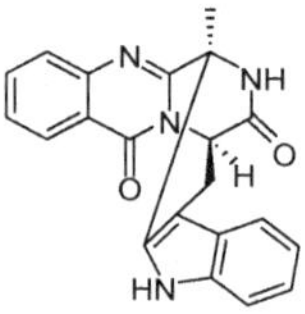

130

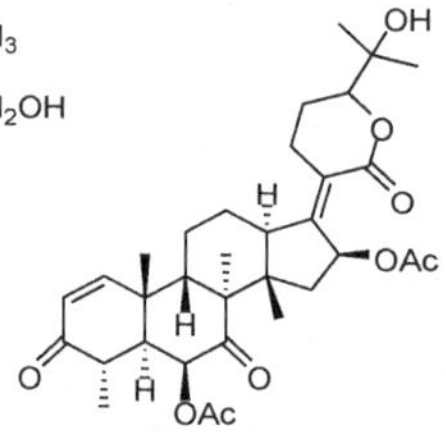

131

14.3.1.2 MNPs from Marine Soft Coral-Derived *Penicillium* sp.

Two 2,5-piperazinedione derivatives, janthinolides A and B (**132–133**), were isolated from the fungus *P. janthinellum* isolated from a soft coral *Dendronephthya* sp. [42].

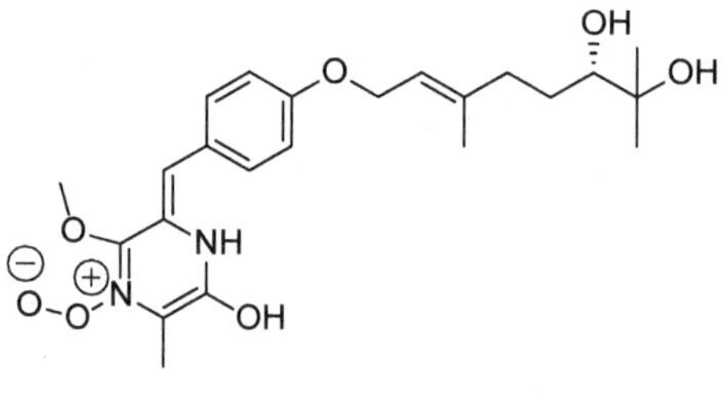

132

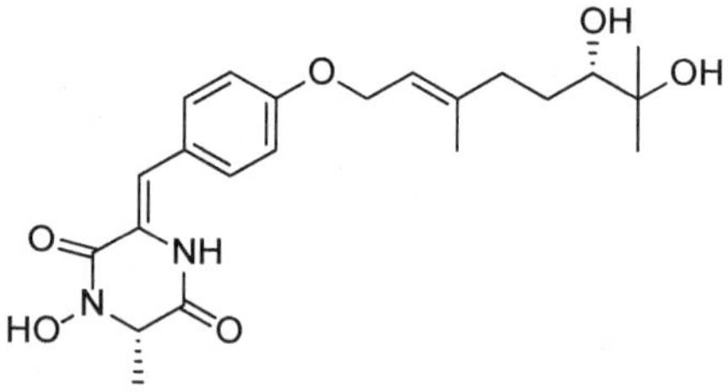

133

14.3.1.3 MNPs from Other Marine Soft Coral-Derived Fungi

Three hirsutane sesquiterpenoids hirsutanol E (**134**) and hirsutanols A and F were isolated from the marine fungus *Chondrostereum* sp. isolated from the soft coral *Sarcophyton tortuosum*. Hirsutanol A showed potent cytotoxicity [43]. Five hydroanthraquinone derivatives, tetrahydroaltersolanols C–F (**135–138**) and dihydroaltersolanol A (**139**), and five alterporriol-type anthranoid dimers, alterporriols N–R (**140–144**), were isolated from a soft coral-derived fungus *Alternaria* sp. [44]. Five triquinane-type sesquiterpenoids, chondrosterins A–E (**145–149**), were isolated from a marine fungus *Chondrostereum* sp. that was isolated from a soft coral *Chondrostereum* sp. Chondrosterin A showed significant cytotoxicity [45]. Three hirsutane sesquiterpenoids chondrosterins F–H (**150–152**) were isolated from the marine fungus *Chondrostereum* sp. that was collected from the soft coral *S. tortuosum*. Incarnal had potent cytotoxicity [46]. An antibacterial chlorinated benzophenone derivative, (±)-pestalachloride D (**153**), was isolated from a marine fungus *Pestalotiopsis* sp. associated with a soft coral *Sarcophyton* sp. [47]. Two sulfur-containing benzofuran derivatives, eurothiocins A and B (**154** and **155**), were isolated from the fungus *Eurotium rubrum* SH-823 obtained from a soft coral *Sarcophyton* sp. Eurothiocins A and B exhibited more potent inhibitory effects against α-glucosidase activity than the clinical α-glucosidase inhibitor acarbose [48].

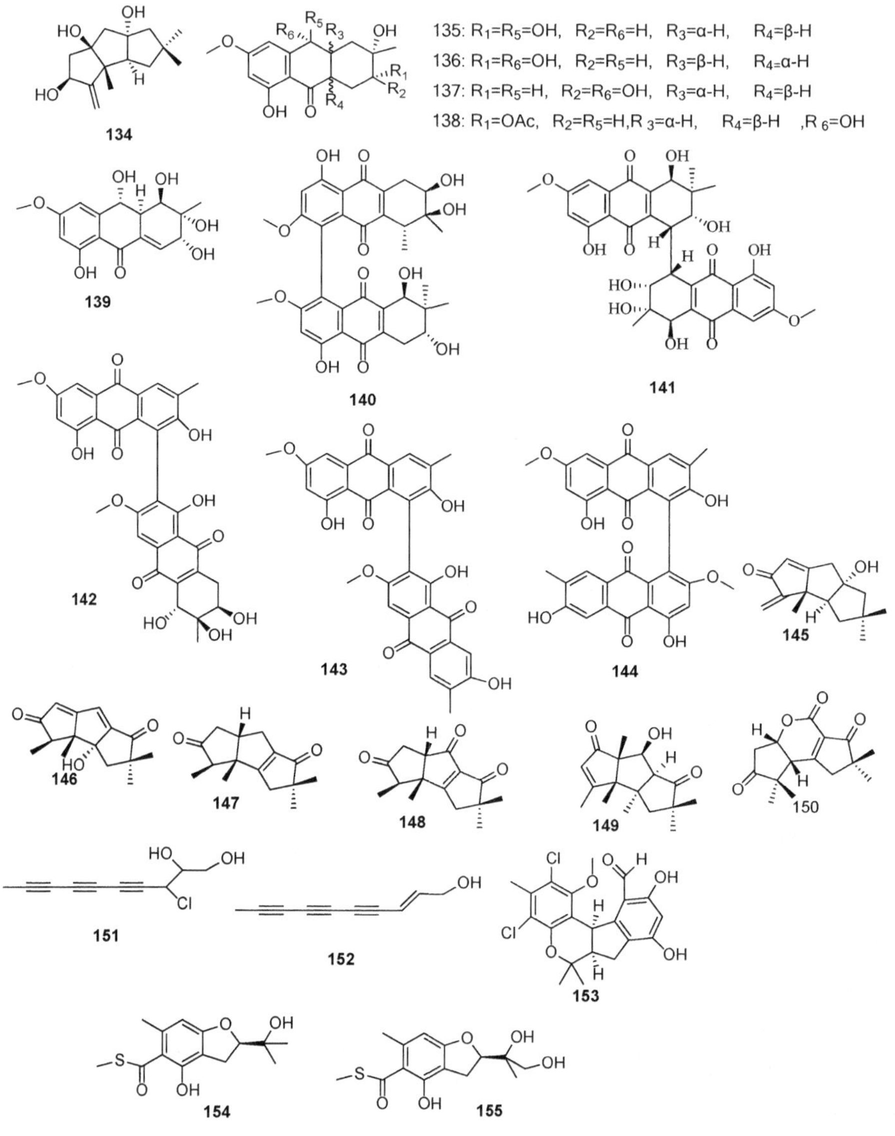

14.3.2 MNPs from Soft Coral-Derived Bacteria and Actinomycetes

Seven novel maleimide derivatives, aqabamycins A–G (**156–162**), were isolated from a marine bacterium *Vibrio* sp. associated with the soft coral *S. polydactyla* together with three natural products 3-nitro-1*H*-indazole (**163**), indazole-3-carbaldehyde (**164**), and 1,4-dithiane (**165**) [49]. Aqabamycins C and E–G displayed different antibacterial activity against *B. subtilis*, *M. luteus*, *E. coli*, and *Proteus vulgaris*. Aqabamycins A and D–G showed strong cytotoxicity against mouse lymphocytic leukemia L1210 cell line. Aqabamycin E had cytotoxicity toward breast cancer MDA-MB321, MCF-7, and colorectal adenocarcinoma Colo-320 cell lines with MIC values of 25/20/50 and 30/50/30 μM, respectively [49]. A depsipeptide thiocoraline (**166**) was isolated from a strain of soft coral symbiotic actinomycetes *Micromonosporad* L-13-ACM2-092. Thiocoraline showed strong cytotoxicity against tumor cell lines P388, A549, HT-29, and MEL-28 with IC_{50} values of 0.002, 0.002, 0.01, and 0.002 μM [4]. Two chlorinated polyketides, strepchloritides A–B (**167–168**), were isolated from the soft coral-associated actinomycetes strain *Streptomyces* sp. OUCMDZ-1703 [50].

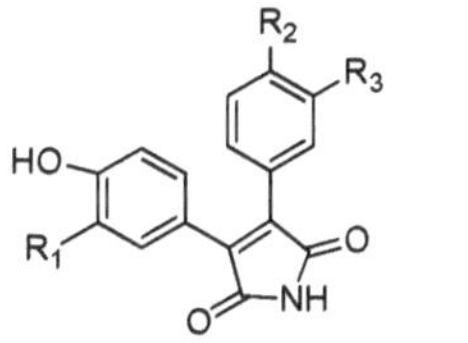

156:R_1=R_2=R_3=H
157:R_1=H,R_2=OH,R_3=NO_2
158:R_1=NO_2,R_2=R_3=H
159:R_1=R_3=NO_2,R_2=OH

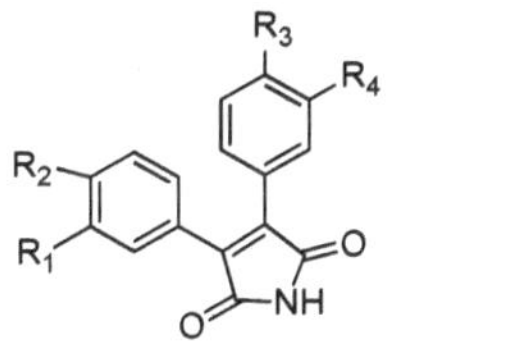

160:R_1=NO_2,R_2=OH,R_3=R_4=H
160a:R_1=R_2=H,R_3=OH,R_4=NO_2
161:R_1=R_4=NO_2,R_2=R_3=OH

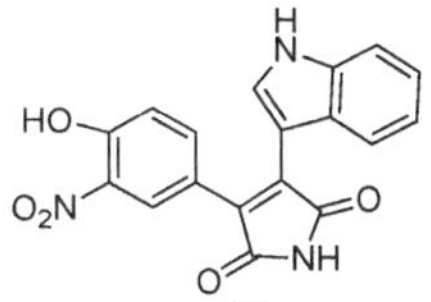

162

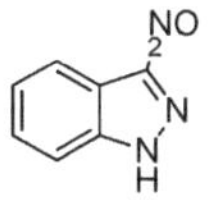

163

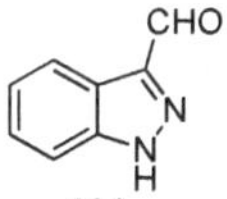

164

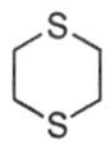

165

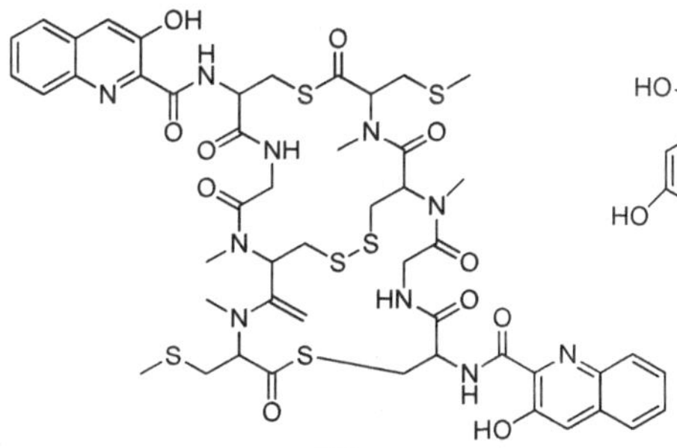

166

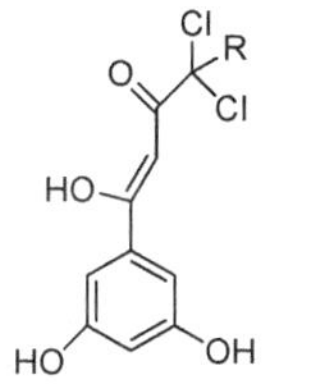

167:R=H
168:R=Cl

14.4 Summary

According to the literature statistics, until 2015 year, over 170 new compounds had been isolated from microbes associated with gorgonians and soft corals, and the studies were mainly focused on the secondary metabolites of coral-associated fungi (Tables 14.1 and 14.2). These compounds included poly ketones, pyrones, half terpenes, cyclic peptides, macrocyclic lactones, mycotoxins, alkaloids, sterols, and so on, and some of them showed strong bioactives, such as cytotoxic, antimalarial, antibacterial, antifungal, and antifouling activity, which indicate that coral-associated microbes are important resources for finding new drugs.

Table 14.1 Marine natural products from marine gorgonian-derived microorganisms (1991–2015)

Compound	Producing strain	Source	Bioactivity	References
1–3	*Aspergillus* sp.	*Dichotella gemmacea*		[5]
4–5	*Aspergillus* sp.	*Dichotella gemmacea*	4: cytotoxicity	[6]
6–10	*Aspergillus sydowii* PSU-F154	*Annella* sp.		[7]
11–14	*Aspergillus* sp. SCSGAF 0076	*Melitodes squamata*	13: antifouling activity	[8]
15–18	*Aspergillus* sp. SCSGAF0093			[9]
19–22	*Aspergillus* sp.	*Dichotella gemmacea*		[10]
23–27	*Aspergillus* sp. XS-20090B15			[11]
28	*Aspergillus* sp.	*Dichotella gemmacea*		[12]
29–39	*Aspergillus* sp.			[13]
40–41	*Penicillium* sp. PSU-F44	*Annella* sp.		[14]
42–50	*Penicillium* sp. PSU-F40	*Annel* sp.		[15]
51–55	*Penicillium citrinum* PSU-F51	*Annella* sp.	54: antibacterial activity Cytotoxicity	[16]
56–62	*Penicillium commune* 518	Wang et al. 2012	56,69,62: antimicrobial activities	[17]
63–65	*Penicillium oxalicum* SCSGAF 0023	*Dichotella gemmacea*	63,64: cytotoxicity	[18]
66–67	*Penicillium oxalicum* SCSGAF 0023	*Dichotella gemmacea*	67: cytotoxicity	[19]
68–71	*Penicillium pinophilum* XS-20090E18			[20]
72–75	*Nigrospora* sp. PSU-F5	*Annella* sp.	72: antibacterial activity	[21]
76–79	*Nigrospora* sp. PSU-F18	*Annella* sp.	76: antibacterial activity	[22]
80	*Xylaria* sp. PSU-F100	*Annella* sp.		[23]
81–85	*Fusarium* spp. PSU-F14 and PSU-F135	*Annella* sp.		[24]
86–87	*Nigrospora* sp. PSU-F11 and PSU-F12	*Annella* sp.	86: cytotoxicity	[25]
88–90	*Cochliobolus lunatus*	*Dichotella gemmacea*	88: antifouling activity Antibacterial	[26]
91–93	*Curvularia* sp. PSUF22	*Annella* sp.		[27]
94–95	*Trichoderma aureoviride* PSU-F95	*Annella* sp.		[28]

(continued)

Table 14.1 (continued)

Compound	Producing strain	Source	Bioactivity	References
96	*Scopulariopsis* sp.	*Carijoa* sp.	Antibacterial activity	[29]
97–100	*A. versicolor*	*D. gemmacea*		[30]
101–105	*Eurotium* sp. XS-200900E6			[31]
106	*Acremonium polychromum* PSU-F125	*Annella* sp.		[32]
107–110	*Nigrospora oryzae* SCSGAF 0111	*Verrucella umbraculum*		[33]
111	*Cladosporium* sp.	*Anthogorgia ochracea (GXWZ-07)*	Cytotoxicity	[34]
112–114	*Talaromyces* sp.	*Subergorgia suberosa*	114: antifouling activities	[35]
115	*Bacillus amyloliquefaciens*	*Junceella juncea*	Antibacterial activity	[36]
116–117	Actinomycete *Streptomyces* sp.	*Pacifigorgia* sp.	116: cytotoxic activity	[3]

Table 14.2 Marine natural products from marine soft coral-derived microorganisms (1997–2015)

Compound	Producing strain	Source	Bioactivity	References
118–119	*Aspergillus terreus*	*Sinularia kavarattiensis*	118: cytotoxicity	[37]
120	*Aspergillus* sp.	*Sarcophyton tortuosum*		[38]
121–123	*Aspergillus versicolor* LCJ-5-4	*Cladiella* sp.	123: antifungal activity	[39]
124–129	*Aspergillus versicolor* LCJ-5-4		127: antioxidative activity	[40]
130–131	*Aspergillus fumigatus* KMM 4631	*Sinularia* sp.		[41]
132–133	*Penicillium janthinellum*	*Dendronephthya* sp.		[42]
134	*Chondrostereum* sp.	*Sarcophyton tortuosum*	Cytotoxicity	[43]
135–144	*Alternaria* sp.			[44]
145–149	*Chondrostereum* sp.	*Chondrostereum* sp.	145: cytotoxicity	[45]
150–152	*Chondrostereum* sp.	*Sarcophyton tortuosum*	150,151,152: cytotoxicity	[46]
153	*Sarcophyton* sp.		Antibacterial activity	[47]
154–155	*Eurotium rubrum* SH-823	*Sarcophyton* sp.	154,155: inhibitory	[48]
156–165	*Vibrio* sp.	*Sinularia polydactyla*	158,162: cytotoxicity antibacterial	[49]
166	*Micromonosporad* L-13-ACM2–092		Cytotoxicity	[4]
167–168	*Streptomyces* sp. OUCMDZ-1703			[50]

Acknowledgment The authors are grateful for the financial support provided by the National Natural Science Foundation of China (41376160), Strategic Leading Special Science and Technology Program of Chinese Academy of Sciences (XDA100304002), and Regional Innovation Demonstration Project of Guangdong Province Marine Economic Development (GD2012-D01-002).

References

1. Li ZY. Advances in marine microbial symbionts in the China Sea and related pharmaceutical metabolites. Mar Drugs. 2009;7:113–29.
2. Konig GM, Kehraus S, Seibert SF, Abdel-Lateff A, Muller D. Natural products from marine organisms and their associated microbes. Chembiochem. 2006;7:229–38.
3. Tapiolas DM, Roman M, Fenical W, Shout TJ, Clardy J. Octalactin-A and octalactin-B -cytotoxic 8-membered-ring lactones from a marine bacterium, *Streptomyces* sp. J Am Chem Soc. 1991;113:4682–3.
4. Romero F, Espliego F, Baz JP, DeQuesada TG, Gravalos D, DelaCalle F. Thiocoraline, a new depsipeptide with antitumor activity produced by a marine Micromonospora.I. Taxonomy, fermentation, isolation, and biological activities. J Antibiot. 1997;50:734–7.
5. Wei MY, Wang CY, Liu QA, Shao CL, She ZG, Lin YC. Five sesquiterpenoids from a marine-derived fungus *Aspergillus* sp. isolated from a gorgonian *Dichotella gemmacea.* Mar Drugs. 2010;8:941–9.
6. Shao CL, Wan CY, Wei MY, Gu YC, She ZG, Qian PY, et al. Aspergilones A and B, two benzylazaphilones with an unprecedented carbon skeleton from the gorgonian-derived fungus *Aspergillus* sp. Bioorg Med Chem Lett. 2011;21:690–3.
7. Trisuwan K, Rukachaisirikul V, Kaewpet M, Phongpaichit S, Hutadilok-Towatana N, Preedanon S, et al. Sesquiterpene and xanthone derivatives from the sea fan-derived fungus *Aspergillus sydowii* PSU-F154. J Nat Prod. 2011;74:1663–7.
8. Bao J, Zhang XY, Xu XY, He F, Nong XH, Qi SH. New cyclic tetrapeptides and asteltoxins from gorgonian-derived fungus *Aspergillus* sp. SCSGAF 0076. Tetrahedron. 2013;69:2113–7.
9. Xu XY, He F, Zhang XY, Bao J, Qi SH. New mycotoxins from marine-derived fungus *Aspergillus* sp. SCSGAF0093. Food Chem Toxicol. 2013;53:46–51.
10. Chen M, Shao CL, Fu XM, Xu RF, Zheng JJ, Zhao DL, et al. Bioactive indole alkaloids and phenyl ether derivatives from a marine-derived *Aspergillus* sp. fungus. J Nat Prod. 2013;76:547–53.
11. Chen M, Shao CL, Fu XM, Kong CJ, She ZG, Wang CY. Lumazine peptides penilumamides B-D and the cyclic pentapeptide asperpeptide A from a gorgonian-derived *Aspergillus* sp. fungus. J Nat Prod. 2014;77:1601–6.
12. Chen M, Shao CL, Kong CJ, She ZG, Wang CY. A new anthraquinone derivative from a gorgonian-derived fungus *Aspergillus* sp. Chem Nat Compd. 2014;50:617–20.
13. Chen M, Wang KL, Liu M, She ZG, Wang CY. Bioactive steroid derivatives and butyrolactone derivatives from a gorgonian-derived *Aspergillus* sp. fungus. Chem Biodivers. 2015;12:1398–406.
14. Trisuwan K, Rukachaisirikul V, Sukpondma Y, Phongpaichit S, Preedanon S, Sakayaroj J. Lactone derivatives from the marine-derived fungus *Penicillium* sp. PSU-F44. Chem Pharm Bull. 2009;57:1100–2.
15. Trisuwan K, Rukachaisirikul V, Sukpondma Y, Phongpaichit S, Preedanon S, Sakayaroj J. Furo 3,2-h isochroman, furo 3,2-h isoquinoline, isochroman, phenol, pyranone, and pyrone derivatives from the sea fan-derived fungus *Penicillium* sp. PSU-F40. Tetrahedron. 2010;66:4484–9.
16. Khamthong N, Rukachaisirikul V, Phongpaichit S, Sita P, Jariya S. Bioactive polyketides from the sea fan-derived fungus Penicillium citrinum PSU-F51. Tetrahedron. 2012;68:8245–50.

17. Wang JF, Lei PP, Wang Y, Wang H, Li J, Zhuang YB, Zhu WM. Antimicrobial aromatic polyketides from gorgonian-associated fungus, *Penicillium commune* 518. Chin J Chem. 2012;30:1236–42.
18. Sun YL, He F, Liu KS, Zhang XY, Bao J, Wang YF, et al. Cytotoxic dihydrothiophene-condensed chromones from marine-derived fungus *Penicillium oxalicum*. Planta Med. 2012;78:1957–61.
19. Bao J, Luo JF, Qin XC, Xu XY, Zhang XY, Tu ZC, et al. Dihydrothiophene-condensed chromones from a marine-derived fungus *Penicillium oxalicum* and their structure-bioactivity relationship. Bioorg Med Chem Lett. 2014;24:2433–6.
20. Zhao DL, Shao CL, Wang KL, Guan FF, Shi T, Wang CY. Azaphilone and diphenyl ether derivatives from a gorgonian-derived strain of the fungus *Penicillium pinophilum*. J Nat Prod. 2015;78:2310–4.
21. Trisuwan K, Rukachaisirikul V, Sukpondma Y, Preedanon S, Phongpaichit S, Rungjindamai N, Sakayaroj J. Epoxydons and a pyrone from the marine-derived fungus *Nigrospora* sp. PSU-F5. J Nat Prod. 2008;71:1323–6.
22. Trisuwan K, Rukachaisirikul V, Sukpondma Y, Preedanon S, Phongpaichit S, Sakayaroj J. Pyrone derivatives from the marine-derived fungus *Nigrospora* sp. PSU-F18. Phytochemistry. 2009;70:554–7.
23. Rukachaisirikul V, Khamthong N, Sukpondma Y, Pakawatchal C, Phongpaichit S, Sakayaroj J. An 11 cytochalasin derivative from the marine-derived fungus *Xylaria* sp. PSU-F100. Chem Pharm Bull. 2009;57:1409–11.
24. Trisuwan K, Khamthong N, Rukachaisirikul V, Phongpaichit S, Preedanon S, Sakayaroj J. Anthraquinone, cyclopentanone, and naphthoquinone derivatives from the sea fan-derived fungi *Fusarium* spp. PSU-F14 and PSU-F135. J Nat Prod. 2010;73:1507–11.
25. Rukachaisirikul V, Khamthong N, Sukpondma Y, Phongpaichit S, Hutadilok-Towatana N, Graidist P, et al. Cyclohexene, diketopiperazine, lactone and phenol derivatives from the sea fan-derived fungi Nigrospora sp. PSU-F11 and PSU-F12. Arch Pharm Res. 2010;33:375–80.
26. Shao CL, Wu HX, Wang CY, Liu QA, Xu Y, Wei MY, et al. Potent antifouling resorcylic acid lactones from the gorgonian-derived fungus *Cochliobolus lunatus*. J Nat Prod. 2011;74:629–33.
27. Trisuwan K, Rukachaisirikul V, Phongpaichit S, Preedanon S, Sakayaroj J. Modiolide and pyrone derivatives from the sea fan-derived fungus *Curvularia* sp. PSU-F22. Arch Pharm Res. 2011;34:709–14.
28. Khamthong N, Rukachaisirikul V, Tadpetch K, Kaewpet M, Phongpaichit S, Preedanon S, et al. Tetrahydroanthraquinone and xanthone derivatives from the marine-derived fungus Trichoderma aureoviride PSU-F95. Arch Pharm Res. 2012;35:461–8.
29. Shao CL, Xu RF, Wei MY, She ZG, Wang CY. Structure and absolute configuration of fumiquinazoline L, an alkaloid from a gorgonian-derived *Scopulariopsis* sp. fungus. J Nat Prod. 2013;76:779–82.
30. Chen M, Fu XM, Kong CJ, Wang CY. Nucleoside derivatives from the marine-derived fungus *Aspergillus versicolor*. Nat Prod Res. 2014;28:895–900.
31. Chen M, Shao CL, Wang KL, Xu Y, She ZG, Wang CY. Dihydroisocoumarin derivatives with antifouling activities from a gorgonian-derived *Eurotium* sp. fungus. Tetrahedron. 2014;70:9132–8.
32. Khamthong N, Rukachaisirkul V, Pakawatchai C, Saithong S, Phongpaichit S, Preedanon S, et al. Acremonoside, a phenolic glucoside from the sea fan-derived fungus *Acremonium polychromum* PSU-F125. Phytochem Lett. 2014;10:50–4.
33. Dong JJ, Bao J, Zhang XY, Xu XY, Nong XH, Qi SH. Alkaloids and citrinins from marine-derived fungus *Nigrospora oryzae* SCSGAF 0111. Tetrahedron Lett. 2014;55:2749–53.
34. Cao F, Yang Q, Shao CL, Kong CJ, Zheng JJ, Liu YF, et al. Bioactive 7-Oxabicyclic [6.3.0] lactam and 12-membered macrolides from a gorgonian-derived *Cladosporium* sp. fungus. Mar Drugs. 2015;13:4171–8.
35. Chen M, Han L, Shao CL, She ZG, Wang CY. Bioactive diphenyl ether derivatives from a gorgonian-derived fungus *Talaromyces* sp. Chem Biodivers. 2015;12:443–50.

36. Gao CH, Tian XP, Qi SH, Luo XM, Wang P, Zhang S. Antibacterial and antilarval compounds from marine gorgonian-associated bacterium *Bacillus amyloliquefaciens* SCSIO 00856. J Antibiot. 2010;63:191–3.
37. Parvatkar RR, D'Souza C, Tripathi A, Naik CG. Aspernolides A and B, butenolides from a marine-derived fungus *Aspergillus terreus*. Phytochemistry. 2009;70:128–32.
38. Li HJ, Cai YT, Chen YY, Lam CK, Lan WJ. Metabolites of marine fungus *Aspergillus* sp. collected from soft coral *Sarcophyton tortuosum*. Chem Res Chin Univ. 2010;26:415–9.
39. Zhuang Y, Teng X, Wang Y, Liu PP, Li GQ, Zhu WM. New quinazolinone alkaloids within rare amino acid residue from coral-associated fungus *Aspergillus versicolor* LCJ-5-4. Org Lett. 2011;13:1130–3.
40. Zhuang Y, Teng X, Wang Y, Liu PP, Wang H, Li J, et al. Cyclopeptides and polyketides from coral-associated fungus *Aspergillus versicolor* LCJ-5-4. Tetrahedron. 2011;67:7085–9.
41. Afiyatullov SS, Zhuravleva OI, Antonov AS, Kalinovsky AI, Pivkin MV, Menchinskaya ES, et al. New metabolites from the marine-derived fungus *Aspergillus fumigatus*. Nat Prod Commun. 2012;7:497–500.
42. Xue CM, Li T, Deng ZW, Fu HZ, Lin WH. Janthinolide A-B, two new 2, 5-piperazinedione derivatives from the endophytic *Penicillium janthinellum* isolated from the soft coral *Dendronephthya* sp. Pharmazie. 2006;61:1041–4.
43. Li HJ, Lan WJ, Lam CK, Yang F, Zhu XF. Hirsutane sesquiterpenoids from the marine-derived fungus *Chondrostereum* sp. Chem Biodivers. 2011;8:317–24.
44. Zheng CJ, Shao CL, Guo ZY, Chen JF, Deng DS, Yang KL, et al. Bioactive hydroanthraquinones and anthraquinone dimers from a soft coral-derived *Alternaria* sp. fungus. J Nat Prod. 2012;75:189–97.
45. Li HJ, Xie YL, Xie ZL, Chen Y, Lam CK, Lan WJ. Chondrosterins A-E, triquinane-type sesquiterpenoids from soft coral-associated fungus *Chondrostereum* sp. Mar Drugs. 2012;10:627–38.
46. Li HJ, Chen T, Xie YL, Chen WD, Zhu XF, Lan WJ. Isolation and structural elucidation of Chondrosterins F-H from the marine fungus *Chondrostereum* sp. Mar Drugs. 2013;11:551–8.
47. Wei MY, Li D, Shao CL, Deng DS, Wang CY. (±)-Pestalachloride D, an antibacterial racemate of chlorinated benzophenone derivative from a soft coral-derived fungus *Pestalotiopsis* sp. Mar Drugs. 2013;11:1050–60.
48. Liu ZM, Xia GP, Chen SH, Liu YY, Li HX, She ZG. Eurothiocin A and B, sulfur-containing benzofurans from a soft coral-derived fungus *Eurotium rubrum* SH-823. Mar Drugs. 2014;12:3669–80.
49. Al-Zereini W, Yao CBFF, Laatsch H, Anke H. Aqabamycins A-G: novel nitro maleimides from a marine Vibrio species: I. Taxonomy, fermentation, isolation and biological activities. J Antibiot. 2010;63:297–301.
50. Fu P, Kong F, Wang YF, Wang Y, Liu PP, Zuo GY, et al. Antibiotic metabolites from the coral-associated actinomycete *Streptomyces* sp. OUCMDZ-1703. Chin J Chem. 2013;31:100–4.

Chapter 15
Natural Products from Sponges

Bing-Nan Han, Li-Li Hong, Bin-Bin Gu, Yang-Ting Sun, Jie Wang, Jin-Tang Liu, and Hou-Wen Lin

Contents

Abstract The sponge is one of the oldest multicellular invertebrates in the world. Marine sponges represent one of the extant metazoans of 700–800 million years. They are classified in four major classes: Calcarea, Demospongiae, Hexactinellida, and Homoscleromorpha. Among them, three genera, namely, *Haliclona*, *Petrosia*, and *Discodemia* have been identified to be the richest source of biologically active compounds. So far, 15,000 species have been described, and among them, more than 6000 species are found in marine and freshwater systems throughout tropical, temperate, and polar regions. More than 5000 different compounds have been

B.-N. Han · L.-L. Hong · B.-B. Gu · Y.-T. Sun · J. Wang · J.-T. Liu · H.-W. Lin (✉)
Marine Drugs Research Center, Department of Pharmacy, State Key Laboratory of Oncogenes and Related Genes, Ren Ji Hospital, School of Medicine, Shanghai Jiao Tong University, Shanghai, People's Republic of China

Z. Li (ed.), *Symbiotic Microbiomes of Coral Reefs Sponges and Corals*,
https://doi.org/10.1007/978-94-024-1612-1_15

isolated and structurally characterized to date, contributing to about 30% of all marine natural products. The chemical diversity of sponge products is high with compounds classified as alkaloids, terpenoids, peptides, polyketides, steroids, and macrolides, which integrate a wide range of biological activities, including antibacterial, anticancer, antifungal, anti-HIV, anti-inflammatory, and antimalarial. There is an open debate whether all natural products isolated from sponges are produced by sponges or are in fact derived from microorganisms that are inhaled though filter-feeding or that live within the sponges. Apart from their origin and chemoecological functions, sponge-derived metabolites are also of considerable interest in drug development. Therefore, development of recombinant microorganisms engineered for efficient production of sponge-derived products is a promising strategy that deserves further attention in future investigations in order to address the limitations regarding sustainable supply of marine drugs.

Keywords Sponge · Sponge holobiont · Natural products · Alkaloids · Peptides · Polyketides · Macrolides · Terpenoids · Steroids · Bioactivity

15.1 Introduction

Considering that oceans comprise over 70% of the earth's surface and harbor a tremendous variety of flora and fauna, marine habitat represents an unexplored source of new bioactive molecules. Although still quite young by many standards, since the 1950s, this field of marine natural products has undergone exponential growth and proven to be a productive source for structurally diverse secondary metabolites. Due to long evolutionary processes favoring the accumulation of strongly bioactive compounds, sponges (Porifera) and their associated microorganisms have become the largest contributors of marine natural products. Seemingly primitive and morphologically defenseless organisms like sponges developed ingenious survival strategies which rely heavily on the accumulation of defensive products protecting them from a multitude of stress factors that involve overgrowth by fouling organisms, attacking by predators, and invasion by pathogenic microorganisms. Sponges are classified in four major classes: Calcarea, Demospongiae, Hexactinellida, and Homoscleromorpha. Among them, three genera, namely, *Haliclona*, *Petrosia*, and *Discodemia* have been identified to be the richest source of biologically active compounds. The chemical diversity of sponge products is high with compounds classified as alkaloids, terpenoids, peptides, polyketides, steroids, and macrolides, which integrate a wide range of biological activities, including antibacterial, anticancer, antifungal, anti-HIV, anti-inflammatory, and antimalarial.

15.2 Bioactive Alkaloids from Marine Sponges

Biologically significant alkaloids, as a special and important class of bioactive natural products, are widely distributed over terrestrial and marine organisms. Recent studies have demonstrated that marine invertebrates and microorganisms are abundant sources of these secondary metabolites. Among these natural products, imidazole-, oxazole-, and thiazole-containing alkaloids are often found to show diverse significant biological activities, including antitumor, antibacterial, antiviral, antimalarial, immunosuppressive activities, etc.

The following review summarizes the latest progress on the isolation, structure identification of a diverse 209 alkaloids from 66 marine sponges with potent biological activities within the literature coverage from 1986 to 2016.

In the year of 1993, xestocyclamine A (**1**) was isolated from Papua New Guinea collections of the sponge *Xestospongia* sp. Pure xestocyclamine A exhibited an IC_{50} = 4 μg/mL (10.1 μM) against PKC ϵ and also exhibits activity in a whole cell IL-1 release assay with an IC_{50} of 1 μM [1].

In the year of 1994, madangamine A (**2**) was isolated from the marine sponge *Xestospongia ingens*, which showed in vitro cytotoxicity against murine leukemia P388 (ED_{50} 0.93 μg/mL) [2].

In 1996, sceptrine (**3**) and ageliferine (**4**) were isolated from *Xestospongia* sp. and *Agelas novaecaledoniae* collected at Baic de Prony, exhibiting a high affinity for somatostatin (IC_{50} = 0.27 μM and 2.2 μM) [3]. Within the same year, 4,5-dibromopyrrole-2-carbamide (**5**) was isolated from the marine sponge *Agelas mauritiana* collected off Hachijo-jima Island, Japan. 4,5-dibromopyrrole-2-carbamide (**5**) promoted larval metamorphosis of the ascidian *Ciona savignyi* at a concentration of 2.5 μg/mL (9.36 μM) [4]. (±)-xestospongin D (**6**) was isolated from the Singapore marine sponge *Niphates* sp. collected from south of the Beting Bemban Besar reef. (**6**) was found to inhibit growth of certain human cancer cell lines comprising the NCI panel (leukemia subpanel, mean GI_{50} 3.62 ± 2.02 μM; breast subpanel, mean GI_{50} 4.53 ± 1.98 μM) as well as the murine P388 lymphocytic leukemia (ED_{50} 1.7 μg/mL) (3.67 μM) [5].

In the year of 1998, the chemical investigation of Micronesian sponge *Oceanapia* sp. from Truk Lagoon, afforded two active pyridoacridine alkaloids: the known compound kuanoniamine D (**7**) as well as the new N-deacyl derivative of the kuanoniamines (**8**). The IC_{50} of N-deacyl derivative of the kuanoniamines (**8**) was 1.2 μg/mL (3.77 μM) against HeLa cells and 2.0 μg/mL against MONO-MAC 6 cells. In receptor binding assays, kuanoniamine D (**7**) showed potent affinity to A1 adenosine receptors with Ki value of 2.94 μM [6].

In 1999, halitulin (**9**) was isolated from the sponge *Haliclona tulearensis* collected in Sodwana Bay, Durban, South Africa. It was found to be cytotoxic against several tumor cell lines: P388, A549, HT29, and MEL28 in concentration of

12–25 ng/mL [7]. Discorhabdin Q (16, 17-dehydrodiscorhabdin B) (**10**) was isolated from cytotoxic extracts of the sponge *Latrunculia purpurea* and numerous collections of *Zyzzya massalis*, *Z. fuliginosa*, and Z. spp from Australia and Fiji. In the NCI 60-cell line antitumor screen, discorhabdin Q (**10**) exhibited moderate cytotoxicity (mean panel GI_{50} = 0.5 μg/mL) [8]. In the same year, five new steroidal alkaloids, plakinamines C and D (**11**), and three related compounds were isolated from the Vanuatu sponge *Corticium* sp. collected off Porth Havannah, Vanuatu, South Pacific. Two new compounds (**12** and **13**) performed good in vitro cytotoxicity against human bronchopulmonary non-small-cell lung carcinoma cells (NSCLC-N6) with IC_{50} values of 3.3–5.7 μg/mL, while plakinamine D (**11**) was cytotoxic with IC_{50} < 3.3 μg/mL [9].

In 2000, topsentins B1 (**14**) was isolated from the marine sponge *Rhaphisia lacazei*, collected in the Mediterranean Sea which showed antiproliferative activity against human bronchopulmonary cancer cells with an IC_{50} of 6.3 μg/mL [10]. Dragmacidin F (**15**), possessing an unprecedented carbon skeleton, was isolated from a marine sponge of the genus *Halicortex* collected off the southern coast of Ustica Island (Italy), which showed good in vitro antiviral activity toward HIV-1 (EC_{50} = 0.91 μM) [11].

In the year of 2001, makaluvamine P (**16**) was isolated from the sponge *Zyzzya* cf. *fuliginosa* collected in the waters off the Vanuatu Islands. Makaluvamine P (**16**) was found to inhibit the growth of KB tumor cells with 64% on at 3.2 μg/mL (9.5 μM) [12].

In 2002, the sponge *Stylissa massa* collected from the shallow waters around Helgoland afforded eight known alkaloids, among which 10 E-hymenialdisine (**17**) and 10 Z-hymenialdisine (**18**) were active in the initial Raf/MEK-1/MAPK signaling cascade assay (IC_{50} = 3 and 6 nM) [13]. Arenosclerins A–C (**19–21**) and haliclonacyclamine E (**22**), isolated from the marine sponge *Arenosclera brasiliensis*, exhibited certain cytotoxity against human HL-60 (leukemia), L929 (fibrosarcoma), B16 (melanoma), and U138 (colon) cancer cell lines at concentrations between 1.5 and 7.0 μg/mL [14]. In addition, isonaamidine E (**23**) was isolated from two sponges, *Leucetta chagosensis* and *Leucetta* cf. *chagosensis*, collected from the Great Barrier Reef and the Fiji Islands, which was found to be cytotoxic toward several tumor cell lines (GI_{50} values was 1.3 μg/mL) [15].

In 2003, manadomanzamines A (**24**) and B (**25**) were isolated from an Indonesian sponge *Acanthostrongylophora* sp. (Haplosclerida: Petrosiidae). Manadomanzamines A (**24**) and B (**25**) exhibited strong activity against *Mycobacterium tuberculosis* (Mtb) with MIC values of 1.9 and 1.5 μg/M (2.4 μM) [16].

In 2004, seven pyrrole alkaloids isolated from *Agelas* sponges were tested for interactions with the cellular calcium homeostasis. Among them, brominated pyrrole alkaloids reduced voltage-dependent calcium elevation in PC12 cells, and dibromosceptrin (**26**) was the most potent alkaloid with a half maximal of 2.8 μM [17].

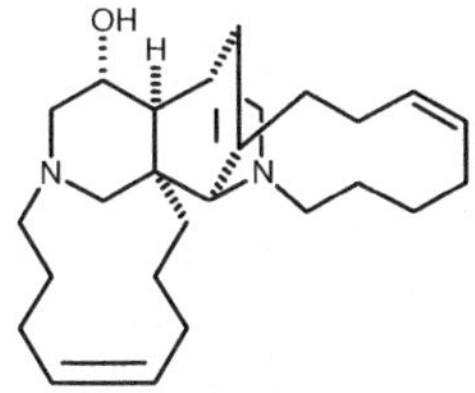

Xestocyclamine A (**1**)

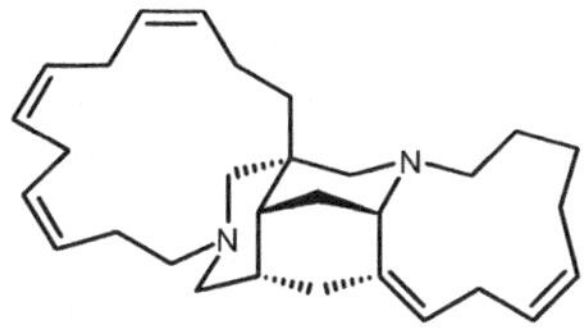

Madangamine A (**2**)

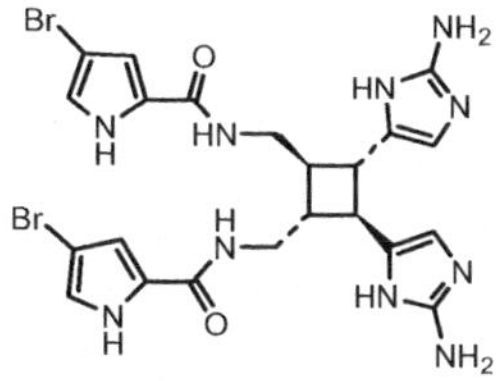

Sceptrine (**3**)

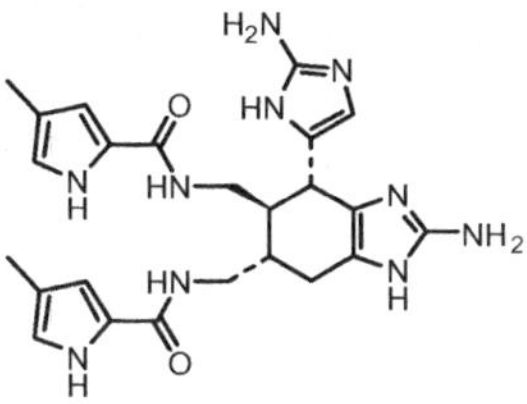

Ageliferine (**4**)

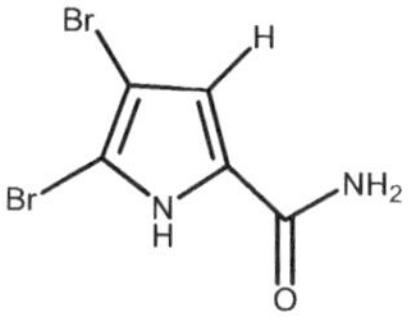

4,5-dibromopyrrole-2-carbamide (**5**)

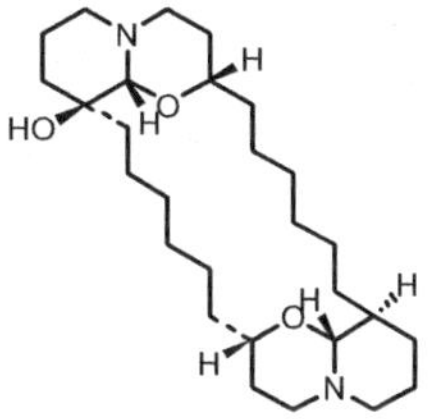

(±)-XestosponginD (**6**)

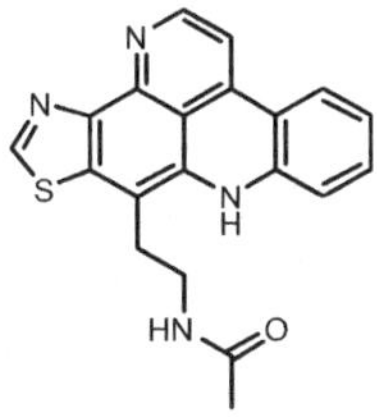

Kuanoniamine D (**7**)

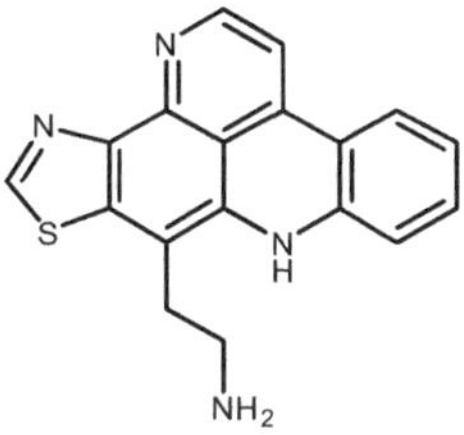

N-deacyl derivative of the kuanoniamines (**8**)

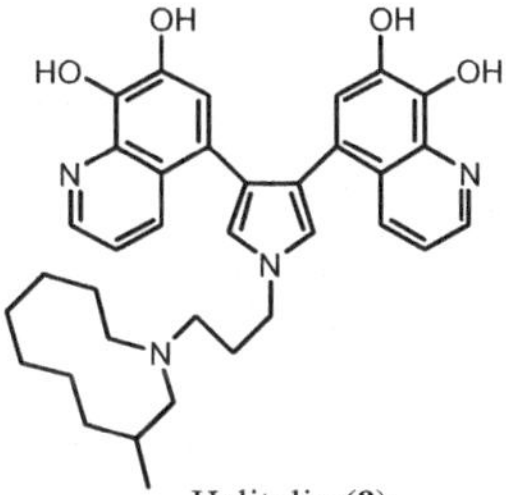

Halitulin (**9**)

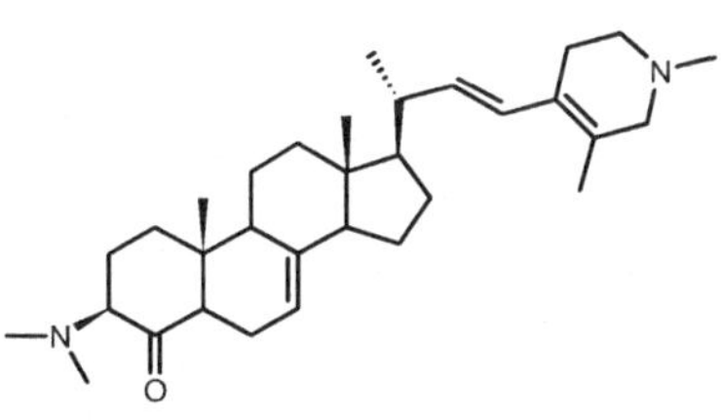

N,N-dimethyl-4-oxo-3-epi-plakinamine B (**12**)

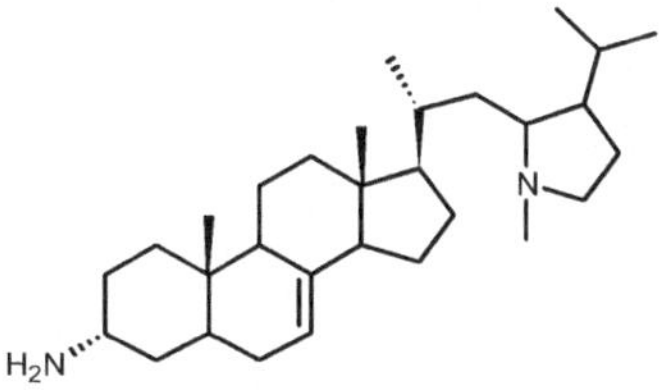

(**13**)

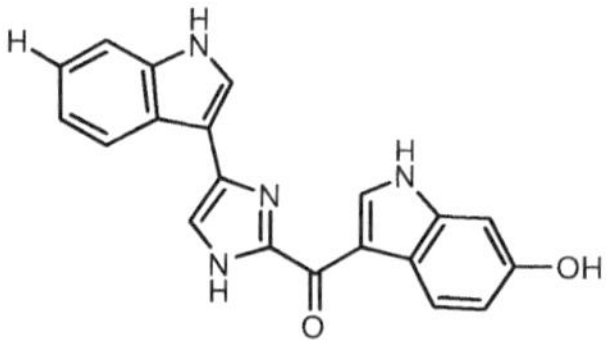

Topsentins B1 (**14**)

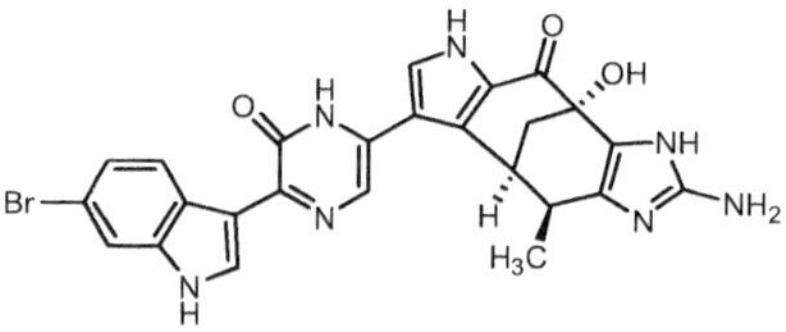

Dragmacidin F (**15**)

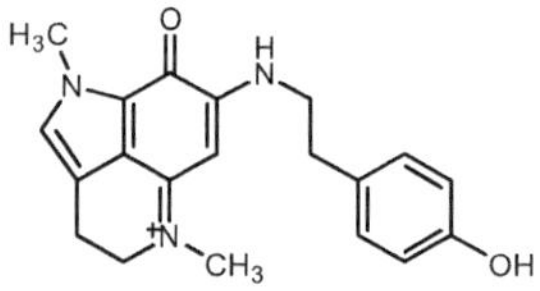

Makaluvamine P (**16**)

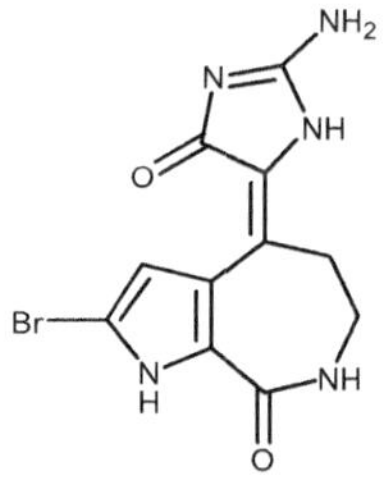

10E-hymenialdisine (**17**)

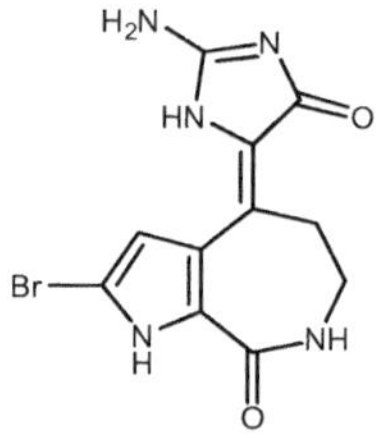

10Z-hymenialdisine (**18**)

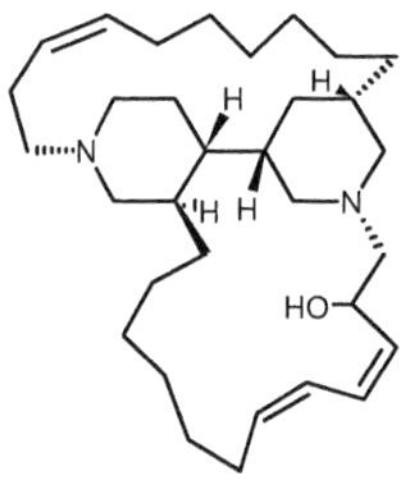

Arenosclerin A (**19**)

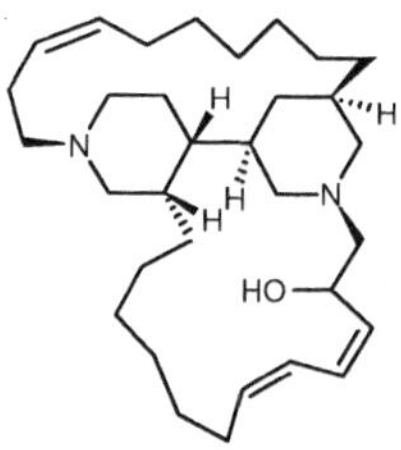

Arenosclerin B (**20**)

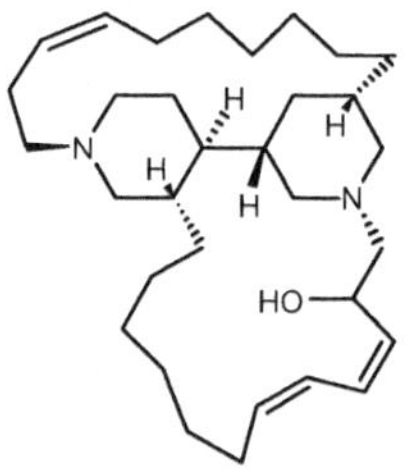

Arenosclerin C (**21**)

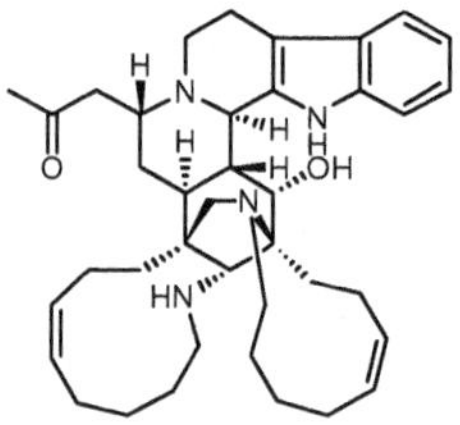

Manadomanzamines A (**24**)

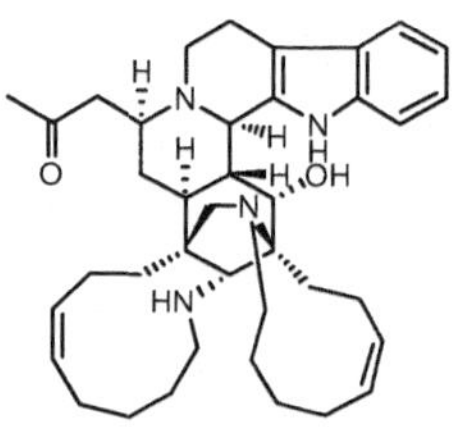

Manadomanzamines B (**25**)

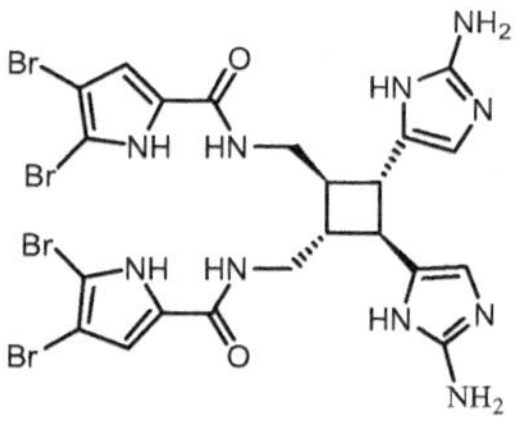

Dibromosceptrin (**26**)

In 1994, manzamine A (**27**), 8-hydroxymanzamine A (**28**), and 8-methoxymanzamine A (**29**) were derived from an unidentified sponge, *Pachypellina* sp. All the compounds showed antitumor and anti-HSV-II activities [18], and the compounds **27** and **28** also showed potent anti-inflammatory, antifungal, and anti-HIV-1 activities [19]. In 2004, compounds manzamine J (**30**), 8-hydroxymanzamine J (**31**), manzamine A N-oxide (**32**), manzamine E (**33**), 6-hydroxymanzamine E (**34**), manzamine F (**35**), and ircinol A (**36**) isolated from a common Indonesian sponge of the genus *Acanthostrongylophora* showed diverse activities against malaria, mycobacterium tuberculosis, leishmania, HIV-1, and AIDS opportunistic infections [20]. In 2006, a structurally related compound manzamine Y (**37**) was obtained from the genus *Acanthostrongylophora.* Compounds **27**, **28**, **33**, **34**, **35**, and **37** displayed the cytotoxicity against *Plasmodium falciparum* and Vero cells [21]. Compounds **27**, **28**, and **33** also showed neuritogenic activity against Neuro-2a cells with IC_{50} values of 3.3, 3.2, and 5.7 μM [22]. In 2009, the new analogues zamamidine C (**38**), 3,4-dihydro-6-hydroxy-10,11-epoxymanzamine A (**39**), and 3,4-dihydromanzamine J N-oxide (**40**) were purified from an Okinawan marine sponge *Amphimedon* species. All the compounds showed cytotoxicity against the three human tumor cell lines P388 murine leukemia L1210, human epidermoid carcinoma KB cells, and murine leukemia, and compounds **38** and **40** also possessed inhibitory activities against *T.b. brucei* (IC_{50} = 0.27, 4.44, μg/mL, respectively) and *P. falciparum* (IC_{50} = 0.58, 7.02, μg/mL, respectively) in vitro [23].

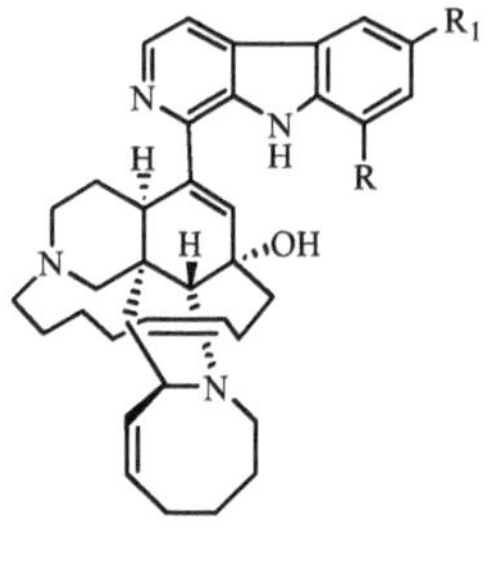

27 R = H R1= H
28 R = OH R1 = H
29 R = OCH3 R1 = H
37 R = H R1 = OH

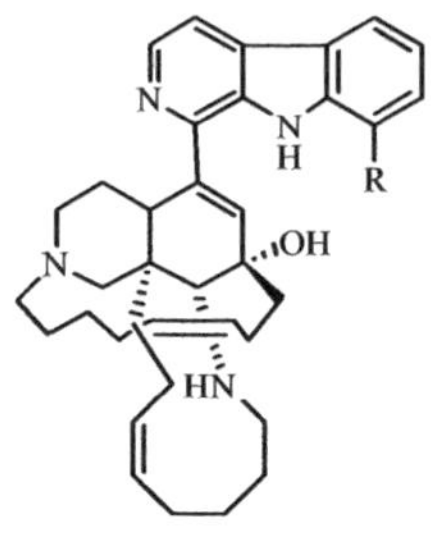

30 R = H
31 R = OH

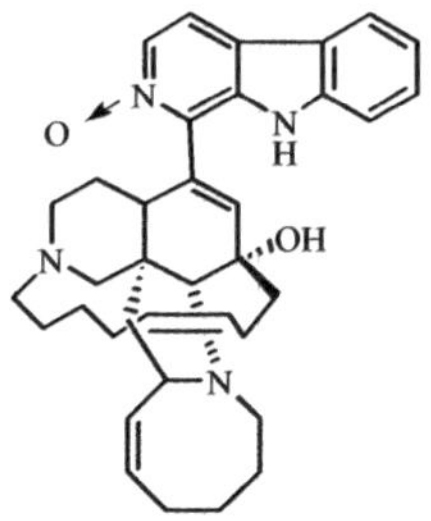

32

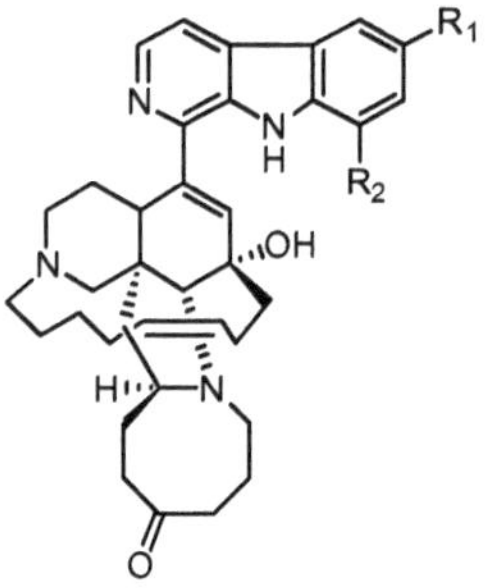

33 R1 = H, R2 = H
34 R1 = OH, R2 = H
35 R1 = H, R2 = OH

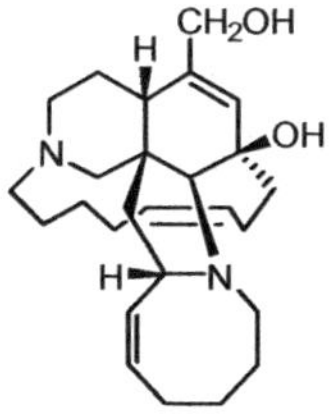

36

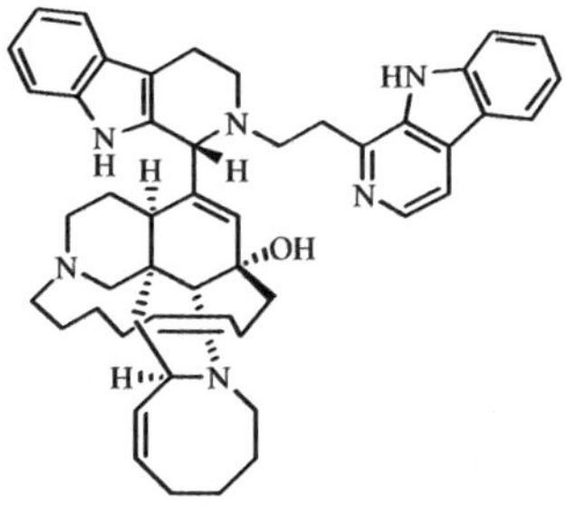

38 zamamidine C

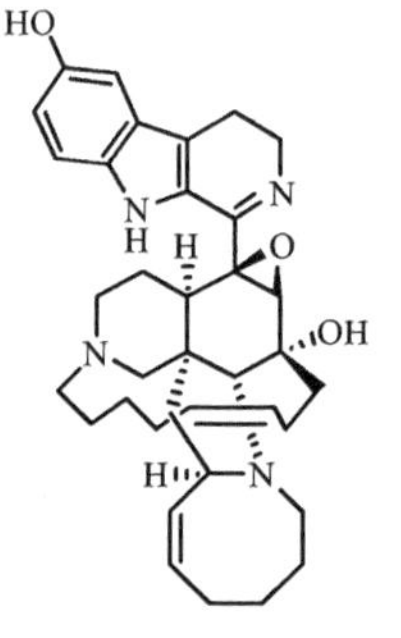

39 3,4-dihydro-6-hydroxy-10,11-epoxymanzamine A

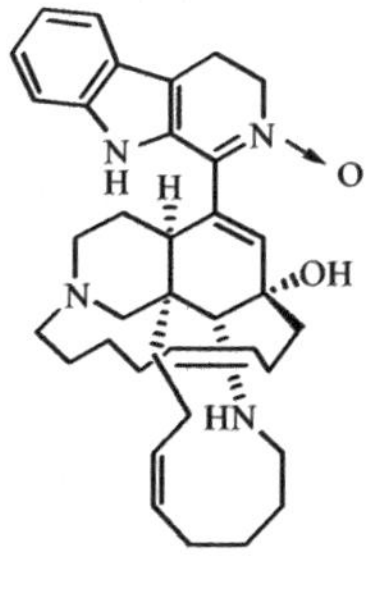

40 3,4-dihydromanzamine J N-oxide

In the year 2004, three new pyrroloiminoquinone alkaloids, 3-dihydro-7,8-dehydrodiscorhabdin C (**41**), 14-bromo-3-dihydro-7,8-dehydrodiscorhabdin C (**45**), discorhabdin V (**46**), and three known compounds 14-bromodiscorhabdin C (**44**), 14-bromo-3-dihydrodiscorhabdin C (**42**), and 3-dihydrodiscorhabdin C (**43**) were yielded from the sponge *Tsitsikamma pedunculata* with the cytotoxicity activity against human colon tumor (HCT-116) cancer cell line at IC_{50} values of 0.197, 0.222, 1.266, 0.077, 0.645, and 0.323 μM, respectively. Then tsitsikammamine A (**47**) and tsitsikammamine B (**48**) were isolated from the sponge *Tsitsikamma favus* with cytotoxicity activity inhibiting against HCT-116 cancer cell line at IC_{50} of 1.414 and 2.382 μM. Two new compounds 1-methoxydiscorhabdin D (**49**) and 1-aminodiscorhabdin D (**50**) and five known metabolites, damirone B (**51**), makaluvic acid A (**52**), makaluvamine C (**53**), discorhabdin G (**54**), and discorhabdin N (**55**), obtained from the sponge *Latrunculia bellae* exhibited cytotoxic activity against HCT-116 cell line with IC_{50} values of 0.232, 0.119, 3.102, 1.089, 0.327, and 2.249 μM, respectively. Discorhabdin A (**56**) and discorhabdin D (**57**) isolated from the sponge *Strongylodesma algoaensis* displayed cytotoxicity activity against HCT-116 cancer cell line which IC_{50} values are 0.007 and 0.595 μM, respectively [24].

In 2008, compounds discorhabdin G (**54**), B (**58**), L (**59**), and W (**60**) isolated from *Latrunculia* species sponges were found cytotoxic toward the P388 murine leukemia cell line with IC_{50} values of 0.1–1.08 μM [25]. In 2012, bispyrroloiminoquinone alkaloids, tsitsikammamine C (**61**), and makaluvamines J (**62**), G (**63**), and L (**64**) were obtained from the Australian marine sponge *Zyzzya* sp. Tsitsikammamine C displayed potent antimalarial activity with IC_{50} values of 13 and 18 nM against chloroquine-sensitive (3D7) and chloroquine-resistant (Dd2) *Plasmodium falciparum*, respectively. Compounds **62–64** displayed potent growth inhibitory activity (IC_{50} < 100 nM) against both *P. falciparum* lines and only moderate cytotoxicity against HEK293 cells (IC_{50} = 1–4 μM) [26].

Fascaplysin (**65**) isolated from the sponge *Thorectandra* sp. in 2005 displayed inhibitory activity in the Cdc25B assay with an IC_{50} value of 1.0 μg/mL [27]. Compounds (*R*)-6″-debromohamacanthin A (**66**) and *cis*-3,4-dihydrohamacanthin B (**67**) isolated from the sponge *Spongosorites* sp. were cytotoxic against A549 (human lung cancer), SK-OV-3 (human ovarian cancer), SK-MEL-2 (human skin cancer), XF498 (human CNS cancer), and HCT 15 (human colon cancer) cell lines with IC_{50} values ranging from 2.83 to 3.85 μg/mL [28]. In 2006, structurally similar compounds deoxytopsentin (**68**), hamacanthin A (**69**), and hamacanthin B (**70**) were isolated from the sponge *Spongosorites* sp., all of which showed antimicrobial activity against various strains of bacteria and fungi, among which deoxytopsentin (**68**) and hamacanthin A (**69**) also exhibited significant antibacterial activity against methicillin-resistant *Staphylococcus aureus* and moderate cytotoxicity against cancer cell lines [29]. Compounds cortistatins A (**71**), B (**72**), C (**73**), and D (**74**) from the sponge *Corticium simplex* displayed potent antiproliferative activities against HUVECs which IC_{50} values are from 0.0018 to 1.1 μM. The compound **71** also exhibited activities against normal human dermal fibroblast (NHDF), epidermoid carcinoma cells (KB3-1), human chronic myelogenous leukemia cells (K562), and murine neuroblastoma cells (Neuro-2A) with IC_{50} values of 6.0, 7.0, 7.0, and 6.0 μM [30]. The marine sponge *Dactylia* sp. yielded one alkaloid ircinamine B (**75**), which was cytotoxic against P388 cell line (IC_{50} = 0.28 μM) [31].

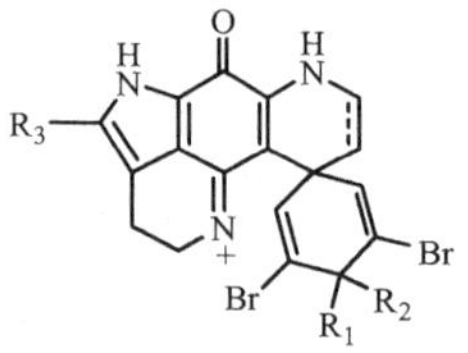

41 $R_1 + R_2 = O$, $R_3 = Br$
42 $R_1 = H$, $R_2 = OH$, $R_3 = Br$
43 $R_1 = H$, $R_2 = OH$, $R_3 = H$
44 $R_1 = H$, $R_2 = OH$, $R_3 = H$ Δ^7
45 $R_1 = H$, $R_2 = OH$, $R_3 = Br$ Δ^7

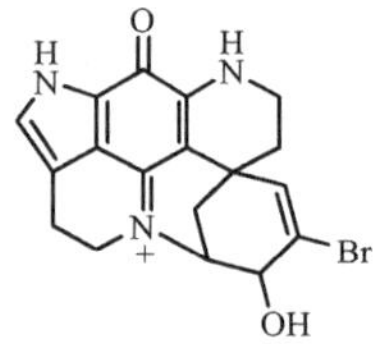

46 discorhabdin V

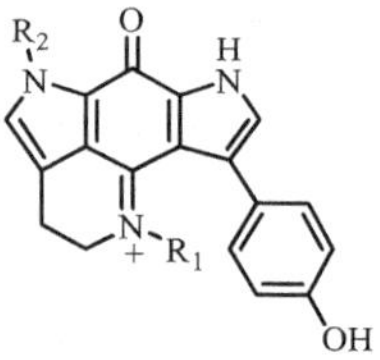

47 $R_1 = R_2 = H$
48 $R_1 = H$, $R_2 = Me$

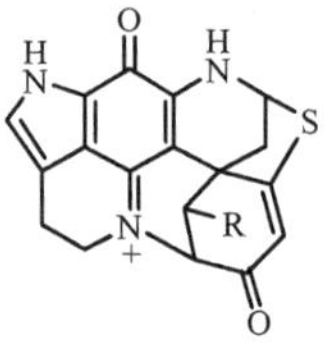

49 R = OMe
50 $R = NH_2$
55 R = NH COOH

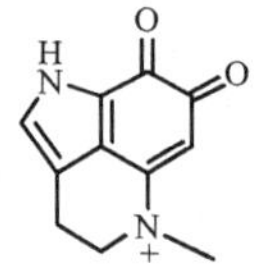

51 damirone B

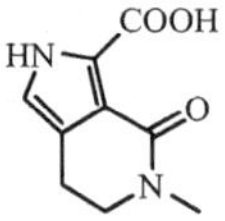

52 makaluvic acid A

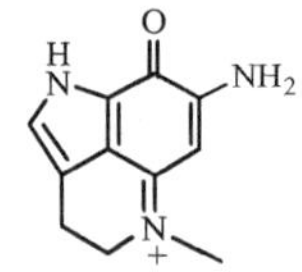

53 makaluvamine C

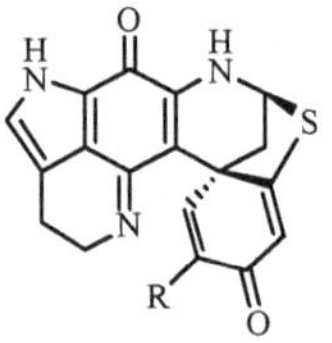

54 discorhabdin G R = H
58 discorhabdin B R = Br

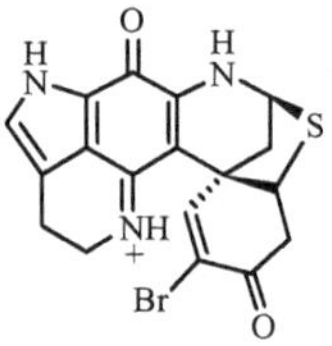

56 discorhabdin A

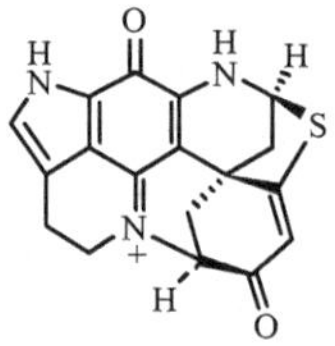

57 discorhabdin D

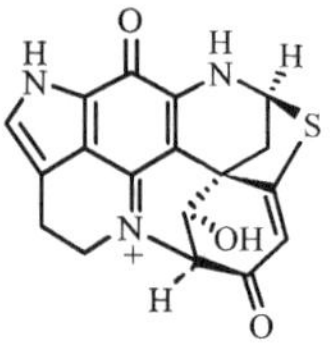

59 discorhabdin L

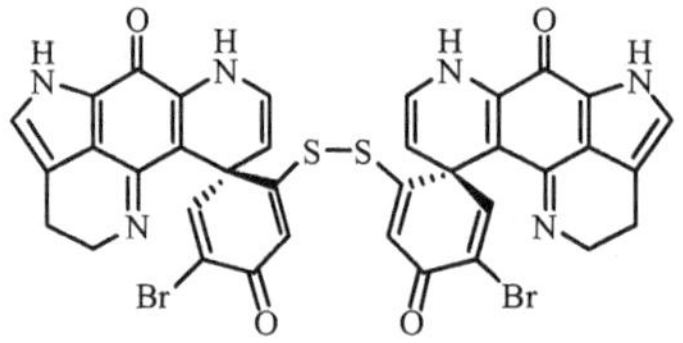

60 discorhabdin W

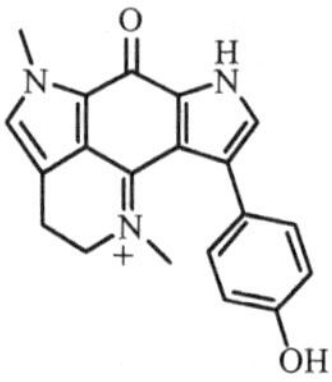

61 tsitsikammamine C

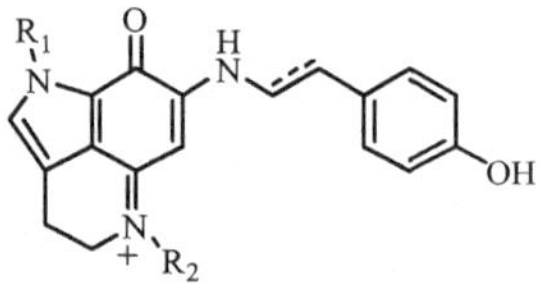

62 $R_1 = H$, $R_2 = CH3$
63 $R_1 = CH_3$, $R_2 = CH3$ Δ^7
64 $R_1 = H$, $R_2 = CH3$ Δ^7

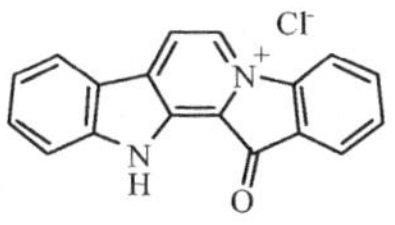

65 fascaplysin

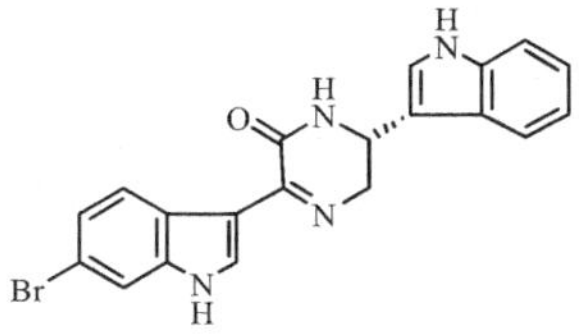

66 (R)-6"-Debromohamacanthin A

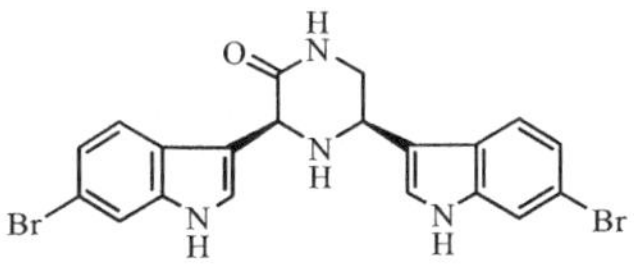

67 cis-3,4-Dihydrohamacanthin B

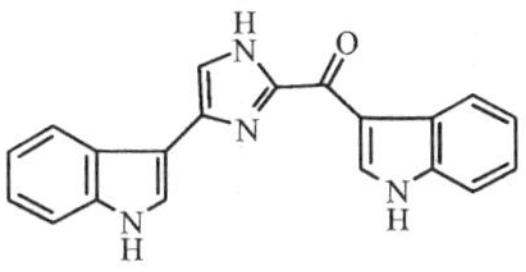

68 deoxytopsentin

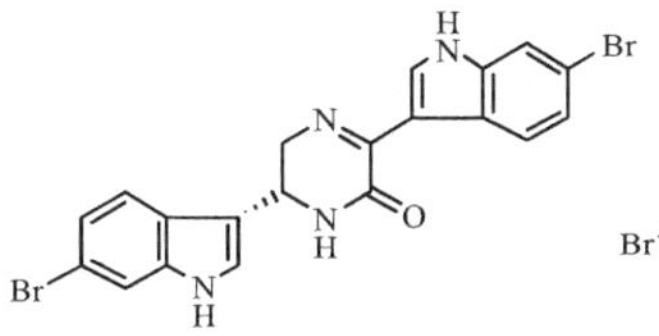

69 hamacanthin A

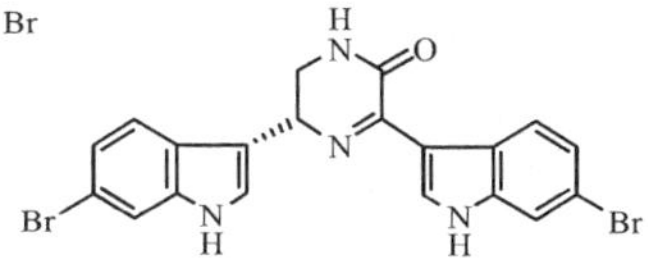

70 hamacanthin B

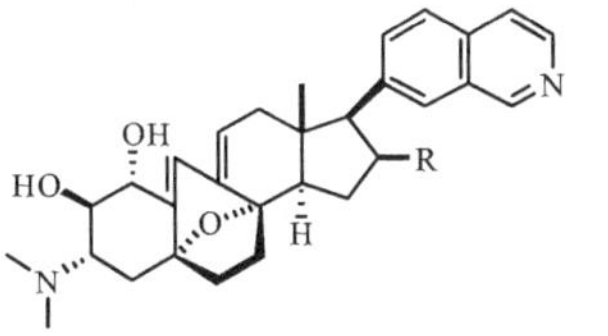

71 R = H
72 R = OH

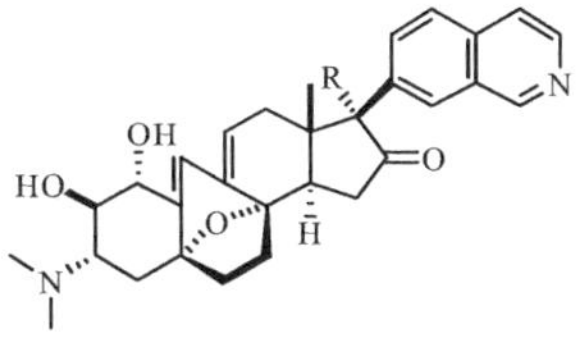

73 R = H
74 R = OH

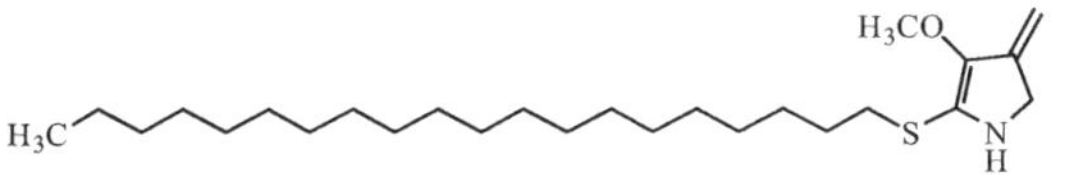

75 Ircinamine B

In 2007, psammaplysenes C (**76**) and D (**77**) were identified from the sponge *Psammoclemma* sp. as the P2X7 receptor antagonists both for the treatment of inflammatory disease with IC_{50} value of 7 μM [32]. Three bis-piperidine alkaloids, haliclonacyclamine F (**78**) and arenosclerins D (**79**) and E (**80**), were isolated from the marine sponge *Pachychalina alcaloidifera*, which displayed cytotoxic activity against SF295 (human CNS), MDA-MB-435 (human breast), and HL-60 (leukemia) cancer cell lines with IC_{50} values ranging from 1.0 to 4.5 μg/mL [33]. In 2010, structurally related compound neopetrosiamine A (**81**) was isolated from the marine sponge *Neopetrosia proxima* collected off the west coast of Puerto Rico, which showed inhibitory activity against MALME-3M melanoma cancer, CCRFCEM leukemia, and MCF-7 breast cancer with IC_{50} values of 1.5, 2.0, and 3.5 μM, respectively, also showing the antiplasmodial activity against *Plasmodium falciparum* (IC_{50} = 2.3 μM) [34]. The compound haliclonacyclamine A (**82**) was isolated from the *Haliclona* sponge *Haliclona* sp. at Solomon Islands. In vitro assay of haliclonacyclamine A against the chloroquine-sensitive 3D7 and chloroquine-resistant strain *P. falciparum* FcB1 gave, respectively, IC_{50} of 0.33 and 0.052 mg/mL. The cytotoxicity of **82** was measured on breast cancer cells MCF-7 with IC_{50} value of 2.6 mg/mL [35].

16b-Hydroxycrambescidin 359 (**83**), batzelladines L and M (**84**, **85**), ptilomycalin A (**86**), crambescidine 800 (**87**), batzelladine C (**88**), and dehydrobatzelladine C (**89**) were isolated from the Jamaican sponge *Monanchora unguifera*, all of which exhibited antimalarial activity against *Plasmodium falciparum* D6 clone and W2 clone with IC_{50} values ranging from 73 to 270 ng/mL. Moreover, their activities of antitumor, anti-tuberculosis, HIV-1, antimicrobial and antimalarial were evaluated. Among them, batzelladine L (**84**) showed the most potent activity against *Mycobacterium tuberculosis* with a MIC of 1.68 mg/mL, ptilomycalin A (**86**), and crambescidine 800 (**87**) exhibited potent activities against human HIV-1 virus with EC_{50}/EC_{90} values of 0.011/0.046 and 0.04/0.12 μM, respectively [36]. In 2009, structurally similar compounds norbatzelladine A (**90**), dinorbatzelladine A (**91**), dinordehydrobatzelladine B (**92**), and dihomodehydrobatzelladine C (**93**) were obtained from marine sponge *Monanchora arbuscula* (de Laubenfels, 1953) collected in Martinique; norbatzelladine L (**94**) and clathriadic acid (**95**) were yielded from marine sponge *Clathria calla* (de Laubenfels, 1934) collected in Guadeloupe. Compounds **90–94** possessed potent antitumor cytotoxic activities against MDA-MB-231, A549, and HT29 with GI_{50} values ranging from 0.7 to 7.9 μM. Compounds **90–95** showed antimalarial activity with IC_{50} values ranging from 0.2 to 2.3 μM [37]. Similar compounds, monalidine A (**96**); batzelladines D (**97**), F (**98**), and L (**84**); and norbatzelladine L (**99**), were yielded from the marine sponge *Monanchora arbuscula*, collected off the southeastern coast of Brazil in 2015, displaying the activities against *Trypanosoma cruzi* and *Leishmania infantum* with IC_{50} values ranging from 2.0 to 8.0 μM [38]. Four brominated pyrrole-imidazole alkaloids (**100–103**) from the Caribbean sponges *Stylissa caribica* and *Agelas wiedenmayeri* were tested for interactions with cellular calcium homeostasis using PC12 cells, massadine (100, EC_{50}: 5.32 μM), and stylissadines A (101, 4.48 μM) and B (102, 4.67 μM), and tetrabromostyloguanidine (103, 15.6 μM) reduced voltage-dependent calcium entry in PC12 cells as measured with Fura II as calcium indicato [39]. 8,8′-Dienecyclostellettamine (**104**) isolated from the marine sponge *Amphimedon compressa* showed potent antibacterial activities against six clinic bacteria, *Candida albicans*, *Escherichia coli*, *Pseudomonas aerug*, *Cryptococcus neoformans*, MRS, and *Aspergillus fumigatus*, with IC_{50} values of 0.4, 1.3, 2.1, 2.5, 0.25, and 0.3 μg/mL, respectively [40]. In 2009, an analogue njaoaminiums B (**105**) isolated from the marine sponge *Reniera* sp., collected off the coasts of Pemba Island, Tanzania, showed cytotoxicity against the three human tumor cell lines MDA-MB-231, A549, and HT29 cell lines with GI_{50} values of 4.8, 4.1, and 4.2 μM [41].

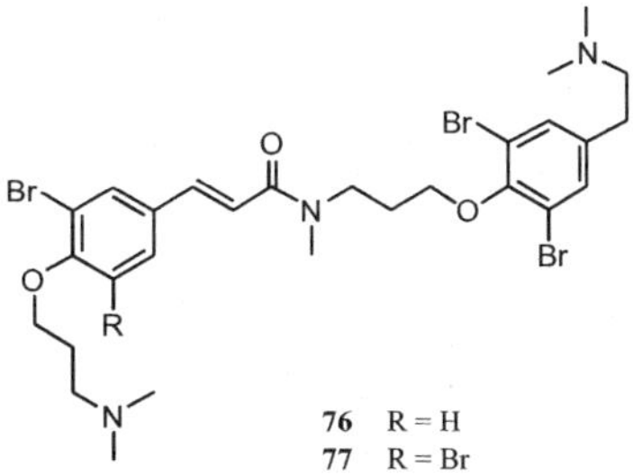

76 R = H
77 R = Br

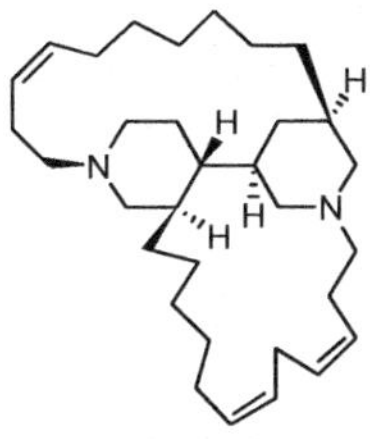

78 haliclonacyclamine F

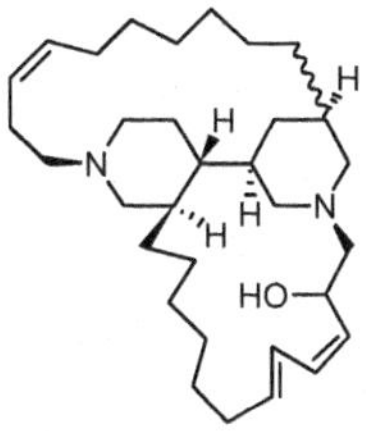

79 arenosclerins D

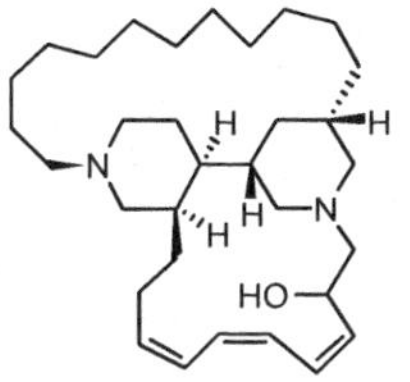

80 arenosclerins E

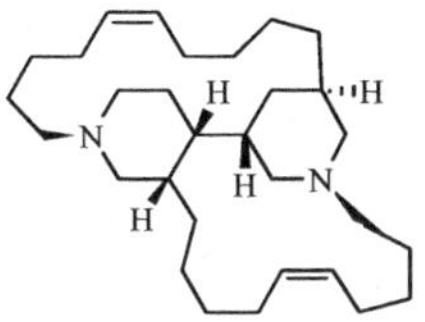

81 neopetrosiamine A

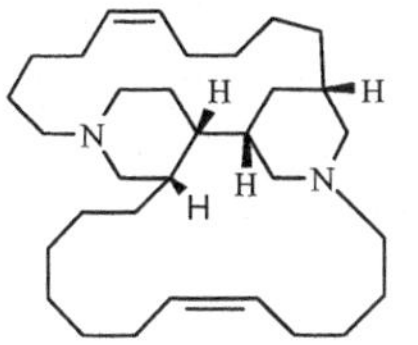

82 neopetrosiamine A

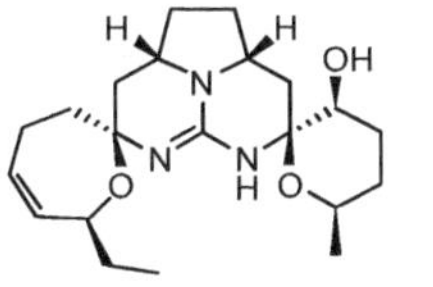

83 16b-hydroxycrambescidin 359

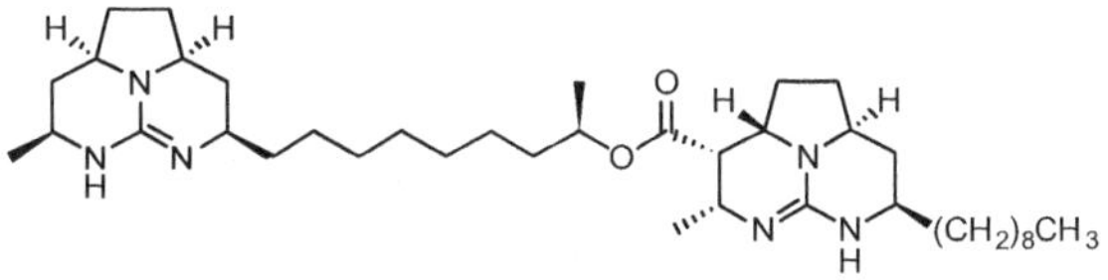

84 batzelladines L

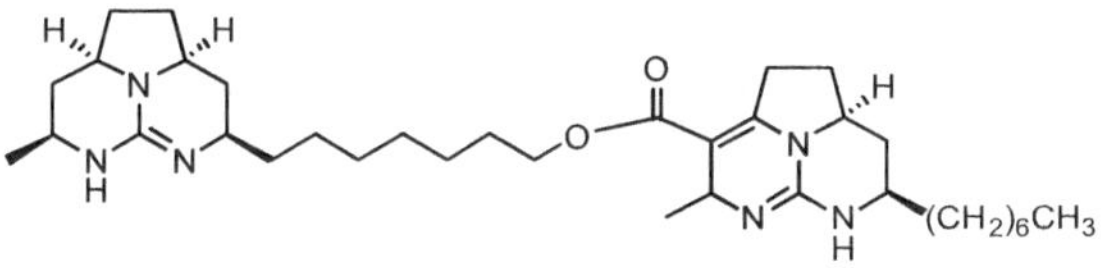

85 batzelladines M

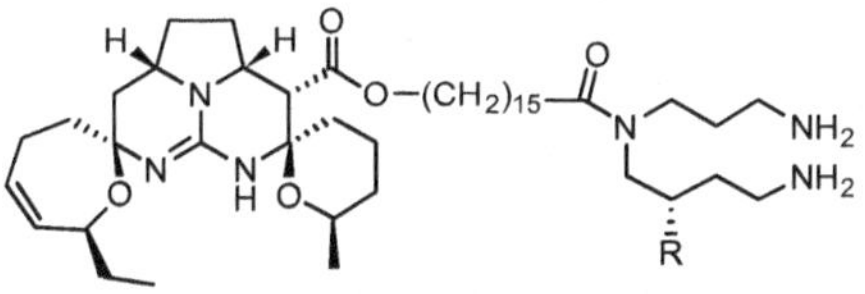

86 ptilomycalin A R = H
87 crambescidine 800 R = OH

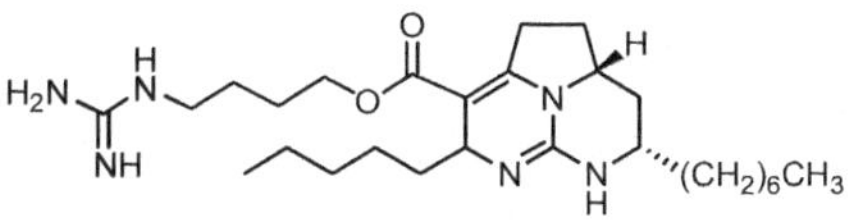

88 batzelladine C

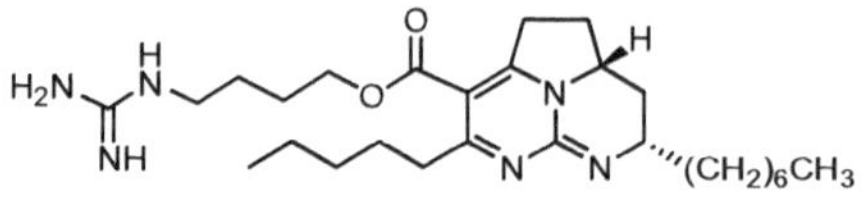

89 dehydrobatzelladine C

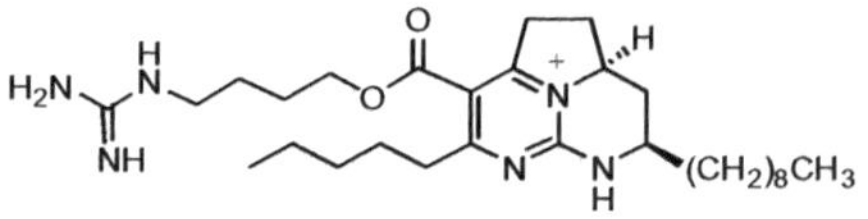

93 dihomodehydrobatzelladine C

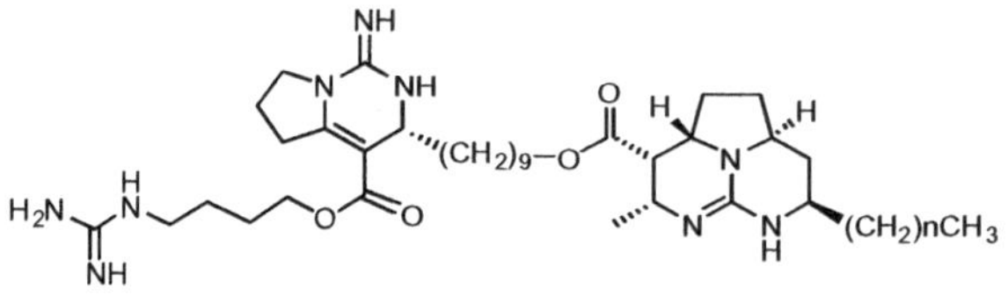

90 norbatzelladine A n = 7
91 dinorbatzelladine A n = 6

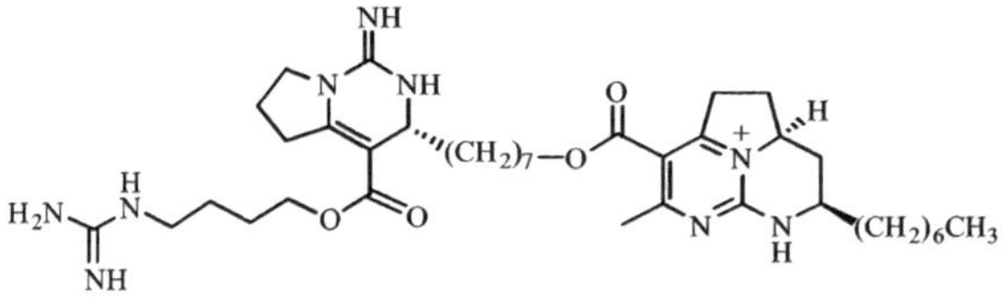

92 dinordehydrobatzelladine B

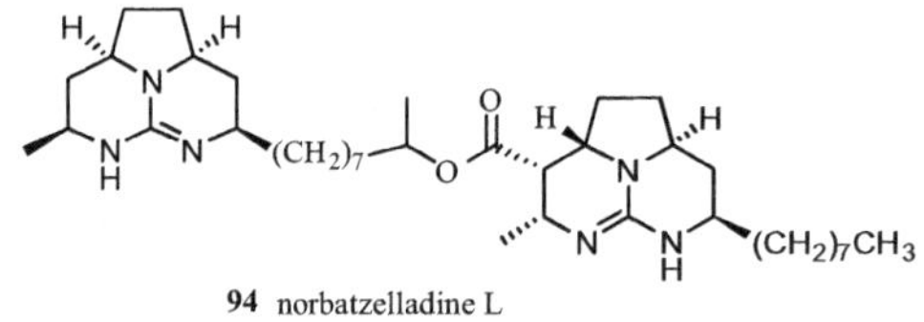

94 norbatzelladine L

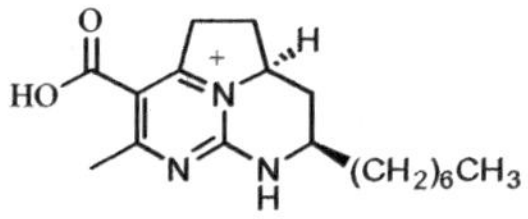

95 clathriadic acid

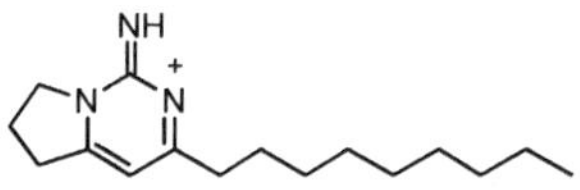

96 monalidine A

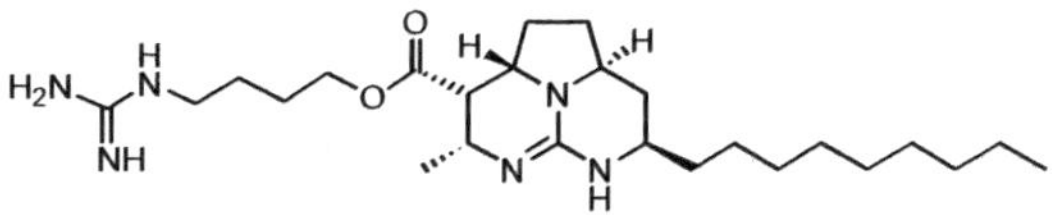

97 batzelladines D

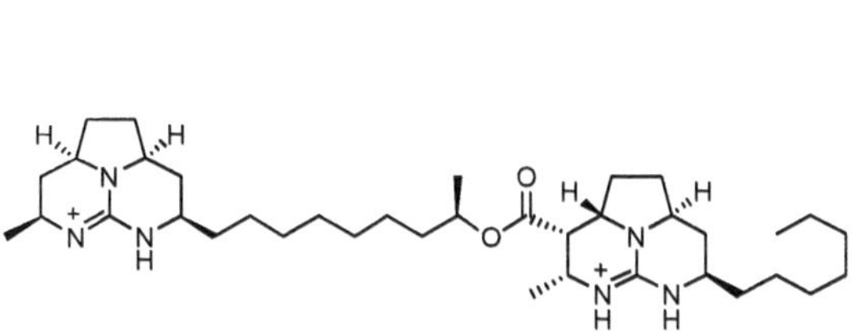

98 batzelladines F

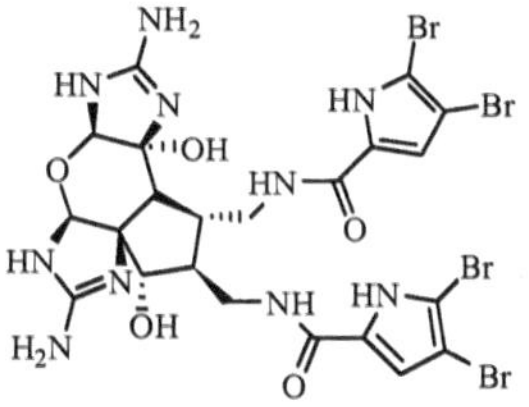

100 massadine

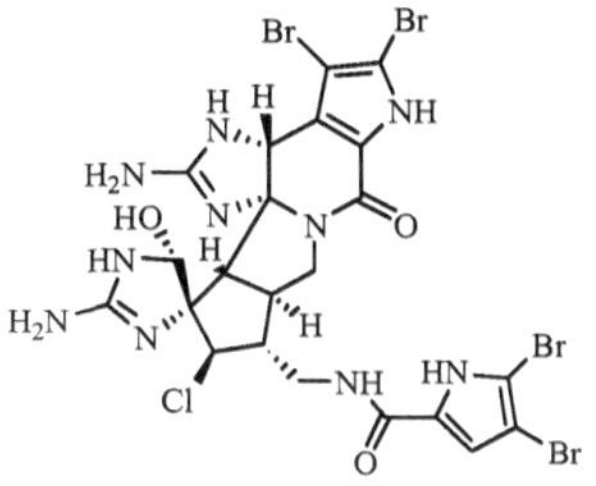

103 tetrabromostyloguanidine

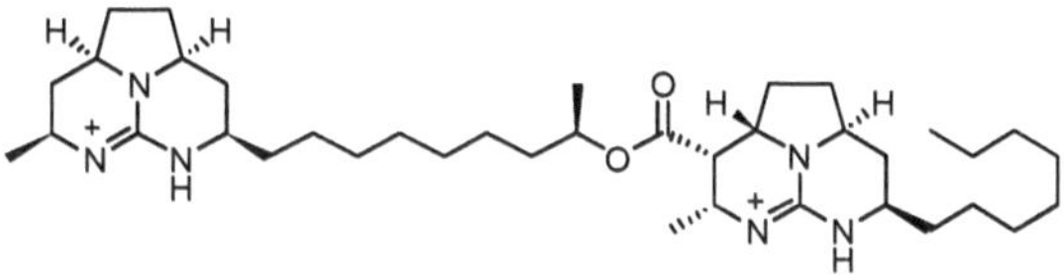

99 batzelladines I

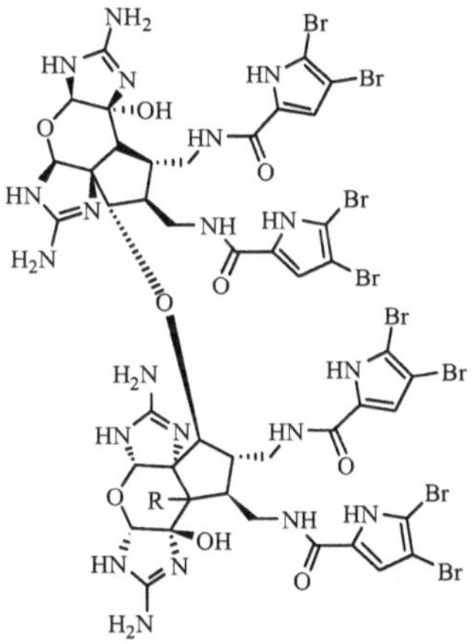

101 R = βH stylissadines A
102 R = αH stylissadines B

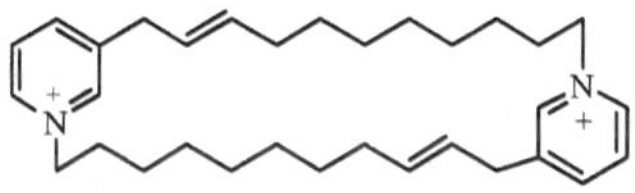

104 8,8'-dienecyclostellettamine

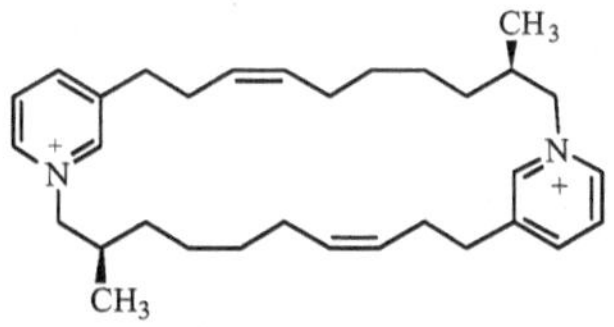

105 Njaoaminium B

In 2008, compounds (+)-aplysinillin (**106**) and dienone (**107**) yielded from the marine sponge *Aplysinella* sp. collected from the Federated States of Micronesia were evaluated for their cancer cell growth inhibition against the MCF-7 cancer cell line with IC_{50} values of 1.19 ± 0.10 and 0.89 ± 0.11 μM, respectively [42]. Trachycladindoles A–F (**108–113**) were yielded from a southern Australian marine sponge, *Trachycladus laevispirulifer*, and they showed cytotoxicity against lung (A549), colorectal (HT29), and breast (MDA-MB-231) cell lines with GI_{50} values ranging from 0.3 to 12.2 μM [43].

In 2009, nagelamides Q (**114**) and R (**115**), isolated from Okinawan marine sponges of the genus *Agelas*, showed antimicrobial activity against *Trichophyton mentagrophytes* with MIC values of 6 μg/mL [44]. Benzosceptrin C (**116**), isolated from an Okinawan marine sponge of the genus *Agelas*, displayed antimicrobial activity against *Micrococcus luteus* and *Cryptococcus neoformans* with MIC values of 6 μg/mL, respectively [45].

In 2010, chemical investigation of the Australian marine sponge *Ecionemia geodides* found a new pyridoacridine alkaloid, ecionines A (**117**) along with the previously isolated marine natural product meridine (**118**). The compounds exhibited moderate cytotoxicity against a panel of human bladder cancer cell lines, including the increasingly metastatic TSU-Pr1 series (TSU-Pr1, TSU-Pr1-B1, and TSU-Pr1-B2) and the superficial bladder cancer cell line 5637, with IC_{50} values ranging from 3 to 7 μM [46]. Bastadin 26 (**119**) isolated from Australian marine sponge *Ianthella flabelliformis* showed potent affinity for the guinea pig δ-opioid receptors with IC_{50} value of 206 nM and a Ki value of 100 nM [47]. Eleven DOPA-derived pyrrole alkaloids, named baculiferins A–C, E–H, and K–N (**120–130**), were isolated from the Chinese marine sponge *Iotrochota baculifera* and found to be potent inhibitors against the HIV-1 IIIB virus in both MT4 and MAGI cell lines with IC_{50} values ranging from 0.1 to 8.4 μM [48]. Monanchocidin (**131**), a guanidine alkaloid with an unprecedented skeleton system possessing cytotoxicity against human leukemia THP-1 with IC_{50} value of 5.1 μM, was isolated from the sponge *Monanhora pulchra* [49]. Two bromotyrosine alkaloids, ceratinadins A and B (**132–133**), were isolated from an Okinawan marine sponge *Pseudoceratina* sp., and they showed antifungal activity against *Cryptococcus neoformans* (MIC, 4 and 8 μg/mL, respectively) and *Candida albicans* (MIC, 2 and 4 μg/mL, respectively) [50]. Psammaplysin F (**134**) yielded from the Australian marine sponge *Hyattella* sp. inhibited the growth of two different strains of the parasite *Plasmodium falciparum* (Dd2 and 3D7) with IC_{50} values of 1.4 and 0.87 μM [51]. In 2011, psammaplysin F (**134**), psammaplysins G (**135**), and psammaplysin H (**136**) were yielded from marine sponge *Pseudoceratina* sp., all of which inhibited the growth of the 3D7 line of *Plasmodium falciparum* with IC_{50} values of 1.92, 5.22, and 0.41 μM, respectively. The compound psammaplysin F also showed cytotoxicity against HepG2 cell line with IC_{50} value of 3.7 μM [52].

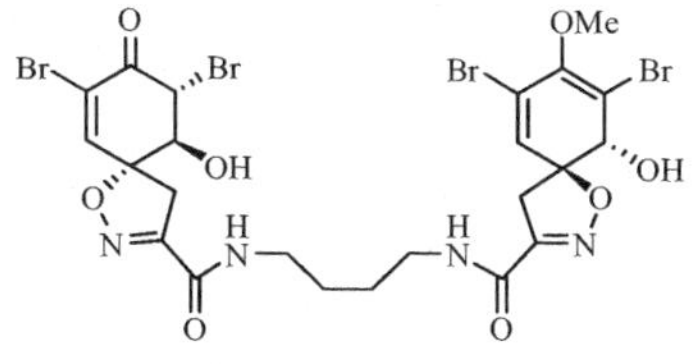

106 (+)-aplysinillin

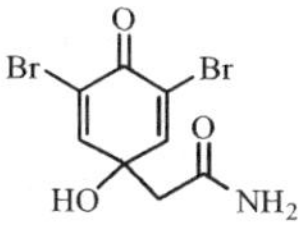

107 dienone

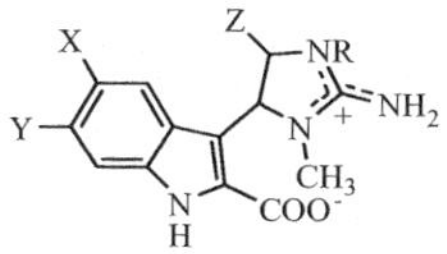

	X	Y	Z	R
108	Br	H	H	H
109	Br	H	H	Me
110	Br	OH	H	H
111	Br	OH	H	Me
112	Br	H	OH	Me
113	Br	OH	OH	Me

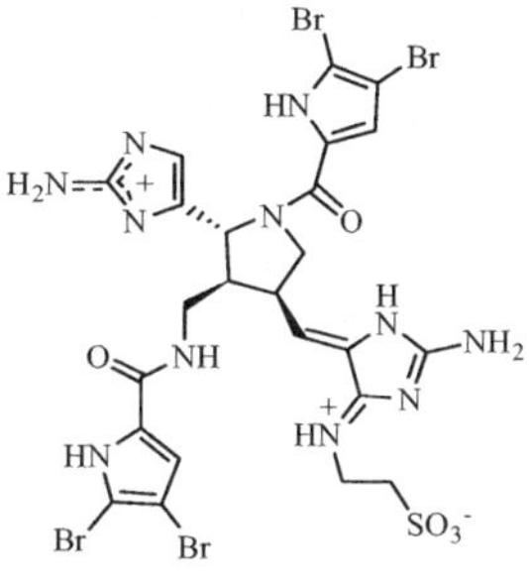

114 Nagelamides Q

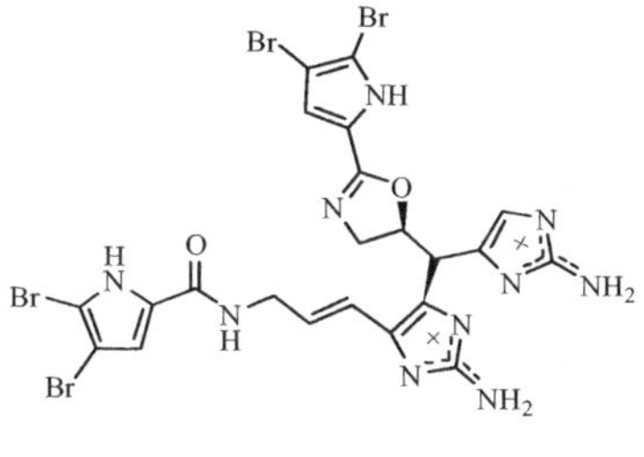

115 Nagelamides R

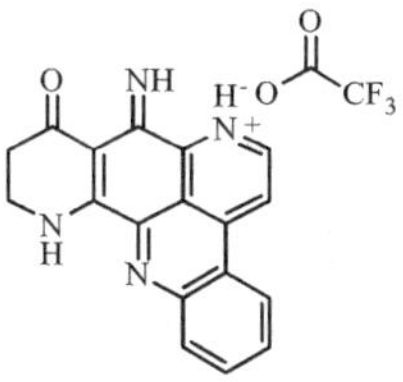

117 ecionines A

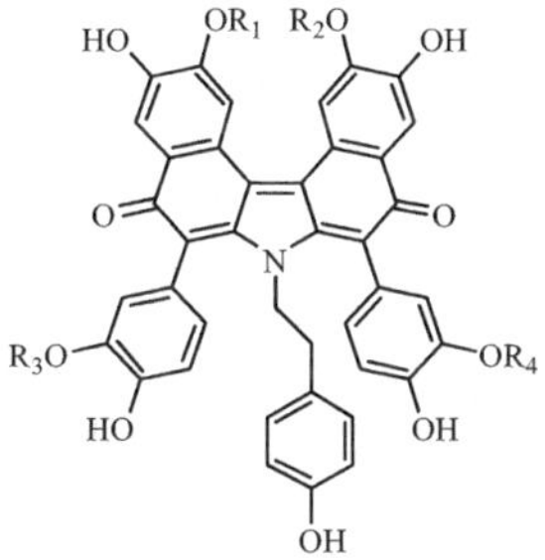

120	$R_1 = SO_3H$,	$R_2 = R_3 = R_4 = H$
121	$R_3 = SO_3H$,	$R_1 = R_2 = R_4 = H$
122	$R_3 = R_4 = SO_3H$,	$R_1 = R_2 = H$
123	$R_1 = R_3 = SO_3H$,	$R_2 = R_4 = H$
124	$R_1 = R_4 = SO_3H$,	$R_2 = R_3 = H$
125	$R_1 = R_3 = R_4 = O_3H$,	$R_2 = H$
126	$R_1 = R_2 = R_3 = SO_3H$,	$R_4 = H$

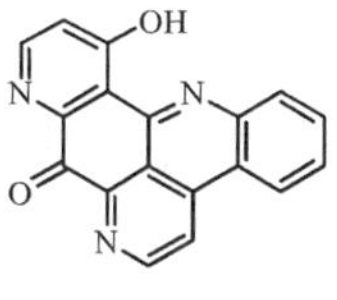

118 meridine

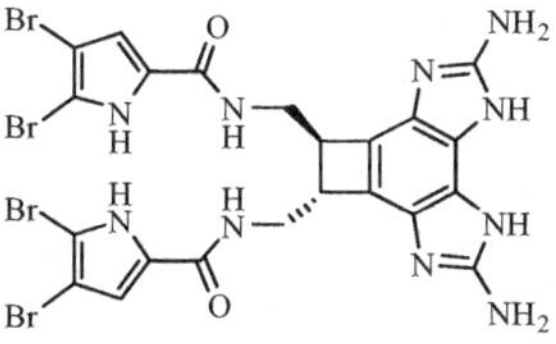

116 benzosceptrin C

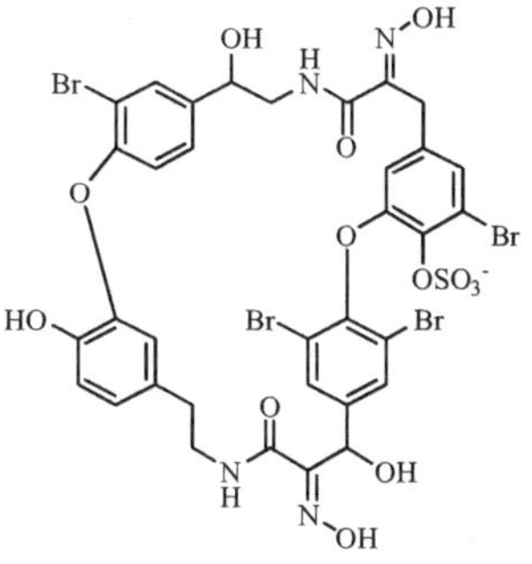

119 bastadin 26

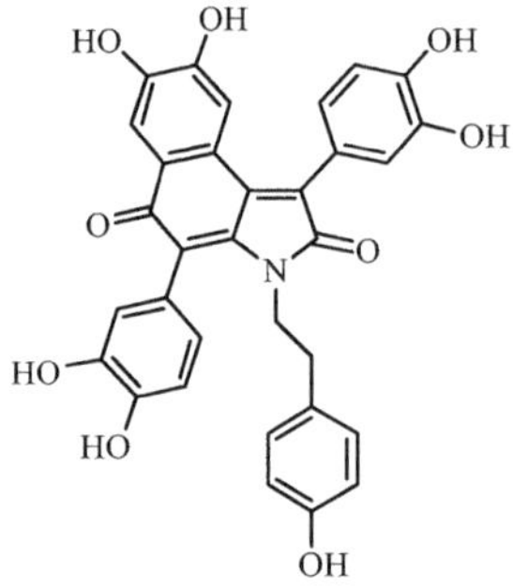

127 baculiferin K

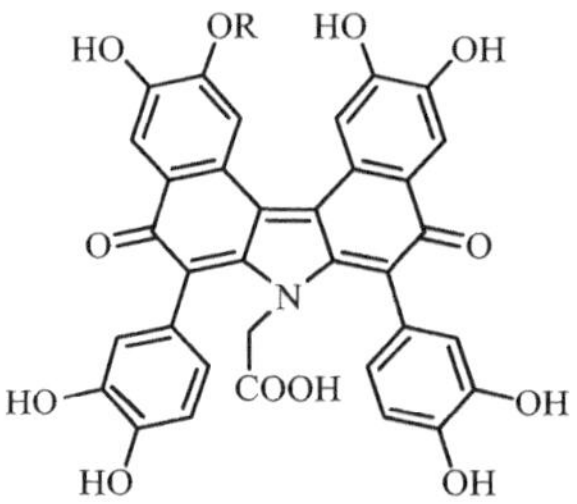

128 R = H
129 R = SO_3H

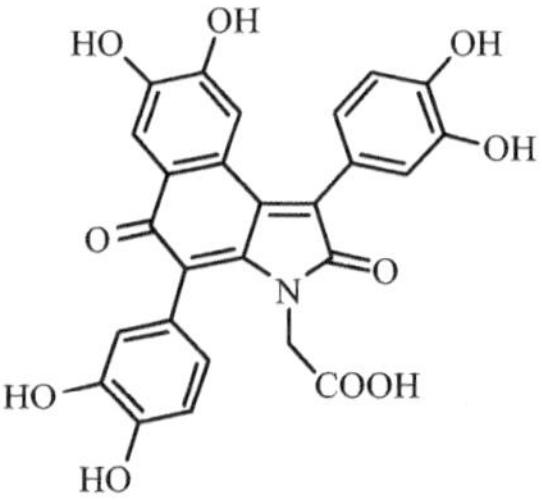

130 baculiferin N

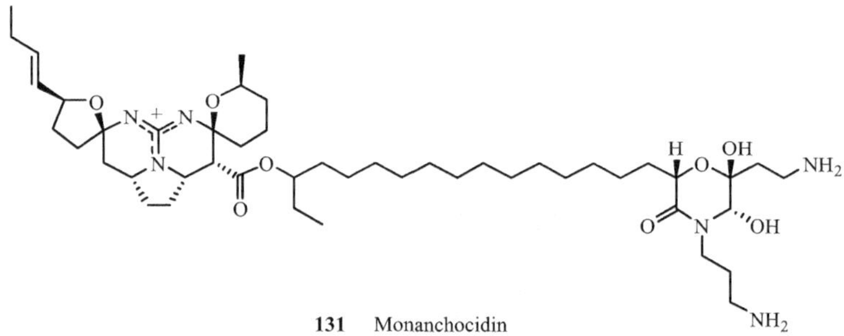

131 Monanchocidin

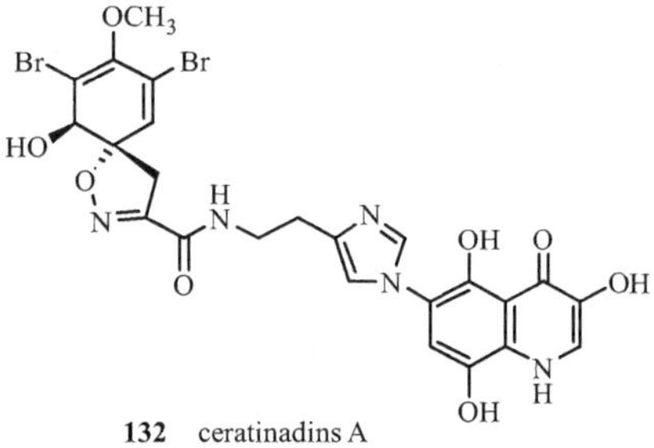

132 ceratinadins A

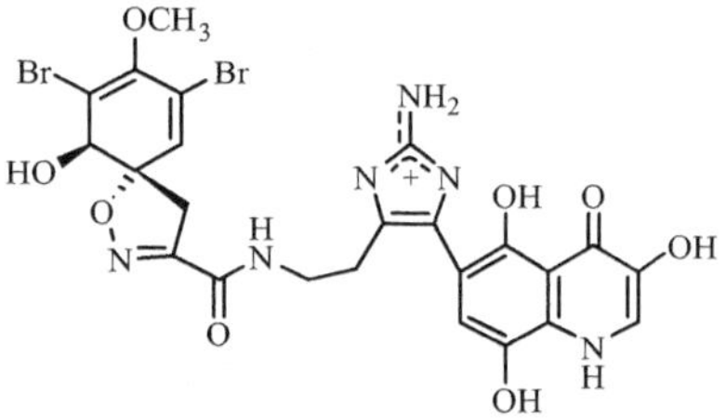

133 ceratinadins B

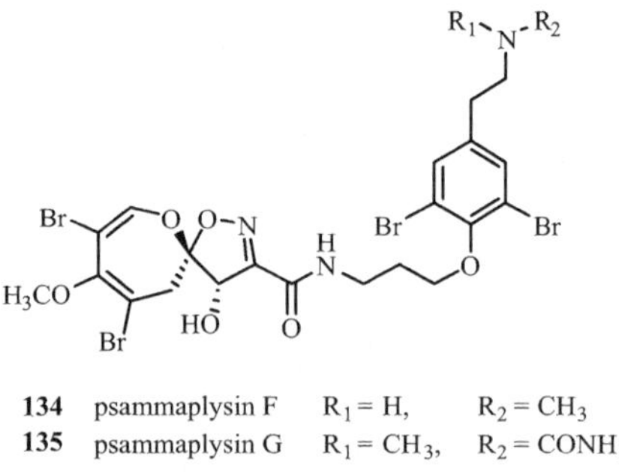

134 psammaplysin F $R_1 = H$, $R_2 = CH_3$
135 psammaplysin G $R_1 = CH_3$, $R_2 = CONH_2$

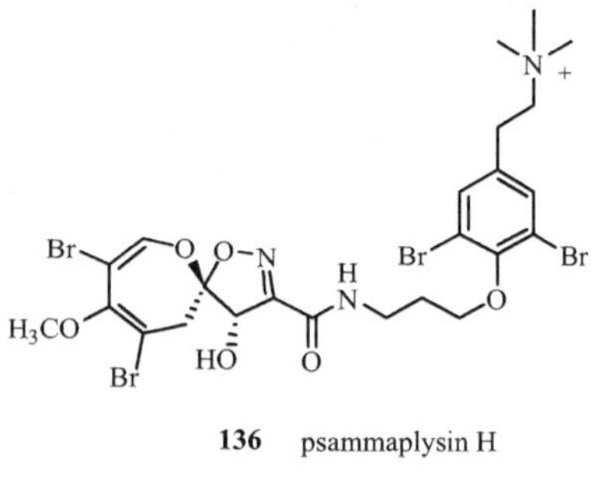

136 psammaplysin H

In 2011, 8b*β*-hydroxyptilocaulin (**137**) and ptilocaulin (**138**) isolated from *Monanchora arbuscula* colonies collected off the northeastern Brazilian coast presented cytotoxicity against HL-60 cell line with IC_{50} values of 7.89 and 5.77 μM, respectively [53]. Polycyclic guanidine alkaloids monanchocidins A–E (**139–143**) isolated from the Far Eastern marine sponge *Monanchora pulchra* showed potent cytotoxic activities against HL-60 human leukemia cells with IC_{50} values of 540, 200, 110, 830, and 650 nM, respectively [54]. Monanchomycalins A (**144**) and B (**145**) isolated from the marine sponge *Monanchora pulchra* showed cytotoxic activities against HL-60 human leukemia cells with IC_{50} values of 120 and 140 nM, respectively [55]. Compounds spermatinamine (**146**) yielded from the Australian marine sponge *Pseudoceratina* sp. inhibited secretion of the *Yersinia* outer protein YopE and the enzyme activity of YopH with IC_{50} value of 6 μM [56].

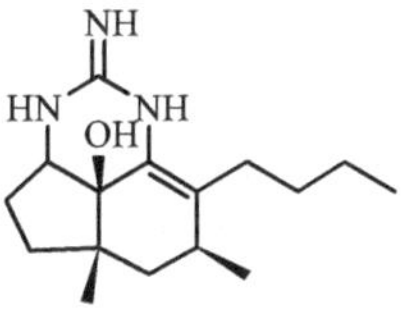

137 8b*β*-hydroxyptilocaulin

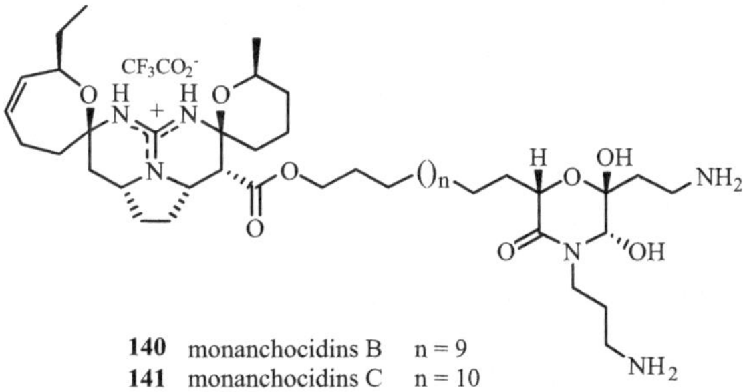

140 monanchocidins B n = 9
141 monanchocidins C n = 10

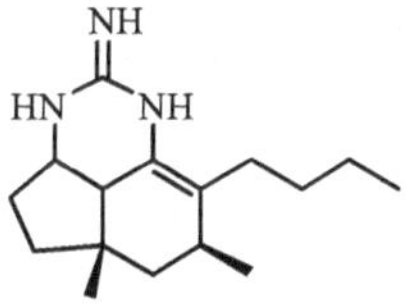

138 ptilocaulin

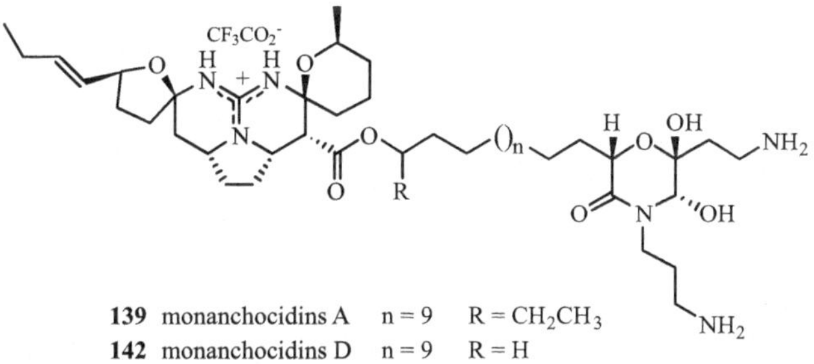

139 monanchocidins A n = 9 R = CH_2CH_3
142 monanchocidins D n = 9 R = H
143 monanchocidins E n = 8 R = CH_2CH_3

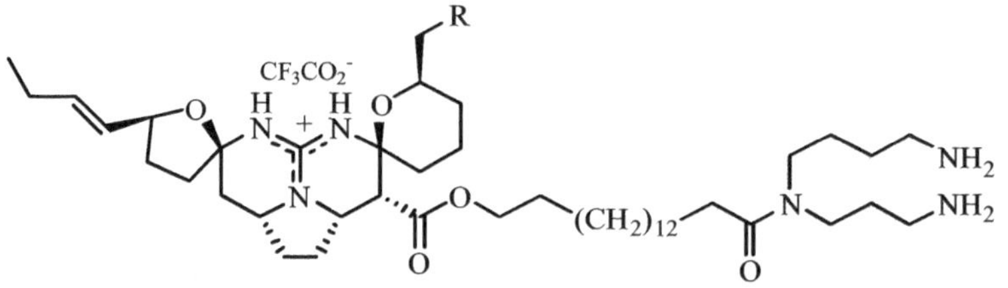

144 monanchomycalins A R = CH_2CH_3
145 monanchomycalins B R = H

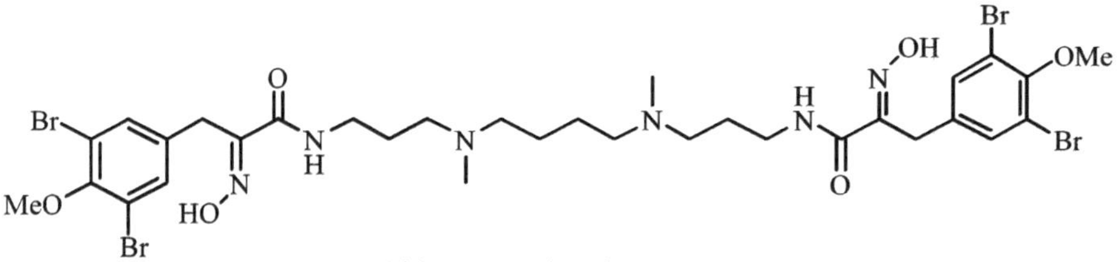

146 spermatinamine

In 2012, 12-N-methyl stevensine (**147**), Z-hymenialdisine (**148**), and Z-debromohymenialdisine (**149**) were obtained from a collection of Indonesian marine sponge *Stylissa* species off Derawan Islands, Berau, NE Kalimantan, which showed significant activity against mouse lymphoma cell line L5187Y with EC_{50} values of 3.5, 1.8, and 2.1 µg/mL, respectively [57]. Densanins A (**150**) and B (**151**) were isolated from the sponge *Haliclona densaspicula* and displayed relatively potent inhibitory effects on lipopolysaccharide-induced nitric oxide production in BV2 microglial cells with IC_{50} values of 1.05 and 2.14 µM, respectively [58]. Ingamine A (**152**), 22(S)-hydroxyingamine A (**153**), and dihydroingenamine D (**154**) were isolated from marine sponge *Petrosid Ng5 Sp5* (family Petrosiidae) obtained from the open repository of the National Cancer Institute, USA. All compounds showed strong antiplasmodial activity against chloroquine-sensitive (D6) and chloroquine-resistant (W2) strains of *Plasmodium falciparum* with IC_{50} values of 57–220 ng/mL. Compounds **152–154** also displayed weak antimicrobial and moderate antileishmanial activities against *Leishmania donovani* promastigotes [59]. Six halogenated alkaloids named purpuroines A, C–D, and F–H (**155–160**) were isolated from the marine sponge *Iotrochota purpurea*. Bioassay for the regulation of tyrosine kinases revealed compounds **155–160** possessing inhibitory activities against the kinase LCK and PLK1 with IC_{50} values ranging from 0.94 to 11.88 µg/mL [60]. Nakijinamines A (**161**) obtained from an Okinawan marine sponge *Suberites* sp. exhibited antimicrobial activity against *Candida albicans*

(IC_{50} = 0.25 μg/mL), *Cryptococcus neoformans* (IC_{50} = 0.5 μg/mL), *Trichophyton mentagrophytes* (IC_{50} = 0.25 μg/mL), and *Micrococcus luteus* (MIC 2 μg/mL) [61]. Two alkaloids, (−)-ageloxime D (**162**) and ageloxime B (**163**), were isolated from the marine sponge *Agelas mauritiana*, and both showed activity against *Cryptococcus neoformans* with IC_{50} values of 5.94 and 4.96 μg/mL, respectively. Compound **163** also exhibited antibacterial activity against *Staphylococcus aureus* (IC_{50} = 7.21 μg/mL) and methicillin-resistant *S. aureus* (IC_{50} = 9.20 μg/mL) [62].

In 2013, thiaplakortones A–D (**164–167**) obtained from the Australian marine sponge *Plakortis lita* displayed significant growth inhibition against chloroquine-sensitive (3D7) and chloroquine-resistant (Dd2) *Plasmodium falciparum* (IC_{50} values <651 nM) and only moderate cytotoxicity against HEK293 cells (IC_{50} values >3.9 μM). **164** was the most active natural product, with IC_{50} values of 51 and 6.6 nM against 3D7 and Dd2 lines, respectively [63]. Calyculin A (**168**) was isolated from the marine sponge *Discodermia calyx* collected off Shikine-jima Island, Japan, which exhibited potent cytotoxicity as well as tumor promotion activity, attributed to its strong and specific inhibition of Ser/Thr protein phosphatases 1 (PP1, IC_{50} = 1.4 nM) and 2A (PP2A, IC_{50} = 2.6 nM) [64]. Two alkaloids, pyrinodemins G and H (**169**, **170**), were isolated from an Okinawan marine sponge *Amphimedon* sp., and they showed cytotoxicity against P388 murine leukemia cells (IC_{50} 9.6 and 2.5 μg/mL, respectively) in vitro [65]. Three dimeric bromopyrrole alkaloids, nagelamides X–Z (**171–173**), were isolated from a marine sponge *Agelas* sp., and they exhibited antimicrobial activity with IC_{50} values ranging from 2.0 to 8.0 μg/mL [66]. Unique bromopyrrole alkaloids, nagelamides U and W (**174**, **175**), were isolated from a marine sponge *Agelas* sp., exhibiting inhibitory activity against *Candida albicans* (IC_{50} 4 μg/mL, each) [67]. Spongiacidin C (**176**) isolated from the marine sponge *Stylissa massa* inhibited USP7 most strongly with an IC_{50} of 3.8 μM among several USP family members tested [68]. *N*-containing metabolites (**177**, **178**) were isolated from the South China Sea sponge *Agelas clathrodes* and showed moderate cytotoxicity against cancer cell line SGC7901 [69]. 2-Methoxy-3-oxoaaptamine (**179**), 2,3-dihydro-2,3-dioxoaaptamine (**180**), demethyl(oxy)aaptamine (**181**), 3-aminodemethyl(oxy)aaptamine (**182**), and 3-(methylamino) demethyl(oxy)aaptamine (**183**) were isolated from a marine sponge of *Aaptos* sp., among which **179** was presented antimycobacterial activity against *Mycobacterium smegmatis* in both active-growing and dormancy-inducing hypoxic conditions with a minimum inhibitory concentration (MIC) of 6.25 μg/ml, and compounds **180–183** showed antimycobacterial activities under hypoxic condition selectively, with MIC values of 1.5–6.25 μg/ml [70].

In 2015, compounds 10-methoxy-2-methylimidazo[4,5,1-ij] pyrido[2,3,4-de] quinolone (**184**), 3-(phenethylamino) demethyl(oxy)aaptamine (**185**), demethyl(oxy) aaptamine (**181**), aaptamine (**186**), and 3-(methylamino)demethyl(oxy)aaptamine (**183**) were isolated from the South China Sea sponge *Aaptos aaptos*, exhibiting cytotoxic activities against HeLa, K562, MCF-7, and U937 cell lines with IC_{50} values in the range of 0.90–12.32 μM [71]. Netamines Q (**187**) isolated from the Madagascar sponge *Biemna laboutei* exhibited antiplasmodial activities with IC_{50} values of 8.37 μM [72]. Indole alkaloids **188**, penaresin (**189**), indolecarbaldehyde (**190**), and plakohypaphorine D (**191**) were isolated from the sponge *Plakortis* sp. collected from Zampa in Okinawa, all of which showed cytotoxicity against P388 cells with IC_{50} values of 0.6, 5, 0.1, and 3.2 μg/mL, respectively [73]. The investigation of South China Sea nudibranch *Jorunna funebris* and its sponge-prey *Xestospongia* sp. led to the isolation of fennebricin C (**192**), fennebricin D (**193**), renieramycin J (**194**), fennebricin A (**195**), renierone (**196**), and *N*-formyl-1,2-dihydrorenierone (**197**). All of the compounds showed the inhibitory activities of NF-κB signaling pathway with IC_{50} values ranging from 1.0 to 9.7 μM, and fennebricin A (**195**) also exhibited growth inhibition against both A549 and HL-60 cell lines with IC_{50} values of 6.2 and 2.5 μM [74]. Crambescin A2 392 (**198**), crambescin A2 406 (**199**), crambescin A2 420 (**200**), and Scheme 575948 (**201**) were obtained from the marine sponge *Pseudaxinella reticulata* collected off the Bahamas. These compounds showed antifungal activity against the human pathogens *Cryptococcus neoformans* var. *gattii* with MIC_{50} values of 1.2, 0.85, and 1.1, 2.5 μM [75]. Ceratinine H (**202**), psammaplysin E (**203**), ceratinophenol A (**204**) were isolated from a new collection of the Red Sea marine sponge *Pseudoceratina arabica*. Compounds **202** and **203** showed potent antiproliferative activities against HeLa cells with IC_{50} values of 2.56 and 2.19 μM; **203** and **204** showed potent antimigratory activity with IC_{50} values of 0.31 and 10.4 μM, respectively [76]. (10E,12Z)-Haliclonadiamine (**205**), halichondriamines A (**206**) and B (**207**), haliclonadiamine (**208**), and papuamine (**209**) were isolated from the Okinawan marine sponge *Halichondria panicea*. These compounds exhibited antimycobacterial activities with inhibition zones of 7–16 mm at 10 μg/disc and also showed apparent activity against the proliferation of the cancer cell line Huh-7 with IC_{50} values from 3.6 to 7.8 μM [77].

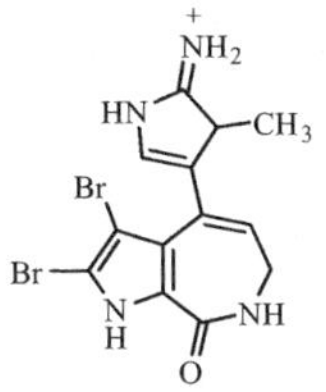

147 12-Nmethyl stevensine

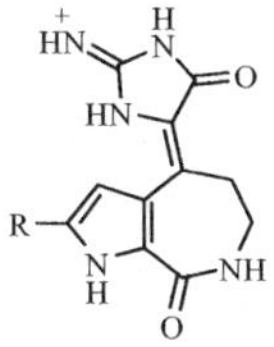

148 R = Br
149 R = H

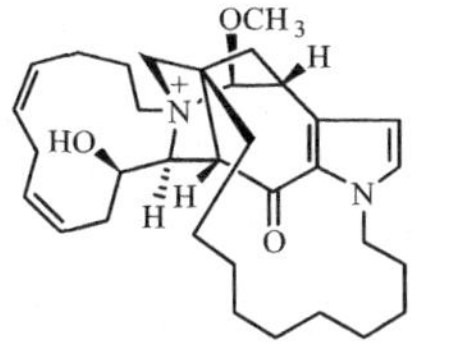

150 densanins A

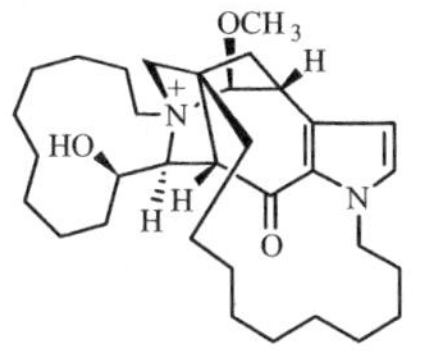

151 densanins B

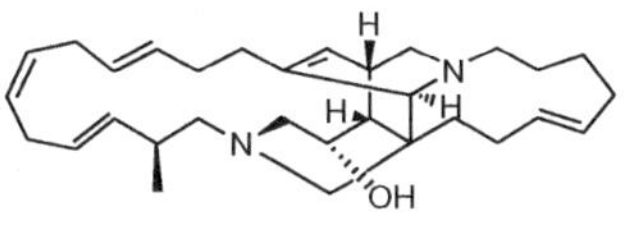

152 R = H
153 R = OH

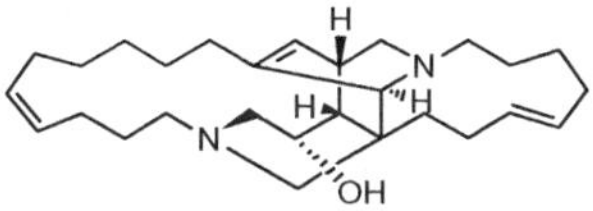

154 dihydroingenamine D

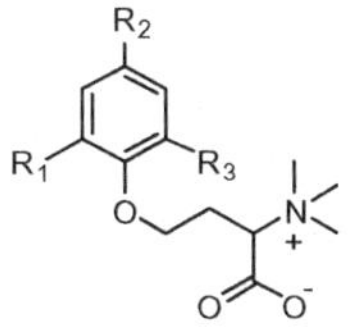

155 $R_1 = R_2 = R_3 = Br$
156 $R_1 = R_3 = Cl$, $R_2 = Br$
157 $R_1 = R_2 = Br$, $R_3 = I$

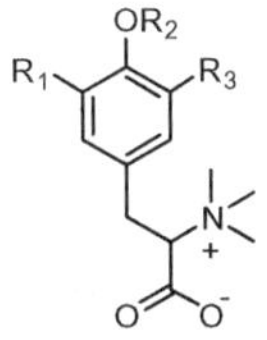

158 $R_1 = R_3 = I$, $R_2 = Me$
159 $R_1 = R_3 = I$, $R_2 = H$
160 $R_1 = Br$, $R_2 = Me$, $R_3 = I$

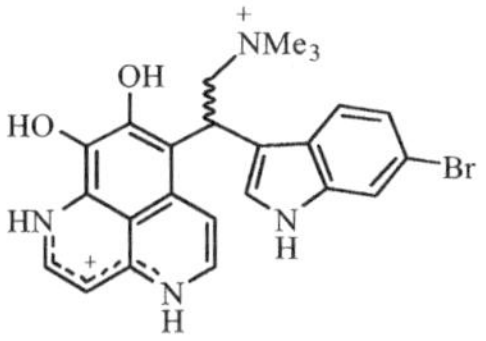

161 nakijinamines A

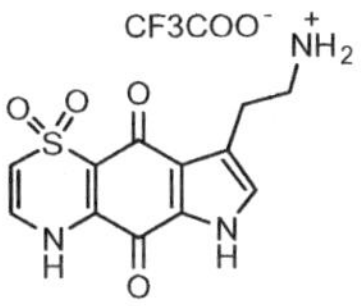

164 Thiaplakortones A

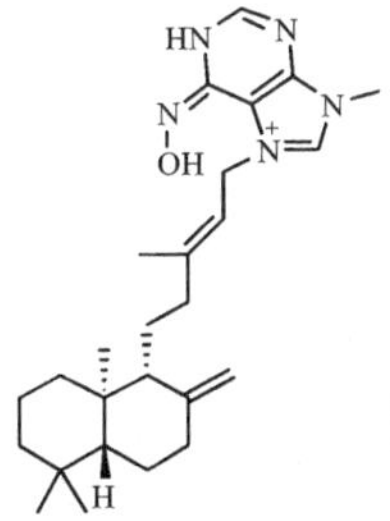

162 (-)-ageloxime D

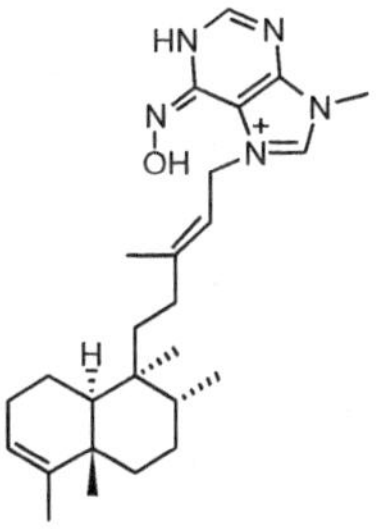

163 ageloxime B

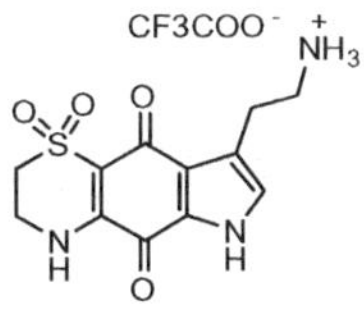

165 Thiaplakortones B

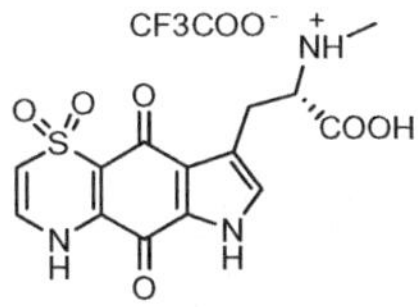

166 Thiaplakortones C

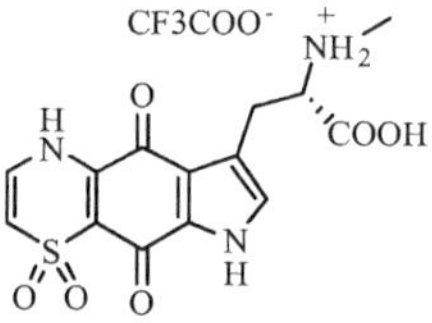

167 Thiaplakortones D

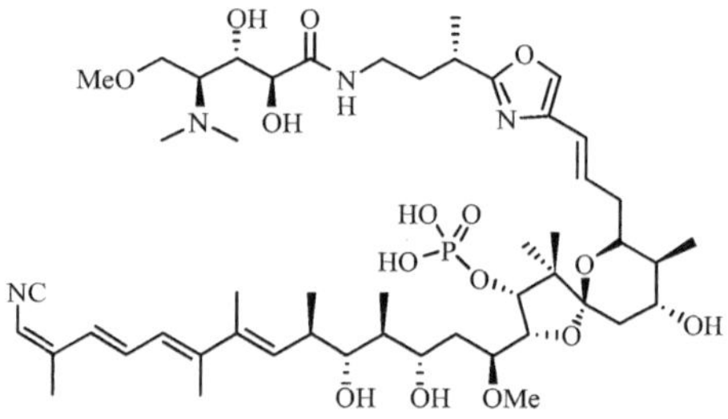

168 Calyculin A

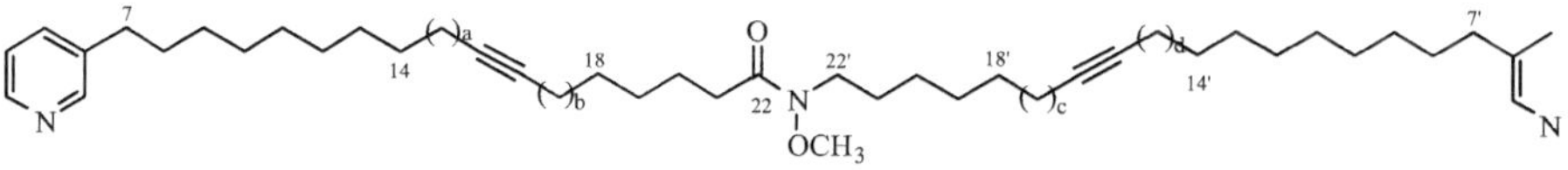

169 Pyrinodemin G
(a+b=1, c+d=1)

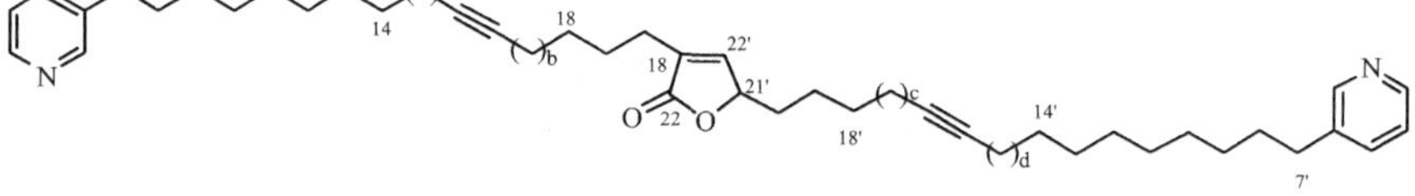

170 Pyrinodemin H
(a+b=1, c+d=1)

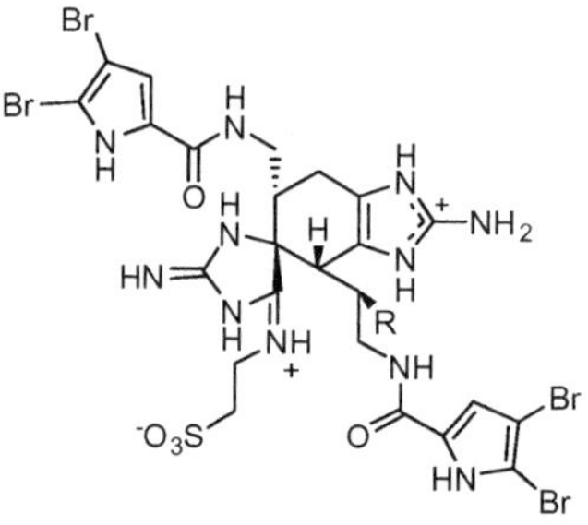

171 nagelamides X R = OH
172 nagelamides Y R = H

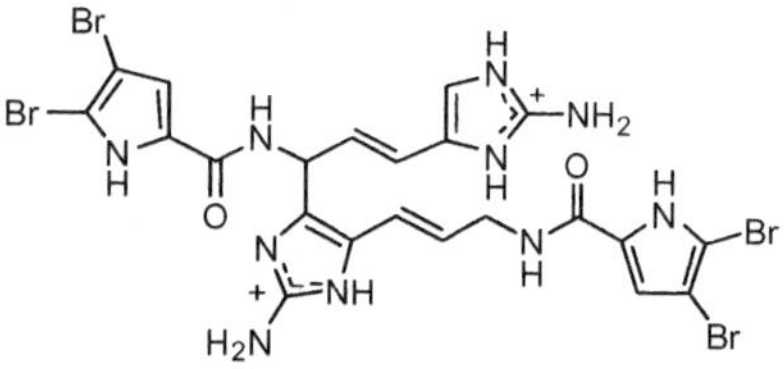

173 nagelamides Z

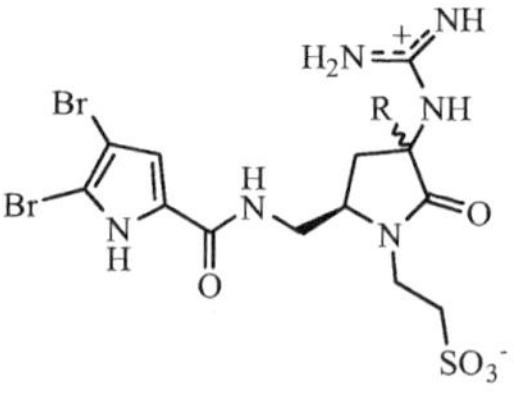

174 Nagelamides U R=¦Â-H

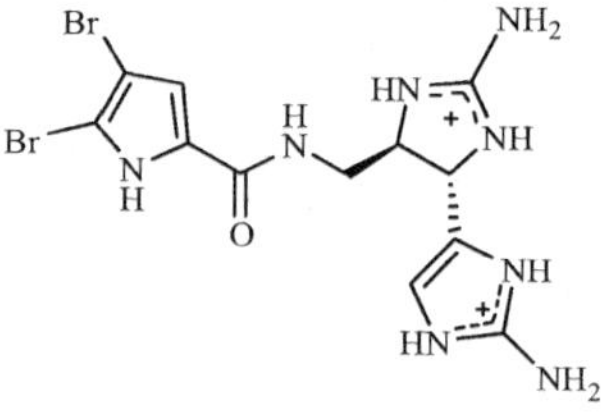

175 Nagelamides W

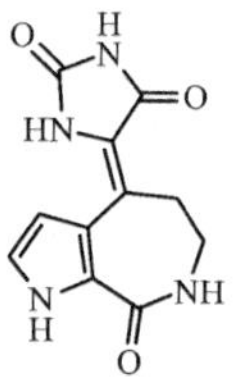

176 Spongiacidin C

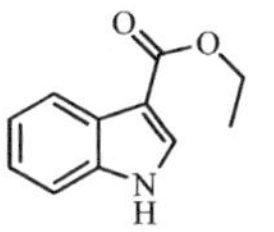

177

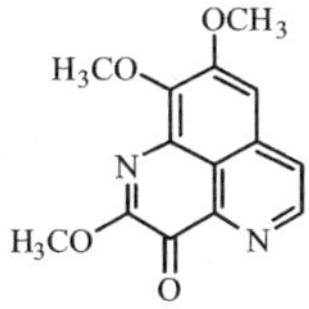

179 2-Methoxy-3-oxoaaptamine

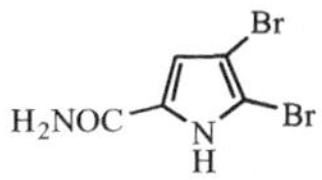

178

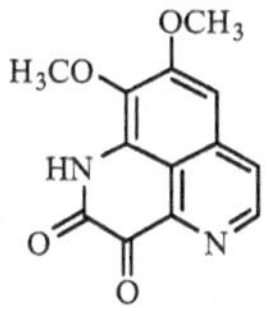

180 2,3-Dihydro-2,3-dioxoaaptamine

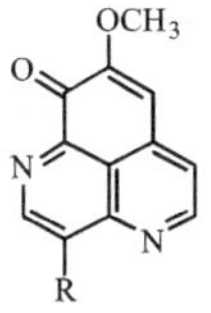

181 Demethyl(oxy)aaptamine R=H
182 3-Aminodemethyl(oxy)-aaptamine R=NH2
183 3-(Methylamino)demethyl-(oxy)aaptamine R=NHCH3

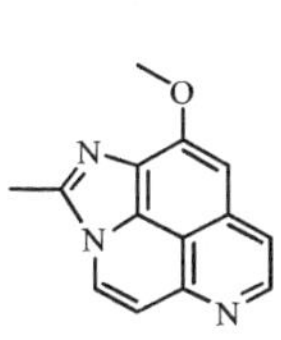

184

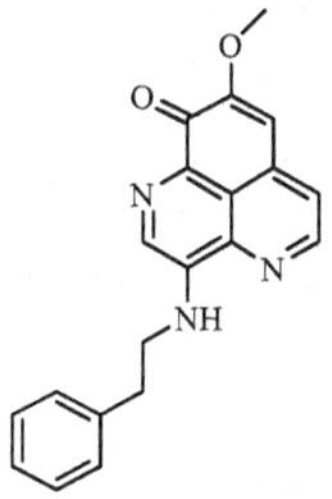

185 3-(phenethylamino) demethyl(oxy)aaptamine

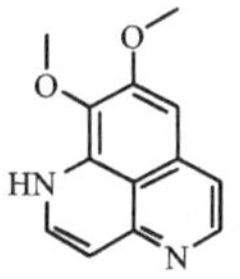

186 aaptamine

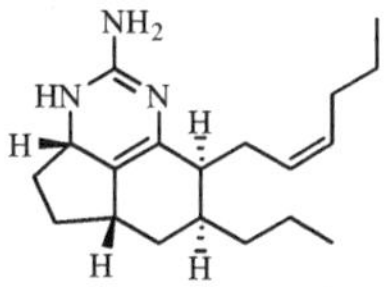

187 Netamine Q

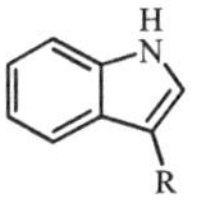

188 R = C ≡ C-CONH2
189 R = CH = CH-COOH
190 R = C(O)H

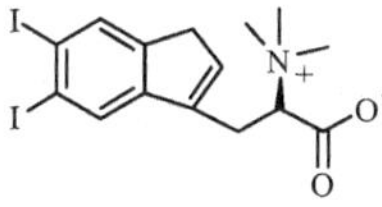

191 plakohypaphorine D

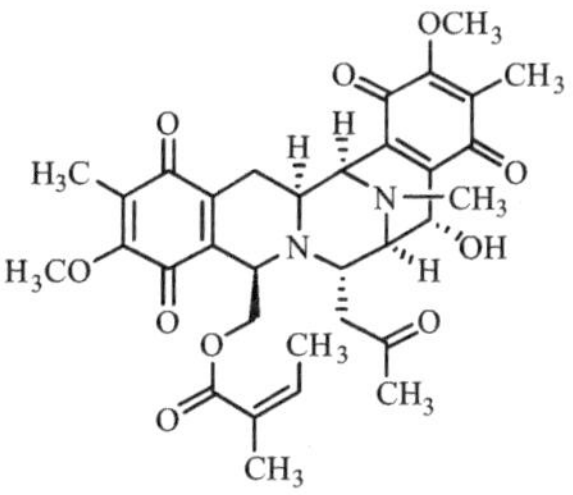

192 Fennebricin C

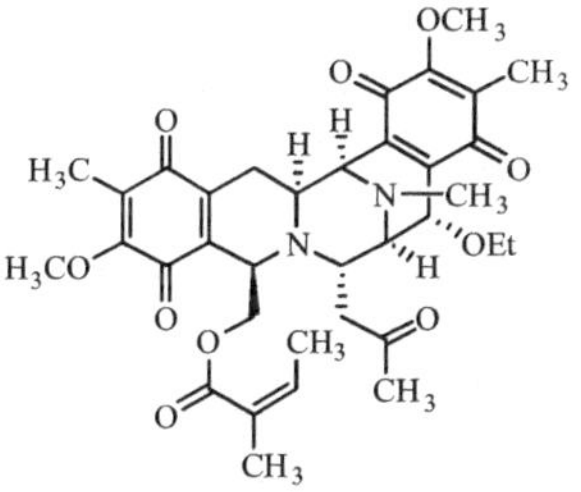

193 Fennebricin D

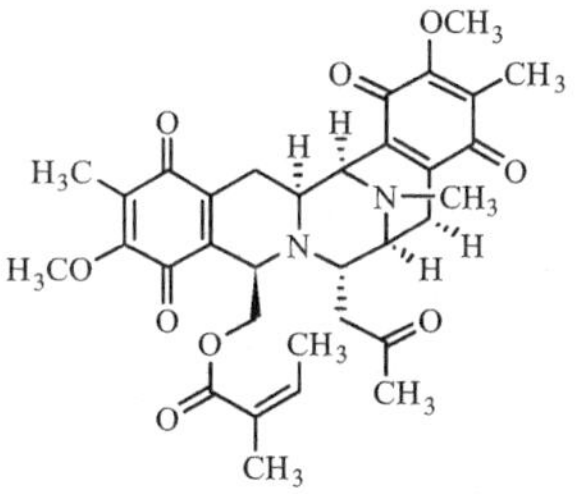

194 Renieramycin J

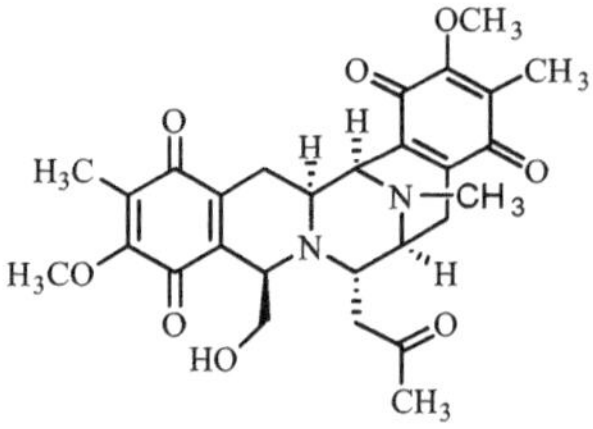

195 Fennebricin A

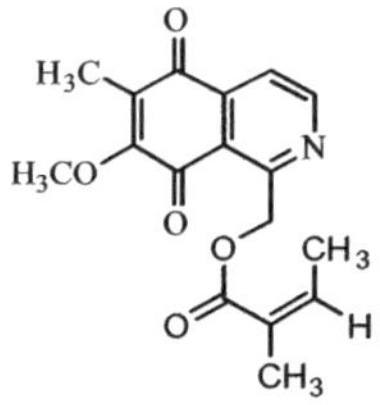

196 renierone

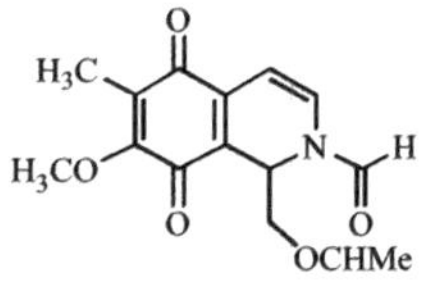

197 N-formyl-1,2-dihydrorenierone

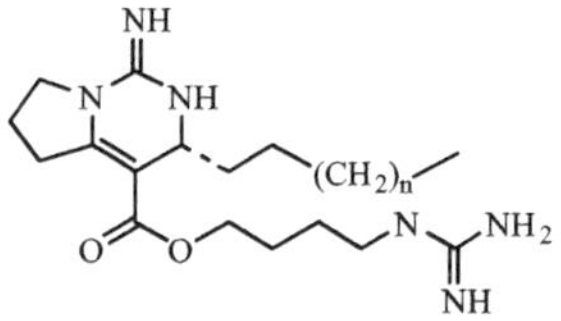

198	crambescin A2 392	n = 4
199	crambescin A2 406	n = 5
200	crambescin A2 420	n = 6
201	Sch 575948	n = 2

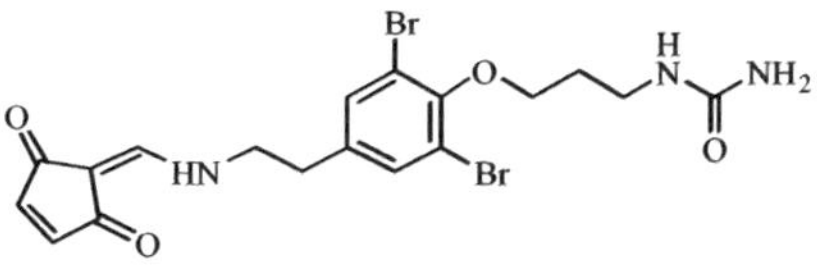

202 Ceratinine H

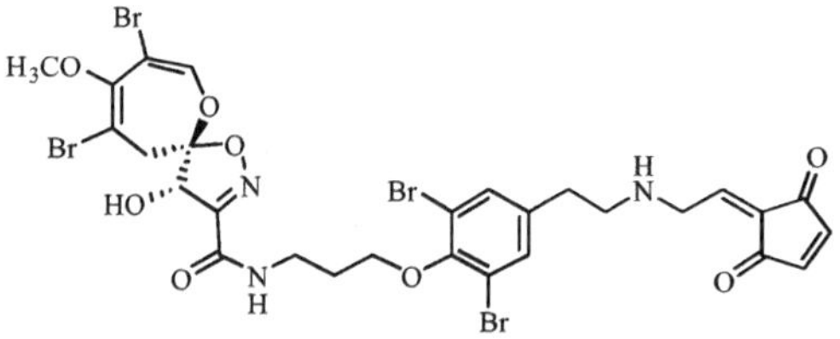

203 psammaplysin E

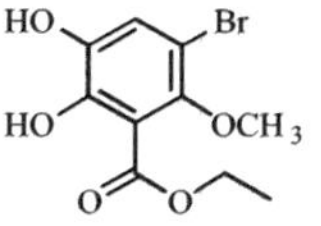

204 ceratinophenol A

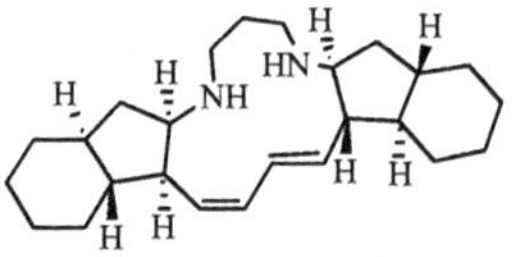

205 (10E,12Z)-Haliclonadiamine

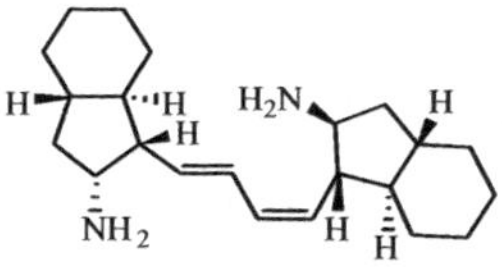

206 Halichondriamine A

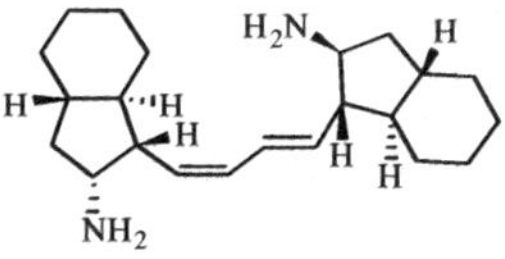

207 Halichondriamine B

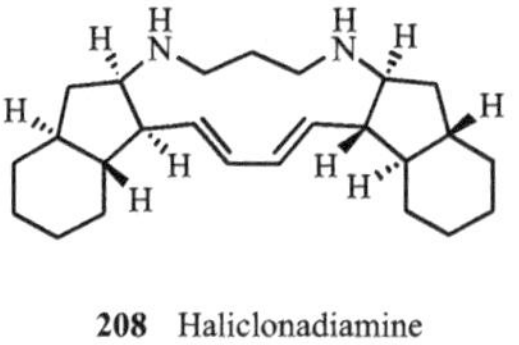

208 Haliclonadiamine

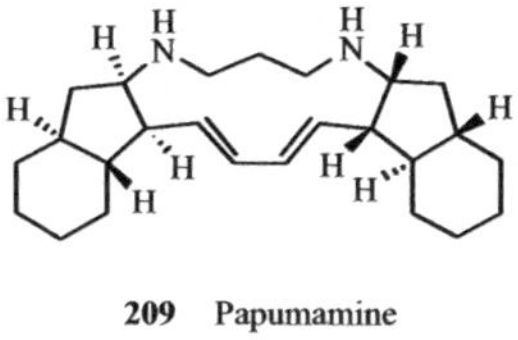

209 Papumamine

15.3 Bioactive Peptides

Marine sponges are shown to have a large variety of resources of bioactive peptides. From the structure, they possess linear peptides, cyclic peptides, and depsipeptides which of them have highly modified structural features of nonproteinogenic amino acid or hydroxy acid group, while others have those with minimal differences from the common ribosomal peptides.

This review summarizes the isolation, structure identification of a diverse 109 peptides from 27 marine sponges with a variety of potent biological activities within the literature coverage from 1991 to 2016.

15.3.1 Linear Peptides

Nazumamide A (**210**) was isolated from the marine sponge *Theonella* sp. in 1991 and displayed inhibition of thrombin with an IC_{50} of 4.63 μM [78].

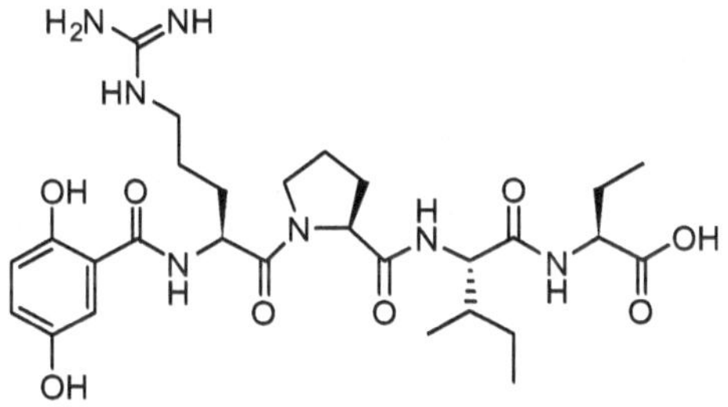

210 nazumamide A

In 1995, the known metabolite hemiasterlin (**211**) and the novel metabolites hemiasterlin A (**212**), hemiasterlin B (**213**), and criamide B (**214**) were isolated from a specimen of *Cymbasrefa* sp. at Motupore and Madang in Papua New Guinea. Hemiasterlin (**211**) displayed significant in vitro cytotoxicity against murine leukemia P388, human breast cancer MCF-7, human glioblastoma/astrocytoma U373, and human ovarian carcinoma HEY with ED_{50} values of 0.087, 170, 22.81, and 2.66 nM, respectively. Interesting, compared to hemiasterlin, hemiasterlin A (**212**) with absence of N-methyl motif attached on the indole ring showed higher activity against human glioblastoma/astrocytoma U373 but less active against human ovarian carcinoma HEY with ED_{50} values of 2.93 and 14.84 nM, respectively. Hemiasterlin B (**213**) closely related to **212** exhibited less cytotoxicity against murine leukemia P388, human breast cancer MCF-7, and human ovarian carcinoma HEY with ED_{50} values of 14.06, 0.13, and 0.032 μM, respectively. Similar compound criamide B (**214**) was observed to display potent in vitro activity against murine leukemia P388, human breast cancer MCF-7, human glioblastoma/astrocytoma U373, human ovarian carcinoma HEY, human colon LOVO, and human lung A549 cell lines with ED_{50} values of 0.011, 9.97, 0.4, 0.28, 0.22, and 0.43 μM, respectively [79].

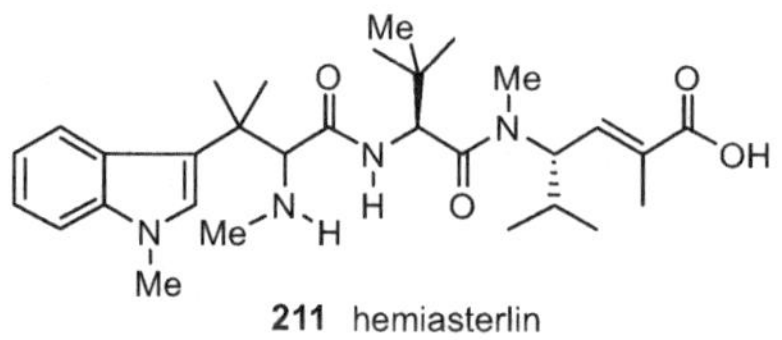

211 hemiasterlin

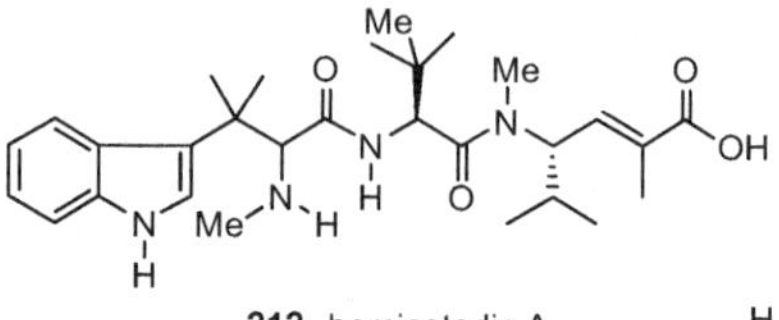

212 hemiasterlin A

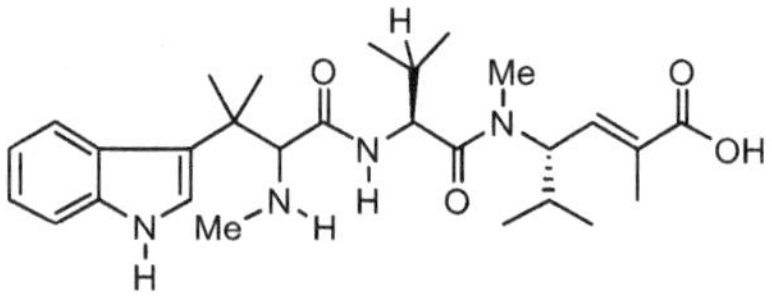

213 hemiasterlin B

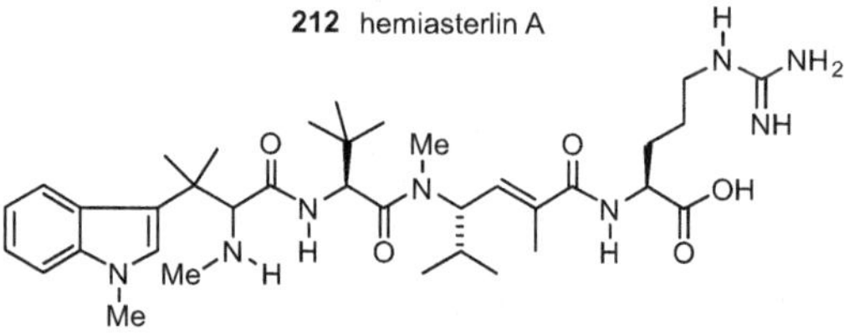

214 criamide B

Halicylindramides D (**215**), a tridecapeptide, was isolated from the marine sponge *Halichondria cylindrata* in Japan in 1996, which was cytotoxic against P388 murine leukemia cells with an IC_{50} value of 1.25 μM [80].

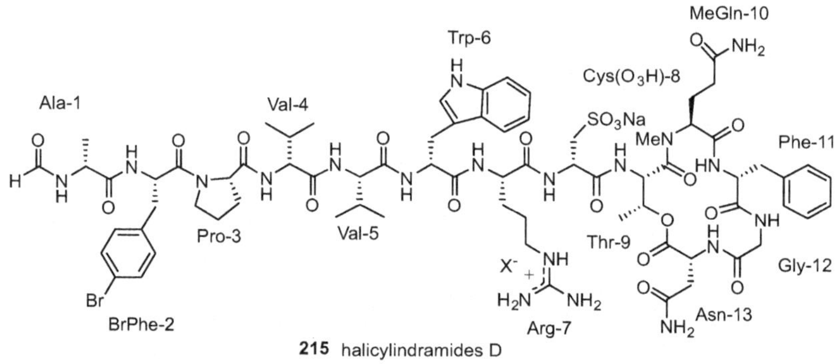

215 halicylindramides D

Koshikamide A_1 (**216**) isolated from a marine sponge, *Theonella* sp. collected from southwestern Japan in 1999, showed potent cytotoxicity against P388 leukemia cells with an IC_{50} value of 1.69 μM [81].

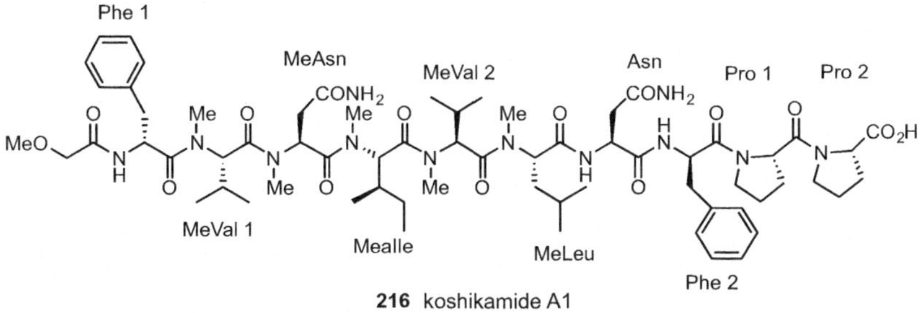

216 koshikamide A1

In 1999 six peptides, pseudotheonamides A1 (**217**), A2 (**218**), B2 (**219**), C (**220**), D (**221**), and dihydrocyclotheonamide A (**222**), were derived from the marine sponge *Theonella swinhoei* collected off Hachijo-jima Island, which exhibited selective serine protease inhibitory activities: inhibition of thrombin with IC_{50} values of 1.0, 3.0, 1.3, 0.19, 1.4, and 0.33 μM, respectively, while they inhibited trypsin with IC_{50} values of 4.5, >10, 6.2, 3.8, >10, and 6.7 μM, respectively [82].

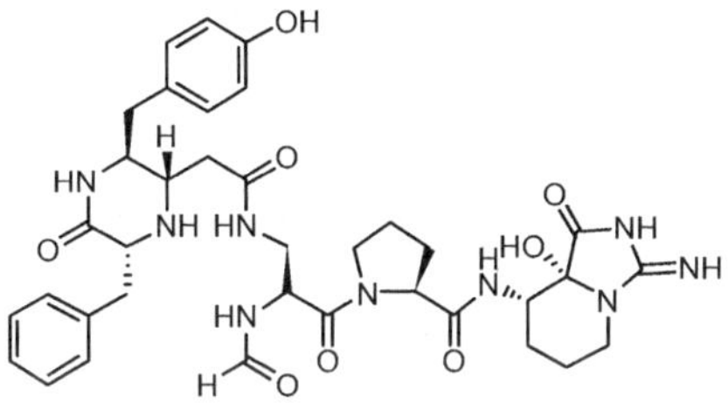

217 pseudotheonamides A_1

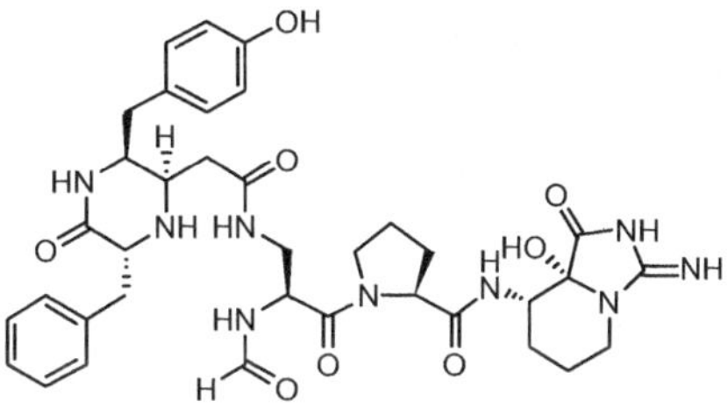

218 pseudotheonamides A_2

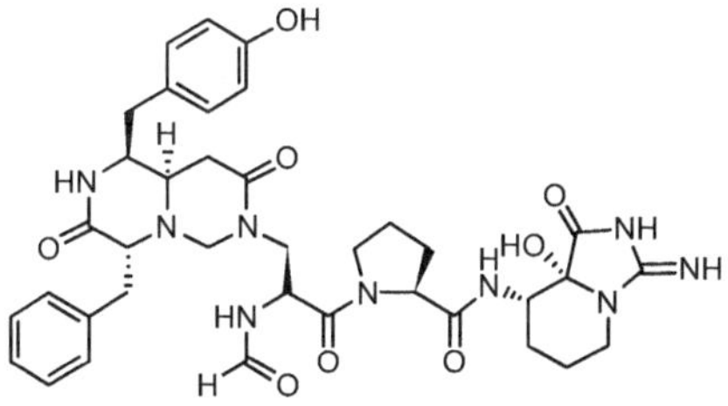

219 pseudotheonamides B_2

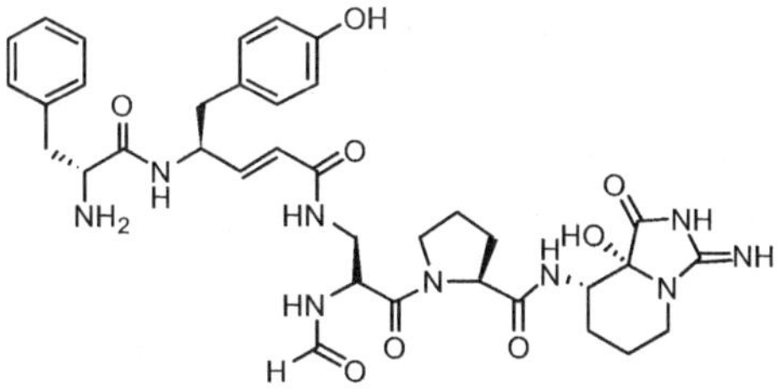

220 theonamide C

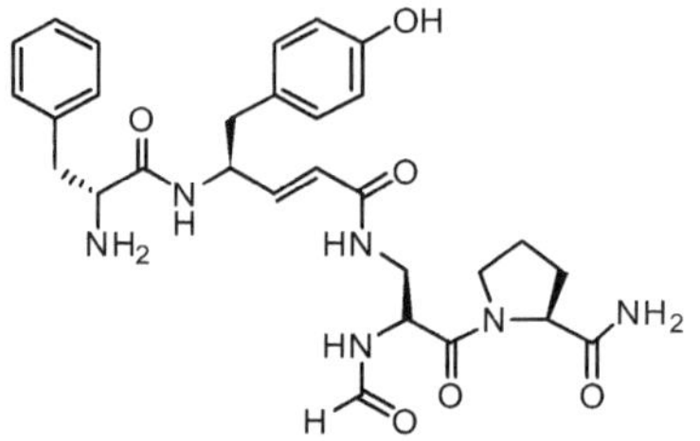

221 pseudotheonamide D

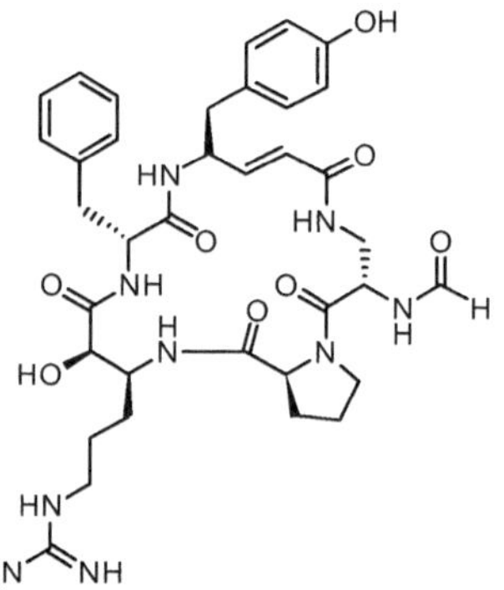

222 dihydrocyclotheonamide A

Miraziridine A (**223**) was isolated from the marine sponge *Theonella* aff. *mirabilis* in 2000 during the collection cruise on R/V *Toyoshio-maru* of Hiroshima University to the Amami and Tokara Islands, which was reported as a cathepsin B inhibitor with an IC_{50} value of 2.1 μM [83].

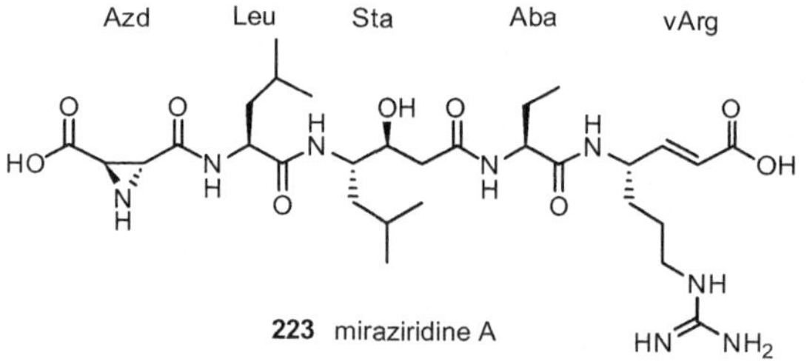

223 miraziridine A

In 2002, dysinosin A (224) was identified as a novel inhibitor of factor VIIa and thrombin, with Ki value of 0.108 and 0.452 μM, respectively, from a new genus and species of Australian sponge of the family Dysideidae [84].

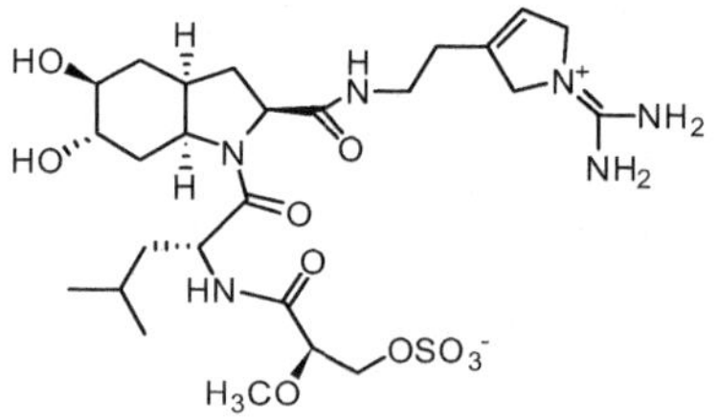

224 dysinosin A

In 2004, dysinosins B–D (**225–227**) were isolated from the sponge *Lamellodysidea chlorea* collected off Low Isles, Queensland, Australia, which inhibited factor VIIa at a Ki of 0.090, 0.124, and 1.320 μM, respectively, and thrombin at a Ki of 0.170, 0.550, and >5.1 μM, respectively [85].

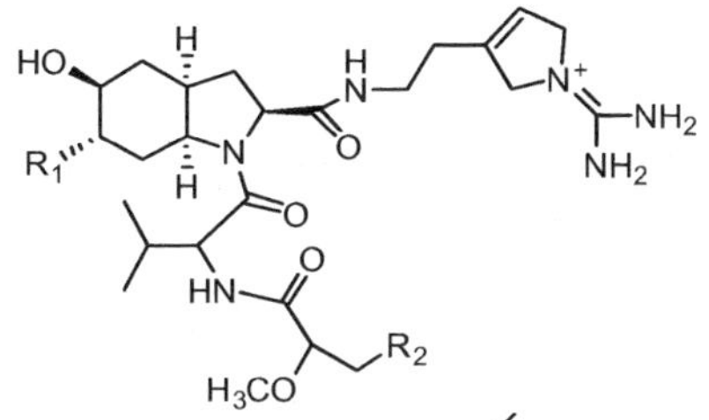

225 dysinosin B R_1 = 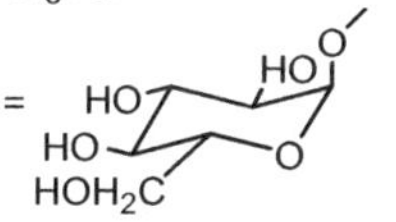 $R_2 = OSO_3^-$

226 dysinosin C $R_1 = OH$ $R_2 = OSO_3^-$

227 dysinosin D $R_1 = OH$ $R_2 = OH$

Marine sponge *Haliclona* sp. collected at Sulawesi Island, Indonesia, in 2004, led to the isolation of kendarimide A (**228**), which reversed MDR in KB-C2 cells mediated by P-glycoprotein (P-gp) at a concentration of 6 μM [86].

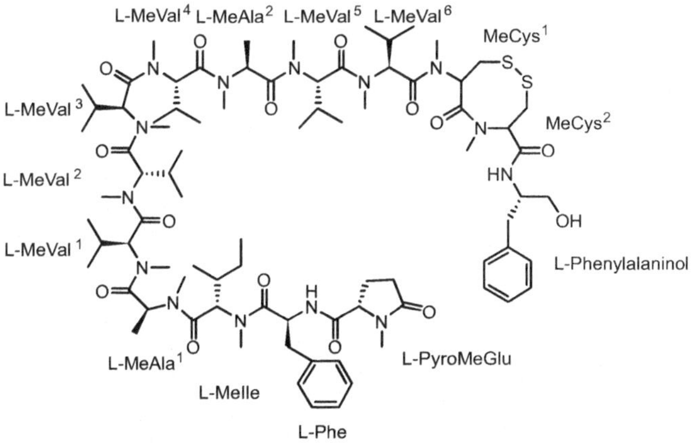

228 kendarimide A

In 2005, some highly cytotoxic polypeptides with 48 amino acid residues, such as polytheonamides A (**229**), B (**230**), and C (**231**) were isolated from the marine sponge *Theonella swinhoei* collected from Hachijo-jima Island. They were tested against P388 murine leukemia cells with IC_{50} values of 15.5×10^{-6}, 13.51×10^{-6}, and 13.48×10^{-6} μM, respectively [87, 88], and polytheonamide B (**230**) exhibited cytotoxicity against HeLa human uterine cervix carcinoma cells with an IC_{50} value of 0.58 nM [89], L1210 murine lymphocytic leukemia cells with an $IC_{50} < 0.8$ nM [90], and Neuro-2a mouse neuroblastoma cells with an $IC_{50} < 0.2$ nM [91].

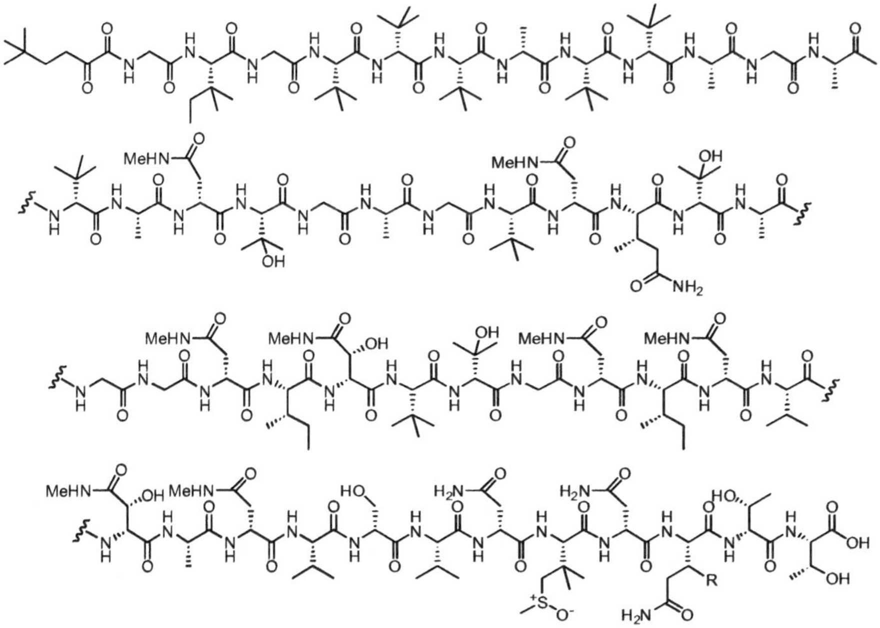

229 polytheonamides A R = H
230 polytheonamides B R = H
231 polytheonamides C R = Me

In 2005, the chemical investigation of marine sponge *Theonella* sp. collected off Shimo-koshiki-jima Island, Kagoshima, led to isolation of the koshikamide A_2 (**232**), which exhibited moderate cytotoxicity against P388 cells with an IC_{50} value of 4.6 μM [92].

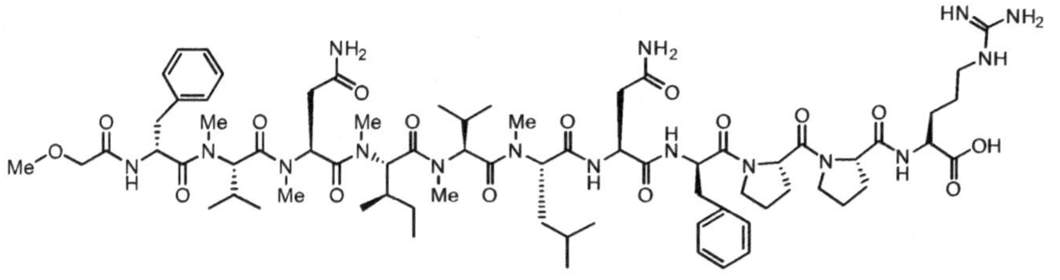

232 koshikamide A2

A chlorinated peptide, sintokamide A (**233**), was isolated from the marine sponge *Dysidea* sp. collected in Indonesia in 2008, which was found to be an inhibitor of N-terminus transactivation of the androgen receptor in prostate cancer cells with an IC_{50} at 9.8 μM [93].

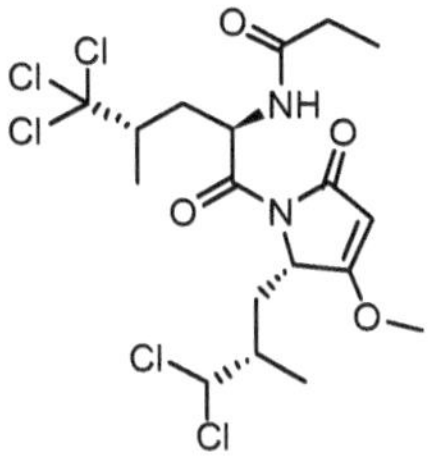

233 sintokamide A

Yaku'amides A (**234**) and B (**235**) were isolated from the marine sponge *Ceratopsion* sp. collected at Yakushinsone in the East China Sea in 2008, which exhibited potent cell growth inhibitory activity against P388 murine leukemia cells with IC_{50} values of 8.54 and 2.42 nM, respectively. Interestingly, the profile of growth inhibitory activity of Yaku'amides A (**234**) was clearly unique and unusual compared with other anticancer drugs when against a panel of 39 human cancer cell lines [94].

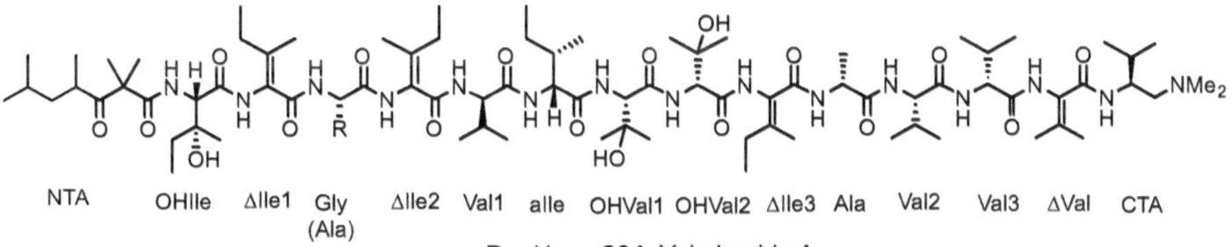

R = H **234** Yaku'amide A
R = CH3 **235** Yaku'amide B

15.3.2 Cyclic Peptides

In 1996, aciculitins A–C (236–238) were isolated from the lithistid sponge *Aciculites orientalis* (Negros, Siquijor, Philippines). Aciculitins A–C were cytotoxic to the human colon tumor cell line HCT-116 with an IC_{50} of 0.5 μg/mL and inhibited the growth of *Candida albicans* at a loading of 2.5 μg/disk in the standard disk assay [95].

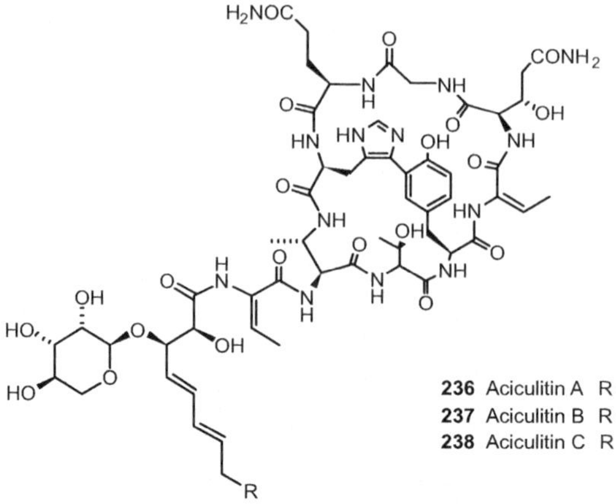

236 Aciculitin A R = C_5H_{11}
237 Aciculitin B R = C_6H_{13}
238 Aciculitin C R = C_7H_{15}

Through bioassay-guided separation, arenastatin A (239) was isolated from the marine sponge *Dysidea arenaria* (Okinawan, Japan) in 1995, which exhibited extremely potent cytotoxicity with IC_{50} of 5 pg/ml against KB cells [96].

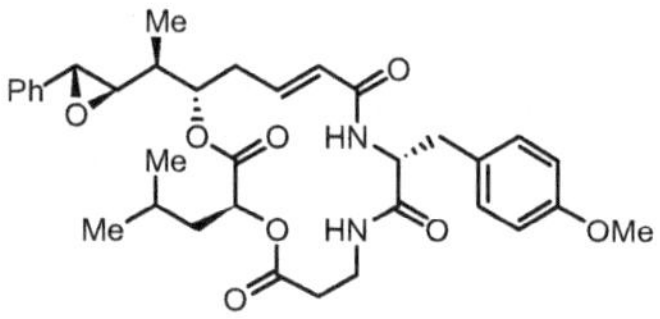

239 Arenastatin A

In 1998, a chlorate cyclic depsipeptide, cyclolithistide A (240), was isolated from the marine sponge *Theonella swinhoei*. It is important to note that cyclolithistide A exhibited significant antifungal activity against *Candida albicans* (ATCC 24433) in the agar disk diffusion assay. At a dose of 20 μg/disk, the inhibition activity was comparable to 90% of the standard, nystatin at a dose of 100 μg/disk [97].

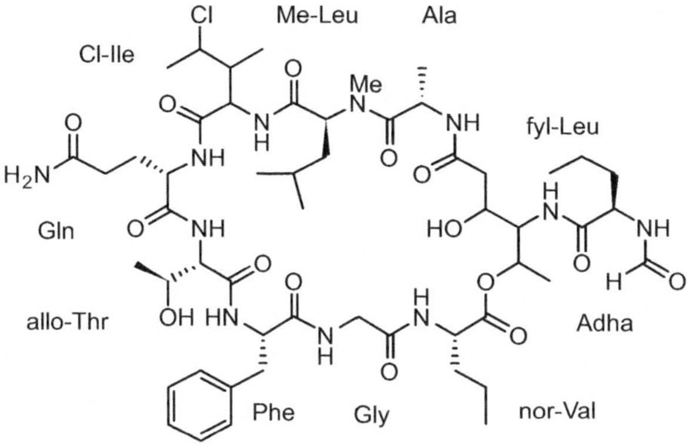

240 Cyclolithistide A

Cyclotheonamide A (241), cyclotheonamide E (242), and cyclotheonamides E2 (243) and E3 (244) were isolated from the marine sponge *Theonella swinhoei* (Tanegashima Island, Tokyo, in July 1993) as a series of potent serine protease inhibitors. In the inhibitory assays against thrombin and trypsin, cyclotheonamide A, E, E2, and E3 exhibited significant inhibition activities with IC_{50} values of 23 and 16 nM, 2.9 and 30 nM, 13 and 55 nM, and 9.5 and 52 nM, respectively [98].

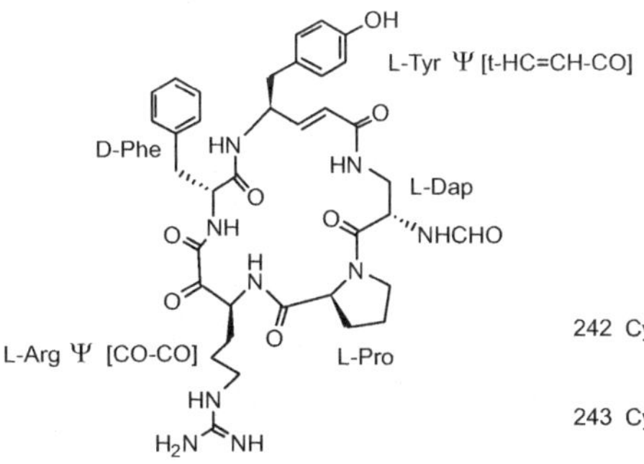

241 cyclotheonamide A

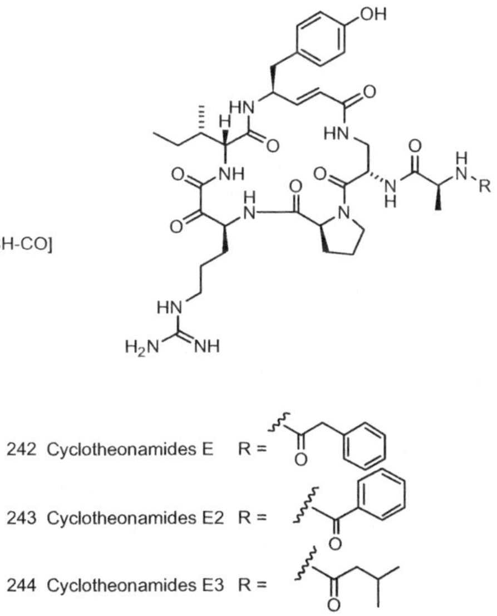

An anti-HIV cyclodepsipeptide, homophymine A (245), was isolated from the marine sponge *Homophymia* sp. (New Caledonian, in 1992), which effectively inhibited the HIV-1 infection with an IC_{50} of 75 nM. Direct cytotoxicity of 17 against the host cells was observed with a TC_{50} (toxic concentration) of 1.19 μM

[99]. Nine cyclodepsipeptides, homophymines B–E (246–249) and A1–E1 (250–254), were also isolated from the polar extracts of the sponge *Homophymia* sp. (New Caledonian, in 1992). Homophymines displayed very potent antiproliferative activity (IC_{50} in the nM range) against a panel of human cancer cell lines [100].

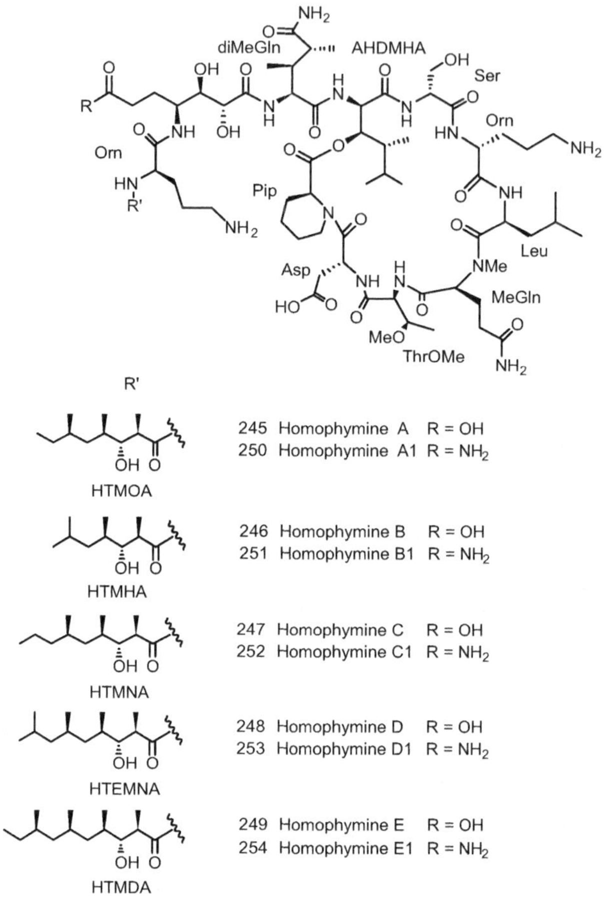

In 2009, jaspamide (255) and jaspamides B–P (256–269) were isolated from the marine sponge *Jaspis splendens*. All tested jaspamide derivatives exhibited antiproliferative activities with IC_{50} values ranging from 0.01 to 33 μM against human breast adenocarinoma (MCF-7) and colon carcinoma (HT29) cell lines [101]. Jaspamides B and C exhibited cytotoxicity against the human NSCLC-N6 cancer cell line with IC_{50} values of 3.3 and 1.1 μg/mL, respectively [102]. Jaspamides Q and R together with jaspamide exhibited potent activities against mouse lymphoma (L5178Y) cell lines with IC_{50} values in the ng/mL range (<0.1 μg/mL, <0.16 μM) [103].

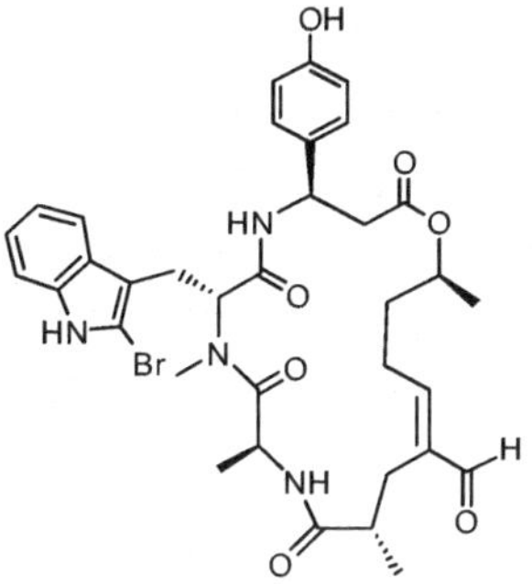

261 Jaspamide G

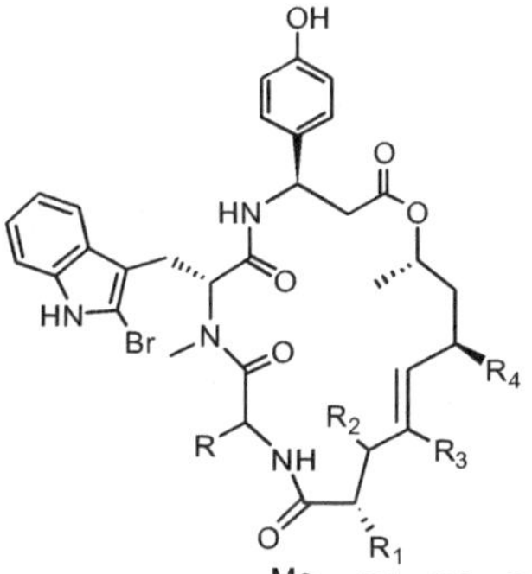

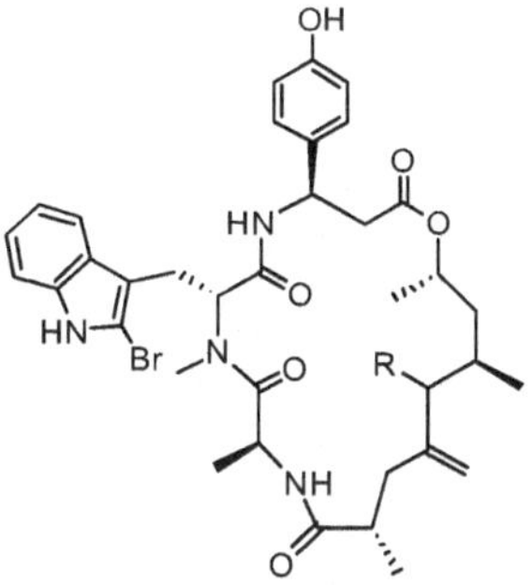

255 Jaspamide R = Me R1 = R3 = R4 = Me, R2 = H

258 Jaspamide D R = ···Et R1 = R3 = R4 = Me, R2 = H

259 Jaspamide E R = CH_2OH R1 = R3 = R4 = Me, R2 = H

260 Jaspamide F R = CH_3 R3 = R4 = Me, R1 = R2 = H

262 Jaspamide H R = CH_3 R1 = R3 = Me, R2 = R4 = H

256 Jaspamide B R = =O

257 Jaspamide C R = —OH

263 Jaspamide J R = CH_3 R3 = R4 = Me, R1 = R2 = H

264 Jaspamide K R = CH_3 R1 = R3 = R4 = Me, R2 = OH

265 Jaspamide L R = CH_3 R1 = R4 = Me, R2 = H, R3 = CH_2OH

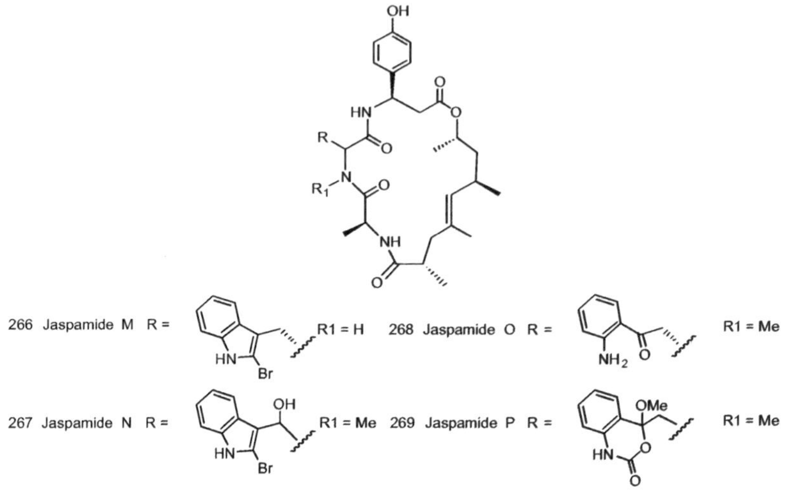

Kapakahines A–C (270–272) were isolated from the sponge *Cribrochalina olemda* (Pohnpei, Federated States of Micronesia, in April 1992, and recollected in August 1993). Kapakahines A, B, and C showed moderate cytotoxicity against P388 murine leukemia cells at IC_{50} values of 5.4, 5.0, and 5.0 μg/mL, respectively [104].

In 1995, five cyclic peptides, keramamides E (273), F (274), G (275), H (276), and J (277), containing an oxazole or a thiazole ring, were isolated from the marine sponge *Theonella* sp. (Okinawan). Keramamide E exhibited cytotoxicity against L1210 murine leukemia cells and human epidermoid carcinoma KB cells with IC_{50} values of 1.60 and 1.55 μg/mL, respectively, while keramamides G, H, and J showed weak cytotoxicity (IC_{50} ~10 μg/mL) [105]. Keramamide F showed cytotoxicity against human epidermoid carcinoma KB cells and murine lymphoma L1210 cells with IC_{50} values of 1.4 and 2.0 μg/mL, respectively [106].

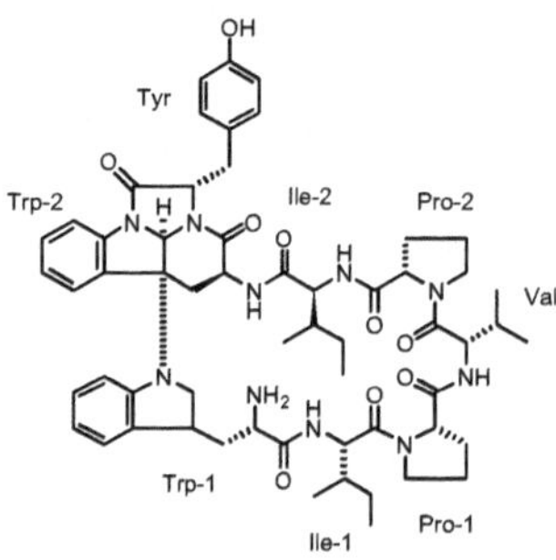

270 Kapakahine A

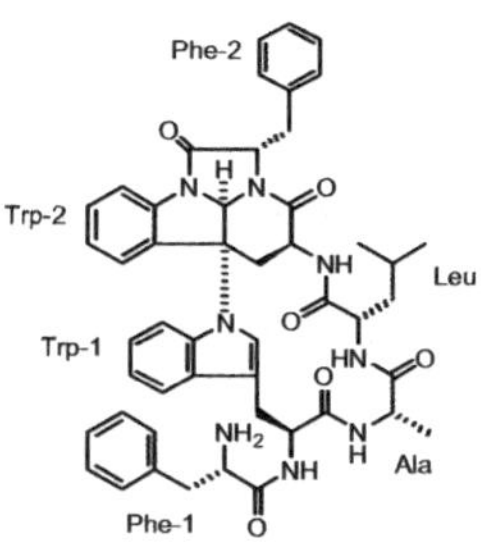

271 Kapakahine B

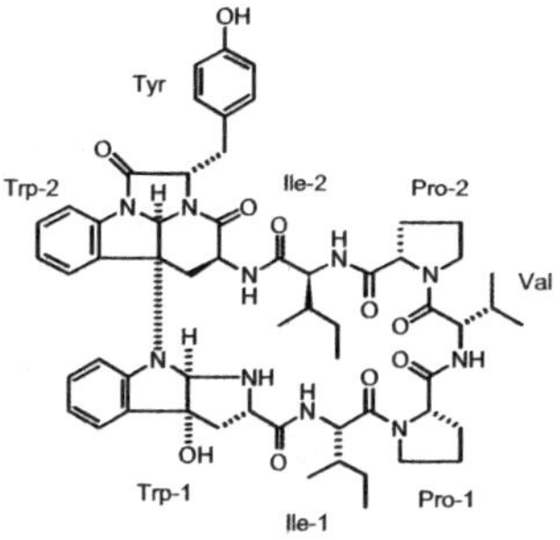

272 Kapakahine C

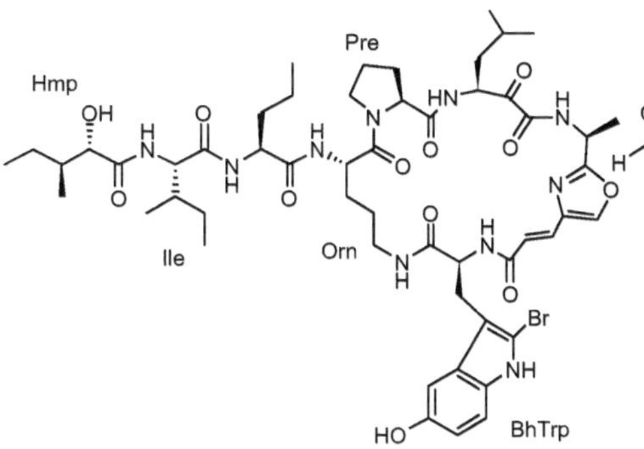

273 Keramamides E

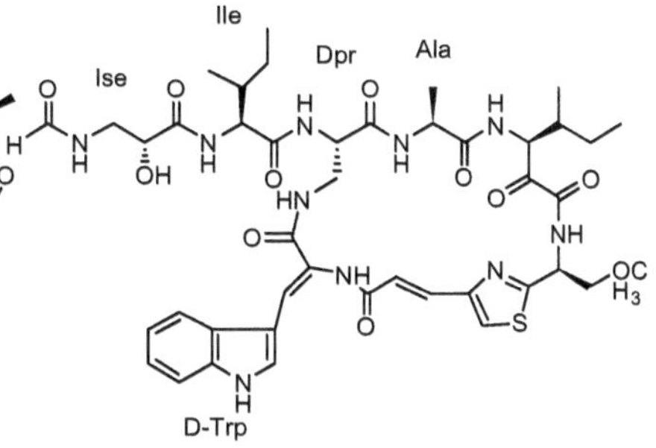

274 Keramamide F

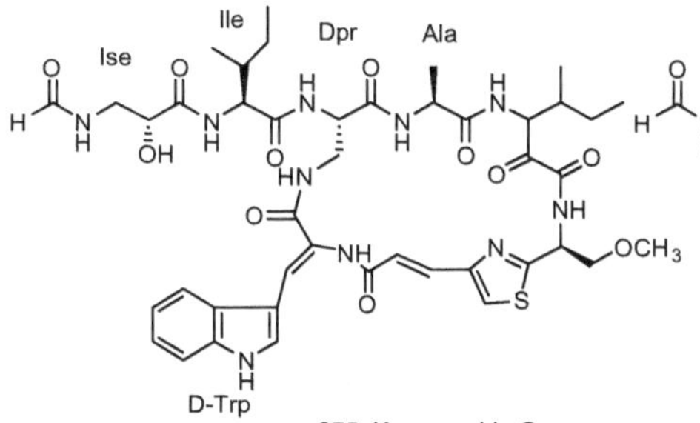

275 Keramamide G

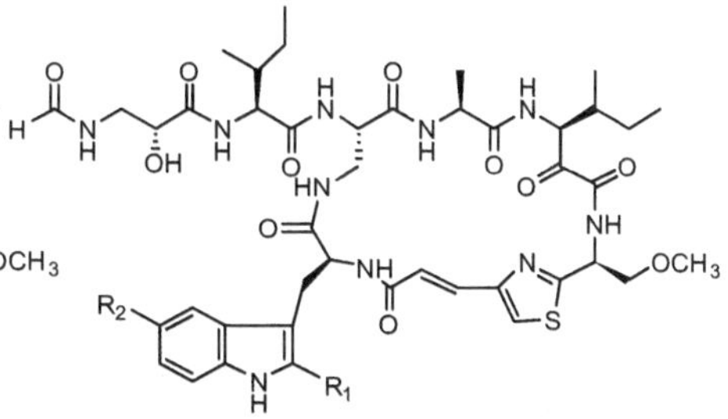

276 Keramamide H R1 = Br, R2 = OH
277 Keramamide J R1 = R2 = H

Four cyclic peptides, microsclerodermins F–I (278–281), were isolated from *Microscleroderma* sp. (Palau, Koror, in 1997). All four microsclerodermins showed very similar cytotoxicity against the HCT-116 cell line with IC_{50}'s of 1.0 μg/mL (280), 1.1 μg/mL(281), 1.8 μg/mL (278), and 2.4 μg/mL (279) [107].

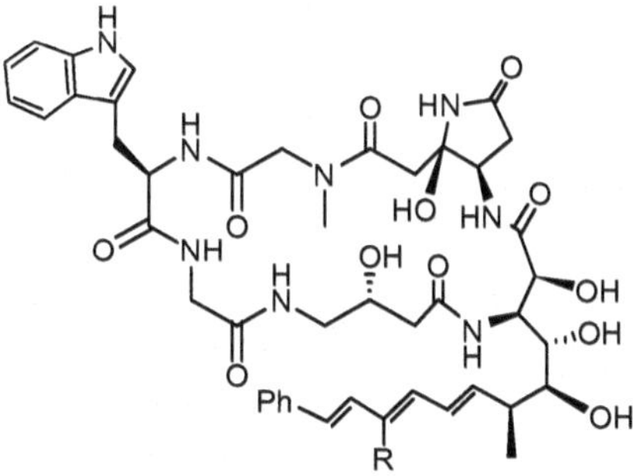

278 Microsclerodermins F R = H
280 Microsclerodermins H R = CH_3

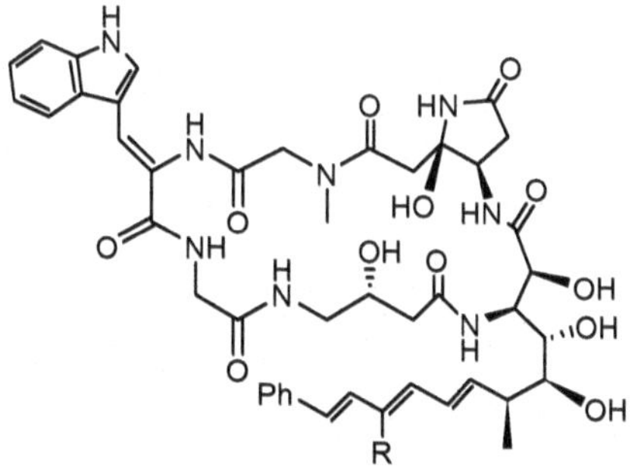

279 Microsclerodermins G R = H
281 Microsclerodermins I R = CH_3

Motuporin (282), a cyclic pentapeptide, was isolated from the marine sponge *Theonella swinhoei* (Papua, New Guinea, 1992). Motuporin inhibited protein phosphatase-1 in a standard phosphorylase phosphatase assay at a concentration of < 1 nM, making it one of the most potent PPl inhibitors known, and it also displayed considerable in vitro cytotoxicity against murine leukemia (P388: IC_{50} 6 μg/mL), human lung (A549: IC_{50} 2.4 μg/mL), ovarian (HEY: IC_{50} 2.8 μg/mL), colon (LoVo: IC_{50} 2.3 μg/mL), breast (MCF-7: IC_{50} 12.4 μg/mL), and brain (U373MG: IC_{50} 2.4 μg/mL) cancer cell lines [108].

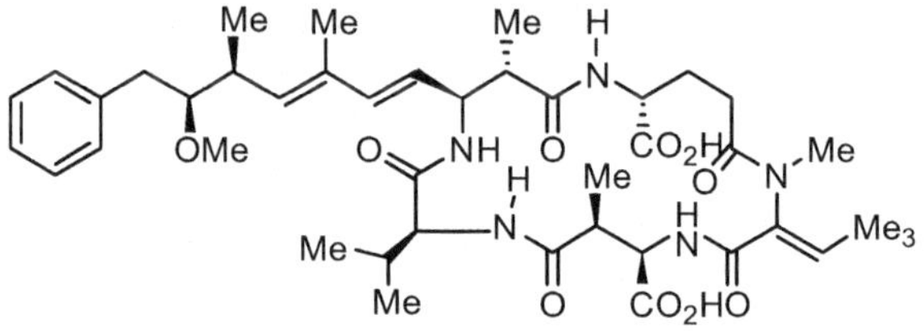

282 Motuporin

Neosiphoniamolide A (283), a potent antifungal cyclodepsipeptide, was isolated from the sponge *Neosiphonia supertes* (New Caledonia, 1989). Neosiphoniamolide A inhibited the growth of the fungi *Piricularia oryzae* and *Helminthosprium gramineum* with IC_{90} values of 5 ppm [109].

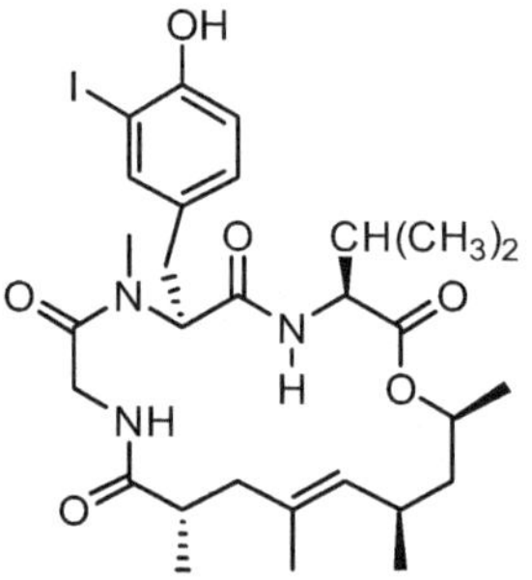

283 Neosiphoniamolide A

In 1991, orbiculamide A (284), a cyclic peptide isolated from themarine ponge *Theonella* sp., exhibited cytotoxic activity against P388 murine leukemia cells (IC_{50} 4.7 μg/mL) [110].

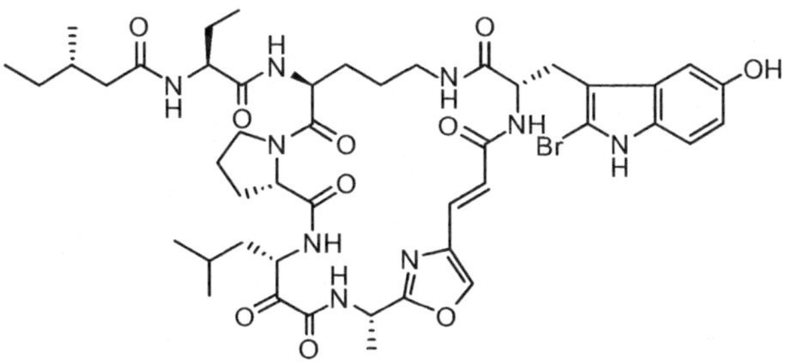

284 Orbiculamide A

The chemical investigation of marine sponge *Theonella swinhoei* (Malaita Island, Solomon Islands, in July 2004) resulted in the isolation of nine cyclopeptides, perthamides C–K (285–263). Perthamides were proved to inhibit TNF-α and IL-8 release in primary human keratinocytes cells and therefore could represent potentially leads for the treatment of psoriasis [111, 112].

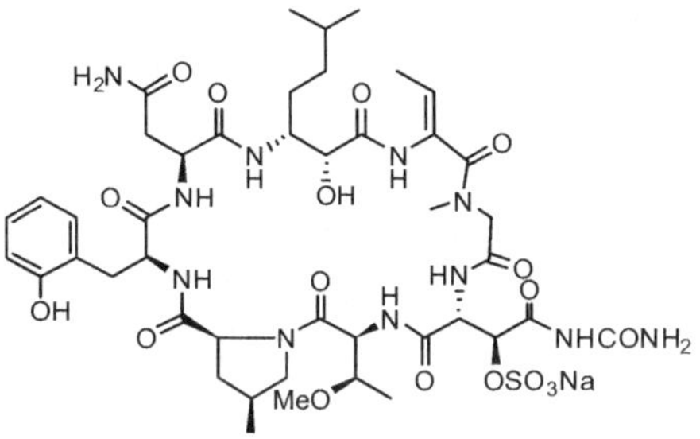

285 Perthamide C

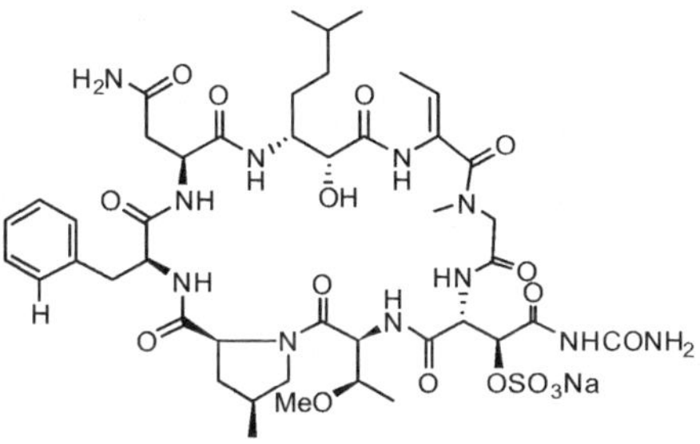

286 Perthamide D

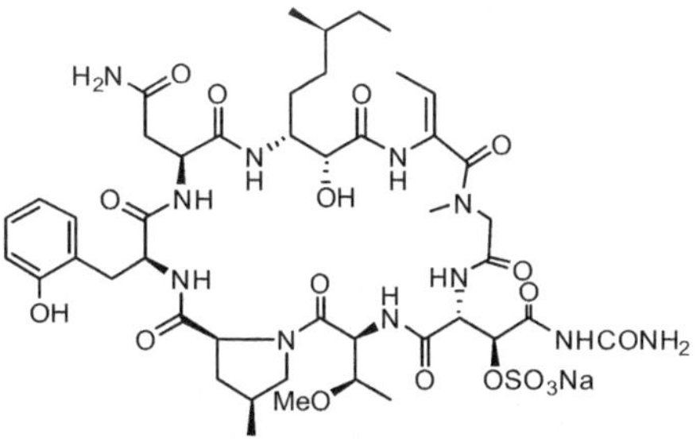

287 Perthamide E

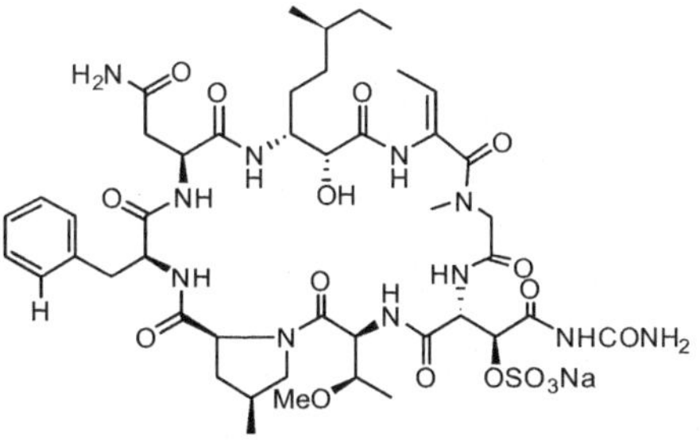

289 Perthamide F

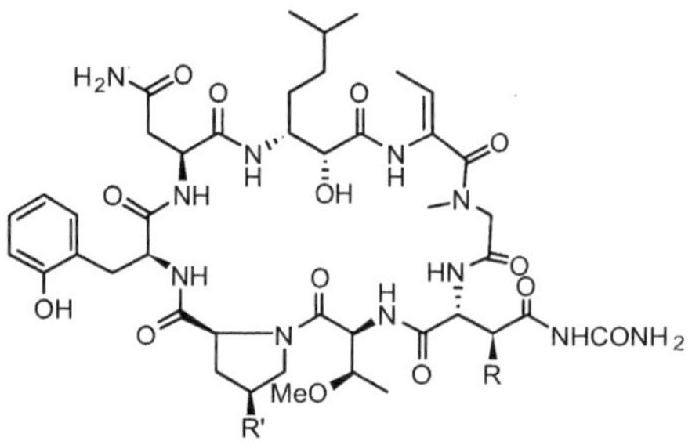

290 Perthamide G R=OSO_3Na R'=H
291 Perthamide H R=OH R'=CH_3

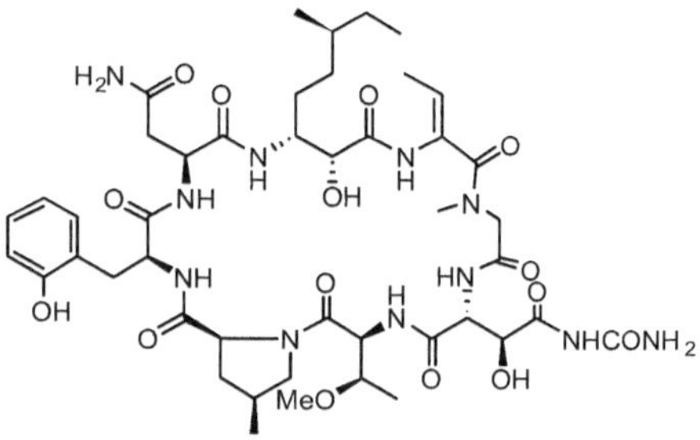

292 Perthamide I

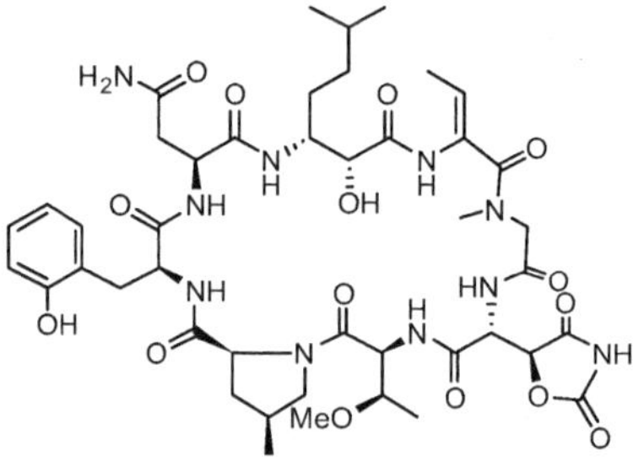

293 Perthamide J

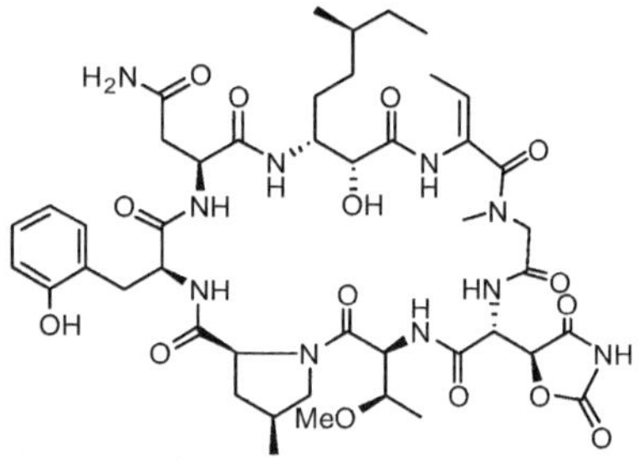

294 Perthamide K

A proline-rich cyclic octapeptide, hymenistatin 1 (295), was isolated from the sponge *Hymeniacidon* sp. (Palau, Western Pacific Ocean, 1985), which was found to be active against the P388 leukemia cell line (ED_{50} 3.5 μg/mL) [113].

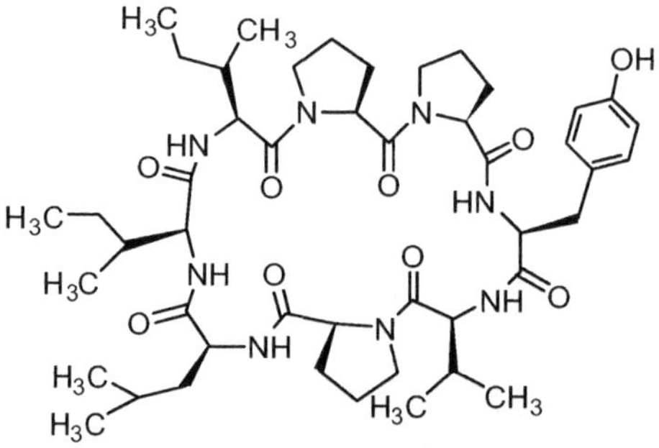

295 Hymenistatin 1

Axinastatin 1 (296), a proline-rich cyclic peptide, was isolated from the marine sponge *Axinella* sp. (Palau, Western Pacific, in 1985), with P388 lymphocytic leukemia inhibitory activity (ED_{50} 0.21 μg/mL) [114].

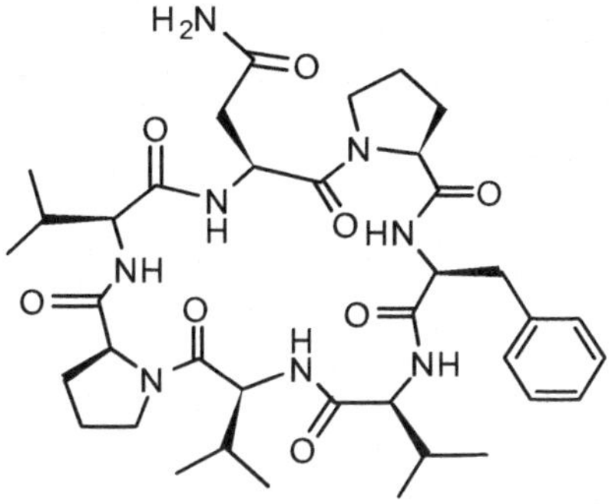

296 Axinastatin 1

In 1993, axinastatin 4 (297), another proline-rich cycloheptapeptide, was isolated from marine sponge *Axinella* cf. (collected in western Indian Ocean, The Republic of The Comoros), which showed comparable cell growth inhibitory activity against a series of human cancer cell lines (P388 lymphocytic leukemia cell line, ED_{50} = 0.057 μg/mL) [115].

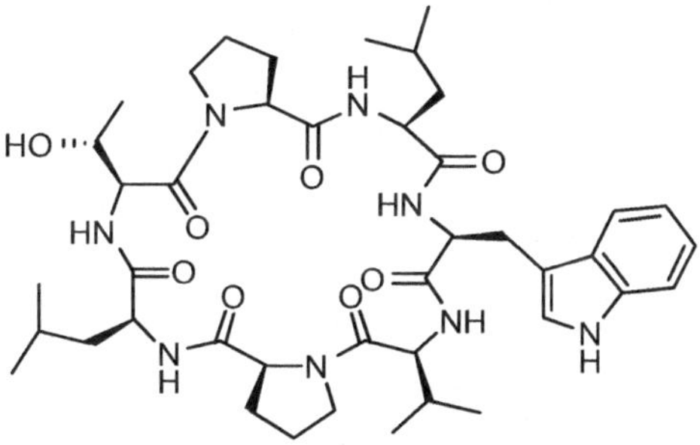

297 Axinastatin 4

In 1993, a cyclic heptapeptide, hymenamide B (298), with a prolylproline segment was isolated from the marine sponge *Hymeniacidon* sp. (Okinawan, Japan). Hymenamide B showed cytotoxicity against murine lymphoma L1210 cells and human epidermoid carcinoma KB cells with IC_{50} value of 3.2 and 6.0 μg/mL in vitro, respectively [116].

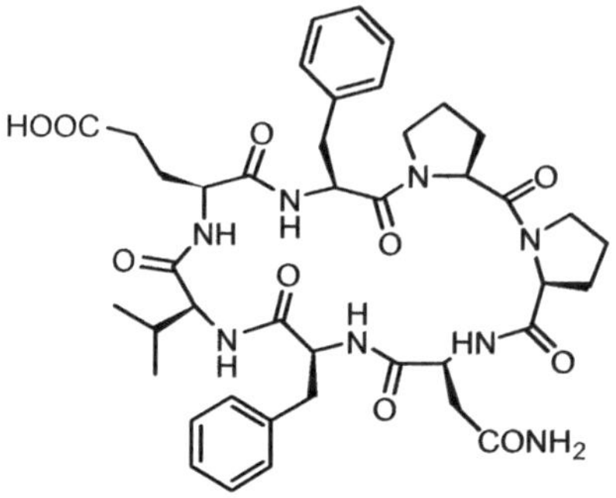

298 Hymenamide B

A cycloheptapeptide designated phakellistatin 1 (299) was isolated from two Indo-Pacific sponges, *Phakellia costata* (Truk Archipelago, 1985–1987) and *Stylotella aurantium* (Palau Archipelago, in 1985), which appeared moderate anti-tumor activity (P388 murine leukemia ED_{50} 7.5 μg/ml) [117].

In 1995, cyclic heptapeptide phakellistatin 2 (300) was isolated from the marine sponge *Phakellia carteri*, showed cell growth inhibitory activity of ED_{50} 0.34 μg/mL against the P388 lymphocytic leukemia cell line [118, 119]. Phakellistatin 4 (301) isolated from *Phakellia costata*, showed GI_{50} values of about 0.6 μM in different human cancer cell lines [119]. Phakellistatin 5 (302), a metabolite of marine sponge *Phakellia costada* collected from the Federated States of Micronesia (Chuuk), exhibited significant cell growth inhibitory activity to the P388 murine lymphocytic leukemia and the human cancer cell lines representing ovarian (OVCAR-3), CNS (SF295), lung (NCI-H460), prostate (DU-145), colon (KM20L2), and melanoma (SK-MEL-5) cancers, with GI_{50} values ranging from 0.14 to 0.74 μg/mL [120].

In 2003, cyclodecapeptide designated phakellistatin 12 (303) was isolated as a trace (1.7×10^{-6}% yield) constituent of the Western Pacific Ocean (Federated States of Micronesia-Chuuk) sponge *Phakellia* sp. with activity against P388 lymphocytic leukemia ED_{50} 2.8 μg/mL [121].

A cyclic heptapeptide phakellistatin 13 (304) isolated from the sponge *Phakellia fusca* Thiele, collected off Yongxing Island of China, in 1998, exhibited potent cytotoxicity against the human hepatoma BEL-7404 cell line with an ED_{50} < 2 μg/mL [122].

Phakellistatins 15–18 (305–309), together with five known cyclopeptides, phakellistatin 13, hymenistatin 1, and hymenamides G, H, and J, were isolated from the South China Sea sponge *Phakellia fusca*, 2007. The new cyclopeptides 74–78 were tested for cytotoxic activity in vitro. Phakellistatin 15 exhibited cytotoxicity against cancer cell line P388 with an IC_{50} value of 8.5 μM. Phakellistatin 16 showed cytotoxicity against cancer cell lines P388 and BEL-7402 with IC_{50} values of 5.4 and 14.3 μM, respectively. Phakellistatins 17 and 18 showed no cytotoxicity against the cancer cell lines P388 and BEL-7402 [123]. The synthetic cyclic peptides of phakellistatins were chemically but not biologically identical with the natural products [116–119].

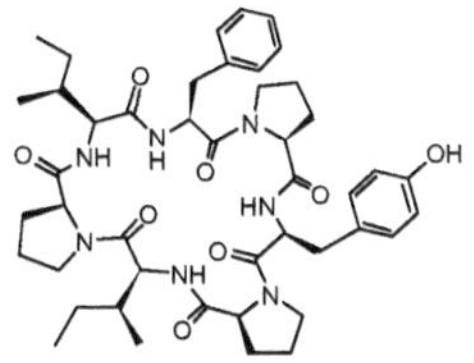

299 Phakellistatin 1

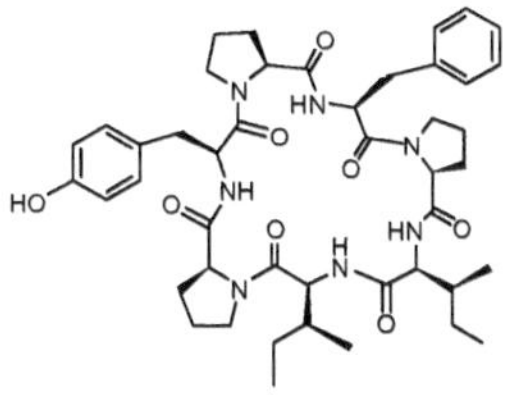

300 Phakellistatin 2

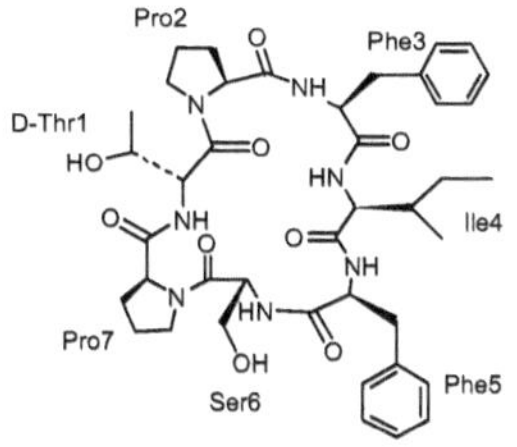

301 Phakellistatin 4

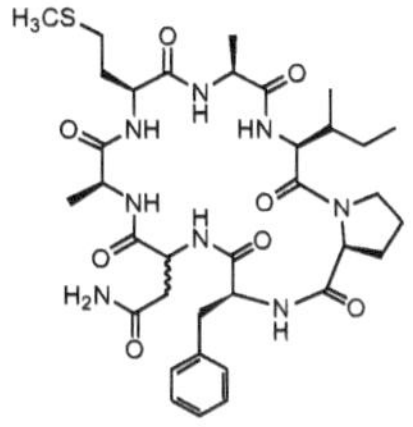

302 Phakellistatin 5

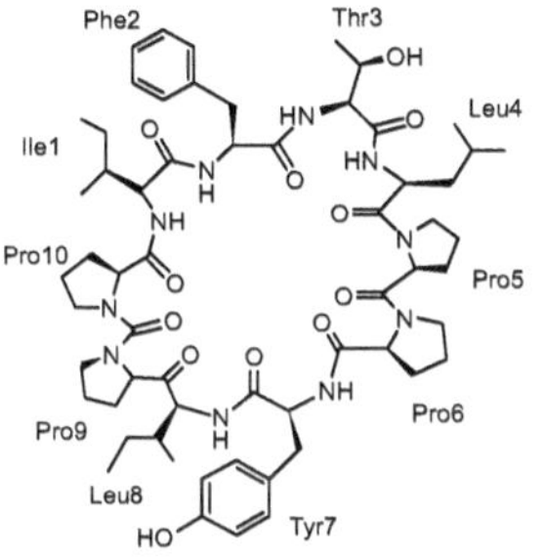

303 Phakellistatin 12

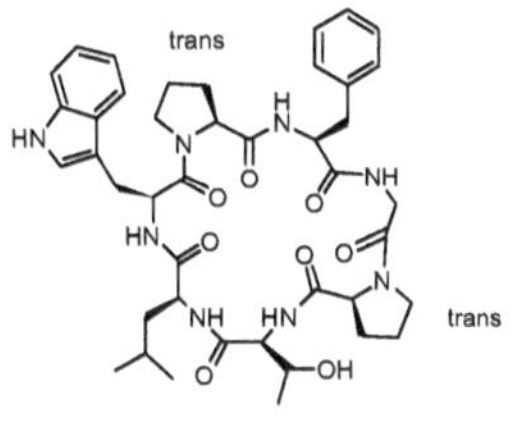

304 Phakellistatin 13

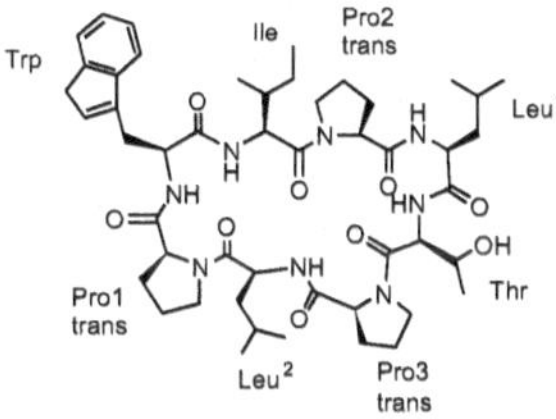

305 Phakellistatins 15

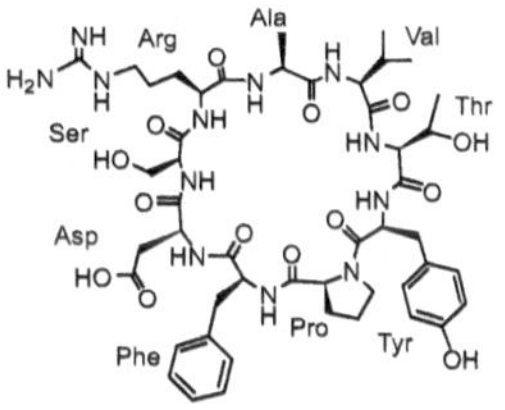

306 Phakellistatins 16 3a trans-Pro (major conformer)
307 Phakellistatins 16 3b cis-Pro (major conformer)

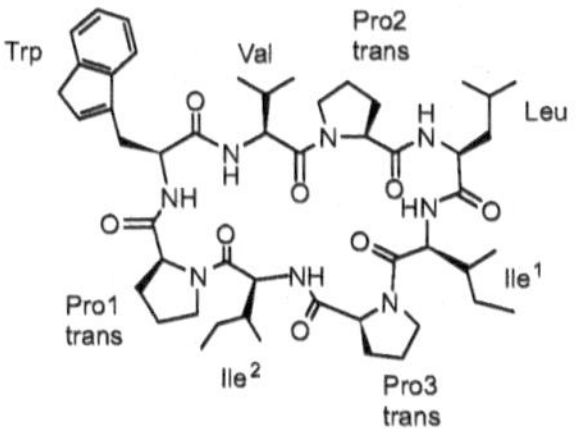

308 Phakellistatins 17

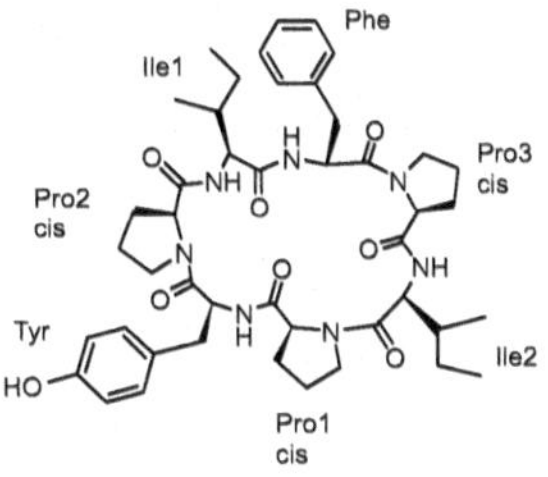

309 Phakellistatins 18

Reniochalistatins A–E (310–314) were isolated and characterized from the marine sponge *Reniochalina stalagmitis* (Yongxing Island, South China Sea, 2009). The cyclic octapeptide reniochalistatin E showed biological activity in various cytotoxicity assays employing different tumor cell lines (RPMI-8226, MGC-803, HL-60, HepG2, and HeLa), against myeloma RPMI-8226 and gastric MGC-803 cells with IC_{50} values of 4.9 and 9.7 μM, respectively, but with no activity against leukemia HL-60 and hepatoma HepG2 (IC_{50} > 20.0 μM) and cervical HeLa (IC_{50} 17.3 μM) cells [124].

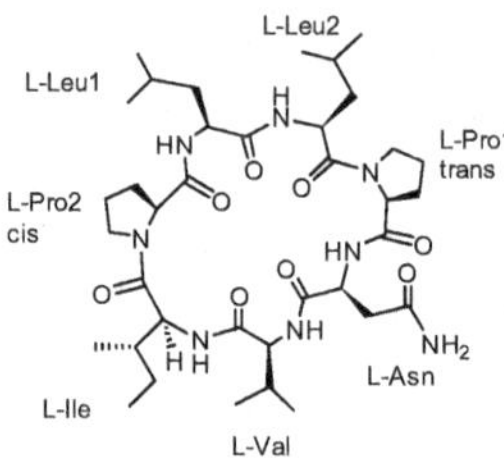

310 Reniochalistatin A

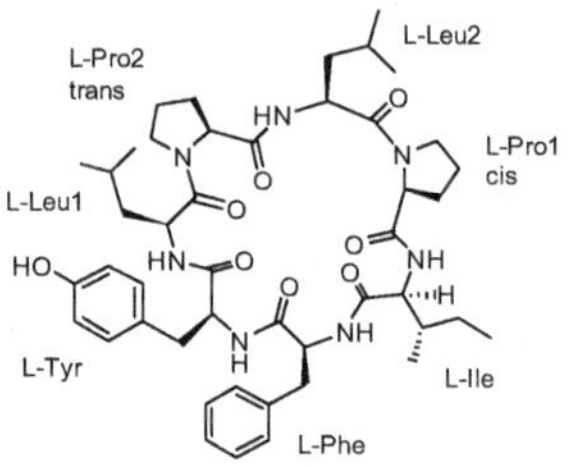

311 Reniochalistatin B

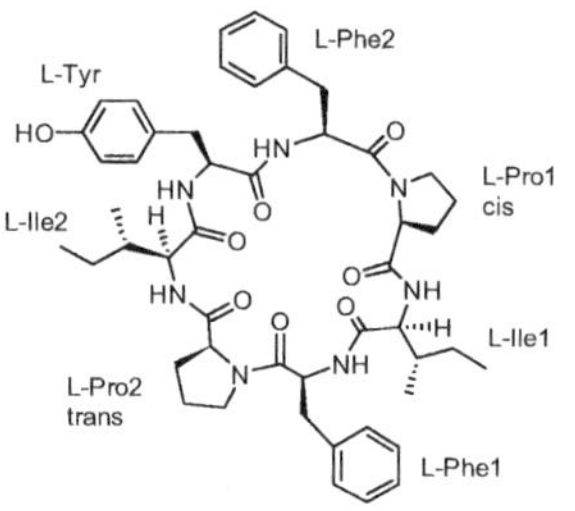

312 Reniochalistatin C

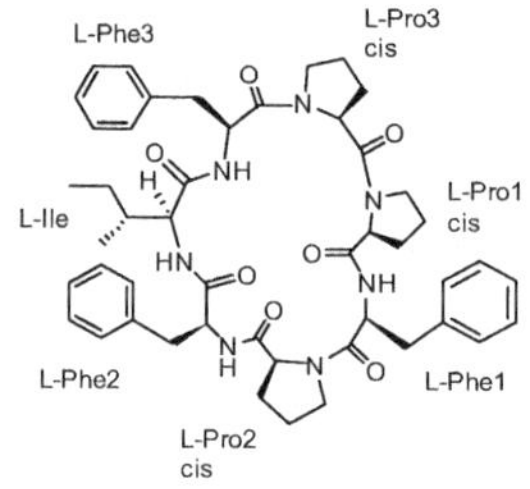

313 Reniochalistatin D

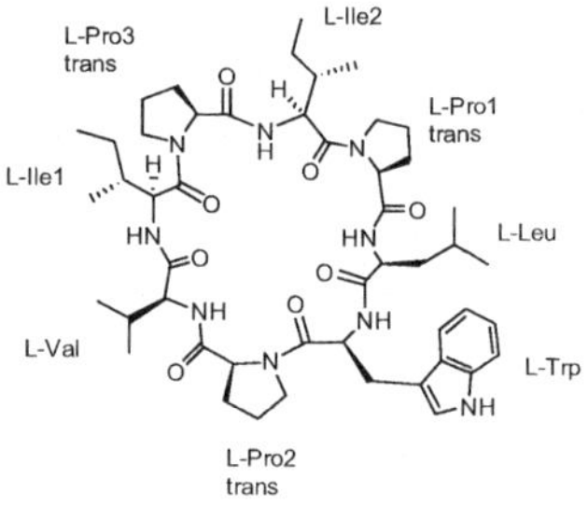

314 Reniochalistatin E

From 2008 to 2015, the chemical investigation of Indonesian sponge *Callyspongia aerizusa* collected from three different locations in Indonesia as indicated: Makassar, S. Sulawesi; Lembeh, N. Sulawesi; and Ambon, Maluku) afforded 13 cyclic peptides callyaerins A–M; callyaerins A (315) and B (316) showed potent anti-TB (*Mycobacterium tuberculosis*) activity with MIC_{90} values of 2 and 5 μM, respectively. Callyaerin A showed strong anti-TB activity, but not cytotoxic to THP-1 (the human monocytic cell line) or MRC-5 (the human fetal lung fibroblast cell line) cells (IC_{50} > 10 μM), which indicated the potential of these compounds as promising anti-TB agents. Callyaerins E (317) and H (318) exhibited strong activity against the L5178Y cell line with ED_{50} values of 0.39 and 0.48 μM, respectively. On the other hand, callyaerin A also showed strong antifungal activity toward *C. albicans*. Callyaerin G (319) was found to be cytotoxic toward the mouse lymphoma cell line (L5178Y) and HeLa cells with ED_{50}(s) of 0.53 and 5.4 μg/mL, respectively [98].

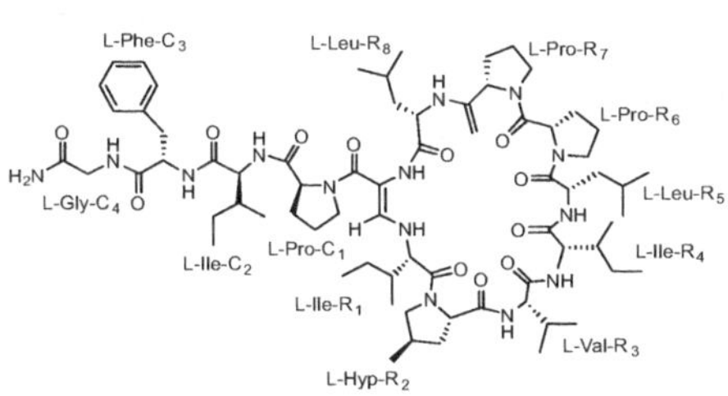

315 Callyaerin A

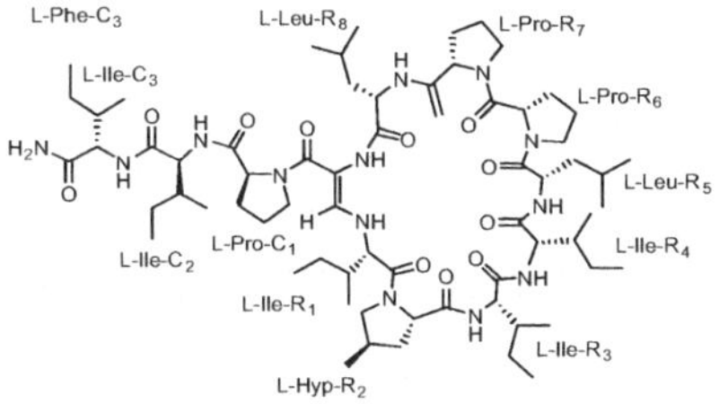

316 Callyaerin B

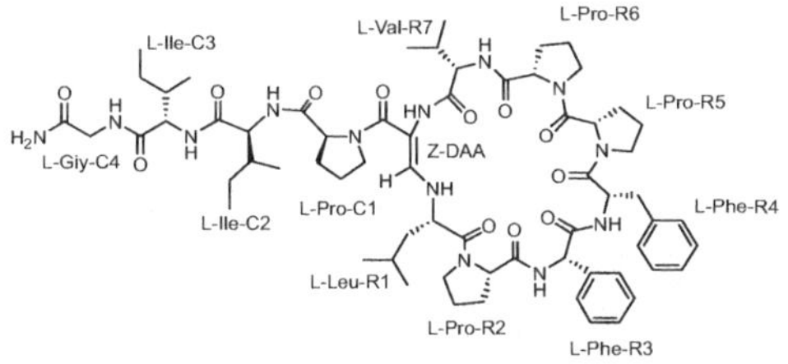

317 Callyaerin E

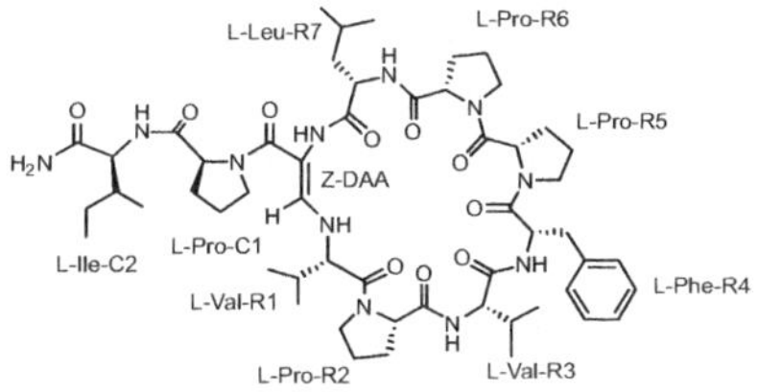

318 Callyaerin H

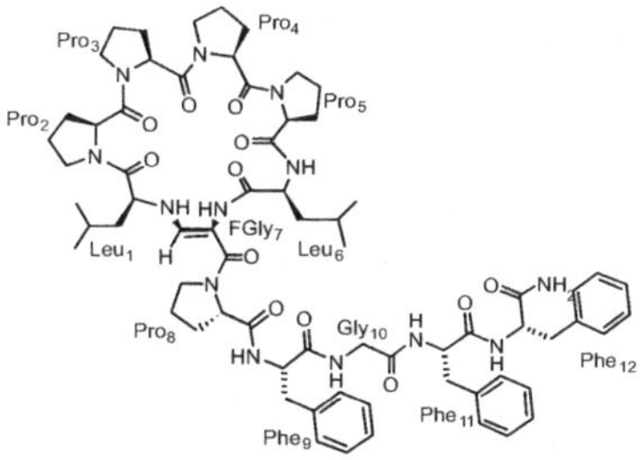

319 Callyaerin G

15.4 Bioactive Polyketides

Marine aliphatic polyketides are a class of compounds that present diverse and interesting biological properties. Due to the versatility of their biosynthetic production mechanism, these compounds exhibit remarkable diversity, both in terms of structural complexity and biological activity. Polyketides are constructed as highly oxygenated stereo chemically enriched scaffolds, sometimes with the characteristic presence of macrocyclic lactones, cyclic five- or six-membered ethers, or polyethers that act as a conformational constraint. This review summarizes the isolation, structure identification of a diverse 104 polyketides from 23 genera of marine sponges with a variety of potent biological activities within the literature coverage from 1968 to 2016.

Phormidolide A, first isolated from the cyanobacterium *Phormidium* sp., which is toxic to brine shrimp (LC_{50} = 1.5 μM) [125], together with two new cytotoxic macrolides named phormidolides B (**320**) and C (**321**), were identified from a sponge of the Petrosiidae family, collected off the coast of Pemba (Tanzania), in 2014 [126]. Cytotoxic activities tested using three human tumor cell lines, lung (A549), colon (HT29), and breast (MDA-MB-231), manifested that phormidolides B and C have significant cytotoxic activities against these three cell lines with IC_{50} around 0.5–1.4 μM [126]. The fact of original discovery from a cyanobacteria species suggested that phormidolide A (**322**) is actually metabolite synthesized by symbiotic cyanobacterium of the sponge, supporting the relevance of symbiotic bacteria as sources of bioactive polyketides and peptides in sponges.

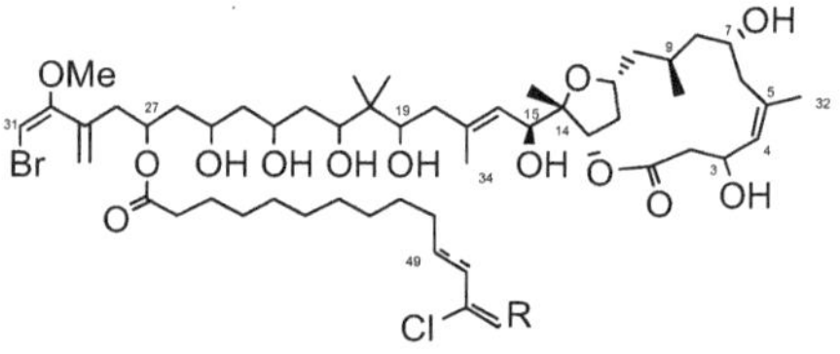

320 Phormidolide B R=Cl ¦¤^{49}
321 Phormidolide C R=Br

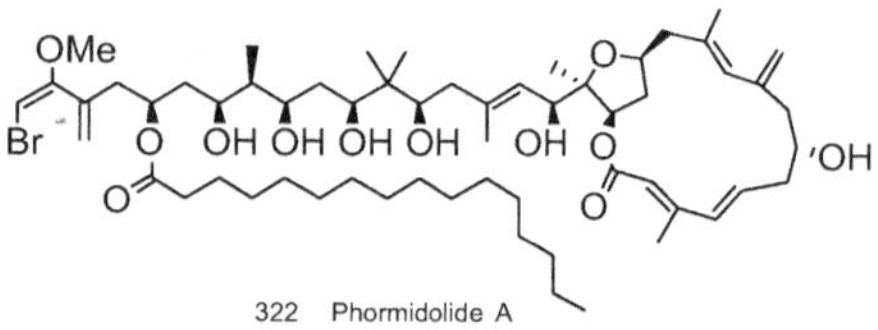

322 Phormidolide A

Plakilactones and gracilioethers are oxygenated polyketides of the plakortin family isolated from the marine sponge *Agelas gracilis*, collected in southern Japan, in 2009 [127] and *Plakinastrella mamillaris*, collected at the Fiji Islands, in 2012–2013 [128–130]. In bioassay-guided fractionation of the lipophilic sponge *Agelas gracilis* extract, Fusetani et al. obtained three new antimalarial compounds against *Plasmodium falciparum*, gracilioethers A–C (**323–325**), with IC_{50} values of 0.5–10 μg/mL, whereas gracilioether B also showed antileishmanial activity [127]. A few years later, the Zampella group isolated several plakilactone- and gracilioether-polyketides, together with the previously known gracilioethers A–C compounds from another sponge *Plakinastrella mamillaris*. Among them, gracilioether B, gracilioether C, and plakilactone C (**326**) demonstrated activation of PPARγ in a dose-dependent manner with relative EC_{50} values of ≈ 5, 10, and 2 μM, respectively, and further mechanism study demonstrated that gracilioether B and plakilactone C covalently bind to the PPARγ substrate domain through a Michael addition reaction involving a cysteine residue and the *α*,*β*-unsaturated ketone group in their side chains, whereas gracilioether C is a noncovalent agonist for PPARγ [128]. In additon, gracilioether H (**327**) inhibited chloroquine-resistant CR FC29 strain in vitro with the antiplasmodial activity of IC_{50} 3.26 μM [129].

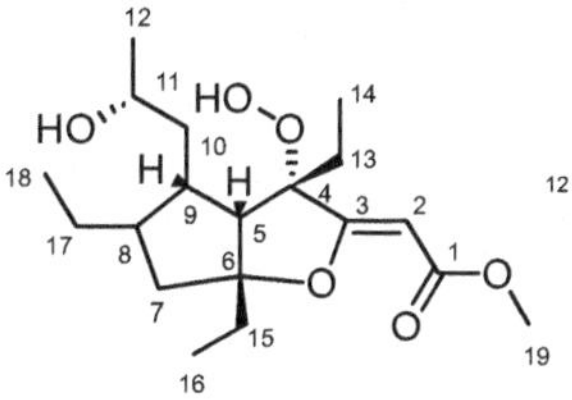

323 Gracilioether A

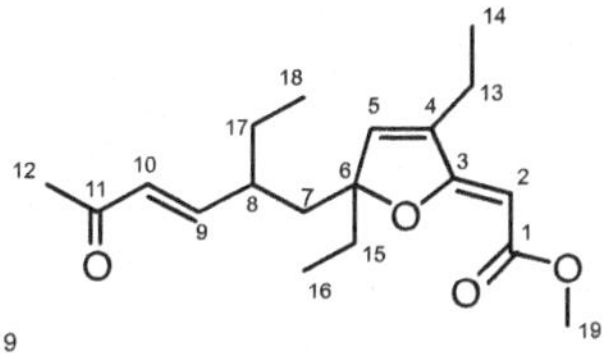

324 Gracilioether B

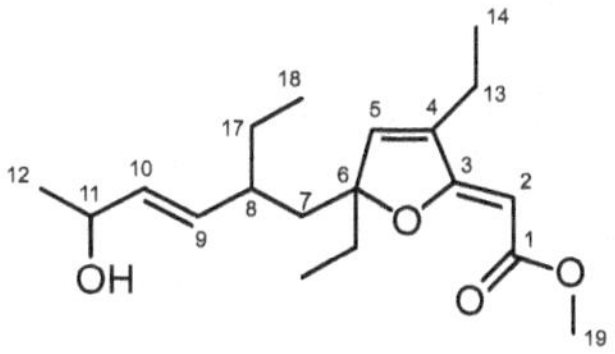

325 Gracilioether C

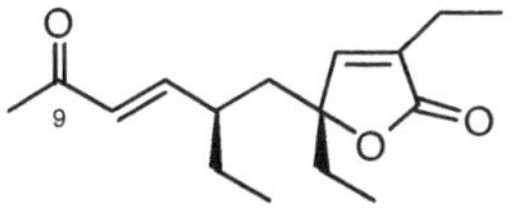

326 Plakilactone C

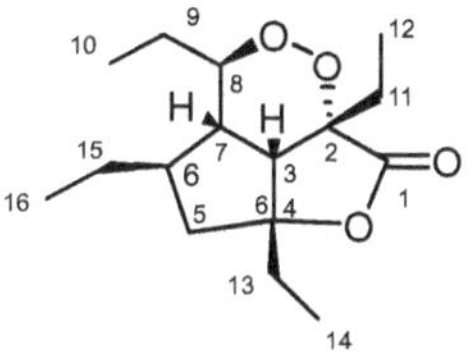

327 Gracilioether H

Smenamide A (**328**) and B (**329**), hybrid peptide/polyketide compounds consisting of a dolapyrrolidinone unit isolated from a Caribbean sponge *Smenospongia aurea*, are collected by SCUBA along the coast of Little Inagua (Bahamas Islands), in 2013 [131]. Structures of smenamides revealed the products of the cyanobacterial metabolism, and 16S rRNA metagenomic analysis detected *Synechococcus spongiarum* as the only cyanobacterium present in *S. aurea*. Smenamides A and B show potent cytotoxic activity at nanomolar levels on lung cancer Calu-1 cells with IC_{50} values of 48 nM and 49 nM, respectively. The clear pro-apoptotic mechanism of action of smenamide A makes smenamides promising results to antitumor drug design.

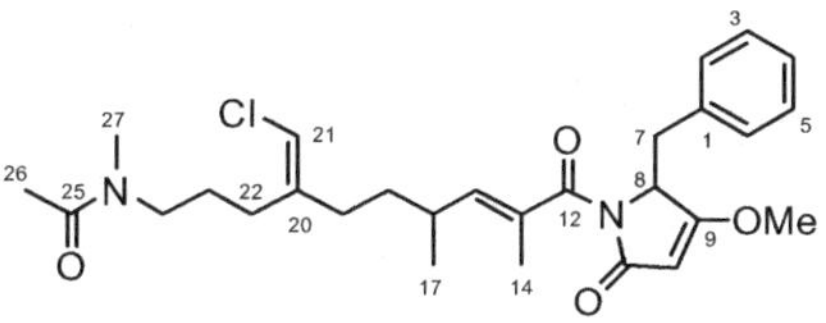

328 Smenamide A

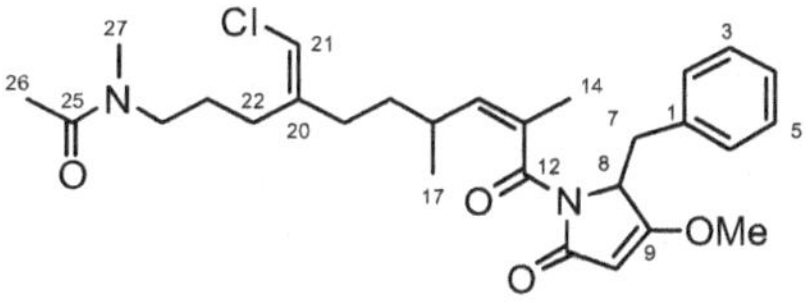

329 Smenamide B

Plakortide R–U (**330–333**), endoperoxide polyketides isolated from the marine sponge *Plakinastrella mamillaris*, collected at Fiji Islands, in 2013 [132]. Pharmacological analysis demonstrated that plakortide U showed the best antiplasmodial activity in vitro against chloroquine-resistant FcM29 strain (IC_{50} 0.80 μM), while the remaining compounds showed a moderate antiplasmodial activity (IC_{50} range: 5–50 μM).

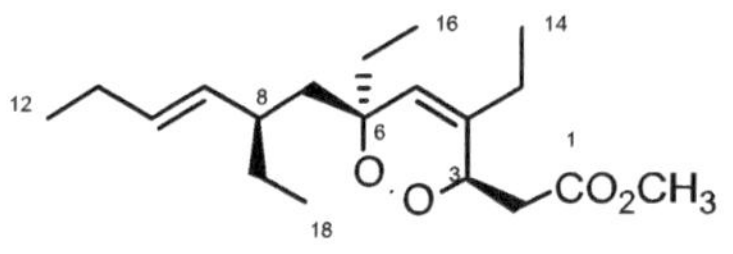

330 Plakortide R

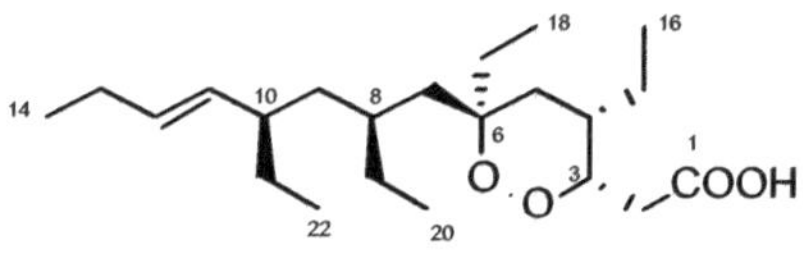

331 Plakortide S

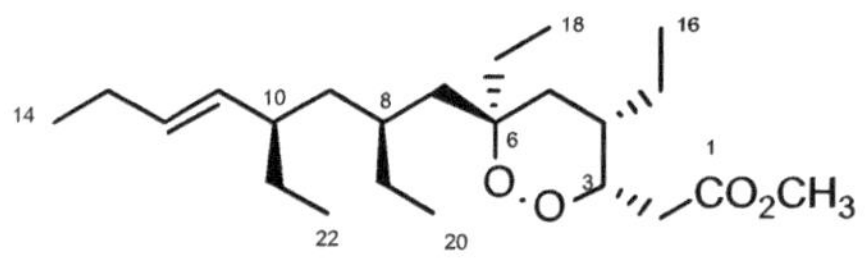

332 Plakortide T

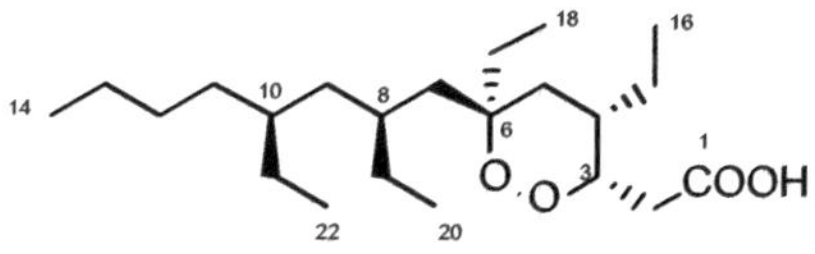

333 Plakortide U

Manzamenones L–N (**334–336**), dimeric fatty-acid derivatives, consisting of an octahydroindenone with three carboxy groups and two hexadecanyl chains, isolated from an Okinawan marine sponge of the genus *Plakortis*, collected in Okinawan, in 2012 [133]. Antimicrobial activity tests against several bacteria and fungi showed that manzamenone M had moderate antimicrobial activities against *Escherichia coli*, *Staphylococcus aureus*, *Candida albicans*, and *Cryptococcus neoformans* (MIC or IC_{50}, 8–32.0 μg/mL), and manzamenone L did not exhibit activity (MIC or IC_{50}, >32.0 μg/mL). From the point of view of structure-activity relationships between manzamenone L and manzamenone M, a free carboxylic acid at C-5 position might be important for the activities. Manzamenone N showed moderate antimicrobial activities against *E. coli*, *C. albicans*, and *C. neoformans* (MIC or IC_{50}, 4–32.0 μg/mL).

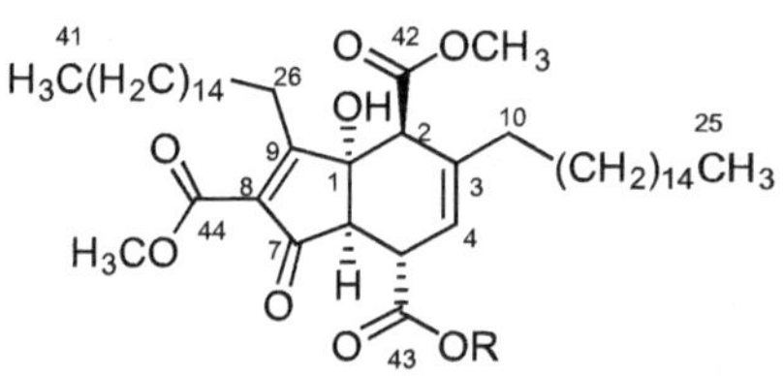

334 Manzamenones L R=CH_3
335 Manzamenones M R=H

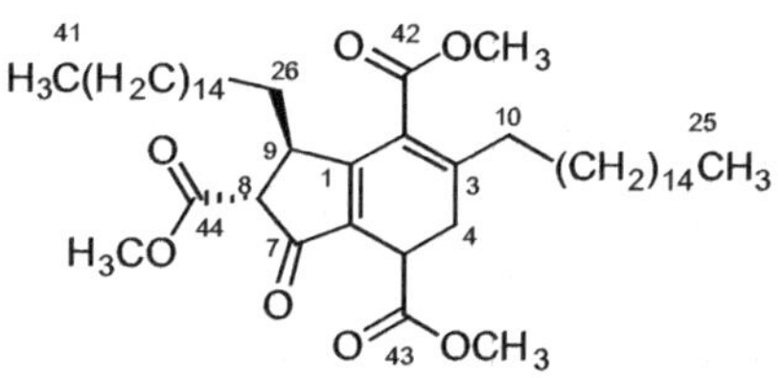

336 Manzamenones N

Tedanolide macrolides, which includes tedanolide (**337**) [134] isolated from *Tedania ignis*, collected in Caribbean, in 1984; 13-deoxytedanolide [135] isolated from *Mycale adhaerens*, collected by SCUBA off Hiburi Island (−10 to −15 m) of the Uwa Sea, 750 km southwest of Tokyo, in 1991; tedanolide C (**338**) [136] isolated from *Ircinia* sp., collected in Milne Bay (S 10°14.278′ E 150°54.782′), Papua New Guinea, in 2006; and precandidaspongiolides A (**339**) and B (**340**) and candidaspongiolides A (**341**) and B (**342**) from *Candidaspongia* sp., collected in Papua New Guinea, in 2011(isolated as an inseparable mixture of two isomers in equilibrium) [137], exhibited highly cytotoxic in the subnanomolar to nanomolar range against various cancer cell lines. Cell-flow cytofluorometry analysis revealed that tedanolide caused accumulation of cells in the S phase at concentration as low as 0.01 μg/mL [134]. 13-Deoxytedanolide (**343**) showed remarkable cytotoxicity against P388 murine leukemia cells with IC_{50}'s of 94 pg/mL. 13-Deoxytedanolide also performed highly in vivo antitumor activity against P388: *T*/*C* = 189% at a dose of 0.125 mg/kg [135]. Tedanolide C exhibited potent cytotoxicity against HCT-116 cells in vitro with IC_{50} value of 9.53×10^{-8} M and caused a strong S-phase arrest [136]. Precandidaspongiolides A and B showed excellent selectivity against melanoma cell lines in the NCI 60-cell line screen, and the LC_{50} values for precandidaspongiolides A/B against melanoma cell lines were significantly lower than other tumor cell lines (seven of the nine melanoma cell lines in the panel had nanomolar LC_{50} values around 19–174 nM); further, precandidaspongiolides A/B were evidenced as P-gp substrates [137]. Studies of SARs of 13-deoxytedanolide [135], precandidaspongiolides A and B, and candidaspongiolides A and B [137] reported that the southern hemisphere of 13-deoxytedanolide comprised the pharmacophore and the epoxide-bearing side chain of 13-deoxytedanolide was essential for the activity [138]; the hemiketal of precandidaspongiolide B and candidaspongiolide B was not essential for the activity, while potency was affected when the primary alcohol of precandidaspongiolides A and candidaspongiolide A was substituted, and the C-7 acetylation of candidaspongiolides A and B increased potency [137]. The similarities between the myriaporones [139] and the candidaspongiolides [137] afforded further evidence in support of microbial symbionts as producers of candidaspongiolides, 13-Deoxytedanolide [140] and candidaspongiolide A [141]. However, the underlying reasons for the candidaspongiolides' melanoma selectivity have yet to be determined.

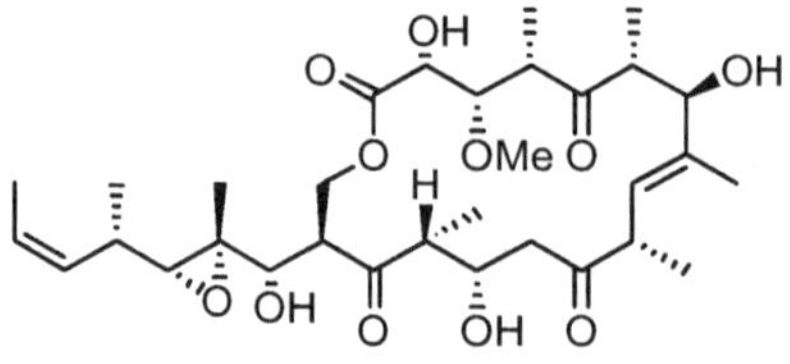

337 Tedanolide

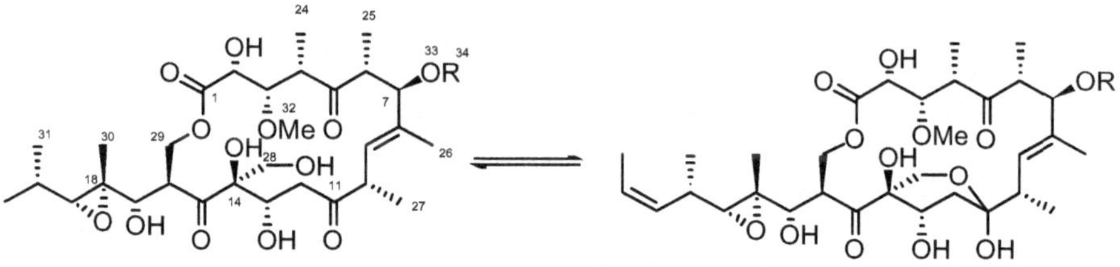

338 Precandidaspongiolides A R=H
339 Precandidaspongiolides B R=Ac

340 Candidaspongiolide A R=H
341 Candidaspongiolide B R=Ac

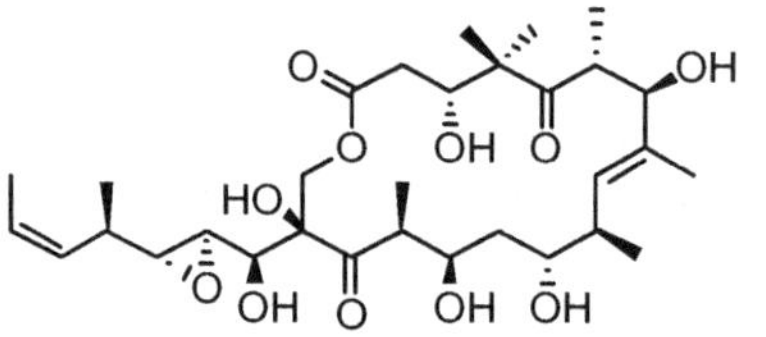

342 Tedanolide C

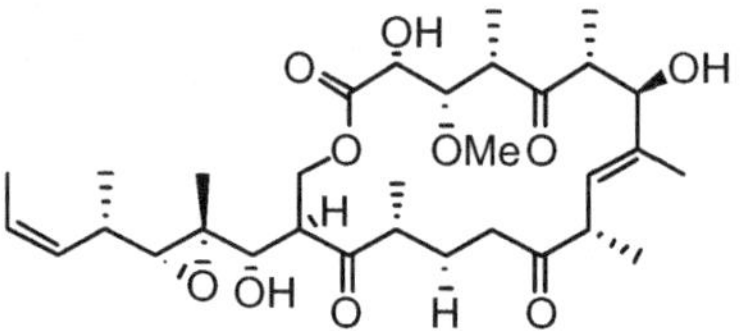

343 13-deoxytedanolide

Lehualides are polyketide derivatives isolated from sponge of the genus *Plakortis*, which have a long alkyl chain with varying degrees of saturation and incorporate α- or γ-pyrone moieties, coupled with thioacetate or thiol functionalities [142, 143]. Lehualide B (**344**) showed moderate cytotoxicity in vitro against an ovarian cancer cell line (IGROV-ET) with a GI_{50} value of 0.83 μM. Lehualide D (**345**) exhibited moderate cytotoxicity to ovarian (IGROV-ET) and leukemia (K562) cell lines with GI_{50} values of 0.73 and 0.23 μM, respectively (the sample of sponge was collected from waters between Lehua Rock and Niihau Island, Hawaii, in July 2003) [143]. Lehualides F (**346**) and G (**346**) exhibited IC_{50} values for cytotoxicity against the human promyelocytic leukemia (HL-60) cell line of 6.2 and 5.4 μM, respectively (the *Plakortis* sponge specimen was collected from a cave off the coast of 'Eua Island, Tonga) [142].

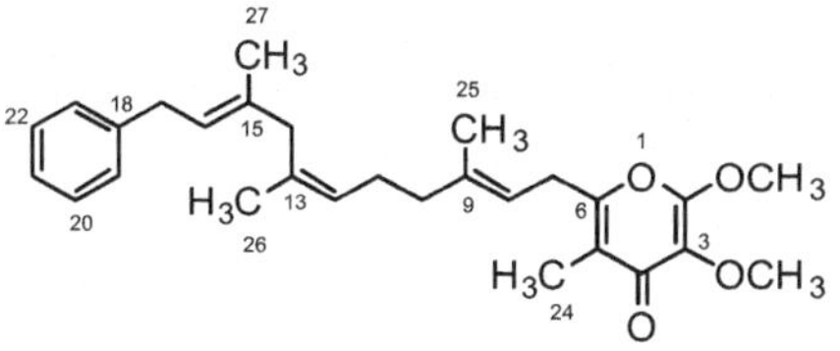

344 Lehualide B

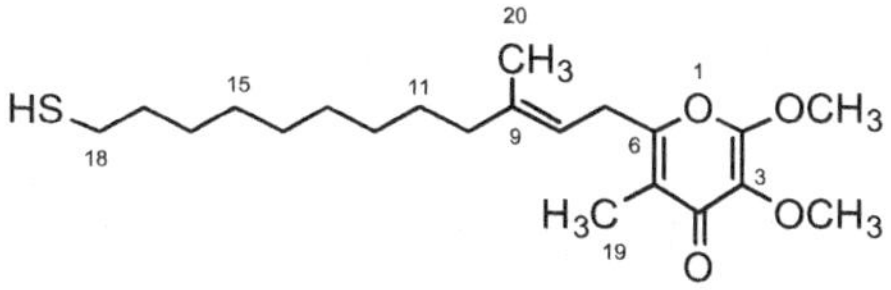

345 Lehualide D

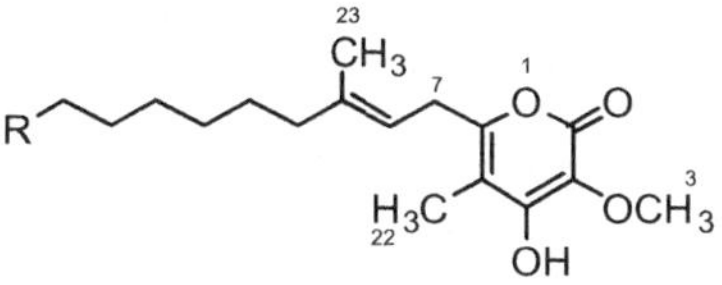

346 Lehualides F R= 

347 Lehualides G R= 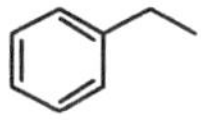

Franklinolides A–C (**348–350**) are the first examples of polyketide phosphodiesters isolated from an aqueous EtOH extract of a sponge sample CMB-01989, collected during deepwater (−105 m) scientific trawling operations in the Great Australian Bight, a massive *Geodia* sp. thinly encrusted with a *Halichondria* sp., in 2010 [144]. SAR studies, using in vitro cytotoxicity and cell proliferation assays against stomach (AGS), colon (HT29), and human brain (SH-SY5Y) cancer cell lines, and a noncancerous control cell line, demonstrated that franklinolides A was the dominant cytotoxic agent (GI_{50} range from 0.1 to 0.3 μM). SAR analysis defined the relative importance of key structural features: (1) a very significant 30- to >300-fold decrease in cytotoxicity following hydrolysis of franklinolides A to bitungolide A [145] (the de-3-*O*-methyl-2-phosphoglyceric acid derivative of franklinolides A); (2) 2- to 7-fold decrease on isomerization of franklinolides A to the 12*E*,14*E* isomer franklinolides B; and (3) a 30- to 50-fold decrease on isomerization of franklinolides A to the 12*E*,14*Z* isomer franklinolides C.

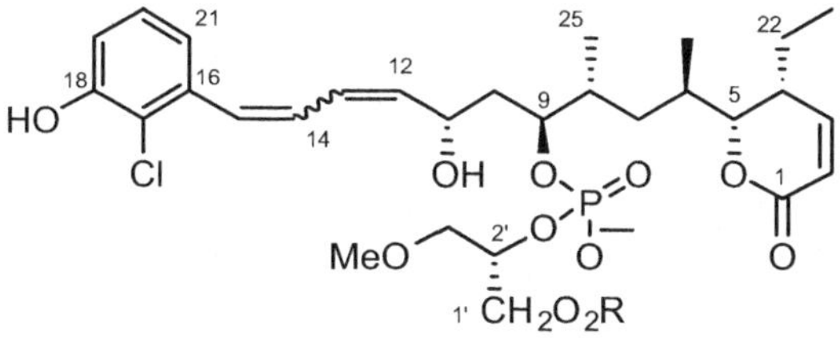

348 Franklinolides A 12*Z*,14*Z* and R=H
349 Franklinolides B 12*E*,14*E* and R=H
350 Franklinolides C 12*E*,14*Z* and R=H

Halenaquinone-type polyketides (**351–358**) were isolated from marine sponges of the genus *Xestospongia* (collected in the South Pacific [146], in the Benga Lagoon, Fiji Islands [147], in Fiji (coll. Nos. 89,109 and 91,007) or Vanuatu (coll. no. 90033) [148], in Kerama Islands, Okinawa [149] and in Okinawan [150]) and *Adocia* (collected from the Eten Island area of Truk Lagoon in January 1984 and November 1985 at 5–10 m depths [151] and Manado, Indonesia, on May 15, 1993 [152]), which showed a number of biological activities [146, 147, 151, 152, 148–150]. In vitro assays and preliminary SAR studies (in 2010) showed that among the compounds 1–8, halenaquinone appeared as the most PLA_2 inhibitor of the series with an IC_{50} of 3.7 μM. Incorporating a dioxothiazine unit in compounds 2 and 3 led to a 30- to 40-fold decrease in potency when compared with the most active pentacyclic polyketide 1, and among pentacyclic polyketide compounds, the presence of a secondary alcohol at C-3 rather than a ketone abolished PLA_2 inhibition (4 and 5). These results highlighted that the C-3 oxidation state and the presence of quinone ring E were important for the anti-PLA_2 activity [27, 146]. Similar conclusions were reported regarding the SAR of protein tyrosine kinase inhibition exhibited by halenaquinone (1) [147]. Farnesyltransferase (FTase) inhibitory experiments showed that the presence of dioxothiazine substitution (IC_{50} 1 1.57 μM vs. 2 1.48 μM and 3 3.75 μM) led to little variation in human (FH) FTase inhibitory activity. Interestingly tetrahydrohalenaquinones A (4) and B (5) did not show such activity. Furthermore, quinol sulfates 6 (IC_{50} 16.11 μM) and 7 (IC_{50} 6.71 μM) exhibited modest activity suggesting that the presence of a quinone moiety was essential for FTases inhibitory activity. Sub-micromolar inhibition of farnesyltransferase enzyme of orhalquinone 8 (IC_{50} 0.40 μM), highlighting this scaffold as a significant modification for enhancing activity [146]. Antiplasmodial activities against FcB1 and 3D7 *Plasmodium falciparum* strains revealed that compound 2, 3, and 8 were the most active of the series with values of IC_{50} 1.08, 3.89, and 9.22, respectively.

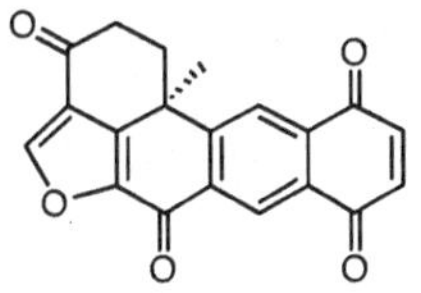

351 Halenaquinone

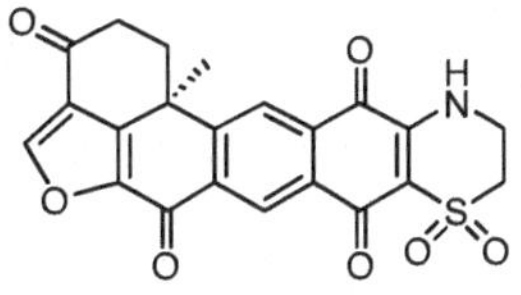

352 3-Ketoadociaquinone A

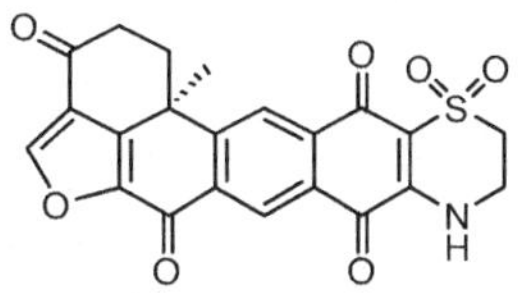

353 3-Ketoadociaquinone B

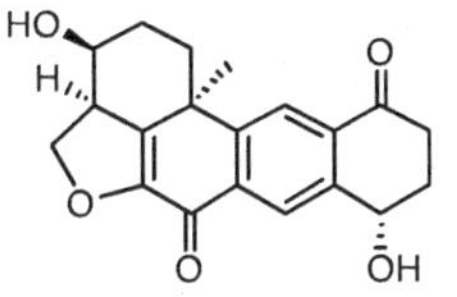

354 Tetrahydrohalenaquinone A

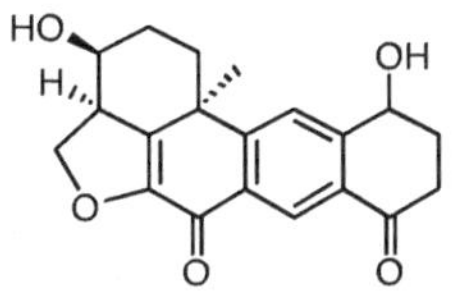

355 Tetrahydrohalenaquinone B

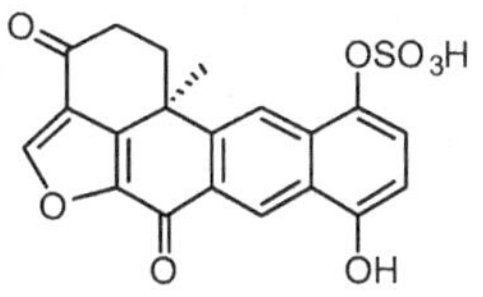

356 Halenaquinol sulfate

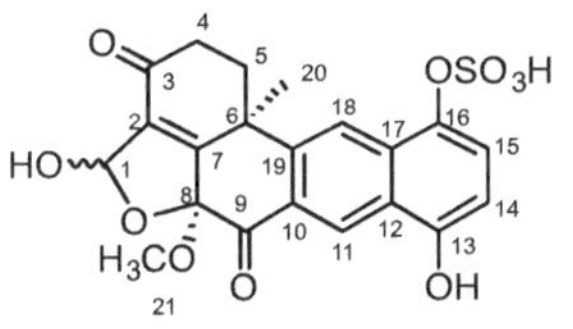

357 Xestosaprol C methylacetal

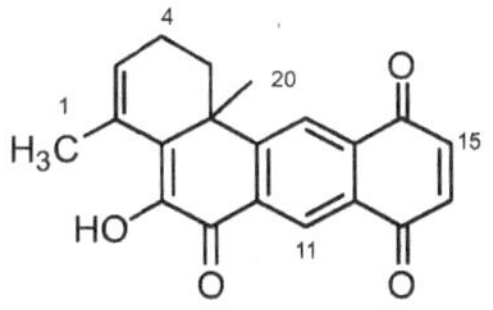

358 Orhalquinone

Marine sponges from family Plakinidae own a great number of simple endoperoxide or peroxyketal polyketides possessing five- or six-membered 1,2-dioxygenated rings (1,2-dioxolane or 1,2-dioxane, respectively). Plakortin (**359**), dihydroplakortin (**360**), 3-epiplakortin (**361**), and plakortide Q (**362**) isolated from marine sponges of *Plakortis halichondroides* (collected at Hookers Reef, Panama) in 1978 [152] and *P. simplex* (collected at the Caribbean Sea) in 1999 [153]. All compounds exhibited a strong in vitro antimalarial activity against D10 (chloroquine-sensitive) and W2 (chloroquine-resistant) strains of *Plasmodium falciparum*, with a more potent activity on the W2 strain (IC_{50} ~ 180 ng/mL), lacking of cytotoxicity [154]. Plakortide I is purified from an unidentified sponge of the genus *Plakortis* and collected at Discovery Bay, Jamaica, in 2002 and represented the first report of an endoperoxide with an α,β-unsaturated ketone moiety in the "western" alkyl side chain and exhibited significant antimalarial activity against the W2 strain with an IC_{50} value of 570 ng/mL and a selectivity index of >8.4 [155]. Plakortides M (**363**) and N (**364**) were isolated from the sponge *P. halichondroides*, collected in Puerto Rico, in 2003, which exhibited potent cytotoxic activity against a number of cancer cell lines in the NCI human cancer screening program but with less selectivity [156]. Peroxyketal polyketides peroxyplakoric A_3 (**365**) and B_3 (**366**) esters isolated from *Plakortis* sp., collected at Zamami Island, Okinawa Prefecture, in 1993, showed IC_{50} = 50 ng/mL against *P. falciparum* with a selective toxicity index (about 200) [157]. SAR studies played crucial roles in both "western" alkyl side chain and the conformational behavior of the dioxane ring of these compounds according to the interaction with the Fe(II)-heme [158, 159].

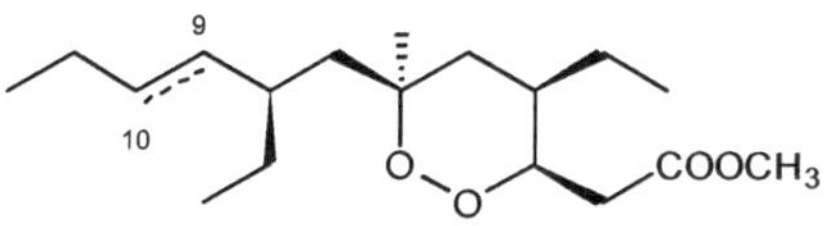

359 Plakortin ¦¤^{9,10}
360 Dihydroplakortin

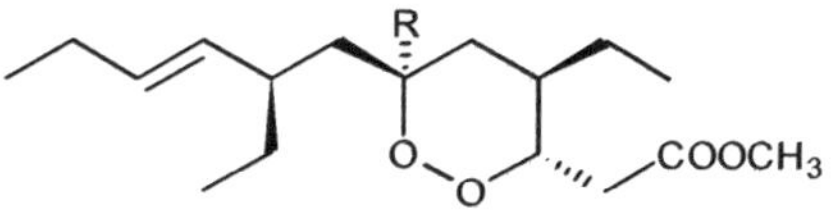

361 3-Epiplakortin R=H
362 Plakortide Q R=CH_2CH3

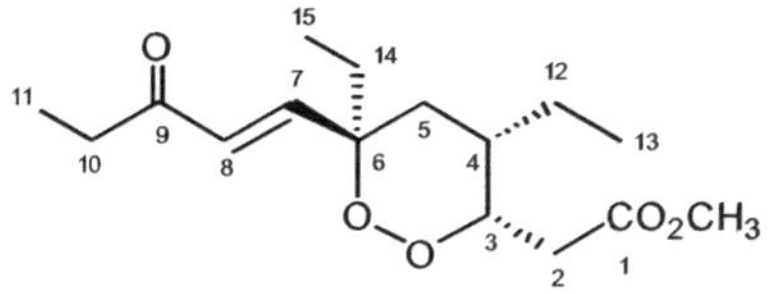

363 Plakortide I

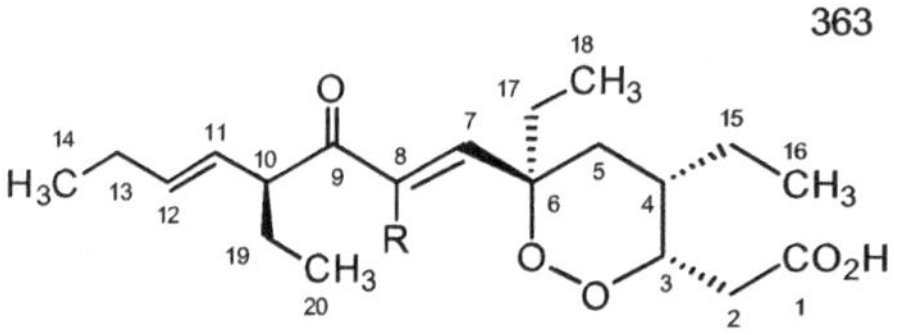

364 Plakortides M R=H
365 Plakortides N R=CH_2CH_3

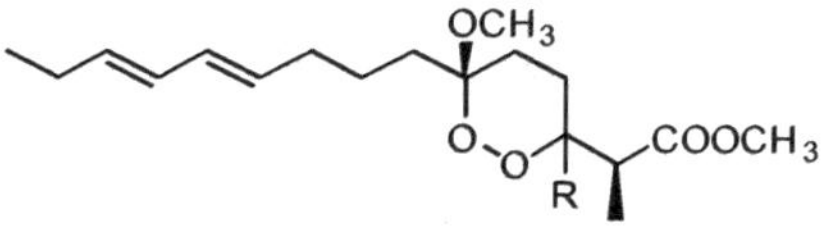

366 peroxyplakoric A_3 R= ···H
367 peroxyplakoric B_3 R= ◄H

Pederins, incorporating the pederin skeleton, which have now been isolated from five genera of the marine sponges: *Mycale* sp. and *Stylinos* sp. (order Poecilosclerida), *Trachycladus* sp. (order Axinellida), and *Theonella* sp. and *Discodermia* sp. (order Lithistida) [44–7 {Clardy, J.; He, H. U. S. Patent 1995, 5,476,953}] [160–171]. It is remarkable to note that the first compound of this class of toxic polyketides, pederin, was isolated in 1953 from the beetle *Paederus fuscipes* [161]. The presence of pederin class of polyketides in such taxonomically distinct organisms indicated the possible microbial origin of these compounds. Pederins exhibited multiple interesting pharmacological activities. Pederin (**368**), the chemical defense agent of the blister beetle, and mycalamides A (**369**) and B (**370**) were reported to disrupt protein synthesis [161, 162, 164]. Furthermore, mycalamides A and B together with mycalamides C (**371**) and D (**372**) showed potent activity against the P388 murine leukemia cell line, giving IC_{50} values of 3.0, 0.7, 95.0, and 35.0 ng/mL, respectively [164, 168]. Theopederins A–L (**373–384**) were markedly cytotoxic against P388 murine leukemia cells with the activity of nM level [165, 167, 171]. 13-Des-*O*-methyl-onnamide A (**385**), dihydroonnamide A (**386**), onnamide B (**387**), 17-

oxoonnamide B (**388**), onnamide C (**389**), onnamide D (**390**), and onnamide A (**391**) were highly cytotoxic against the P388 cell line with IC_{50} values of 0.15, 0.04, 0.13, 0.10.0.07, 0.02, and 0.01 μg/mL, respectively [166]. Onnamide F (**392**) was active against fungi *Saccharomyces cerevisae* with a value of LD_{99} 1.4 μg/mL. No activity was observed against *Bacillus subtilis* or *Eschericha coli*, indicating a selective toxicity for eukaryotes [170]. Icadamides A (**393**) and B (**394**) were reported to show in vitro cytotoxicity against HCT-116 human colon carcinoma cell line with IC_{50} value of 63 nM and 0.17 nM, respectively. Among pederin type of compounds, only icadamide B (**395**) was studied for in vivo antitumor activity and exhibited activity against intraperitoneally and subcutaneously implanted tumors such as P388 mouse leukemia, M109 mouse lung tumors, and asbestos-induced pulmonary squamous cell carcinoma. Icadamide C displayed potent cytotoxic activity against a small panel of five human solid tumor cell lines (A549, SK-OV-3, SK-MEL-2, XF-498, and HCT-15) with ED_{50} values of less than 0.1 μg/mL [160].

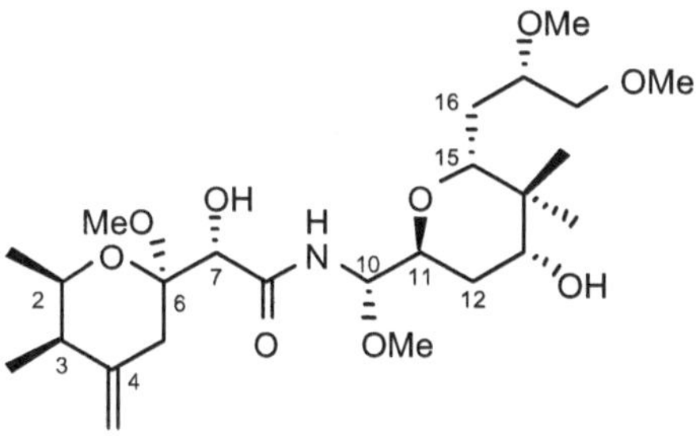

368 Pederin

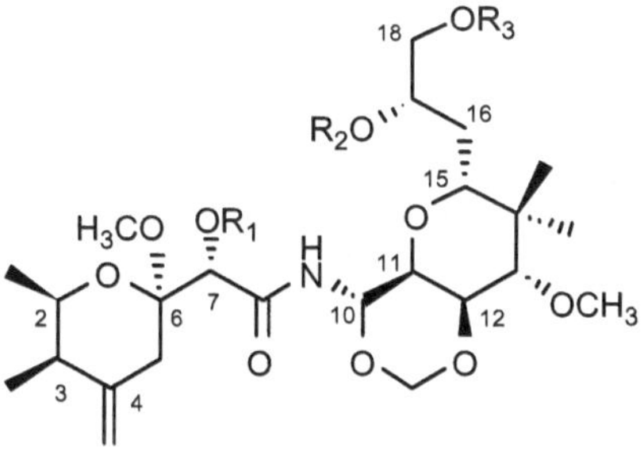

369 Mycalamides A R1=R2=R3=H
370 Mycalamides B R1=R3=H,R2=CH3

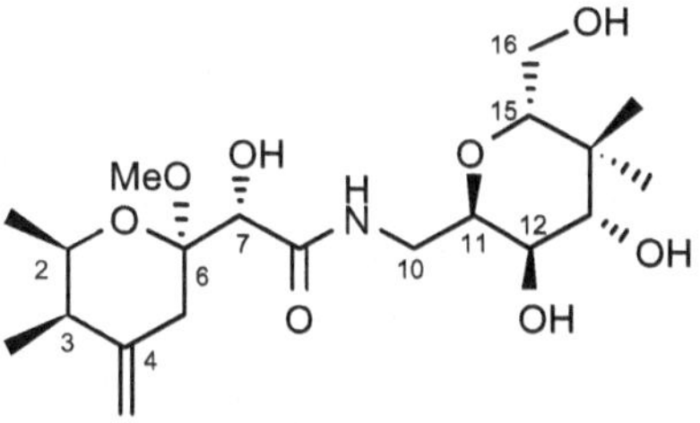

371 Mycalamides C

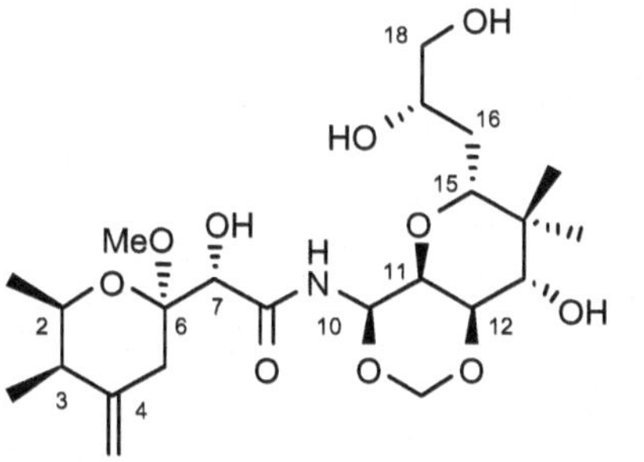

372 Mycalamides D

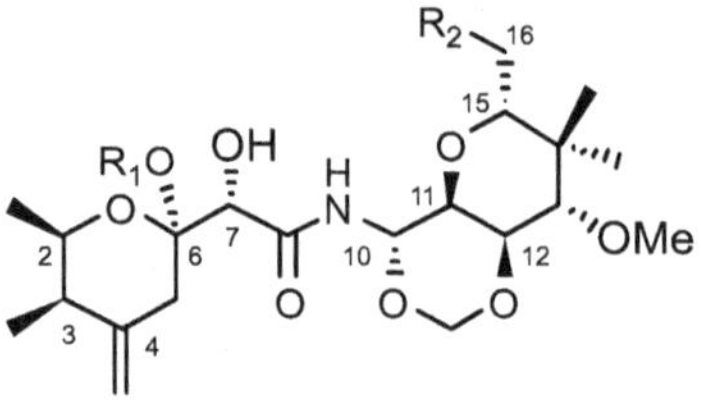

373 Theopederins A R_1=Me, R_2= 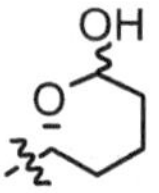

374 Theopederins B R_1=Me, R_2= 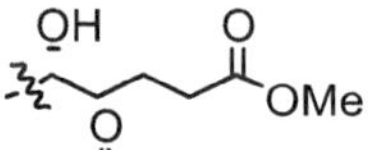

375 Theopederins C R_1=Me, R_2= 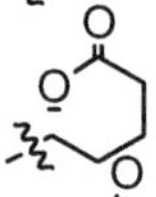

376 Theopederins D R_1=Me, R_2= 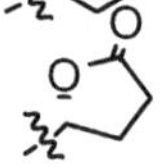

377 Theopederins E R_1=Me, R_2= OH

378 Theopederins F R_1=Me, R_2= 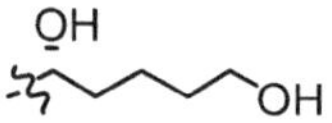

379 Theopederins G R_1=H, R_2= 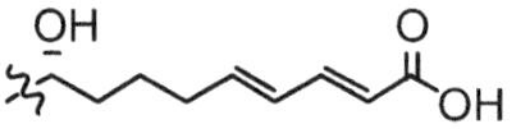

380 Theopederins H R_1=H, R_2= 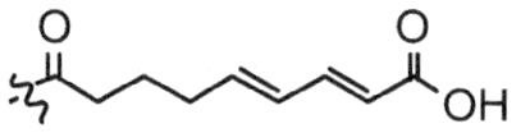

381 Theopederins I R_1=H, R_2= 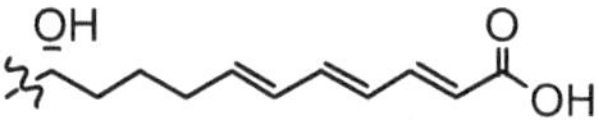

382 Theopederins J R_1=H, R_2= 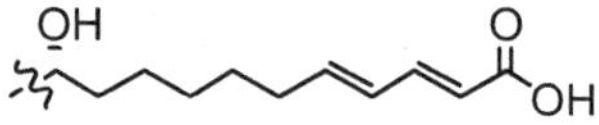

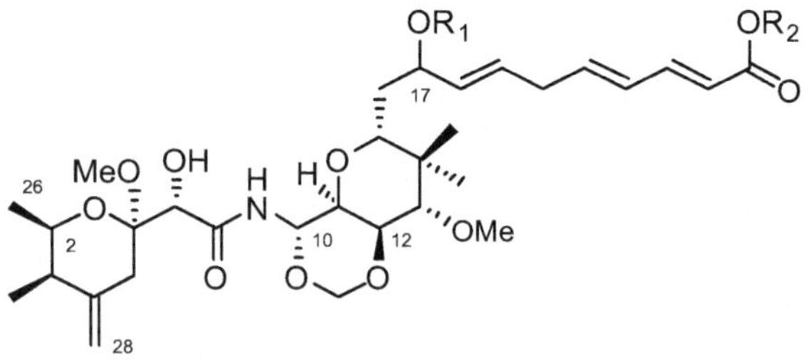

383 Theopederins K R_1=Me R_2=H
384 Theopederins L R_1=H, R_2=H

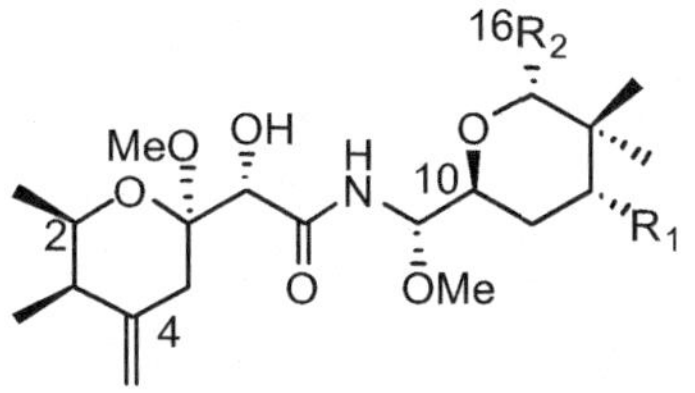

391 Onnamide A R1=OH, R2= 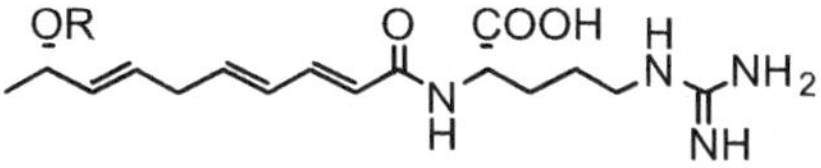

392 Onnamide F R1=OMe, R2= 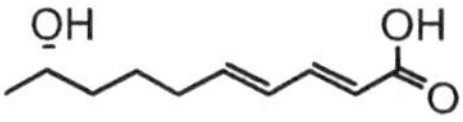

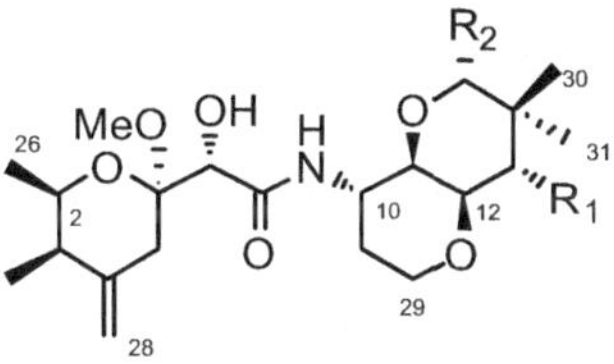

385 13-Des-O-methyl-onnamide A R_1=OH, R_2= 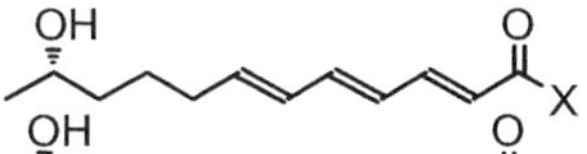

386 Dihydroonnamide A R_1=OMe, R_2= 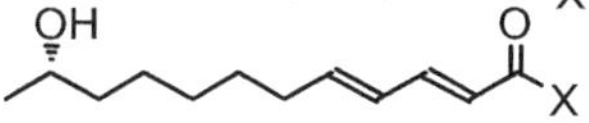

387 Onnamide B R_1=OMe, R_2= 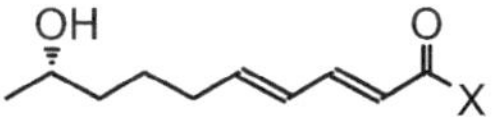

388 I7-Oxoonnamide B R_1=OMe, R_2= 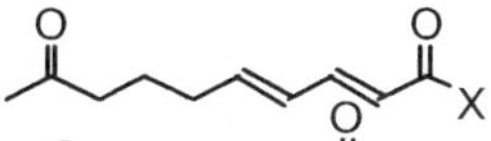

389 Onnamide C R_1=OMe, R_2= 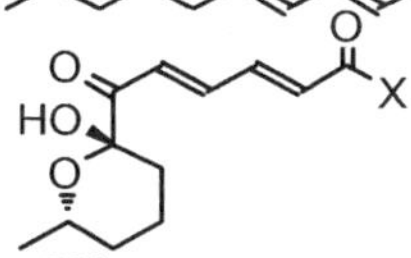

390 Onnamide D R_1=OMe, R_2= 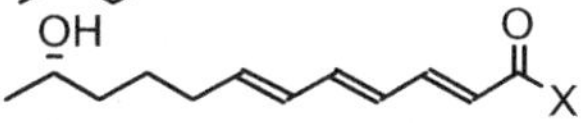

393 Icadamide A R_1=OMe, R_2= 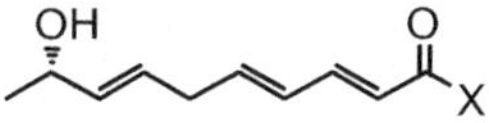

394 Icadamide C R_1=OMe, R_2= 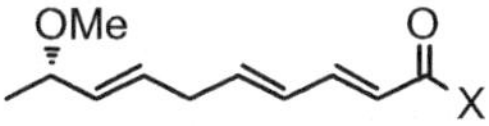

395 Icadamide B R_1=OMe, R_2= 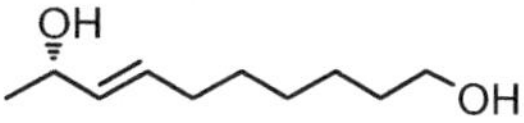

X= 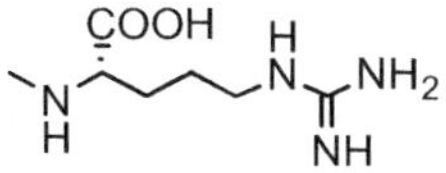

The marine sponge *Discodermia calyx* (order Lithistida, family Theonellidae) was reported to contain the calyculins (**396–413**) [172, 173, 78, 174–176], unique polyketides bearing nitrogen and phosphorus functions. These macrolides exhibited a variety of biological activities including antitumor and smooth muscle contractile, which are attributed to inhibition of protein phosphatase 1 and 2A (all these compounds exhibited nM scale of inhibition activity). SAR studies showed that the 17-phosphate, 13-hydroxyl, and the hydrophobic tetraene moieties were all necessary for binding to the phosphatase 1 and 2A [176].

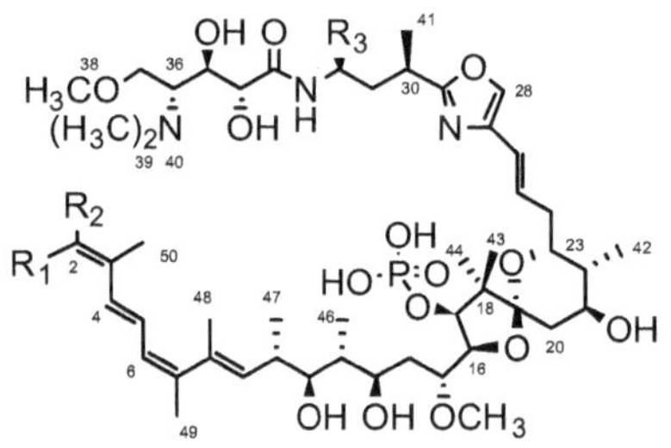

Calyculins A-D(396-399)

396	R_1=CN	R_2=H	R_3=H
397	R_1=H	R_2=CN	R_3=H
398	R_1=CN	R_2=H	R_3=CH3
399	R_1=H	R_2=CN	R_3=CH3

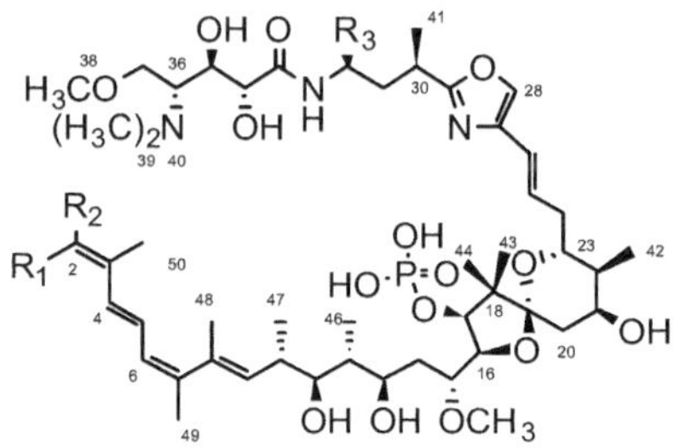

Calyculins A-H(400-407)

	R1	R2	R3		
404	CN	H	H	400	6*E* insomer of 5
405	H	CN	H	401	6*E* insomer of 6
406	CN	H	CH3	402	6*E* insomer of 7
407	H	CN	CH3	403	6*E* insomer of 8

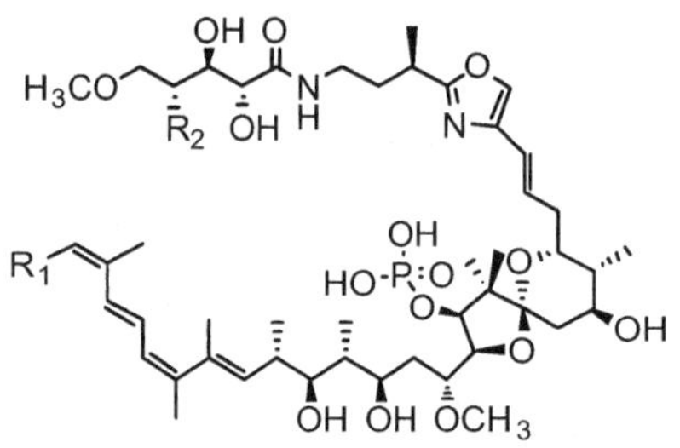

408	Des-N-methylcalyculin A	R_1=CN	R_2=NHMe
409	Calyculinamide A	R_1=$CONH_2$	R_2=NMe_2
410	Calyculinamide F	R_1=$CONH_2$	R_2=$N(CH_3)_2$

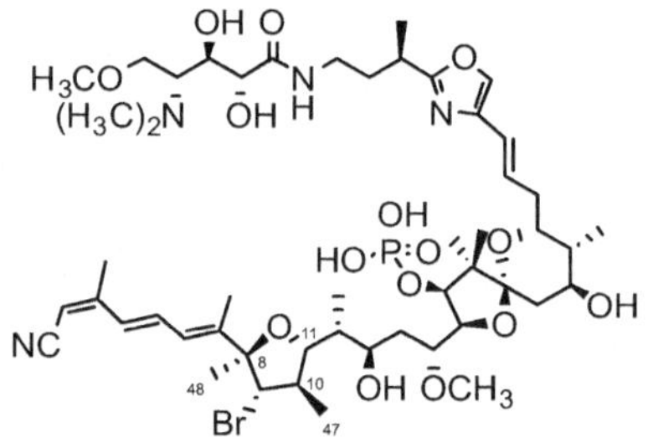

411 Calyculin J

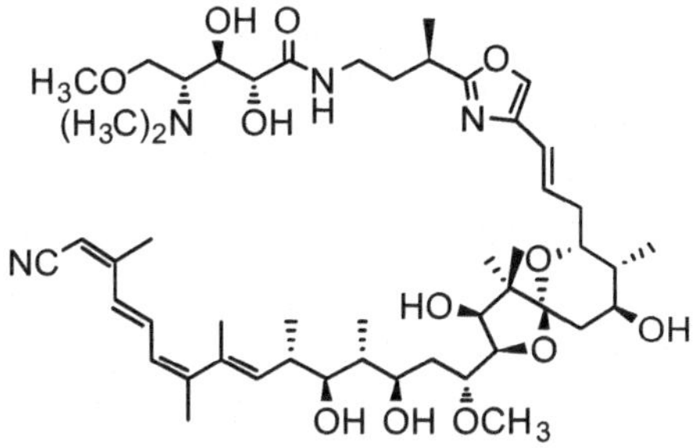

412 Dephosphonocalyculin A

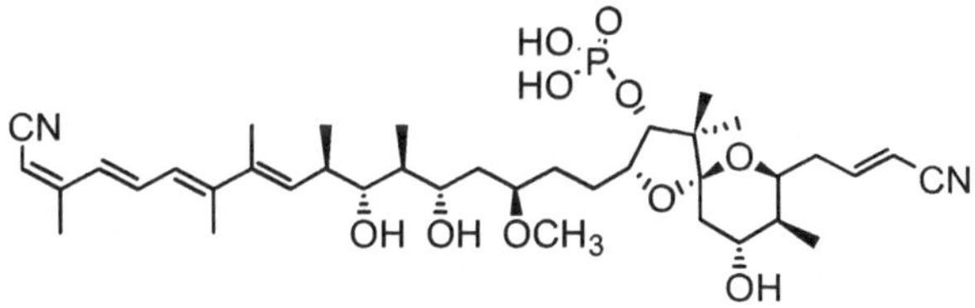

413 Hemicalyculin A

The sponge-derived polyketide macrolides fijianolides are an important subset of 20-membered ring containing compounds isolated from marine sponge *Cacospongia mycofijiensis*, *Hyatella* sp., and *Fasciospongia rimosa* [177–180]. Fijianolide B extremely potent (KB[a] IC_{50} = 29 nM, MDA-MB-435 IC_{50} = 5.7 nM), while fijianolide A (**414**) (HT29 IC_{50} = 21 μM, KB IC_{50} > 39 μM, MDA-MB-435 IC_{50} = 2 μM) was also very active but at a reduced potency [181, 182]. Neolaulimalide (**415**) showed cytotoxicity against P388, A549, HT29, and MEL28 cell lines at 0.01–0.05 μg/mL [180]. Fijianolide B (**416**) together with fijianolides E (**417**) and G (**418**) were also shown to disrupt interphase and mitotic division, and fijianolide B was more potent than fijianolides E and G [177]. An in vivo evaluation of fijianolide B using tumor-bearing severe combined immuno-deficiency mice demonstrated significant inhibition of growth of HCT-116 tumor cells over 28 days [177].

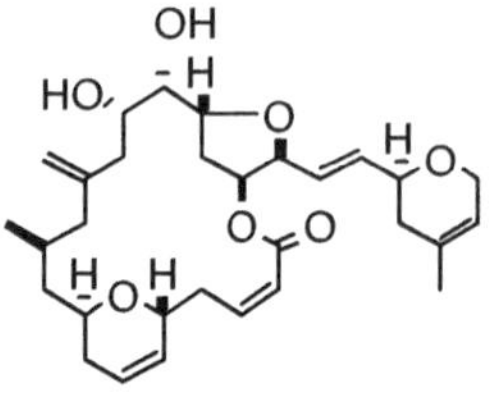

414 Fijianolide A

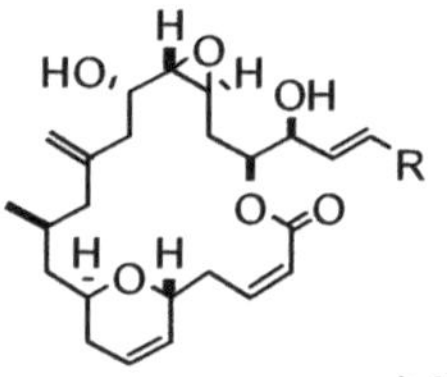

416 Fijianolide B R=

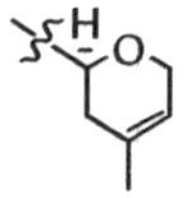

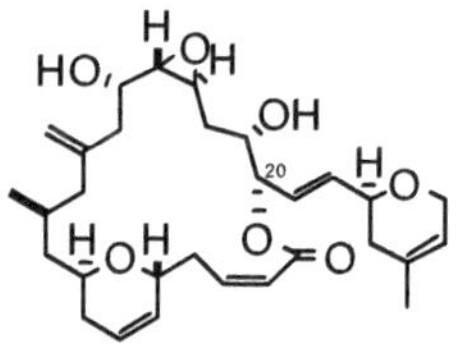

415 Neolaulimalide

417 Fijianolide E R=

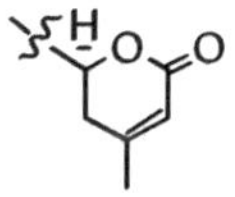

418 Fijianolide G R=

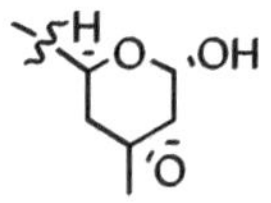

Spiculoic acids and zyggomphic acids are indane-type polyketides, which have integrated phenylacetic acid, butyrate, and propionate units, isolated from marine sponge *Plakortis zyggompha* and *P. angulospiculatus* [183–185]. The vast majority of polyketides made in nature are assembled from acetate and propionate building blocks, whereas these spiculoic acids and zyggomphic acids incorporated the intact of butyrate units. Furthermore, the location of the olefin functionality formed by reduction of the β ketone and dehydration after condensation of the phenylacetic acid starter unit with the first butyrate is unusual. Normally, the dehydration step in polyketide biosynthesis would yield an α,β-unsaturated ester. In the biosyntheses of these polyketides, dehydration occurs in the opposite direction, leading to conjugation between the olefin and the phenyl ring [184]. Spiculoic acid A (**419**) showed in vitro cytotoxicities against the breast MCF-7, breast MDA-MB-231, lung carcinoma A549, and colon carcinoma HT29 cell lines with IC_{50} values of 8.0, 2.4, 4.6, and 8.1 μg/mL, respectively [183, 184]. Zyggomphic acid (**420**) exhibited in vitro antitumor activities against the breast MDA-MB-231, lung carcinoma A549, and colon carcinoma HT29 cell lines with IC_{50} values of 1.2, 3.3, and 3.6 μg/mL, respectively [183].

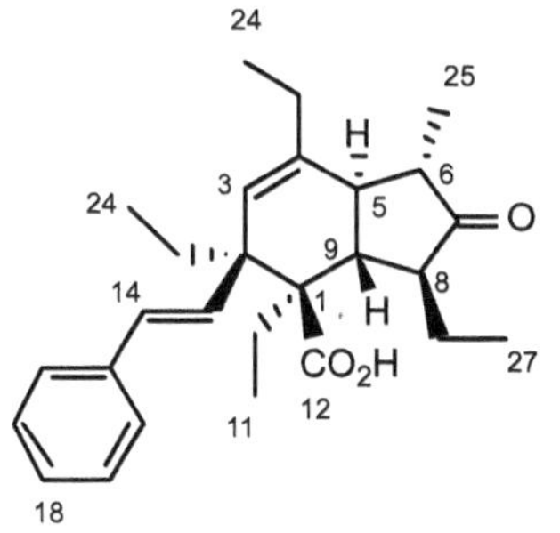

419 Spiculoic acid A

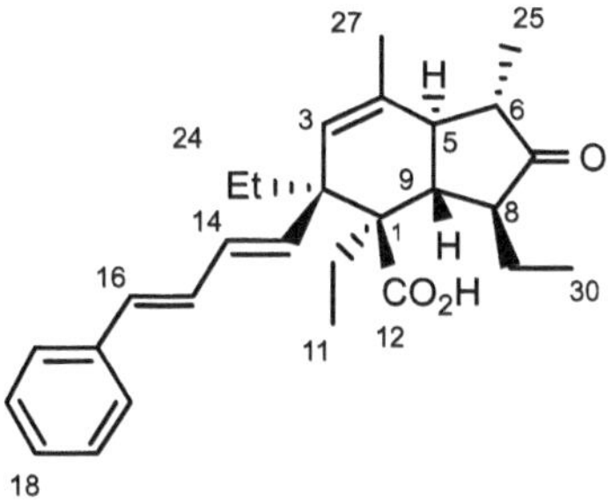

420 Zyggomphic acid

Dihalenaquinolides A (**421**) and B (**422**), novel pentacyclic polyketide dimers, were isolated from marine sponge *Petrosia elastica* collected in Nan-wan, Taiwan, during June 1998 [186]. Dihalenaquinolide A inhibited the growth of PC-3 tumor cells at 10 μg/mL, while compound dihalenaquinolide B was inactive [186].

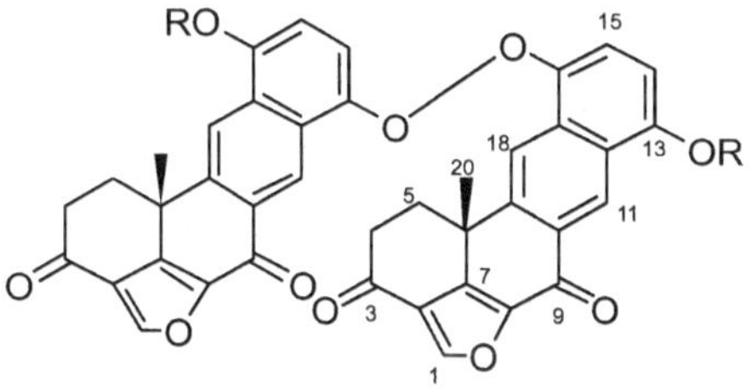

421 Dihalenaquinolides A R=CH_3
422 Dihalenaquinolides B R=CH_2CH_3

Callystatin A (**423**) is a novel polyketide with a terminal α,β-unsaturated δ-lactone, isolated from the marine sponge *Callyspongia truncata*, collected at Goto Islands, Nagasaki Prefecture, in 1997 and exhibited potent cytotoxicity against KB cells at IC_{50} 0.01 ng/ml. Through analogue, syntheses and the assessment of their biological potencies against KB cells manifested the ketonic carbonyl, the 19-hydroxyl, and the three asymmetric methyl groups located in the β-hydroxyketone part of callystatin A contributing to the cytotoxic potency, respectively. Moreover, the α,β-unsaturated δ-lactone portion served as a conclusive functional group for the cytotoxic activity [187].

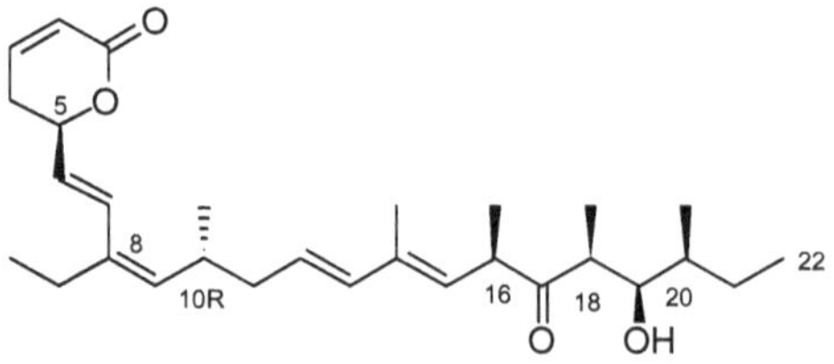

423 Callystatin A

15.5 Bioactive Macrolides

Natural products possessing a macrocyclic lactone moiety are considered to be "macrolides," and most of them are likely to belong to polyketides from a biogenetic viewpoint. Because so many macrolides have been reported in recent years, the selection of compounds derived from 36 marine sponges may seem arbitrary and depends on the potency of their biological activities reported from year 1984 to year 2014.

In 1984, tedanolide (**424**) was isolated from *Tedania ignis* (Caribbean), which exhibited highly cytotoxic with ED_{50} 2.5 × 10^{-4} μg/mL in KB (human carcinoma of the nasopharynx) and 1.6 × 10^{-5} μg/mL in PS (lymphocytic leukemia) [134]. Eight antitumor compounds including norhalichondrins A–C (**425–427**), homohalichondrins A–C (**428–430**), and halichondrins B (**431**) and C (**432**) were found from *Halichondria okadai* Kadota (Miura Peninsula, Tokyo) in 1986, among which halichondrin B exhibited remarkable in vivo antitumor activity [188]. In 1987, halichondramide (**433**), dihydrohalichondramide (**434**), and isohalichondramide (**435**) were isolated from the Pacific sponge *Halichondria* sp. (Kwajelein Island), among which halichondramide showed significant activity against *Candida albicuns* at 0.01 μg/disk in the standard disk assay [189], and dihydrohalichondramide and isohalichondramide had antifungal activity and inhibited cell division in the fertilized sea urchin egg assay [190]. In 1989, mycalolides A–C (**436–438**) were isolated from a sponge *Mycale* sp. (Kii Peninsula, Japan) and demonstrated antifungal activities against many pathogenic fungi and cytotoxic against B-16 melanoma cells with IC_{50}s of 0.5–1.0 ng/mL [191]. The chemical investigation of Okinawan marine sponge *Jaspis* sp. in 1993 yielded jaspisamides A–C (**439–441**) which exhibited cytotoxicities against L1210 murine leukemia cells in vitro, with IC_{50} values of <0.001, <0.001, and < 0.001 μg/mL, and against human epidermoid carcinoma KB cells in vitro with IC_{50} values of 0.015, 0.006, and 0.013 μg/mL, respectively [192]. In 1988, an Indonesian sponge *Hyattella* sp. was collected offshore from Manado (northern Sulawesi, Indonesia) yielded laulimalide (**442**) and isolaulimalide (**443**), and laulimalide displayed potent cytotoxicity, IC_{50} = 15 ng/mL, against the KB cell line [179]. Swinholide A (**444**) was first isolated from a Red Sea sponge *Theonella swinhoei* (Gulf of Eilat, Israel, in 1985), demonstrating in vitro antifungal activity [193]. It was re-isolated from the Okinawan marine sponge *Theonella swinhoei* in 1989 and exhibited potent cytotoxic activity (IC_{50} 0.04 μg/mL) for KB cell [194]. Its absolute configuration was elucidated by means of the X-ray diffraction method and chemical derivations in 1990 [195]. Then, in 1990, swinholide B (**445**), swinholide C (**446**), and isoswinholide A (**447**) were also found from the Okinawan marine sponge *Theonella swinhoei*, among which, swinholide B and swinholide C exhibited potent cytotoxicity almost equivalent to that of swinholide A toward KB cell lines (IC_{50} 0.041 and 0.052 μg/mL, respectively), while isoswinholide A showed weaker cytotoxicity (IC_{50} 1.1 μg/mL) [196].

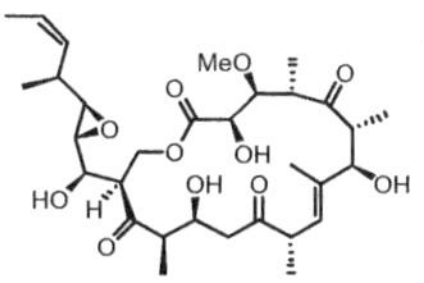

424 Tedanolide

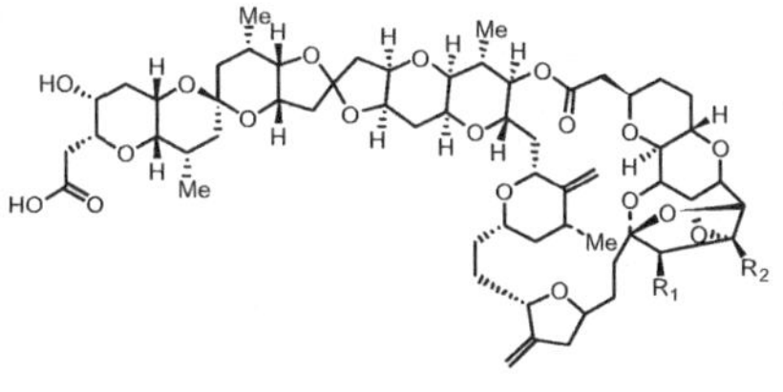

425 Norhalichondrin A: R_1, R_2 = OH
426 Norhalichondrin B: R_1, R_2 = H
427 Norhalichondrin C: R_1 =H, R_2 = OH

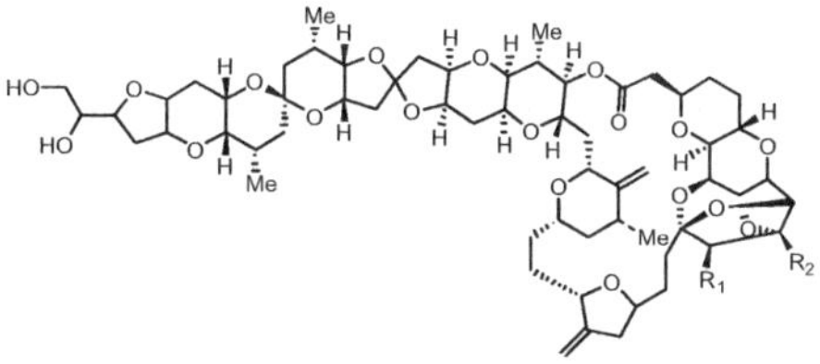

428 Homohalichondrin A: R_1, R_2 = OH
429 Homohalichondrin B: R_1, R_2 = H
430 Homohalichondrin C: R_1 = H, R_2 = OH

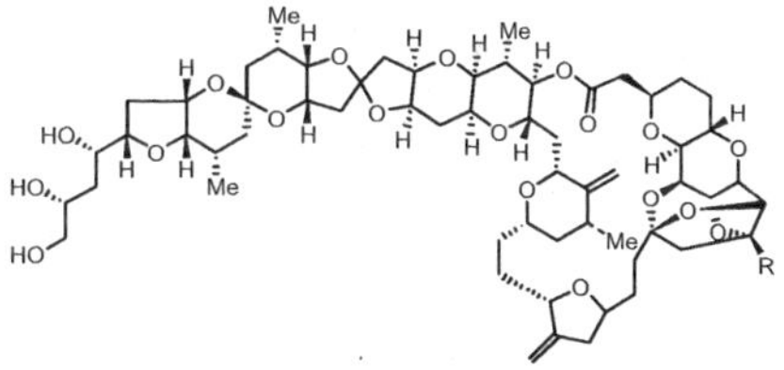

431 Halichondrins B: R = H
432 Halichondrins C: R = OH

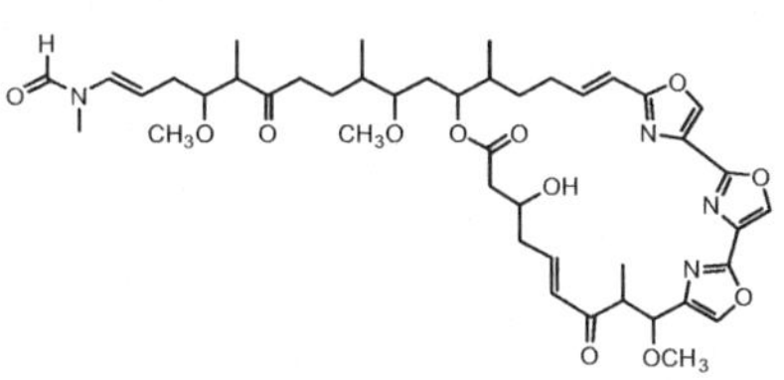

433 Halichondramide

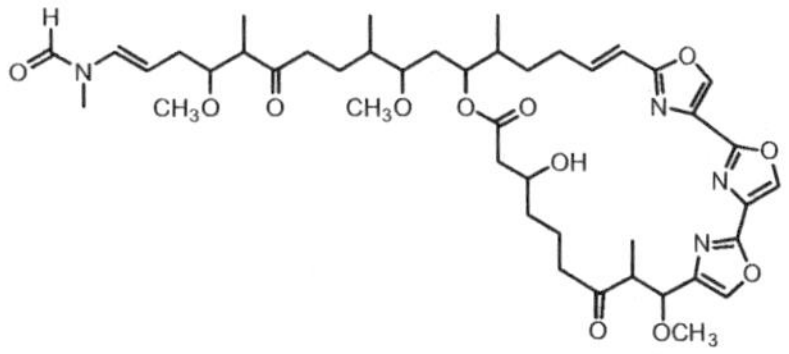

434 Dihydrohalichondramide

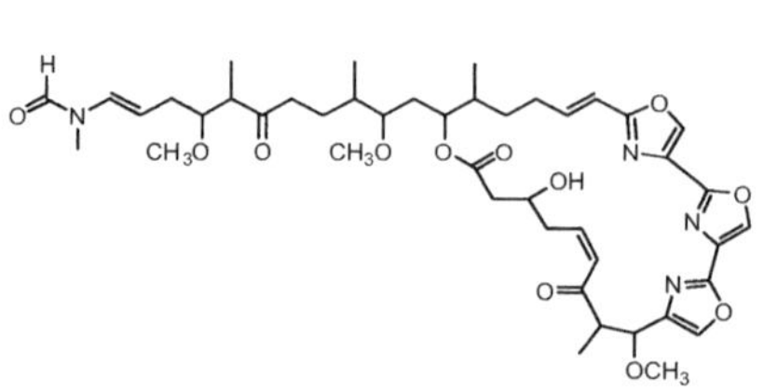

435 Isohalichondramide

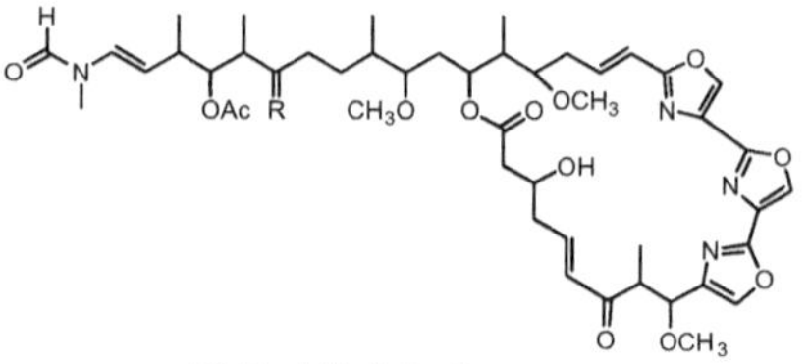

436 Mycalolide A: R = O

437 Mycalolide B: R = 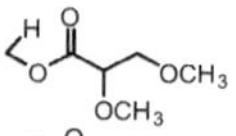

438 Mycalolide C: R = 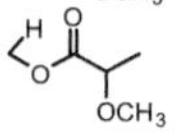

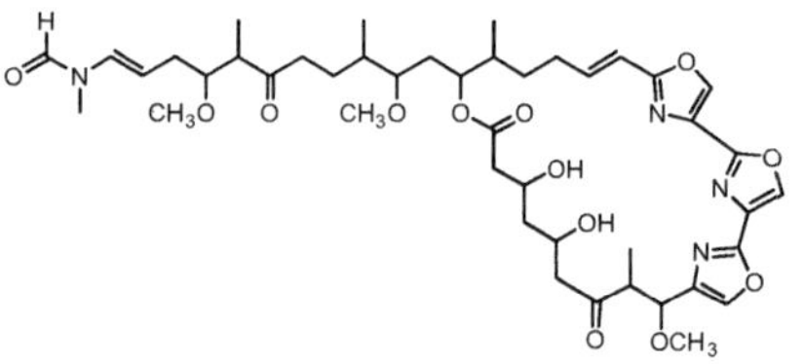

439 Jaspisamide A

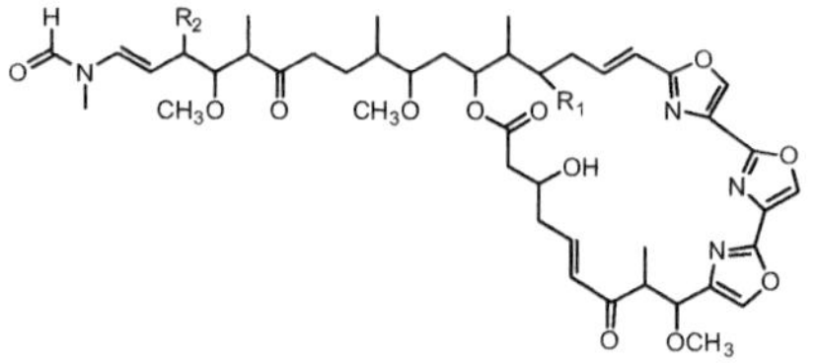

440 Jaspisamide B: R_1 = OH, R_2 = H
441 Jaspisamide C: R_1 = H, R_2 = CH_3

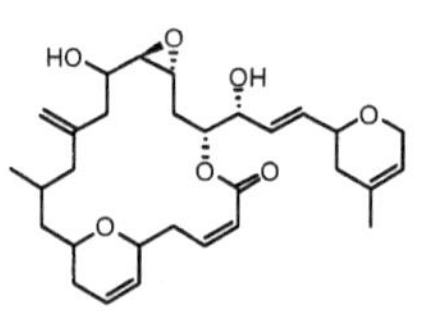

442 Laulimalide

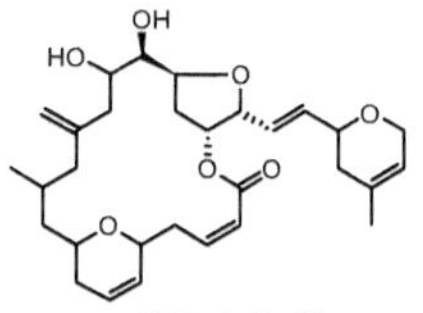

443 Isolaulimalide

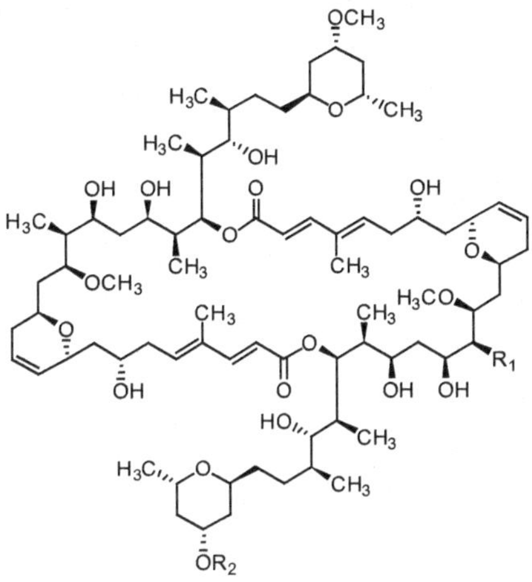

444 Swinholide A : R1 = R_2 = CH_3
445 Swinholide B : R_1 = H, R_2 = CH_3
446 Swinholide C : R_1 = CH_3, R_2 = H

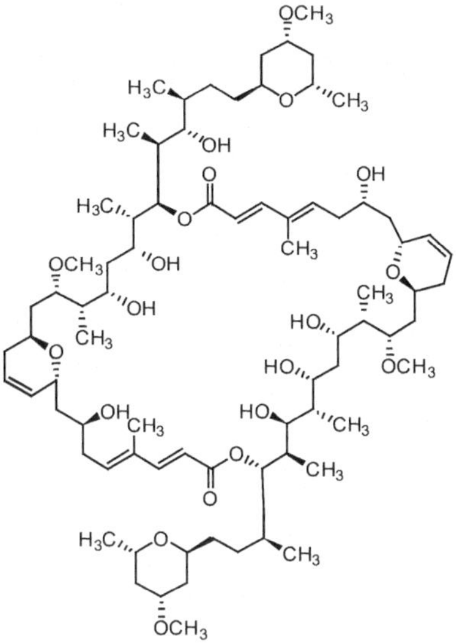

447 Isoswinholide A

In 1993, altohyrtins A–C (**448–450**) and 5-desacetylaltohyrtin A (**451**) were found in sponge *Hyrtios altum*, which were of great interest for cytotoxic activities against KB cells with IC_{50} values of 0.01, 0.02, 0.4, and 0.3 ng/mL, respectively [94, 197, 198]. A highly potent polyether macrolide antimitotic agent designated halistatin **1** (**452**) was isolated from *Phakellia carteti* (Grand Comore Island, Republic of Comoros), which showed strong cytotoxicity to L1210 murine leukemia cells with IC_{50} values of 0.5 nM. In the further mechanism study, it was shown to cause accumulation of cells arrested in mitosis, inhibited tubulin polymerization, and inhibits binding of radiolabeled vinblastine and GTP to tubulin [199]. Cinachyrolide A (**453**) was isolated from *Cinachyra* sp. which was highly cytotoxic against L1210 murine leukemia cells with an IC_{50} of <0.6 ng/mL [200]. In 1994, superstolides A (**454**) and B (**455**) were obtained from the deepwater marine sponge *Neosiphonia superstes* (New Caledonia) and demonstrated highly cytotoxic against human bronchopulmonary non-small-cell lung carcinoma NSCLC-N6-L 16 cells (IC_{50} 0.04 and 0.039 μg/mL), murine leukemia P388 cells (IC_{50} both of 0.003 μg/mL), and human nasopharyngeal carcinoma KB cells (IC_{50} 0.02 and 0.005 μg/mL) [201, 202]. Reidispongiolides A (**456**) and B (**457**) were isolated from *Rekiivpmgia werulea* n.gen. n.sp. (South of New Caledonia), which exhibited potent cytotoxicity against various human carcinoma cells with IC_{50} values of 0.01–0.16 μg/mL [203]. Lasonolide A (**458**) was isolated from *Forcepia* sp. (British Virgin Islands), as a potent cytotoxin against the A549 human lung carcinoma and P388 murine leukemia cell lines with IC_{50} values of 40 and 2 ng/mL, respectively. Further, it inhibited cell adhesion in the EL-4.IL-2 cell line with an IC_{50} of 19 ng/mL [204]. Isohomohalichondrin B (**459**) was found in New Zealand sponge *Lissodendoryx* sp. and showed significant cytotoxic activity against the P388 cell lines and selective cytotoxicity in the NCI'S primary screen [205].

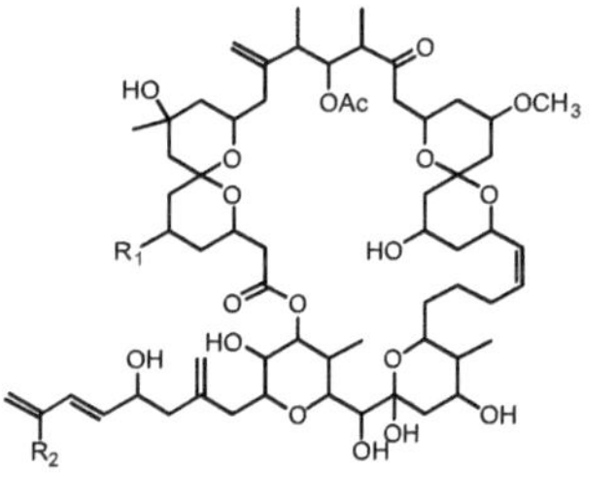

448 Altohyrtin A: R_1 = OAc, R_2 = Cl
449 Altohyrtin B: R_1 = OAc, R_2 = Br
450 Altohyrtin C: R_1 = OAc, R_2 = H
451 5-Desacetylaltohyrtin A: R_1 = OH, R_2 = Cl

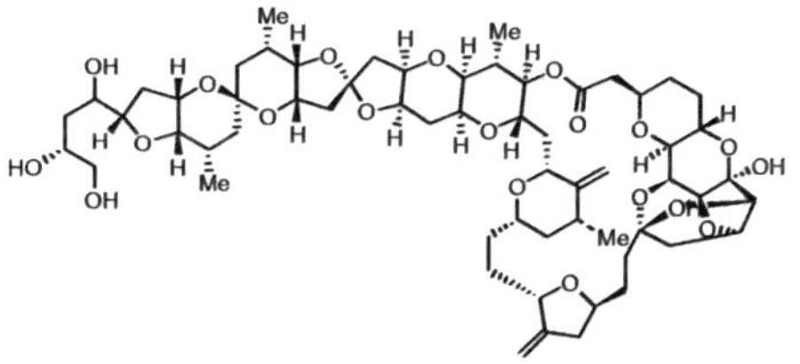

452 Halistatin 1

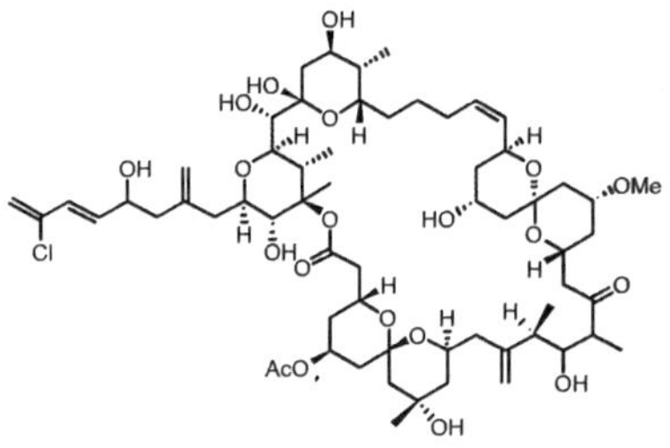

453 Cinachyrolide A

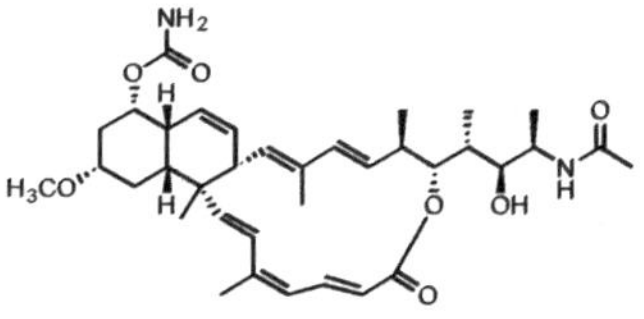

454 Superstolide A

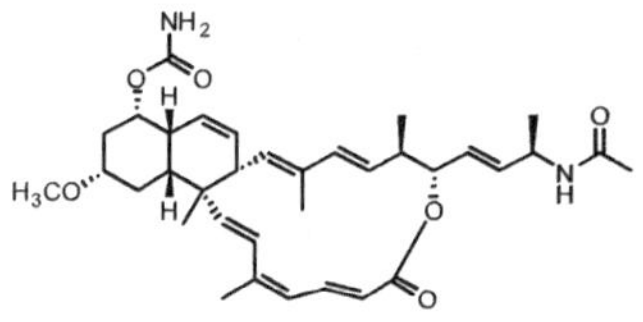

455 Superstolide B

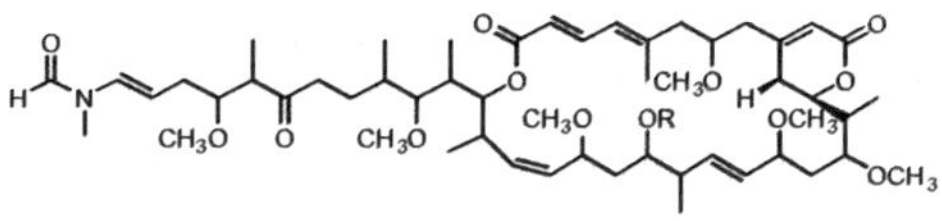

456 Reidispongiolide A: R = CH_3
457 Reidispongiolide B: R = H

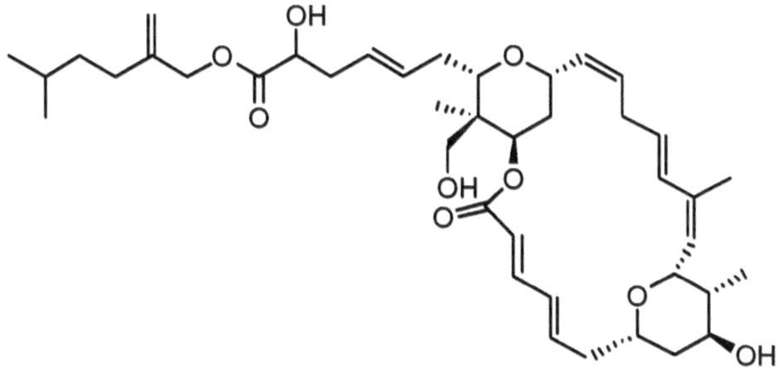

458 Lasonolide A

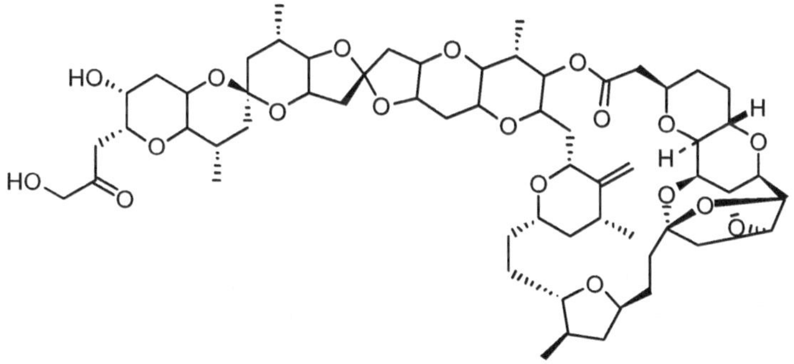

459 Isohomohalichondrin B

The chemical investigation of Indian Ocean marine sponge *Phorbas* sp. (Muiron Island, Australia, in 1995) yielded phorboxazoles A (**460**) and B (**461**), which exhibited in vitro antifungal activity against *Candida albicans* at 0.1 μg/disk and extraordinary cytostatic activity [206]. In 1996, theonezolide A (**462**), a novel polyketide macrolide, isolated from the Okinawan marine sponge *Theonella* sp. (Okinawa, Japan), caused a marked platelet shape change at low concentrations (0.2–0.6 μM) [207]. Leucascandrolide A (**463**), a doubly O-bridged 18-membered macrolide of a new type, i.e., possessing little C_1-branching vs. extensive 1,3-dioxygenation and a peculiar side chain, was isolated from a calcareous sponge of a new genus, *Leucuscundra caveoluta* from the Coral Sea. It showed strong cytotoxic activity in vitro on KB cells and less marked action on P388 cells, as well as very strong inhibition of *Candida albicans* [208]. Compared with leucascandrolide A, leucascandrolide B (**464**) was found from the same sponge sample which showed only marginal cytotoxicity on tumor cell lines, with an IC_{50} of 5 μg/mL on KB cells and > 10 μg/mL on P388 murine leukemia cells and no activity on *Candida albicans* [209]. A marine sponge *Fasciospongia rimosa* (Okinawa, Japan) yielded zampanolide (**465**) which showed potent cytotoxicity (IC_{50} 1–5 ng/mL) against P388, A549, HT29, and MEL28 cell lines [210]. In 1997, study of a deepwater sponge *Lissodendoryx* sp. (Kaikoura Peninsula, New Zealand) yielded neonorhalichondrin

B (**466**), neohomohalichondrin B (**467**), 55-methoxyisohomohalichondrin B (**468**), 53-methoxyneoisohomohalichondrin B (**469**), and 53-epi-53-methoxyneoisohomohalichondrin B (**470**), and further antitumor assay demonstrated that they were highly cytotoxic against P388 cells with IC_{50} values of 0.4, 0.8, 10, and 0.1 ng/mL except 53-epi-53-methoxyneoisohomohalichondrin B, respectively [211]. Two novels, highly potent, cytotoxic macrolides and salicylihalamides A (**471**) and B (**472**), were isolated from the sponge *Haliclona* sp. (Rottnest Island, Australia). COMPARE pattern-recognition analyses of the NCI 60-cell mean-graph screening profiles of salicylihalamide A did not reveal any significant correlations to the profiles of known antitumor compounds in the NCI's "standard agent database," thus supporting the conclusion that the salicylihalamides represent a potentially important new class for antitumor lead optimization and in vivo investigations [212]. In 1998, thiomycalolides A (**473**) and B (**474**) were obtained from *Mycale* sp. (Kii Peninsula, Japan), which exhibited highly cytotoxic against P388 murine leukemia cells with an IC_{50} value of 18 ng/mL each [213].

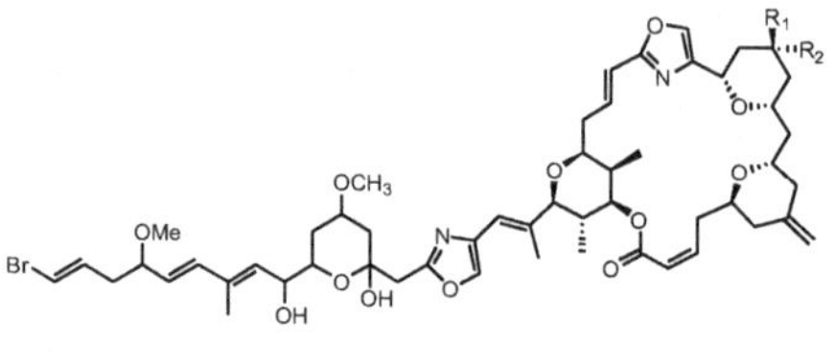

460 Phorboxazole A: R_1 = OH, R_2 = H
461 Phorboxazole B: R_1 = H, R_2 = OH

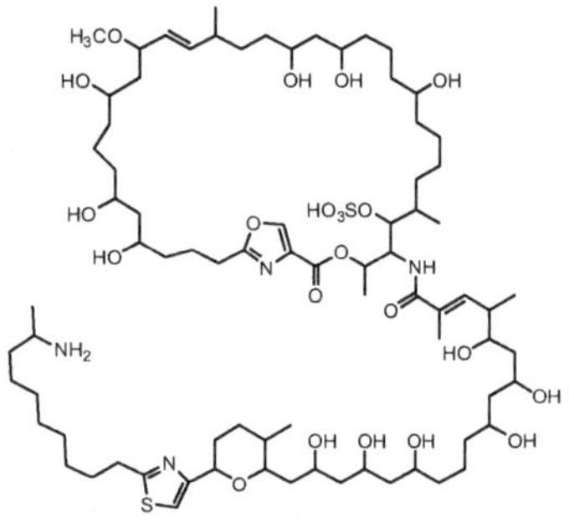

462 Theonezolide A

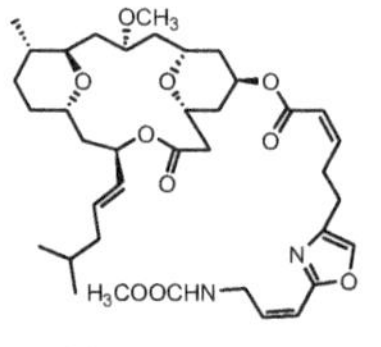

463 Leucascdndrolide A

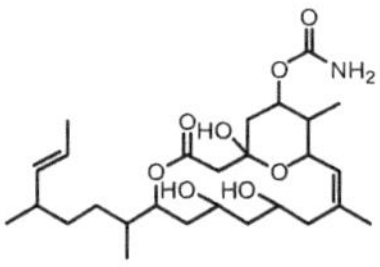

464 Leucascdndrolide B

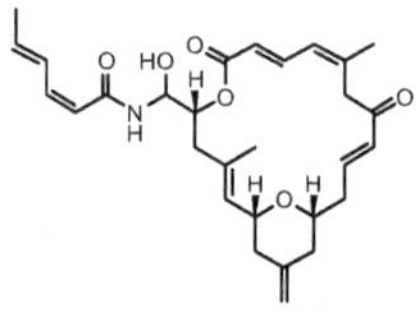

465 Zampanolide

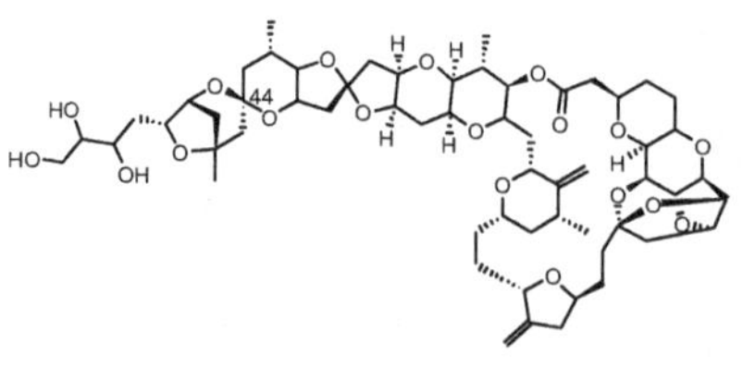

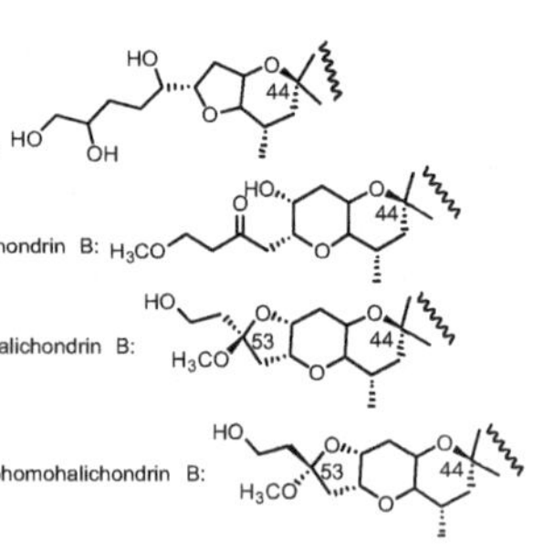

466 Neonorhalichondrin B
467 Neohomohalichondrin B:
468 55-Methoxyisohomohalichondrin B:
469 53-Methoxyneoisohomohalichondrin B:
470 53-Epi-53-methoxyneoisohomohalichondrin B:

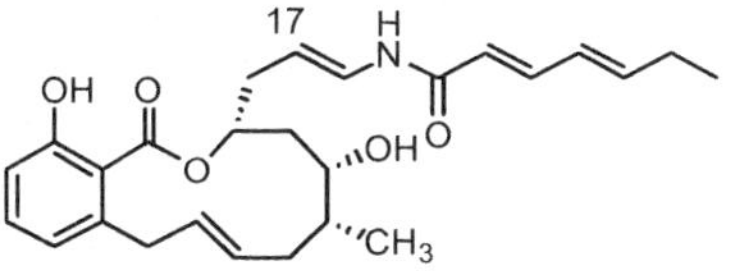

471 Salicylihalamide A
472 Salicylihalamide B: 17Z

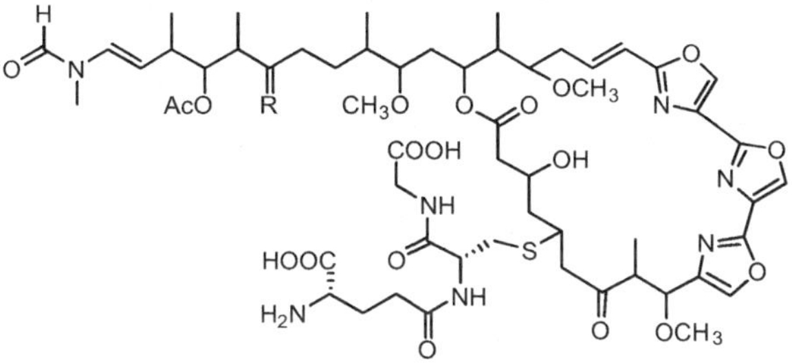

473 Thiomycalolide A: R = O

474 Thiomycalolide B: R = 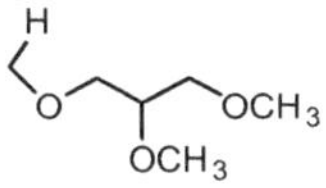

Chemical investigation of an Okinawan sponge *Ircinia* sp. resulted in isolation of haterumalides NA, NB, NC, ND, and NE (**475–479**), among which, haterumalides NA exhibited cytotoxicity against P388 cells, with an IC_{50} of 0.32 μg/mL, and moderate acute toxicity against mice, with an LD_{99} of 0.24 g/kg which was found in 1999 [214]. In 2000, antiproliferative bioassay-guided fractionation of an aqueous extract of the marine sponge *Chondropsis* sp. (Wollongong, Australia) provided chondropsins A (**480**) and B (**481**). Testing of chondropsin A in the NCI 60-cell screen revealed a mean-graph profile that did not correlate significantly with the profile of any compound class represented in the NCI standard agents database [215]. In 2001, 73-deoxychondropsin A (**482**) and chondropsin C (**483**) were isolated from two different collections of marine sponges belonging to the genus *Ircinia* (*Ircinia ramose*, Australia; *Ircinia* sp., Philippines) and exhibited IC_{50}'s of approximately 0.8 and 0.2 ng/mL toward the LOX and MOLT-4 cell lines, respectively [216]. Chondropsin D (**484**) exhibited IC_{50}'s of approximately 10 and 250 ng/mL toward the LOX and MOLT-4 cell lines, respectively [217]. In 2000, peloruside A (**485**) was found to be cytotoxic to P388 murine leukemia cells at approximately 10 ng/mL (18 nM) which was found in sponge *Mycale* sp. [218]. In 2010, peloruside B (**486**), a natural congener of peloruside A, was isolated from the New Zealand marine sponge *Mycale hentscheli*. Peloruside B was found to promote microtubule polymerization and arrest cells in the G2/M phase of mitosis similar to paclitaxel, and its bioactivity was comparable to that of peloruside A [219].

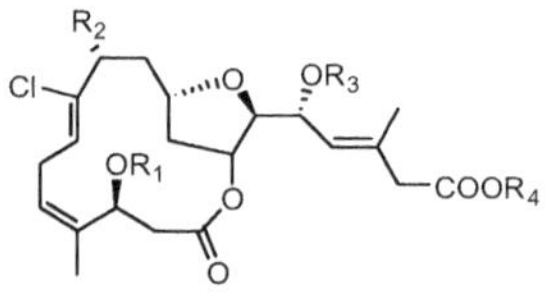

475 Haterumalide NA: R_1 = Ac, R_2 = H, R_3 = H, R_4 = H
476 Haterumalide NB: R_1 = Ac, R_2 = H, R_3 = H, R_4 = ^{n}Bu
477 Haterumalide NC: R_1 = Ac, R_2 = OH, R_3 = H, R_4 = ^{n}Bu
478 Haterumalide ND: R_1 = Ac, R_2 = OH, R_3 = H, R_4 = H
479 Haterumalide NE: R_1 = H, R_2 = H, R_3 = H, R_4 = H

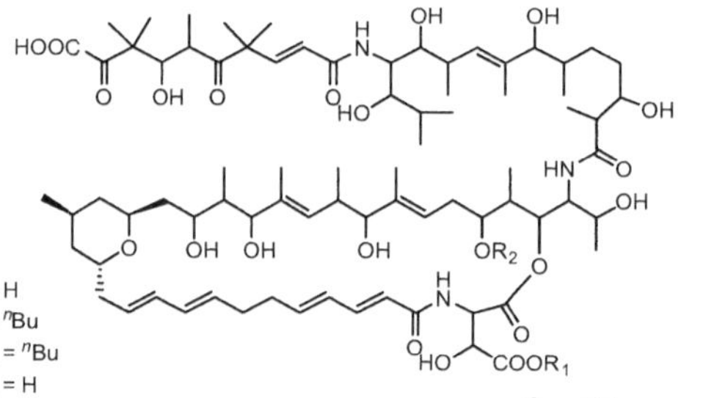

480 Chondropsin A: R_1 = 1, R_2 = 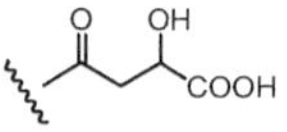

481 Chondropsin B: R_1 = R_2 = H

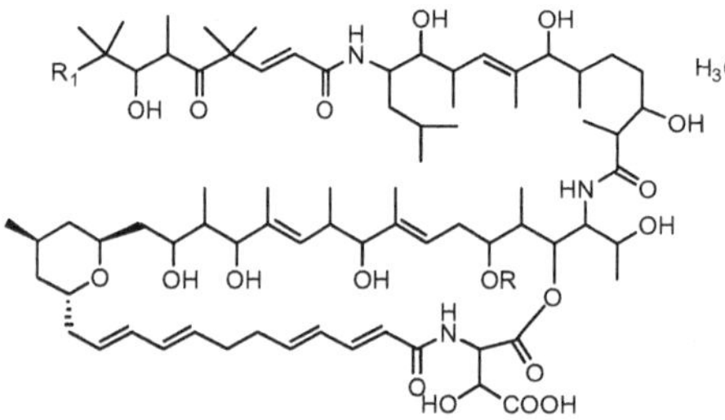

R = 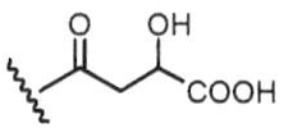

482 73-Deoxychondropsin A: R_1 = $COOCH_3$
483 Chondropsin C: R_1 = H

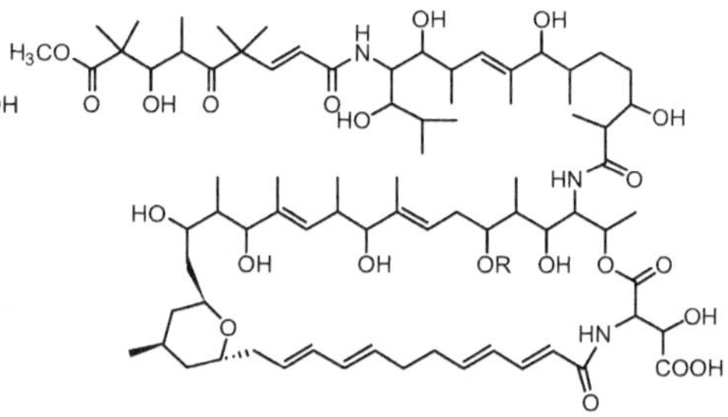

R = 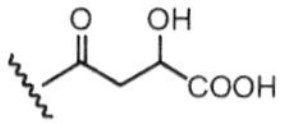

484 Chondropsin D

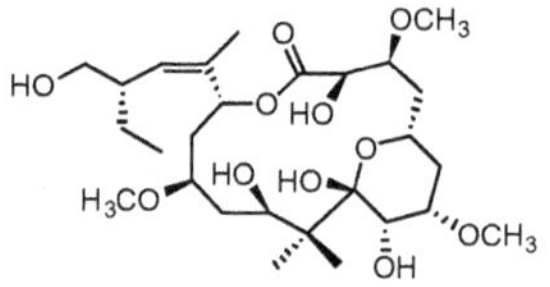

485 Peloruside A

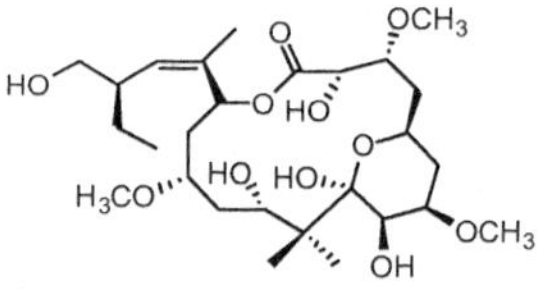

486 Peloruside B

In 2001, dactylolide (**487**) from *Dactylospongia* sp. showed cytotoxic activity against the L1210 and SK-OV-3 tumor cell lines (63% and 40% inhibition at 3.2 μg/mL) [220]. In 2002, 30,32-dihydroxymycalolide A (**488**) was obtained from *Mycale izuensis*, with cytotoxic activity against HeLa cells (IC_{50} value of 2.6 ng/mL) [221]. In 2005, 13-deoxytedanolide (**489**), a highly antitumor macrolide from the marine sponge *Mycale adhaerens*, exhibited cytotoxic activity against P388 murine leukemia cells with IC_{50} of 0.064 ng/mL [138]. In 2006, tedanolide C (**490**) isolated from *Ircinia* sp. (Milne Bay, Papua New Guinea) exhibited potent cytotoxicity against HCT-116 cells in vitro with IC_{50} value of 9.53×10^{-2} μM and caused a strong S-phase arrest [136]. Leiodolides A (**491**) and B (**492**) were found in marine sponge

Leiodermatium (Palau) from deep water, and leiodolide A showed significant cytotoxicity (average GI_{50} = 2.0 μM) in the National Cancer Institute's 60 cell line panel with enhanced activity against HL-60 leukemia and OVCAR-3 ovarian cancer cell lines [222]. The Red Sea sponge *Theonella swinhoei* (Hurghada, Egypt) yielded swinholide I (**493**) and hurghadolide A (**494**) which were in vitro cytotoxic against human colon adenocarcinoma (HCT-116) with IC_{50} values of 5.6 and 365 nM, respectively. Furthermore, they could disrupt the actin cytoskeleton in the range of 70 and 7.3 nM, respectively. In addition, they were both active against *Candida albicans* [223].

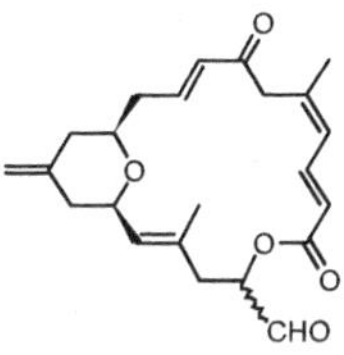
487 Dactylolide

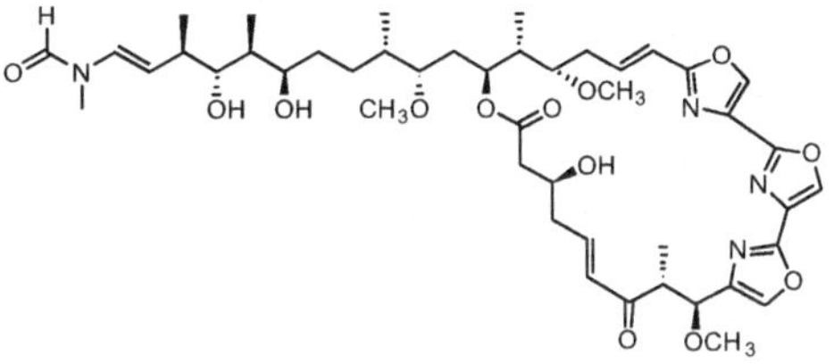
488 30,32-Dihydroxymycalolide A

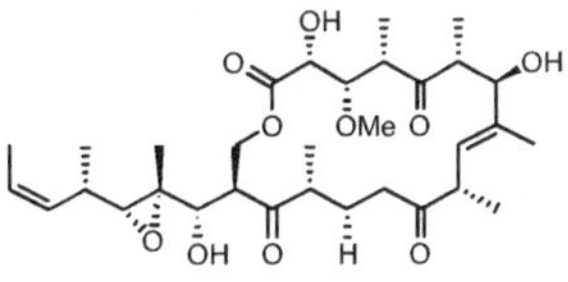
489 13-Deoxytedanolide

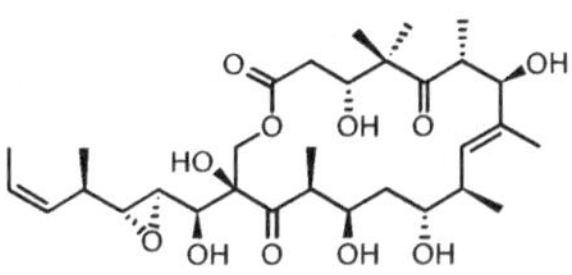
490 Tedanolide C

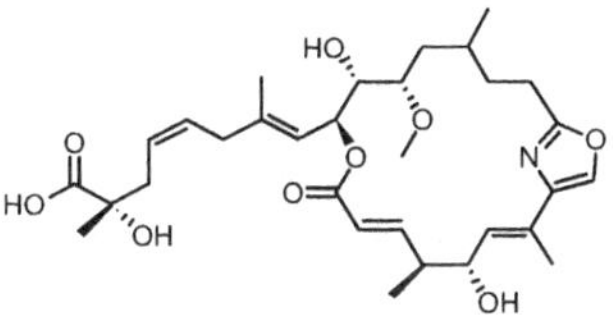
491 Leiodolides A

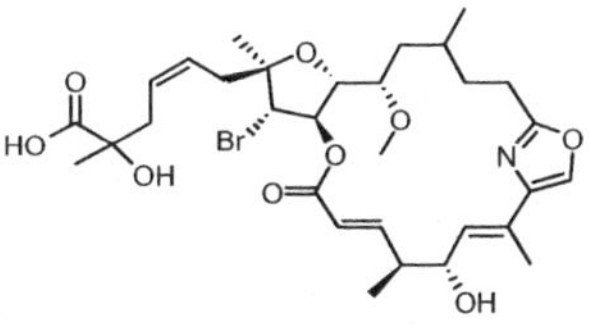
492 Leiodolides B

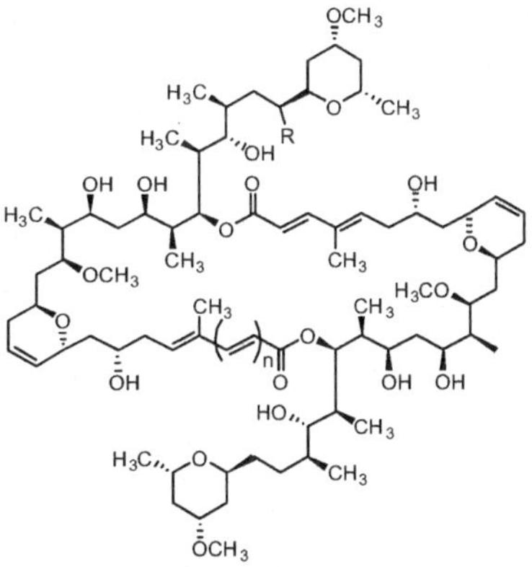
493 Swinholide I : n = 1, R = OH
494 Hurghadolide A: n = 0, R = H

In 2007, neopeltolide (**495**) was isolated from a deepwater sponge of the family Neopeltidae (Jamaica), and it showed potent in vitro anti-proliferation activity against A549 human lung adenocarcinoma, the NCI-ADR-RES human ovarian sarcoma, and the P388 murine leukemia cell lines, with IC_{50}'s of 1.2, 5.1, and 0.56 nM, respectively. Neopeltolide also inhibited the growth of the fungal pathogen *Candida albicans* with a minimum inhibitory concentration of 0.62 μg/mL [224]. The study of marine sponge *Poecillastra* sp. (Grand Bahama Island, Bahamas) yielded poecillastrins B (**496**) and C (**497**), which were of interest for cytotoxicity against a human melanoma tumor cell line (LOX) with an IC_{50} value of less than 1 μg/mL [225]. In 2008, mirabilin (**498**) was isolated from the marine sponge *Siliquariaspongia mirabilis* (Federated States of Micronesia), which prevented the tumor cell line HCT-116 from growing with an IC_{50} value of 0.27 ± 0.09 μM [226]. Three nitrogenous macrolides designated salarin A (**499**), B (**500**) and tulearin A (**501**) were isolated from the Madagascar *Fascaplysinopsis* sp. sponge. Both salarins carry an acetylcarbamate moiety, and in addition, salarin A contains a triacylamine group and salarin B contains a methoxymethylketone lactam. Tulearin A was featured with a naturally rare carbamate ester. They were found to be toxic to brine shrimp larvae, and salarin A and tulearin A were also cytotoxic to leukemia cells [227].

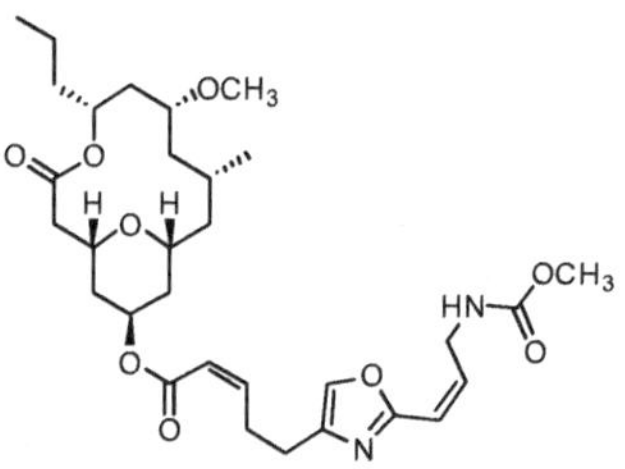

495 Neopeltolide

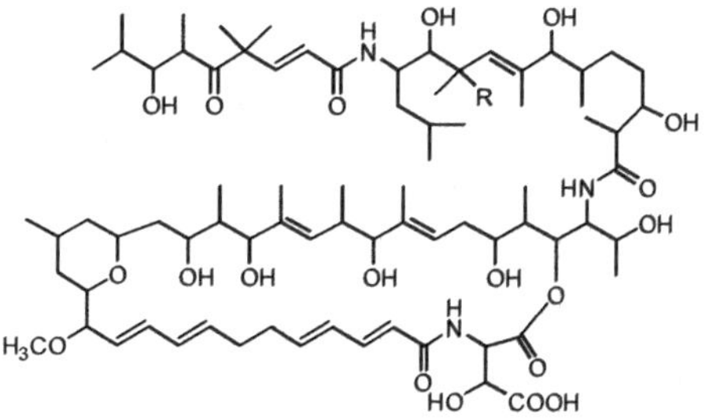

496 Poecillastrin B: R = CH_3
497 Poecillastrin C: R = H

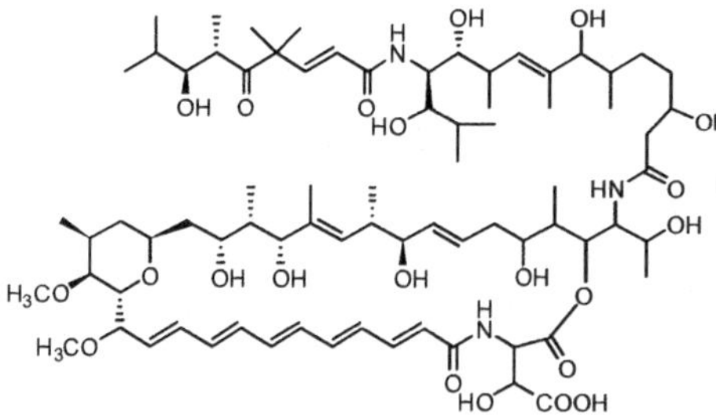

498 Mirabilin

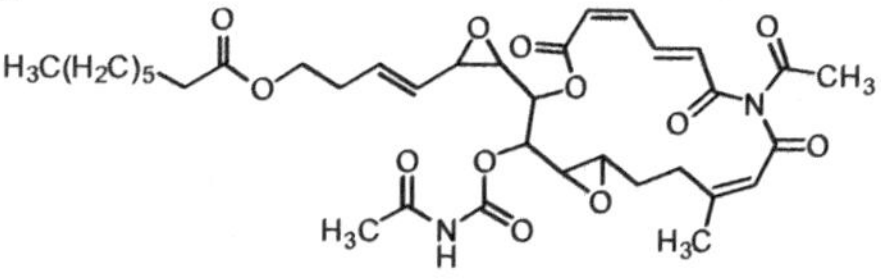

499 Salarin A

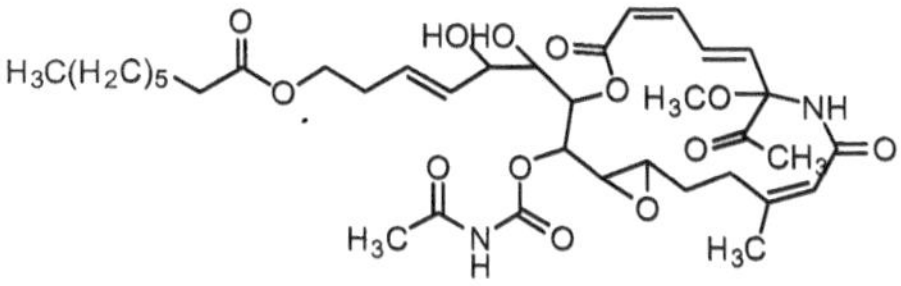

500 Salarin B

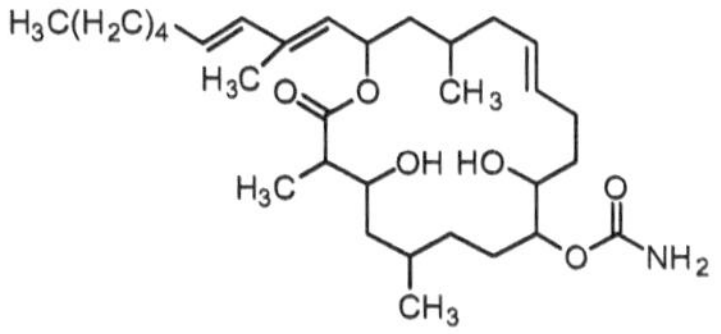

501 Tulearin A

In the year of 2009, the deepwater marine sponge *Lissodendoryx* sp. (Kaikoura Peninsula, New Zealand) was found to contain halichondrin B-1140, halichondrin B-1092, halichondrin B-1020, and halichondrin B-1076 (**502–505**), which exhibited highly cytotoxicity against the P388 cell lines with IC_{50} values of 2.0, 0.76, 1.1, and 1.1 ng/mL, respectively [228]. Leiodermatolide (**506**) isolated from the marine sponge *Leiodermatium* sp. (Fort Lauderdale, Florida,2011) was found to exhibit potent and selective antimitotic activity (IC_{50} < 10 nM) against a range of human cancer cell lines by inducing G2/M cell cycle arrest [229]. Kabiramides B–D, G, J, and K (**507–512**) were isolated from the sponge *Pachastrissa nux*, showed moderate to strong antimalarial and cytotoxic activities, except for kabiramide G, which possessed only potent cytotoxicity [230]. In 2013, kabiramides L (**513**) and I (**514**) were obtained from the same sponge sample *Pachastrissa nux*, and both exhibited a moderate antiplasmodial activity against *Plasmodium falciparum* K1 with IC_{50}s of 2.6 and 4.5 μM, respectively [231]. In 2014, callyspongiolide (**515**), a structurally unique polyketide-derived macrolide, was isolated from the marine sponge *Callyspongia* sp. collected in Indonesia, and it showed strong cytotoxicity against human Jurkat J16 T and Ramos B lymphocytes [232].

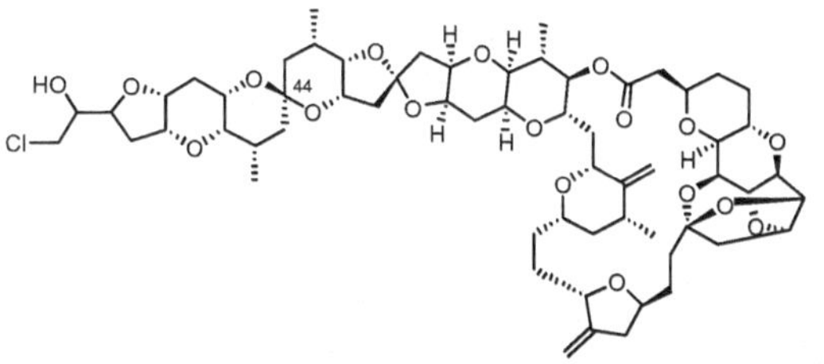

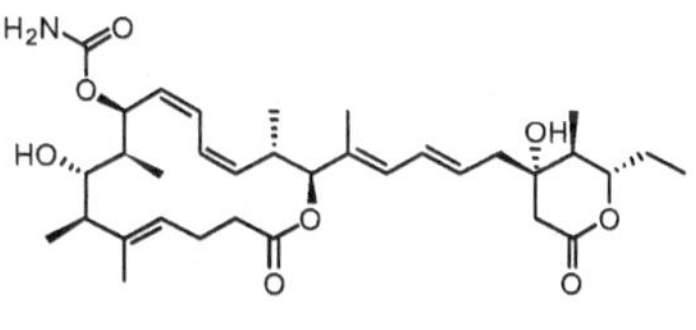

506 Leiodermatolide

502 Halichondrin B-1140

503 Halichondrin B-1092:

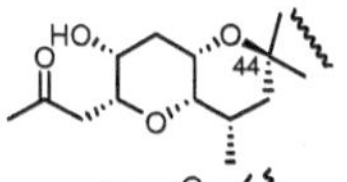

504 Halichondrin B-1020:

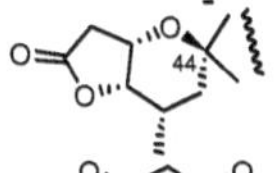

505 Halichondrin B-1076:

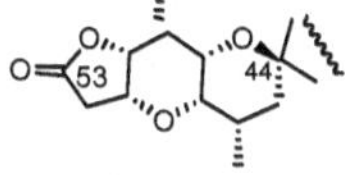

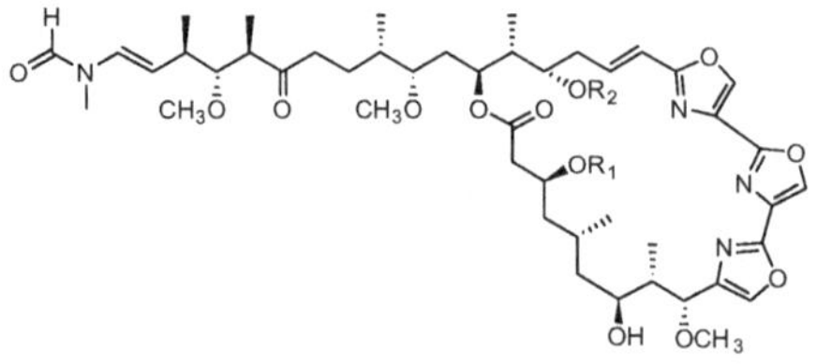

507 Kabiramide B: R_1 = $CONH_2$, R_2 = H
508 Kabiramide C: R_1 = $CONH_2$, R_2 = CH_3
509 Kabiramide D: R_1 = H, R_2 = CH_3

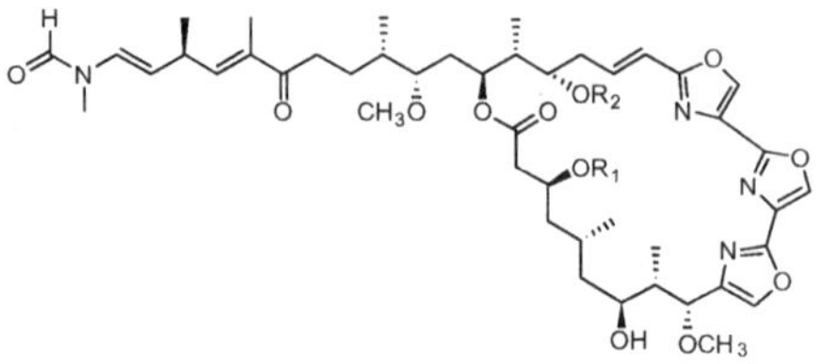

510 Kabiramide G: R_1 = $CONH_2$, R_2 = CH_3
511 Kabiramide J: R_1 = $CONH_2$, R_2 = H
512 Kabiramide K: R_1 = H, R_2 = CH_3
513 Kabiramide L: R_1 = H, R_2 = H

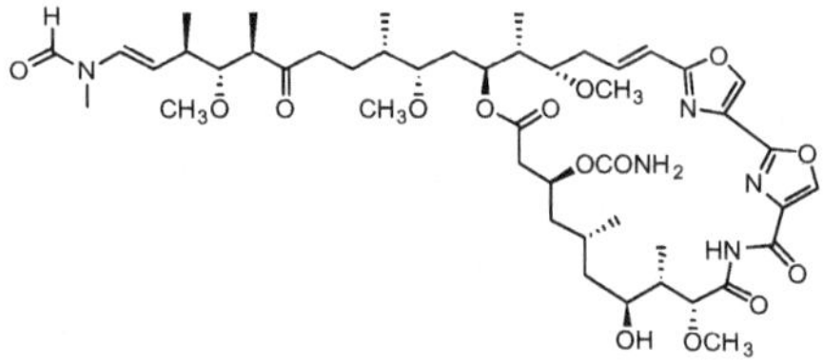

514 Kabiramide I

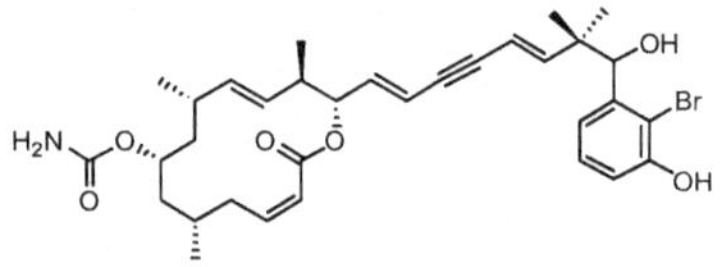

515 Callyspongiolide

15.6 Bioactive Terpenoids

Marine organisms produce a wide array of fascinating terpenoid structures distinguished by characteristic structural features. Since sponges are one of the prime resources of sesquiterpenes,diterpenes,sesterterpenes,and triterpenes, we here will survey 170 terpenoids from 32 sponges with various biological activities such as antitumor, anti-inflammation, antifouling, fungicide, as well as pesticide.

15.6.1 Sesquiterpenes

In 1996, chemical investigation of sponge *Acanthella eavernos*a (Hachijo-jima Island, Tokyo), yielded Isocyanate and isothiocyanate derivatives **516** and **517**, both of which were highly active in antifouling assay with EC_{50} value of 0.05 μg/mL [233]. In 2008 year, three new sesquiterpene quinones isohyatellaquinone (**518**), 7, 8-dehydrocyclospongiaquinone-2 (**519**), and 9-epi-7, 8-dehydrocyclospongiaquinone-2 (**520**), along with the known quinones mamanuthaquinone (**521**) and ilimaquinone (**522**), were isolated from *Dactylospongia elegans* (Coral Gardens dive site at the Inner Gneerings reef, Australia, in 2007). All compounds were active against the breast cancer (BC) and small cell lung cancer (NCI-H187) cell lines an IC_{50} range of 1.50–12.4 μg/mL except compound **519** [234].

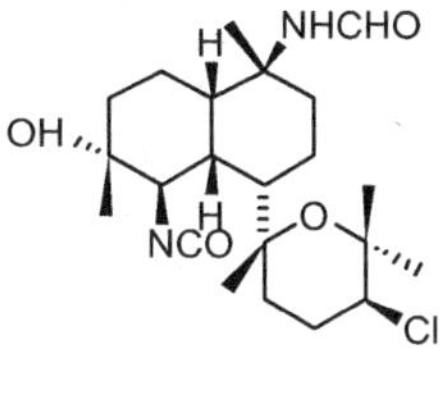

516

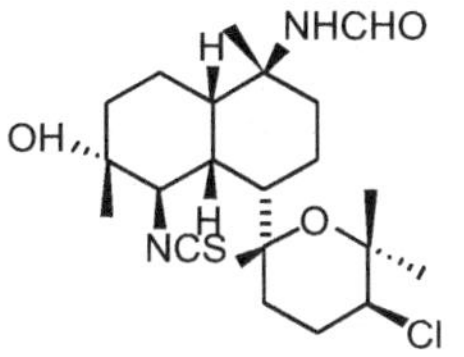

517

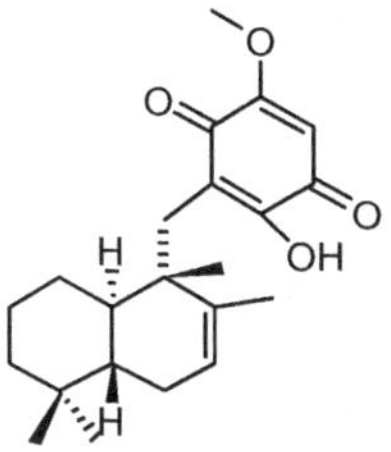

518

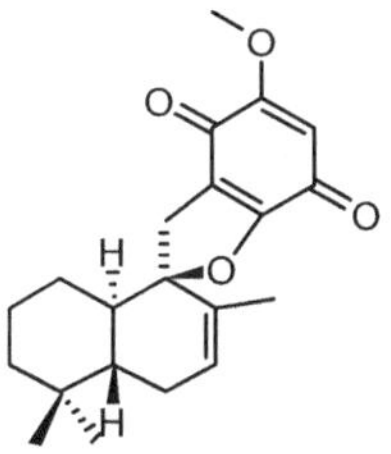

519

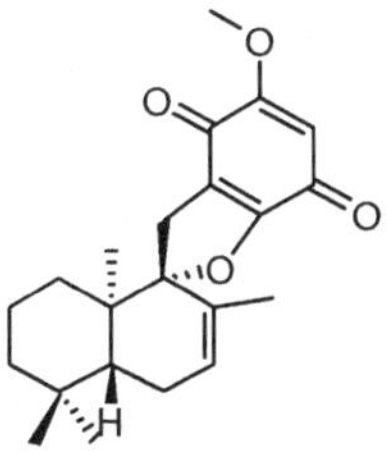

520

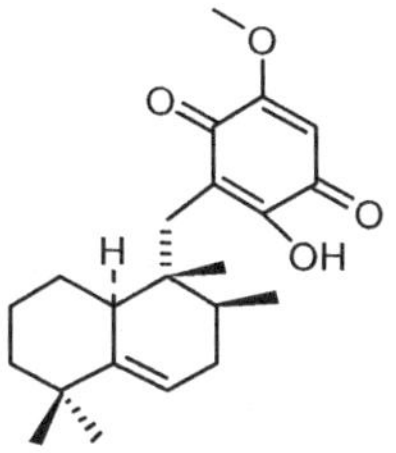

521

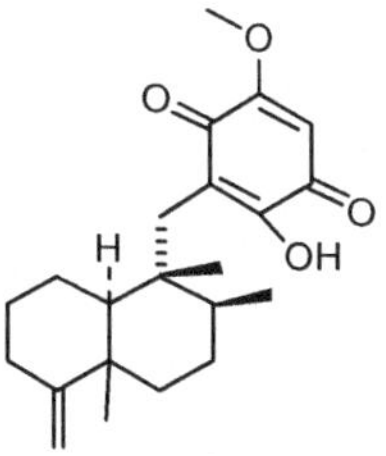

522

In 2011, nakijinol B (**523**), nakijinol B diacetate (**524**), smenospongines B (**525**) and C (**526**) were found in the marine sponge *Dactylospongia elegans* (Pugh Shoal, northeast of Truant Island, in November 1990), along with two known compounds [ilimaquinone (**527**) and 5-epi-ilimaquinone as a 1:1 mixture, dactyloquinone B (**528**)]. All compounds tended out to be active from 1.8 to 46 μM but lacking selectivity for tumor versus normal cell lines (SF-268, H460, MCF-7, HT29, and CHO-K1). And the 1:1 mixture of ilimaquinone (**527**) and 5-epi-ilimaquinone was found to be the most cytotoxic with GI_{50} values ranging from 1.8 to 5.4 μM [235].

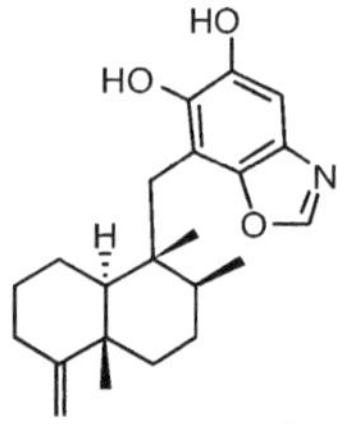

523

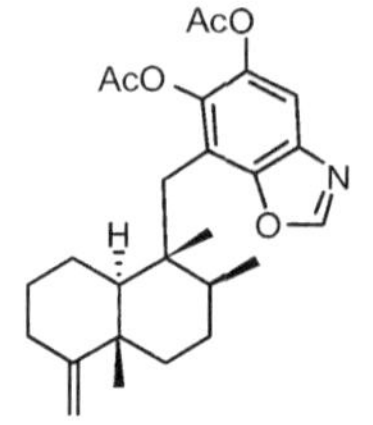

524

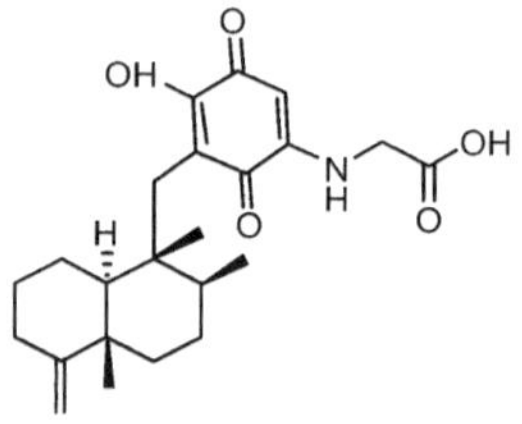

525

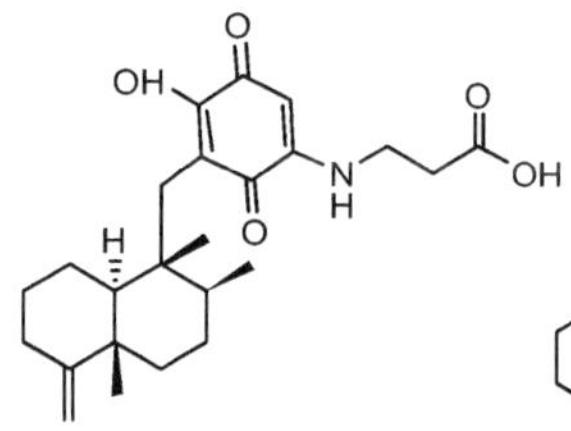

526

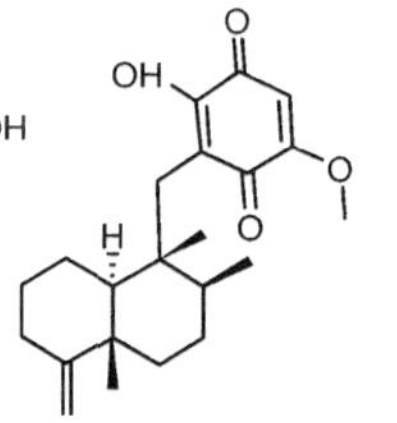

527

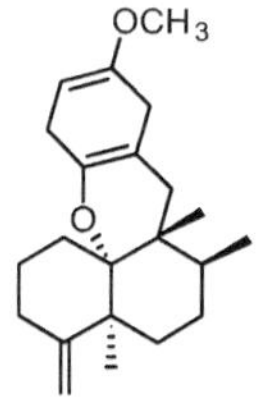

528

In 2011, *Halichondria* sp. (Unten Port, Okinawa) was the source of three new dimeric sesquiterpenoids, halichonadins G–I (**529–531**), of which Halichonadins G (**529**) and I (**531**) were active against murine lymphoma L1210 cells and human epidermoid carcinoma KB cells with an IC_{50} range of 3.4–6.9 μg/mL [236].

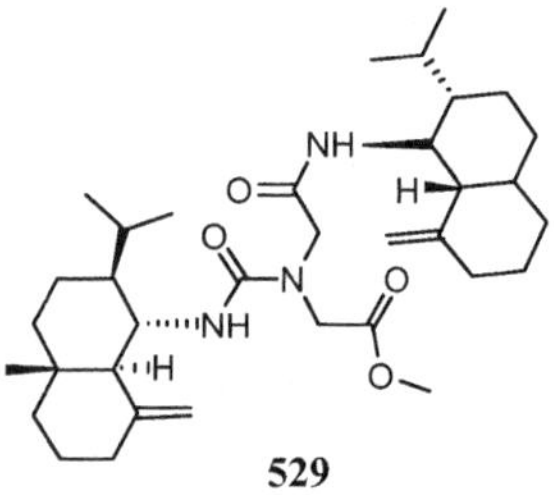

529

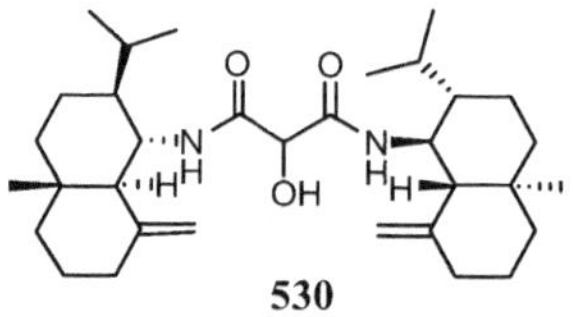

530

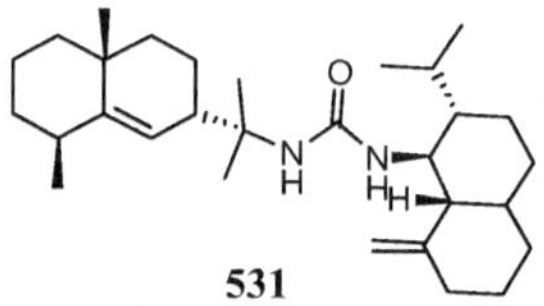

531

In 2012, a chemical study of sponge *Dysidea avara* (Xisha islands, South China Sea) resulted in discovery of dysidavarones A–D (**532–535**), of which dysidavarone A (**532**) inhibited protein tyrosine phosphatase 1B (PTP1B) with IC_{50} value of 9.98 μM. These four compounds are the first example of sesquiterpenes featured with the unprecedented "dysidavarane" carbon skeleton [237].

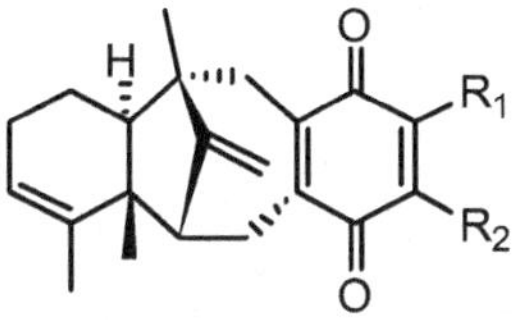

532 R_1=OEt R_2=H
533 R_1=H R_2=OEt

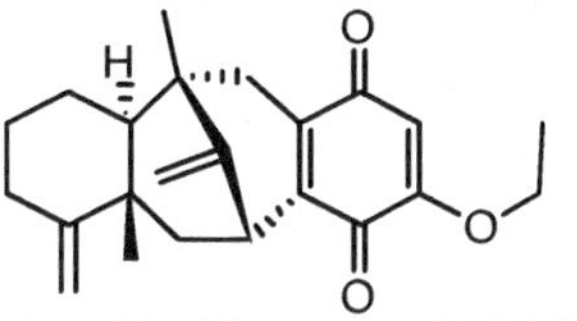

Dysidavarone D (**534**)

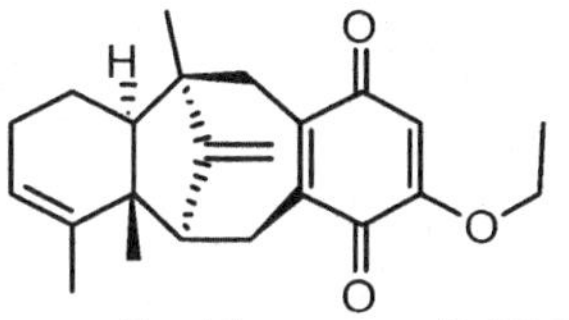

Dysidavarone D (**535**)

In 2013, chemical investigation of *Euryspongia* sp. (Iriomote Island, Okinawa, Japan) revealed the discovery of Euryspongins A–C (**536–538**). The compounds **536–538** were not active against protein tyrosine phosphatase 1B (PTP1B) while compound **539** (the dehydrated product of **536**) tended to inhibite PTP1B with IC_{50} value of 3.6 μM, highlighting the importance of the absence of an OH group at C-4 for activity [238].

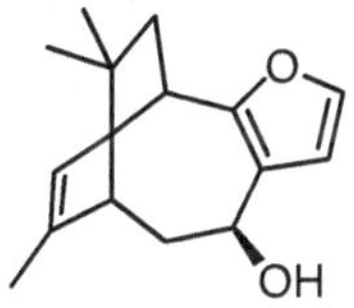

Euryspongin A (**536**)

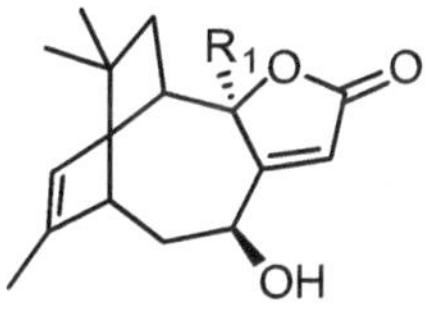

Euryspongin B (**537**) R1=H
Euryspongin C (**538**) R1=OCH_3

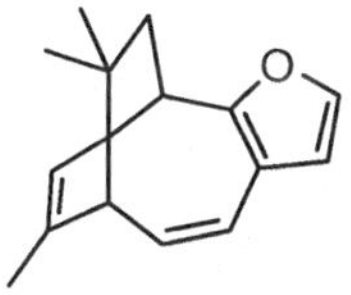

Dehydroeuryspongin A (**539**)

15.6.2 Diterpenes

In 1998, four new norditerpenes (**540–543**) were isolated from the marine sponge *Diacarnus* cf. *spinopoculum* (Solomon Islands). Compounds **540** and **541** showed moderate activity against the NCI's 60 cell line [239]. In 2002, an inseparable mixture, sarcotins K (**544**) and L (**545**) were reported from the marine sponge *Sarcotragus* sp. (Cheju Island, Korea, in July 1998). The mixture was evaluated for cytotoxicity against five human tumor cell lines, but only showed activity against SK-MEL-2 cell line with an ED_{50} value of 6.2 μg/mL [240].

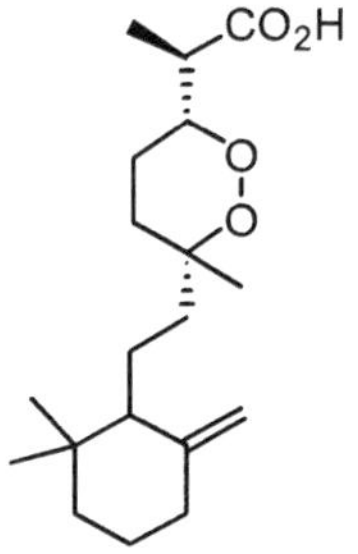

540

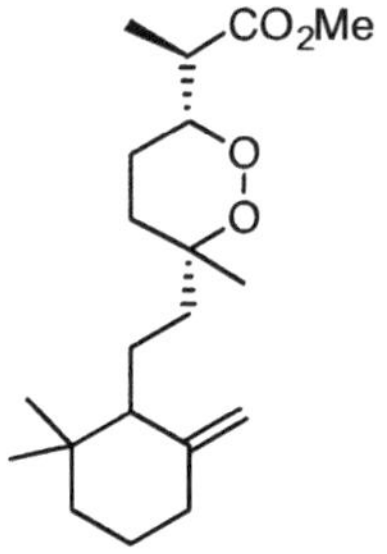

541

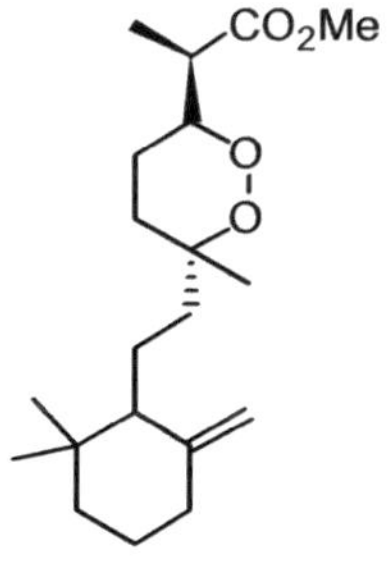

542

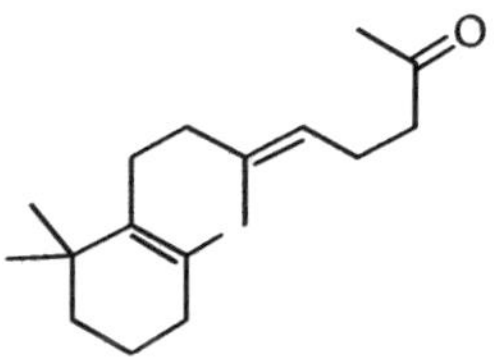

543

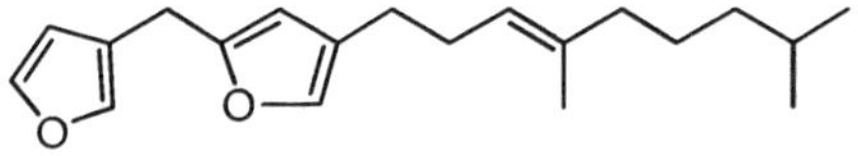

544

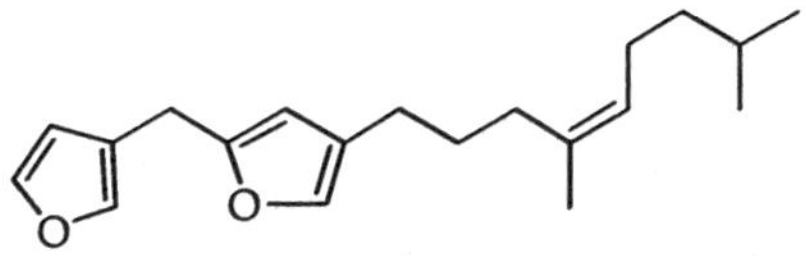

545

In 2009, seven new spongian-class diterpenes (**546–552**) were discovered from the sponge *Dysidea* cf. *arenaria* (Okinawa Island), of which compounds **547**, **551**, and **552** exhibited cytotoxicity against NBT-T2 rat bladder epithelial cells with IC_{50} values of 1.9, 1.8, and 4.2 μg/mL, respectively [241]. A new isonitrile diterpene, namely 8-isocyanoamphilecta-11(20), 15-diene (**553**), along with three known isonitriles (**554–556**), were reported in the same year from the sponge *Ciocalapata* sp. (vicinity of Koh-Tao, Thailand, in April 2002). The four isonitriles extinctively supressed *Plasmodium falciparum* K1 with mean IC_{50} values from 0.09 to1.07 μmol/L [242].

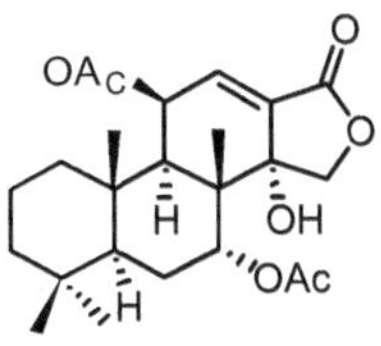

546

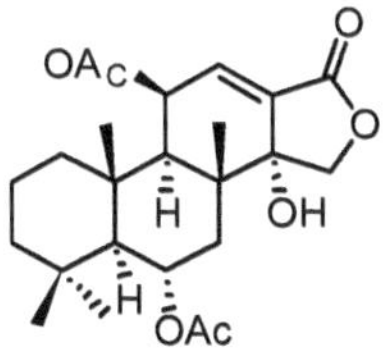

547

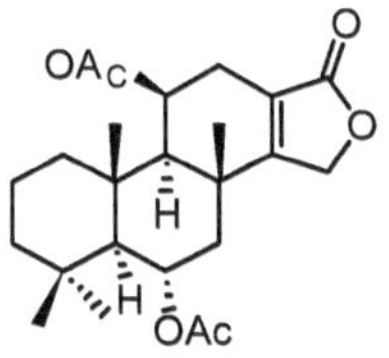

548

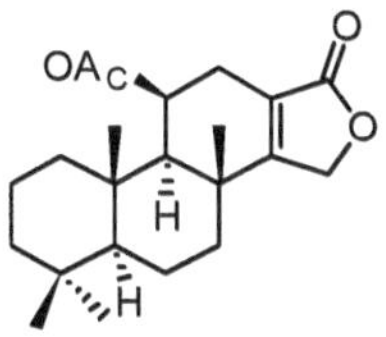

549

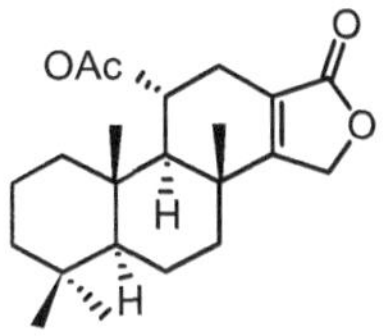

550

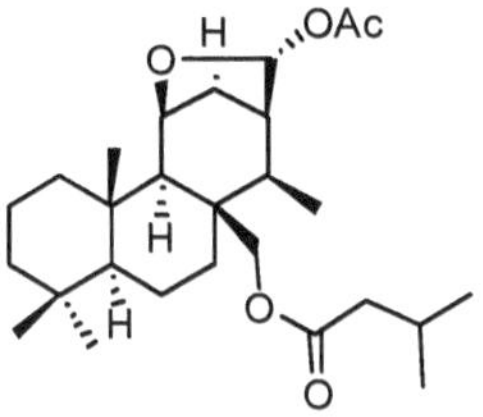

551

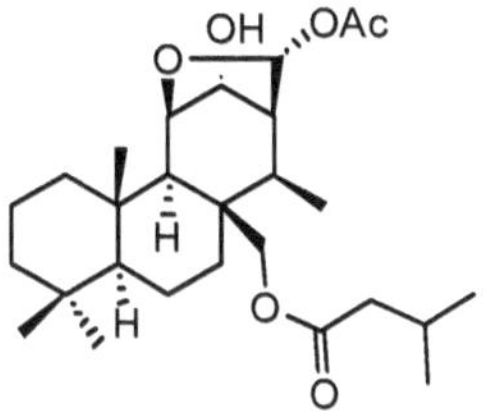

552

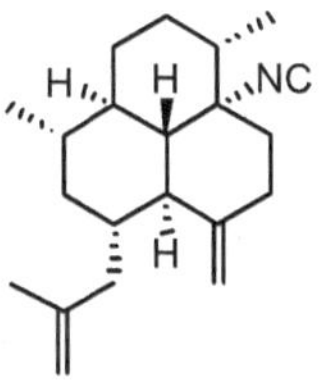

553

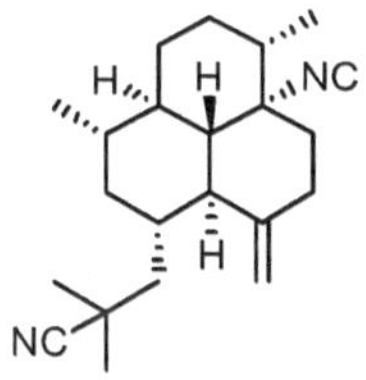

554

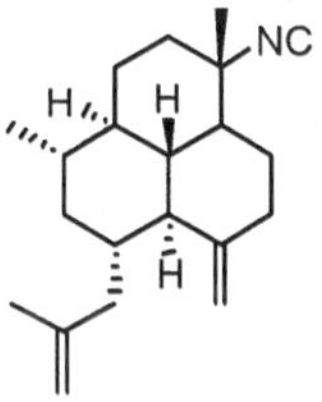

555

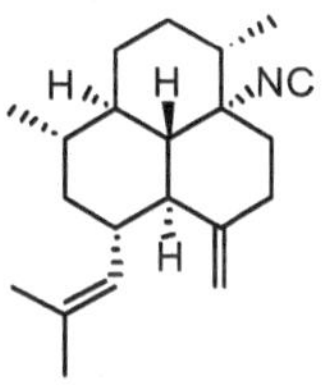

556

In 2011, four novel 9-*N*-methyladeninium diterpenoids, agelasine M (**557**), 2-oxo-agelasine B (**558**), gelasineA (**559**), and gelasine B (**560**), together with the known agelasine B (**561**) and F (**562**), were isolated from the marine sponge *Agelas* sp. (Kimbe Bay, Papua New Guinea, in November 2007). Compounds **557** and **562** exhibited potent cytotoxicity against Jurkat cells with IC_{50} values of 3.0 and 3.6 μg/mL while compounds **561** and **562** were strongly active against *Trypanosoma brucei* with IC_{50} values of 8.4 and 3.3 μg/mL [243].

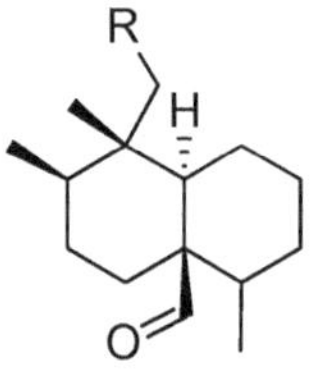

557

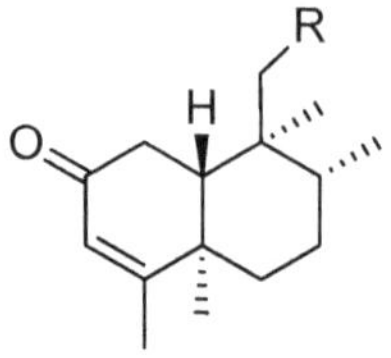

558

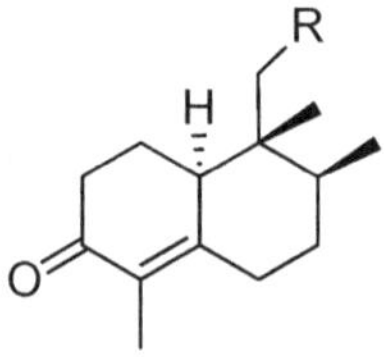

559

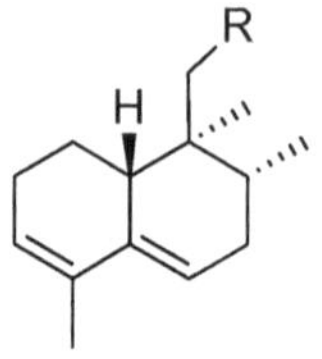

560

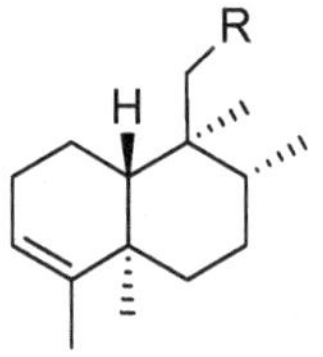

561

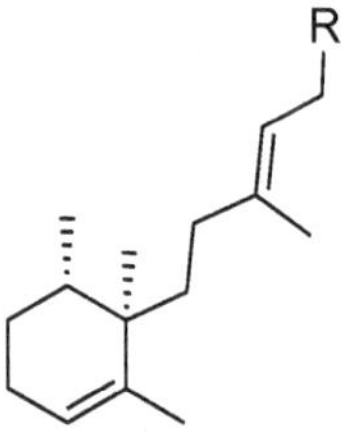

562

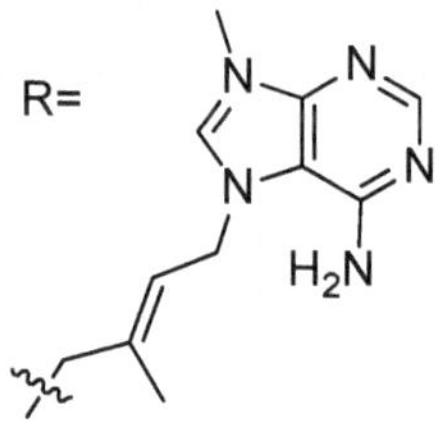

In 2012, two new spongian-class diterpenes (**563**, **564**), along with two known compounds (**565**, **566**), were found in the marine sponge *Chromodoris* sp. (Okinawa Island), and they showed moderate cytotoxity against NBT-T2 rat bladder epithelial cells with IC_{50} values of 5.6, 12, 3.4, and 3.8 μg/mL, respectively [244].

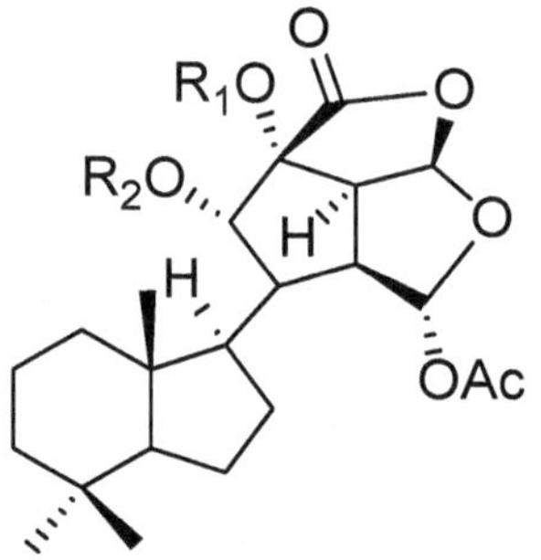

563: R1=R2=H
565: R1=R2=Ac
566: R1=H, R2=Ac

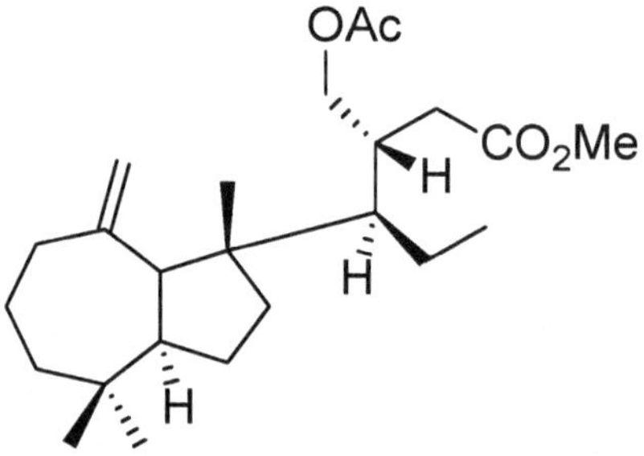

564

In 2015, chemical investigation of the New Zealand marine sponge *Hamigera tarangaensis* (Cavalli Island, New Zealand, in December 2003) resulted in the isolation of nine new nitrogenous hamigeran diterpenoids, namely hamigeran M(**567**), hamigerans N – Q(**568**–**571**), 19-epi-hamigeran Q (**572**), and 18-epi-hamigerans N,

P, and Q (**573**–**575**). Hamigeran M (**567**) exhibited potent cytotoxic activity against HL-60 promyelocytic leukemic cell line at 6.9 μM. And in this work, the structure of hamigeran D (**576**) was revised [245].

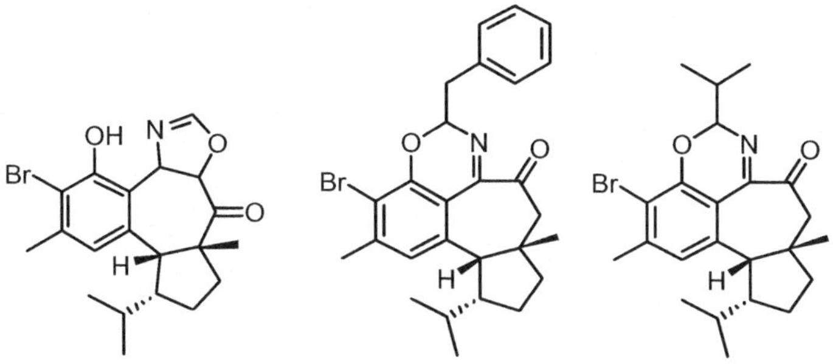

567 **568** 18 *S** **573** 18 *R** **569** 18 *S**

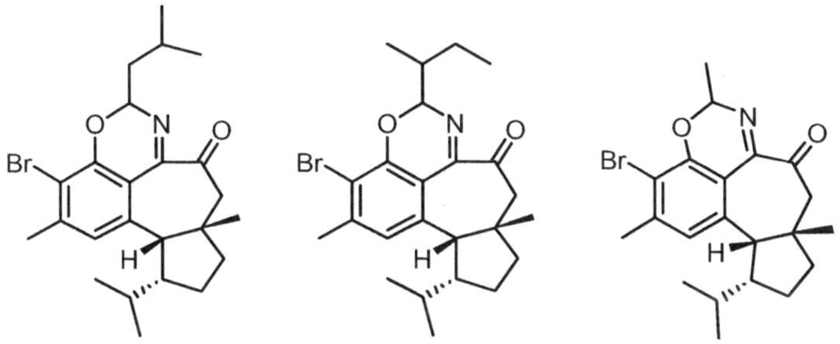

570 18 *S** **574** 18 *R** **571** 18 *S** **572** 18 *R**, 19-*epi* **575** 18 *R** **576** 18 *S*

In 2015, a new meroditerpene, 26-O-ethylstrongylophorine-14 (**577**), was isolated from the Okinawan marine sponge *Strongylophora strongilata* (Iriomote Island, Okinawa, Japan, in 2010) together with six known strongylophorines (**578**–**583**). Compounds **577, 578**, and **580** inhibited the activity of protein tyrosine phosphatase 1B (PTP1B) with an IC_{50} value of 8.7, 8.5, 9.0 μM, respectively [246]. A chemical investigation of the marine sponge *Petrosia corticata* (North Sulawesi, Indonesia, in 2007) resulted in the characterization of two new strongylophorine derivatives in 2015, 26-O-methylstrongylophorine-16 (**584**) and 26-O-ethylstrongylophorine-16 (**585**), along with six known congeners (**586**–**592**). Most of these compounds showed potent inhibitory activity against the chymotrypsin-like activity of the proteasome in the low concentration (μg/mL) [247].

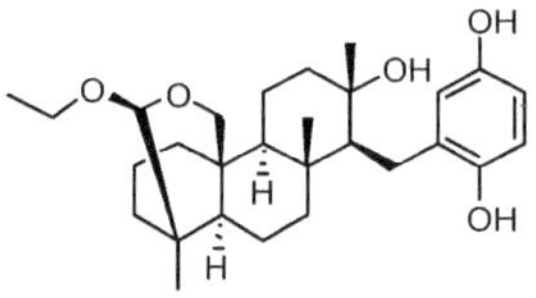

577

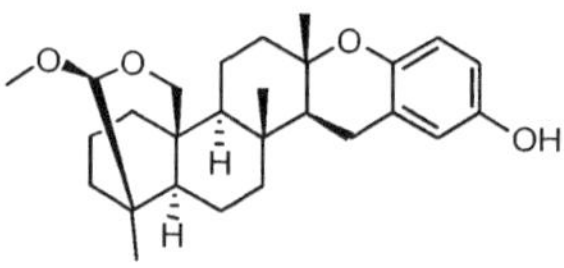

578

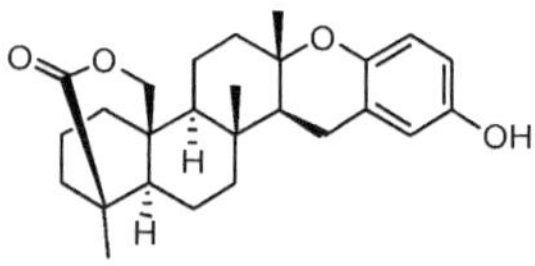

579

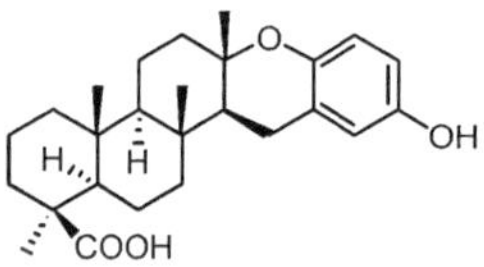

580

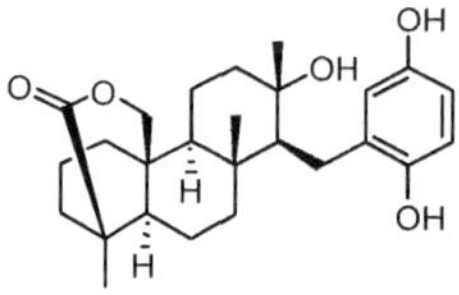

581

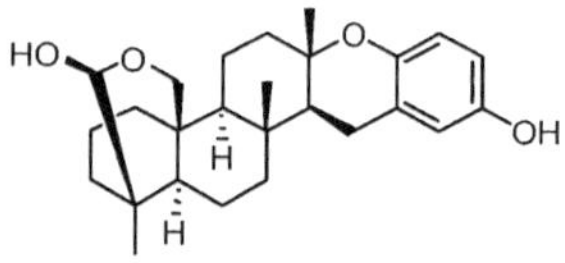

582

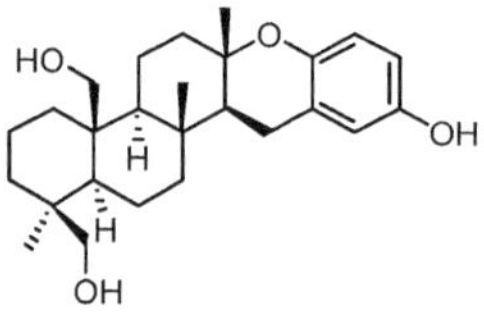

583

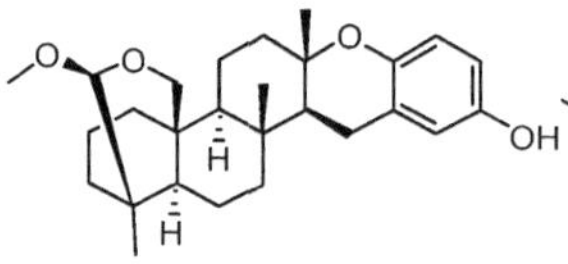

584

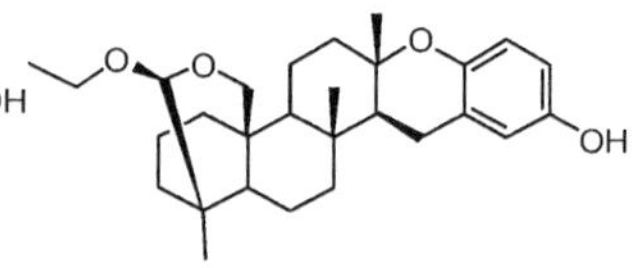

585

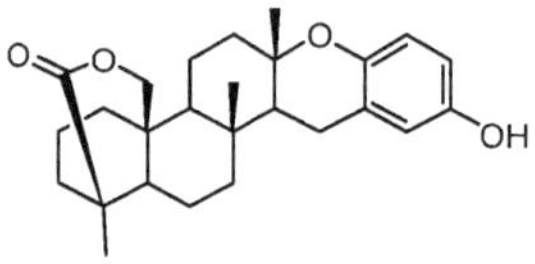

586

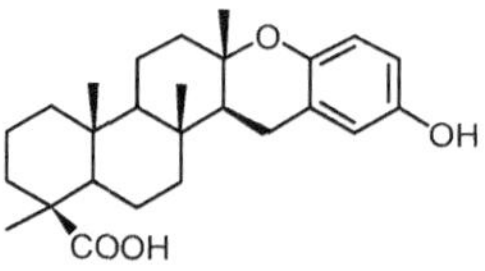

587

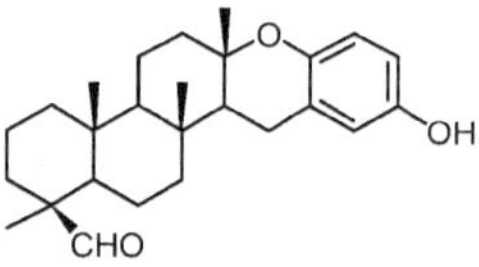

588

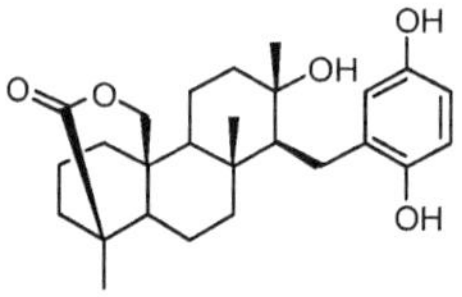

589

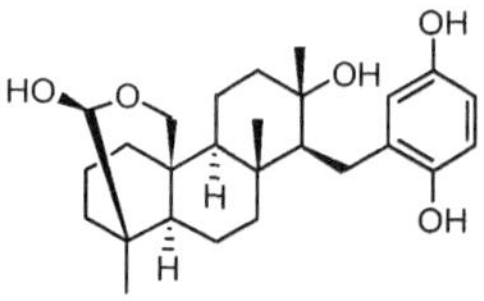

590

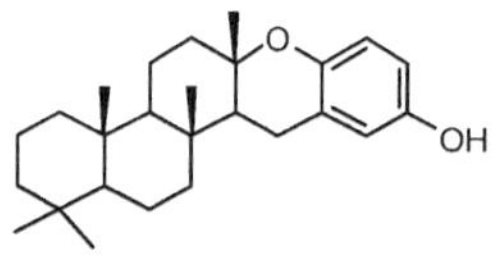

591

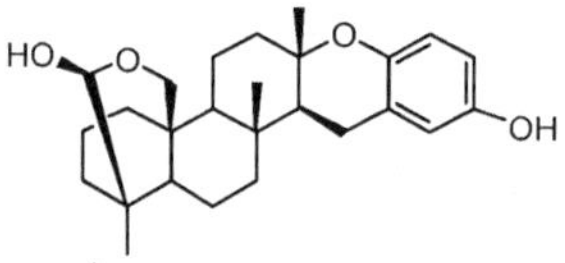

592

15.6.3 Sesterterpenes

In 1991, three new norsesterterpene peroxides, phyllofenone B (**593**), phyllofolactone A (**594**), and phyllofolactone B (**595**), were isolated from the sponge *Latrunculia sp.* (Jervis Bay, Australia). Among them, phyllofenone B was active against the P388 cell line with an IC_{50} value of 5 μg/mL [248]. In 1991, a study of the Adriatic sponge *Fasciospongia cavernosa* (Yongxing Island, South China Sea, in April 1988) yielded 25-Deoxycacospongionolide B (**596**), which showed a high cytotoxicity with an EC_{50} value of 0.74 μg/mL in the Arremia salina bioassay [249].

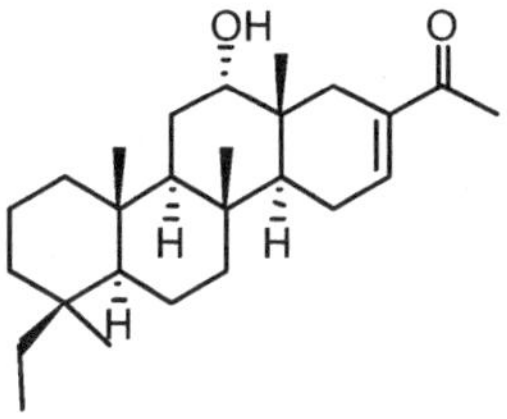

593

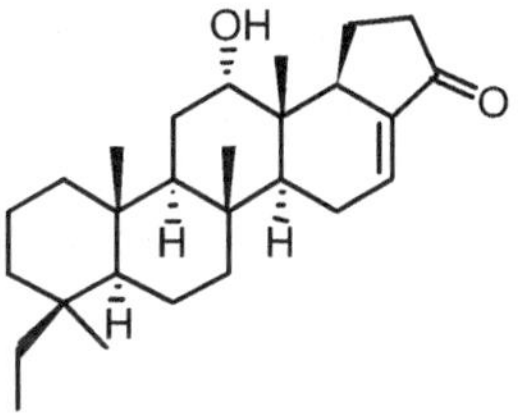

594

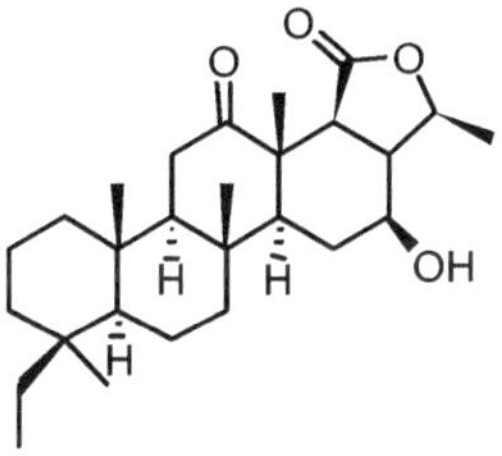

595

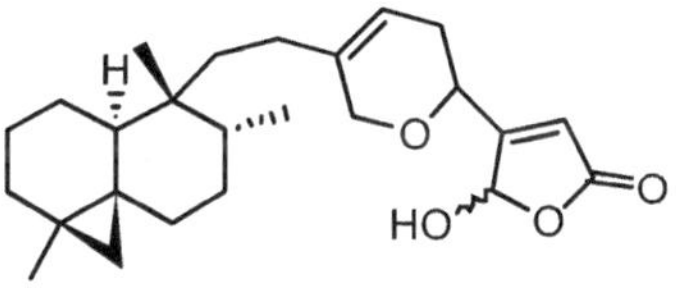

596

In 1998, chemical analysis of *Hyrtios erecta* (Amami-Oshima, Japan, in 1993) resulted in the discovery of novel scalarane sesterterpenes (**597–601**), and their stereo structures were characterized by means of spectral analyses, X-ray crystallography, and chemical reactions. Compound **597** showed potent in vitro and in vivo antitumor activities. In addition, the structure-activity relationship was also discussed using computer-assisted structure matching of **597** and aragusterols [250].

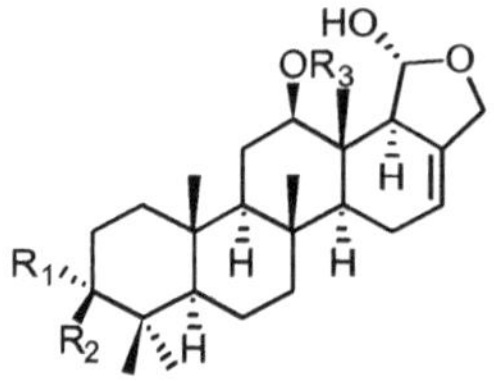

597 R1, R2=O;R3=H
598 R1=R3=H;R2=OH
599 R1=H;R2=OH;R3=Ac
600 R1=R2=R3=H

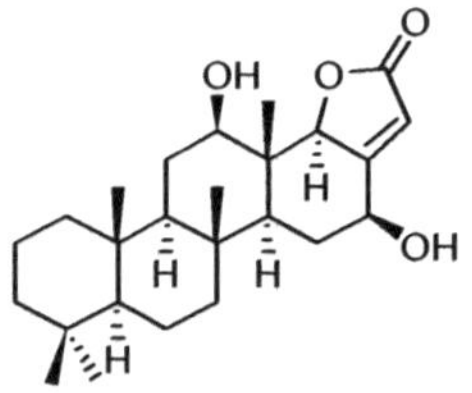

601

In 2002, barangcadoic acid A (**602**) and rhopaloic acids D-G (**603–606**) found in the marine sponge *Hippospongia sp*, were reported to possess RCE protease inhibitory activity [251]. In 2010, a chemical investigation of the South China Sea sponge *Phyllospongia foliascens* (Yongxing Island, South China Sea, in June 2007) resulted in discovery of a new scalarane sesterterpene, phyllofolactone M (**607**) [252].

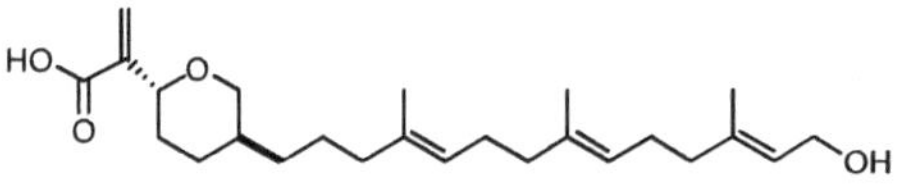

602

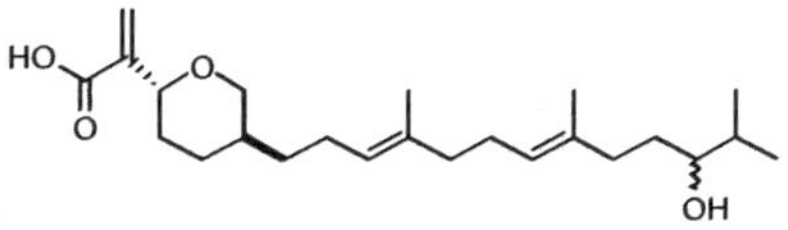

603

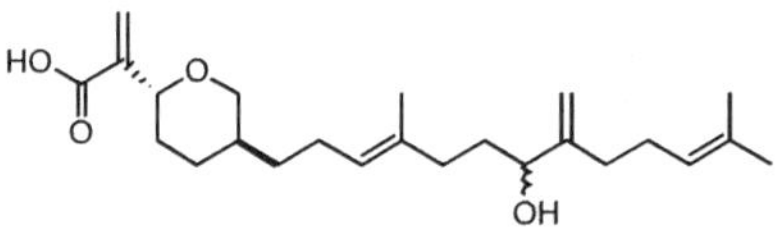

604

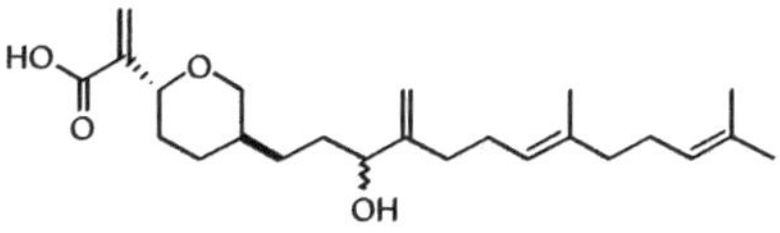

605

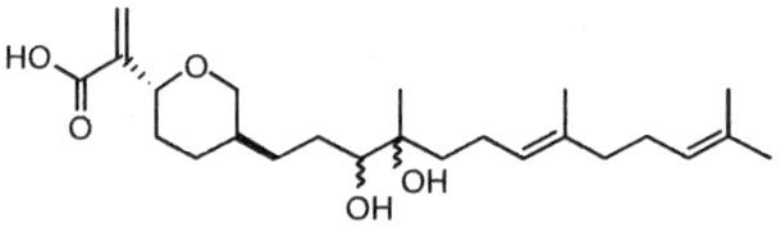

606

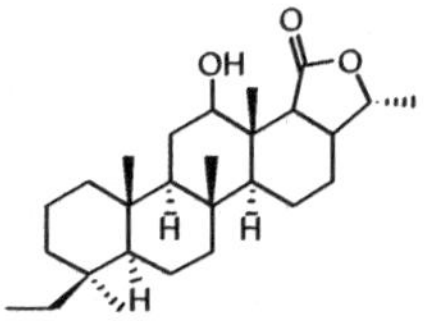

607

In 2011, a study of the sponge *Coscinoderma* sp. (Chuuk Island, Federated States of Micronesia, in June 2006) yielded eight new sesterterpenes (**607–615**), which displayed moderate cytotoxicity toward the K562 cell line and inhibitory activities of isocitrate lyase, sortase A, and Na^+/K^+-ATPase [253].

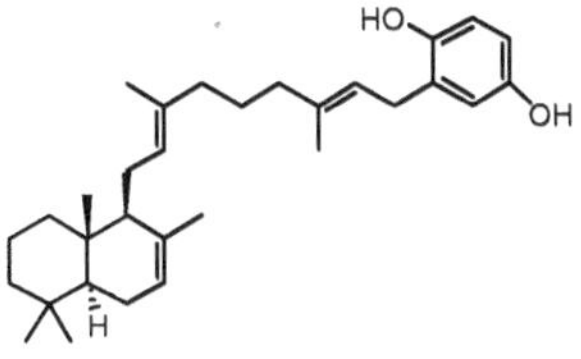

608

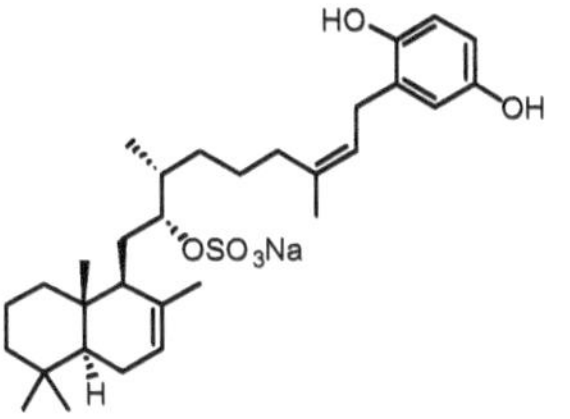

609

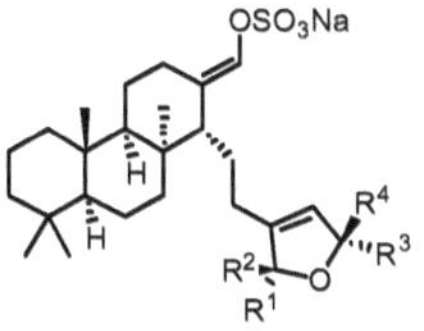

610 R^1=H, R^2=H,R^3R^4=O
611 R^1R^2=O, R^3=H, R^4=OCH_3
612 R^1R^2=O, R^3=OCH_3,R^4=H
613 R^1=OCH3, R^2=H,R^3=OCH3,R^4=H

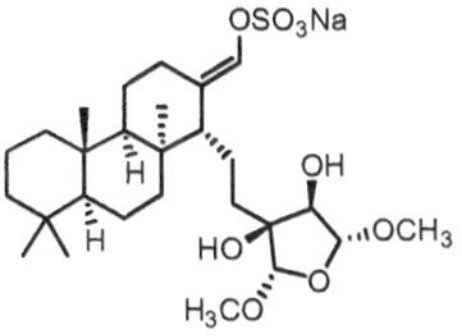

614

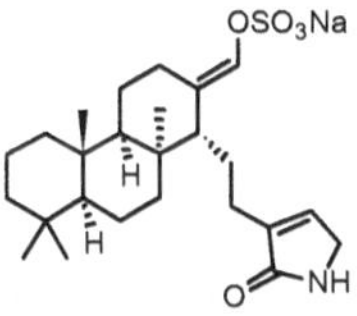

615

In 2011, five new sesterterpenes (**616–620**) were isolated from the sponge *Hyatella* sp. (Soheuksan-do, West Sea, Korea, on June 18, 2007). Compounds **616–619** contained oxidized furan moieties, while compound **620** possessed a corresponding lactam. Compound **619** exhibited moderate antibacterial activity with MIC value of 1.56 μg/mL [254].

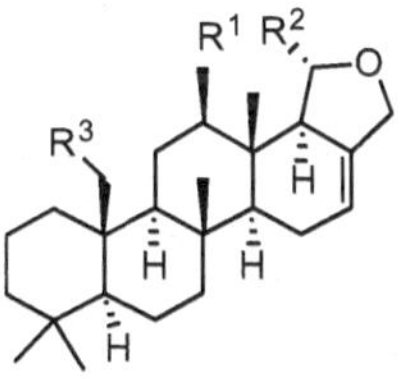

616 R^1=OAc,R^2=OH,R^3=OAc
617 R^1=OH,R^2=OH,R^3=OAc
618 R^1=OAc,R^2=OCH_3,R^3=H

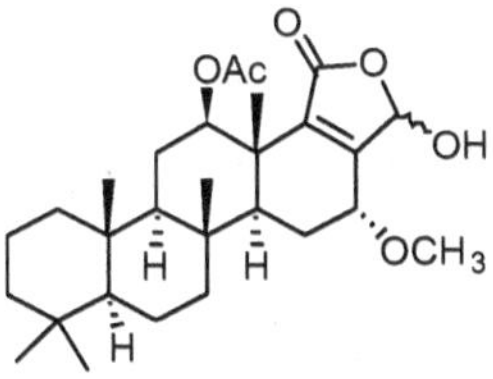

619

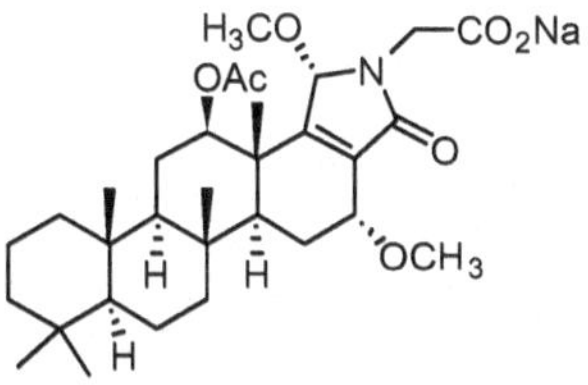

620

In 2011, hippolides A–H (**621–630**) were obtained from the sponge *Hippospongia lachne* (Yongxing Island, South China Sea, in June 2007), and their absolute configurations were established by the modified Mosher's method and CD data. Hippolide A exhibited cytotoxicity against A549, HeLa, and HCT-116 cell lines with IC_{50} values of 5.22×10^{-2},4.80×10^{-2} and 9.78 μM, respectively [255].

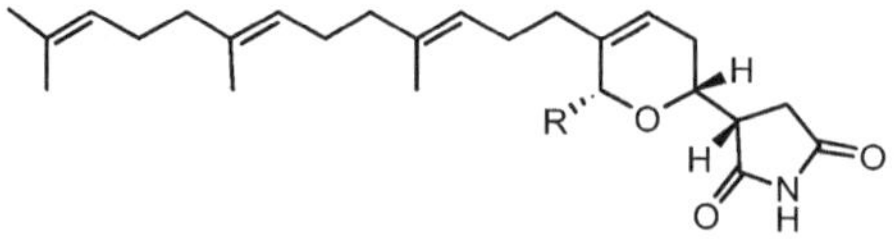

621 R=OH
622 R=OCH_3

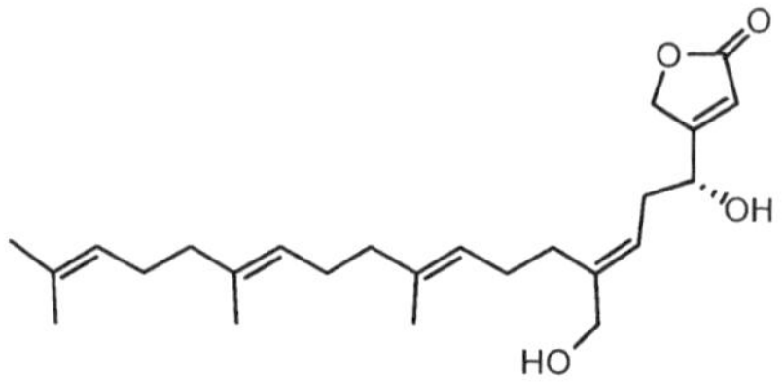

623

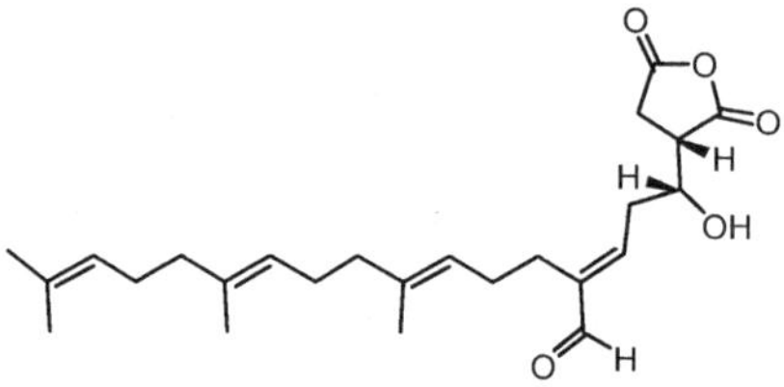

624

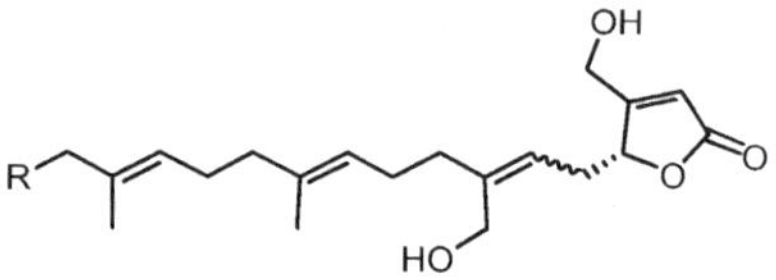

625 6*E*,R= 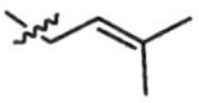

626 6*Z*,R= 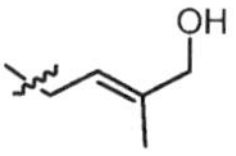

627 6*Z*,R= 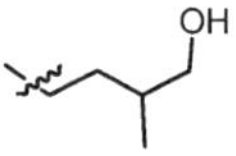

628 6*Z*,R= 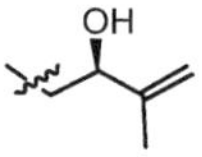

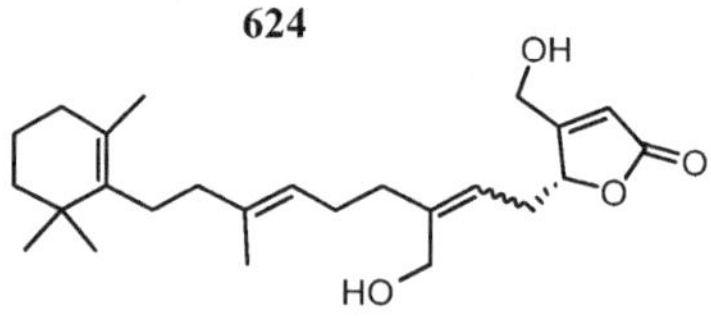

629 6*E*
630 6*Z*

In 2011, two new sesterterpenoids, phorbasones A (**631**) and B (**632**), were found from the Korean marine sponge *Phorbas* sp., and their complete structures were elucidated by spectral data and chemical reactions. Phorbasone A exhibited a positive effect on the calcium deposition activity in C3H10T1/2 cells [256].

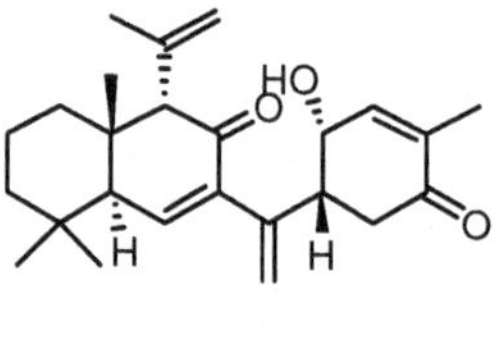

631 Phorbasone A

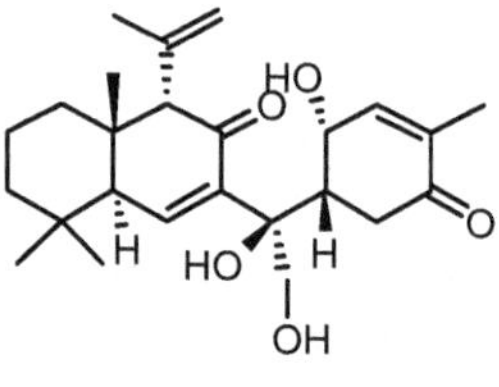

632 Phorbasone B

In 2012, phorone A (**633**) and isophorbasone A (**634**) featured with two new carbon skeletons were identified, along with ansellone B (**635**) and phorbasone A acetate (**636**), from a Korean marine sponge, *Phorbas* sp. Ansellone B (**635**) and phorbasone A acetate (**636**) exhibited potent inhibitory activity on nitric oxide production in RAW 264.7 LPS-activated mouse macrophage cells with IC_{50} values of 4.5 and 2.8 μM, respectively [257].

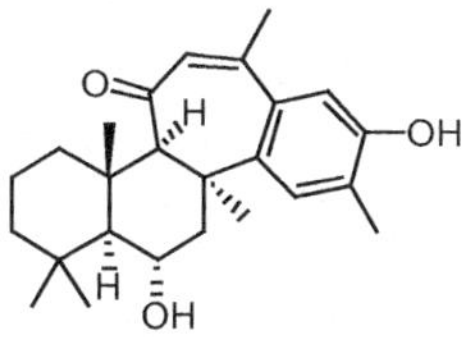

633 phorone A

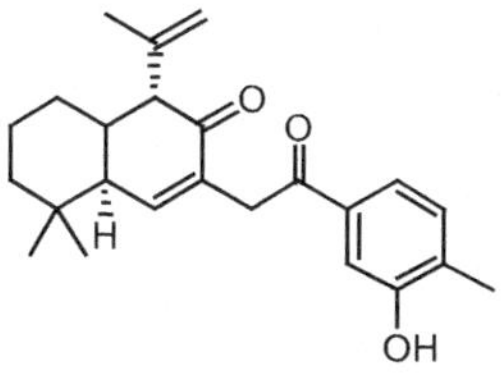

634 isophorbasone A

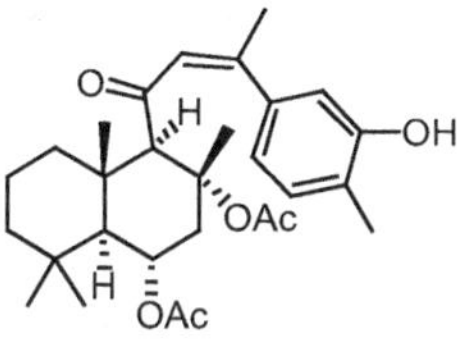

635

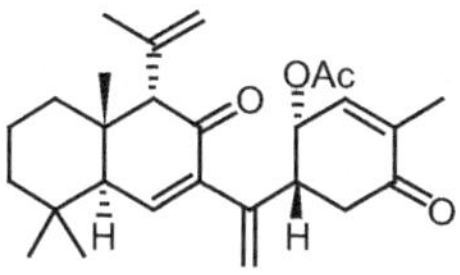

636

In 2013, four new sesterterpenoids, ansellone B (**637**), phorbadione (**638**), secoepoxyansellone A (**639**), and alotaketal C (**640**) were isolated from the sponge *Phorbas* sp. (Howe Sound, British Columbia). Ansellone B (**637**) possessed an unprecedented heterocyclic skeleton with an oxocane ring, while secoepoxyansellone A (**639**) was the first example of the degraded "secoansellane" sesterterpenoid carbon skeleton. Alotaketal C (**640**) activated cAMP signaling in HEK293 cells with an EC_{50} of 6.5 μM [258].

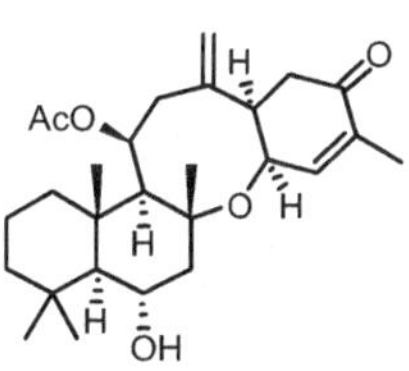

637

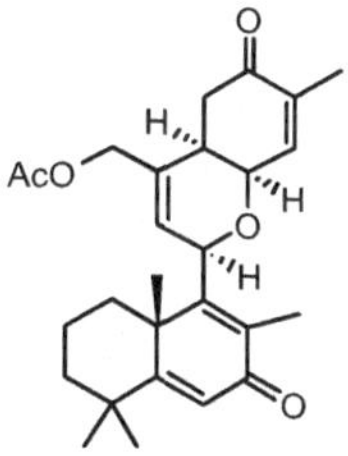

638

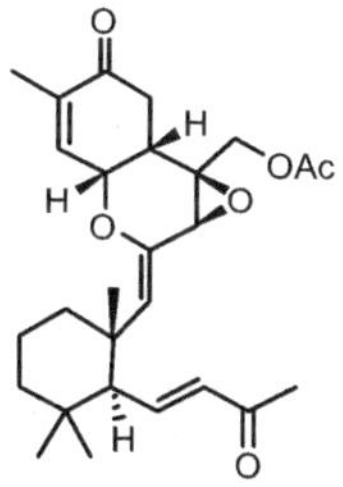

639

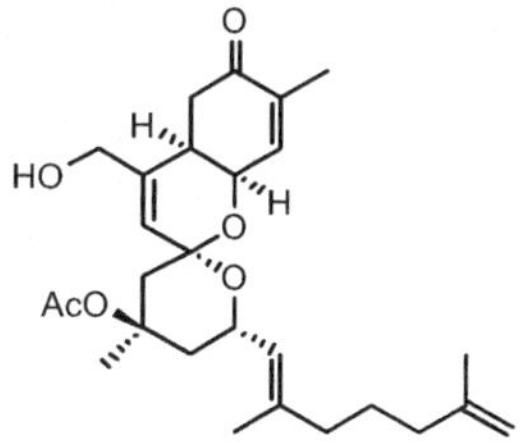

640

Meanwhile, three novel scalarane sesterterpenes, 12-deacetoxy-23-hydroxyscalaradial (**641**), 12-dehydr-oxy-23-hydroxyhyrtiolide (**642**), and 12-O-acetyl-16-deacetoxy-23-acetoxyscalarafuran (**643**), along with four known derivatives (**644–647**), were isolated from *Psammocinia* sp. (South Sea of Korea). They exhibited cytotoxicity against intractable human cancer cell lines A498, ACHN, MIA-paca, and PANC-1, with mean IC_{50} values in the range of 0.4–48 μM [259].

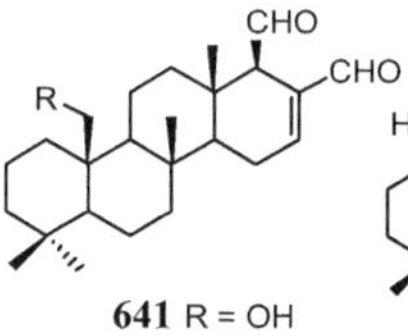

641 R = OH
647 R = H

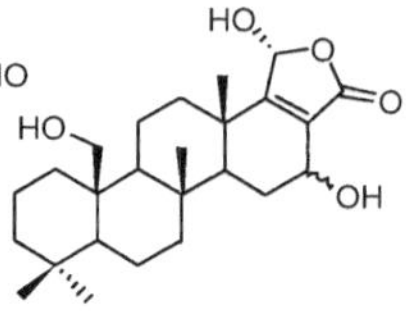

642

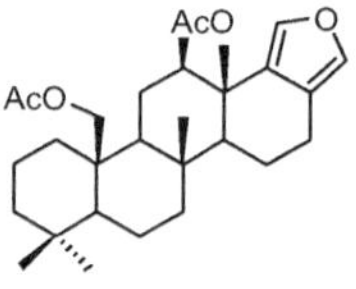

643

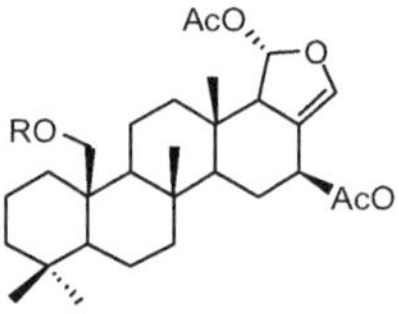

644 R = H
646 R = Ac

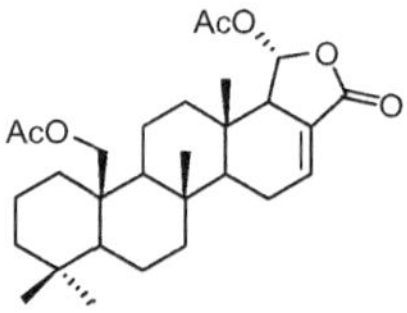

645

Phorbaketals D–K (**648–656**) with a spiroketal-modified benzopyran moiety, along with phorbins A–C (**657–659**), were characterized from the sponge *Monanchora* sp. (Gageo Island, southwestern Korea, in July 2009), and the absolute configurations of them were established with the modified Mosher's method and CD spectroscopic data analysis. Phorbin A (**657**) showed potent cytotoxicity against MIA-paca and PANC-1 human pancreatic cancer cell lines, similar to or better than the positive control [260].

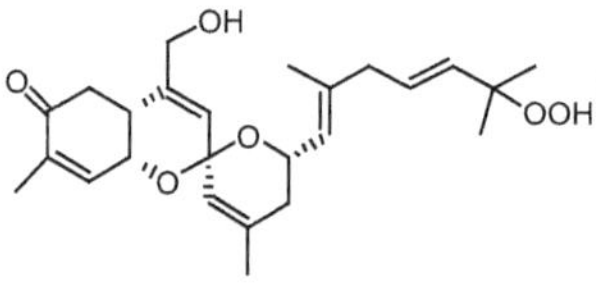

648

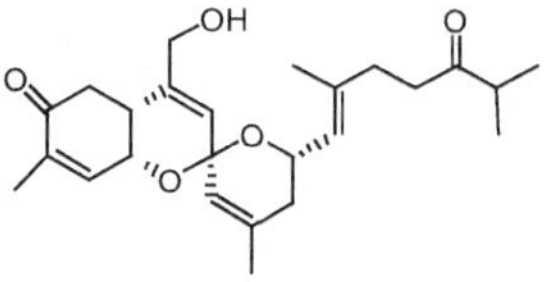

649

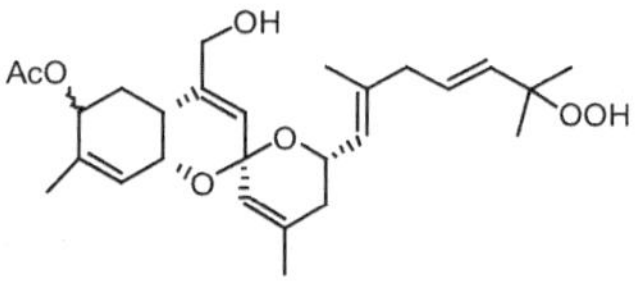

650 (R)-OAc
651 (S)-OAc

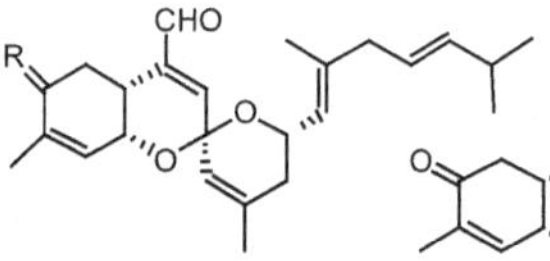

652 R=H,OH(R)
653 R=O
654 R=H,OAc(R)
655 R=H,OAc(S)

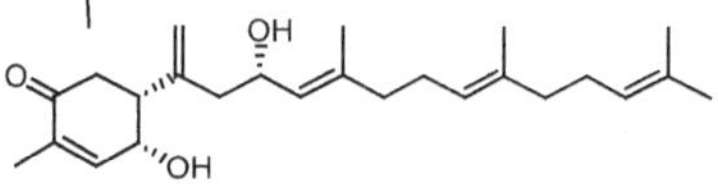

656

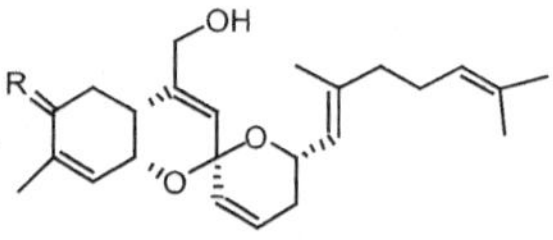

657 R=O
658 R=H,OH(R)
659 R=H,OH(S)

In 2014, five new scalarane derivatives (**660–664**) acting as inhibitors of TDP-43 nuclear factor were discovered from *Hyrtios* sp. and *Petrosaspongia* sp. (Fiji Islands) [261].

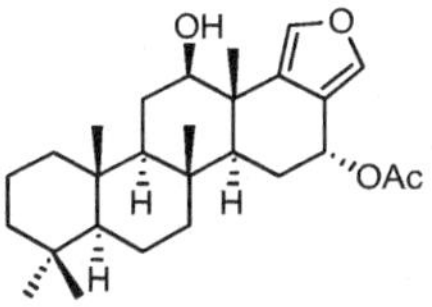

660

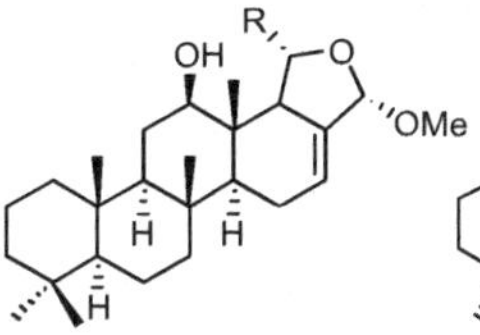

661

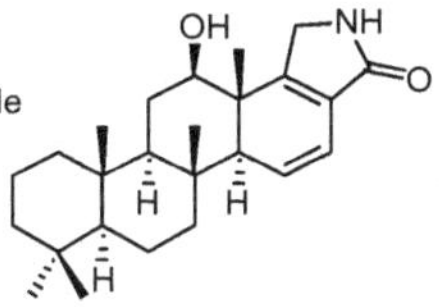

662

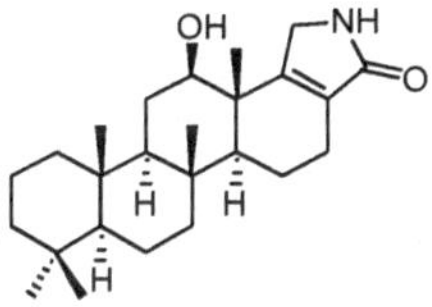

663

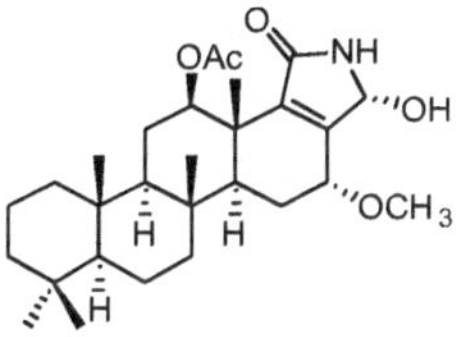

664

In 2015, phyllospongins A–E (**665–668**) were identified from *Phyllospongia lamellosa* (Hurghada, Egypt).These scalarane sesterterpenes exhibited potent cytotoxic activity against HCT-116, HePG2, MCF-7 cell lines with an IC_{50} range of 0.29–2.14 μM. Phyllospongins D showed cytotoxicity against HCT-116 as potent as doxorubicin. Phyllospongins E showed cytotoxicity against MCF-7 comparable to doxorubicin [262].

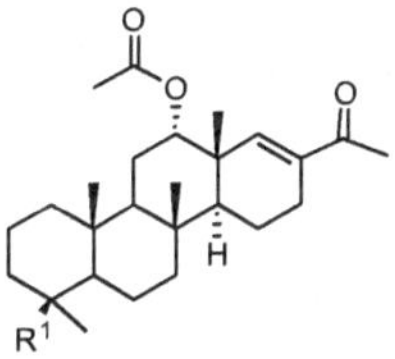

Phyllospongin A (**664**, R^1=CH_3,R^2= -CH_3COO)
Phyllospongin B (**665**,R^1=CH_2CH_3,R^2= -CH_3COO)

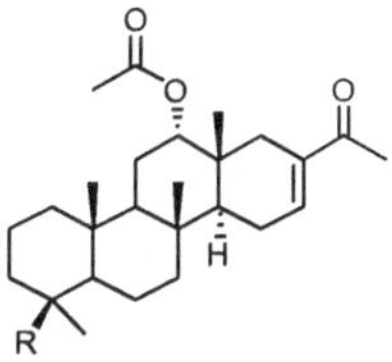

Phyllospongin C (**666**,R=CH3)

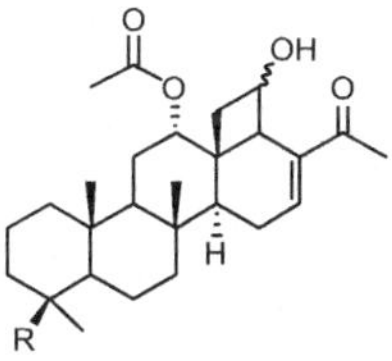

Phyllospongin D (**667**,R=CH_3,-OH)
Phyllospongin E (**668**,R=CH_3,-OH)

In 2015, phorone B (**669**) and ansellone C (**670**) structurally close related to the sesterterpenes of the phorone and ansellone classes were isolated from the marine sponge *Clathria gombawuiensis* (Gageo-do, Korea, on September 9–11, 2006). The two compounds showed temperate cytotoxic activity against A549 and K562 cell lines with mean IC_{50} values in the range 3.9–5.4 μM. The cytotoxicity of **1** may be related to the presence of a free phenolic –OH group, as the corresponding O-methoxy derivative is inactive [263].

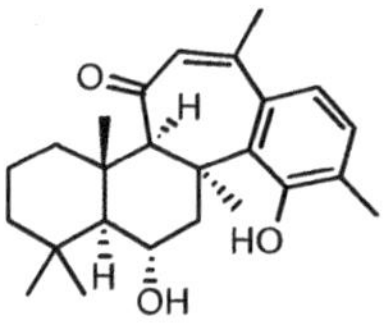

669

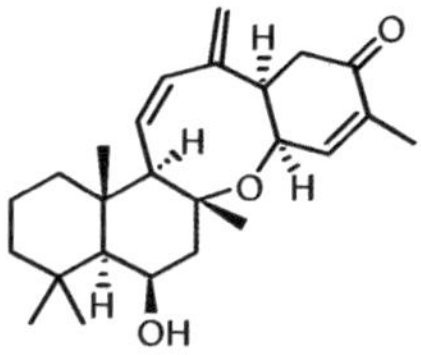

670

15.6.4 Triterpenes

In 2010, nine new isomalabaricane derivatives, globostelletins A–I (**671**–**679**), were isolated from *Rhabdastrella globostellata* (Hainan Island, South China Sea, in June 2006), along with five known compounds (**680**–**684**). Of which, globostelletins C and D (**673**–**674**) were a pair of inseparable geometrical isomers with a ratio of 1:1. Compounds **673**–**674** and **678**–**682** exhibited activity against A2780 cell lines with low μM (IC_{50} < 10 μM) [264].

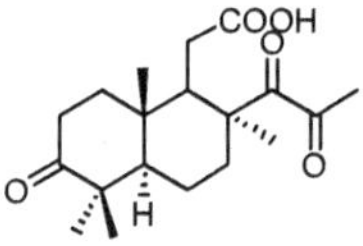

671

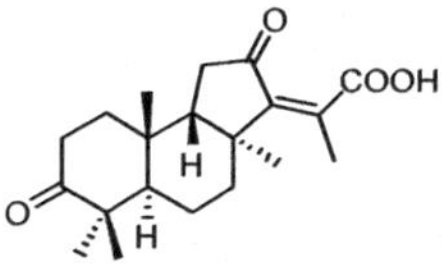

672

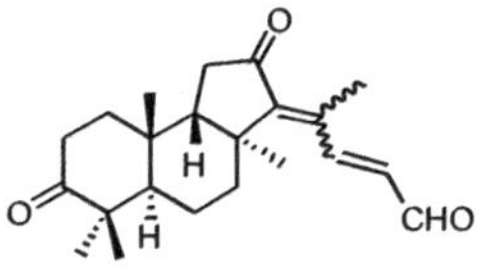

673 13*E*
674 13*Z*

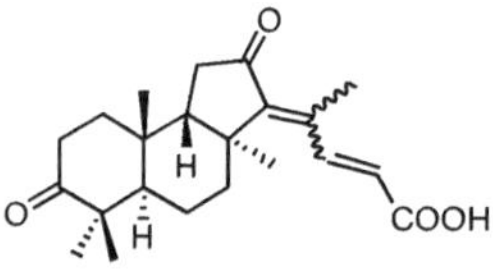

675 13*E*
676 13*Z*

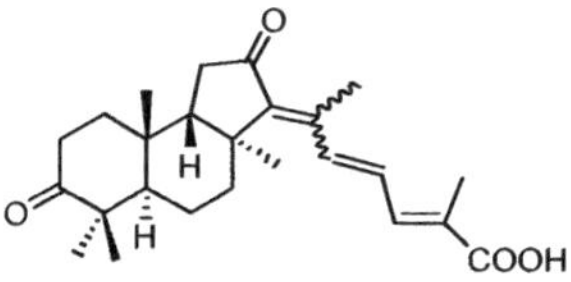

677 13*E*
680 13*Z*

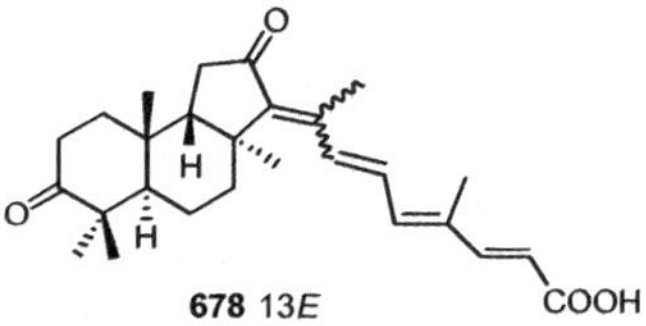

678 13*E*
679 13*Z*

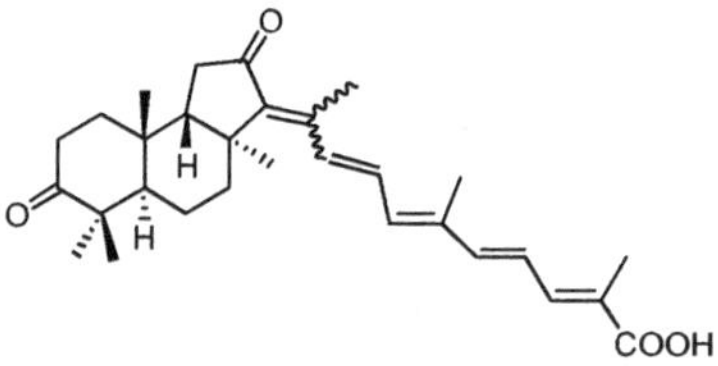

681 13*E*
682 13*Z*

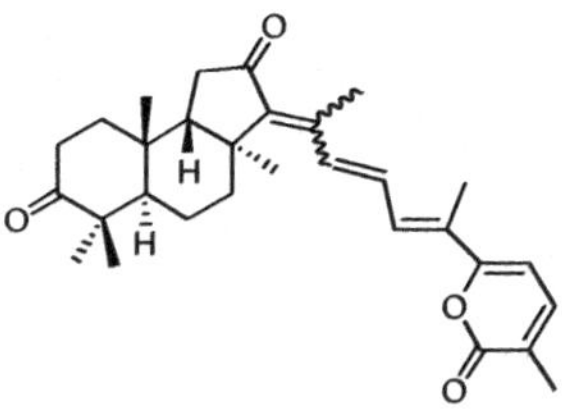

683 13*E*
684 13*Z*

15.7 Bioactive Sterols

The basic role of sterols is the maintenance of optimal fluidity of cell membranes, although these compounds also serve as precursors for the production of diverse steroid classes such as the polyhydroxylated marine sterols. In recent years, research in the marine sterol field has progressed at an impressive pace. Here we intend to provide a description of the characteristic features of 107 sterols with potent biological activities from 29 marine sponges covering literatures from 1993 to 2015.

15.7.1 Alkaloidal Sterols

In 2014, the chemical study of marine sponge *Corticium niger* collected from the Philippines resulted in the discovery of two new steroidal alkaloids plakinamines N (**685**) and O (**686**), which were tested for antiproliferative activity and showed remarkable inhibitory effects against all of the colon cell lines with mean GI_{50} values of 11.5 and 2.4 μM [265].

In 2007, a tropical sponge *Phorbas amaranthus* (Key Largo, Florida) was found to contain a 24-imidazolyl steroidal alkaloid amaranzole A (**687**). The structure was elucidated on the basis of MS, NMR, exciton-coupled CD spectrum, and comparison with model compounds [266]. In the same year, steroidal alkaloid 4-acetoxy-plakinamine B (**688**) bearing a stigmastane skeleton was found in the Thai sponge *Corticium* sp. The compound inhibited acetylcholinesterase with IC_{50} value of 3.75 μM [267]. A series of steroidal alkaloids cortistatins J–L (**689–691**) were isolated from the Indonesian marine sponge *Corticium simplex*. Cortistatin J exhibited potent cytostatic antiproliferative activity against human umbilical vein endothelial cells (HUVECs) at 8 nM [268].

In 2006, four steroidal isoquinoline alkaloids cortistatins A–D (**692–695**) were obtained from the marine sponge *Corticium simplex* (Flores Is., Indonesia). These compounds exhibited highly selective antiproliferative activity against HUVECs with IC_{50} values of 0.0018, 1.1, 0.019, and 0.15 μM, respectively [269].

In 2003, four steroidal alkaloids, plakinamine I–K (**696–698**) and dihydroplakinamine K (**699**), were obtained from a Philippine sponge *Corticium niger*. These compounds exhibited significant in vitro cytotoxicity against the human colon tumor cell line HCT-116 with IC_{50} values of 10.6, 6.1, 1.4, and 1.4 μM, respectively [270].

In 2002, four plakinamine-type steroidal alkaloids were isolated from a marine sponge *Vanuatuan Corticium*, of which plakinamine G (**700**) and tetrahydroplakinamine A (**701**) were quite active against rat glioma cells (IC_{50}'s 6.8 and 1.4 μg/mL, respectively) [271].

In the year of 1999, chemical study of *Corticium* sp. from Vanuatu yielded plakinamines C (**702**) and D (**703**) and three other steroidal alkaloids (**704–706**), all of which showed potent cytotoxicity against human bronchopulmonary non-small-cell lung carcinoma cells with IC_{50} values of <3.3–5.7 μg/mL [272].

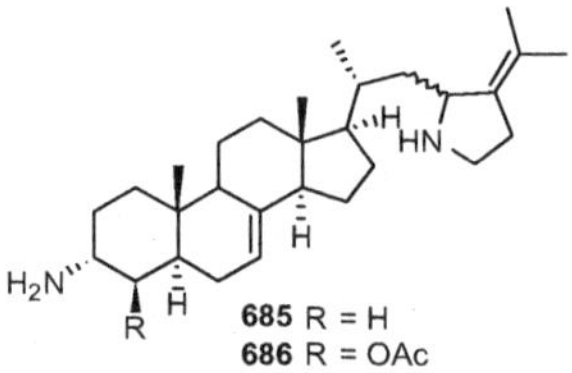

685 R = H
686 R = OAc

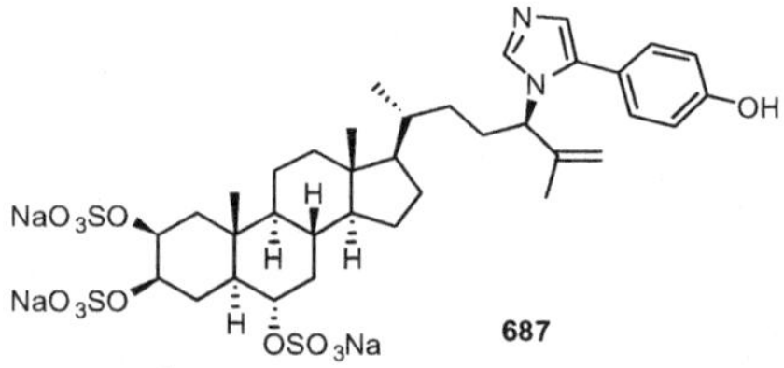

687

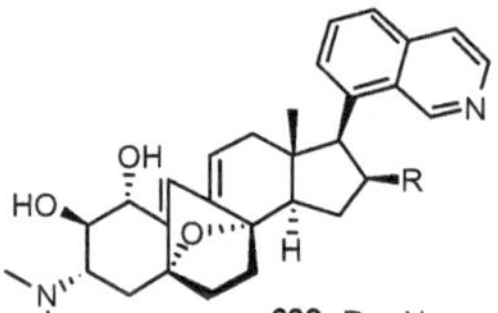

692 R = H
693 R = OH

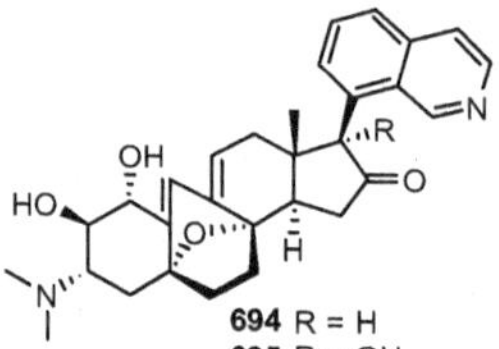

694 R = H
695 R = OH

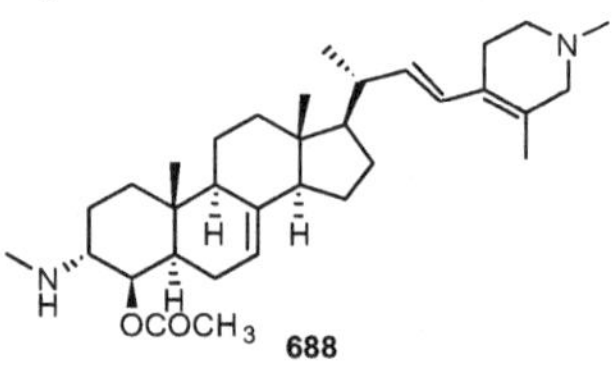

688

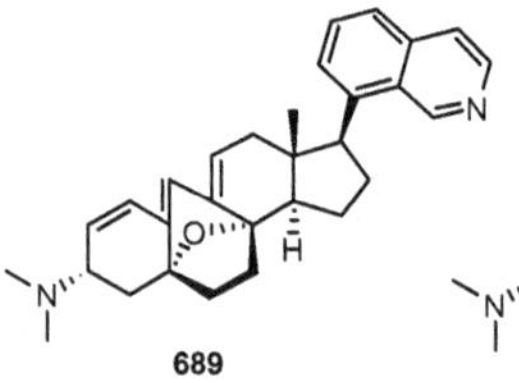

689

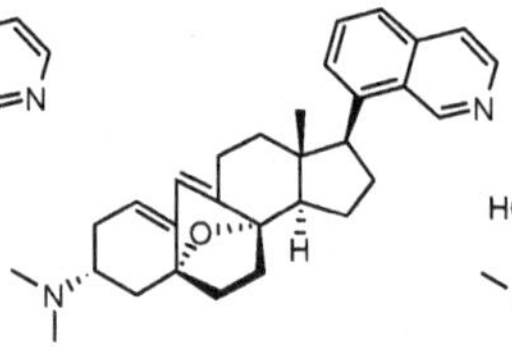

690

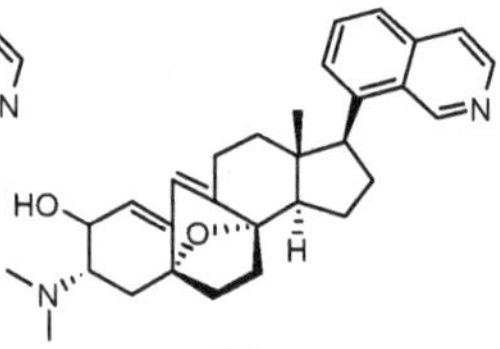

691

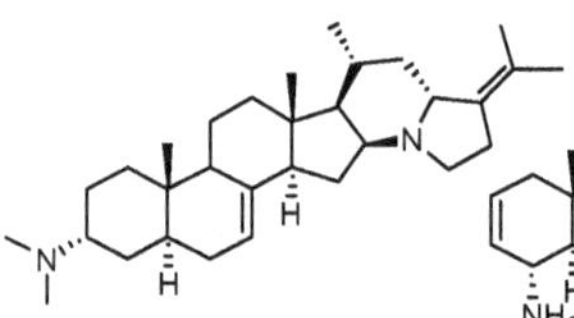

696

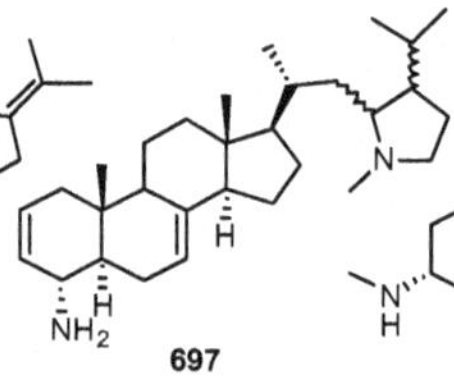

697

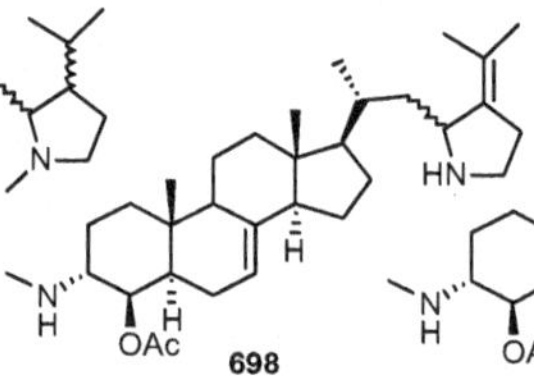

698

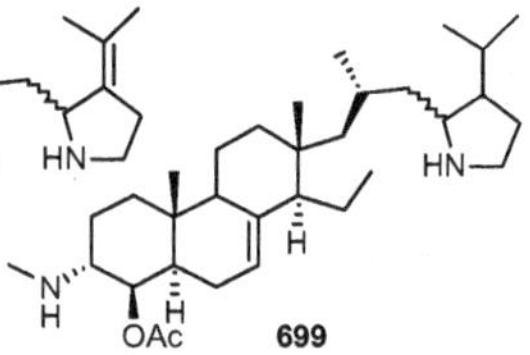

699

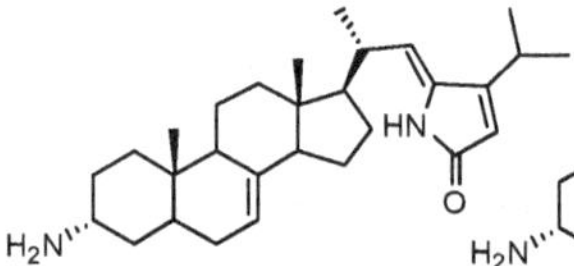

700

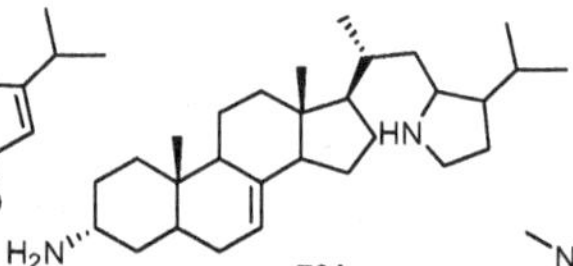

701

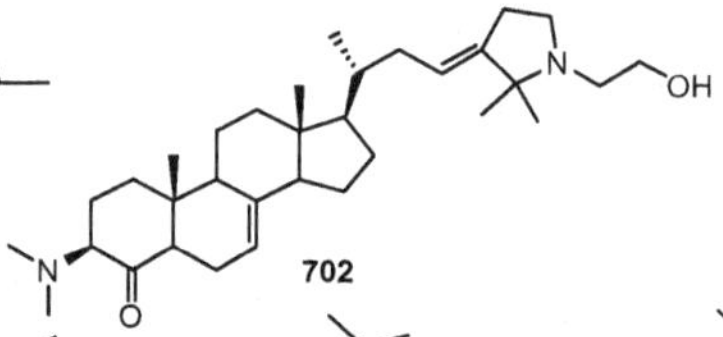

702

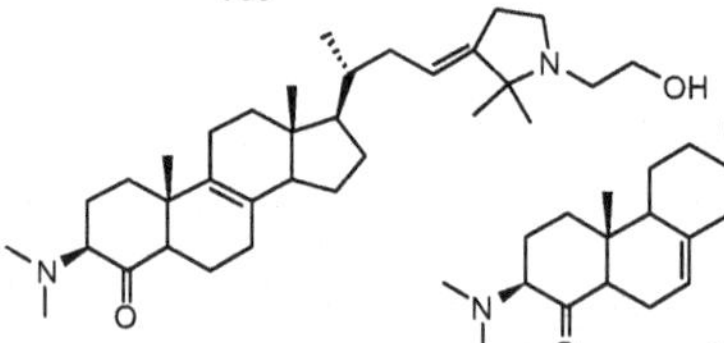

703

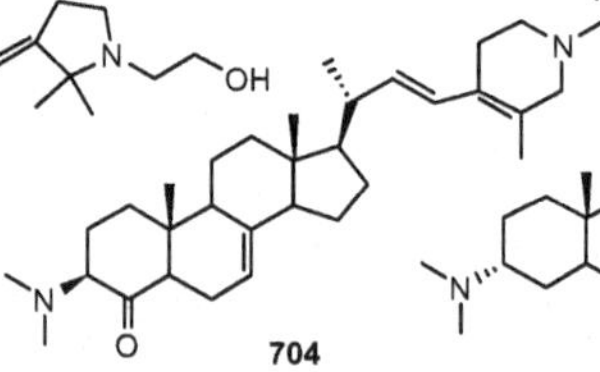

704

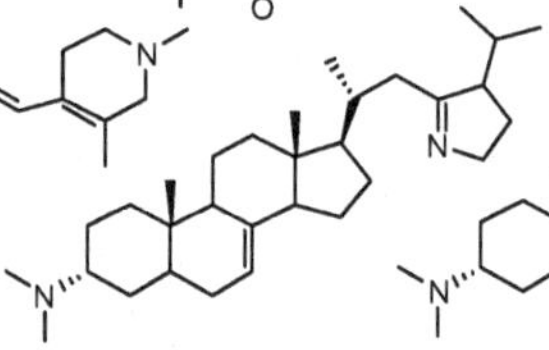

705

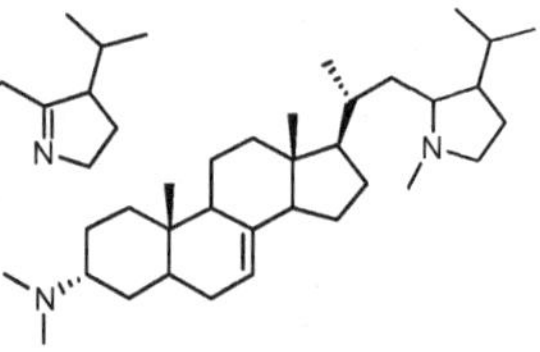

706

15.7.2 *Sulfated Sterols*

In 2012, two new dimeric sterols manadosterols A (**707**) and B (**708**) were found in the sponge *Lissodendryx fibrosa* in Indonesia, which are potent inhibitors of the Ubc13-Uev1a complex with IC_{50} values of 0.09 and 0.13 μM and therefore have potential as anticancer agents [273].

In 2011, shishicrellastatin A and B (**709** and **710**), two dimeric steroid derivatives, were reported from the marine sponge *Crella* (*Yvesia*) *spinulata*. Both of them exhibited bioactivity against cathepsin B with an IC_{50} value of 6.9 μM each [274].

In 2009, three sulfated sterol dimers, fibrosterol sulfates A–C (**711–713**), were isolated from *Lissodendoryx* (*Acanthodoryx*) *fibrosa* collected in the Philippines. Fibrosterol sulfate B inhibited protein kinase Cζ with IC_{50} value of 5.6 μM [275].

In the year of 2008, a species of *Spheciospongia* sp. (Cagayan de Oro, Philippines) was the source of sterol sulfates, spheciosterol sulfates A–C (**714–716**), all of which potent inhibited protein kinase Cζ with IC_{50} values of 1.59, 0.53, 0.11, and 1.21 μM, respectively [276].

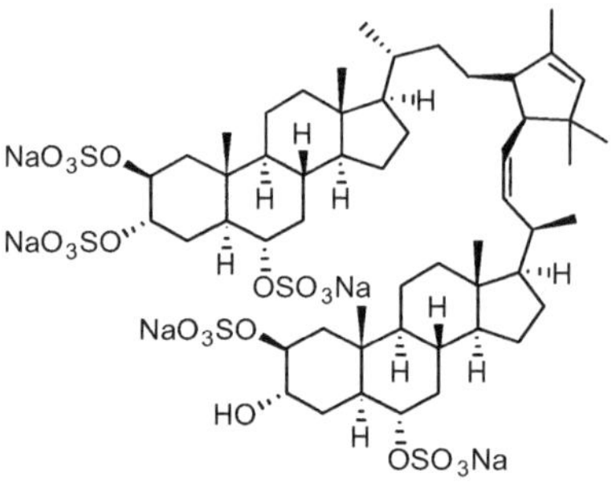

707

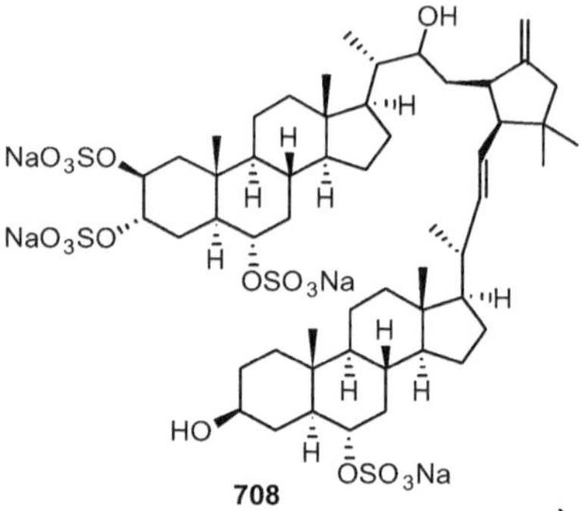

708

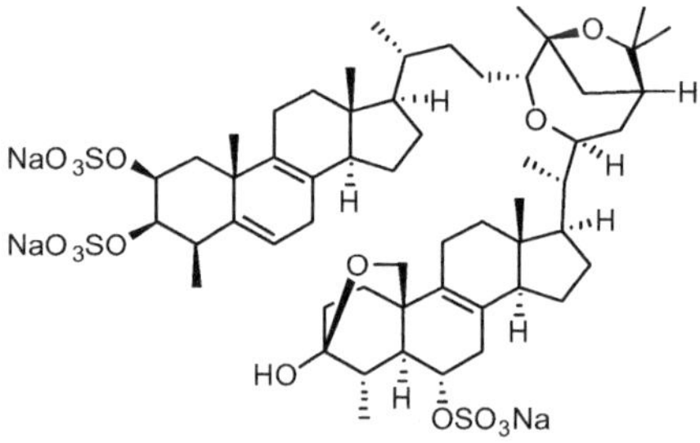

709

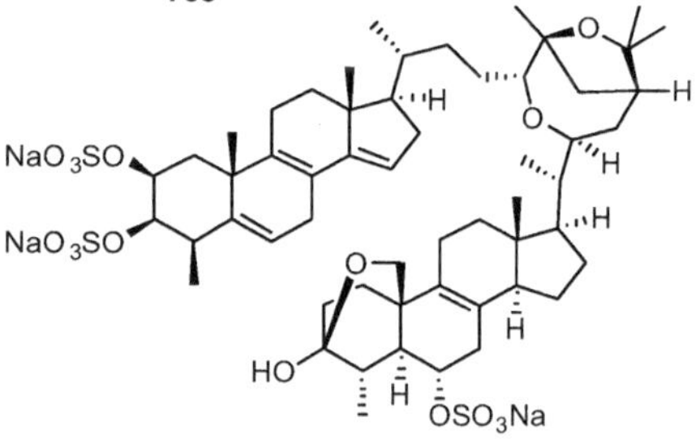

710

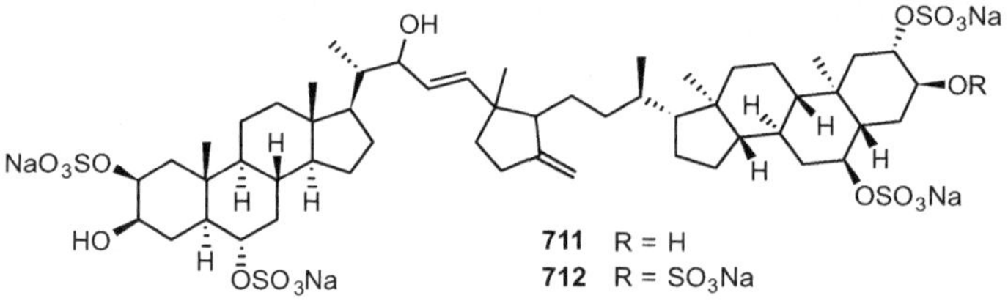

711 R = H
712 R = SO_3Na

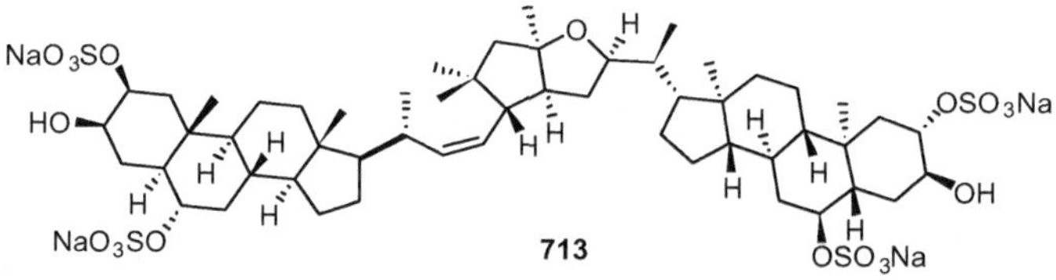

713

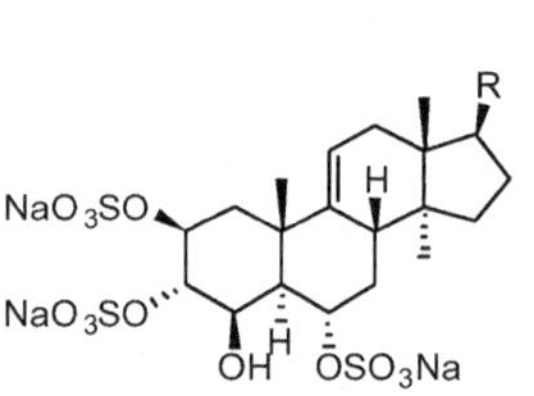

714 R = 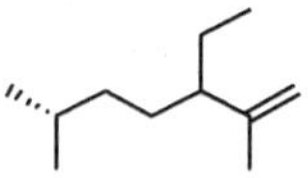

715 R = 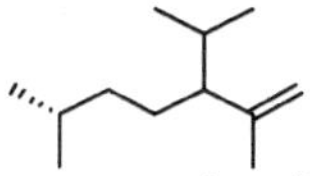

716 R = 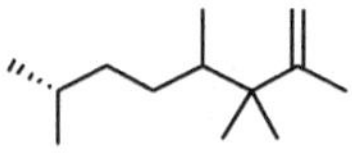

In 2008, three new marine polar steroids, chlorotopsentiasterol sulfate D (**717**), topsentiasterol sulfate F (**718**), and iodotopsentiasterol D (**719**), were isolated from the marine sponge *Topsentia* sp. (Vang Fong Bay, Vietnam). Chlorotopsentiasterol sulfate D proved to be an effective inhibitor of *endo*-1,3-β-D-glucanase [277].

In 2007, sulfated sterol 24ξ,25-dimethyl-3*α*-hydroxyl-cholest-5-ene-2*β*-ol sodium sulfate (**720**) which showed cytotoxicity to four human cancer cell lines was obtained from *Halichondria rugosa* [278].

In 2007, an undescribed marine sponge *Euryspongia* was the source of eurysterols A (**721**) and B (**722**), with the former being cytotoxicity against human colon carcinoma (HCT-116) cells with IC_{50} value of 2.9 μg/mL, also antifungal to amphotericin-B-resistant *Candida albicans* [279].

In 2003, a sterol sulfate Scheme 572423 (**723**), along with the known halistanol sulfate, was isolated from a *Topsentia* species collected in the Bahamas. The compounds were found as $P2Y_{12}$ inhibitors with IC_{50} of 0.48 and 2.2 μM, respectively [280].

In 2001, a Philippine sponge of the genus *Xestospongia* yielded two sulfated sterols, ibisterol B (**724**) and C (**725**) and an epoxysteroid (**726**) that were found to be inhibitors of HIV-1 integrase [281].

In 1998, marine sponge *Crella* sp. from Vanuatu Island yielded crellastatin A (**727**), a new nonsymmetric dimeric steroid, which exhibited cytotoxic activity against NSCLC-N6 cells with the IC_{50} value of 1.5 μg/mL [282].

In 1996, halistanol disulfate B (**728**), a sterol sulfate, was isolated from the MeOH extract of a South African sponge *Pachastrella* sp. The compound had an IC_{50} of 2.1 μM for inhibition of endothelin converting enzyme [283].

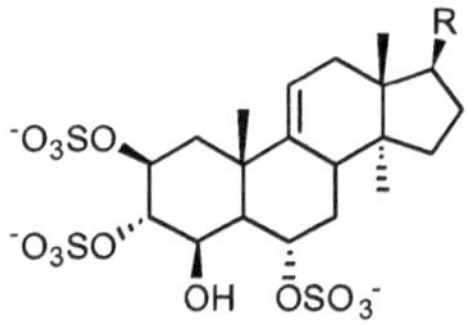

717 R = 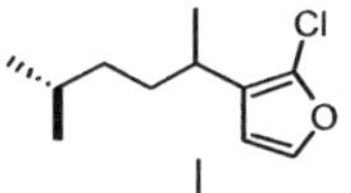

718 R = 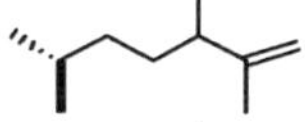

719 R = 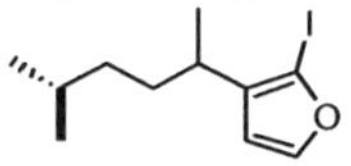

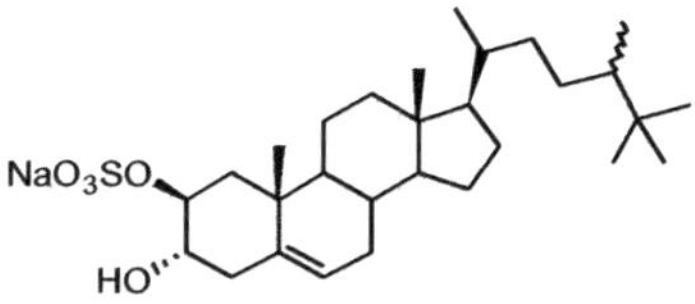

720

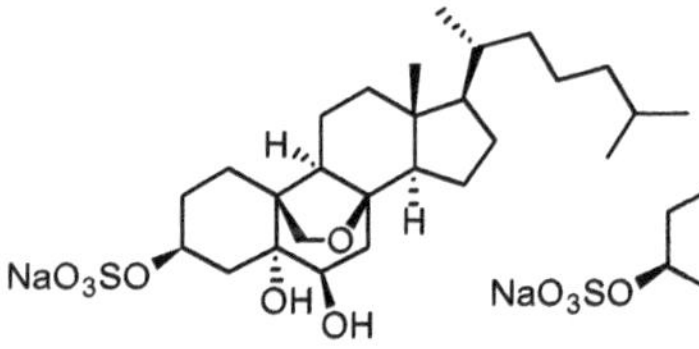

721

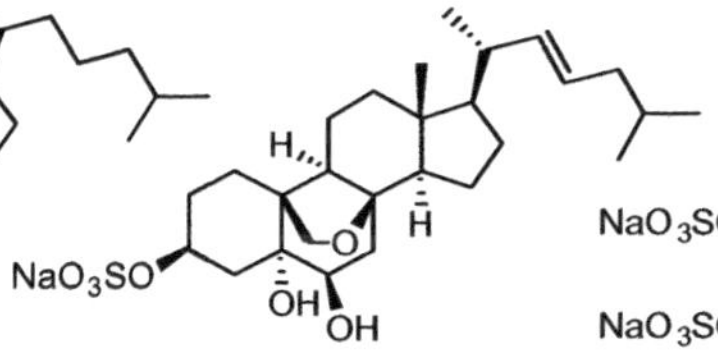

722

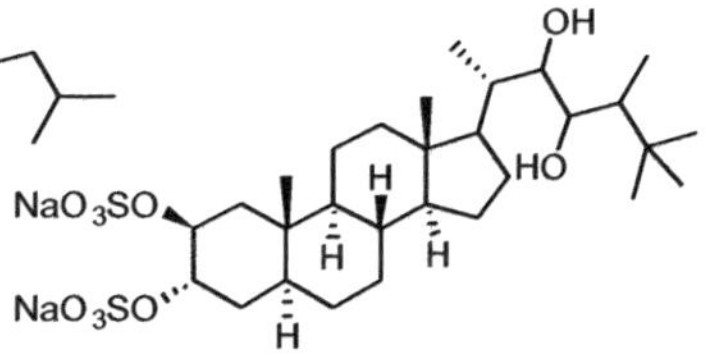

723

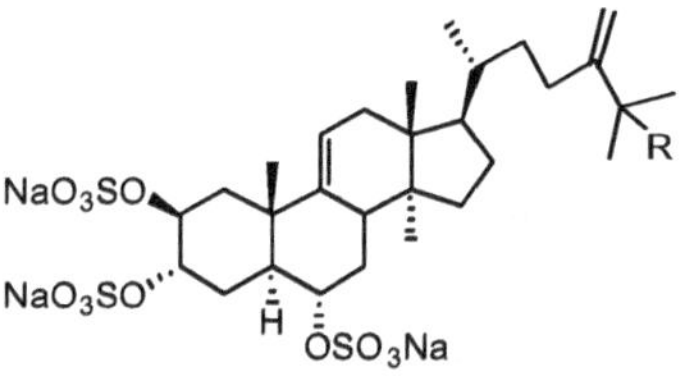

724 R = H
725 R =Me

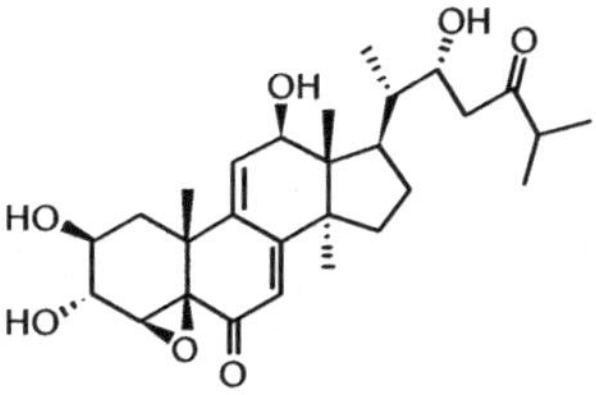

726

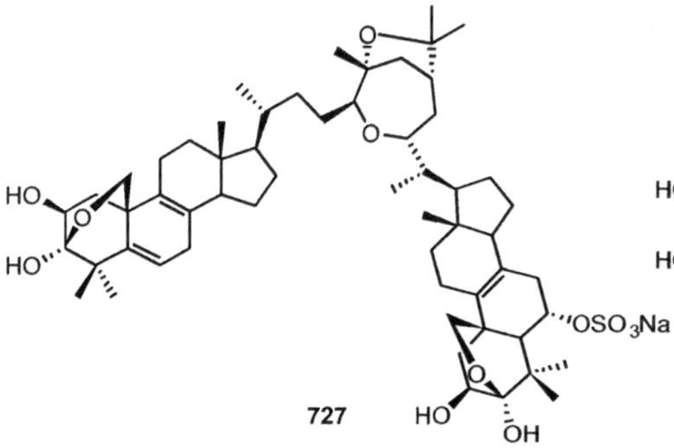

727

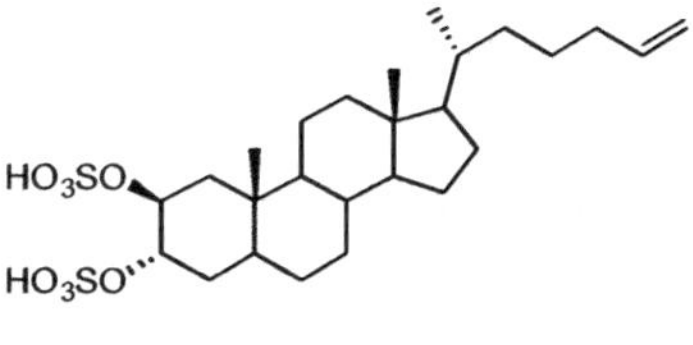

728

15.7.3 *Glycoside Sterols*

In 2010, steroidal glycosides, namely, pandarosides E–J, and their methyl esters were isolated from a Caribbean sponge *Pandaros acanthifolium*, with all except pandaroside H exhibiting antiprotozoal activity. Methyl ester of pandaroside G (**730**) potently inhibited the growth of *Trypanosoma brucei rhodesiense* (IC_{50} = 0.038 μM) and *Leishmania donovani* (IC_{50} = 0.051 μM) especially [284].

In 2000, a potent penasterol disaccharide eryloside F (**731**) inhibited human platelet aggregation in vitro as a thrombin receptor antagonist, which was obtained from the sponge *Erylus formosus*. The IC_{50} values of the compound inhibited SFLLRN and U-46619-induced platelet aggregation were at 0.3 and 1.7 μg/mL, respectively [285].

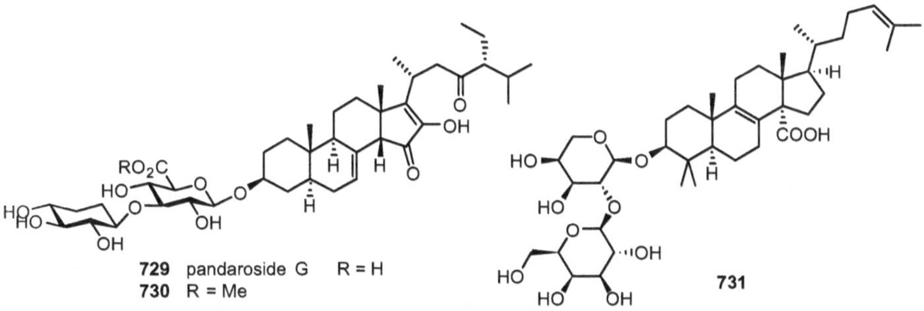

15.7.4 *Others*

In 2014, cinanthrenol A (**732**), an estrogenic steroid containing phenanthrene nucleus, was isolated from the marine sponge *Cinachyrella* sp. Absolute configuration of the compound was established as 16S, 17S, and 19S by the modified Mosher's method. Cinanthrenol A bound to estrogen receptor in a competitive manner against estradiol with an IC_{50} value of 10 nM. Moreover, it has cytotoxicity against P388 and HeLa cells as well, with IC_{50} 4.5 and 0.4 μg/mL, respectively [286].

Also in this year, extracts of the sponge *Theonella swinhoei*, which collected at the Xisha island, yielded two novel sterols swinhoeisterols A and B (**733** and **734**) with an unprecedented 6/6/5/7 tetracyclic systems. The absolute configurations of the compounds were assigned by X-ray diffraction, TDDFT/ECD calculations, and modified Mosher's method. Swinhoeisterols A exhibit a potent inhibitory activity (IC_{50} = 2.9 μM) against the histone acetyltransferase (h)p300 [287].

In 2013, aragusterol I (**735**), 21-*O*-octadecanoyl-xestokerol A (**736**), and 7*β*-hydroxypetrosterol (**737**), three cyclopropanated sterols, were isolated from the Vietnamese marine sponge *Xestospongia testudinaria* [288]. The compound 21-O-octadecanoyl-xestokerol A showed antifouling activity with EC_{50} values similar to that of the antifoulant marine pollutant tributyltin oxide.

In 2008, aminosteroids clionamine D (**738**) which has an unprecedented spiro bislactone side chain were isolated from South African specimens of the sponge *Cliona celata* [289].

In 2004, the Okinawan sponge *Terpios hoshinota* yielded nakiterpiosin (**739**) and nakiterpiosinone (**740**), both exhibiting cytotoxicity potentially against mouse lymphocytic leukemia cells (P388) with IC_{50} values of 0.01 mg/mL, respectively [290].

In 2001, *Stelletta hiwasaensis* from Japan was found to produce orostanal (**741**), abeo-sterol derivative, which induced apoptosis in human acute promyelotic leukemia cell with IC_{50} value of 1.7 μM [291].

In 1999, glaciasterol B 3-monoacetate (**742**), a new 9,11-secosterol which exhibited potent toxic activity (LC_{50} = 0.54 μg/ml) to brine shrimp, was isolated from the Tyrrhenian sponge *Fasciospongiu cavernosa* [292].

In 1996, an Okinawan marine sponge *Xestospongia* sp. was found to contain aragusteroketals A (**743**) and C (**744**); both of them exhibited potent cytotoxic activity with the same IC_{50} value of 4 ng/ml against KB cells [293].

In 1995, pentacyclic steroid xestobergsterol C (**745**), possessing a cis C/D ring junction, which exhibited cytotoxicity against L-1210 murine leukemia cells with IC_{50} values of 4.1 μg/mL, was obtained from the Okinawan marine sponge *Ircinia* sp. [294].

In 1993, an Okinawan marine sponge of the genus *Xestospongia* was the source of aragusterol A–D (**746–749**) [295–297], all of which except aragusterol C strongly inhibited the proliferation of KB cells at IC_{50} values of 0.042, 3.3, and 0.041 μg/mL, respectively.

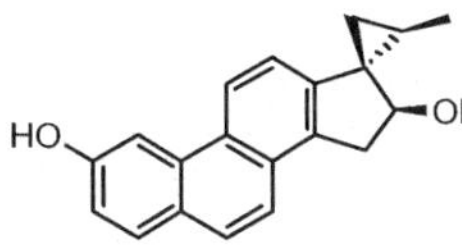

732

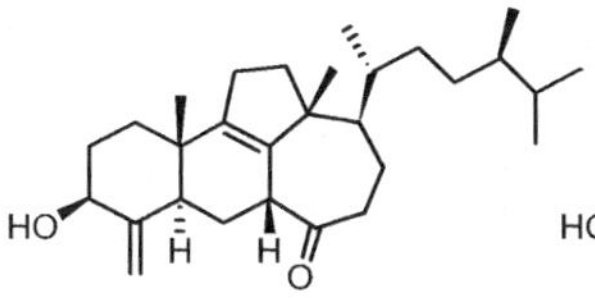

733

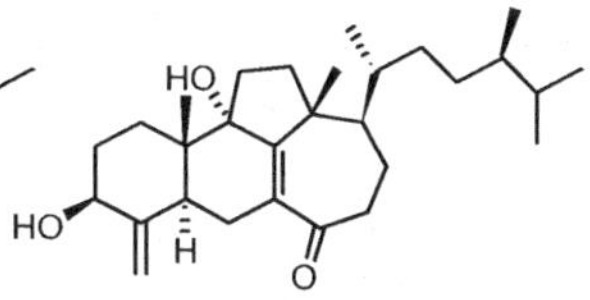

734

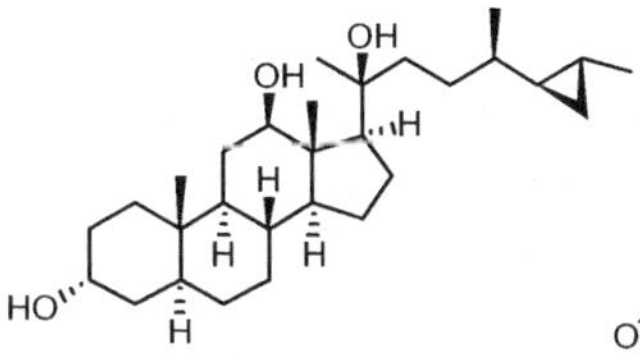

735

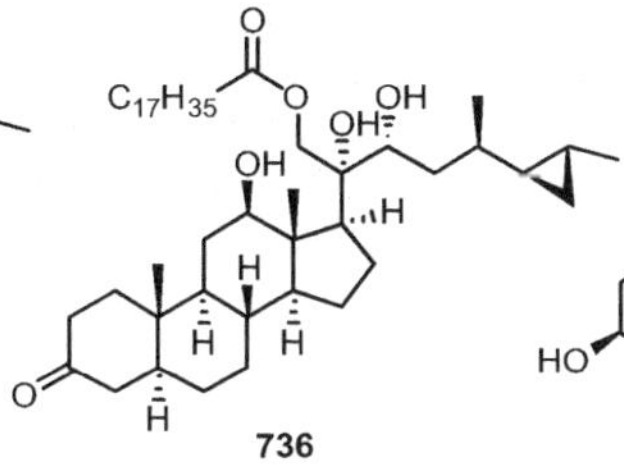

736

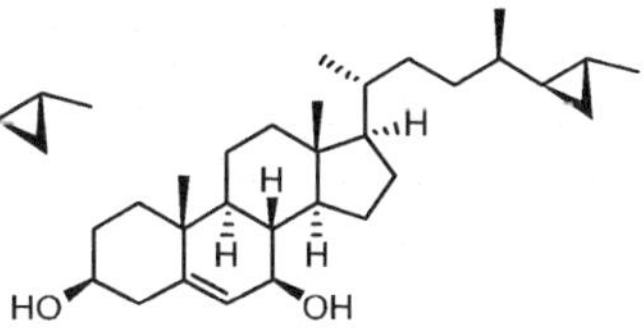

737

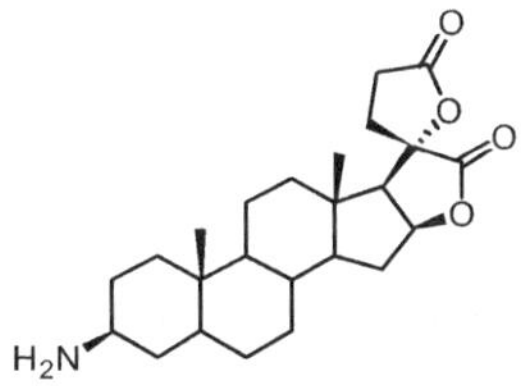

738

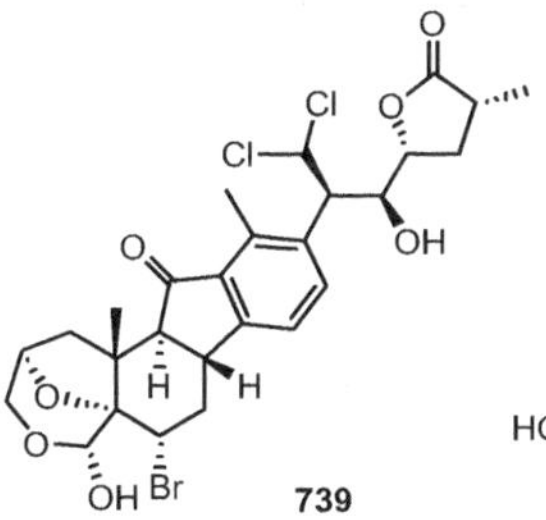

739

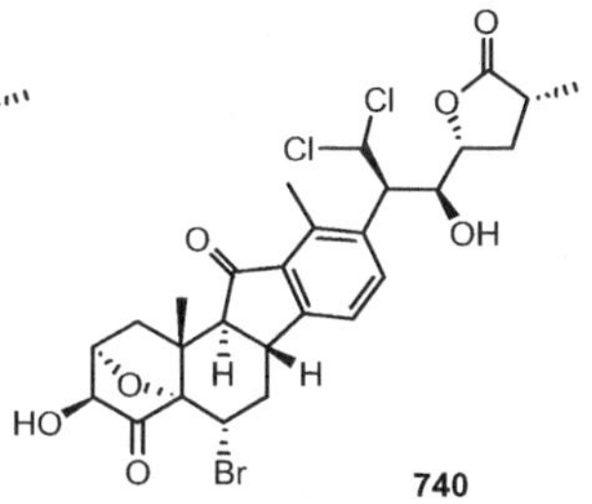

740

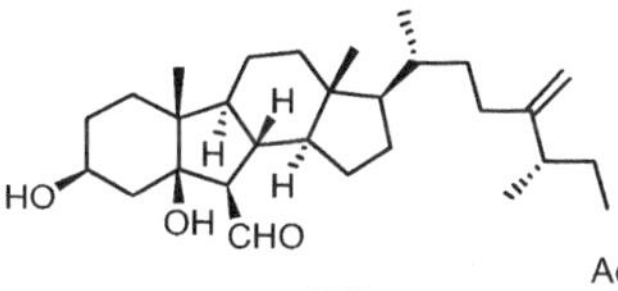

741

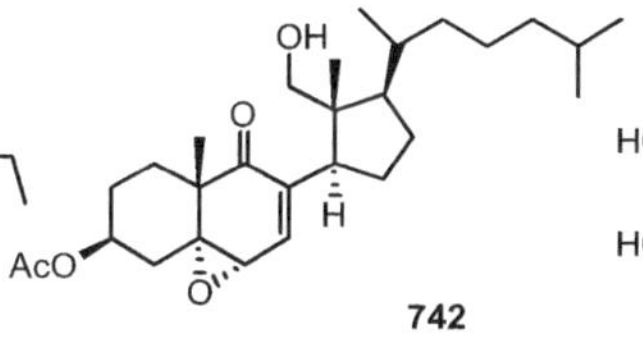

742

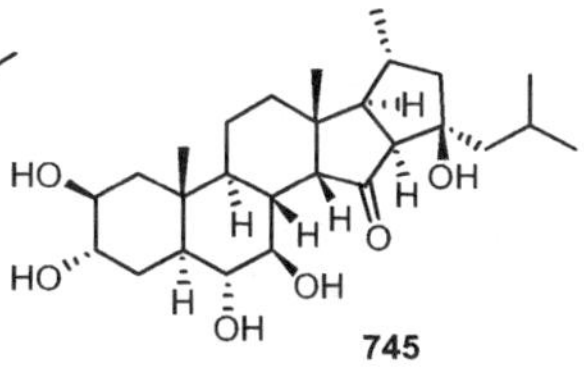

745

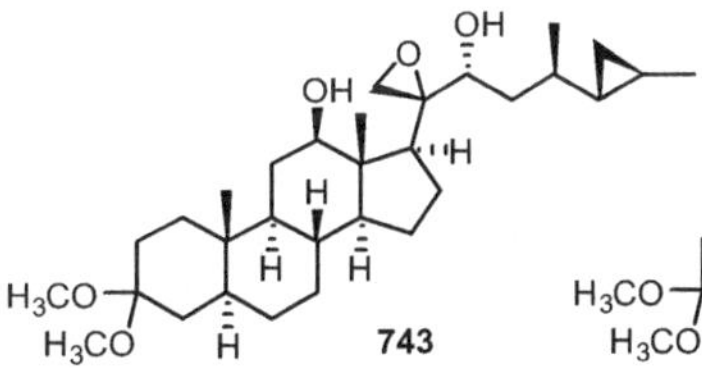

743

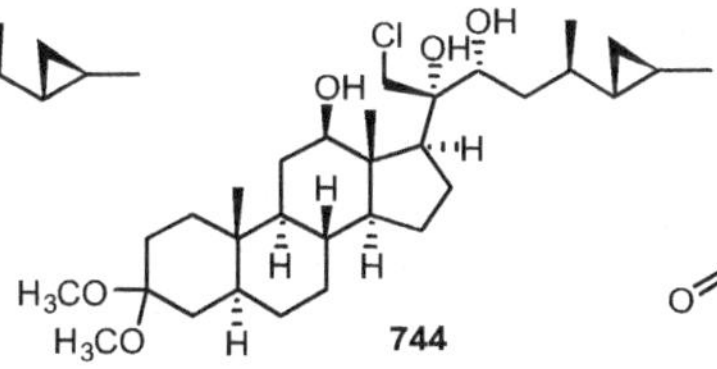

744

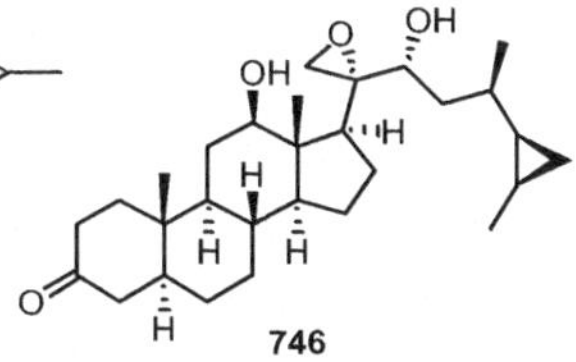

746

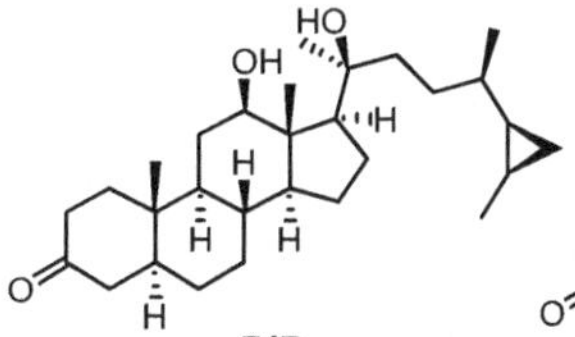

747

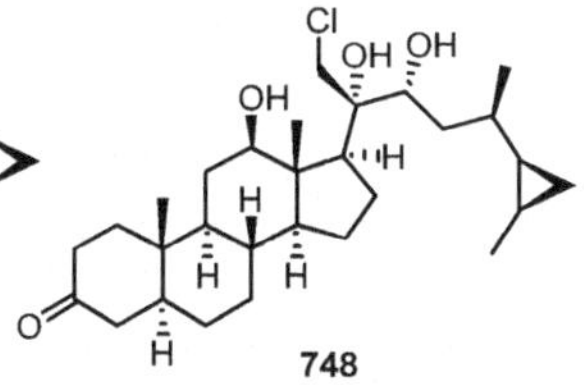

748

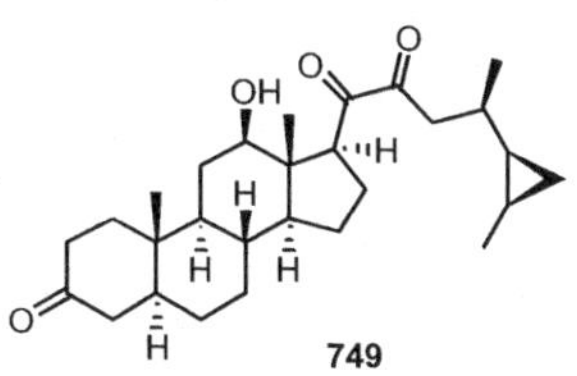

749

In 2015, a steroidal ketone (**750**), bearing an ergosta-22,25-diene side chain, was isolated from the South China Sea marine sponge *Xestospongia testudinaria*. The compound exhibited potent activity against protein tyrosine phosphatase 1B (PTP1B) with an IC_{50} value of 4.27 μM [298].

In the same year, antibacterial compound gelliusterol E (**751**) was obtained from the Red Sea sponge *Callyspongia* aff. *implexa* [299]. Gelliusterol E inhibited the formation and growth of gram-negative bacterium *Chlamydia trachomatis* in a dose-dependent manner with an IC_{50} value of 2.3 μM.

In 2012, Okinawan sponge *Dysidea* sp. was the source of dysideasterols F–H (**752–754**), all of which inhibited human epidermoid carcinoma cells strongly with a similar cytotoxic effect with IC_{50} values of 0.15–0.3 μM [300].

In 2010, an undescribed species of *Topsentia* was the source of three isopropyl steroids, topsentinols K (**755**), L (**756**) and K trisulfate (**757**); the latter was an inhibitor of BACE1 dose-dependently with an IC_{50} value of 1.2 μM [301].

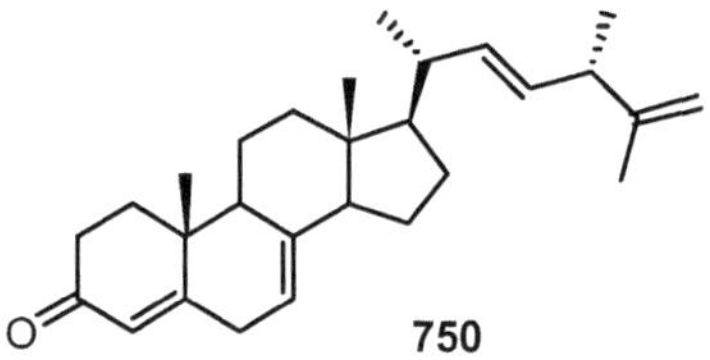

750

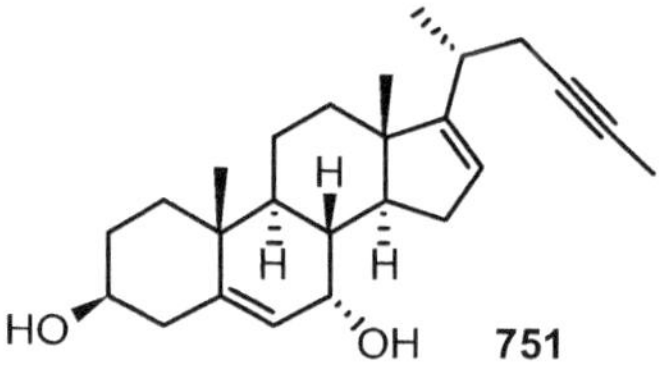

751

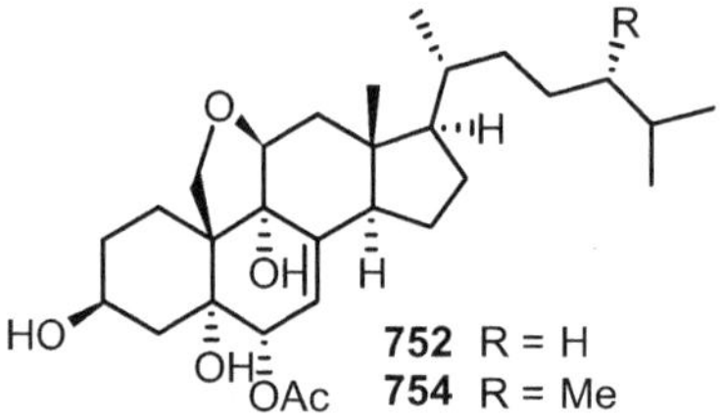

752 R = H
754 R = Me

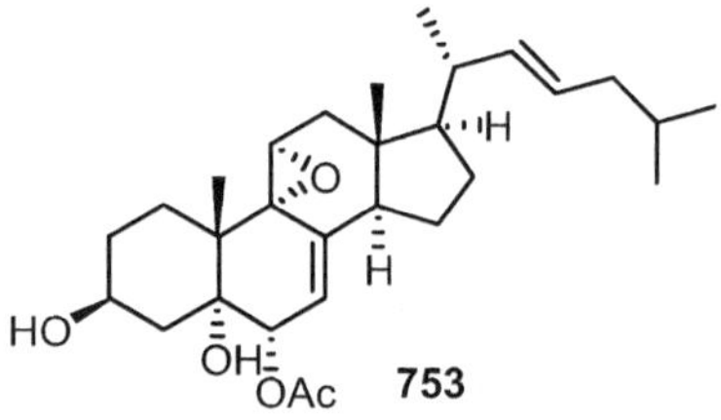

753

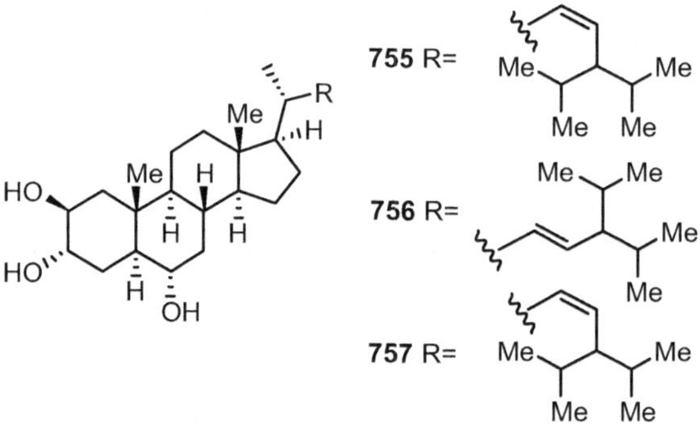

In 2009, five norselic acids A–E (**758–762**) were obtained from the sponge *Crella* sp. collected in Antarctica, all of which inhibited the growth of the Leishmania parasite at low micromolar levels [302].

Also in 2009, Australian marine sponge *Psammoclema* sp. contained a series of trihydroxysterols (**763–766**), of which **4** were cytotoxic against a panel of cancer cell lines [303].

In 2008, *Ircinia aruensis* collected in Naozhou Island yielded six epoxysterols. Compound **767** exhibited cytotoxicity against four cancer cell lines (7402, H-460, LOVO, and MCF) with IC_{50} values of 4.3, 2.8, 5.1, and 3.5 μg/mL, respectively [304].

In 2005, *Homaxinella* sp. (Korea) contained a series of highly degraded sterols demethylincisterols A_1–A_4 (**768–771**) and butoxyderivatised sterols homaxisterols A_1–A_4 (**772–775**), all of which but homaxisterols A_4 inhibited a panel of five human solid tumor cell lines, and especially **770** displayed significant cytotoxicity [305].

In 2004, Italian sponge *Cliona nigricans* provided two polychlorinated steroids clionastatins A (**776**) and B (**777**), which exhibited potential antitumor activity in three cell lines with IC_{50} values ranging from 0.8 to 2.0 μg/mL [306].

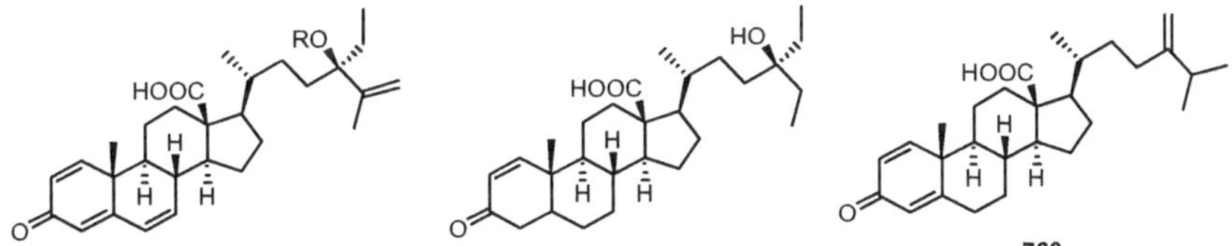

758 R = H
762 R = Ac

759

760

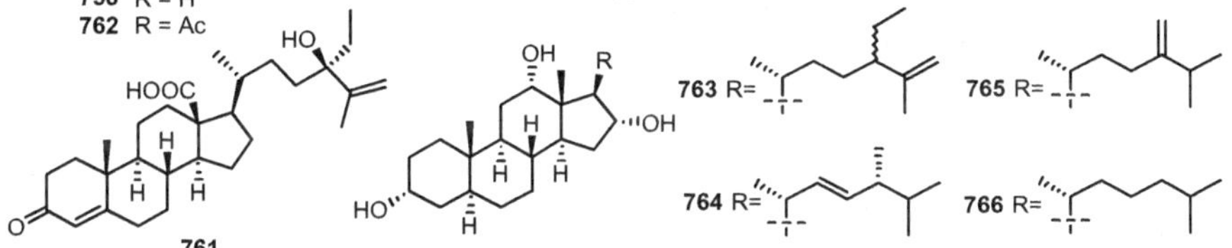

761

763 R=

764 R=

765 R=

766 R=

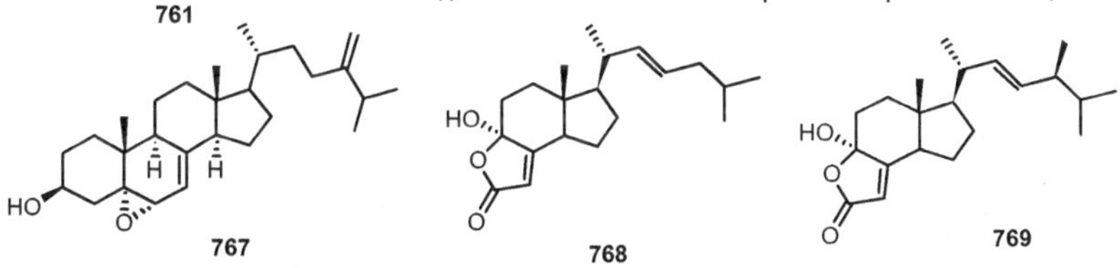

767

768

769

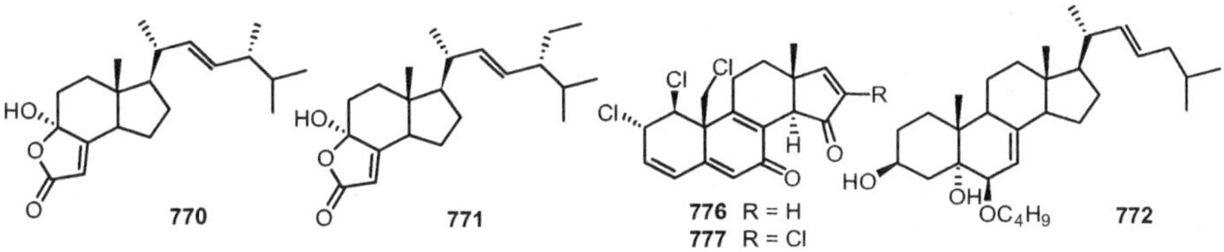

770

771

776 R = H
777 R = Cl

772

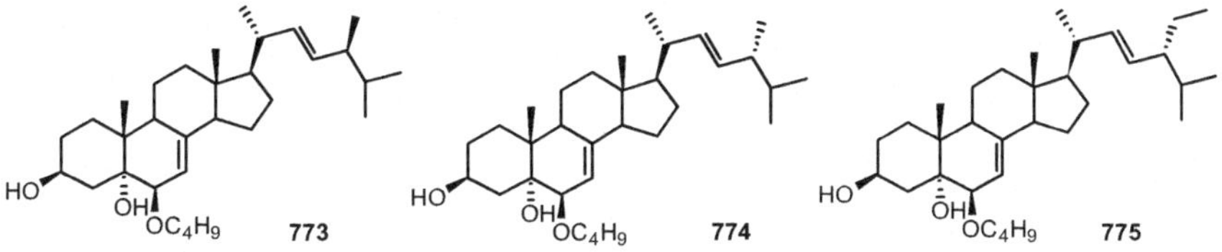

773

774

775

In 2004, four sterols (**778–781**) were isolated from the marine sponge *Axinella* cf. *bidderi* (Yemen, Indian Ocean), which possess, respectively, the cholestene and the cholestane skeleton with a cyclic enol ether linkage between C-18 and C-22. These sterols showed cytotoxic activity against prostate, ovary, pancreas, colon, and lung cell lines with GI_{50} values ranging from 0.6 to 8.3 μg/mL [307].

In 2002, three new sterols (**782–784**), isolated from the marine sponge *Polymastia tenax*, were found to have potent antiproliferative activity toward A549, HT29, H-116, MS-1, and PC-3 tumor cells in the range 0.5–10 μg/mL [308].

In 2001, an undescribed species of *Gellius* collected in the Caribbean coast of Panama was the source of four acetylenic sterols, gelliusterols A–D (**785–788**); **785**, **786**, and **787** were found to be cytotoxic to a panel of cancer cell lines [309].

In 2000, northern Australia *Dysidea* sp. yielded three polyoxygenated sterols (**789–791**) that inhibited the binding of IL-8 to the human recombinant IL-8 receptor-type A with the IC_{50} values of 20, 5.5, and 4.5 μM, respectively [310].

A deepwater marine sponge *Scleritoderma* sp. cf. *paccardi* which collected from Turneffe Islands (Belize) in 1985 yielded a sterol ether (**792**) named 24(*R*)-methyl-5α-cholest-7-enyl 3β-methoxymethyl ether. IC_{50} of the compound inhibited the murine P388 tumor cell line was 2.3 μg/mL [311].

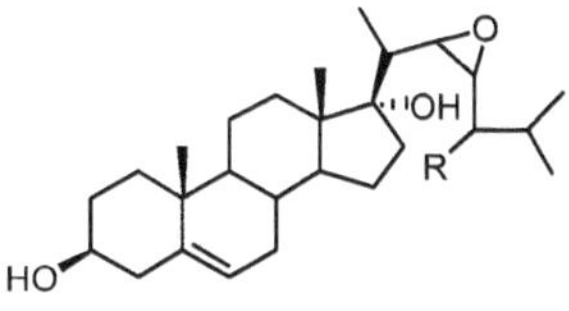

778 R = Me
779 R = H

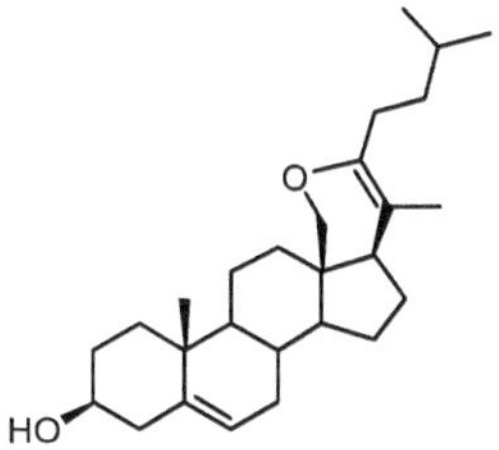

780

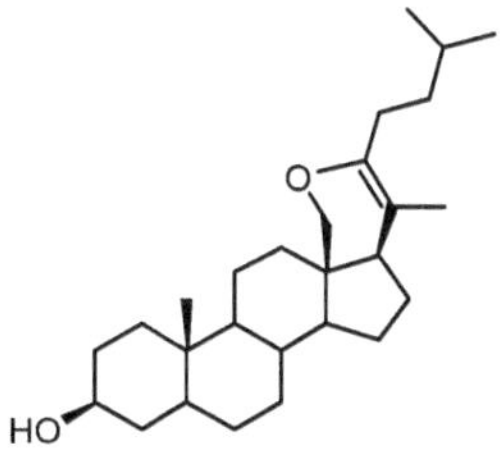

781

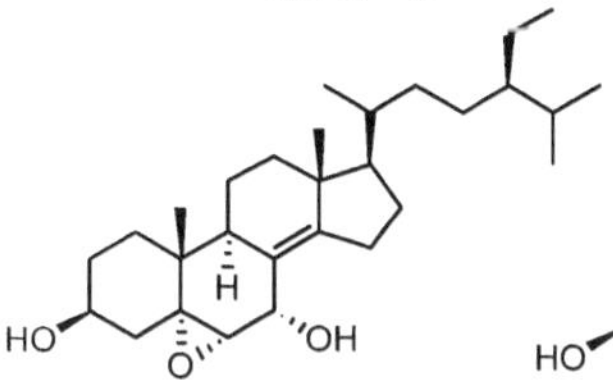

782

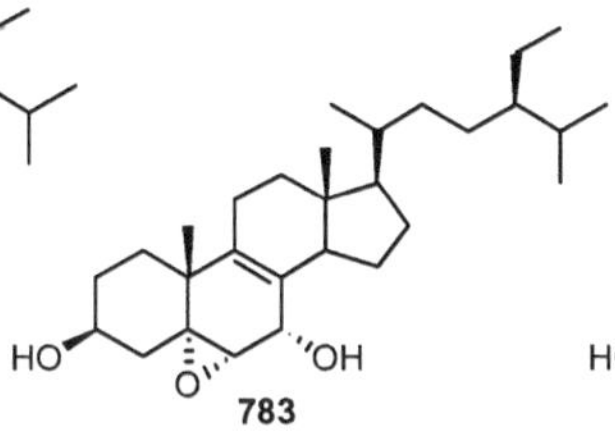

783

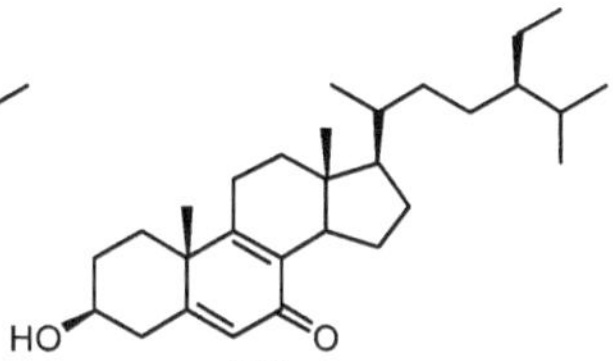

784

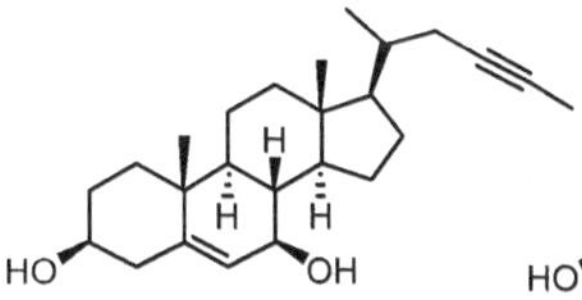

785

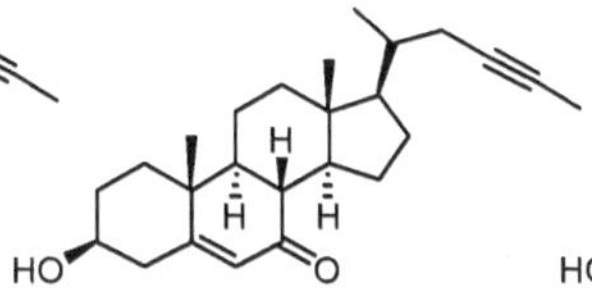

786

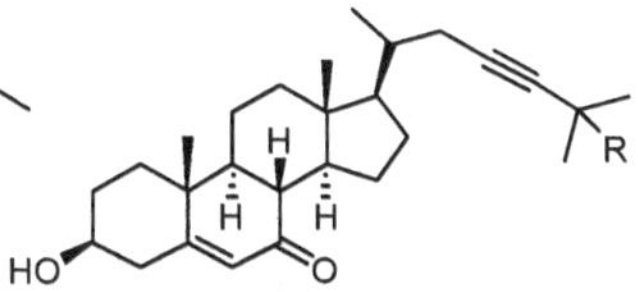

787 R = H
788 R = OH

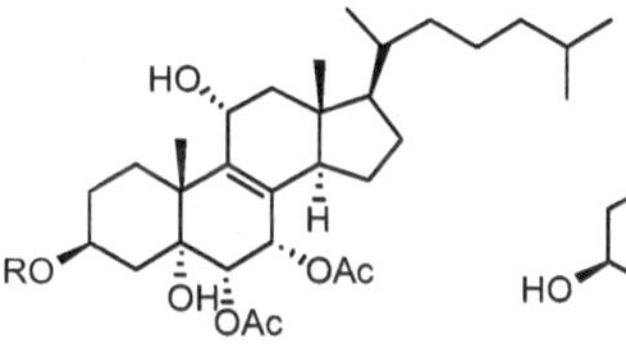

789 R = H
790 R = Ac

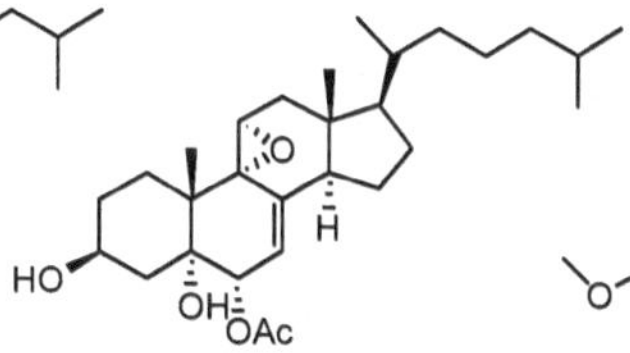

791

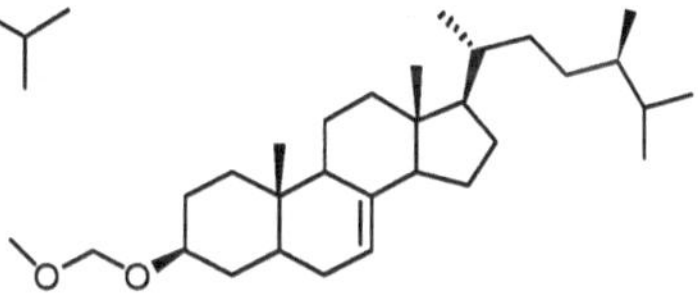

792

15.8 Bioactive Potentials from Diverse Sponges Derived Natural Products

For the past decades, marine sponges have been considered as a very fertile field for the discovery of bioactive natural chemical substances with respect to the diversity of their primary and secondary chemical components and metabolites [312]. It was proved that marine sponges produce an enormous array of antitumor, antiviral, anti-inflammatory, antibiotic, and other bioactive molecules that have the potential for therapeutic use. Studies showed that different components affect the targeted disease by different mechanisms. Natural chemical products that can act as inhibitors of transcription factors may be effective against both malignant neoplasms and viral diseases. Most bioactive metabolites from sponges proved to be inhibitors of certain enzymes, which often mediate or produce mediators of intracellular or intercellular messengers involved in the pathogenesis of a disease [313]. Here, we introduce several kinds of recently confirmed bioactive lead compounds isolated from marine sponge which pharmacological mechanisms were studied.

Cytarabine (Fig. 15.1) is derived from the related marine natural products spongothymidine and spongouridine, nucleosides with a modified sugar moiety. These nucleosides were isolated from the Caribbean sponge *Crypotethia crypta* by Bergmann and Feeney in 1951 [314]. Cytarabine is mainly used in the treatment of acute myeloid leukemia and acute lymphocytic leukemia (ALL), where it is the backbone of induction chemotherapy [315]. Cytosine arabinoside interferes with the synthesis of DNA. Its mode of action is due to its rapid conversion into cytosine arabinoside triphosphate, which damages DNA when the cell cycle holds in the S phase (synthesis of DNA). The cells requiring DNA replication for mitosis are therefore most affected [315]. Cytosine arabinoside was also considered as inhibitor of both DNA and RNA polymerases and nucleotide reductases needed for DNA synthesis. As an antiviral agent, cytarabine was often used to inhibit deoxycytidine utilization. Due to the rapid deamination in the body into the inactive uracil derivative, cytarabine therefore is often given by continuous intravenous infusion [316].

Renieramycin G (Fig. 15.2) was isolated from marine sponges *Xestospongia* and *Cribrochalina* [317]. A recent study demonstrated that Renieramycin G induced apoptosis via the p53-dependent pathway and inhibited the progression and metastasis of non-small cell lung cancer [318]. Renieramycin G possesses an amide carbonyl group at the C-21 position demonstrated minimal antiproliferative activity in human colorectal cancer (HCT-116) and human lung adenocarcinoma (A549) cells owing to both compounds possessing an amide carbonyl group at the C-21 position.

Fig. 15.1 Cytarabine

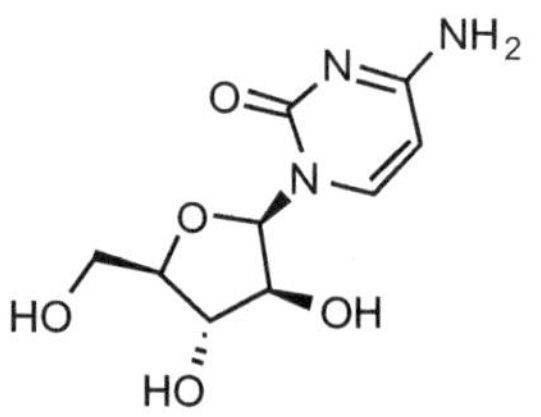

Fig. 15.2 Renieramycin G

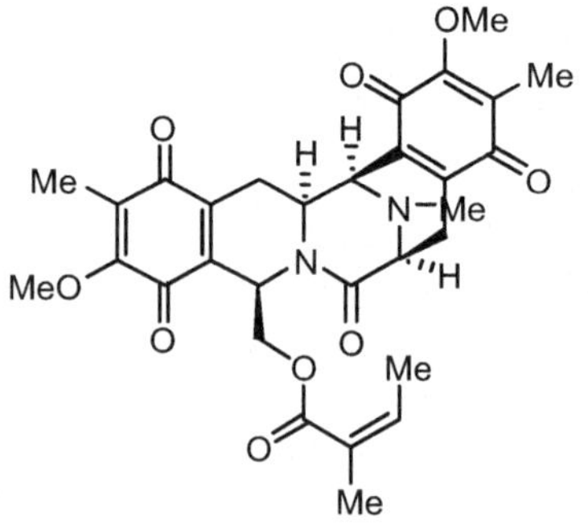

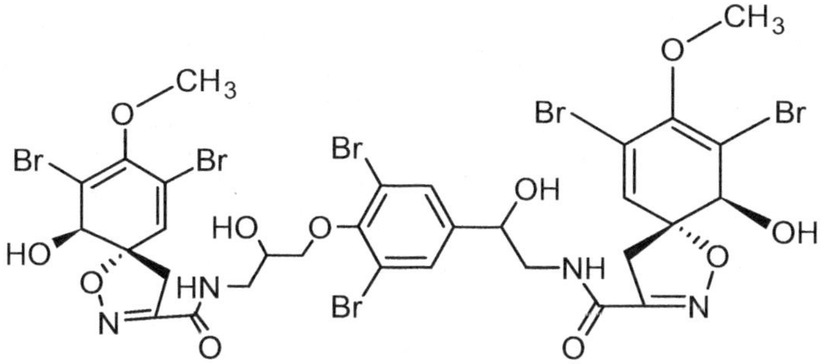

Fig. 15.3 Isofistularin-3

The minimal antiproliferative activity for members of the tetrahydroisoquinoline family possessing an amide carbonyl group at the C-21 position seems consistent compared to their C-21 cyano- or carbinolamine-containing relatives.

The brominated alkaloid Isofistularin-3 (Iso-3), a new DNA methyltransferase (DNMT)1 inhibitor, was from the marine sponge *Aplysina aerophoba* [319] (Fig. 15.3). Docking analysis confirmed DNMT inhibition data in vitro and revealed binding of Iso-3 within the DNA binding site of DNMT1 [320]. Subsequent increased expression of tumor suppressor gene aryl hydrocarbon receptor (AHR) could be correlated to decreased methylation of CpG sites within the essential Sp1 regulatory region of its promoter. Iso-3 induced growth arrest of cancer cells in G0/G1 with increasing p21 and p27 expression and reducing cyclin E1, PCNA and c-myc levels. Fluorescent and transmission electron microscopy revealed that the reduced proliferation was accompanied by morphological changes as autophagy. Furthermore, Iso-3 strongly synergized with tumor necrosis factor-related apoptosis-inducing ligand (TRAIL) in RAJI [combination index (CI) = 0.22] and U937 cells (CI = 0.21) and increased TRAIL-induced apoptosis via a mechanism involving reduction of survivin expression but not of Bcl-2 family proteins nor X-linked inhibitor of apoptosis protein (XIAP) [321]. Treatment of Iso-3 decreased FLIPL expression and triggered activation of endoplasmic reticulum (ER) stress with increased GRP78 expression, which eventually induced TRAIL receptor death receptor (DR)5 surface expression. Importantly, as a potential candidate for further anticancer drug development, Iso-3 reduced the viability, colony, and in vivo tumor forming potential without affecting the viability of PBMCs from healthy donors or zebrafish development [321].

Ageladine A (Fig. 15.4) was derived from the sponge *Agelas nakamurai* [322], exhibiting in vitro and in vivo antiangiogenic activity associated with its MMP inhibition. However, subsequent study confirmed that it is resulted from the selective inhibition of kinases such as yeast Sps1/Ste20-related kinase 4, dual specificity tyrosine-phosphorylation-regulated kinase 1A, and tyrosine kinase 2 [323]. Ageladine A and its synthetic analogues feature highly selective angiogenesis inhibition, at concentrations of which no cytotoxicity are shown in the National Cancer

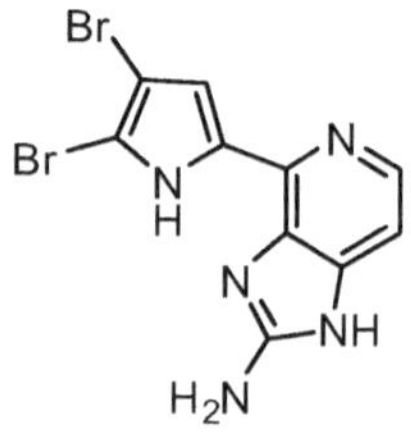

Fig. 15.4 Ageladine A

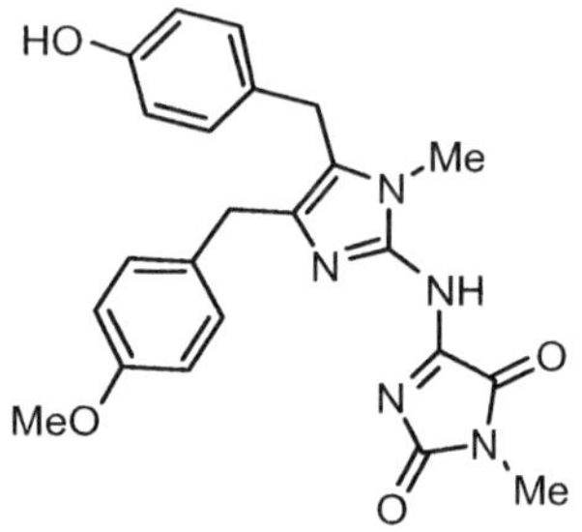

Fig. 15.5 Naamidine A

Institute (NCI) panel of 60 human cancer cell lines. Moreover, ageladine A is a fluorescent dye that is pH-sensitive and is of interest for imaging [324].

Naamidine A (Fig. 15.5) was found in marine sponge *Leucetta chagosensis*. Bioactivity study showed that this compound inhibits the EGF signaling pathway and is more specific for the EGF-mediated mitogenic activity than for the insulin-mediated mitogenic activity [325]. In 2009, LaBarbera et al. illustrated that naamidine A triggers biomarkers of apoptosis like externalization of phosphatidylserine, cleavage and activation of caspases-3, −8, and − 9, and disruption of the mitochondrial membrane potential indicating that the cell death caused by naamidine A in epidermoid carcinoma cells (A-431) is a consequence of apoptosis instead of cytotoxicity. It is also reported that naamidine A inhibits the growth of tumor xenograft by activating caspase-3 manifesting apoptosis activity in vivo. Besides, naamidine A-caused apoptosis does not depend on functional p53 and is independent of extracellular signal-regulated kinase 1/2 [326].

Araguspongine C (Fig. 15.6) is a group of macrocyclic oxaquinolizidine alkaloids derived from the marine sponge *Xestospongia* species [327]. Araguspongine C prevented the proliferation of varied breast cancer cell lines in vitro dose-dependently. Characterized by vacuole formation and upregulation of autophagy markers such as Atg3, Atg7, Atg16L, and LC3A/B suppressing c-Met and HER2 receptor tyrosine kinase activation [328], araguspongine C induces autophagic cell death in HER2-overexpressing BT-474 breast cancer cells. What's more, docking research and cell-free Z-LYTE assays revealed that araguspongine C owing the direct interaction potentially with the receptor tyrosine kinases c-Met and HER2 at their kinase domains. Especially, araguspongine C treatment causes the suppression of PI3K/Akt/mTOR signaling cascade in breast cancer cells by autophagy [328].

Fig. 15.6 Araguspongine C

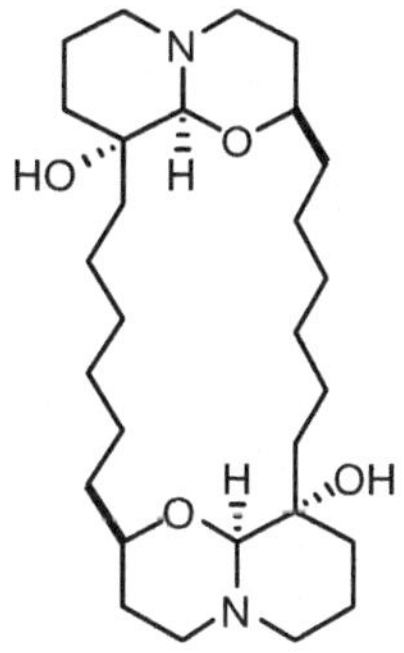

Fig. 15.7 Psammaplysene A

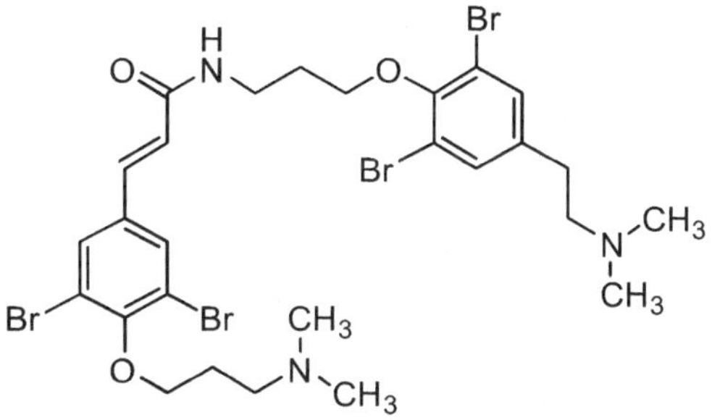

Psammaplysene A (Fig. 15.7) is derived from the marine sponge *Psammaplysilla* sp. Schroeder et al. demonstrated that psammaplysene A was an active inhibitor of a transcription factor FOXO1 nuclear export, whereas psammaplysene B shows less potential [329]. Extensive study illustrated that human endometrial cancer cells treated for 24 h with 1 μM psammaplysene A causes cell death leading to morphology changes with apoptosis and PARP cleavage, revealing that the cell death aroused from psammaplysene A is a result of apoptosis instead of cytotoxicity. Psammaplysene A bringing about the doubling of cells in the G2/M phase and FOXO1 silencing in ECC-1 cells reduces psammaplysene A-induced apoptosis [330].

Agelasines from a marine sponge *Agelas clathrodes* have structurally unique compounds which possess mono or bi-cyclic diterpenoids with a 9-methyladeninium chromophore. The cytotoxicity of agelasine B (Fig. 15.8) and its mechanism have been extensively studied which are of great interest that its higher toxicity in cancer cells (IC50 = 3.22, 2.99, and 6.86 μM MCF-7, SKBr3, and PC-3 cells, respectively) than in normal cells (fibroblasts, IC50 = 32.91 μM) where agelasine B upregulated the intracellular concentration of Ca^{2+} and caused fast Ca^{2+} release via the endoplasmic reticulum (ER). This research indicated that sarcoplasmic-ER Ca^{2+}-ATPase activity is inhibited by agelasine B. What's more, intracellular Ca^{2+} accumulation in the mitochondria is tied to apoptosis. In addition, this marine sponge toxin induces DNA fragmentation and severely enhances caspase-8 activity in MCF-7 cells [331].

Fig. 15.8 Agelasine B

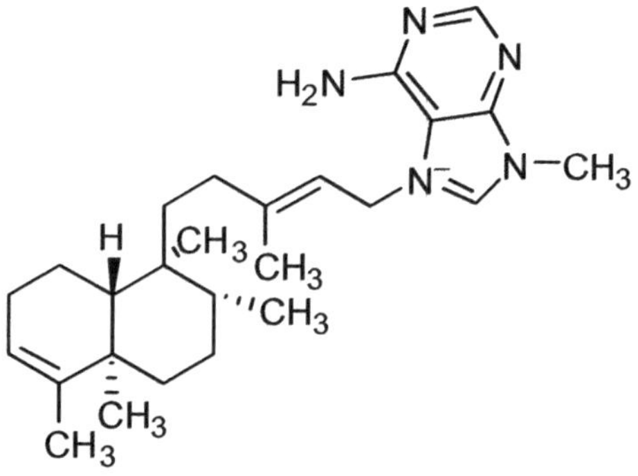

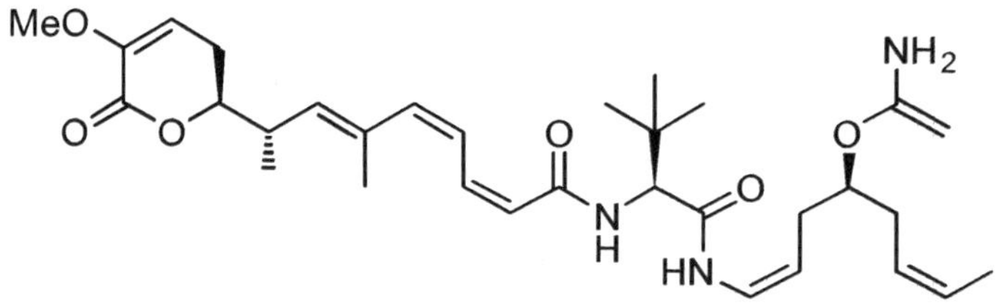

Fig. 15.9 PM060184

Therefore, agelasine B is potential for treating breast cancer for less toxicity in normal breast cells.

PM060184 (Fig. 15.9) belongs to a new family of tubulin-binding agents originally isolated from the marine sponge *Lithoplocamia lithistoides* [332]. It was published that PM060184 presents the highest known affinities among tubulin-binding agents and that it targets tubulin dimers at a new binding site. PM060184 has a potent antitumor activity in a panel of different tumor xenograft models. Moreover, PM060184 is able to overcome P-gp-mediated resistance in vivo, an effect that could be related to its high binding affinity for tubulin. PM060184 is an inhibitor of tubulin polymerization that reduces microtubule dynamicity in cells by 59% [333]. PM060184 suppresses microtubule shortening and growing at a similar extent. This action affects cells in interphase and mitosis. In the first case, the compound induced a disorganization and fragmentation of the microtubule network and the inhibition of cell migration. In the second case, it induced the appearance of multipolar mitosis and lagging chromosomes at the metaphase plate. These effects correlated with a nonclassical apoptosis pathway, which caused prometaphase arrest and induction of caspase-dependent apoptosis or appearance of cells in a multinucleated interphase-like state. Taken together, PM060184 represents a new tubulin-binding agent with promising potential as an anticancer agent [334].

Eribulin mesylate (E7389) (Fig. 15.10) is a microtubule dynamics inhibitor with antitumor activity, which is effective against not only a broad range of human cancer

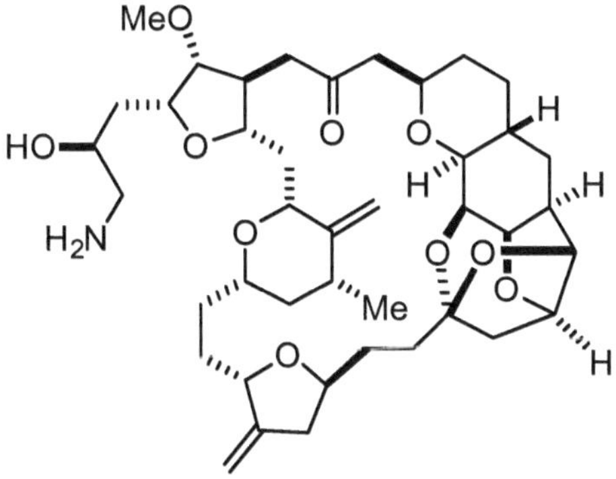

Fig. 15.10 Eribulin mesylate

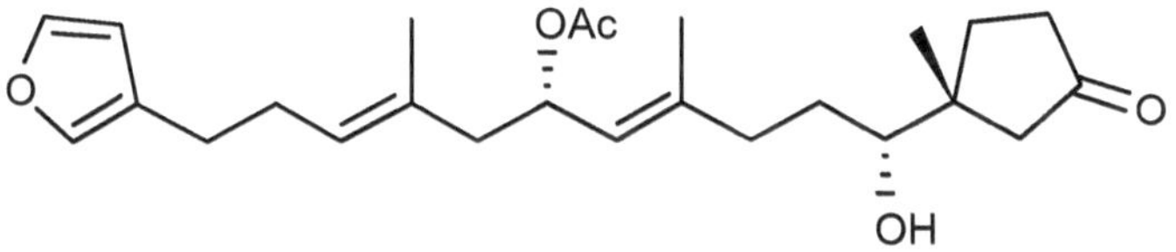

Fig. 15.11 10-Acetylirciformonin B

cell lines but also human tumor xenograft models derived from melanoma, colon, breast, ovarian, and pancreatic cancer [336]. This non-taxane molecule is a structurally simplified synthetic analogue of halichondrin B derived from the marine sponge *Halichondria okadai* [335]. Different from other tubulin-targeted agents like taxanes, epothilones, and vinca alkaloid, which affect both growth and shortening of microtubules, eribulin only affects growth by binding to the microtubules and suppressing microtubule polymerization without affecting shortening, thereby sequestering tubulin into nonfunctional aggregates [336, 337]. The prohibited formation of mitotic spindles leads to G2/M cell cycle arrest and apoptosis as a result of prolonged mitotic blockage. Eribulin remains to be active in taxane-resistant cell lines with β-tubulin mutations and those which overexpress P-gp according to in vitro studies. It's reported to be manageably safe in a 21-day-cycle administration, neutropenia being the main dose limit among all toxicities, with an MTD of 1.4 mg/m^2 in phase I studies, but seems to be both effective and safe in several phase II studies. Therefore, it has been applied to patients with locally advanced or metastatic breast cancer previously treated with at least two chemotherapeutic regimens for advanced disease [335–337].

10-Acetylirciformonin B (Fig. 15.11) classified as furanoterpenoid which is isolated from marine sponge *Ircinia* sp. can restrain the growth of leukemia HL-60 cells, with an IC50 value of 1.7 μg/mL obtained at 48 h of treatment due to its unique structure of the linear C22-sesterterpenoid. What's more, its anticancer activity was activated by inducting of DNA damage and apoptosis. To be detailed, DNA damage was mediated by the phosphorylation of histone H2AX, p-CHK2

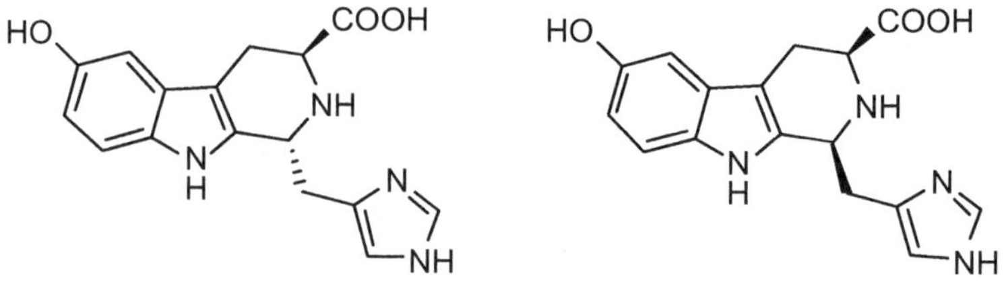

Fig. 15.12 Hyrtioreticulins A and B

Fig. 15.13 Peloruside

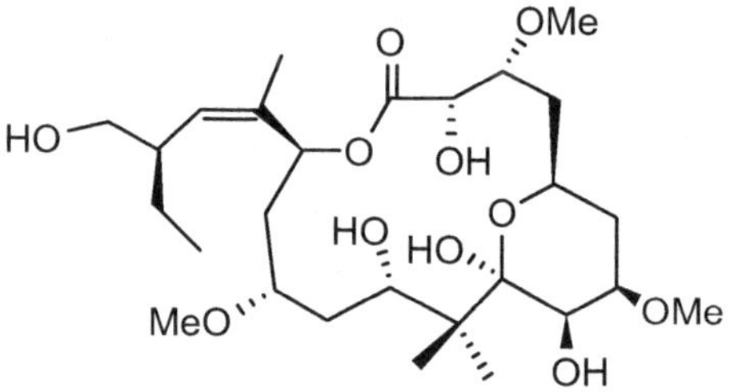

(checkpoint kinase), sensitive markers of DNA double-strand breaks (DSBs), and apoptosis which was triggered by caspase-8, caspase-9, and caspase-3, resulting to PARP cleavage, the downregulation of Bcl-xL and the upregulation of Bax [338].

Hyrtioreticulins A and B (Fig. 15.12) belong to indole alkaloids derived from the marine sponge *Hyrtios reticulatus*. These alkaloids have inhibition of E1-ubiquitin intermediate formation from 0.75 to 11 μg/mL in IC50 values [339]. Moreover, the structures are approximately the same except for their stereochemistry at C-1, in which hyrtioreticulin A is *trans*-configured and hyrtioreticulin B is *cis*, respectively, demonstrating that the *trans* configuration reinforces inhibitory activity against E1, the ubiquitin-activating enzyme required for ubiquitination in the ubiquitin-proteasome pathway involving in a large variety of cellular events such as cell cycle control, transcription, and development [340]. Deregulation of this pathway, therefore, can cause numerous diseases like cancer. Consequently, the ubiquitin pathway plays a significant role in anticancer drugs. In that context, hyrtioreticulins A and B catch more attention on the bioactivity on new anticancer therapeutics.

Peloruside A (Fig. 15.13) is a microtubule-stabilizing agent isolated from a New Zealand marine sponge [341]. Peloruside prevents growth of a panel of cancer cell lines at low nanomolar concentrations, including cell lines that are resistant to paclitaxel. Three xenograft studies in athymic nu/nu mice were performed to assess the efficacy of peloruside compared with standard anticancer agents such as paclitaxel, docetaxel, and doxorubicin. In the first study, peloruside A, 5 and 10 mg/kg (QD × 5) caused growth inhibition (%TGI of 84% and 95%, respectively), on the growth of H460 non-small cell lung cancer xenografts, whereas standard treatments with paclitaxel (8 mg/kg, QD × 5) and docetaxel (6.3 mg/kg, Q2D × 3) were much less

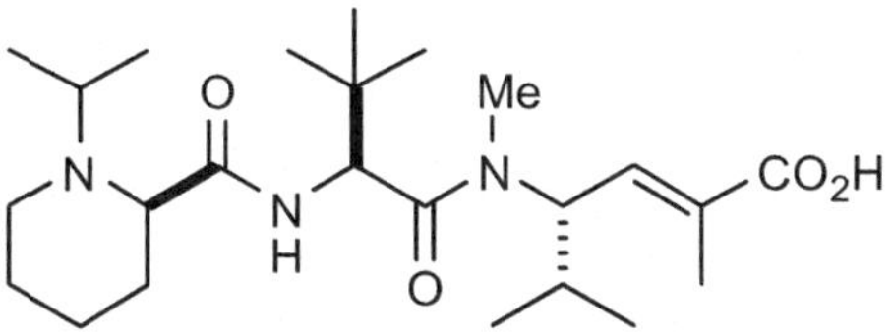

Fig. 15.14 E7974

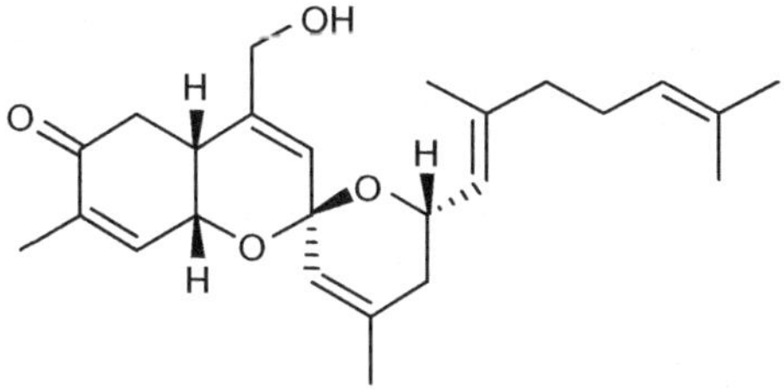

Fig. 15.15 Phorbaketal A

effective (%TGI of 50% and 18%, respectively). In a second xenograft study using A549 lung cancer cells and varied schedules of dosing, activity of peloruside was again superior compared with the taxanes with inhibitions ranging from 51% to 74%, compared with 44% and 50% for the two taxanes [342]. A third xenograft study in a P-glycoprotein-overexpressing NCI/ADR-RES breast tumor model showed that peloruside was better tolerated than either doxorubicin or paclitaxel. Peloruside is highly effective in preventing the growth of lung and P-glycoprotein-overexpressing breast tumors in vivo and that further therapeutic development is necessary [343].

E7974 (Fig.15.14) is a synthetic analogue of the marine sponge natural product hemiasterlin. Hemiasterlin, a potent cytotoxic tripeptide, was originally isolated from marine sponges [344]. E7974 acts via a tubulin-based antimitotic mechanism. E7974 inhibits polymerization of purified tubulin in vitro with IC_{50} values similar to those of vinblastine. In cultured human cancer cells, E7974 induces G_2/M arrest and marked disruption of mitotic spindle formation. Consistent with this observation, E7974 induces caspase-3 activation and PARP cleavage, typical biochemical markers of apoptosis. Only a short cellular exposure to E7974 is sufficient to induce maximum mitotic arrest, suggesting that E7974's antitumor effects in vivo may persist even after blood levels of the drug decrease after drug administration. Investigation of interactions of E7974 with purified tubulin using two synthetic tritiated photoaffinity analogues of E7974 indicated that E7974 seems to share a unique, predominantly α-tubulin-targeted mechanism with other hemiasterlin-based compounds, suggesting the hemiasterlins evolved to mainly target α-tubulin, not β-tubulin subunits unlike many tubulin-targeted natural products [345].

Phorbaketal A (Fig. 15.15) is a metabolite of the marine sponge *Phorbas* sp. [346]. This tricyclic sesterterpenoid has significant inhibitory effect on the production of nitric oxide (NO) and inflammatory cytokines such as tumor necrosis factor-alpha, interleukin (IL)-1beta, IL-6, and monocyte chemotactic protein-1 which is induced by

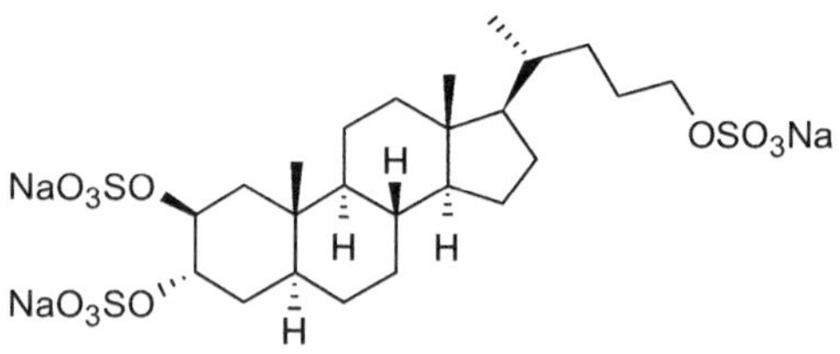

Fig. 15.16 Solomonsterol A

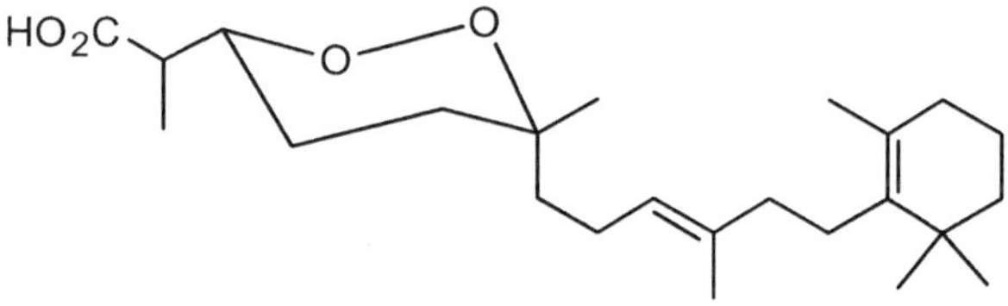

Fig. 15.17 Epimuqubilin A

LPS and the expression of inducible NO synthase in RAW 264.7 cells. Additionaly, it inhibited the transcription of a crucial signaling molecule named nuclear factor-kappaB (NF-κB) in inflammation, whereas the expression of heme oxygenase-1 (HO-1) proved to be upregulated in LPS-stimulated RAW 264.7 cells [347].

Solomonsterol A (Fig. 15.16), a selective pregnane X receptor (PXR) agonist found in the marine sponge *Theonella swinhoei*, shows anti-inflammatory activity and immune dysfunction and attenuates systemic inflammation in the rheumatoid arthritis mouse model. It is reported that solomonsterol A had an effect on protecting from the development of arthritis according to arthritis score, CRP, and plasma cytokines. In addition, anti-collagen antibodies (CAIA) could reduce the expression of inflammatory markers including TNFα, IFNγ, and IL-17 and chemokines MIP1α and RANTES in draining lymph nodes in the rheumatoid arthritis mouse model which are induced by injecting transgenic mice harboring a humanized PXR [348].

Epimuqubilin A (Fig. 15.17), a norsesterterpene peroxide isolated from marine sponge *Latrunculia* sp., inhibits nitric oxide production in LPS-stimulated RAW 264.7 cells (IC(50) = 7.6 μM). At both the mRNA and protein levels, cyclooxygenase-2 (COX-2) and inducible nitric oxide synthase (iNOS) are suppressed in a dose-dependent manner. Mitogen-activated protein kinases (MAPKs), one major upstream signaling pathway involved in the transcription of both COX-2 and iNOS, were not affected by treatment of epimuqubilin A. However, the compound blocked the phosphorylation of inhibitor κB (IκB) kinase (IKKβ), resulting in the stabilization of IκBα and inhibition of NF-κB p65 nuclear translocation and DNA binding. Levels of phosphorylated IKKα were not affected. This is an unique mechanistic relationship that suggests epimuqubilin A warrants further exploration as a potential therapeutic agent [349].

Ascididemin (ASC) is a marine alkaloid and belongs to the group of pyridoacridine alkaloids, mostly being isolated from sponges and tunicates (Fig. 15.18). Ascididemin was isolated in 1988 from the tunicate *Didemnum* sp. and showed

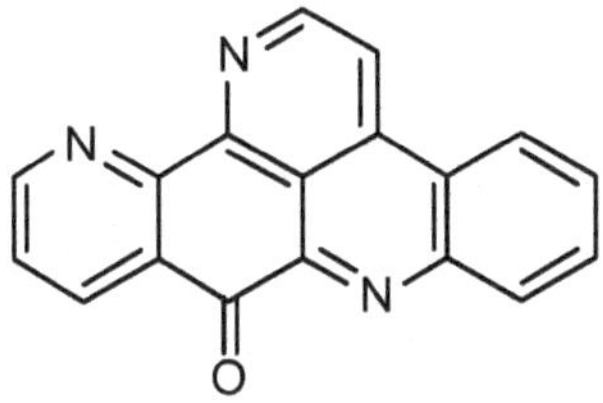

Fig. 15.18 Ascididemin

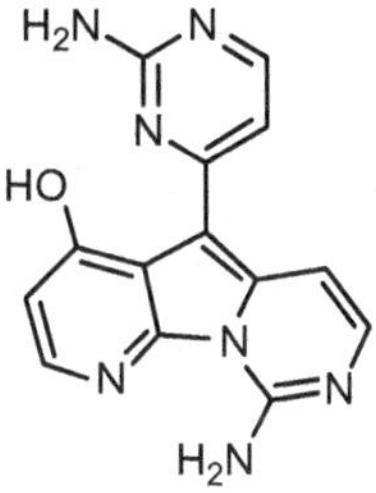

Fig. 15.19 Variolin B

remarkable in vitro cytotoxicity against a variety of cancer cells including multidrug resistant cells lines [350, 351]. Similar to other planar pyridoacridine derivatives, ascididemin interacts with DNA and recognizes triplex and quadruplex structures especially G-quadruplexes. Different mechanisms of action for this alkaloid have been proposed including topoisomerase I and II toxins. However, experiments with cell lines resistant to these toxins indicated that topo I and II are not potential targets for ascididemin (ASC) in the cell but may be by reactive oxygen species (ROS) to cleave DNA. Own work contributes to its action on the apoptotic signaling in Jurkat leukemia cells. Interestingly, ACS induces a mitochondrial pathway that requires the activation of caspase-2 upstream of mitochondria. Caspase-2 activation was not blocked by the overexpression of Bcl-2 proteins such as Bcl-xL and was responsible for caspase-9 activation. As a possible link between caspase-2 and mitochondrial activation, Bid was found to be cleaved by ASC as a specific caspase-2 inhibitor inhibits the ASC-induced cleavage of Bid [351]. In addition, JNK was activated by ASC upstream of mitochondria via reactive oxygen species. Caspase-2 activation provides a possible link between the DNA damaging activity and the induction of apoptosis. To this end, ASC might be a valuable chemical tool to induce DNA damage and apoptotic signaling events.

Variolin B (VAR-B) (Fig. 15.19) is a natural product isolated from the sponge *Kirkpatrickia variolosa*, found in Antarctica. VAR-B has been shown potent pro-apoptotic activity. In different human cancer cell lines, both compounds inhibited colony formation, caused cell cycle arrest, and induced apoptosis at concentrations ranging from 0.1 to 2 μM. Although variolins induced an increase in the levels of p53 with an increase in p21, their cytotoxicities did not appear to be dependent on p53 status as their potency was comparable in cells with wild-type p53 or in sub-lines with inactivated p53. Both VAR-B and dVAR-B prevent the cells from enter-

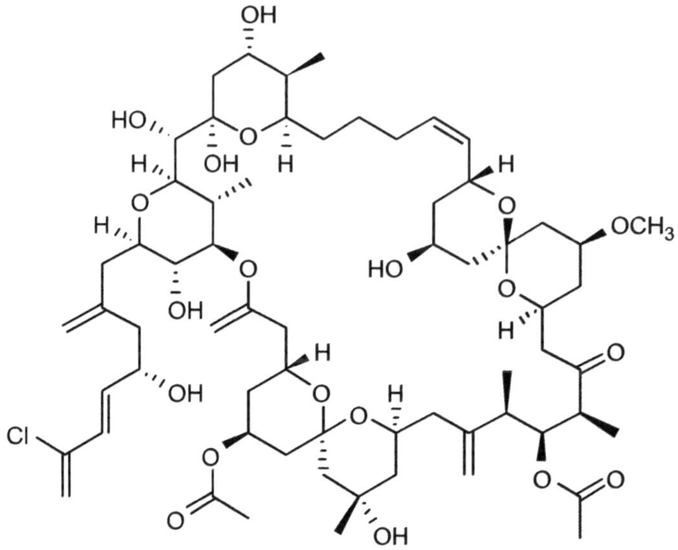

Fig. 15.20 Spongistatin 1

ing S phase, blocking cells in G1 and cause an accumulation of cells in G2. The apoptosis induced by VAR-B and dVAR-B occurred very rapidly in some cell lines (e.g., Jurkat leukemia cells) and was already evident 4 h after the beginning of treatment. Although intercalation of dVAR-B in DNA has been demonstrated, neither VAR-B nor dVAR-B produces detectable breaks in DNA, which are consistent with the in vitro biochemical assays that also demonstrated that dVAR-B is not topoisomerase I or II poison. Instead, each of these variolins appears to inhibit cyclin-dependent kinases (CDKs) in the lM range. CDK1-cyclin B, CDK2-cyclin A, and CDK2/cyclin E complexes were inhibited in a range of concentrations lower than those required to inhibit the activity of CDK4/cyclin D or CDK7/cyclin H complexes. Variolins are a new class of CDK inhibitors that activate apoptosis in a p53-independent fashion, and thus they may be effective against tumors with p53 mutations or deletions [352].

Spongistatin is a macrocyclic lactone that has been isolated from the marine sponges *Spirastrella spinispirulifera* and *Hyrtios* sp. by the group of Pettit. Spongistatin 1 (Fig. 15.20) showed interesting apoptotic features in various tumor cells. In leukemic cell lines, it triggered caspase-dependent apoptosis through the release of cytochrome c, Smac/Diablo, and Omi/HtrA2 from the mitochondria into the cytosol. Spongistatin 1 caused degradation of the anti-apoptotic XIAP, which suggested it might be a promising drug for the treatment of chemoresistance due to overexpression of XIAP. Moreover spongistatin 1 induces apoptosis more efficiently in human primary leukemic cells of children suffering from acute leukemia at low nanomolar concentrations than clinically applied conventional drugs used in micromolar concentrations. In addition normal healthy peripheral blood cells were significantly less affected by spongistatin 1 [353]. Besides leukemic cells, spongistatin 1 showed promising apoptotic potential in mammary cancer cells including the treatment-resistant cell line MCF-7 lacking caspase-3. Regarding the apoptotic signaling pathways of spongistatin 1, two interesting features can be reported. First, spongistatin 1-induced cell death involves the pro-apoptotic proteins AIF and endonuclease G. Both proteins translocate from mitochondria to the nucleus and contribute to spongistatin 1-mediated apoptosis as shown via gene silencing. Second, spongistatin 1 acts as a tubulin-depolymerizing agent and is able to free the pro-apoptotic Bcl-2 family member Bim from its sequestration both by the microtubular complex and by the anti-apoptotic protein Mcl-1. Silencing of Bim by siRNA leads to a diminished translocation of AIF and endonuclease G to the nucleus and subsequently reduces rate of apoptosis. By using spongistatin 1 as a chemical tool, Bim has been suggested to be an important factor upstream of mitochondria by executing a central role in the caspase-independent apoptotic signaling pathway induced by spongistatin 1. These different apoptotic features indicate that the apoptosis signaling is cell line-specific. Finally, spongistatin 1 affects highly invasive pancreatic tumor cells by not only inhibiting their invasion and migration but also by inducing anoikis in these cells. Bcl-2 seems to be a major target for spongistatin 1 in these processes. Besides tumor cells, spongistatin inhibits angiogenic activity of endothelial cells via inhibition of PKC-a [354].

References

1. Rodriguez J, Schatzman RC, Lou L, Crews P. An alkaloid protein kinase C inbibitor, xestocyclamine A, from the marine sponge *Xestospongia* sp. J Am Chem Soc. 1993;115:10436–7.
2. Peters BM, Kurz L, Schatzman RC, Mccarley D, Lou L, et al. Novel marine sponge alkaloids:V. an alkaloid protein kinase C inhibitor, xestocyclamine A, from the marine sponge *Xestospongia* sp. ChemInform. 1994;25:10436–7.
3. Vassas A, Bourdy G, Paillard JJ, Lavayre J, Païs M, Quirion JC, et al. Naturally occurring somatostatin and vasoactive intestinal peptide inhibitors. Isolation of alkaloids from two marine sponges. Planta Med. 1996;62:28–30.
4. Tsukamoto S, Kato H, Hiroshi-Hirota A, Fusetani N. Mauritiamine, a new antifouling oroidin dimer from the marine sponge *Agelas mauritiana*. J Nat Prod. 1996;59:501–3.
5. Pettit GR, Orr B, Herald DL, Doubek DL, Tackett L, Schmidt JM, et al. Isolation and X-ray crystal structure of racemic Xestospongin D from the Singapore marine sponge *Niphates* sp. Bioorg Med Chem Lett. 1996;6:1313–8.
6. Eder C, Schupp P, Proksch P, Wray V, Steube K, Müller CE, et al. Soest: bioactive pyridoacridine alkaloids from the micronesian sponge *Oceanapia* sp. J Nat Prod. 1998;61:301–5.
7. Kashman Y, Koren-Goldshlager G, Gravalos MDG, Schleyer M. Halitulin, a new cytotoxic alkaloid from the marine sponge *Haliclona tulearensis*. Tetrahedron Lett. 1999;40:997–1000.
8. Mariegeneviève-Dijoux WRG, Hallock YF, Ii JHC, Boyd MR. A new discorhabdin from two sponge genera. J Nat Prod. 1999;62:636–7.
9. Marino SD, Iorizzi M, Zollo F, Roussakis C, Debitus C. Plakinamines C and D and three other new steroidal alkaloids from the sponge *Corticium* sp. J Nat Prod. 1999;30:636–7.

10. Casapullo A, Bifulco G, Bruno I, Riccio R. New bisindole alkaloids of the topsentin and hamacanthin classes from the Mediterranean marine sponge *Rhaphisia lacazei*. J Nat Prod. 2000;63:447–51.
11. Cutignano A, Bifulco G, Bruno I, Casapullo A, Gomez-Paloma L, Riccio R, et al. Dragmacidin F: a new antiviral bromoindole alkaloid from the Mediterranean sponge *Halicortex* sp. Cheminform Abstr. 2000;31:41.
12. Casapullo A, Cutignano A, Bruno I, Bifulco G, Debitus C, Gomezpaloma L, et al. Makaluvamine P, a new cytotoxic pyrroloiminoquinone from *Zyzzya cf. fuliginosa*. J Nat Prod. 2001;64:1354–6.
13. Tasdemir D, Mallon RG, Michael L, Feldberg S, Kim K, Collins D, et al. Ireland: Aldisine alkaloids from the Philippine sponge *Stylissa massa* are potent inhibitors of mitogen-activated protein kinase kinase-1 (MEK-1). J Med Chem. 2002;45:529–32.
14. Torres YR, Berlinck RG, Nascimento GG, Fortier SC, Pessoa C, de Moraes MO, et al. Antibacterial activity against resistant bacteria and cytotoxicity of four alkaloid toxins isolated from the marine sponge *Arenosclera brasiliensis*. Farmacologia Marinha. 2002;40:885–91.
15. Gross H, Kehraus S, König GM, Woerheide G, Wright AD. New and biologically active imidazole alkaloids from two sponges of the genus Leucetta. J Nat Prod. 2002;65:1190–3.
16. Hamann MT. Manadomanzamines A and B: a novel alkaloid ring system with potent activity against mycobacteria and HIV-1. J Am Chem Soc. 2003;125:13382–6.
17. Bickmeyer U, Drechsler C, Kock M, Assmann M. Brominated pyrrole alkaloids from marine Agelas sponges reduce depolarization-induced cellular calcium elevation. Toxicon. 2004;44:45–51.
18. Ichiba T, Corgiat JM, Scheuer PJ, Kelly-Borges M. 8-Hydroxymanzamine A, a β-carboline alkaloid from a sponge *Pachypellina* sp. J Nat Prod. 1994;57:168–70.
19. Yousaf M, Hammond NL, Peng J, Wahyuono S, McIntosh KA, Charman WN, et al. New manzamine alkaloids from an Indo-Pacific sponge. Pharmacokinetics, oral availability, and the significant activity of several manzamines against HIV-I, AIDS opportunistic infections, and inflammatory diseases. J Med Chem. 2004;47:3512–7.
20. Rao KV, Kasanah N, Wahyuono S, Tekwani BL, Schinazi RF, Hamann MT, et al. Three new manzamine alkaloids from a common Indonesian sponge and their activity against infectious and tropical parasitic diseases. J Nat Prod. 2004;67:1314–8.
21. Rao KV, Donia MS, Peng J, Garcia-Palomero E, Alonso D, Martinez A, et al. Manzamine B and E and ircinal A related alkaloids from an Indonesian Acanthostrongylophora sponge and their activity against infectious, tropical parasitic, and Alzheimer's diseases. J Nat Prod. 2006;69:1034–40.
22. Zhang B, Higuchi R, Miyamoto T, Soest RWV. Neuritogenic activity-guided isolation of a free base form manzamine A from a marine sponge, *Acanthostrongylophora aff. ingens*. Chem Pharm Bull. 2008;56:866–9.
23. Yamada M, Takahashi Y, Kubota T, Fromont J, Ishiyama A, Otoguro K, et al. 3,4-dihydro-6-hydroxy-10,11-epoxymanzamine A, and 3,4-dihydromanzamine J N-oxide, new manzamine alkaloids from sponge *Amphimedon* sp. Tetrahedron. 2009;65:2313–7.
24. Antunes EM, Beukes DR, Kelly M, Samaai T, Barrows LR, Marshall KM, et al. Cytotoxic pyrroloiminoquinones from four new species of south African latrunculid sponges. J Nat Prod. 2004;67:1268–76.
25. Grkovic T, Ding Y, Li XC, Webb VL, Ferreira D, Copp BR, et al. Enantiomeric discorhabdin alkaloids and establishment of their absolute configurations using theoretical calculations of electronic circular dichroism spectra. J Org Chem. 2008;73:9133–6.
26. Davis RA, Buchanan MS, Duffy S, Avery VM, Charman SA, Charman WN, et al. Antimalarial activity of pyrroloiminoquinones from the Australian marine sponge *Zyzzya* sp. J Med Chem. 2012;55:5851–8.
27. Cao S, Foster C, Lazo JS, Kingston DG. Sesterterpenoids and an alkaloid from a *Thorectandra* sp. as inhibitors of the phosphatase Cdc25B. Bioorg Med Chem. 2005;13:5094–8.

28. Bao B, Sun Q, Yao X, Hong J, Lee CO, Sim CJ, et al. Cytotoxic bisindole alkaloids from a marine sponge *Spongosorites* sp. J Nat Prod. 2005;68:711–5.
29. Mar W, Kim S, Kim JY, Lee TH, Kim JG, Shin D, et al. Antimicrobial activity and cytotoxicity of bis (indole) alkaloids from the sponge *Spongosorites* sp. Biol Pharm Bull. 2006;29:570–3.
30. Aoki S, Watanabe Y, Sanagawa M, Setiawan A, Kotoku N, Kobayashi M, et al. Cortistatins A, B, C, and D, anti-angiogenic steroidal alkaloids, from the marine sponge *Corticium simplex*. J Am Chem Soc. 2006;128:3148–9.
31. Sato S, Kuramoto M, Ono N. Ircinamine B, bioactive alkaloid from marine sponge *Dactylia* sp. Tetrahedron Lett. 2006;47:7871–3.
32. Buchanan MS, Carroll AR, Addepalli R, Avery VM, Hooper JN, Quinn RJ, et al. Psammaplysenes C and D, cytotoxic alkaloids from *Psammoclemma* sp. J Nat Prod. 2007;70:1827–9.
33. Oliveira JH, Nascimento AM, Kossuga MH, Cavalcanti BC, Pessoa CO, Moraes MO, et al. Cytotoxic alkylpiperidine alkaloids from the Brazilian marine sponge *Pachychalina alcaloidifera*. J Nat Prod. 2007;70:538–43.
34. Wei X, Nieves K, Rodriguez AD. Neopetrosiamine A, biologically active bis-piperidine alkaloid from the Caribbean Sea sponge *Neopetrosia proxima*. Bioorg Med Chem Lett. 2010;20:5905–8.
35. Mani L, Petek S, Valentin A, Chevalley S, Folcher E, Aalbersberg W, et al. The in vivo antiplasmodial activity of haliclonacyclamine A, an alkaloid from the marine sponge *Haliclona* sp. Nat Prod Res. 2011;25:1923–30.
36. Hua HM, Peng J, Dunbar DC, Schinazi RF, Castro Andrews AG, Cuevas C, et al. Batzelladine alkaloids from the caribbean sponge *Monanchora unguifera* and the significant activities against HIV-1 and AIDS opportunistic infectious pathogens. Tetrahedron. 2007;63:11179–88.
37. Laville R, Thomas OP, Berrué F, Marquez D, Vacelet J, Amade P, et al. Bioactive guanidine alkaloids from two Caribbean marine sponges. J Nat Prod. 2009;72:1589–94.
38. Santos MF, Harper PM, Williams DE, Mesquita JT, Pinto EG, Costa-Silva TA, et al. Antiparasitic guanidine and pyrimidine alkaloids from the marine sponge *Monanchora arbuscula*. J Nat Prod. 2015;78:1101–12.
39. Bickmeyer U, Grube A, Klings KW, Köck M. Disturbance of voltage-induced cellular calcium entry by marine dimeric and tetrameric pyrrole–imidazole alkaloids. Toxicon. 2007;50:490–7.
40. Xu NJ, Sun X, Yan XJ. A new cyclostellettamine from sponge *Amphimedon compressa*. Chin Chem Lett. 2007;18:947–50.
41. Laville R, Genta-Jouve G, Urda C, Fernández R, Thomas OP, Reyes F, et al. Njaoaminiums A, B, and C: cyclic 3-alkylpyridinium salts from the marine sponge *Reniera* sp. Molecules. 2009;14:4716–24.
42. Ankudey FJ, Kiprof P, Stromquist ER, Chang LC. New bioactive bromotyrosine-derived alkaloid from a marine sponge *Aplysinella* sp. Planta Med. 2008;74:555–9.
43. Capon RJ, Peng C, Dooms C. Trachycladindoles A-G: cytotoxic heterocycles from an Australian marine sponge, *Trachycladus laevispirulifer*. Org Biomol Chem. 2008;6:2765–71.
44. Araki A, Kubota T, Aoyama K, Mikami Y, Fromont J, Kobayashi JI, et al. Nagelamides Q and R, novel dimeric bromopyrrole alkaloids from sponges *Agelas* sp. Org Lett. 2009;11:1785–8.
45. Kubota T, Araki A, Yasuda T, Tsuda M, Fromont J, Aoyama K, et al. Benzosceptrin C, a new dimeric bromopyrrole alkaloid from sponge *Agelas* sp. Tetrahedron Lett. 2009;50:7268–70.
46. Barnes EC, Said NABM, Williams ED, Hooper JNA, Davis RA. Ecionines A and B, two new cytotoxic pyridoacridine alkaloids from the Australian marine sponge, *Ecionemia geodides*. Tetrahedron. 2010;66:283–7.
47. Carroll AR, Kaiser SM, Davis RA, Moni RW, Hooper JN, Quinn RJ, et al. A bastadin with potent and selective δ-opioid receptor binding affinity from the Australian sponge *Ianthella flabelliformis*. J Nat Prod. 2010;73:1173–6.
48. Fan G, Li Z, Shen S, Zeng Y, Yang Y, Xu M. Baculiferins A-O, O-sulfated pyrrole alkaloids with anti-HIV-1 activity, from the Chinese marine sponge *Iotrochota baculifera*. Bioorg Med Chem. 2010;18:5466–74.

49. Guzii AG, Makarieva TN, Denisenko VA, Dmitrenok PS, Kuzmich AS, Dyshlovoy SA, et al. Monanchocidin: A new apoptosis-inducing polycyclic guanidine alkaloid from the marine sponge *Monanchora pulchra*. Org Lett. 2010;12:4292–5.
50. Kon Y, Kubota T, Shibazaki A, Gonoi T, Kobayashi J. Ceratinadins A-C, new bromotyrosine alkaloids from an Okinawan marine sponge *Pseudoceratina* sp. Bioorg Med Chem Lett. 2010;20:4569–72.
51. Yang X, Davis RA, Buchanan MS, Duffy S, Avery VM, Camp D, et al. Antimalarial bromotyrosine derivatives from the Australian marine sponge *Hyattella* sp. J Nat Prod. 2010;73:985–7.
52. Xu M, Andrews KT, Birrell GW, Tran TL, Camp D, Davis RA, et al. Psammaplysin H, a new antimalarial bromotyrosine alkaloid from a marine sponge of the genus *Pseudoceratina*. Bioorg Med Chem Lett. 2011;21:846–8.
53. Ferreira EG, Wilke DV, Jimenez PC, Oliveira JR, Pessoa ODL, Silveira ER, et al. Guanidine alkaloids from *Monanchora arbuscula*: chemistry and antitumor potential. Chem Biodivers. 2011;8:1433–45.
54. Makarieva TN, Tabakmaher KM, Guzii AG, Denisenko VA, Dmitrenok PS, Shubina LK, et al. Monanchocidins B–E: polycyclic guanidine alkaloids with potent antileukemic activities from the sponge *Monanchora pulchra*. J Nat Prod. 2011;74:1952–8.
55. Makarieva TN, Tabakmaher KM, Guzii AG, Denisenko VA, Dmitrenok PS, Kuzmich AS, et al. Monanchomycalins A and B, unusual guanidine alkaloids from the sponge *Monanchora pulchra*. Tetrahedron Lett. 2012;53:4228–31.
56. Yin S, Davis RA, Shelper T, Sykes ML, Avery VM, Elofsson M, et al. Quinn: Pseudoceramines A-D, new antibacterial bromotyrosine alkaloids from the marine sponge *Pseudoceratina* sp. Org Biomol Chem. 2011;9:6755–60.
57. Fouad MA, Debbab A, Wray V, Müller WEG, Proksch P. New bioactive alkaloids from the marine sponge *Stylissa* sp. Tetrahedron. 2012;68:10176–9.
58. Hwang BS, Jeong EJ, Sim CJ, Rho JR. Densanins A and B, new macrocyclic pyrrole alkaloids isolated from the marine sponge *Haliclona densaspicula*. Org Lett. 2012;14:6154–7.
59. Ilias M, Ibrahim MA, Khan SI, Jacob MR, Tekwani BL, Walker LA, et al. Pentacyclic ingamine alkaloids, a new antiplasmodial pharmacophore from the marine sponge *Petrosid Ng5 Sp5*. Planta Med. 2012;78:1690–7.
60. Shen S, Liu D, Wei C, Proksch P, Lin W. Purpuroines A-J, halogenated alkaloids from the sponge *Iotrochota purpurea* with antibiotic activity and regulation of tyrosine kinases. Bioorg Med Chem. 2012;20:6924–8.
61. Takahashi Y, Tanaka N, Kubota T, Ishiyama H, Shibazaki A, Gonoi T, et al. Heteroaromatic alkaloids, nakijinamines, from a sponge *Suberites* sp. Tetrahedron. 2012;68:8545–50.
62. Yang F, Hamann MT, Zou Y, Zhang MY, Gong XB, Xiao JR, et al. Antimicrobial metabolites from the Paracel Islands sponge *Agelas mauritiana*. J Nat Prod. 2012;75:774–8.
63. Davis RA, Duffy S, Fletcher S, Avery VM, Quinn RJ. Thiaplakortones A-D: antimalarial thiazine alkaloids from the Australian marine sponge *Plakortis lita*. J Org Chem. 2013;78:9608–13.
64. Kimura M, Wakimoto T, Abe I. Allos-hemicalyculin A, a photochemically converted calyculin from the marine sponge *Discodermia calyx*. Tetrahedron Lett. 2013;54:114–6.
65. Kubota T, Kura KI, Fromont J, Kobayashi JI. Pyrinodemins G–I, new bis-3-alkylpyridine alkaloids from a marine sponge *Amphimedon* sp. Tetrahedron. 2013;69:96–100.
66. Tanaka N, Kusama T, Takahashi-Nakaguchi A, Gonoi T, Fromont J, Kobayashi JI. Nagelamides X–Z, dimeric bromopyrrole alkaloids from a marine sponge *Agelas* sp. Org Lett. 2013;15:3262–5.
67. Tanaka N, Kusama T, Takahashi-Nakaguchi A, Gonoi T, Fromont J, Kobayashi JI, et al. Nagelamides U–W, bromopyrrole alkaloids from a marine sponge *Agelas* sp. Tetrahedron Lett. 2013;54:3794–6.
68. Yamaguchi M, Miyazaki M, Kodrasov MP, Rotinsulu H, Losung F, Mangindaan RE, et al. Spongiacidin C, a pyrrole alkaloid from the marine sponge *Stylissa massa*, functions as a USP7 inhibitor. Bioorg Med Chem Lett. 2013;23:3884–6.

69. Yang F, Ji RH, Li J, Gan JH, Lin HW. N-containing metabolites from the marine sponge *Agelas clathrodes*. Nat Prod Commun. 2013;8:1713–4.
70. Arai M, Han C, Yamano Y, Setiawan A, Kobayashi M. Aaptamines, marine spongean alkaloids, as anti-dormant mycobacterial substances. J Nat Med. 2014;68:372–6.
71. Gan JH, Hu WZ, Yu HB, Yang F, Cao MX, Shi HJ, et al. Three new aaptamine derivatives from the South China Sea sponge *Aaptos aaptos*. J Asian Nat Prod Res. 2015;17:1231–8.
72. Gros E, Martin MT, Sorres J, Moriou C, Vacelet J, Frederich M, et al. Netamines O–S, five new tricyclic guanidine alkaloids from the Madagascar sponge *Biemna laboutei* and their antimalarial activities. Chem Biodivers. 2015;12:1725–33.
73. Hanif N, Yamada K, Kitamura M, Kawazoe Y, Voogd NJ, Uemura D, et al. New indole alkaloids from the sponge *Plakortis* sp. Chem Nat Compd. 2015;51:1130–3.
74. Huang RY, Chen WT, Kurtán T, Mándi A, Ding J, Li J, et al. Bioactive isoquinolinequinone alkaloids from the South China Sea nudibranch *Jorunna funebris* and its sponge-prey *Xestospongia* sp. Future Med Chem. 2015;8:17–27.
75. Jamison MT, Molinski TF. Antipodal crambescin A2 homologues from the marine sponge *Pseudaxinella reticulata*. Antifungal structure-activity relationships. J Nat Prod. 2015;78:557–61.
76. Shaala LA, Youssef DTA, Badr JM, Sulaiman M, Khedr A, Sayed KA, et al. Bioactive alkaloids from the Red Sea marine Verongid sponge *Pseudoceratina arabica*. Tetrahedron. 2015;71:7837–41.
77. Abdjul DB, Yamazaki H, Kanno S, Takahashi O, Kirikoshi R, Ukai K, et al. Haliclonadiamine derivatives and 6-epi-Monanchorin from the marine sponge *Halichondria panicea* collected at Iriomote Island. J Nat Prod. 2016;79:1149–54.
78. Matsunaga S, Fujiki H, Sakata D, Fusetani N. Calyculins E, F, G, and H, additional inhibitors of protein phosphatases 1 and 2a, from the marine sponge *discodermia calyx*. Tetrahedron. 1991;47:2999–3006.
79. Coleman JE, Silva ED, Kong F, Andersen RJ, Allen TM. Cytotoxic peptides from the marine sponge *Cymbastela* sp. Tetrahedron. 1995;51:10653–62.
80. Li HY, Matsunaga S, Fusetani N. Halicylindramides D and E, antifungal peptides from the marine sponge *Halichondria cylindrata*. J Nat Prod. 1996;59:163–6.
81. Fusetani N, Warabi K, Nogata Y, Nakao Y, Matsunaga S, Van Soest RR, et al. Koshikamide A 1, a new cytotoxic linear peptide isolated from a marine sponge, *Theonella* sp. Tetrahedron Lett. 1999;40:4687–90.
82. Nakao Y, Masuda A, Matsunaga S, Fusetani N. Pseudotheonamides, serine protease inhibitors from the marine sponge *Theonella swinhoei*. J Am Chem Soc. 1999;121:2425–31.
83. Nakao Y, Fujita M, Warabi K, Matsunaga S, Fusetani N. Miraziridine A, a novel cysteine protease inhibitor from the marine sponge *Theonella aff. mirabilis* 1. J Am Chem Soc. 2000;122:10462–3.
84. Carroll AR, Pierens GK, Fechner G, Almeida Leone P, Ngo A, Simpson M, et al. Dysinosin A: a novel inhibitor of factor VIIa and thrombin from a new genus and species of Australian sponge of the family *Dysideidae*. J Am Chem Soc. 2002;124:13340–1.
85. Carroll AR, Buchanan MS, Edser A, Hyde E, Simpson M, Quinn RJ, et al. Dysinosins BD, inhibitors of factor VIIa and thrombin from the Australian sponge *Lamellodysidea chlorea*. J Nat Prod. 2004;67:1291–4.
86. Aoki S, Cao L, Matsui K, Rachmat R, Akiyama SI, Kobayashi M. Kendarimide A, a novel peptide reversing P-glycoprotein-mediated multidrug resistance in tumor cells, from a marine sponge of *Haliclona* sp. Tetrahedron. 2004;60:7053–9.
87. Hamada T, Matsunaga S, Yano G, Fusetani N. Polytheonamides A and B, highly cytotoxic, linear polypeptides with unprecedented structural features, from the marine sponge, *Theonella swinhoei*. J Am Chem Soc. 2005;127:110–8.
88. Hamada T, Sugawara T, Matsunaga S, Fusetani N. Polytheonamides, unprecedented highly cytotoxic polypeptides from the marine sponge *Theonella swinhoei* 2. Structure elucidation. Tetrahedron Lett. 1994;35:609–12.

89. Iwamoto M, Shimizu H, Muramatsu I, Oiki S. A cytotoxic peptide from a marine sponge exhibits ion channel activity through vectorial-insertion into the membrane. FEBS Lett. 2010;584:3995–9.
90. Hamada T, Sugawara T, Matsunaga S, Fusetani N. Polytheonamides, unprecedented highly cytotoxic polypeptides, from the marine sponge *theonella swinhoei*: 1. Isolation and component amino acids. Tetrahedron Lett. 1994;35:719–20.
91. Hamada T, Matsunaga S, Fujiwara M, Fujita K, Hirota H, Schmucki R, et al. Solution structure of polytheonamide B, a highly cytotoxic nonribosomal polypeptide from marine sponge. J Am Chem Soc. 2010;132:12941–5.
92. Araki T, Matsunaga T, Fusetani N. Koshikamide A2, a cytotoxic linear undecapeptide isolated from a marine sponge of *Theonella* sp. Biosci Biotechnol Biochem. 2005;69:1318–22.
93. Sadar MD, Williams DE, Mawji NR, Patrick BO, Wikanta T, Chasanah E, et al. Sintokamides A to E, chlorinated peptides from the sponge *Dysidea* sp. that inhibit transactivation of the N-terminus of the androgen receptor in prostate cancer cells. Org Lett. 2008;10:4947–50.
94. Ueoka R, Ise Y, Ohtsuka S, Okada S, Yamori T, Matsunaga S, et al. Yaku'amides A and B, cytotoxic linear peptides rich in dehydroamino acids from the marine sponge *Ceratopsion* sp. J Am Chem Soc. 2010;132:17692–4.
95. Bewley CA, He H, Williams DH, Faulkner DJ. Aciculitins A-C: cytotoxic and antifungal cyclic peptides from the lithistid sponge *Aciculites orientalis*. J Am Chem Soc. 1996;118:4314–21.
96. Kobayashi M, Wang W, Ohyabu N, Kurosu M, Kitagawa I. Improved total synthesis and structure-activity relationship of arenastatin A, a potent cytotoxic spongean depsipeptide. Chem Pharm Bull. 1995;43:1598–600.
97. Clark DP, Carroll J, Naylor S, Crews P. An antifungal cyclodepsipeptide, Cyclolithistide A, from the sponge *Theonella swinhoei*. J Org Chem. 1998;63:8757–64.
98. Nakao Y, Oku N, Matsunaga S, Fusetani N. Cyclotheonamides E2 and E3, new potent serine protease inhibitors from the marine sponge of the genus *Theonella*. J Nat Prod. 1998;61:667–70.
99. Zampella A, Sepe V, Luciano P, Bellotta F, Monti MC, D'Auria MV, et al. Homophymine A, an anti-HIV cyclodepsipeptide from the sponge *Homophymia* sp. J Org Chem. 2008;73:5319–27.
100. Zampella A, Sepe V, Bellotta F, Luciano P, D'Auria MV, Cresteil T, et al. Homophymines B-E and A1-E1, a family of bioactive cyclodepsipeptides from the sponge *Homophymia* sp. Org Biomol Chem. 2009;7:4037–44.
101. Gala F, D'Auria MV, Marino S, Sepe V, Zollo F, Smith CD, et al. Jaspamides M-P: new tryptophan modified jaspamide derivatives from the sponge *Jaspis splendens*. Tetrahedron. 2009;65:51–6.
102. Zampella A, Giannini C, Debitus C, Roussakis C, D'Auria MV. New Jaspamide derivatives from the marine sponge *Jaspis splendans* collected in Vanuatu. J Nat Prod. 1999;62:332–4.
103. Ebada SS, Wray V, Voogd NJ, Deng Z, Lin W, Proksch P. Two new jaspamide derivatives from the marine sponge *Jaspis splendens*. Mar Drugs. 2009;7:435–44.
104. Yeung BKS, Nakao Y, Kinnel RB, Carney JR, Yoshida WY, Scheuer PJ, et al. The kapakahines, cyclic peptides from the marine sponge *Cribrochalina olemda*. J Org Chem. 1996;61:7168–73.
105. Shigemori H, Itagaki F, Takao T, Shimonishi Y, Kobayashi JI. Keramamides E, G, H, and J, new cyclic peptides containing an oxazole or a thiazole ring from a *Theonella* sponge. Pept Chem. 1995;32:17–20.
106. Itagaki F, Shigemori H, Ishibashi M, Nakamura T, Sasaki T, Kobayashi J, et al. Keramamide F, a new thiazole-containing peptide from the Okinawan marine sponge *Theonella* sp. J Org Chem. 1992;57:5540–2.
107. Qureshi A, Colin PL, Faulkner DJ. Microsclerodermins F-I, antitumor and antifungal cyclic peptides from the lithistid sponge *Microscleroderma* sp. Tetrahedron. 2000;56:3679–85.
108. Silva ED, Williams DE, Andersen RJ, Klix H, Holmes CFB, Allen TM, et al. Motuporin, a potent protein phosphatase inhibitor isolated from the Papua New Guinea sponge *Theonella swinhoei Gray*. Tetrahedron Lett. 1992;33:1561–4.

109. D'Auria MV, Gomez-Paloma L, Minale L, Zampella A, Debitus C, Perez J, et al. Neosiphoniamolide A, a novel cyclodepsipeptide, with antifungal activity from the marine sponge *Neosiphonia superstes*. J Nat Prod. 1995;58:121–3.
110. Fusetani N, Sugawara T, Matsunaga S, Hirota H. Orbiculamide A: a novel cytotoxic cyclic peptide from a marine sponge *Theonella* sp. J Am Chem Soc. 1991;113:7811–2.
111. Festa C, Marino S, Sepe V, D'Auria MV, Bifulco G, Andres R, et al. Perthamides C-F, potent human antipsoriatic cyclopeptides. Tetrahedron. 2011;67:7780–6.
112. Festa C, Marino S, Sepe V, D'Auria MV, Monti MC, Bucci M, et al. Anti-inflammatory cyclopeptides from the marine sponge *Theonella swinhoei*. Tetrahedron. 2012;68:2851–7.
113. Pettit GR, Clewlow PJ, Dufresne C, Doubek DL, Cerny RL, Rutzler K, et al. Antineoplastic agents. 193. Isolation and structure of the cyclic peptide hymenistatin 1. Can J Chem. 1990;68:708–11.
114. Pettit GR, Herald CL, Boyd MR, Leet JE, Dufresne C, Doubek DL, et al. Antineoplastic agents. 219. Isolation and structure of the cell growth inhibitory constituents from the western Pacific marine sponge *Axinella* sp. J Med Chem. 1991;34:3339–40.
115. Pettit GR, Gao F, Cerny R. Antineoplastic agents. 279. Isolation and structure of axinastatin 4 from the western Indian Ocean marine sponge *Axinella cf. carteri*. Heterocycles. 1993;35:711–8.
116. Kobayashi J, Tsuda M, Nakamura T, Mikami Y, Shigemori H. Hymenamides A and B, new proline-rich cyclic heptapeptides from the Okinawan marine sponge *Hymeniacidon* sp. Tetrahedron. 1993;49:2391–402.
117. Pettit GR, Cichacz Z, Barkoczy J, Dorsaz AC, Herald DL, Williams MD, et al. Isolation and structure of the marine sponge cell growth inhibitory cyclic peptide phakellistatin 1. J Nat Prod. 1993;56:260–7.
118. Pettit GR, Rhodes MR, Tan R. Antineoplastic agents. 400. Synthesis of the Indian Ocean marine sponge cyclic heptapeptide Phakellistatin 2. J Nat Prod. 1999;62:409–14.
119. Pettit GR, Xu JP, Cichacz Z, Schmidt JM, Dorsaz AC, Boyd MR, et al. Antineoplastic agents. 303. Isolation and structure of the human cell growth inhibitory phakellistatin 4 from the western Pacific sponge *Phakellia costata*. Heterocycles. 1995;40:501–6.
120. Pettit GR, Xu JP, Cichacz ZA, Williams MD, Dorsaz AC, Brune DC, et al. Antineoplastic agents 315. Isolation and structure of the marine sponge cancer cell growth inhibitor phakellistatin 5. Bioorg Med Chem Lett. 1994;4:2091–6.
121. Yi Y, Li W. Cyclopeptide compound phakellistatin 12 having antitumor activity. CN1396177A. 2003.
122. Jiang QF, Zhou YJ, Yao JZ, Lu JG, Zhu J, Yao B, et al. Total synthesis of a marine cyclic peptide: phakellistatin-13. Huaxue Xuebao. 2007;65:253–6.
123. Zhang HJ, Yi YH, Yang GJ, Hu MY, Cao GD, Yang F, et al. Proline-containing cyclopeptides from the marine sponge *Phakellia fusca*. J Nat Prod. 2010;73:650–5.
124. Zhan KX, Jiao WH, Yang F, Li J, Wang SP, Li YS, et al. Reniochalistatins A-E, cyclic peptides from the marine sponge *Reniochalina stalagmitis*. J Nat Prod. 2014;77:2678–84.
125. Lorente A, Gil A, Fernǻndez R, Cuevas C, Albericio F, Álvarez M, et al. Phormidolides B and C, cytotoxic agents from the sea: enantioselective synthesis of the macrocyclic core. Chemistry. 2015;21:150–6.
126. Williamson RT, Boulanger A, Vulpanovici A, Roberts MA, Gerwick WH. Structure and absolute stereochemistry of Phormidolide, a new toxic metabolite from the marine *Cyanobacterium Phormidium* sp. J Org Chem. 2002;67:7927–36.
127. Ueoka R, Nakao Y, Kawatsu S, Yaegashi J, Matsumoto Y, Matsunaga S, et al. Gracilioethers A–C, antimalarial metabolites from the marine sponge *Agelas gracilis*. J Org Chem. 2009;74:4203–7.
128. Festa C, Lauro G, Marino S, D'Auria MV, Monti MC, Casapullo A, et al. Plakilactones from the marine sponge *Plakinastrella mamillaris*. Discovery of a new class of marine ligands of peroxisome proliferator-activated receptor γ. J Med Chem. 2012;55:8303–17.

129. Festa C, Marino S, D'Auria MV, Deharo E, Gonzalez G, Deyssard C, et al. Gracilioethers E–J, new oxygenated polyketides from the marine sponge *Plakinastrella mamillaris*. Tetrahedron. 2012;68:10157–63.
130. Festa C, D'Amore C, Renga B, Lauro G, Marino S, D'Auria MV, et al. Oxygenated polyketides from *Plakinastrella mamillaris* as a new chemotype of PXR agonists. Mar Drugs. 2013;11:2314–27.
131. Teta R, Irollo E, Della Sala G, Pirozzi G, Mangoni A, Costantino V, et al. Smenamides A and B, chlorinated peptide/polyketide hybr,ids containing a dolapyrrolidinone unit from the Caribbean sponge *Smenospongia aurea*. Evaluation of their role as leads in antitumor drug research. Mar Drugs. 2013;11:4451–63.
132. Festa C, Marino S, D'Auria MV, Taglialatela-Scafati O, Deharo E, Petek S, et al. New antimalarial polyketide endoperoxides from the marine sponge *Plakinastrella mamillaris* collected at Fiji Islands. Tetrahedron. 2013;69:3706–13.
133. Kubota T, Ishiguro Y, Takahashi-Nakaguchi A, Fromont J, Gonoi T, Kobayashi JI, et al. Manzamenones L–N, new dimeric fatty-acid derivatives from an Okinawan marine sponge *Plakortis* sp. Bioorg Med Chem Lett. 2013;23:244–7.
134. Schmitz FJ, Gunasekera SP, Yalamanchili G, Hossain MB, Van der Helm D. Tedanolide: a potent cytotoxic macrolide from the Caribbean sponge *Tedania ignis*. J Am Chem Soc. 1984;106:7251–2.
135. Fusetani N, Sugawara T, Matsunaga S, Hirota H. Bioactive marine metabolites: IIIV. Cytotoxic metabolites of the marine sponge *Mycale adhaerens Lambe*. J Org Chem. 1991;56:4971–4.
136. Chevallier C, Bugni TS, Feng X, Harper MK, Orendt AM, Ireland CM, et al. Tedanolide C: a potent new 18-membered-ring cytotoxic macrolide isolated from the Papua New Guinea marine sponge *Ircinia* sp. J Org Chem. 2006;71:2510–3.
137. Whitson EL, Pluchino KM, Hall MD, McMahon JB, McKee TC. New Candidaspongiolides, tedanolide analogues that selectively inhibit melanoma cell growth. Org Lett. 2011;13:3518–21.
138. Nishimura S, Matsunaga S, Yoshida S, Nakao Y, Hirota H, Fusetani N. Structure–activity relationship study on 13-deoxytedanolide, a highly antitumor macrolide from the marine sponge *Mycale adhaerens*. Biorg Med Chem. 2005;13:455–62.
139. Cheng JF, Lee JS, Sakai R, Jares-Erijman EA, Silva MV, Rinehart KL, et al. Myriaporones 1–4, cytotoxic metabolites from the Mediterranean Bryozoan *Myriapora truncate*. J Nat Prod. 2007;70:332–6.
140. Nishimura S, Matsunaga S, Yoshida M, Hirota H, Yokoyama S, Fusetani N, et al. 13-Deoxytedanolide, a marine sponge-derived antitumor macrolide, binds to the 60S large ribosomal subunit. Biorg Med Chem. 2005;13:449–54.
141. Trisciuoglio D, Uranchimeg B, Cardellina JH, Meragelman TL, Matsunaga S, Fusetani N, et al. Induction of apoptosis in human cancer cells by candidaspongiolide, a novel sponge polyketide. J Natl Cancer Inst. 2008;100:1233–46.
142. Barber JM, Quek NCH, Leahy DC, Miller JH, Bellows DS, Northcote PT, et al. Lehualides E–K, cytotoxic metabolites from the tongan marine sponge *Plakortis* sp. J Nat Prod. 2011;74:809–15.
143. Sata N, Abinsay H, Yoshida WY, Horgen FD, Sitachitta N, Kelly M, et al. Lehualides A–D, metabolites from a Hawaiian sponge of the genus *Plakortis*. J Nat Prod. 2005;68:1400–3.
144. Zhang H, Conte MM, Capon RJ. Franklinolides A-C from an Australian marine sponge complex: phosphodiesters strongly enhance polyketide cytotoxicity. Angew Chem. 2010;49:9904–6.
145. Sirirath S, Tanaka J, Ohtani II, Ichiba T, Rachmat R, Ueda K, et al. Bitungolides A–F, new polyketides from the Indonesian sponge *Theonella cf. swinhoei*. J Nat Prod. 2002;65:1820–3.
146. Longeon A, Copp BR, Roué M, Dubois J, Valentin A, Petek S, et al. New bioactive halenaquinone derivatives from South Pacific marine sponges of the genus *Xestospongia*. Biorg. Med. Chem. 2010;18:6006–11.

147. Lee RH, Slate DL, Moretti R, Alvi KA, Crews P. Marine sponge polyketide inhibitors of protein tyrosine kinase. Biochem Biophys Res Commun. 1992;184:765–72.
148. Alvi KA, Rodriguez J, Diaz MC, Moretti R, Wilhelm RS, Lee RH, et al. Protein tyrosine kinase inhibitory properties of planar polycyclics obtained from the marine sponge *Xestospongia cf. carbonaria* and from total synthesis. J Org Chem. 1993;58:4871–80.
149. Kobayashi JI, Hirase T, Shigemori H, Ishibashi M, Bae MA, Tsuji T, et al. New Pentacyclic compounds from the Okinawan marine sponge *Xestospongia sapra*. J Nat Prod. 1992;55:994–8.
150. Kobayashi M, Shimizu N, Kitagawa I, Kyogoku Y, Harada N, Uda H. Absolute stereostructures of halenaquinol and halenaquinol sulfate, pentacyclic hydroquinones from the okinawan marine sponge *xestospongia sapra*, as determined by theoretical calculation of CD spectra. Tetrahedron Lett. 1985;26:3833–6.
151. Schmitz FJ, Bloor SJ. Xesto- and halenaquinone derivatives from a sponge, *Adocia sp.*, from Truk lagoon. J Org Chem. 1988;53:3922–5.
152. Cao S, Foster C, Brisson M, Lazo JS, Kingston DGI. Halenaquinone and xestoquinone derivatives, inhibitors of Cdc25B phosphatase from a *Xestospongia* sp. Biorg Med Chem. 2005;13:999–1003.
153. Cafieri F, Fattorusso E, Taglialatela-Scafati O, Ianaro A. Metabolites from the sponge *Plakortis simplex*. Determination of absolute stereochemistry of plakortin isolation and stereostructure of three plakortin related compounds. Tetrahedron. 1999;55:7045–56.
154. Fattorusso E, Parapini S, Campagnuolo C, Basilico N, Taglialatela-Scafati O, Taramelli D, et al. Activity against plasmodium falciparum of cycloperoxide compounds obtained from the sponge *Plakortis simplex*. J Antimicrob Chemother. 2002;50:883–8.
155. Hu JF, Gao HF, Kelly M, Hamann MT. Plakortides I–L, four new cyclic peroxides from an undescribed Jamaican sponge *Plakortis* sp. (Homosclerophorida, Plakinidae). Tetrahedron. 2001;57:9379–83.
156. Sol Jiménez M, Garzón SP, Rodríguez AD. Plakortides M and N, bioactive polyketide endoperoxides from the Caribbean marine sponge *Plakortis halichondrioides*. J Nat Prod. 2003;66:655–61.
157. Kobayashi M, Kondo K, Kitagawa I. Antifungal peroxyketal acids from an Okinawan marine sponge of *Plakortis* sp. Chem Pharm Bull. 1993;41:1324–6.
158. Fattorusso C, Persico M, Calcinai B, Cerrano C, Parapini S, Taramelli D, et al. Manadoperoxides A–D from the Indonesian sponge *Plakortis cfr. simplex*. Further insights on the structure–activity relationships of simple 1,2-dioxane antimalarials. J Nat Prod. 2010;73:1138–45.
159. Fattorusso C, Campiani G, Catalanotti B, Persico M, Basilico N, Parapini S, et al. Endoperoxide derivatives from marine organisms: 1,2-Dioxanes of the plakortin family as novel antimalarial agents. J Med Chem. 2006;49:7088–94.
160. Shinde PB, Mansoor TA, Luo X, Hong J, Lee CO, Jung JH, et al. Cytotoxic polyketides from the marine sponge *Discodermia calyx*. ChemInform. 2007;38:990–4.
161. Matsumoto T, Yanagiya M, Maeno S, Yasuda S. A revised structure of pederin. Tetrahedron Lett. 1968;9:6297–300.
162. Perry NB, Blunt JW, Munro MHG, Pannell LK. Mycalamide A, an antiviral compound from a New Zealand sponge of the genus *Mycale*. J Am Chem Soc. 1988;110:4850–1.
163. Sakemi S, Ichiba T, Kohmoto S, Saucy G, Higa T. Isolation and structure elucidation of onnamide A, a new bioactive metabolite of a marine sponge, *Theonella* sp. J Am Chem Soc. 1988;110:4851–3.
164. Perry NB, Blunt JW, Munro MHG, Thompson AM. Antiviral and antitumor agents from a New Zealand sponge, *Mycale* sp. 2. Structures and solution conformations of mycalamides a and B. J Org Chem. 1990;55:223–7.
165. Fusetani N, Sugawara T, Matsunaga S. Bioactive marine metabolites. 41. Theopederins A-E, potent antitumor metabolites from a marine sponge, *Theonella* sp. J Org Chem. 1992;57:3828–32.

166. Matsunaga S, Fusetani N, Nakao Y. Eight new cytotoxic metabolites closely related to onnamide A from two marine sponges of the genus *Theonella*. Tetrahedron. 1992;48:8369–76.
167. Tsukamoto S, Matsunaga S, Fusetani N, Toh-E A. Theopederins F-J: five new antifungal and cytotoxic metabolites from the marine sponge, *theonella swinhoei* 1. Tetrahedron. 1999;55:13697–702.
168. Simpson JS, Garson MJ, Blunt JW, Munro MHG, Hooper JNA. Mycalamides C and D, cytotoxic compounds from the marine sponge *Stylinos n.* species. J Nat Prod. 2000;63:704–6.
169. West LM, Northcote PT, Hood KA, Miller JH, Page MJ. Mycalamide D, a new cytotoxic amide from the New Zealand marine sponge *Mycale Species*. J Nat Prod. 2000;63:707–9.
170. Vuong D, Capon RJ, Lacey E, Gill JH, Heiland K, Friedel T, et al. Onnamide F: a new nematocide from a southern Australian marine sponge. *Trachycladus laevispirulifer*. J Nat. 2001;64:640–2.
171. Paul GK, Gunasekera SP, Longley RE, Pomponi SA. Theopederins K and L. highly potent cytotoxic metabolites from a marine sponge *Discodermia Species*. J Nat Prod. 2002;65:59–61.
172. Kato Y, Fusetani N, Matsunaga S, Hashimoto K, Fujita S, Furuya T, et al. Bioactive marine metabolites: XVI. Calyculin A. A novel antitumor metabolite from the marine sponge *Discodermia calyx*. J Am Chem Soc. 1986;108:2780–1.
173. Kato Y, Fusetani N, Matsunaga S, Hashimoto K, Koseki K. Isolation and structure elucidation of calyculins B, C, and D, novel antitumor metabolites, from the marine sponge *Discodermia calyx*. Am Chem Soc. 1988;53:3930–2.
174. Matsunaga S, Wakimoto T, Fusetani N. Isolation of four new calyculins from the marine sponge *Discodermia calyx* 1. J Org Chem. 1997;62:2640–2.
175. Matsunaga S, Wakimoto T, Fusetani N, Suganuma M. Isolation of dephosphonocalyculin a from the marine sponge, *Discodermia calyx*. Tetrahedron Lett. 1997;38:3763–4.
176. Wakimoto T, Matsunaga S, Takai A, Fusetani N. Insight into binding of Calyculin A to protein phosphatase 1: isolation of hemicalyculin A and chemical transformation of Calyculin A. Chem Biol. 2002;9:309–19.
177. Johnson TA, Tenney K, Cichewicz RH, Morinaka BI, White KN, Amagata T, et al. Sponge-derived fijianolide polyketide class: further evaluation of their structural and cytotoxicity properties. J Med Chem. 2007;50:3795–803.
178. Quinoa E, Kakou Y, Crews P. Fijianolides, polyketide heterocycles from a marine sponge. J Org Chem. 1988;53:3642–4.
179. Corley DG, Herb R, Moore RE, Scheuer PJ, Paul VJ. Laulimalides. New potent cytotoxic macrolides from a marine sponge and a nudibranch predator. J Org Chem. 1988;53:3644–6.
180. Tanaka JI, Higa T, Bernardinelli G, Jefford CW. New cytotoxic macrolides from the sponge *Fasciospongia rimosa*. Chem Lett. 1996;25:255–6.
181. Mooberry SL, Tien G, Hernandez AH, Plubrukarn A, Davidson BS. Laulimalide and isolaulimalide, new paclitaxel-like microtubule-stabilizing agents. Cancer Res. 1999;59:653–60.
182. Gallagher BM Jr, Fang FG, Johannes CW, Pesant M, Tremblay MR, Zhao H, et al. Synthesis and biological evaluation of (−)-laulimalide analogues. Bioorg Med Chem Lett. 2004;14:575–9.
183. Berrue F, Thomas OP, Laville R, Prado S, Golebiowski J, Fernandez R, et al. The marine sponge *Plakortis zyggompha*: a source of original bioactive polyketides. Tetrahedron. 2007;63:2328–34.
184. Huang XH, van Soest R, Roberge M, Andersen RJ. Spiculoic acids A and B, new polyketides isolated from the Caribbean marine sponge *Plakortis angulospiculatus*. Org Lett. 2004;6:75–8.
185. Berrué F, Thomas OP, Fernández R, Amade P. Iso-, Nor-, and dinor-spiculoic acids A, polyketides from the marine sponge *Plakortis zyggompha*. J Nat Prod. 2005;68:547–9.
186. Ching Shen Y, Sai Prakash CV, Guh JH. New pentacyclic polyketide dimeric peroxides from a Taiwanese marine sponge *Petrosia elastica*. Tetrahedron Lett. 2004;45:2463–6.
187. Murakami N, Sugimoto M, Kobayashi M. Participation of the β-hydroxyketone part for potent cytotoxicity of callystatin A, a spongean polyketide. Biorg Med Chem. 2001;9:57–67.

188. Hirata Y, Uemura D. Halichondrins – antitumor polyether macrolides from a marine sponge. Pure Appl Chem. 1986;58:701–10.
189. Kernan MR, Faulkner DJ. Halichondramide, an antifungal macrolide from the sponge *halichondria* sp. Tetrahedron Lett. 1987;28:2809–12.
190. Kernan MR, Molinski TF, Faulkner DJ. Macrocyclic antifungal metabolites from the Spanish dancer nudibranch Hexabranchus sanguineus and sponges of the genus *Halichondria*. J Org Chem. 1988;53:5014–20.
191. Fusetani N, Yasumuro K, Matsunaga S, Hashimoto K. Mycalolides A – C, hybrid macrolides of ulapualides and halichondramide, from a sponge of the genus *Mycale*. Tetrahedron Lett. 1989;30:2809–12.
192. Kobayashi JI, Murata O, Shigemori H, Sasaki T, Jaspisamides A-C. New Cytotoxic Macrolides from the Okinawan Sponge *Jaspis* sp. J Nat Prod. 1993;56:787–91.
193. Carmely S, Kashman Y. Structure of swinholide-a, a new macrolide from the marine sponge *theonella swinhoei*. Tetrahedron Lett. 1985;26:511–4.
194. Kobayashi M, Tanaka JI, Katori T, Matsuura M, Kitagawa I. Structure of swinholide A, a potent cytotoxic macrolide from the Okinawan marine sponge *theonella swinhoei*. Tetrahedron Lett. 1989;30:2963–6.
195. Kitagawa I, Kobayashi M, Katori T, Yamashita M, Tanaka J, Doi M, et al. Absolute stereostructure of swinholide A, a potent cytotoxic macrolide from the Okinawan marine sponge *Theonella swinhoei*. J Am Chem Soc. 1990;112:3710–2.
196. Kobayashi M, Tanaka JI, Katori T, Kitagawa I. Marine natural products. XXIII.: Three new cytotoxic dimeric macrolides, Swinholides B and C and Isoswinholide A, congeners of Swinholide A, from the Okinawan marine sponge *Theonella swinhoei*. Chem Pharm Bull. 1990;38:2960–6.
197. Kobayashi M, Aoki S, Sakai H, Kawazoe K, Kihara N, Sasaki T, et al. Altohyrtin A, a potent anti-tumor macrolide from the Okinawan marine sponge *Hyrtios altum*. Tetrahedron Lett. 1993;34:2795–8.
198. Kobayashi M, Aoki S, Sakai H, Kihara N, Sasaki T, Kitagawa I. Altohyrtins B and C and 5-desacetylaltohyrtin A, potent cytotoxic macrolide congeners of altohyrtin A, from the Okinawan marine sponge *Hyrtios altum*. Chem Pharm Bull. 1993;41:989–91.
199. Pettit GR, Tan R, Gao F, Williams MD, Doubek DL, Boyd MR, et al. Isolation and structure of halistatin 1 from the eastern Indian Ocean marine sponge *Phakellia carteri*. J Org Chem. 1993;58:2538–43.
200. Fusetani N, Shinoda K, Matsunaga S. Bioactive marine metabolites. 48. Cinachyrolide A: a potent cytotoxic macrolide possessing two spiro ketals from marine sponge *Cinachyra* sp. J. Am. Chem Soc. 1993;115:3977–81.
201. D'Auria MV, Debitus C, Paloma LG, Minale L, Zampella A. Superstolide A: a potent cytotoxic macrolide of a new type from the new Caledonian deep water marine sponge *Neosiphonia superstes*. J Am Chem Soc. 1994;116:6658–63.
202. D'Auria MV, Paloma LG, Minale L, Zampella A, Debitus C. A novel cytotoxic macrolide, Superstolide B, related to Superstolide A, from the new Caledonian marine sponge *Neosiphonia superstes*. J Nat Prod. 1994;57:1595–7.
203. D'Auria MV, Paloma LG, Minale L, Zampella A, Verbist JF, Roussakis C, et al. Reidispongiolide A and B, two new potent cytotoxic macrolides from the new caledonian sponge *Reidispongia coerulea*. Tetrahedron. 1994;50:4829–34.
204. Horton PA, Koehn FE, Longley RE, McConnell OJ. Lasonolide A, a new cytotoxic macrolide from the marine sponge *Forcepia* sp. J Am Chem Soc. 1994;116:6015–6.
205. Litaudon M, Hart JB, Blunt JW, Lake RJ. Munro, M.h. Isohomohalichondrin B, a new antitumour polyether macrolide from the New Zealand deep-water sponge *Lissodendoryx* sp. Tetrahedron Lett. 1994;35:9435–8.
206. Searle PA, Molinski TF. Phorboxazoles A and B: potent cytostatic macrolides from marine sponge *Phorbas species*. J Am Chem Soc. 1995;117:8126–31.

207. Rho MC, Park YH, Sasaki S, Ishibashi M, Kondo K, Kobayashi JI, et al. The mode of rabbit platelet shape change and aggregation induced by theonezolide-A, a novel polyketide macrolide, isolated from the Okinawan marine sponge *Theonella* sp. Can J Physiol Pharmacol. 1996;74:193–9.
208. D'Ambrosio M, Guerriero A, Pietra F, Debitus C. Leucascandrolide A, a new type of macrolide: the first powerfully bioactive metabolite of calcareous sponges (Leucascandra caveolata, a new genus from the coral sea). Helv Chim Acta. 1996;79:51–60.
209. D'Ambrosio M, Tatò M, Pocsfalvi G, Debitus C, Pietra F. Leucascandrolide B, a new 16-membered, extensively methyl-branched polyoxygenated macrolide from the Calcareous sponge *Leucascandra caveolata* from northeastern waters of New Caledonia. Helv Chim Acta. 1999;82:347–53.
210. Tanaka JI, Higa T. Zampanolide, a new cytotoxic marcrolide from a marine sponge. Tetrahedron Lett. 1996;37:5535–8.
211. Litaudon M, Hickford SJH, Lill RE, Lake RJ, Blunt JW, Munro MHG. Antitumor polyether macrolides: new and hemisynthetic Halichondrins from the New Zealand deep-water sponge *Lissodendoryx* sp. J Org Chem. 1997;62:1868–71.
212. Erickson KL, Beutler JA, Cardellina JH, Boyd MR. Salicylihalamides A and B, novel cytotoxic macrolides from the marine sponge *Haliclona* sp. J Org Chem. 1997;62:8188–92.
213. Matsunaga S, Nogata Y, Fusetani N. Thiomycalolides: new cytotoxic trisoxazole-containing macrolides isolated from a marine sponge *Mycale* sp. J Nat Prod. 1998;61:663–6.
214. Takada N, Sato H, Suenaga K, Arimoto H, Yamada K, Ueda K, et al. Isolation and structures of haterumalides NA, NB, NC, ND, and NE, novel macrolides from an Okinawan sponge *Ircinia* sp. Tetrahedron Lett. 1999;40:6309–12.
215. Cantrell CL, Gustafson KR, Cecere MR, Pannell LK, Boyd MR. Chondropsins A and B: novel tumor cell growth-inhibitory macrolide lactams from the marine sponge *Chondropsis* sp. J Am Chem Soc. 2000;122:8825–9.
216. Rashid MA, Gustafson KR, Boyd MR. New chondropsin macrolide lactams from marine sponges in the genus Ircinia. Tetrahedron Lett. 2001;42:1623–6.
217. Rashid MA, Cantrell CL, Gustafson KR, Boyd MR. Chondropsin D, a new 37-membered-ring macrolide lactam from the marine sponge *Chondropsis Species*. J Nat Prod. 2001;64:1341–4.
218. West LM, Northcote PT, Battershill CN. Peloruside A: a potent cytotoxic macrolide isolated from the New Zealand marine sponge *Mycale* sp. J Org Chem. 2000;65:445–9.
219. Singh AJ, Xu CX, Xu X, West LM, Wilmes A, Chan A, et al. Peloruside B, a potent antitumor macrolide from the New Zealand marine sponge *Mycale hentscheli*: isolation, structure, total synthesis, and bioactivity. J Org Chem. 2010;75:2–10.
220. Cutignano A, Bruno I, Bifulco G, Casapullo A, Debitus C, Gomez-Paloma L, et al. Dactylolide, a new cytotoxic macrolide from the Vanuatu sponge *Dactylospongia* sp. Eur J Org Chem. 2001;2001:775–8.
221. Phuwapraisirisan P, Matsunaga S, van Soest RWM, Fusetani N. Isolation of a new mycalolide from the marine sponge *Mycale izuensis*. J Nat Prod. 2002;65:942–3.
222. Sandler JS, Colin PL, Kelly M, Fenical W. Cytotoxic macrolides from a new species of the deep-water marine sponge *Leiodermatium*. J Org Chem. 2006;71:7245–51.
223. Youssef DTA, Mooberry SL. Hurghadolide A and Swinholide I, potent actin-microfilament disrupters from the Red Sea sponge *Theonella swinhoei*. J Nat Prod. 2006;69:154–7.
224. Wright AE, Botelho JC, Guzmán E, Harmody D, Linley P, McCarthy PJ, et al. Neopeltolide, a macrolide from a lithistid sponge of the family *Neopeltidae*. J Nat Prod. 2007;70:412–6.
225. Takada K, Choi BW, Rashid MA, Gamble WR, Cardellina JH, Van QN, et al. Structural assignment of Poecillastrins B and C, macrolide lactams from the deep-water Caribbean sponge *Poecillastra Species*. J Nat Prod. 2007;70:428–31.
226. Plaza A, Baker HL, Bewley CA. Mirabilin, an antitumor macrolide lactam from the marine sponge *Siliquariaspongia mirabilis*. J Nat Prod. 2008;71:473–7.
227. Bishara A, Rudi A, Aknin M, Neumann D, Ben-Califa N, Kashman Y, et al. Salarins A and B and tulearin A: new cytotoxic sponge-derived macrolides. Org Lett. 2008;10:153–6.

228. Hickford SJH, Blunt JW, Munro MHG. Antitumour polyether macrolides: four new halichondrins from the New Zealand deep-water marine sponge *Lissodendoryx* sp. Biorg Med Chem. 2009;17:2199–203.
229. Paterson I, Dalby SM, Roberts JC, Naylor GJ, Guzmán EA, Isbrucker R, et al. Leiodermatolide, a potent antimitotic macrolide from the marine sponge *Leiodermatium* sp. Angew Chem Int Ed. 2011;50:3219–23.
230. Sirirak T, Kittiwisut S, Janma C, Yuenyongsawad S, Suwanborirux K, Plubrukarn A. Kabiramides J and K, trisoxazole macrolides from the sponge *Pachastrissa nux*. J Nat Prod. 2011;74:1288–92.
231. Sirirak T, Brecker L, Plubrukarn A. Kabiramide L, a new antiplasmodial trisoxazole macrolide from the sponge *Pachastrissa nux*. Nat Prod Res. 2012;27:1213–9.
232. Pham CD, Hartmann R, Böhler P, Stork B, Wesselborg S, Lin W, et al. Callyspongiolide, a cytotoxic macrolide from the marine sponge *Callyspongia* sp. Org Lett. 2014;16:266–9.
233. Hirota H, Tomono Y, Fusetani N. Terpenoids with antifouling activity against barnacle larvae from the marine sponge *Acanthella cavernosa*. Tetrahedron. 1996;52:2359–68.
234. Yong KWL, Jankam A, Hooper JNA, Suksamrarn A, Garson MJ. Stereochemical evaluation of sesquiterpene quinones from two sponges of the genus *Dactylospongia* and the implication for enantioselective processes in marine terpene biosynthesis. Tetrahedron. 2008;64:6341–8.
235. Ovenden SPB, Nielson JL, Liptrot CH, Willis RH, Tapiolas DM, Wright AD, et al. Sesquiterpene benzoxazoles and sesquiterpene quinones from the marine sponge *Dactylospongia elegans*. J Nat Prod. 2011;74:65–8.
236. Suto S, Tanaka N, Fromont J. Kobayashi, J.i. Halichonadins G-J, new sesquiterpenoids from a sponge *Halichondria* sp. Tetrahedron Lett. 2011;52:3470–3.
237. Jiao WH, Huang XJ, Yang JS, Yang F, Piao SJ, Gao H, et al. Dysidavarones A-D, new sesquiterpene quinones from the marine sponge *Dysidea avara*. Org Lett. 2012;14:202–5.
238. Yamazaki H, Nakazawa T, Sumilat DA, Takahashi O, Ukai K, Takahashi S, Namikoshi M. Euryspongins A-C, three new unique sesquiterpenes from a marine sponge *Euryspongia* sp. Bioorg Med Chem Lett. 2013;23:2151–4.
239. Sperry S, Valeriote FA, Corbett TH, Crews P. Isolation and cytotoxic evaluation of marine sponge-derived norterpene peroxides. J Nat Prod. 1998;61:241–7.
240. Liu Y, Hong J, Lee CO, Im KS, Kim ND, Choi JS, et al. Cytotoxic pyrrolo and furanoterpenoids from the sponge *Sarcotragus species*. J Nat Prod. 2002;65:1307–14.
241. Agena M, Tanaka C, Hanif N, Yasumoto-Hirose M, Tanaka J. New cytotoxic spongian diterpenes from the sponge *Dysidea cf. arenaria*. Tetrahedron. 2009;65:1495–9.
242. Wattanapiromsakul C, Chanthathamrongsiri N, Bussarawit S, Yuenyongsawad S, Plubrukarn A, Suwanborirux K. 8-Isocyanoamphilecta-11(20),15-diene, a new antimalarial isonitrile diterpene from the sponge *Ciocalapata* sp. Can J Chem. 2009;87:612–8.
243. Calcul L, Tenney K, Ratnam J, McKerrow JH, Crews P. Structural variations to the 9-N-methyladeninium diterpenoid hybrid commonly isolated from *Agelas* sponges. Aust J Chem. 2011;64:846.
244. Uddin MH, Hossain MK, Nigar M, Roy MC, Tanaka J. New cytotoxic spongian-class rearranged diterpenes from a marine sponge. Chem Nat Compd. 2012;48:412–5.
245. Dattelbaum JD, Singh AJ, Field JJ, Miller JH, Northcote PT. The nitrogenous hamigerans: unusual amino acid-derivatized aromatic diterpenoid metabolites from the New Zealand marine sponge *Hamigera tarangaensis*. J Org Chem. 2015;80:304–12.
246. Lee JS, Abdjul DB, Yamazaki H, Takahashi O, Kirikoshi R, Ukai K, et al. Strongylophorines, new protein tyrosine phosphatase 1B inhibitors, from the marine sponge *Strongylophora strongilata* collected at Iriomote Island. Bioorg Med Chem Lett. 2015;25:3900–2.
247. Noda A, Sakai E, Kato H, Losung F, Mangindaan REP, de Voogd NJ, Yokosawa H, et al. Strongylophorines, meroditerpenoids from the marine sponge *Petrosia corticata*, function as proteasome inhibitors. Bioorg Med Chem Lett. 2015;25:2650–3.
248. Zeng L, Fu X, Su J, Pordesimo EO, Traeger SC, Schmitz FJ. Novel bishomoscalarane sesterterpenes from the sponge *Phyllospongia foliascens*. J Nat Prod. 1991;54:421–7.

249. De Rosa S, Puliti R, Crispino A, De Giulio A, De Sena C, Iodice C, et al. 25-Deoxycacospongionolide B and cacospongionolide C, two new terpenoids from the sponge *Fasciospongia cavernosa*. Tetrahedron. 1995;51:10731–6.
250. Tsuchiya N, Sato A, Hata T, Sato N, Sasagawa K, Kobayashi T. Cytotoxic scalarane sesterterpenes from a sponge, *Hyrtios erecta*. J Nat Prod. 1998;61:468–73.
251. Craig KS, Williams DE, Hollander I, Frommer E, Mallon R, Collins K, et al. Novel sesterterpenoid and norsesterterpenoid RCE-protease inhibitors isolated from the marine sponge *Hippospongia* sp. Tetrahedron Lett. 2002;43:4801–4.
252. Zhang HJ, Yi YH, Yang F, Chen WS, Lin HW. Sesterterpenes and a new sterol from the marine sponge *Phyllospongia foliascens*. Molecules. 2010;15:834–41.
253. Bae JM, Jeon JE, Lee YJ, Lee HS, Sim CJ, Oh KB, et al. Sesterterpenes from the tropical sponge *Coscinoderma* sp. J Nat Prod. 2011;74:1805–11.
254. Jeon JE, Bae JM, Lee KJ, Oh KB, Shin JH. Scalarane Sesterterpenes from the sponge *Hyatella* sp. J Nat Prod. 2011;74:847–51.
255. Piao SJ, Zhang HJ, Lu HY, Yang F, Jiao WH, Yi YH, et al. Hippolides A-H, acyclic manoalide derivatives from the marine sponge *Hippospongia lachne*. J Nat Prod. 2011;74:1248–54.
256. Rho JR, Hwang BS, Joung S, Byun MR, Hong JH, Lee HY. Phorbasones A and B, sesterterpenoids isolated from the marine sponge *Phorbas* sp. and induction of osteoblast differentiation. Org Lett. 2011;13:884–7.
257. Wang W, Lee Y, Lee TG, Mun B, Giri AG, Lee J, et al. Phorone A and isophorbasone A, sesterterpenoids isolated from the marine sponge *Phorbas* ***sp***. Org Lett. 2012;14:4486–9.
258. Daoust J, Chen M, Wang M, Williams DE, Chavez MAG, Wang YA, et al. Sesterterpenoids isolated from a northeastern Pacific *Phorbas* sp. J Org Chem. 2013;78:8267–73.
259. Hahn D, Won DH, Mun B, Kim H, Han C, Wang W, et al. Cytotoxic scalarane sesterterpenes from a Korean marine sponge *Psammocinia* sp. Bioorg Med Chem Lett. 2013;23:2336–9.
260. Wang W, Mun B, Lee Y, Venkat Reddy M, Park Y, Lee J, et al. Bioactive sesterterpenoids from a Korean sponge *Monanchora* sp. J Nat Prod. 2013;76:170–7.
261. Festa C, Cassiano C, D'Auria MV, Debitus C, Monti MC, De Marino S. Scalarane sesterterpenes from *Thorectidae* sponges as inhibitors of TDP-43 nuclear factor. Org Biomol Chem. 2014;12:8646–55.
262. Hassan MHA, Rateb ME, Hetta M, Abdelaziz TA, Sleim MA, Jaspars M, et al. Scalarane sesterterpenes from the Egyptian Red Sea sponge *Phyllospongia lamellosa*. Tetrahedron. 2015;71:577–83.
263. Woo J-K, Kim C-K, Ahn C-H, Oh D-C, Oh K-B, Shin J. Additional sesterterpenes and a nortriterpene saponin from the sponge *Clathria gombawuiensis*. J Nat Prod. 2015;78:218–24.
264. Li J, Xu B, Cui J, Deng Z, de Voogd NJ, Proksch P, et al. Globostelletins A–I, cytotoxic isomalabaricane derivatives from the marine sponge *Rhabdastrella globostellata*. Biorg Med Chem. 2010;18:4639–47.
265. Sunassee SN, Ransom T, Henrich CJ, Beutler JA, Covell DG, McMahon JB, et al. Steroidal alkaloids from the marine sponge *Corticium niger* that inhibit growth of human colon carcinoma cells. J Nat Prod. 2014;77:2475–80.
266. Morinaka BI, Masuno MN, Pawlik JR, Molinski TF. Amaranzole A, a new N-imidazolyl steroid from *Phorbas amaranthus*. Org Lett. 2007;9:5219–22.
267. Langjae R, Bussarawit S, Yuenyongsawad S, Ingkaninan K, Plubrukarn A. Acetylcholinesterase-inhibiting steroidal alkaloid from the sponge *Corticium* sp. Steroids. 2007;72:682–5.
268. Aoki S, Watanabe Y, Tanabe D, Setiawan A, Arai M, Kobayashi M, et al. Novel abeo-9(10-19)-androstane-type steroidal alkaloids with isoquinoline unit, from marine sponge *Corticium simplex*. Tetrahedron Lett. 2007;48:4485–8.
269. Aoki S, Watanabe Y, Sanagawa M, Setiawan A, Kotoku N, Kobayashi M. Cortistatins A, B, C, and D, anti-angiogenic steroidal alkaloids, from the marine sponge *Corticium simplex*. J Am Chem Soc. 2006;128:3148–9.

270. Ridley CP, Faulkner DJ. New cytotoxic steroidal alkaloids from the Philippine sponge *Corticium niger*. J Nat Prod. 2003;66:1536–9.
271. Borbone N, De Marino S, Iorizzi M, Zollo F, Debitus C, Esposito G, et al. Minor steroidal alkaloids from the marine sponge *Corticium* sp. J Nat Prod. 2002;65:1206–9.
272. Marino SD, Iorizzi M, Zollo F, Roussakis C, Debitus C. Plakinamines C and D and three other new steroidal alkaloids from the sponge *Corticium* sp. Eur JOrg Chem. 1999;1999:697–701.
273. Ushiyama S, Umaoka H, Kato H, Suwa Y, Morioka H, Rotinsulu H, et al. Manadosterols A and B, sulfonated sterol dimers inhibiting the Ubc13-Uev1A interaction, isolated from the marine sponge *Lissodendryx fibrosa*. J Nat Prod. 2012;75:1495–9.
274. Murayama S, Imae Y, Takada K, Kikuchi J, Nakao Y, van Soest RW, et al. Shishicrellastatins, inhibitors of cathepsin B, from the marine sponge *Crella (Yvesia) spinulata*. Bioorg Med Chem. 2011;19:6594–8.
275. Whitson EL, Bugni TS, Chockalingam PS, Conception GP, Feng X, Jin G, et al. Fibrosterol sulfates from the Philippine sponge *Lissodendoryx (Acanthodoryx) fibrosa*: sterol dimers that inhibit PKC. J Org Chem. 2009;74:5902–8.
276. Whitson EL, Bugni TS, Chockalingam PS, Concepcion GP, Harper MK, He M, et al. Spheciosterol sulfates, PKCzeta inhibitors from a philippine sponge *Spheciospongia* sp. J Nat Prod. 2008;71:1213–7.
277. Guzii AG, Makarieva TN, Denisenko VA, Dmitrenok PS, Burtseva YV, Krasokhin VB, et al. Topsentiasterol sulfates with novel iodinated and chlorinated side chains from the marine sponge *Topsentia* sp. Tetrahedron Lett. 2008;49:7191–3.
278. Zhang HJ, Sun JB, Lin HW, Wang ZL, Tang H, Cheng P, et al. A new cytotoxic cholesterol sulfate from marine sponge *Halichondria rugosa*. Nat Prod Res. 2007;21:953–8.
279. Boonlarppradab C, Faulkner DJ. Eurysterols A and B, cytotoxic and antifungal steroidal sulfates from a marine sponge of the genus *Euryspongia*. J Nat Prod. 2007;70:846–8.
280. Yang SW, Buivich A, Chan TM, Smith M, Lachowicz J, Pomponi SA, et al. A new sterol sulfate, Sch 572423, from a marine sponge, *Topsentia* sp. Bioorg Med Chem Lett. 2003;13:1791–4.
281. Lerch ML, Faulkner DJ. Unusual polyoxygenated sterols from a Philippines sponge *Xestospongia* sp. Tetrahedron. 2001;57:4091–4.
282. D'Auria MV, Giannini C, Zampella A, Minale L, Debitus C, Roussakis C. Crellastatin A: a cytotoxic bis-steroid sulfate from the Vanuatu marine sponge *Crella* sp. J Org Chem. 1998;63:7382–8.
283. Patil AD, Freyer AJ, Breen A, Carte B, Johnson RK. Halistanol disulfate B, a novel sulfated sterol from the sponge *Pachastrella* sp.:inhibitor of endothelin converting enzyme. J Nat Prod. 1996;59:606–8.
284. Regalado EL, Tasdemir D, Kaiser M, Cachet N, Amade P, Thomas OP. Antiprotozoal steroidal saponins from the marine sponge *Pandaros acanthifolium*. J Nat Prod. 2010;73:1404–10.
285. Stead P, Hiscox S, Robinson PS, Pike NB, Sidebottom PJ, Roberts AD, et al. Eryloside F, a novel penasterol disaccharide possessing potent thrombin receptor antagonist activity. Bioorg Med Chem Lett. 2000;10:661–4.
286. Machida K, Abe T, Arai D, Okamoto M, Shimizu I, de Voogd NJ, et al. Cinanthrenol A, an estrogenic steroid containing phenanthrene nucleus, from a marine sponge *Cinachyrella* sp. Org Lett. 2014;16:1539–41.
287. Gong J, Sun P, Jiang N, Riccio R, Lauro G, Bifuico G, et al. New steroids with a rearranged skeleton as (h)P300 inhibitors from the sponge *Theonella swinhoei*. Org Lett. 2014;16:2224–7.
288. Nguyen XC, Longeon A, Pham VC, Urvois F, Bressy C, Trinh TT, et al. Antifouling 26,27-cyclosterols from the Vietnamese marine sponge *Xestospongia testudinaria*. J Nat Prod. 2013;76:1313–8.
289. Keyzers RA, Daoust J, Davies-Coleman MT, Van Soest R, Balgi A, Donohue E. Autophagy-modulating aminosteroids isolated from the sponge *Cliona celata*. Org Lett. 2008;10:2959–62.

290. Teruya T, Nakagawa S, Koyama T, Arimoto H, Kita M, Uemura D. Nakiterpiosin and Nakiterpiosinone, novel cytotoxic C-Nor-D-Homosteroids from the Okinawan sponge *Terpios hoshinota*. Tetrahedron. 2004;60:6989–93.
291. Miyamoto T, Kodama K, Aramaki Y, Higuchi R. Soest, Orostanal R.W.M.V. A novel abeo-sterol inducing apoptosis in leukemia cell from a marine sponge, *Stelletta hiwasaensis*. Tetrahedron Lett. 2001;42:6349–51.
292. Rosa SD, Giulio AD, Crisping A, Iodice C, Tommonaro G. New 9,11-sSecosterol from the Tyrrhenian sponge *Fasciospongia Cavernosa*. Nat Prod Lett. 1999;13:15–20.
293. Kobayashi M, Chen YJ, Higuchi K, Aoki S, Kitagawa I. Aragusteroketals A and C, two novel cytotoxic steroids from a marine sponge of *Xestospongia* sp. Chem Pharm Bull. 1996;44:1840–2.
294. Kobayashi J, Shinonaga H, Shigemori H. Xestobergsterol C, a new pentacyclic steroid from the Okinawan marine sponge *Ircinia* sp. and absolute stereochemistry of xestobergsterol A. J Nat Prod. 1995;58:312–8.
295. Iguchi K, Fujita M, Nagaoka H, Mitome H, Yamada Y. Aragusterol a: A potent antitumor marine steroid from the okinawan sponge of the genus, *Xestospongia*. Tetrahedron Lett. 1993;34:6277–80.
296. Iguchi K, Shimura H, Taira S, Yokoo C, Matsumoto K, Yamada Y. Aragusterol B and D, new 26,27-cyclosterols from the Okinawan marine sponge of the genus *Xestospongia*. J Org Chem. 1994;59:7499–502.
297. Shimura H, Iguchi K, Yamada Y, Nakaike S, Yamagishi T, Matsumoto K, et al. A novel halogenated marine steroid from an Okinawan sponge, *Xestospongia* sp., possessing potent antitumor activity. Experientia. 1994;50:134–6.
298. He WF, Xue DQ, Yao LG, Li J, Liu HL, Guo YW. A new bioactive steroidal ketone from the South China Sea sponge *Xestospongiatestudinaria*. J Asian Nat Prod Res. 2015;18:1–5.
299. Abdelmohsen UR, Cheng C, Reimer A, Kozjak-Pavlovic V, Ibrahim AK, Rudel T, et al. Antichlamydial sterol from the Red Sea sponge *Callyspongia aff. implexa*. Planta Med. 2015;81:382–7.
300. Govindam SV, Choi BK, Yoshioka Y, Kanamoto A, Fujiwara T, Okamoto T, et al. Novel cytotoxic polyoxygenated steroids from an Okinawan sponge *Dysidea* sp. Biosci Biotechnol Biochem. 2012;76:999–1002.
301. Dai J, Sorribas A, Yoshida WY, Kelly M, Williams PG. Topsentinols, 24-isopropyl steroids from the marine sponge *Topsentia* sp. J Nat Prod. 2010;73:1597–600.
302. Ma WS, Mutka T, Vesley B, Amsler MO, McClintock JB, Amsler CD, et al. Norselic acids A-E, highly oxidized anti-infective steroids that deter mesograzer predation, from the Antarctic sponge *Crella* sp. J Nat Prod. 2009;72:1842–6.
303. Holland IP, McCluskey A, Sakoff JA, Gilbert J, Chau N, Robinson PJ, et al. Steroids from an Australian sponge *Psammoclema* sp. J Nat Prod. 2009;72:102–6.
304. Xu S, Liao X, Du B, Zhou X, Huang Q, Wu C. A series of new 5,6-epoxysterols from a Chinese sponge *Ircinia aruensis*. Steroids. 2008;73:568–73.
305. Mansoor TA, Hong J, Lee CO, Bae SJ, Im KS, Jung JH. Cytotoxic sterol derivatives from a marine sponge *Homaxinella* sp. J Nat Prod. 2005;68:331–6.
306. Fattorusso E, Taglialatela-Scafati O, Petrucci F, Bavestrello G, Calcinai B, Cerrano C, et al. Polychlorinated androstanes from the burrowing sponge *Cliona nigricans*. Org Lett. 2004;6:1633–5.
307. Funel C, Berrue F, Roussakis C, Fernandez Rodriguez R, Amade P. New cytotoxic steroids from the Indian Ocean sponge *Axinella cf. bidderi*. J Nat Prod. 2004;67:491–4.
308. Santafe G, Paz V, Rodriguez J, Jimenez C. Novel cytotoxic oxygenated C29 sterols from the Colombian marine sponge *Polymastia tenax*. J Nat Prod. 2002;65:1161–4.
309. Gallimore WA, Kelly M, Scheuer PJ. Gelliusterols A–D, new acetylenic sterols from a sponge, *Gellius species*. J Nat Prod. 2001;64:741–4.

310. Leone PdA, Redburn J, Hooper JNA, Quinn RJ. Polyoxygenated dysidea sterols that inhibit the binding of [I 125] IL-8 to the human recombinant IL-8 receptor type A. J. Nat. Prod. 2000;63:694–7.
311. Gunasekera SP, Kelly-Borges M, Longley RE. A new cytotoxic sterol methoxymethyl ether from a deep water marine sponge *Scleritoderma* sp. *cf. paccardi*. J Nat Prod. 1996;59:161–2.
312. Belarbi E. Producing drugs from marine sponges. Biotechnol Adv. 2003;21:585–98.
313. Blunt JW, Copp BR, Keyzers RA, Munro M, Prinsep MR. Marine natural products. Nat Prod Rep. 2013;30:237–323.
314. Bergmann W, BURKE DC. Contributions to the study of marine products. XXXIX. The nucleosides of sponges. III. 1 Spongothymidine and spongouridine2. J Org Chem. 1955;20:1501–7.
315. Pigneux A, Perreau V, Jourdan E, Vey N, Dastugue N, Huguet F, et al. Adding lomustine to idarubicin and cytarabine for induction chemotherapy in older patients with acute myeloid leukemia: the BGMT 95 trial results. Haematologica. 2007;92:1327–34.
316. Lancet JE, Roboz GJ, Cripe LD, Michelson GC, Fox JA, Leavitt RD, et al. A phase 1b/2 study of vosaroxin in combination with cytarabine in patients with relapsed or refractory acute myeloid leukemia. Haematologica. 2015;100:231–7.
317. Oku N, Matsunaga S, van Soest RW, Fusetani N, Renieramycin J. A highly cytotoxic tetrahydroisoquinoline alkaloid, from a marine sponge *Neopetrosia* sp. J Nat Prod. 2003;66:1136–9.
318. Halim H, Chunhacha P, Suwanborirux K, Chanvorachote P. Anticancer and antimetastatic activities of Renieramycin M, a marine tetrahydroisoquinoline alkaloid, in human non-small cell lung cancer cells. Anticancer Res. 2011;31:193–201.
319. Teeyapant R, Woerdenbag HJ, Kreis P, Hacker J, Wray V, Witte L, et al. Antibiotic and cytotoxic activity of brominated compounds from the marine sponge *Verongia aerophoba*. Z Naturforsch C. 2003;48:939–45.
320. Song J, Rechkoblit O, Bestor TH, Patel DJ. Structure of DNMT1-DNA complex reveals a role for autoinhibition in maintenance DNA methylation. Science. 2011;331:1036–40.
321. Florean C, Schnekenburger M, Lee JY, Kim KR, Mazumder A, Song S, et al. Discovery and characterization of Isofistularin-3, a marine brominated alkaloid, as a new DNA demethylating agent inducing cell cycle arrest and sensitization to TRAIL in cancer cells. Oncotarget. 2016;7:24027–49.
322. Meketa ML, Weinreb SM. Total synthesis of ageladine A, an angiogenesis inhibitor from the marine sponge *Agelas nakamurai*. Org Lett. 2006;8:1443–6.
323. Shengule SR, Loa-Kum-Cheung WL, Parish CR, Blairvacq M, Meijer L, Nakao Y, et al. A one-pot synthesis and biological activity of ageladine A and analogues. J Med Chem. 2011;54:2492–503.
324. Bickmeyer U. The alkaloid Ageladine A, originally isolated from marine sponges, used for pH-sensitive imaging of transparent marine animals. Mar Drugs. 2012;10:223–33.
325. Copp BR, Fairchild CR, Cornell L, Casazza AM, Robinson S, Ireland CM. Naamidine A is an antagonist of the epidermal growth factor receptor and an in vivo active antitumor agent. J Med Chem. 1998;41:3909–11.
326. LaBarbera DV, Modzelewska K, Glazar AI, Gray PD, Kaur M, Liu T, et al. The marine alkaloid naamidine A promotes caspase-dependent apoptosis in tumor cells. Anti-Cancer Drugs. 2009;20:425–36.
327. Jaimovich E, Mattei C, Liberona JL, Cardenas C, Estrada M, Barbier J, et al. Xestospongin B, a competitive inhibitor of IP3-mediated Ca2+ signalling in cultured rat myotubes, isolated myonuclei, and neuroblastoma (NG108-15) cells. FEBS Lett. 2005;2005(579):2051–7.
328. Akl MR, Ayoub NM, Ebrahim HY, Mohyeldin MM, Orabi KY, Foudah AI, et al. Araguspongine C induces autophagic death in breast cancer cells through suppression of c-Met and HER2 receptor tyrosine kinase signaling. Mar Drugs. 2015;13(1):288–311.
329. Schroeder FC, Kau TR, Silver PA, Clardy J. The psammaplysenes, specific inhibitors of FOXO1a nuclear export. J Nat Prod. 2005;68:574–6.

330. Berry E, Hardt JL, Clardy J, Lurain JR, Kim JJ. Induction of apoptosis in endometrial cancer cells by psammaplysene A involves FOXO1. Gynecol Oncol. 2009;112:331–6.
331. Pimentel AA, Felibertt P, Sojo F, Colman L, Mayora A, Silva ML, et al. The marine sponge toxin agelasine B increases the intracellular Ca(2+) concentration and induces apoptosis in human breast cancer cells (MCF-7). Cancer Chemother Pharmacol. 2012;69:71–83.
332. Martin MJ, Coello L, Fernandez R, Reyes F, Rodriguez A, Murcia C, et al. Isolation and first total synthesis of PM050489 and PM060184, two new marine anticancer compounds. J Am Chem Soc. 2013;135:10164–71.
333. Pera B, Barasoain I, Pantazopoulou A, Canales A, Matesanz R, Rodriguez-Salarichs J, et al. New interfacial microtubule inhibitors of marine origin, PM050489/PM060184, with potent antitumor activity and a distinct mechanism. ACS Chem Biol. 2013;8:2084–94.
334. Martinez-Diez M, Guillen-Navarro MJ, Pera B, Bouchet BP, Martinez-Leal JF, Barasoain I, et al. PM060184, a new tubulin binding agent with potent antitumor activity including P-glycoprotein over-expressing tumors. Biochem Pharmacol. 2014;88:291–302.
335. Cortes J, Vahdat L, Blum JL, Twelves C, Campone M, Roche H, et al. Phase II study of the halichondrin B analog eribulin mesylate in patients with locally advanced or metastatic breast cancer previously treated with an anthracycline, a taxane, and capecitabine. J Clin Oncol. 2010;28:3922–8.
336. Gitlitz BJ, Tsao-Wei DD, Groshen S, Davies A, Koczywas M, Belani CP, et al. A phase II study of halichondrin B analog eribulin mesylate (E7389) in patients with advanced non-small cell lung cancer previously treated with a taxane: a California cancer consortium trial. J Thorac Oncol. 2012;7:574–8.
337. Spira AI, Iannotti NO, Savin MA, Neubauer M, Gabrail NY, Yanagihara RH, et al. A phase II study of eribulin mesylate (E7389) in patients with advanced, previously treated non-small-cell lung cancer. Clin Lung Cancer. 2012;13:31–8.
338. Su JH, Chang WB, Chen HM, El-Shazly M, Du YC, Kung TH, et al. 10-acetylirciformonin B, a sponge furanoterpenoid, induces DNA damage and apoptosis in leukemia cells. Molecules. 2012;17:11839–48.
339. Yamanokuchi R, Imada K, Miyazaki M, Kato H, Watanabe T, Fujimuro M, et al. Hyrtioreticulins A-E, indole alkaloids inhibiting the ubiquitin-activating enzyme, from the marine sponge *Hyrtios reticulatus*. Bioorg Med Chem. 2012;20:4437–42.
340. Bagola K, von Delbruck M, Dittmar G, Scheffner M, Ziv I, Glickman MH, et al. Ubiquitin binding by a CUE domain regulates ubiquitin chain formation by ERAD E3 ligases. Mol Cell. 2013;50:528–39.
341. Hood KA, West LM, Rouwe B, Northcote PT, Berridge MV, Wakefield SJ, et al. Peloruside A, a novel antimitotic agent with paclitaxel-like microtubule- stabilizing activity. Cancer Res. 2002;62:3356–60.
342. Wilmes A, O'Sullivan D, Chan A, Chandrahasen C, Paterson I, Northcote PT, La Flamme AC, et al. Synergistic interactions between peloruside A and other microtubule-stabilizing and destabilizing agents in cultured human ovarian carcinoma cells and murine T cells. Cancer Chemother Pharmacol. 2011;68:117–26.
343. Meyer CJ, Krauth M, Wick MJ, Shay JW, Gellert G, De Brabander JK, et al. Peloruside A inhibits growth of human lung and breast tumor xenografts in an athymic nu/nu mouse model. Mol Cancer Ther. 2015;14:1816–23.
344. Chevallier C, Richardson AD, Edler MC, Hamel E, Harper MK, Ireland CM. A new cytotoxic and tubulin-interactive milnamide derivative from a marine sponge *Cymbastela* sp. Org Lett. 2003;5:3737–9.
345. Kuznetsov G, TenDyke K, Towle MJ, Cheng H, Liu J, Marsh JP, et al. Tubulin-based antimitotic mechanism of E7974, a novel analogue of the marine sponge natural product hemiasterlin. Mol Cancer Ther. 2009;8:2852–60.
346. Rho JR, Hwang BS, Sim CJ, Joung S, Lee HY, Kim HJ. Phorbaketals A, B, and C, sesterterpenoids with a spiroketal of hydrobenzopyran moiety isolated from the marine sponge *Phorbas* sp. Org Lett. 2009;11:5590–3.

347. Seo YJ, Lee KT, Rho JR, Choi JH. Phorbaketal A, isolated from the marine sponge *Phorbas* sp., exerts its anti-inflammatory effects via NF-kappaB inhibition and heme oxygenase-1 activation in lipopolysaccharide-stimulated macrophages. Mar Drugs. 2015;13:7005–19.
348. Mencarelli A, D'Amore C, Renga B, Cipriani S, Carino A, Sepe V, et al. Solomonsterol A, a marine pregnane-X-receptor agonist, attenuates inflammation and immune dysfunction in a mouse model of arthritis. Mar Drugs. 2014;12:36–53.
349. Park EJ, Cheenpracha S, Chang LC, Pezzuto JM. Suppression of cyclooxygenase-2 and inducible nitric oxide synthase expression by epimuqubilin A via IKK/IkappaB/NF-kappaB pathways in lipopolysaccharide-stimulated RAW 264.7 cells. Phytochem Lett. 2011;4:426–31.
350. Delfourne E, Bastide J. Marine pyridoacridine alkaloids and synthetic analogues as antitumor agents. Med Res Rev. 2003;23:234–52.
351. Dirsch VM, Kirschke SO, Estermeier M, Steffan B, Vollmar AM. Apoptosis signaling triggered by the marine alkaloid ascididemin is routed via caspase-2 and JNK to mitochondria. Oncogene. 2004;23:1586–93.
352. Simone M, Erba E, Damia G, Vikhanskaya F, Di Francesco AM, Riccardi R, et al. Variolin B and its derivate deoxy-variolin B: new marine natural compounds with cyclin-dependent kinase inhibitor activity. Eur J Cancer. 2005;41:2366–77.
353. Schyschka L, Rudy A, Jeremias I, Barth N, Pettit GR, Vollma AM. Spongistatin 1: a new chemosensitizing marine compound that degrades XIAP. Leukemia. 2008;22:1737–45.
354. Rothmeier AS, Ischenko I, Joore J, Garczarczyk D, Furst R, Bruns CJ, et al. Investigation of the marine compound spongistatin 1 links the inhibition of PKCalpha translocation to nonmitotic effects of tubulin antagonism in angiogenesis. FASEB J. 2009;23:1127–37.

Chapter 16
Natural Products from Corals

Guoqiang Li, Pinglin Li, and Xuli Tang

Contents

Abstract Cnidarians (mainly corals) are one of the most important resources of marine natural products (MNPs). Corals have rich species diversity, with more than 10 orders having 7000 species been found. The orders Alcyonacea (soft corals) and Gorgonacea (sea fans) which are referred to as ahermatypes, or non-reef building corals, have contributed with promising bioactive marine compounds. Till now, there have been more than 5800 compounds obtained from corals all over the world, attributing almost 20% of the total MNPs. These secondary metabolites are mostly representative as terpenoids and steroids. More than a hundred of the coral-derived natural products have great potential for the development of new pharmaceuticals and antifoulants. In this section, the research progress on coral-derived MNPs during the past two decades was included, mainly focusing on five points: (1) coral-

---

G. Li (✉) · P. Li (✉) · X. Tang
Key Laboratory of Marine Drugs, Chinese Ministry of Education, School of Medicine and Pharmacy, Ocean University of China, Qingdao, People's Republic of China
e-mail: liguoqiang@ouc.edu.cn; lipinglin@ouc.edu.cn

Z. Li (ed.), *Symbiotic Microbiomes of Coral Reefs Sponges and Corals*,
https://doi.org/10.1007/978-94-024-1612-1_16

produced structures with brand-new skeleton, (2) coral-derived MNPs with significant bioactivities, (3) some representative approaches on total synthesis of coral-derived novel compounds, (4) a short review of the coral-derived compounds in preclinical and clinical development, and (5) Conclusion.

Keywords Coral · Natural products · Bioactivity

16.1 Introduction

Marine invertebrates are rich sources of bioactive compounds. Although sponges are the foremost providers of marine bioactive compounds, corals are also being studied with promising results. Coral is a large, diverse group of marine invertebrates that include over 7500 species. These corals are a rich source of novel, bioactive secondary metabolites. More than 5800 compounds have been obtained from corals all over the world with over 3400 articles published till now. The orders Alcyonacea (soft corals) and Gorgonacea (sea fans), which are referred to as ahermatypes, or non-reef building corals, and do not always have zooxanthellae, have contributed with the highest number of bioactive marine compounds.

Soft corals are generally brightly colored that they have all the colors of the rainbow, but their predominant hues are rust, red, orange, yellow, olive, and purple, and gorgonian octocorals known as sea whips and sea fans are normally erect, flattened, branching, and reminiscent of a fan, or they may be whiplike, bushy, or even encrusting. However, the incidence of predation in these organisms is low due to the toxic compounds they produce. The soft corals and gorgonian are rich sources of secondary metabolites including diterpenes, sesquiterpenes, capnellenes, and steroids. Some of those compounds have great potential for the development of new pharmaceuticals and antifoulants. The soft corals contribute to almost 70% of the coral-derived MNPs, with the genera *Sinularia*, *Sarcophyton*, *Lobophytum*, *Nephthea*, *Paralemnalia*, *Xenia*, *Klyxum*, *Clavularia*, *Cladiella*, *Cespitularia*, and *Carijoa* mostly studied, while gorgonian is representative of the genera *Junceella*, *Ellisella*, *Dichotella*, *Briareum*, *Isis*, *Menella*, *Pseudopterogorgia*, *Leptogorgia*, *Eunicea*, *Subergorgia*, *Muricella*, *Anthogorgia*, *Astrogorgia*, *Echinomuricea*, and *Acanthogorgia*. In this section, the research progress on coral-derived MNPs during the past two decades was included.

16.2 Coral-Produced Novel Compounds

Since 1995 there are over a hundred of novel compounds with the new skeletons isolated and identified from corals, with steroids and terpenoids the most representatives.

16.2.1 Novel Steroids

A new hemiketal steroid, named cladiellin A (**1**) with antioxidant activity, was first isolated from the soft coral *Cladiella* sp. collected from Sanya Bay, Hainan Island, of China [1]. Three unusual new steroid thioesters, parathiosteroids A–C (**2–4**), were isolated from the soft coral *Paragorgia* sp. collected in Madagascar. These compounds displayed cytotoxicity against three human tumor cell lines at the micromolar level [2]. Cladocorans A (**5**) and B (**6**), showing a γ-hydroxybutenolide end group, were isolated from the Mediterranean coral *Cladocora caespitosa* and are novel sesterterpenoids whose structures were revised by the synthesis [3]. Four antifouling secosteroids named isogosterones A–D (**7–10**) have been isolated from a Japanese octocoral *Dendronephthya* sp. [4] (Fig. 16.1).

16.2.2 Novel Diterpenoids

From a rare alcyonacean, *Eleutherobia* species (possibly *E. albiflora*), collected near Bennett's Shoal in Western Australia, a new diterpene glycoside, eleutherobin (**11**), was obtained and found to be the second molecule possessing the unique microtubule-stabilizing properties of paclitaxel [5]. Methyl sarcotroates A (**12**) and B (**13**), two novel diterpenoids having a tetradecahydrocyclopenta[3′,4′]cyclobuta[1′,2′:4,5]cyclonona[1,2-b]oxirene ring system, were isolated from the Hainan soft coral *Sarcophyton trocheliophorum*. Methyl sarcotroate B showed significant inhibitory activity against protein tyrosine phosphatase 1B (PTP1B) [6]. Bielschowskysin (**14**) with highly oxygenated hexacyclic structure was isolated

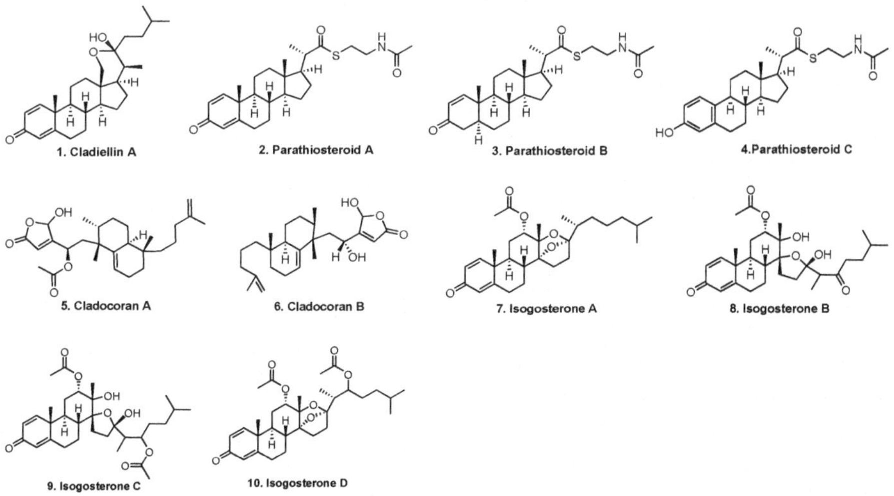

Fig. 16.1 The structures of the novel steroids isolated from corals

from the Caribbean gorgonian *Pseudopterogorgia kallos*. Bielschowskysin exhibited antimalarial activity against *Plasmodium falciparum* as well as strong anticancer activity against two human cancer cell lines [7].

An unprecedented heptacyclic C40 *bis*-diterpenoid, bisgersolanolide (**15**), was isolated from the Caribbean gorgonian *Pseudopterogorgia bipinnata* (Verrill) collected in San Andrés Island, Colombia [8]. The novel diterpenoid aquariolide A (**16**) with an unprecedented highly rearranged carbon skeleton (named aquariane) was isolated from cultured *Erythropodium caribaeorum* [9]. Three new briarane diterpenoids, briareolate esters L–N (**17–19**), were isolated from a gorgonian *Briareum asbestinum*. Briareolate esters L and M are the first natural products having a ten-membered macrocyclic ring with a (*E,Z*)-dieneone [10].

Tortuosenes A (**20**) and B (**21**), having a new C-2/C-20-cyclized cembranoid skeleton, were isolated from the Formosan soft coral *Sarcophyton tortuosum* [11]. The novel aberrarone (**22**) was isolated from the Caribbean Sea whip, *Pseudopterogorgia elisabethae*. Aberrarone showed antimalarial activity against a chloroquine-resistant strain of the protozoan parasite *Plasmodium falciparum* [12]. Three novel compounds, designated kitungolides A–C (**23–25**), were isolated from a soft coral of a new genus belonging to the family Xeniidae collected at Kitungamwe, Kenya. The three new compounds are of a unique heterotricyclic skeleton [13]. Corallolides A (**26**) and B (**27**) with antiparasitic and antituberculosis activity, respectively, were isolated from the Caribbean gorgonian *Pseudopterogorgia bipinnata* collected near Providencia Island [14].

An unprecedented C, C-linked dimeric norcembranoid (sinulochmodin A, **28**), a novel isocembranoid (sinulochmodin B, **29**), and a novel yonarane norditerpenoid (sinulochmodin C, **30**) with an oxo-THF ring were isolated from a Formosan soft coral *Sinularia lochmodes* [15]. Providencin (**31**) is cytotoxin isolated from the Caribbean gorgonian *Pseudopterogorgia kallos*. The highly oxygenated hexacyclic structure is based on a previously undescribed bicyclo[12.2.0]hexadecane ring system [16]. The study of two specimens of *Pseudopterogorgia elisabethae*, collected from a different location at the San Andrés Archipelago, afforded two new norditerpenes, caribenols A (**32**) and B (**33**), with unusual C19 rearranged terpenes [17]. A new xenicane diterpenoid, cristaxenicin A (**34**), was isolated from the deep sea gorgonian *Acanthoprimnoa cristata* [18].

Xenibellols A (**35**) and B (**36**), possessing a unique heterotricyclic skeleton consisting of an unprecedented hexahydrocoumarine system, were isolated from the soft coral *Xenia umbellate* collected at Green Island, Taiwan [19]. A novel compound, plumisclerin A (**37**), was isolated from the Mayotte Island soft coral *Plumigorgia terminosclera*. Plumisclerin A possesses the novel plumisclerane carbon skeleton, including a tricyclo[4,3,1,01,5]decane ring [20]. A novel trispiropentacyclic diterpene, intricarene (**38**), was isolated from the Caribbean gorgonian *Pseudopterogorgia kallos* [21]. Colombiasin A (**39**), belonging to a class of C20 rearranged diterpenes possessing an intricate tetracyclic framework, was isolated from the West Indian gorgonian octocoral *Pseudopterogorgia elisabethae* [22]. Four diterpenes and a norditerpenoid, elisabethin D (**40**), elisabethin D acetate (**41**), 3-epi-elisabanolide (**42**), and elisapterosin A (**43**), all of which possess unusual carbocyclic skeletons, were isolated from the West Indian gorgonian *Pseudopterogorgia elisabethae* [23].

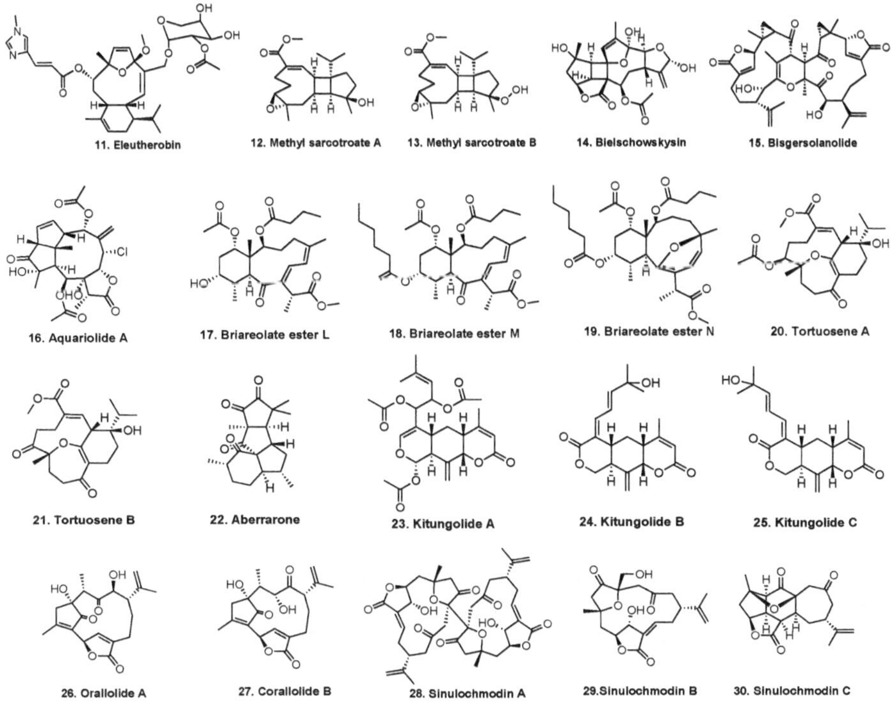

Fig. 16.2 The structures of the novel diterpenoids isolated from corals

Nanolobatolide (**44**), possessing a novel C18 structure, was isolated from the Formosan soft coral *Sinularia nanolobata* [24]. Two antituberculosis diterpenoid alkaloids with the uncommon benzoxazole moiety, pseudopteroxazole (**45**) and *seco*-pseudopteroxazole (**46**), were isolated from the West Indian gorgonian *Pseudopterogorgia elisabethae*. Pseudopteroxazole potently inhibited growth of *Mycobacterium tuberculosis* H37Rv, while *seco*-pseudopteroxazole showed moderate-to-strong inhibitory activity [25]. Two novel prostanoid-related marine oxylipins, tricycloclavulone (**47**) having a tricyclo[5.3.0.01,4]decane ring system and clavubicyclone (**48**) having a bicyclo[3.2.1]octane ring system, were isolated from the Okinawan soft coral *Clavularia viridis* [26]. Zoaramine (**49**), whose core resembles the structure of norzoanthamine alkaloid, was isolated from an Atlantic variety of *Zoanthus* sp. [27] (Fig. 16.2).

16.2.3 Novel Sesquiterpenoids

Two new sesquiterpenoids bearing the fused 3/6/5 tricyclic ring system, shagenes A (**50**) and B (**51**), were isolated from an undescribed soft coral collected from the Scotia Arc in the Southern Ocean [28]. Paesslerins A (**52**) and B (**53**), sesquiterpenoids with an unprecedented tricyclic 2,8,8,10-tetramethyltricyclo[4.3.2.02,5] undecane skeleton, were isolated from the subantarctic soft coral *Alcyonium paessleri* [29]. Two new terpenoids, chabranol (**54**) and capillosanol (**55**), having unprecedented terpenoid skeletons, were isolated from the soft corals *Sinularia capillosa* and *Nephthea chabroli*, respectively [30]. Isishippuric acid B (**56**) bearing a novel 4,5-seco-6-norquadrane skeleton isolated from the gorgonian coral *Isis hippuris* exhibited potent cytotoxicity (ED_{50} < 0.1 μg/mL) toward three cancer cell lines [31] (Fig. 16.3).

16.3 Coral-Derived Bioactive Compounds

In terms of biodiversity, marine bioactive compounds display varied potential applications, in cosmetics, as nutraceuticals, and in agrochemical industries. The ability of corals (soft coral and gorgonian) to produce powerful pharmaceutical compounds has been well documented. Only the compounds displaying an $IC_{50} \leq 10.0$ μg/mL or μM (except where stated otherwise) and $ED_{50} \leq 4.0$ μg/mL were considered for the present study, as these values are commonly used in the surveyed literature to ascertain relevant bioactivity.

16.3.1 Antimicrobial Activity

The diterpenoid alkaloids, pseudopteroxazole (**45**) and seco-pseudopteroxazole (**46**), showed strong growth inhibition on *Mycobacterium tuberculosis* H37Rv (97% inhibition induced at 12.5 μg/mL) [25]. Litosterol (**57**) and nephalsterol C (**58**) inhibit the growth of *M. tuberculosis* (90% and 96%, respectively), with MIC values of 3.13 and 12.5 μg/mL, respectively [32].

Pseudopterosins (PsG (**59**), PsP (**60**), PsQ (**61**), PsS (**62**), PsT (**63**), PsU (**64**), 3-O-Ac-PsU (**65**)) and seco-pseudopterosins (seco-PsJ (**66**), seco-PsK (**67**)) and the inter-converting mixture of non-glycosylated diterpenes (IMNGD (**68**)) were active against *Staphylococcus aureus* and *Enterococcus faecalis* [33].

Erectathiol (**69**), a sesquiterpene isolated from *Nephthea erecta*, exhibited significant activity against *S. enteritidis*, more effective than ampicillin [34]. A lactone cembrane diterpene, sarcophytolide (**70**) isolated from the Red Sea soft coral *Sarcophyton glaucum*, showed activity against *S. aureus* (Gram-positive) with MIC of 0.19 μg/mL, against *P. aeruginosa* (Gram-negative) with MIC of 0.22 μg/mL, and against *Saccharomyces cerevisiae* (yeast) with MIC of 0.13 μg/mL [35]. Craterellin A (**71**) showed antibacterial activity against *Bacillus cereus* with a MIC value of 3.12 μm [36]. Dioxanyalolide (**72**) showed antibacterial activity against *Escherichia coli* at 1.25 μg/mL [37].

(5*S*)-3-[(3*E*,5*S*)-5-hydroxy-3-hepten-6-yn-1-yl]-5-methyl-2(5H)-furanone (**73**) showed antimicrobial activity toward pathogenic bacterial strains, including

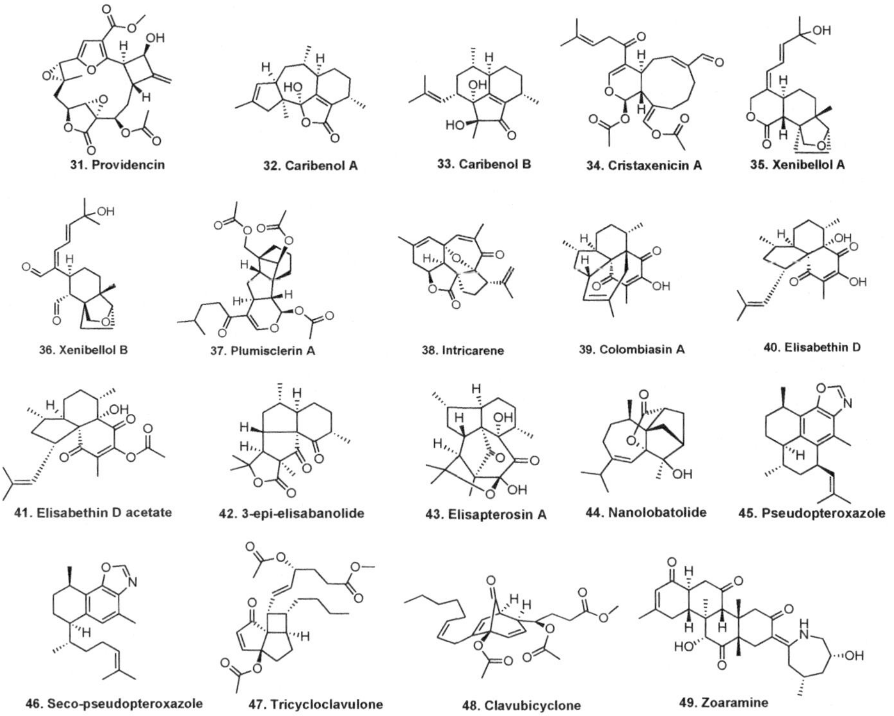

Fig. 16.2 (continued)

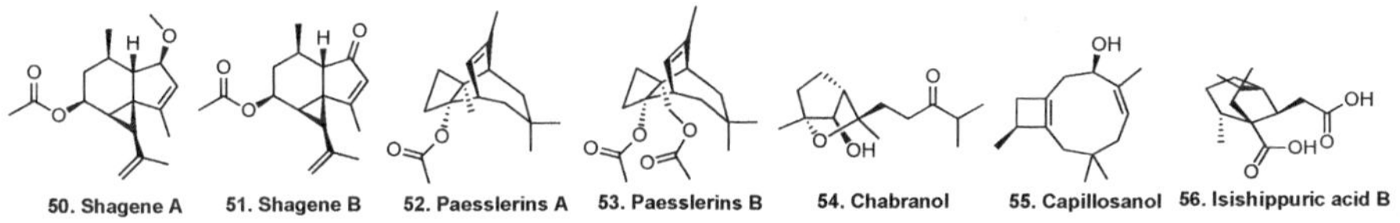

Fig. 16.3 The structures of the novel sesterpenoids isolated from corals

Bacillus cereus, *Salmonella typhi*, *Escherichia coli*, *Staphylococcus aureus*, and *Pseudomonas aeruginosa* at concentrations of 0.20 mg/mL [38]. 15β-hydroxypregna-1,4,20-trien-3-one (**74**) exhibited antibacterial activity against *Pseudomona puido*, with a MIC value of 31 nM [39]. Carijodienone (**75**) significantly inhibited the growth in the ranges 6–8 and 5–10 μg/mL against *B. cereus* and *K. pneumoniae*, respectively [40]. Subergosterone B (**76**) exhibited antibacterial activity against *Bacillus cereus* with MIC value of 1.56 μm [41].

Xeniolide I (**77**) possesses antibacterial activity at 1.25 μg/ml (*Escherichia coli* and *Bacillus subtilis*), and novaxenicin B (**78**) induces apoptosis in transformed mammalian cells at 1.25 μg/ml [42]. The two major cembranes in *Eunicea knighti*, **79** and **80**, showed potent biofilm inhibition in *Staphylococcus aureus* ATCC 25923

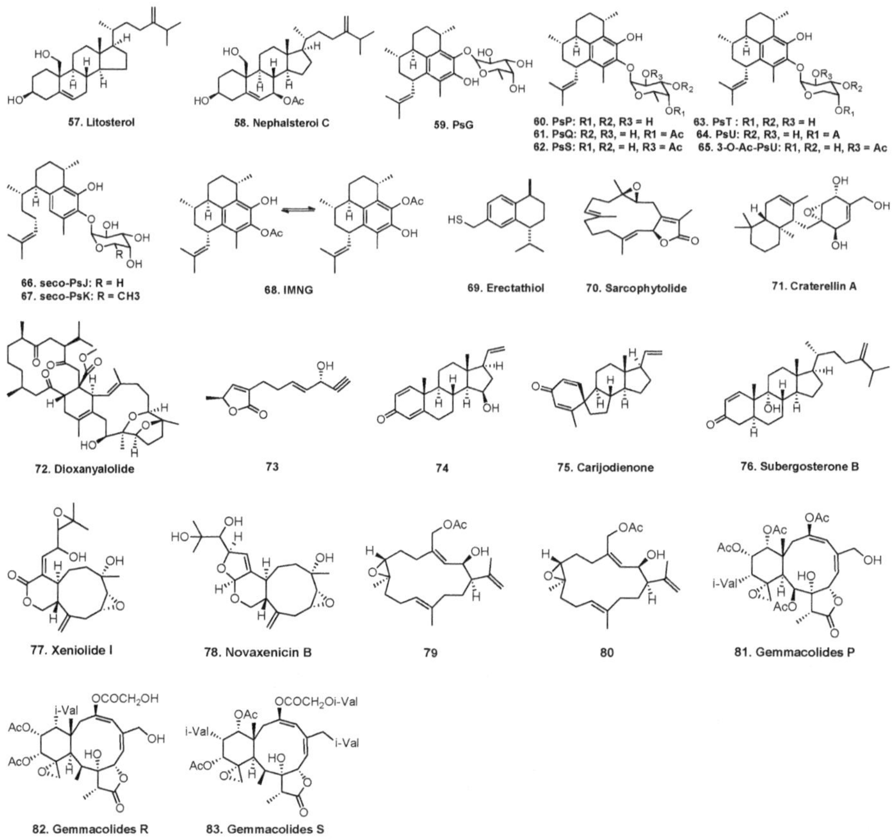

Fig. 16.4 The antimicrobial compounds isolated from corals

(Gram positive) with IC_{50} values of 0.16 ± 0.03 and 0.08 ± 0.01 μM, respectively [43]. Gemmacolides P (**81**), R (**82**), and S (**83**) exhibited potent antifungal activity against *Septoria tritici* with IC_{50} values less than 7 μM [44] (Fig. 16.4).

16.3.2 Anti-inflammatory

Pseudopterosin A (**84**), which was originally isolated from the Caribbean Sea whip *Pseudopterogorgia elisabethae*, possesses potent anti-inflammatory and analgesic activities [45]. At a concentration of 10 μM, tortuosene A (**20**) isolated from the Formosan soft coral *Sarcophyton tortuosum* exhibited significant inhibition (56.0 ± 3.1%) toward N-formylmethionyl-leucyl-phenylalanine/cytochalasin B (fMLP/CB)- induced superoxide anion (O2 •–) generation [11]. At the same concentration, paraminabic acids B (**85**) and C (**86**) isolated from a Formosan soft coral

Paraminabea acronocephala reduced the levels of iNOS to 63.9 ± 6.3% and 53.5 ± 8.6%, respectively [46].

Lobophytone Z (**89**) isolated from a Chinese soft coral *Lobophytum pauciflorum* inhibited NO production with an IC_{50} value of 2.6 μm [47]. At concentration of 10 μM, durumolides A (**88**), K (**89**), and L (**90**) isolated from the soft coral *Lobophytum durum* significantly reduced the levels of the iNOS protein (34.7 ± 7.9%, 0.8 ± 0.6% and 5.7 ± 2.2%, respectively) and COX-2 protein (62.5 ± 4.3%, 47.8 ± 9.0% and 71.6 ± 5.8%, respectively) compare with the control cells (LPS alone) [48]. Sinularin (**91**) isolated from the soft coral *Sinularia flexibilis* have anti-edematous effects on paw edema induced by carrageenan or adjuvant [49].

Using the TPA-induced ear edema model, fuscoside E (**92**) and B (**93**) isolated from Caribbean octocoral *Eunicea fusca* showed strong exhibited inhibition levels of 80.5% and 81.5% (0.5 mg/ear), respectively, comparable to that shown by the indomethacin [50]. Secosterols pinnisterols A (**94**) and C (**95**) from gorgonian *Pinnigorgia* sp. displayed significant inhibitory effects on the release of elastase (IC_{50} = 3.32 and 2.81 μM) and inhibitory effects on the generation of superoxide anions (IC_{50} = 2.33 and 2.50 μM) by human neutrophils [51]. At a concentration of 10 μM, briarenolides M (**96**), P (**97**), S (**98**), and T (**99**) from Formosan octocoral *Briareum* sp. inhibited the accumulation of the pro-inflammatory inducible nitric oxide synthase (iNOS) protein to 49.6, 58.4, 57.4, and 53.5%, respectively, and briarenolides N (**100**), P, and T inhibited the accumulation of the pro-inflammatory cyclooxygenase-2 (COX-2) protein to 53.9, 59.1, and 59.3%, respectively, in LPS (lipopolysaccharide)-stimulated RAW264.7 macrophage cells [52]. Briarenolides U–Y (**101–105**) significantly inhibited the expression of the pro-inflammatory inducible nitric oxide synthase (iNOS) (41.9%, 47.3%, 50.1%, 66.2%, and 54.3%, respectively) and cyclooxygenase-2 (COX-2) (26.1%, 35.6%, 58.1%, 67.2%, and 55.4%, respectively) of the lipopolysaccharide (LPS)-stimulated RAW264.7 macrophage cells [53].

At a concentration of 10 μM, sinulacembranolide A (**106**) inhibited the accumulation of the pro-inflammatory inducible nitric oxide synthase (iNOS, 8.55% ± 3.32%) protein of the lipopolysaccharide (LPS)-stimulated RAW264.7 marcophage cells [54]. Glaucumolides A (**107**) and B (**108**) from a cultured soft coral *Sarcophyton glaucum* displayed strong inhibition of superoxide anion generation and elastase release in human neutrophils stimulated by fMLP/CB. At concentrations of 5, 10, and 20 μM, glaucumolide A significantly reduced the levels of iNOS and COX-2 to 59.4 ± 9.0 and 66.5 ± 4.4%, 31.3 ± 6.5 and 78.3 ± 5.0%, and − 2.6 ± 2.7 and − 0.5 ± 3.2%, respectively. At concentrations of 10 and 20 μM, glaucumolide B significantly reduced the levels of iNOS and COX-2 to 75.9 ± 3.5 and 64.3 ± 6.9% and 43.4 ± 5.0 and 6.0 ± 3.6%, respectively [55]. Hirsutosteroside A (**109**), a new pregnane glycoside from soft corals *Cladiella hirsute*, significantly inhibited the fMLP/CB-induced elastase release with IC_{50} values of 4.1 ± 0.1 μM [59]. New cembranoid isosinulaflexiolide K (**110**) isolated from cultured soft corals *Sinularia sandensis* significantly reduced the levels of iNOS to 30.9 ± 4.1% and the levels of COX-2 to 47.1 ± 3.8%, respectively, at a concentration of 10 μM [56].

Klyflaccisteroids C (**111**) and F (**112**) from the soft coral *K. flaccidum* showed strong inhibitions toward superoxide anion generation with IC_{50} of less than 5 μM [57]. Krempfielin N (**113**) isolated from a Taiwanese soft coral *Cladiella krempfi* significantly inhibited superoxide anion generation and elastase release in human neutrophils induced by *N*-formyl-methionyl-leucyl-phenylalanine/cytochalasin B (FMLP/CB) with IC_{50} of 4.94 ± 1.68 μM [58]. Lobocrasols A (**114**) and B (**115**) isolated from the Vietnamese soft coral *Lobophytum crissum* significantly inhibited TNFα-induced NF-κB transcriptional activity in HepG2 cells, with IC_{50} values of 6.30 ± 0.42 and 6.63 ± 0.11 μM, respectively [59]. Sarcopanol A (**116**) isolated from the Vietnamese soft coral *Sarcophyton pauciplicatum* significantly inhibited TNFα-/INFγ-induced NF-κB transcriptional activity in human keratinocyte (HaCaT) cells, with EC_{50} value of 8.27 ± 3.28 μM [60]. At a concentration of 10 μM, paraminabic acids B (**85**) and C (**86**) from soft coral *Paraminabea acronocephala* reduced the levels of iNOS to 63.9 ± 6.3% and 53.5 ± 8.6%, respectively, whereas paraminabic acid B enhanced the expression of COX-2 (130.5 ± 9.8%) [46].

At a concentration of 1 μM, lochmolin A (**117**) from the soft coral *Sinularia lochmodes* reduced the level of LPS-induced COX-2 to 36.6 ± 3.8% [61]. The new cembranoids, sarcocrassocolides A (**118**), D (**119**), and F–Q (**120–131**) from a soft coral *Sarcophyton crassocaule*, display significant in vitro anti-inflammatory activity in LPS-stimulated RAW264.7 macrophage cells by inhibiting the expression of the iNOS protein [62]. 6-*epi*-yonarasterol B (**132**) from a gorgonian *Echinomuricea* sp. showed significant inhibitory effects on the generation of superoxide anions and the release of elastase (>89% at a concentration of 10 μg/mL) by human neutrophils [63]. The anti-inflammatory activity of a new dilophol diterpene eunicidiol (**133**) was evaluated by measuring their ability to reduce phorbol myristate acetate (PMA)-induced edema in a mouse ear model [64].

A new clovane-related sesquiterpenoid, clovan-2,9-dione (**134**), isolated from the gorgonian *Rumphella antipathies* displayed significant inhibitory effects on the generation of superoxide anion and the release of elastase by human neutrophils with IC_{50} of less than 5 μg/mL [65]. An immunomodulatory assay of marine cembrane compounds, isolated from a soft coral *Lobophytum crassum*, on mouse bone marrow-derived dendritic cells (BMDCs) indicated that lobocrassin B (**135**) did not stimulate TNF-α production from BMDCs but effectively inhibited LPS-induced DC activation by inhibiting the production of TNF-α and exhibited a broad spectrum of inhibiting DC activation mediated by various TLR agonists [66]. At a concentration of 10 μM, klysimplexins J–N (**136–140**), R (**141**), and S (**142**) from the cultured soft coral *Klyxum simplex* significantly reduced the expression of iNOS protein [67] (Fig. 16.5).

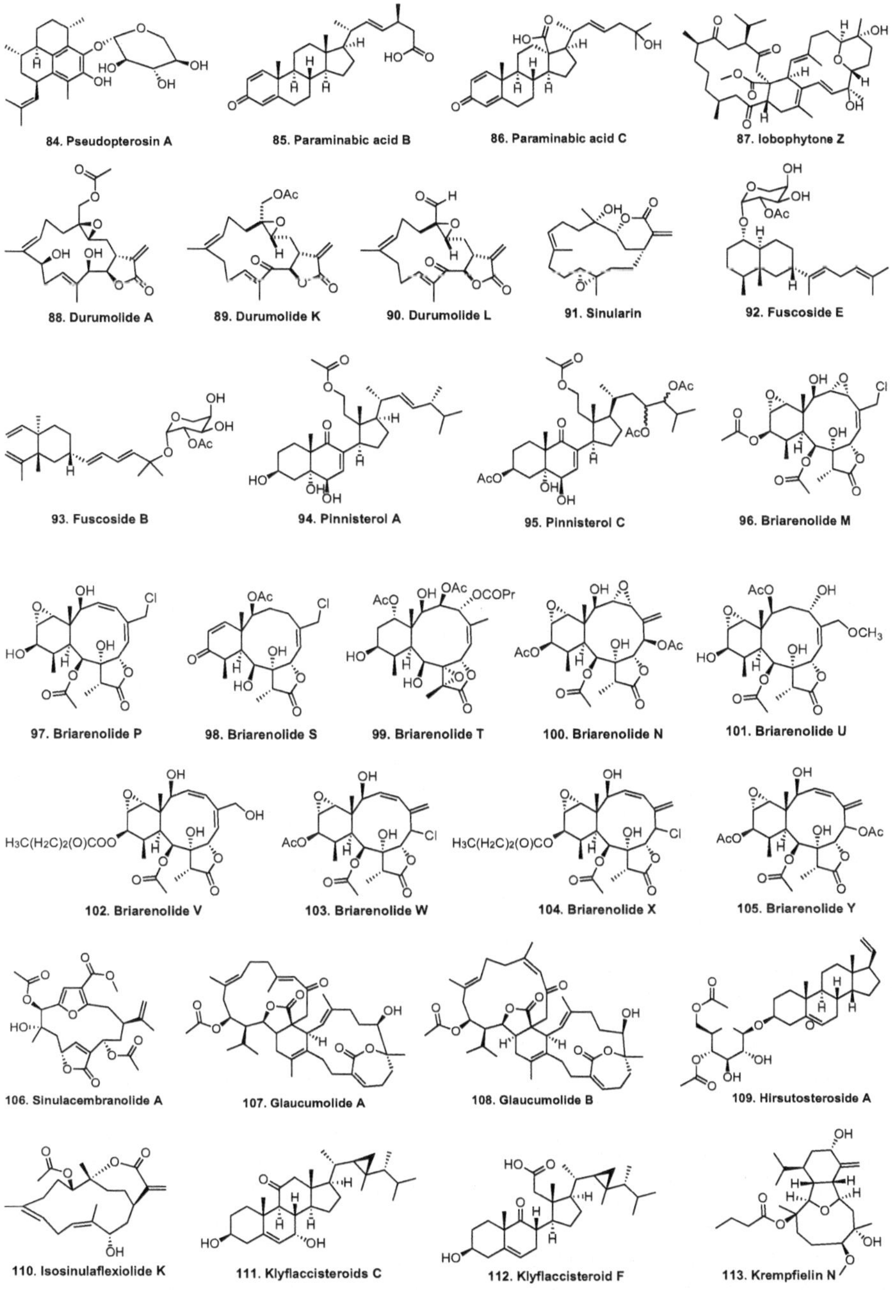

Fig. 16.5 The anti-inflammatory compounds isolated from corals

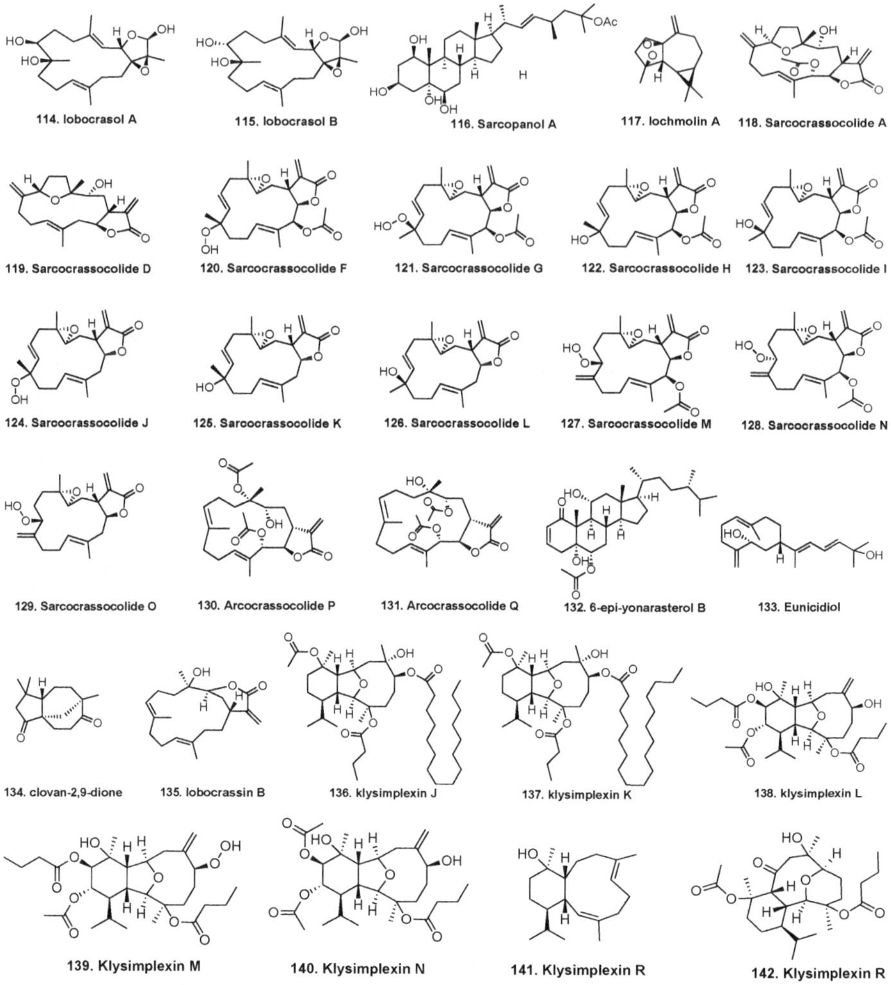

Fig. 16.5 (continued)

16.3.3 Antitumor Activity

Eleutherobin (**11**) isolated from a rare alcyonacean, *Eleutherobia* species (possibly *E. albiflora*), showed significant cytotoxicity against wide cancer cells with IC_{50} range of 10–15 nM in vitro [5]. It was the second molecule to possess the unique microtubule-stabilizing properties of paclitaxel. While in the antimitotic activity in the cell-based assay, desmethyleleutherobin (**143**) and isoeleutherobin A (**144**) isolated from the Caribbean octocoral *Erythropodium caribaeorum* were slightly more potent than eleutherobin, with IC_{50} of 20 and 50 nM, respectively, than the analogues eleutherobin (IC_{50} 100 nM), Z-eleutherobin (**145**) (IC_{50} 250 nM), desacetyleleutherobin (**146**) (IC_{50} 400 nM), sarcodictyin A (**147**) (IC_{50} 2 μM), caribaeoside (**148**) (IC_{50} 20 μM), and caribaeolin (**149**) (IC_{50} 20 μM) [68].

The cembranoid (1S,2S,3E,7E,11E)-3,7,11,15-cembratetraen-17,2-olide (LS-1, **150**) isolated from *Lobophytum cristagalli* inhibited the proliferation of HT-29 cells, and IC_{50} value for LS-1 was about 3.7 μM. LS-1 also showed significantly cytotoxic activity against A549 lung cancer cells and SNU-C5 human colon carcinoma cells with IC_{50} values of 5.1 μM and 6.6 μM, respectively [69]. Subsequent study revealed further that LS-1 induced SNU-C5/5-FU, 5-FU-resistant colon cancer cells, via the activation of the TGF-β pathway with downregulation of CEA. The study suggested that LS-1 may have the potential for treatment of colon cancer, including chemotherapy-resistant colon cancer [69].

Muricenones A (**151**) and B (**152**) isolated from the gorgonian *Muricea* sp. showed a significant and selective activity as inhibitors of the growth of A549 cells with GI_{50} values of 2 and 3 μg/mL, respectively [70]. Lobocrassin B (**135**) isolated from the soft coral *Lobophytum crissum* exhibited modest cytotoxicity against K562, CCRF-CEM, Molt4, and HepG2 cells with IC_{50} of ranging from 0.1 to 8.2 μM [71]. The cytotoxic activity of plumisclerin A (**37**) isolated from the soft coral *Plumigorgia terminosclera* displayed cytotoxic activity against lung (A549), colon (HT29), and breast (MDA-MB-231) cell lines, with GI_{50} values of 4.7, 2.1, and 6.1 μM, respectively [20]. A cembranolide (**153**) isolated from *Lobophytum cristagalli* was a potent (IC_{50} 150 nM) inhibitor of farnesyl protein transferase (FPT) [72].

Sarcophyolide B (**154**) from the soft coral *Sarcophyton elegans* showed significant inhibition against A2780 with IC_{50} value of 2.92 μM, respectively [73]. Pacificins C (**155**), H (**156**), K (**157**), and L (**158**) from the Formosan soft coral *Nephthea pacifica* exhibited cytotoxicity against P-388 cells with ED_{50} ranging from 1.44 to 3.2 μg/mL [74]. Sinuflexolide (**159**) and sinuflexibilin (**160**) from the soft coral *Sinularia flexibilis* exhibited significant cytotoxicity toward A549, HT-29, KB, and P-388 cells with ED_{50} values of less than 1 μg/mL [75]. Paraminabic acid C (**86**) from soft coral *Paraminabea acronocephala* showed potent cytotoxicity against Hep3B, MDA-MB-231, MCF-7, and A-549 cancer cell lines, with IC_{50} values ranging from 2.05 to 2.83 μg/mL [46].

Juncenolide C (**161**) from a red gorgonian *Junceella juncea* exhibited cytotoxicity against human hepa adenocarcinoma (HEPA 59 T/VGH) and oral epidermoid carcinoma (KB-16) cells at concentrations of 6.6 and 7.8 μg/mL, respectively [76]. 4α-Hydroxy-5-episinuleptolide (**162**) from the cultured *Sinularia numerosa* exhibited moderate cytotoxicity against CCRF-CEM cells with an IC_{50} value of 4.21 μg/mL [77]. Juncenolide A (**163**) exhibited cytotoxicity against human colon adenocarcinoma (DLD-1) and oral epidermoid carcinoma (KB-16) cells at a concentration of 3.4 and 5.9 μg/mL, respectively [78]. Chondrosterin A (**164**) from the Taiwanese gorgonian *Junceella juncea* showed cytotoxic activities against A549, CNE2, and LoVo cancer lines with IC_{50} values of 2.45, 4.95, and 5.47 μM, respectively [79].

11-Epi-sinulariolide acetate (11-epi-SA) (**165**) isolated from the cultured soft coral *Sinularia flexibilis* has been examined for potential anti-cell migration and invasion effects on hepatocellular carcinoma cells (HCC) [80]. Further investigation revealed that 11-epi-SA suppressed the phosphorylation of ERK1/2 and p38MAPK and also the expression of the phosphorylation of FAK/PI3K/AKT/mTOR pathways [81]. The steroids stoloniferones O–Q (**166–167**) from the soft coral *Clavularia viridis* showed strong cytotoxicity against HT-29 and P388 cell lines with IC_{50} values ranging from 0.1 to 0.5 μg mL − 1 [82].

Sinulariaoid A (**168**) from the soft coral *Sinularia* sp. had specific cytotoxic activity against HepG2/ADMs with IC_{50} values of 9.70 μM through inducing apoptosis, and its selective toxicity toward HepG2/ADM cells was not related to P-glycoproteins [83]. Eupalmerin acetate (**169**) (EPA, from gorgonian *Eunicea palmeri*) induced phosphorylation of JNK was transient in both U87-MG and U373-MG cells [84].

Waixenicin A (**170**) from the soft coral *Sarcothelia edmondsoni* was found to inhibit cell proliferation through magnesium-dependent block of transient receptor potential melastatin 7 (TRPM7) channels. And Waixenicin A represents the first potent and relatively specific inhibitor of TRPM7 ion channels [85]. The cembrane-based diterpenoids, pavidolides B (**171**) and C (**172**) from the soft coral *Sinularia pavida*, showed selective inhibition against human promyelocytic leukemia cell line HL-60 with IC_{50} of 2.7 and 5.3 μg/mL [86]. Sinuladiterpene B (**173**), a cembrane diterpene from *Sinularia flexibilis*, exhibited cytotoxic activity against WiDr (human colon adenocarcinoma) cell lines with ED_{50} 8.37 mg/mL [87]. Subergorgic acid methyl ester (**174**) from the Taiwanese gorgonian coral *Subergorgia suberosa* exhibited cytotoxicity against HeLa cells with an ED_{50} of 4.3 μg/mL but inactive against the growth of KB cells [88].

Isolated from the Okinawan soft coral *Clavularia koellikeri*, *Ent-trans*-cembranolide (**175**) inhibited the proliferation of NCI-H522 (lung cancer) strongly with an IC_{50} of 0.66 μg/mL and those of LOX-IMVI (melanoma) and MKN74 (stomach cancer) with IC_{50}s of 0.72 and 0.81 μg/mL, respectively. 2-Acetoxystolonidiol (**176**) showed cytotoxic activity against human colorectal adenocarcinoma cells (DLD-1) with an IC_{50} of 5.0 μg/mL [89]. Bielschowskysin (**14**) from the Caribbean gorgonian *Pseudopterogorgia kallos* exhibited antiplasmodial activity (IC_{50} = 10 μg/mL) against *Plasmodium falciparum* and displayed cytotoxicity against the EKVX non-small cell lung cancer (GI_{50} < 0.01 μM) and CAKI-1 renal cancer (GI_{50} = 0.51 μM) [7] (Fig. 16.6).

16.3.4 Antifouling Activity

The antifoulants were identified as potent inhibition with the EC_{50} values lower than the standard requirement (EC50 < 25 μg/mL) as established by the US Navy program. Ophirin (**177**) and calicophirin B (**178**) from the South China Sea gorgonian

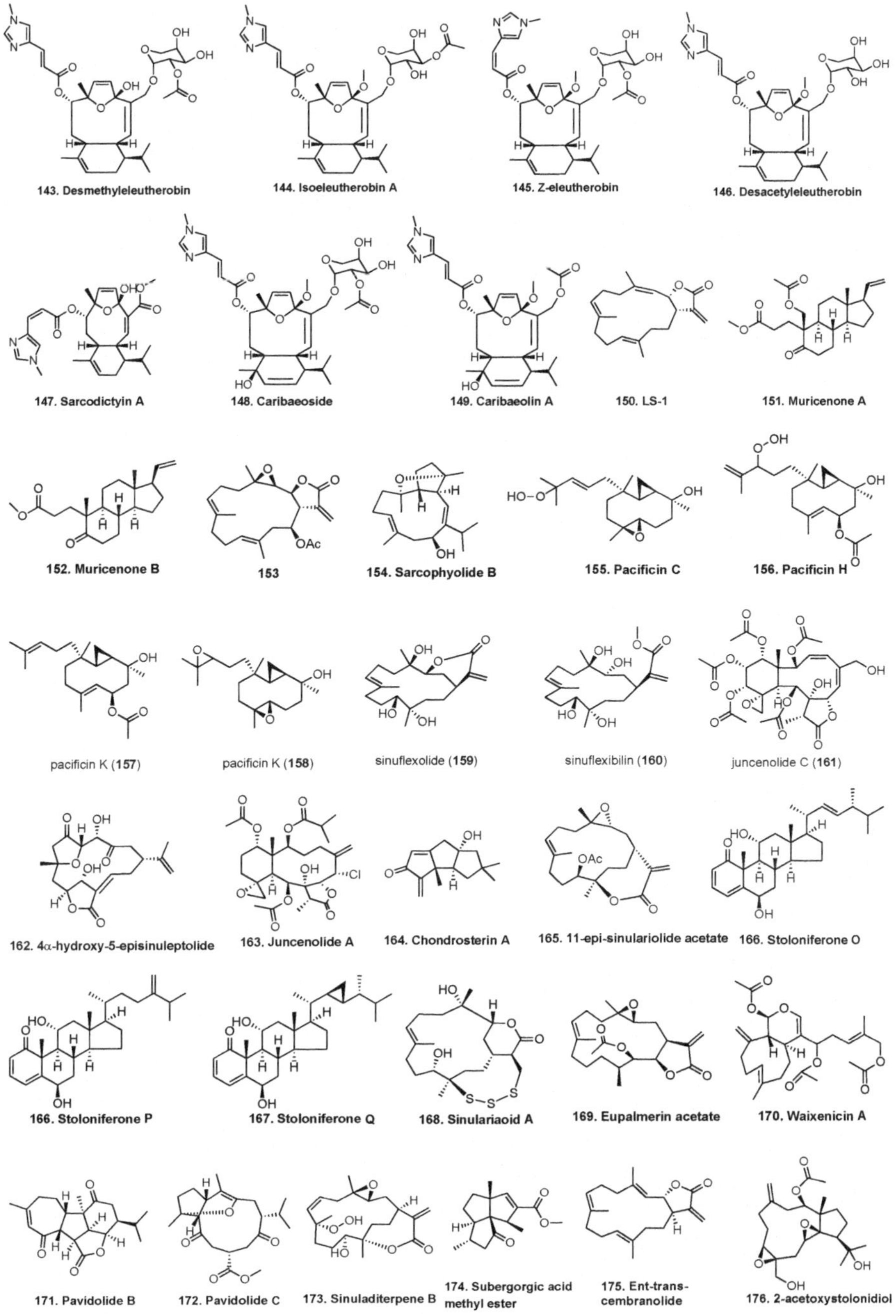

Fig. 16.6 The antitumor compounds isolated from corals

Muricella sibogae showed significant antifouling activity against the green mussel *Perna viridis*, with EC_{50} values of 1.8 and 18.1 μg/mL and LC_{50} values of 33.4 and > 50 μg/mL, respectively [90]. Sarcophytol-A acetate (**179**) from soft coral *Sarcophyton infundibuliforme* exhibited potent antifouling activity against the larval settlement of barnacle *Balanus amphitrite* at nontoxic concentration with EC50 values of 1.75 μg/mL [91]. Briarane diterpenoids; gemmacolides A (**180**), B (**181**), and D (**182**); and juncin ZII (**183**), from the South China Sea gorgonian *Junceella juncea*, had potent antifouling activities at nontoxic concentrations with EC_{50} values of 0.004, 0.005, 2.82, and 0.447 μg mL − 1, respectively, toward barnacle *B. amphitrite* larvae [92].

Sinulariol J (**184**) from the Chinese soft coral *Sinularia rigida* significantly inhibited the larval settlement of *B. amphitrite* with EC50 of 5.65 μg mL − 1 [93]. Anthogonoid A (**185**), antsimplexin A (**186**), and klysimplexin G (**187**) from the Beibu Gulf gorgonian *Anthogorgia caerulea* have antilarval (*B. amphitrite*) settlement activity with ED_{50} values of 0.56, 5.28, and 2.83 μg mL − 1, respectively, and their LD_{50} values are larger than 200 μg mL − 1 [91]. New unusual cholestane derivatives from the South China Sea gorgonian *Subergorgia suberosa*, pentacyclic steroid 16,22-epoxy-20β,23*S*-dihydroxycholest-1-ene-3-one (**188**), exhibited inhibitory effect with EC_{50} values of 5.3 μg/mL, respectively [94]. Junceellolide A (**189**) and junceellonoid D (**190**) from the South China Sea gorgonian *Junceella fragilis* had potent antifouling activities at nontoxic concentrations with EC_{50} values of 5.6 and 10 μM, respectively [95].

Capillosanane A (**191**) from the soft coral *Sinularia capillosa* exhibited potent antifouling activity against *Balanus amphitrite*, with an IC_{50} value of 9.70 μM [96]. Pavidolides C (**172**) and D (**192**) from the soft coral *Sinularia pavida* exhibited moderate inhibition against the larval settlement of barnacle *Balanus amphitrite* with ED_{50} of 4.32 and 2.12 μg/ml and low cytotoxicity (LD50 > 50 μg/mL) [86]. Isogosterones A–D (**7–10**) from an octocoral *Dendronephthya* sp. inhibited the settlement of *B. amphitrite* cyprid larvae with an EC_{50} value of 2.2 μg/mL [4]. 1,7-Guaiazulenequinone (**193**) and ketolactone (**194**) from a Chinese gorgonian *Anthogorgia* species showed significant inhibition against the larval settlement of barnacle *B. amphitrite* with EC_{50} < 7.0 μg/mL [97].

(−)-6α-Hydroxy polyanthellin A (**195**) from the soft coral *Cladiella krempfi* showed activity against the cyprids of the fouling barnacle, *Balanus amphitrite* (EC_{50} 9.02 μg/mL, LC_{50} 36 μg/mL, and therapeutic ratio = 4) [98]. The EC_{50} values of subergorgic acid (**196**) from *Subergorgia suberosa* against *B. amphitrite* larvae were 1.2/3.2 μg mL − 1, respectively, while the LC_{50} of both compounds exceeded 200 μg/mL [99]. A sterol nephthoacetal (**197**) from soft coral *Nephthea* sp. exhibited significant inhibitory effect against *Bugula neritina* larvae with EC50 value of 2.5 μg/mL while having low toxicity with LC_{50} > 25.0 μg/mL [100] (Fig. 16.7).

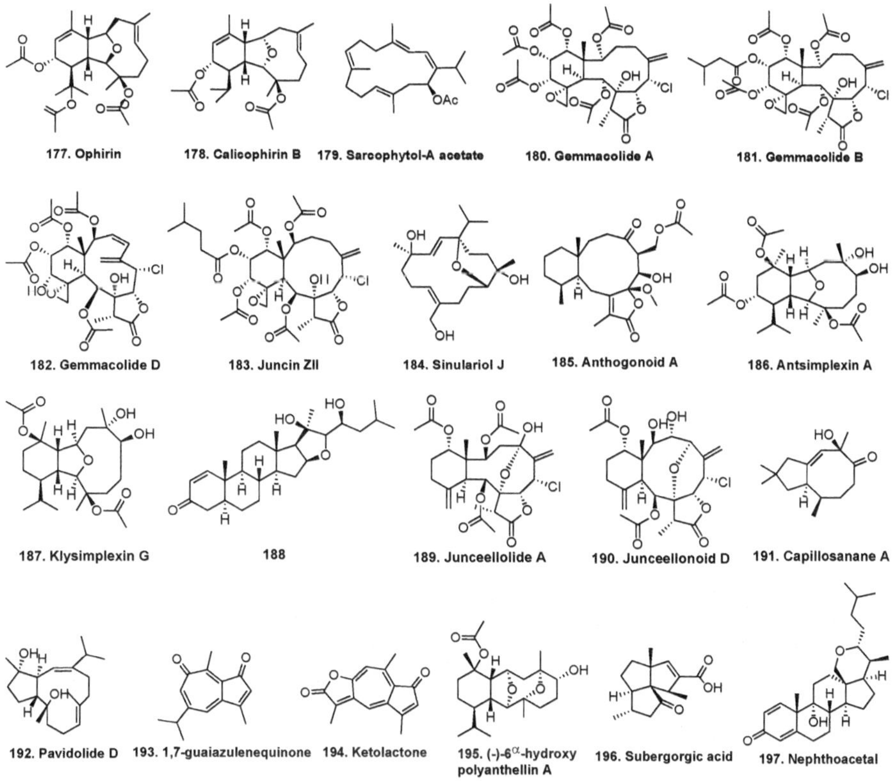

Fig. 16.7 The antifouling compounds isolated from corals

16.3.5 *Antifeedant Activity*

Hydroxycolorenone (**198**) from the Indonesian soft coral *Nephthea chabrolii* inhibited the growth of the larvae of *Spodoptera littoralis* by causing 94% inhibition of larval growth at a concentration at 530 ppm, resulting the EC_{50} and LC_{50} as 8.8 [±0.26 SE] ppm and 453 [±0.43 SE] ppm, respectively [101]. Hicksoanes A–C (**199**–**201**) from the gorgonian *Subergorgia hicksoni* showed antifeeding activity against goldfish at natural concentration (10 μg mL–1) [102].

Ainigmaptilone A (**202**) from the Antarctic gorgonian coral *Ainigmaptilon antarcticus* was evaluated in antipredator assays and found to significantly inhibit predation by *Odontaster validus* [103]. Two seco-steroids 9,11-secogorgosterol (**203**) and 9,11-secodinosterol (**204**) from *Pseudopterogorgia americana* were not deterrent to reef fishes separately but were deterrent in combination, suggesting a synergistic effect [104]. A sesquiterpenoid heterogorgiolide (**205**) and a citerpenoid (6*E*)-2α,9α-epoxyeunicella-6,11(12)-dien-3β-ol (**206**) from the Brazilian gorgonian *Heterogorgia uatumani*, at natural volumetric concentrations (0.5 and 0.8 mg/cm^3), significantly inhibited feeding relative to controls with high statistical significance [105] (Fig. 16.8).

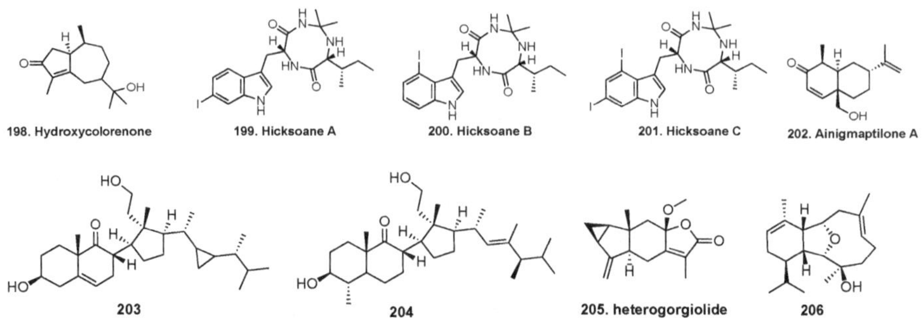

Fig. 16.8 The antifeedant compounds isolated from corals

16.3.6 Other Bioactivities

The tetrahydroxy gorgostane gorgost-5-ene-3β,9α,11α-triol (**207**) from the Caribbean octocoral *Eunicea laciniata* was one of the most potent compounds in a screening using a LXR-SPA binding assay. It inhibited the binding activities of LXRα and LXR-β with IC_{50} values of 0.07 and 0.2 μM, respectively, and was a potent stimulator of LXRα in the HTRF assay (EC50 = 0.05 μM) but did not exhibit any stimulation of LXRβ at 10 μM despite showing good binding activity (IC_{50} = 0.2 μM) [106]. Malonganenones L (**208**) from the South China Sea gorgonian *Echinogorgia pseudossapo* showed moderate inhibitory activity against phosphodiesterases PDE4D with IC_{50} value of 8.5 μM [107].

Shagene A (**50**) from an undescribed genus of octocoral demonstrates 5 μM IC_{50} of the infected macrophage stage of the visceral leishmaniasis causing parasite [27]. Pachycladins A (**209**) and D (**210**), sclerophytin A (**211**), polyanthelin A (**212**), and 3,6-diacetylcladiellisin (**213**) from the Red Sea soft coral *Cladiella pachyclados* showed potent antimigratory activity, comparable to the positive control of 4-hydroxyphenylmethylene hydantoin (PMH) [108]. Aberrarone (**22**) and colombiasin A (**39**) from the Caribbean Sea whip, *Pseudopterogorgia elisabethae*, showed antimalarial *Plasmodium falciparum* (W-2 chloroquine-resistant strain) activity (IC_{50} = 10 μg/mL) [12].

Cycloaplysinopsin C (**214**), a bis(indole) alkaloid isolated from *Tubastrea* sp., inhibited growth of two strains of *Plasmodium falciparum*, one chloroquine-sensitive (F32/Tanzania) and other chloroquine-resistant (FcB1/Colombia) with IC_{50} of 1.48 and 1.2 μg/mL, respectively [109]. Bielschowskysin (**14**) isolated from the Caribbean gorgonian *Pseudopterogorgia kallos* exhibited antiplasmodial activity (IC_{50} = 10 μg/mL) against *Plasmodium falciparum* [7]. Corallolide A (**26**) from the Caribbean gorgonian *Pseudopterogorgia bipinnata* displayed mild antimalarial activity (IC_{50} = 10 μg/mL) against malaria parasite *Plasmodium falciparum* [14]. Cristaxenicin A (**34**) from the deep sea gorgonian *Acanthoprimnoa cristata* showed antiprotozoal activities against *Leishmania amazonensis* and *Trypanosoma congolense* with IC_{50} values of 0.088 and 0.25 μM, respectively [18].

Dolabellanone 9 (**215**) from a Colombian gorgonian of the genus *Eunicea* showed antimalarial activity against the protozoan parasite *Plasmodium falciparum* with IC_{50} = 9.4 μM [110]. Pseudopterosin V (**216**) from the sea whip *Pseudopterogorgia elisabethae* showed the most potent in vitro antimalarial activity, with an IC_{50} of 1 μg/mL against the *Plasmodium falciparum* W2 (chloroquine-resistant) strain [111]. A novel briarane-type diterpene named brianthein A (**217**) isolated from the gorgonian *Briareum excavatum* reversed multidrug resistance (MDR) in human carcinoma cell lines, KB-C2, overexpressing P-glycoprotein (P-gp) [112]. Lobohedleolide (**218**), (7*Z*)-lobohedleolide (**219**), and a new compound, 17-dimethylaminolobohedleolide (**220**), isolated from the soft coral *Lobophytum* sp. exhibited moderate HIV-inhibitory activity (EC50 approximately 3–5 μg/mL) in a cell-based in vitro anti-HIV assay [113]. The NCI's CEM-SS cell line screen is designed to detect gents acting at any stage in the HIV virus reproductive cycle, and rietone (**221**) from the South African soft coral *Alcyonium fauri* showed moderate activity (EC_{50} 1.23/~molar and IC_{50} 9.32/molar) in this bioassay [114] (Fig. 16.9).

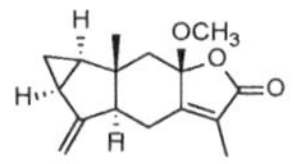

207. Gorgost-5-ene-3β,9α,11β-triol

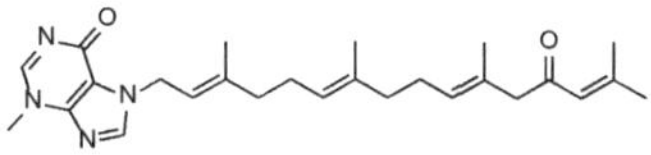

208. Malonganenones L

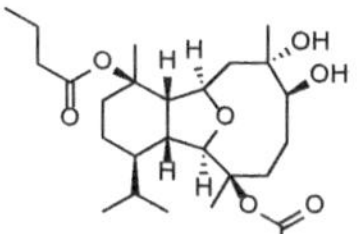

209. Pachycladins A

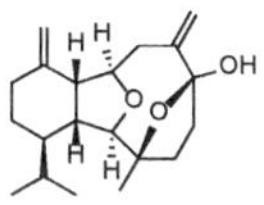

210. Pachycladin D

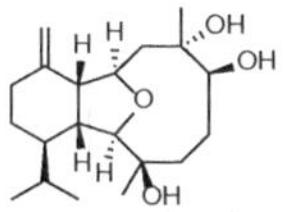

211. Sclerophytin A

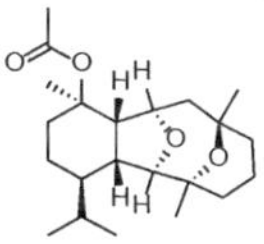

212. Polyanthelin A

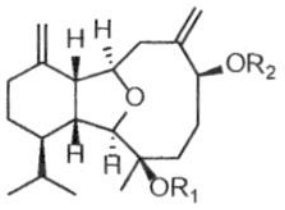

213. 3,6-diacetylcladiellisin

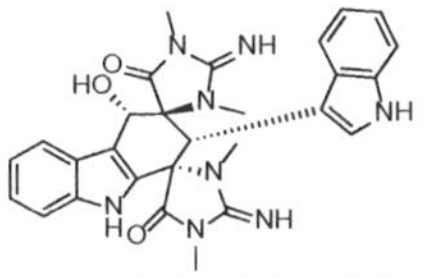

214. Cycloaplysinopsin C

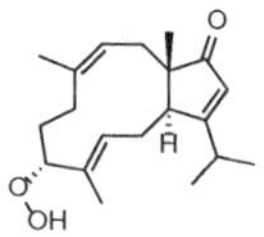

215. Dolabellanone 9

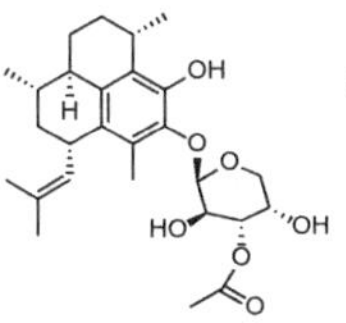

216. pseudopterosin V

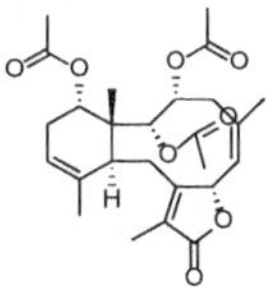

217. Brianthein A

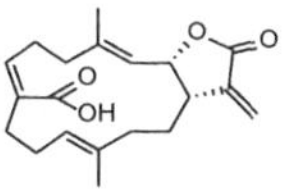

218. Lobohedleolide

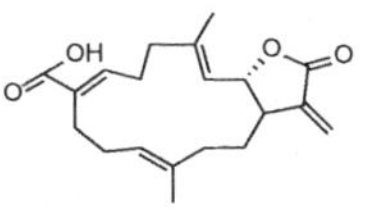

219. (7Z)-lobohedleolide

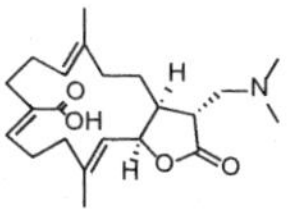

220. 17-dimethylaminolobohedleolide

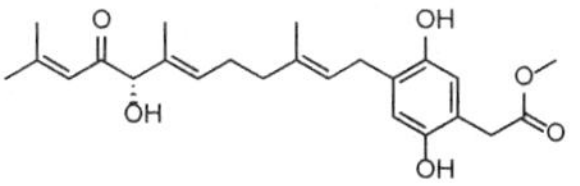

221. Rietone

Fig. 16.9 Other bioactive compounds isolated from corals

16.4 Some Total Synthesis of Coral-Derived Novel Compounds

The pseudopterosin diterpene glycosides have promising anti-inflammatory and analgesic properties. A general synthetic route toward these compounds has accomplished a number of amphilectane and serrulatane, including pseudopterosin A (**84**), pseudopteroxazole (**45**), seco-pseudopterosin A aglycone, erogorgiaene, pseudopterosins G–J aglycone, amphilectolide, and caribenol A (**32**). The key step was a stereoselective Cope rearrangement either promoted by gold catalysis or thermal conditions [115].

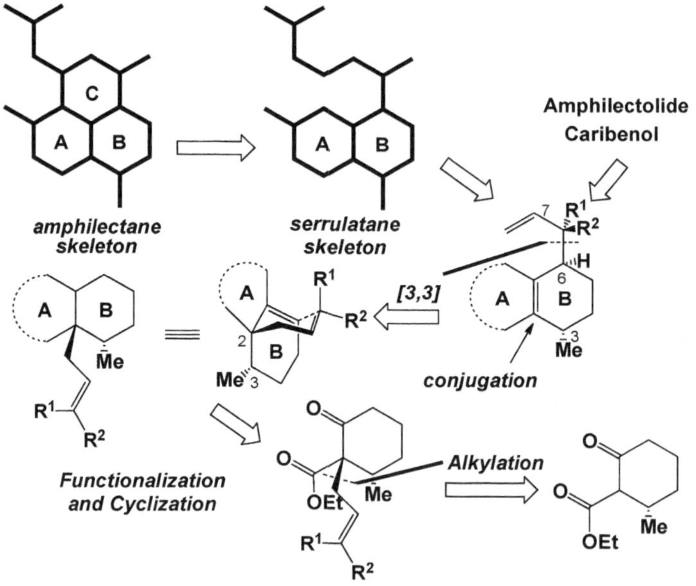

Ileabethoxazole is derived from the serrulatane diterpenoid skeleton by incorporating a subsequent cyclization for the fused cyclopentane ring. *Williams* et al. [116] have specifically focused on the chemistry of ileabethoxazole using a Stille cross-coupling reaction of propargylic stannanes with 5-iodo-1,3-oxazoles to produce 1,1-disubstituted allenes (**11**). An iron-mediated [2 + 2 + 1] carbocyclization yields a novel cyclopentenone for elaboration to **1**.

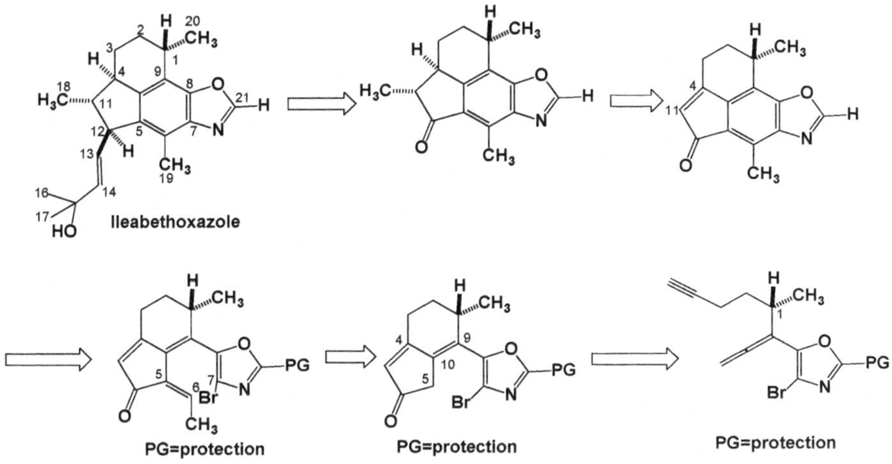

The eunicellin diterpenes (cladiellins) are isolated from a variety of soft corals. Since the isolation of the first member of this class, over 60 additional members have been reported. A challenge to all of the eunicellin diterpenes is the hydroisobenzofuran core [117].

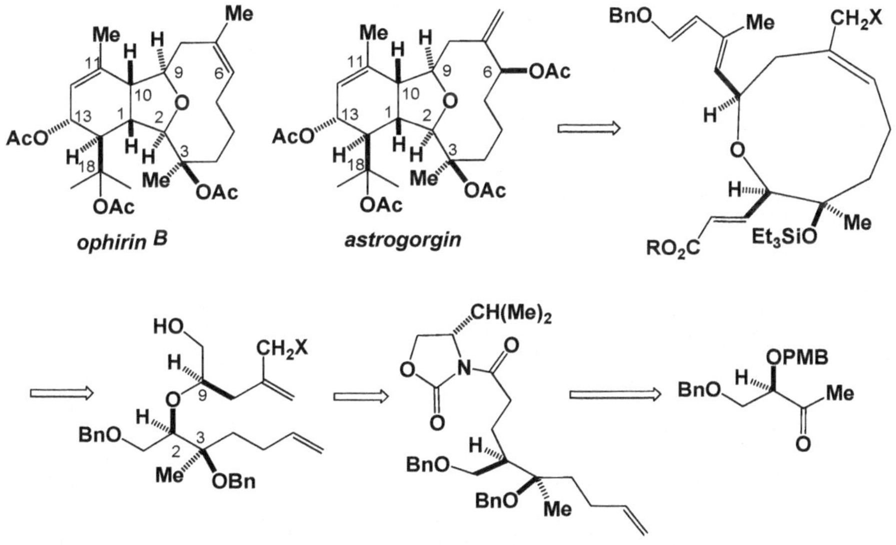

The structurally unusual and compact marine natural product antheliolide A has long remained a formidable challenge to synthetic chemists since the isolation from *Anthelia glauca*. The framework of four-, five-, six-, and nine-membered rings and the particular arrangement of functionality and multiple embedded stereocenters sharply limit the range of chemical reactions that are applicable to the synthesis. Mushti et al. [118] reported the development of a logical synthetic pathway to antheliolide A that efficiently assembles this acetoacetate-diterpenoid composite with excellent stereocontrol.

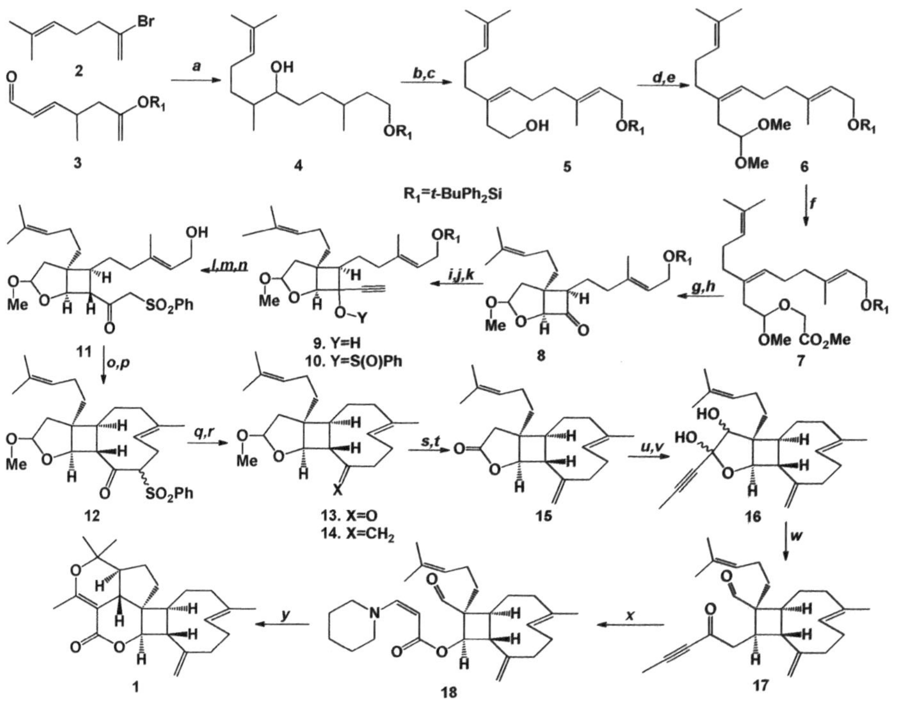

Since clavulone I, about 40 related compounds were isolated from soft coral *Clavularia viridis*. Many of these compounds have antitumor activities. As for the tricycloclavulone, the stereochemistry of the carbon bearing the acetoxy group on the α-chain has not yet been examined. Kazuo et al. [119] firstly finished total synthesis of (+)-tricycloclavulone including a catalytic enantioselective [2 + 2]-cycloaddition reaction using novel chiral copper catalyst.

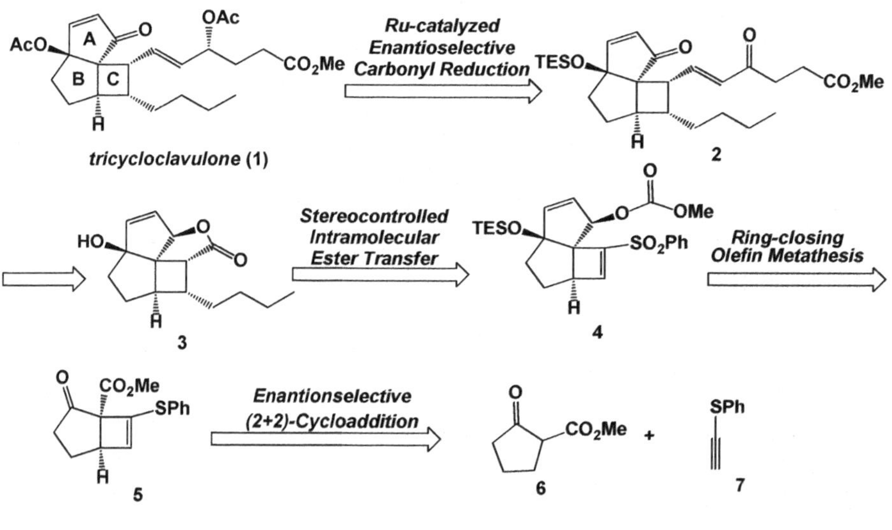

Among the most exciting new tubulin polymerization and microtubule-stabilizing agents are eleutherobin, eleuthosides A and B, and sarcodictyins A and B. Sarcodictyins A and B are rare natural substances originally found by Pietra and his group in the Mediterranean stoloniferan coral *Sarcodictyon roseum*. Their potent antitumor properties and Taxol-like mechanism of action were reported in 1997 by the Pharmacia & Upjohn group. *Nicolaou* et al. [120] reported the two related approaches to these molecules, both utilizing (+)-carvone as starting material.

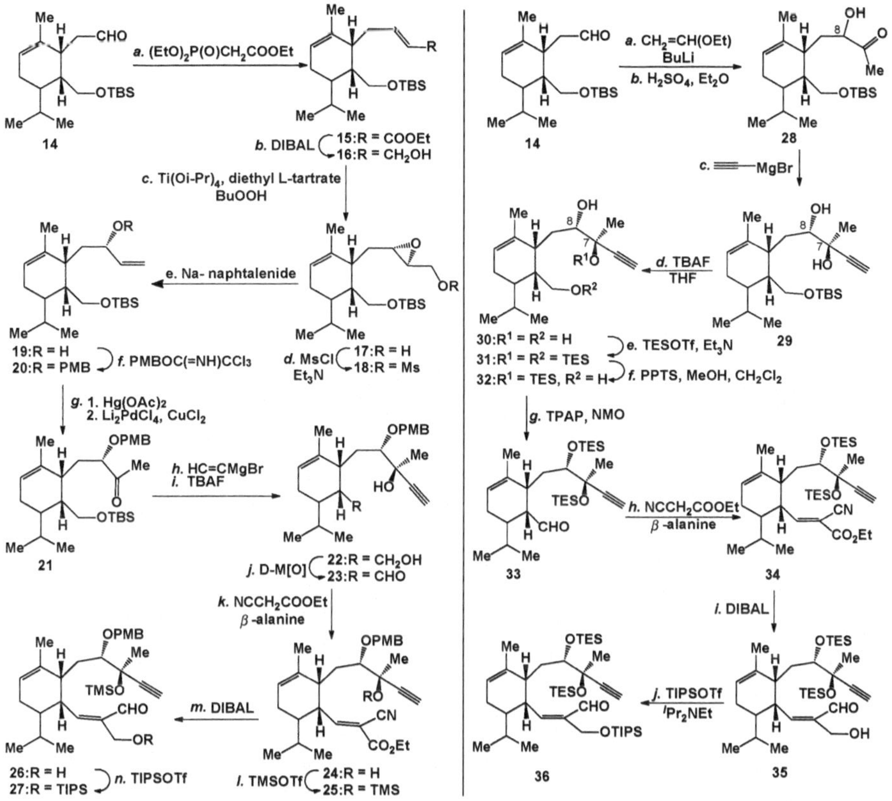

Later, *Nicolaou* et al. [121] reported the first total synthesis of sarcodictyins A and B.

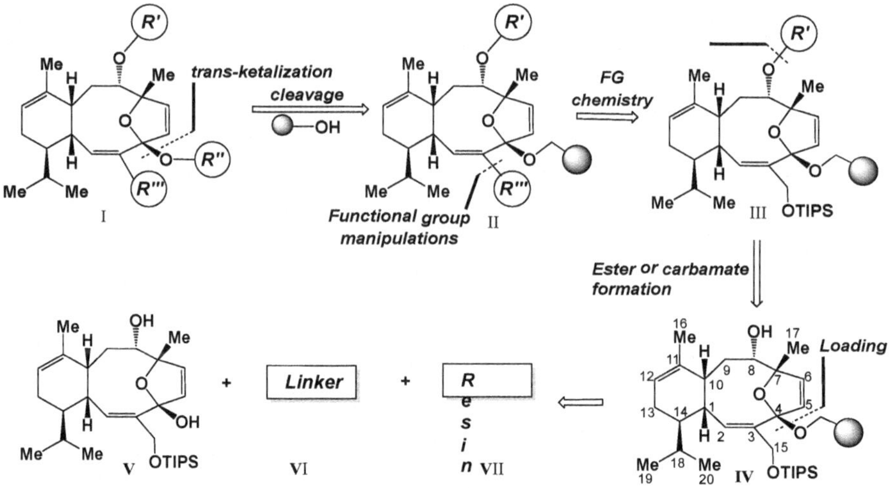

The diverse family of diterpenes derived biosynthetically from (+)-elisabethatriene, such as (−)-colombiasin A, (−)-elisapterosin B, and (+)-erogorgiaene. All these compounds have three stereocenters. These three stereocenters have represented considerable challenges for a synthetic work. *Davies* et al. [22] described a C−H functionalization strategy that has the potential to the stereochemical issues.

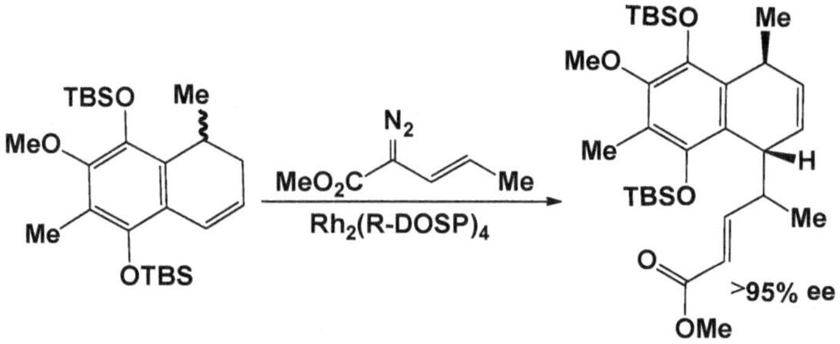

Wang et al. [122] reported enantioselective synthesis of the cytotoxic (−)-sclerophytin A, involving a stereoselective Oshima-Utimoto reaction, a Shibata-Baba indium-promoted radical cyclization, and a novel stereoconvergent epoxide hydrolysis.

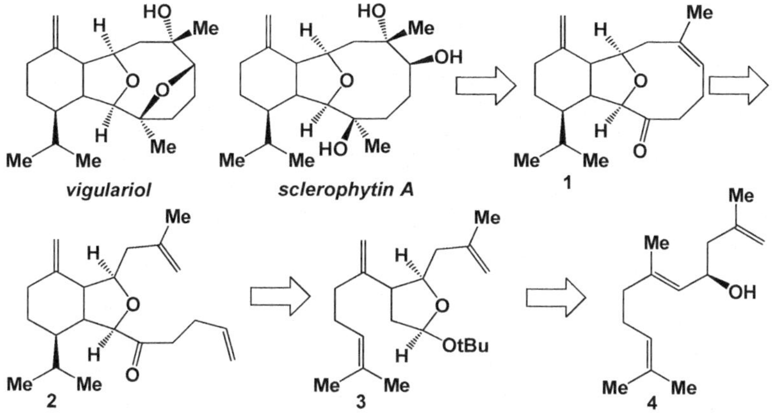

Davoren et al. [123] reported the first total synthesis of solandelactone E by a novel and convergent strategy that required 23 steps. The synthesis features a diastereoselective acetal-directed cyclopropanation of an electron-deficient diene, a sharpless asymmetric dihydroxylation, and a [2,3]-sigmatropic rearrangement of a selenoxide intermediate.

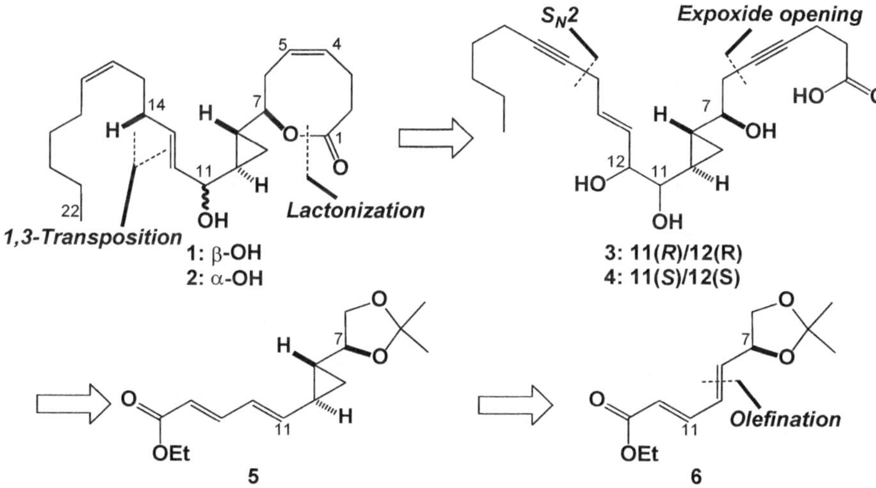

Clavirolide C, a member of the dolabellane diterpenes isolated from the Pacific soft coral *Clavularia viridis*, contains a characteristic trans-bicyclo[9.3.0]tetradecane core. *Brown* et al. [124] showed their interest in this natural product stems from the viewpoint that an efficient and enantioselective total synthesis of clavirolide C.

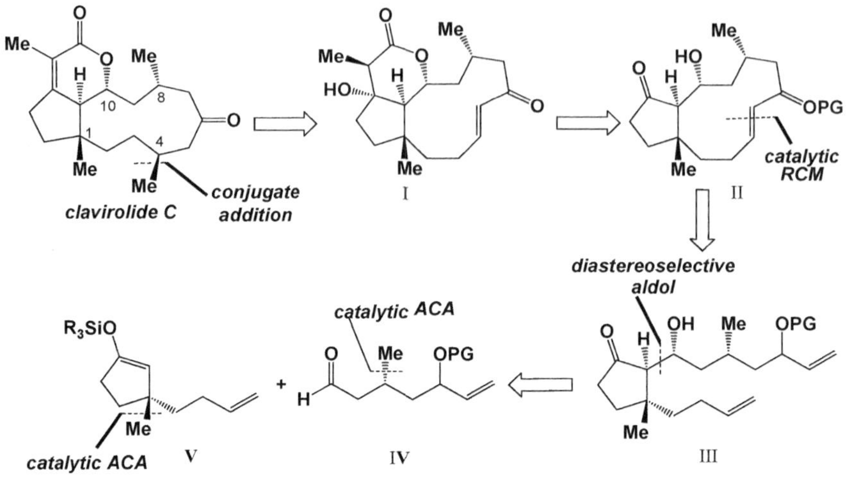

Kündig et al. [125] reported the synthesis work of both enantiomers of acetoxy-tubipofuran using enantioselective and diastereoselective dearomatization sequences starting from the benzaldehyde chromium tricarbonyl complex.

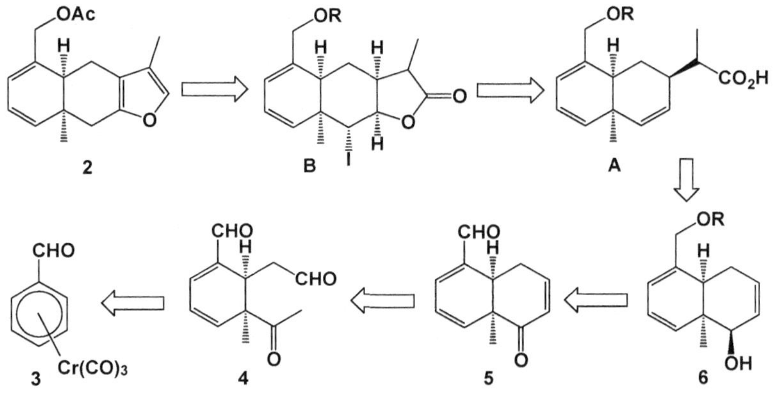

Crimmins et al. [126] reported a highly stereoselective synthesis of 11-acetoxy-4-deoxyasbestinin D in 26 linear steps.

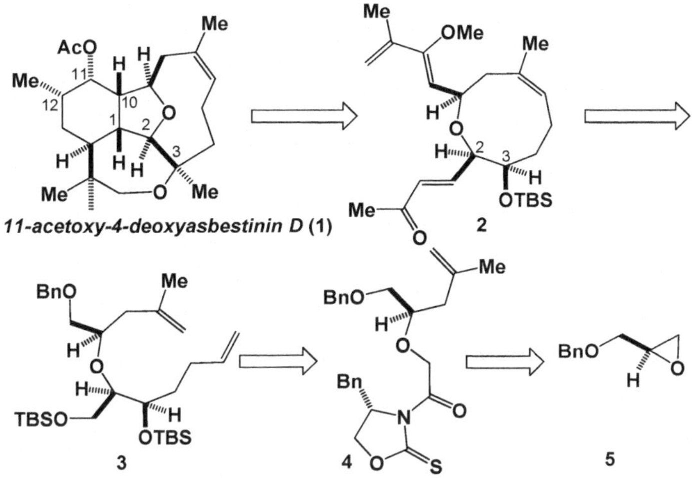

Ichige et al. [127] reported the first asymmetric total synthesis of methyl sarcophytoate, a biscembranoid marine natural product, which was achieved by the thermal Diels-Alder reaction.

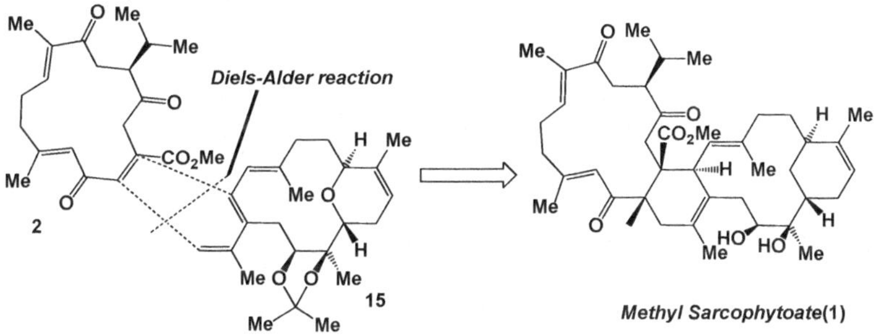

Nicolaou et al. [128] reported total synthesis of eleutherobin and eleuthosides A and B.

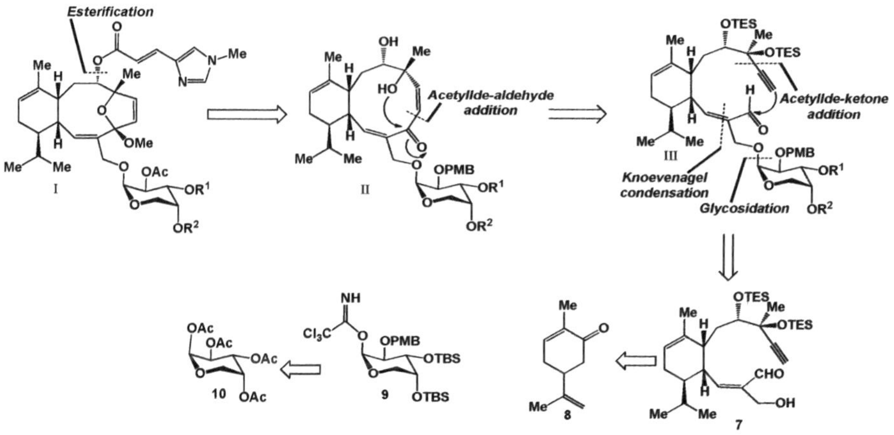

Heckrodt et al. [129] reported total synthesis of elisabethin A by an intramolecular Diels-Alder reaction under biomimetic conditions.

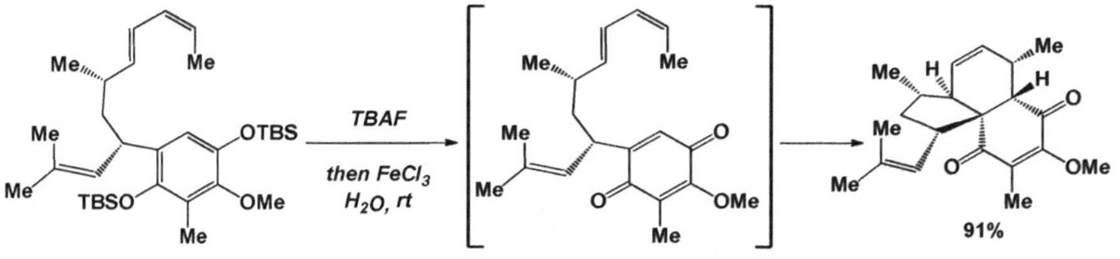

16.5 Coral-Derived Compounds in Preclinical and Clinical Development

The preclinical drug pipeline plays an important role and currently supplying hundreds of novel marine natural compounds with determined mechanisms of action every year, and continue to support the clinical pipeline with potentially valuable compounds. There are currently about 35 coral-derived compounds in preclinical development for anticancer, anti-inflammatory, antibacterial, and antifeedant, with one in phase II status form. The following section will provide detailed information about these compounds [130].

Sarcodictyins A (**147**) and B (**222**) isolated from the Mediterranean corals *Sarcodictyon roseum* and *Eleutherobia aurea* by Pietra's group were originally reported as no biological activity, and then their activity versus tubulin was reported by a group from Pharmacia & Upjohn in 1997 [131]. The synthetic sarcodictyin (**223**) was compared with Taxol and epothilones A and B. The synthetic one (**223**) exhibited polymerizing properties of 46% (for comparison: Taxol, 65%; epothilone A, 72%; epothilone B, 97%). Cytotoxicity studies were conducted with the parental ovarian carcinoma cell line 1A9, and the Taxol-resistant tumor cell lines PTX10 and PTX22 derived from 1A9. **223** showed IC_{50} values of less 6.0 nM against the three cell lines. The apparent inconsistency between tubulin polymerization activity and cytotoxicity for a number of the synthetic compounds may be indicative of the availability of an additional mechanism of action in some cases, which showed the sarcodictyins to be a new class of potential anticancer agents [132].

Eleutherobin (**11**) isolated from a rare alcyonacean, *Eleutherobia* species (possibly *E. albiflora*), possessed significant cytotoxicity against a wide variety of cancer cells with an IC_{50} range of 10–15 nM in vitro. More importantly, eleutherobin was found to stabilize microtubules by competing for the paclitaxel (Taxol) binding site on the microtubule polymer. It was the second molecule to possess the unique microtubule-stabilizing properties of paclitaxel [5]. Cinel et al. (2001) reported that the Ojima pharmacophore proposal implies that changes in the C-11–C-13 regions of eleutherobin (**143–146**) should have an impact on the ability of analogues to stabilize tubulin polymers. The structural changes in caribaeoside alter both the shape and polarity of the diterpene core in the tubulin binding region B of the proposed pharmacophore [133]. In a brief communication, *Britton* et al. [134] completed the first detailed investigation of synthetic transformations of eleutherobin. In a cell-based antimitotic assay, 2′,3′-dihydroeleutherobin (**224**) was 1000-fold less active than eleutherobin.

The proposed mechanism of punaglandins (**225–229**) isolated from the soft coral *Telesto riisei* was reported by *Baker* et al. [135]. The punaglandins can inhibit isopeptidase activity and exhibit antiproliferative effects more potently than A and J series prostaglandins. The punaglandins caused cell death with equal potency and efficacy in RKO cells (competent p53) and RKO-E6 (disrupted p53) cell lines. These data further substantiate that the cytotoxicity mechanism of the punaglandins is p53-independent [136].

Huang et al. [137] investigated the molecular pharmacology of the clavulone II (**230**) derived from the Japanese soft coral *Clavularia viridis*. Working with a human acute promyelocytic leukemia, clavulone II induced downregulation of cyclin D1 expression and G1 arrest of the cell cycle at lower concentrations (1.5 μM), while at higher concentrations (3 μM), clavulone II induced apoptosis with concomitant modulation of caspases and Bcl-2 family proteins.

Two preclinical studies were described for bromovulone III (**231**) isolated from the soft coral *Clavularia viridis* [138]. In 2005, *Chiang* et al. [139] reported that bromovulone III induced apoptosis in hepatocellular carcinoma cells. Then in 2006, this same research group reported that bromovulone III induced apoptosis in human hormone-resistant prostate cancer cells [140].

A dihydroxycapnellene, capnell-9(12)-ene-8b,10a-diol (**232**) from the soft coral *Dendronephthya rubeola*, showed a good antiproliferative activity against the cell line L-929 (murine fibroblasts) and a good cytotoxic activity against the HeLa (human cervix carcinoma) cell line with GI_{50} and CC_{50} values of less 10 μM [84]. *Posadas* et al. [141] reported on the mechanism of action of cavernolide (**233**) from the sponge *Fasciospongia cavernosa*, indicating that cavernolide's potent inhibition of tumor necrosis factor-a, nitric oxide, and prostaglandin E2 in vitro was the result of both human synovial phospholipase A2 (IC_{50} = 8.8 μM) inhibition and inhibition of inducible nitric oxide synthase (iNOS) and cyclooxygenase-2 (COX-2) gene expression in intact cells. Natural sinulamide (**234**) from a soft coral *Sinularia* sp. not only inhibited H,K-ATPase with an IC_{50} value of 5.5 μM but also was cytotoxic against L1210 and P388 with IC_{50} values of 3.1 and 4.5 μg/mL, respectively. It is the first acylated spermine derivative of soft coral origin; however, acylated spermidines are known from soft corals of the same genus [142].

Sinulariolide (**235**) and flexibilide (**236**) from the soft coral *Sinularia flexibilis* showed marked antimicrobial activity and inhibited growth of Gram-positive bacteria, but they were generally inactive against Gram-negative bacteria, yeast, and fungi. Because the cell walls of the fungi and yeast would be the least permeable to diffusion of terpenoid substances. Flexibilide was effective even at concentrations as low as 5 ppm, whereas sinulariolide was effective at concentrations of 10 ppm. These compounds show potential as antibiotics [143]. Lobane diterpene (**237**) of the soft coral *Lobophytum pauciflorum* from Mindoro, Philippines, was found to be active against the Gram-positive *B. subtilis* and the yeast *S. cerevisiae*, causing an inhibition zone of 8 and 7 mm in diameter, respectively. No inhibition, however, was observed for *S. aureus* and *E. coli*. It is remarkable that the observed antibacterial activity of **237** is obviously not caused by general toxicity but is rather due to different modes of action and specific target requirements, which are apparently strongly influenced by the chemical structure of the studied compound [144].

During the preclinical pharmacological research of anti-inflammatory marine compounds, Chao et al. [145] identified the new crassumolides A and C (**238** and **239**) from the soft coral *Lobophytum crassum*. Similarly, Cheng et al. [146] found that new compounds from the soft coral *Lobophytum duru*, durumolides A–C (**240–242**), inhibited both the iNOS and COX-2 proteins in LPS-activated RAW 264.7 cells in vitro (apparent IC_{50} less than 10 μM). Shen et al. [147] showed that frajunolides B and C (**243** and **244**), isolated from the Taiwanese gorgonian *Junceella fragilis*, significantly inhibited superoxide anion and elastase generation from human neutrophils in vitro (apparent IC_{50} greater than 10 μg/mL).

The anti-inflammatory effects of the pseudopterosin A (PSA, **245**) and pseudopterosin E (PSE) from the marine gorgonian *Pseudopterogorgia elisabethae* can in part be attributed to the inhibition of eicosanoid production through a PLA_2- and cyclooxygenase-independent mechanism. The important role of prostaglandins and leukotrienes in diseases such as arthritis, asthma, and inflammatory bowel disease suggests that these novel agents may provide useful chemical templates for developing bioactive drugs with similar or improved pharmacological profiles [148]. *BALA* et al. [149] conducted an extensive bioactivity screen with the sesquiterpene africanene (**246**) from the soft coral *Sinularia leptoclados*.

Ichida et al. [150] used the whole-cell clamp and Fura-2 techniques to study the membrane current and intracellular Ca^{2+} concentration ($[Ca^{2+}]_i$) changes of mouse megakaryocytes in response to palytoxin (PTX, **247**) from *Palythoa* sp. PTX induces a nonselective cation channel in mouse megakaryocytes.

A structure-activity relationship study with stolonidiol (**248**) from the soft coral *Clavularia* sp. demonstrated potent choline acetyltransferase-inducible activity in neuronal cultures in vitro. The mechanism of in vivo analgesic and anti-inflammatory activity of the pseudopterosins results as a consequence of in vivo inhibition of eicosanoid release [151].

Sawant et al. [152] investigated both the marine cembranoid sarcophine (**249**) and a semisynthetic sulfur-containing derivative (**250**) in an in vitro anti-neuroinflammatory assay. Only sarcophine significantly inhibited both generations of superoxide anion and thromboxane B2 (IC_{50} = 1 μM) from activated rat brain macrophages.

Temraz et al. [153] noted that Red Sea soft corals *Sarcophyton glaucum* and *Lobophytum crassum* contained natural product trigonelline (**251**), which increased the electrophysiological excitability of rat cultured dorsal root ganglion neurons.

In screening marine extracts for new eukaryotic translation initiation inhibitors, *Bordeleau* et al. [154] identified the natural product hippuristanol (**252**) from the coral *Isis hippuris*. Hippuristanol was a selective and potent inhibitor of eIF4A RNA-binding activity.

And now, a coral-derived natural compound has been in phase II clinical trials as an anti-inflammatory and wound healing agent. The pseudopterosins (**59–67**) constitute a class of diterpene glycosides isolated from *Pseudopterogorgia elisabethae*. Pseudopterosins A–D were the first of a series. Pseudopterosin A (PsA, **86**) induces topical inflammation in mice, stabilizes cell membranes, prevents the release of prostaglandins and leukotrienes from zymosan-stimulated murine macrophages, and inhibits degranulation of human polymorphonuclear leukocytes and phagosome formation in *Tetrahymena* cells. Pseudopterosins A–D have been licensed to OsteoArthritis Sciences Inc. for medical use as anti-inflammatory drugs [155]. During this test, a simple C-10 O-methyl ether of PsA, methopterosin (VM301, OAS1000, **253**), has been developed from this discovery and has been approved by the US Food and Drug Administration for clinical trials in 1995. In double-blind phase II clinical trials, pseudopterosin increased reepithelization and accelerated the wound healing process [130] (Fig. 16.10).

16.6 Conclusion

Some cnidarian species, especially the soft corals and gorgonians, are promising sources of marine bioactive compounds. Soft corals and gorgonians possess extraordinarily diverse MNPs, with terpenoids the most representatives, showing the remarkable species-specific feature of the corals. There have been almost 3000 new compounds isolated from corals in the last two decades, and dozens of them had unique structures with unprecedented skeletons.

Now the lack of new compounds in the pipelines in some strategic areas (e.g., antibiotics) suggests that the natural product-based discovery program would be more reliable. The coral-derived compounds displayed wide bioactivities including antitumor, anti-inflammatory, antifouling, antifeedant, and antimicrobial activities. During the last two decades, about 170 coral-produced compounds showed signifi-

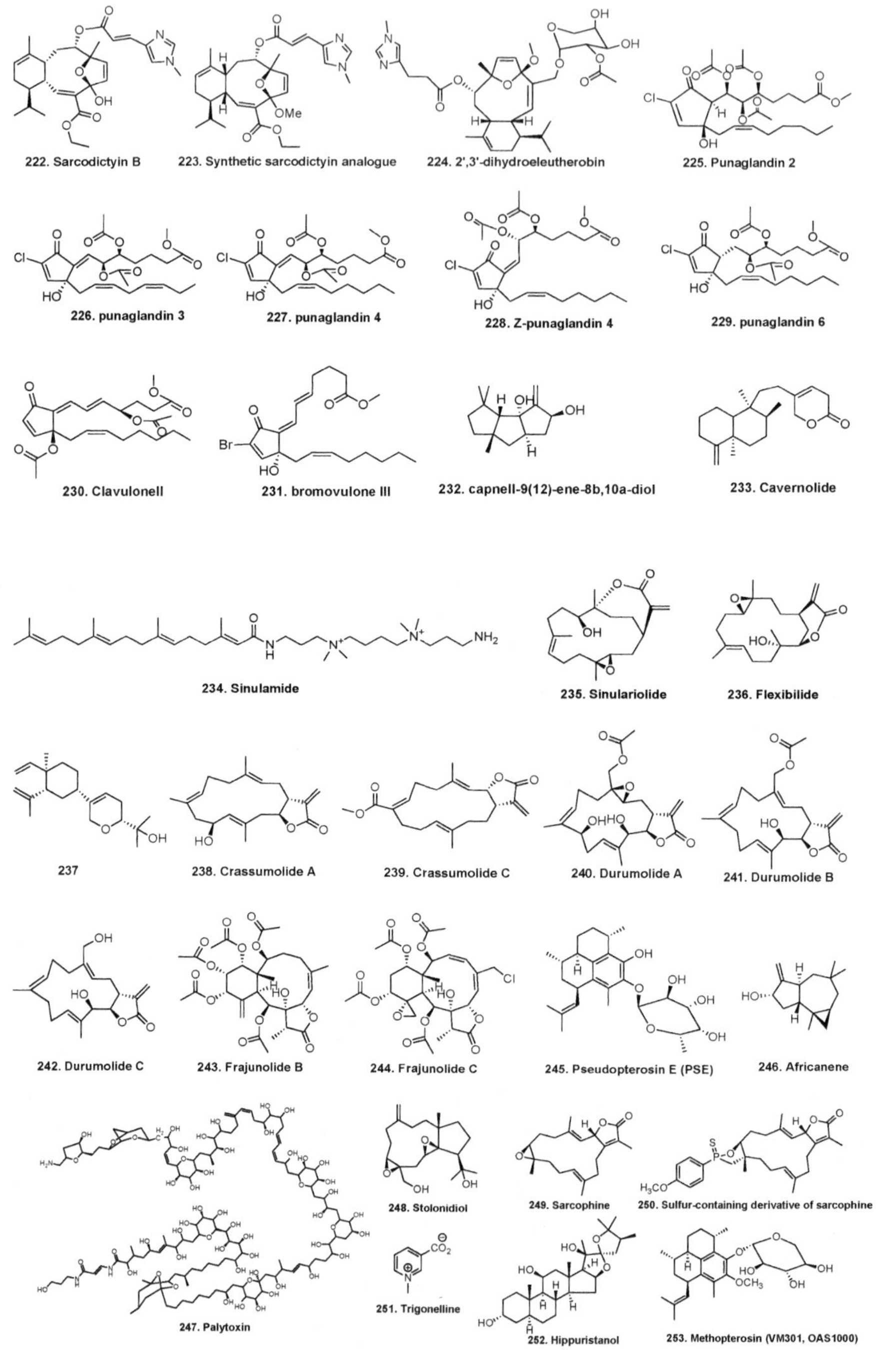

Fig. 16.10 Coral-derived compounds in preclinical and clinical study

cant bioactivities with IC_{50} less than 10 μM. And anti-inflammatory, antitumor, and antimicrobial activities showed the high biotechnological potential of MNPs from corals. There have been more than 50 coral-derived compounds in preclinical progress and 1 compound in phase II study.

References

1. Zhang GW, Ma XQ, Kurihara H, Zhang CX, Yao XS, Su JY, et al. New hemiketal steroid from the soft coral *Cladiella* sp. Org Lett. 2005;7:991–4.
2. Poza JJ, Fernández R, Reyes F, Rodríguez J, Jiménez C. Isolation, biological significance, synthesis, and cytotoxic evaluation of new natural Parathiosteroids A–C and analogues from the soft coral *Paragorgia* sp. J Org Chem. 2008;73:7978–84.
3. (a) Miyaoka H, Yamanishi M, Kajiwara Y, Yamada Y. Total synthesis of cladocorans A and B: a structural revision. J Org Chem 2003;68:3476–9. (b) Fontana A, Ciavatta ML, Cimino G. Cladocoran A and B: two novel γ-hydroxybutenolide sesterterpenes from the Mediterranean coral *Cladocora cespitosa*. J Org Chem 1998;63:2845–49.
4. Tomono Y, Hirota H, Fusetani N, Isogosterones AD. Antifouling 13, 17-Seco steroids from an Octocoral *Dendronephthya* sp. Org Lett. 1999;64:2272–5.
5. Lindel T, Jensen PR, Fenical W, Long BH, Casazza AM, Carboni J, et al. Eleutherobin, a new cytotoxin that mimics paclitaxel (Taxol) by stabilizing microtubules. J Am Chem Soc. 1997;119:8744–5.
6. Liang LF, Kurtan T, Mandi A, Yao LG, Li J, Zhang W, et al. Unprecedented diterpenoids as a PTP1B inhibitor from the Hainan soft coral *Sarcophyton trocheliophorum* Marenzeller. Org Lett. 2012;15:274–7.
7. Marrero J, Rodríguez AD, Baran P, Raptis RG, Sánchez JA, Ortega-Barria E, et al. Bielschowskysin, a gorgonian-derived biologically active diterpene with an unprecedented carbon skeleton. Org Lett. 2004;6:1661–4.
8. Rodríguez AD, Shi JG. Isolation, structure elucidation, and synthesis of bisgersolanolide, a novel heptacyclic bis-diterpenoid from the Gorgonian Octocoral Pseudopterogorgia bipinnata. Org Lett. 1999;1:337–40.
9. Taglialatela-Scafati O, Deo-Jangra U, Campbell M, Roberge M, Andersen RJ. Diterpenoids from cultured *Erythropodium caribaeorum*. Org Lett. 2002;4:4085–8.
10. Gupta P, Sharma U, Schulz TC, Sherrer ES, McLean AB, Robins AJ, et al. Bioactive diterpenoid containing a reversible "spring-loaded"(E, Z)-dieneone michael acceptor. Org Lett. 2011;13:3920–3.
11. Lin KH, Tseng YJ, Chen BW, Hwang TL, Chen HY, Dai CF, et al. Tortuosenes A and B, new diterpenoid metabolites from the Formosan soft coral *Sarcophyton tortuosum*. Org Lett. 2014;16:1314–7.
12. Rodríguez II, Rodríguez AD, Zhao H. Aberrarone: a gorgonian-derived diterpene from *Pseudopterogorgia elisabethae*. J Org Chem. 2009;74:7581–4.
13. Chill L, Rudi A, Benayahu Y, Schleyer M, Kashman Y. Kitungolides A, B, and C, new diterpenes from a soft coral of a new genus. Org Lett. 2004;6:755–8.
14. Ospina CA, Rodríguez AD. Corallolides A and B: bioactive diterpenes featuring a novel carbon skeleton. Org Lett. 2009;11:3786–9.
15. Tseng YJ, Ahmed AF, Dai CF, Chiang MY, Sheu JH. Sinulochmodins AC, three novel terpenoids from the soft coral *Sinularia lochmodes*. Org Lett. 2005;7:3813–6.
16. Marrero J, Rodríguez AD, Baran P, Raptis RG. Isolation and structure of providencin: a highly oxygenated diterpene possessing a unique bicyclo [12.2. 0] hexadecane ring system from the sea plume *Pseudopterogorgia kallos*. Org Lett. 2003;5:2551–4.

17. Wei X, Rodríguez II, Rodríguez AD, Barnes CL. Caribenols A and B, sea whip derived nor-diterpenes with novel tricarbocyclic skeletons. J Org Chem. 2007;72:7386–9.
18. Ishigami ST, Goto Y, Inoue N, Kawazu SI, Matsumoto Y, Imahara Y, et al. Cristaxenicin A, an antiprotozoal xenicane diterpenoid from the deep sea gorgonian *Acanthoprimnoa cristata*. J Org Chem. 2012;77:10962–6.
19. El-Gamal AA, Wang SK, Duh CY. Xenibellols A and B, new diterpenoids from the Formosan soft coral *Xenia umbellata*. Org Lett. 2005;7:2023–5.
20. Martín MJ, Fernandez R, Francesch A, Amade P, de Matos-Pita SS, Reyes F, et al. Plumisclerin A, a diterpene with a new skeleton from the soft coral *Plumigorgia terminosclera*. Org Lett. 2010;12:912–4.
21. Marrero J, Rodríguez AD, Barnes CL. Intricarene, an unprecedented trispiropentacyclic diterpene from the Caribbean Sea plume *Pseudopterogorgia kallos*. Org Lett. 2005;7:1877–80.
22. Davies HM, Dai X, Long MS. Combined CH activation/cope rearrangement as a strategic reaction in organic synthesis: total synthesis of (−)-colombiasin A and (−)-elisapterosin B. J Am Chem Soc. 2006;128:2485–90.
23. Rodríguez AD, Ramírez C, Rodríguez II, Barnes CL. Novel terpenoids from the west Indian Sea Whip *Pseudopterogorgia elisabethae* (Bayer). Elisapterosins A and B: rearranged diterpenes possessing an unprecedented cagelike framework1. J Org Chem. 2000;65:1390–8.
24. Tseng YJ, Wen ZH, Dai CF, Chiang MY, Sheu JH. Nanolobatolide, a new C18 metabolite from the Formosan soft coral *Sinularia nanolobata*. Org Lett. 2009;11:5030–2.
25. Rodríguez AD, Ramírez C, Rodríguez II, González E. Novel antimycobacterial benzoxazole alkaloids, from the west Indian Sea whip *Pseudopterogorgia elisabethae*. Org Lett. 1999;1:527–30.
26. Iwashima M, Terada I, Okamoto K, Iguchi K. Tricycloclavulone and clavubicyclone, novel prostanoid-related marine oxylipins, isolated from the Okinawan soft coral *Clavularia viridis*. J Org Chem. 2002;67:2977–81.
27. Cen-Pacheco F, Norte M, Fernández JJ, Daranas AH. Zoaramine, a zoanthamine-like alkaloid with a new skeleton. Org Lett. 2014;16:2880–3.
28. von Salm JL, Wilson NG, Vesely BA, Kyle DE, Cuce J, Baker BJ. Shagenes A and B, new tricyclic sesquiterpenes produced by an undescribed Antarctic octocoral. Org Lett. 2014;16:2630–3.
29. Rodriguez Brasco MF, Seldes AM, Palermo JA. Paesslerins A and B: novel tricyclic sesquiterpenoids from the soft coral *Alcyonium paessleri*. Org Lett. 2001;3:1415–7.
30. Cheng SY, Huang KJ, Wang SK, Wen ZH, Hsu CH, Dai CF, et al. New terpenoids from the soft corals *Sinularia capillosa* and *Nephthea chabroli*. Org Lett. 2009;11:4830–3.
31. Torihata M, Nakahata T, Kuwahara S. Enantioselective total synthesis of isishippuric acid B via intramolecular Michael reaction. Org Lett. 2007;9:2557–9.
32. Wei X, Nieves K, Rodríguez AD. Bioactive cubitane diterpenoids from a Colombian gorgonian species of the genus *Eunicea*. Pure Appl Chem. 2012;84:1847–55.
33. Correa H, Aristizabal F, Duque C, Kerr R. Cytotoxic and antimicrobial activity of pseudopterosins and seco-pseudopterosins isolated from the octocoral *Pseudopterogorgia elisabethae* of San Andres and Providencia islands (Southwest Caribbean Sea). Mar Drugs. 2011;9:334–44.
34. Sun P, Meng LY, Tang H, Liu BS, Li L, Yi Y, et al. Sinularosides A and B, bioactive 9, 11-secosteroidal glycosides from the South China Sea soft coral *Sinularia humilis* Ofwegen. J Nat Prod. 2012;75:1656–9.
35. Badria FA, Guirguis AN, Perovic S, Steffen R, Müller WE, Schröder HC. Sarcophytolide: a new neuroprotective compound from the soft coral *Sarcophyton glaucum*. Toxicology. 1998;131:133–43.
36. Zheng CJ, Shao CL, Chen M, Niu ZG, Zhao DL, Wang CY. Merosesquiterpenoids and ten-membered macrolides from a soft coral-derived *Lophiostoma* sp. Fungus. Chem Biodivers. 2015;12:1407–14.
37. Yan P, Lv Y, van Ofwegen L, Proksch P, Lin W. Lobophytones A – G, nw isobiscembranoids from the soft coral *Lobophytum pauciflorum*. Org Lett. 2010;12:2484–7.

38. Gomaa MN, Soliman K, Ayesh A, Abd El-Wahed A, Hamza Z, Mansour HM, et al. Antibacterial effect of the red sea soft coral *Sarcophyton trocheliophorum*. Nat Prod Res. 2016;30:729–34.
39. Zhao HY, Shao CL, Li ZY, Han L, Cao F, Wang CY. Bioactive pregnane steroids from a South China Sea gorgonian *Carijoa* sp. Molecules. 2013;18:3458–66.
40. Díaz-Marrero AR, Porras G, Aragón Z, de la Rosa JM, Dorta E, Cueto M, et al. Carijodienone from the octocoral *Carijoa multiflora*. a spiropregnane-based steroid. J Nat Prod. 2011;74:292–5.
41. Sun XP, Cao F, Shao CL, Wang M, Zhang XL, Wang CY. Antibacterial Δ1-3-Ketosteroids from the South China Sea gorgonian coral *Subergorgia rubra*. Chem Biodivers. 2015;12:1068–74.
42. Bishara A, Rudi A, Goldberg I, Benayahu Y, Kashman Y. Novaxenicins A–D and xeniolides I–K, seven new diterpenes from the soft coral *Xenia novaebrittanniae*. Tetrahedron. 2006;62:12092–7.
43. Tello E, Castellanos L, Arévalo-Ferro C, Duque C. Disruption in quorum-sensing systems and bacterial biofilm inhibition by cembranoid diterpenes isolated from the octocoral *Eunicea knighti*. J Nat Prod. 2012;75:1637–42.
44. Li C, La MP, Sun P, Kurtan T, Mandi A, Tang H, et al. Bioactive (3Z, 5E)-11, 20-epoxybriara-3, 5-dien-7, 18-olide diterpenoids from the South China Sea gorgonian *Dichotella gemmacea*. Mar Drugs. 2011;9:1403–18.
45. Look SA, Fenical W, Jacobs RS, Clardy J. The pseudopterosins: anti-inflammatory and analgesic natural products from the sea whip *Pseudopterogorgia elisabethae*. Proc Natl Acad Sci USA. 1986;83:6238–40.
46. Chao CH, Wu YC, Wen ZH, Sheu JH. Steroidal carboxylic acids from soft coral *Paraminabea acronocephala*. Mar Drugs. 2013;11:136–45.
47. Yan P, Deng Z, van Ofwegen L, Proksch P, Lin W. Lobophytones U–Z1, biscembranoids from the Chinese soft coral *Lobophytum pauciflorum*. Chem Biodivers. 2011;8:1724–34.
48. (a) Cheng SY, Wen ZH, Chiou SF, Hsu CH, Wang SK, Dai CF, et al. Durumolides A–E, anti-inflammatory and antibacterial cembranolides from the soft coral *Lobophytum durum*. Tetrahedron. 2008;64:9698–704. (b) Cheng SY, Wen ZH, Wang SK, Chiou SF, Hsu CH, Dai CF, et al. Anti-inflammatory cembranolides from the soft coral *Lobophytum durum*. Bioorg Med Chem 2009;17:3763–9.
49. Huang SY, Chen NF, Chen WF, Hung HC, Lee HP, Lin YY, et al. Sinularin from indigenous soft coral attenuates nociceptive responses and spinal neuroinflammation in carrageenan-induced inflammatory rat model. Mar Drugs. 2012;10:1899–919.
50. Reina E, Puentes C, Rojas J, García J, Ramos FA, Castellanos L, et al. Fuscoside E: a strong anti-inflammatory diterpene from Caribbean octocoral *Eunicea fusca*. Bioorg Med Chem Lett. 2011;21:5888–91.
51. Chang YC, Kuo LM, Hwang TL, Yeh J, Wen ZH, Fang LS, et al. Pinnisterols A–C, new 9, 11-secosterols from a Gorgonian *Pinnigorgia* sp. Mar Drugs. 2016;14:12.
52. Su YD, Wen ZH, Wu YC, Fang LS, Chen YH, Chang YC, et al. Briarenolides M–T, new briarane diterpenoids from a Formosan octocoral *Briareum* sp. Tetrahedron. 2016;72:944–51.
53. Su YD, Wu TY, Wen ZH, Su CC, Chen YH, Chang YC, et al. Briarenolides U–Y, new anti-inflammatory briarane diterpenoids from an Octocoral *Briareum* sp.(Briareidae). Mar Drugs. 2015;13:7138–49.
54. Lin WJ, Wu TY, Su TR, Wen ZH, Chen JJ, Fang LS, et al. Terpenoids from the octocoral *Sinularia gaweli*. Int J Mol Sci. 2015;16:19508–17.
55. Huang CY, Sung PJ, Uvarani C, Su JH, Lu MC, Hwang TL, et al. Glaucumolides A and B, biscembranoids with new structural type from a cultured soft coral *Sarcophyton glaucum*. Sci Rep. 2015;5:15624.
56. Chao CH, Huang TZ, Wu CY, Chen BW, Huang CY, Hwang TL, et al. Steroidal and α-tocopherylhydroquinone glycosides from two soft corals *Cladiella hirsuta* and *Sinularia nanolobata*. RSC Adv. 2015;5:74256–62.
57. Tsai CR, Huang CY, Chen BW, Tsai YY, Shih SP, Hwang TL, et al. New bioactive steroids from the soft coral *Klyxum flaccidum*. RSC Adv. 2015;5:12546–54.

58. Lee YN, Tai CJ, Hwang TL, Sheu JH. Krempfielins N–P, new anti-inflammatory eunicellins from a Taiwanese soft coral *Cladiella krempfi*. Mar Drugs. 2014;12:1148–56.
59. Thao NP, Luyen BTT, Ngan NTT, Song SB, Cuong NX, Nam NH, et al. New anti-inflammatory cembranoid diterpenoids from the Vietnamese soft coral *Lobophytum crassum*. Bioorg Med Chem Lett. 2014;24:228–32.
60. Thao NP, Luyen BTT, Sun YN, Song SB, Van Thanh N, Cuong NX, et al. NF-κB inhibitory activity of polyoxygenated steroids from the Vietnamese soft coral *Sarcophyton pauciplicatum*. Bioorg Med Chem Lett. 2014;24:2834–8.
61. Tseng YJ, Shen KP, Lin HL, Huang CY, Dai CF, Sheu JH. Lochmolins A–G, new sesquiterpenoids from the soft coral *Sinularia lochmodes*. Mar Drugs. 2012;10:1572–81.
62. (a) Lin WY, Lu Y, Su JH, Wen ZH, Dai CF, Kuo YH, et al. Bioactive cembranoids from the Dongsha atoll soft coral *Sarcophyton crassocaule*. Mar Drugs. 2011;9:994–1006. (b) Lin WY, Su JH, Lu Y, Wen ZH, Dai CF, Kuo YH, et al. Cytotoxic and anti-inflammatory cembranoids from the Dongsha atoll soft coral *Sarcophyton crassocaule*. Bioorg Med Chem. 2010;18:1936–41. (c) Lin WY, Lu Y, Chen BW, Huang CY, Su JH, Wen ZH, et al. Sarcocrassocolides M–O, bioactive cembranoids from the Dongsha Atoll soft coral *Sarcophyton crassocaule*. Mar Drugs. 2012;10:617–26.
63. Chung HM, Hong PH, Su JH, Hwang TL, Lu MC, Fang LS, et al. Bioactive compounds from a gorgonian coral *Echinomuricea* sp.(Plexauridae). Mar Drugs. 2012;10:1169–79.
64. Marchbank DH, Berrue F, Kerr RG. Eunicidiol, an anti-inflammatory dilophol diterpene from *Eunicea fusca*. J Nat Prod. 2012;75:1289–93.
65. Chung HM, Su JH, Hwang TL, Li JJ, Chen JJ, Chen YH, et al. Rumphellclovanes C–E, new clovane-type sesquiterpenoids from the gorgonian coral *Rumphella antipathies*. Tetrahedron. 2013;69:2740–4.
66. Lin CY, Lu MC, Su JH, Chu CL, Shiuan D, Weng CF, et al. Immunomodulatory effect of marine cembrane-type diterpenoids on dendritic cells. Mar Drugs. 2013;11:1336–50.
67. Chen BW, Chao CH, Su JH, Tsai CW, Wang WH, Wen ZH, et al. Klysimplexins I–T, eunicellin-based diterpenoids from the cultured soft coral *Klyxum simplex*. Org Biomol Chem. 2011;9:834–44.
68. (a) Cinel B, Roberge M, Behrisch H, van Ofwegen L, Castro CB, Andersen RJ. Antimitotic diterpenes from Erythropodium caribaeorum test pharmacophore models for microtubule stabilization. Org Lett 2000;2:257–60. (b) Roberge M, Cinel B, Anderson HJ, Lim L, Jiang X, Xu L, et al. Cell-based screen for antimitotic agents and identification of analogues of rhizoxin, eleutherobin, and paclitaxel in natural extracts. Cancer Res. 2000;60:5052–8.
69. Hong JY, Boo HJ, Kang JI, Kim MK, Yoo ES, Hyun JW, et al. (1S, 2S, 3E, 7E, 11E)-3, 7, 11, 15-Cembratetraen-17, 2-olide, a cembrenolide diterpene from soft coral *Lobophytum* sp., inhibits growth and induces apoptosis in human colon cancer cells through reactive oxygen species generation. Biol Pharm Bull. 2012;35:1054–63.
70. Ortega MJ, Zubia E, Rodriguez S, Carballo JL, Salva J. Muricenones A and B: new degraded pregnanes from a gorgonian of the genus *Muricea*. Eur J Org Chem. 2002;19:3250–3.
71. Kao CY, Su JH, Lu MC, Hwang TL, Wang WH, Chen JJ, et al. Lobocrassins A–E: new cembrane-type diterpenoids from the soft coral *Lobophytum crassum*. Mar Drugs. 2011;9:1319–31.
72. Coval SJ, Patton RW, Petrin JM, James L, Rothofsky ML, Lin SL, et al. A cembranolide diterpene farnesyl protein transferase inhibitor from the marine soft coral *Lobophytum cristagalli*. Bioorg Med Chem Lett. 1996;6:909–12.
73. Xi Z, Bie W, Chen W, Liu D, Ofwegen LV, Proksch P, et al. Sarcophyolides B–E, new cembranoids from the soft coral *Sarcophyton elegans*. Mar Drugs. 2013;11:3186–96.
74. El-Gamal AA, Wang SK, Dai CF, Chen IG, Duh CY. Prenylbicyclogermacrane diterpenoids from the Formosan soft coral *Nephthea pacifica*. J Nat Prod. 2005;68:74–7.
75. Duh CY, Wang SK, Tseng HK, Sheu JH, Chiang MY. Novel cytotoxic cembranoids from the soft coral *Sinularia flexibilis*. J Nat Prod. 1998;61:844–7.
76. Shen YC, Lin YC, Ko CL, Wang LT. New Briaranes from the Taiwanese Gorgonian *Junceella juncea*. J Nat Prod. 2003;66:302–5.

77. Chen WF, Yin CT, Cheng CH, Lu MC, Fang LS, Wang WH, et al. Norcembranoidal diterpenes from the cultured-type octocoral *Sinularia numerosa*. Int J Mol Sci. 2015;16:3298–306.
78. Shen YC, Lin YC, Chiang MY. Juncenolide A, a new briarane from the Taiwanese gorgonian *Junceella juncea*. J Nat Prod. 2002;65:54–6.
79. Li HJ, Xie YL, Xie ZL, Chen Y, Lam CK, Lan WJ. Chondrosterins A–E, triquinane-type sesquiterpenoids from soft coral-associated fungus *Chondrostereum* sp. Mar Drugs. 2012;10:627–38.
80. Wu YJ, Neoh CA, Tsao CY, Su JH, Li HH. Sinulariolide suppresses human hepatocellular carcinoma cell migration and invasion by inhibiting matrix metalloproteinase-2/-9 through MAPKs and PI3K/Akt signaling pathways. Int J Mol Sci. 2015;16:16469–82.
81. Lin JJ, Su JH, Tsai CC, Chen YJ, Liao MH, Wu YJ. 11-epi-Sinulariolide acetate reduces cell migration and invasion of human hepatocellular carcinoma by reducing the activation of ERK1/2, p38MAPK and FAK/PI3K/AKT/mTOR signaling pathways. Mar Drugs. 2014;12:4783–98.
82. Duh CY, Lo IW, Wang SK, Dai CF. New cytotoxic steroids from the soft coral *Clavularia viridis*. Steroids. 2007;72:573–9.
83. Lei LF, Chen MF, Wang T, He XX, Liu BX, Deng Y, et al. Novel cytotoxic nine-membered macrocyclic polysulfur cembranoid lactones from the soft coral *Sinularia* sp. Tetrahedron. 2014;70:6851–8.
84. Iwamaru A, Iwado E, Kondo S, Newman RA, Vera B, Rodríguez AD, et al. Eupalmerin acetate, a novel anticancer agent from Caribbean gorgonian octocorals, induces apoptosis in malignant glioma cells via the c-Jun NH2-terminal kinase pathway. Mol Cancer Ther. 2007;6:184–92.
85. Zierler S, Yao G, Zhang Z, Kuo WC, Pörzgen P, Penner R, et al. Waixenicin A inhibits cell proliferation through magnesium-dependent block of transient receptor potential melastatin 7 (TRPM7) channels. J Biol Chem. 2011;286:39328–35.
86. Shen S, Zhu H, Chen D, Liu D, van Ofwegen L, Proksch P, et al. Pavidolides A–E, new cembranoids from the soft coral *Sinularia pavida*. Tetrahedron Lett. 2012;53:5759–62.
87. Lo KL, Khalil AT, Kuo YH, Shen YC. Sinuladiterpenes A–F, new cembrane diterpenes from *Sinularia flexibilis*. Chem Biodivers. 2009;6:2227–35.
88. Wang GH, Ahmed AF, Kuo YH, Sheu JH. Two new subergane-based sesquiterpenes from a Taiwanese Gorgonian coral *Subergorgia suberosa*. J Nat Prod. 2002;65:1033–6.
89. Iwashima M, Matsumoto Y, Takenaka Y, Iguchi K, Yamori T. New marine diterpenoids from the Okinawan soft coral *Clavularia koellikeri*. J Nat Prod. 2002;65:1441–6.
90. Zhang XW, Tang XL, Yuan HR, Feng DQ, Su P, Li PL, et al. Two new eunicellin diterpenoids from the South China Sea gorgonian *Muricella sibogae* and their bioactivities. Nat Prod Res. 2015;29:2018–23.
91. (a) Wang CY, Chen AN, Shao CL, Li L, Xu Y, Qian PY. Chemical constituents of soft coral *Sarcophyton infundibuliforme* from the South China Sea. Biochem Syst Eco. 2011;39:853–85. (b) Gao CH, He BJ, Chen YN, Ke K, Lin L, Long B, et al. Two new diterpenoids from the Beibu Gulf Gorgonian *Anthogorgia caerulea*. Zeitschrift für Naturforschung B. 2014;69:116–20.
92. Qi SH, Zhang S, Qian PY, Xu HH. Antifeedant and antifouling briaranes from the South China Sea gorgonian *Junceella juncea*. Chem Nat Compd. 2009;45:49–54.
93. Lai D, Li Y, Xu M, Deng Z, Van Ofwegen L, Qian P, et al. Sinulariols A–S, 19-oxygenated cembranoids from the Chinese soft coral *Sinularia rigida*. Tetrahedron. 2011;67:6018–29.
94. Zhang J, Liang Y, Wang KL, Liao XJ, Deng Z, Xu SH. Antifouling steroids from the South China Sea gorgonian coral *Subergorgia suberosa*. Steroids. 2014;79:1–6.
95. Gribble GW. Biological activity of recently discovered halogenated marine natural products. Mar Drugs. 2015;13:4044–136.
96. Chen D, Chen W, Liu D, van Ofwegen L, Proksch P, Lin W. Asteriscane-type sesquiterpenoids from the soft coral *Sinularia capillosa*. J Nat Prod. 2013;76:1753–63.
97. Chen D, Yu S, van Ofwegen L, Proksch P, Lin W. Anthogorgienes A–O, new guaiazulene-derived terpenoids from a Chinese gorgonian Anthogorgia species, and their antifouling and antibiotic activities. J Agric Food Chem. 2011;60:112–23.

98. Mol VL, Raveendran TV, Parameswaran PS, Kunnath RJ, Rajamohanan PR. α-Hydroxy polyanthellin A-A novel antifouling diterpenoid from the Indian soft coral *Cladiella krempfi* (Hickson). Can J Chem. 2010;89(6):57–60.
99. Qi SH, Zhang S, Yang LH, Qian PY. Antifouling and antibacterial compounds from the gorgonians *Subergorgia suberosa* and *Scripearia gracilis*. Nat Prod Res. 2008;22:154–66.
100. Zhang J, Li LC, Wang KL, Liao XJ, Deng Z, Xu SH. Pentacyclic hemiacetal sterol with antifouling and cytotoxic activities from the soft coral *Nephthea* sp. Bioorg Med Chem Lett. 2013;23:1079–82.
101. Handayani D, Edrada RA, Proksch P, Wray V, Witte L, van Ofwegen L, et al. New oxygenated sesquiterpenes from the Indonesian soft coral *Nephthea chabrolii*. J Nat Prod. 1997;60:716–8.
102. Řezanka T, Hanuš LO, Dembitsky VM, Sigler K. Identification of the eight-membered heterocycles Hicksoanes A–C from the Gorgonian *Subergorgia hicksoni*. Eur J Org Chem. 2008;7:1265–70.
103. Iken KB, Baker BJ. Ainigmaptilones, sesquiterpenes from the Antarctic gorgonian coral *Ainigmaptilon antarcticus*. J Nat Prod. 2003;66:888–90.
104. Epifanio RDA, Maia LF, Pawlik JR, Fenical W. Antipredatory secosterols from the octocoral *Pseudopterogorgia americana*. Mar Ecol-Prog Ser. 2007;329:307–10.
105. Maia LF, Epifanio RDA, Eve T, Fenical W. New fish feeding deterrents, including a novel sesquiterpenoid heterogorgiolide, from the Brazilian gorgonian *Heterogorgia uatumani* (Octocorallia, Gorgonacea). J Nat Prod. 1999;62:1322–4.
106. Jayasuriya H, Herath KB, Ondeyka JG, Guan Z, Borris RP, Tiwari S, et al. Diterpenoid, steroid, and triterpenoid agonists of liver X receptors from diversified terrestrial plants and marine sources. J Nat Prod. 2005;68:1247–52.
107. Sun ZH, Cai YH, Fan CQ, Tang GH, Luo HB, Yin S. Six new tetraprenylated alkaloids from the South China Sea Gorgonian *Echinogorgia pseudossapo*. Mar Drugs. 2014;12:672–81.
108. Hassan HM, Khanfar MA, Elnagar AY, Mohammed R, Shaala LA, Youssef DT, et al. Pachycladins A – E, prostate cancer invasion and migration inhibitory eunicellin-based diterpenoids from the Red Sea soft coral *Cladiella pachyclados*. J Nat Prod. 2010;73:848–53.
109. Meyer M, Delberghe F, Liron F, Guillaume M, Valentin A, Guyot M. An antiplasmodial new (bis) indole alkaloid from the hard coral *Tubastraea* sp. Nat Prod Res. 2009;23:178–82.
110. Wei X, Rodríguez AD, Baran P, Raptis RG. Dolabellane-type diterpenoids with antiprotozoan activity from a Southwestern Caribbean Gorgonian octocoral of the genus *Eunicea*. J Nat Prod. 2010;73:925–34.
111. Rodríguez II, Shi YP, García OJ, Rodríguez AD, Mayer AM, Sánchez JA, et al. New pseudopterosin and seco-pseudopterosin diterpene glycosides from two Colombian isolates of *Pseudopterogorgia elisabethae* and their diverse biological activities. J Nat Prod. 2004;67:1672–80.
112. Aoki S, Okano M, Matsui K, Itoh T, Satari R, Akiyama SI, et al. Brianthein A, a novel briarane-type diterpene reversing multidrug resistance in human carcinoma cell line, from the gorgonian *Briareum excavatum*. Tetrahedron. 2001;57:8951–7.
113. Rashid MA, Gustafson KR, Boyd MR. HIV-inhibitory cembrane derivatives from a Philippines collection of the soft coral Lobophytum species 1. J Nat Prod. 2000;63:531–3.
114. Hooper GJ, Davies-Coleman MT. Sesquiterpene hydroquinones from the South African soft coral *Alcyonium fauri*. Tetrahedron Lett. 1995;36:3265–8.
115. Newton CG, Sherburn MS. Total synthesis of the pseudopterosin aglycones. Nat Prod Rep. 2015;32:865–76.
116. Williams DR, Shah AA. Total synthesis of (+)-ileabethoxazole via an iron-mediated Pauson–Khand [2+ 2+ 1] carbocyclization. J Am Chem Soc. 2014;136:8829–36.
117. (a) Crimmins MT, Brown BH, Plake HR. An intramolecular Diels-Alder approach to the eunicellins: enantioselective total syntheses of ophirin B and astrogorgin. J Am Chem Soc 2006;128:1371–8. (b) Molander GA, St. Jean DJ, Haas J Toward a general route to the eunicellin diterpenes: the asymmetric total synthesis of deacetoxyalcyonin acetate J Am Chem Soc 2004;126:1642–3. (c) Kim H, Lee H, Kim J, Kim S, Kim D. A general strategy for

synthesis of both (6 Z)-and (6 E)-Cladiellin diterpenes: total syntheses of (−)-cladiella-6, 11-dien-3-ol,(+)-polyanthellin A,(−)-cladiell-11-ene-3, 6, 7-triol, and (−)-deacetoxyalcyonin acetate. J Am Chem Soc 2006;128:15851–5. (d) MacMillan DW, Overman LE, Pennington LD. A general strategy for the synthesis of cladiellin diterpenes: enantioselective total syntheses of 6-acetoxycladiell-7 (16), 11-dien-3-ol (deacetoxyalcyonin acetate), cladiell-11-ene-3, 6, 7-triol, sclerophytin A, and the initially purported structure of sclerophytin A. J Am Chem Soc 2001;123:9033–44.

118. Mushti CS, Kim JH, Corey EJ. Total synthesis of antheliolide A. J Am Chem Soc. 2006;128:14050–2.
119. Ito H, Hasegawa M, Takenaka Y, Kobayashi T, Iguchi K. Enantioselective total synthesis of (+)-tricycloclavulone. J Am Chem Soc. 2004;126:4520–1.
120. Nicolaou KC, Xu JY, Kim S, Pfefferkorn J, Ohshima T, Vourloumis D, et al. Total synthesis of sarcodictyins A and B. J Am Chem Soc. 1998;120:8661–73.
121. Nicolaou KC, Winssinger N, Vourloumis D, Ohshima T, Kim S, Pfefferkorn J, et al. Solid and solution phase synthesis and biological evaluation of combinatorial sarcodictyin libraries. J Am Chem Soc. 1998;120:10814–26.
122. Wang B, Ramirez AP, Slade JJ, Morken JP. Enantioselective synthesis of (−)-sclerophytin A by a stereoconvergent epoxide hydrolysis. J Am Chem Soc. 2010;132:16380–2.
123. Davoren JE, Martin SF. Enantioselective synthesis and structure revision of solandelactone E. J Am Chem Soc. 2007;129:510–1.
124. Brown MK, Hoveyda AH. Enantioselective total synthesis of clavirolide C. applications of Cu-catalyzed asymmetric conjugate additions and Ru-catalyzed ring-closing metathesis. J Am Chem Soc. 2008;130:12904–6.
125. Kündig EP, Cannas R, Laxmisha M, Ronggang L, Tchertchian S. Chromium-mediated asymmetric synthesis of both enantiomers of acetoxytubipofuran. J Am Chem Soc. 2003;125:5642–3.
126. Crimmins MT, Ellis JM. Establishing the absolute configuration of the asbestinins: enantioselective total synthesis of 11-acetoxy-4-deoxyasbestinin D. J Am Chem Soc. 2005;127:17200–1.
127. Ichige T, Okano Y, Kanoh N, Nakata M. Total synthesis of methyl sarcophytoate. J Am Chem Soc. 2007;129:9862–3.
128. Nicolaou KC, Ohshima T, Hosokawa S, Van Delft FL, Vourloumis D, Xu JY, et al. Total synthesis of eleutherobin and eleuthosides A and B. J Am Chem Soc. 1998;120:8674–80.
129. Heckrodt TJ, Mulzer J. Total synthesis of elisabethin A: intramolecular Diels-Alder reaction under biomimetic conditions. J Am Chem Soc. 2003;125:4680–1.
130. Mayer AM, Glaser KB, Cuevas C, Jacobs RS, Kem W, Little RD, et al. The odyssey of marine pharmaceuticals: a current pipeline perspective. Trends Pharmacol Sci. 2010;31:255–65.
131. Mayer AM, Lehmann VK. Marine pharmacology in 1998: Marine compounds with antibacterial, anticoagulant, antifungal, antiinflammatory, anthelmintic, antiplatelet, antiprotozoal, and antiviral activities; with actions on the cardiovascular, endocrine, immune, and nervous systems; and other miscellaneous mechanisms of action. Pharmacologist. 2000;43:62–9.
132. Nicolaou KC, Kim S, Pfefferkorn J, Xu J, Ohshima T, Hosokawa S, et al. Synthesis and biological activity of sarcodictyins. Angew Chem Int E. 1998;37:1418–21.
133. Cinel B, Roberge M, Behrisch H, van Ofwegen L, Castro CB, Andersen RJ. Antimitotic diterpenes from erythropodium caribaeorum test pharmacophore models for microtubule stabilization. Org Lett. 2000;2:257–60.
134. Britton R, de Silva ED, Bigg CM, McHardy LM, Roberge M, Andersen RJ. Synthetic transformations of eleutherobin reveal new features of its microtubule-stabilizing pharmacophore. J Am Chem Soc. 2001;123:8632–3.
135. Baker BJ, Scheuer PJ. The punaglandins: 10-chloroprostanoids from the octocoral *Telesto riisei*. J Nat Prod. 1994;57:1346–53.
136. Verbitski SM, Mullally JE, Fitzpatrick FA, Ireland CM. Punaglandins, chlorinated prostaglandins, function as potent Michael receptors to inhibit ubiquitin isopeptidase activity. J Med Chem. 2004;47:2062–70.

137. Huang YC, Guh JH, Shen YC, Teng CM. Investigation of anticancer mechanism of clavulone II, a coral cyclopentenone prostaglandin analog, in human acute promyelocytic leukemia. J Biomed Sci. 2005;12:335–45.
138. Shen YC, Cheng YB, Lin YC, Guh JH, Teng CM, Ko CL. New prostanoids with cytotoxic activity from Taiwanese octocoral *Clavularia viridis*. J Nat Prod. 2004;67:542–6.
139. Chiang PC, Chien CL, Pan SL, Chen WP, Teng CM, Shen YC, et al. Induction of endoplasmic reticulum stress and apoptosis by a marine prostanoid in human hepatocellular carcinoma. J Hepatol. 2005;43:679–86.
140. Chiang PC, Kung FL, Huang DM, Li TK, Fan JR, Pan SL, et al. Induction of Fas clustering and apoptosis by coral prostanoid in human hormone-resistant prostate cancer cells. Eur J Pharmacol. 2006;542:22–30.
141. Posadas I, Terencio MC, De Rosa S, Payá M. Cavernolide: a new inhibitor of human sPLA2 sharing unusual chemical features. Life Sci. 2000;67:3007–14.
142. Sata NU, Sugano M, Matsunaga S, Fusetani N. Sinulamide: an H, K-ATPase inhibitor from a soft coral *Sinularia* sp. Tetrahedron Lett. 1999;40:719–22.
143. Aceret TL, Coll JC, Uchio Y, Sammarco PW. Antimicrobial activity of the diterpenes flexibilide and sinulariolide derived from *Sinularia flexibilis* Quoy and Gaimard 1833 (Coelenterata: Alcyonacea, Octocorallia). Comp Biochem Physiol C: Pharmacol Toxicol Endocrinol. 1998;120:121–6.
144. Edrada RA, Proksch P, Wray V, Witte L, van Ofwegen L. Four new bioactive lobane diterpenes of the soft coral *Lobophytum pauciflorum* from Mindoro, Philippines. J Nat Prod. 1998;61:358–61.
145. Chao CH, Wen ZH, Wu YC, Yeh HC, Sheu JH. Cytotoxic and anti-inflammatory cembranoids from the soft coral *Lobophytum crassum*. J Nat Prod. 2008;71:1819–24.
146. Cheng SY, Wen ZH, Chiou SF, Hsu CH, Wang SK, Dai CF, et al. Durumolides A–E, anti-inflammatory and antibacterial cembranolides from the soft coral *Lobophytum durum*. Tetrahedron. 2008;64:9698–704.
147. Shen YC, Chen YH, Hwang TL, Guh JH, Khalil AT. Four new briarane diterpenoids from the gorgonian coral *Junceella fragilis*. Helv Chim Acta. 2007;90:1391–8.
148. Mayer AM, Jacobson PB, Fenical W, Jacobs RS, Glaser KB. Pharmacological characterization of the pseudopterosins: novel anti-inflammatory natural products isolated from the Caribbean soft coral, *Pseudopterogorgia elisabethae*. Life Sci. 1998;62:PL401–7.
149. BALA SRD, Venkata RD, Bheemasankara RC, Dhananjaya N, Kuttan R, Babu TD. Isolation and structural determination of new sphingolipids and pharmacological activity of africanene and other metabolites from *Sinularia leptoclados*. Chem Pharm Bull. 1999;47:1214–20.
150. Ichida K, Ikeda M, Goto K, Ito K. Characterization of a palytoxin-induced non-selective cation channel in mouse megakaryocytes. Jpn J Pharmacol. 1999;81:200–8.
151. Yabe T, Yamada H, Shimomura M, Miyaoka H, Yamada Y. Induction of choline acetyltransferase activity in cholinergic neurons by stolonidiol: structure-activity relationship. J Nat Prod. 2000;63:433–5.
152. (a) Sawant S, Youssef D, Mayer A, Sylvester P, Wali V, Arant M, et al. Anticancer and anti-inflammatory sulfur-containing semisynthetic derivatives of sarcophine. Chem Pharm Bull 2006;54:1119–123. (b) Mayer AMS, Oh S, Presto E, Glaser KB, Jacobson PB. LPS-primed rat brain microglia: a convenient in vitro model to search for anti-inflammatory marine natural products. Shock. 1997;7(Suppl 2):49.
153. Temraz TA, Houssen WE, Jaspars M, Woolley DR, Wease KN, Davies SN, et al. A pyridinium derivative from Red Sea soft corals inhibited voltage-activated potassium conductances and increased excitability of rat cultured sensory neurones. BMC Pharmacol. 2006;6:10.
154. Bordeleau ME, Mori A, Oberer M, Lindqvist L, Chard LS, Higa T, et al. Functional characterization of IRESes by an inhibitor of the RNA helicase eIF4A. Nat Chem Biol. 2006;2:213–20.
155. Gross H, König GM. Terpenoids from marine organisms: unique structures and their pharmacological potential. Phytochem Rev. 2006;5:115–41.

Chapter 17
Mass Production of Natural Products from Microbes Derived from Sponges and Corals

Shivakumar P. Banakar, Loganathan Karthik, and Zhiyong Li

Contents

Abstract Lack of sufficient pure natural compounds hinders further drug developments. The optimization of fermentation conditions is essential to enhance the yield of metabolites. Microbial genome analysis reveals the presence of a large number of cryptic biosynthetic gene clusters, and different strategies are there to trigger these gene pathways for the extensive study of natural product chemistry. Hence, the advanced technologies play a crucial role to achieve efficient discovery and productivity of novel microbial bioactive compounds. This chapter provides an outline on the mass production of microbial natural products derived from marine sponges and corals.

Keywords Sponge · Corals · Microbes · Natural products · Mass production

---

S. P. Banakar · L. Karthik · Z. Li (✉)
Marine Biotechnology Laboratory, State Key Laboratory of Microbial Metabolism, School of Life Sciences and Biotechnology, Shanghai Jiao Tong University, Shanghai, People's Republic of China
e-mail: zyli@sjtu.edu.cn

Z. Li (ed.), *Symbiotic Microbiomes of Coral Reefs Sponges and Corals*,
https://doi.org/10.1007/978-94-024-1612-1_17

17.1 Introduction

Natural products are unique bioactive compounds, which led to the initiation of drug discovery [1]. Marine invertebrates are unexploited and significant resources in the marine environment to discover novel bioactive compounds. Marine sponges and corals harbor diverse microbial communities, such as actinobacteria, fungi, archaea, and viruses [2, 3]; their bioactive natural products are substantial in the pharmaceutical industries as antimicrobial, anticancer, and immunosuppressants [4]. The developments of innovative technologies have overcomed the hurdles for the discovery and characterization of microbial bioactive natural products. The mass production of microbial bioactive compounds is a significant aspect of achieving effective yield for structural elucidation, bioactivity studies, and pharmaceutical applications.

17.2 Cultured Microbes Derived from Sponges and Corals

17.2.1 Sponges

Sponges inhabit a range of marine and freshwater systems [5], which form a close association with phylogenetically diverse microorganisms [2, 3]. Moreover, the sponges have acquired symbiotic microbial flora through parental sponges, surrounding water, or from other sources [6–8]. Microorganisms derived from the marine sponges are best sources for bioactive natural products [4, 9]. Extensive research of the past two decades on sponge symbiotic microbial communities revealed their phylogenetic diversity and biogeography [10–12] and their vital role in host metabolism and health [13–15].

The cultured actinomycetes derived from the marine sponges are *Dietzia*, *Rhodococcus*, *Streptomyces*, *Salinispora*, *Marinophilus*, *Solwaraspora*, *Salinibacterium*, *Aeromicrobium marinum*, *Williamsia maris*, and *Verrucosispora* [12, 16]. Morphological variants of actinobacteria were isolated from the marine sponge *Haliclona* sp., in the South China Sea, e.g., *Streptomyces*, *Nocardiopsis*, *Micromonospora*, and *Verrucosispora* [17]. Moreover, the marine sponge-associated actinomycetes, like *Rhodococcus* sp. RV157 (*Dysidea avara*) and *Micromonospora* sp. RV43 (*Aplysina aerophoba*), were isolated from Mediterranean sponges, and *Actinokineospora* sp. EG49 (*Spheciospongia vagabunda*) were isolated from the Red Sea sponge, as well as *Nocardiopsis* sp. SBT366 (*Chondrilla nucula*), *Streptomyces* sp. SBT343 (*Petrosia ficiformis*), *Geodermatophilus* sp. SBT350 (*Chondrilla nucula*), *Streptomyces* sp. SBT345 (*Agelas oroides*), *Streptomyces* sp. SBT346 (*Petrosia ficiformis*), and *Micromonospora* sp. SBT373 (*Chondrilla nucula*) [18]. Diversity analysis of cultural actinomycetes associated with 8 species of marine sponges reported the 13 genera, including 5 genera as the first records belong to the 10 families and order *Actinomycetales* from the South China Sea and the Yellow Sea [16]. 180 actinomycete strains including at least 14 new phylotypes within the genera *Micromonospora*, *Verrucosispora*, *Streptomyces*, *Salinispora*, *Solwaraspora*, *Microbacterium*, and *Cellulosimicrobium* were isolated from the Caribbean sponge and sediment samples [19]. The actinomycetes isolated from 15 species of sponges in the South China Sea consisted of 20 genera of 12 families, including the 3 rare genera, such as *Marihabitans*, *Polymorphospora*, and *Streptomonospora* [12]. The marine sponge *Mycale* sp. derived bacterial strains isolation reported from the genera *Actinobacteria*, *Bacteroidetes*, *Gammaproteobacteria*, *Alphaproteobacteria*, and *Firmicutes* [20]. Particularly, 14 new actinobacterial strains were isolated from 3 Mediterranean sponges [21].

Ascomycetous fungi, such as *Sordariomycetes*, *Dothideomycetes*, and *Eurotiomycetes*, are highly dominated in marine sponges [22]. Most of the marine sponges harbored some quite common fungal genera, such as *Acremonium*, *Aspergillus*, *Fusarium*, *Penicillium*, *Phoma*, and *Trichoderma* [23, 24], and few rare genera, such as *Botryosphaeria*, *Epicoccum*, *Paraphaeosphaeria*, and *Tritirachium* [25]. Besides, fungal strains belonging to *Bartalinia* and *Volutella* from *Tethya aurantium* and *Schizophyllum*, *Sporidiobolus*, *Bjerkandera* (*Basidiomycota*), and *Yarrowia* (*Ascomycota*) were isolated from marine sponges [24, 26]. Cultured fungal strains from 10 species of marine sponges in the South China Sea belonged to the predominant genera, viz., *Aspergillus*, *Penicillium*, and *Volutella* and the others, such as *Ascomycete*, *Fusarium*, *Isaria*, *Plectosphaerella*, *Pseudonectria*, *Simplicillum*, and *Trichoderma* [27].

17.2.2 Corals

Corals are sessile marine invertebrates belonging to the phylum *Cnidaria*, living in the compact colonies of many identical individual polyps. Corals are categorized into stony and soft corals. Stony corals are mainly reef-building scleractinian corals, and soft corals include a range of species, like gorgonians and sea pens in the subclass of Alcyonaria or Octocorallia [28]. Corals involve a mutually beneficial symbiosis with photosynthetic dinoflagellate algae *Symbiodinium*. The dynamic relationship between the corals and microorganisms plays a significant role in the coral health [29–34]. Microorganisms associated with corals influence the coral host physiology as well as coral reef ecosystem, like pathogen resistance and biogeochemical cycling of critical nutrients [28, 31]. Fewer reports are available on the isolation of coral-associated microorganisms through the culture-dependent methods [35], whereas the culture-independent studies have revealed the diverse microflora associated with corals [36–44].

Green sulfur bacteria, such as *Alphaproteobacteria*, *Firmicutes* and *Planctomycetales* (*Montastraea annularis*), and *Gammaproteobacteria* and *Betaproteobacteria* (*M. cavernosa*), have been detected in corals [34], and *Alphaproteobacteria* and *Bacteroidetes* were found in the soft coral *Dendronephthya* sp. [36]. Predominant bacterial strains belonging to *Gamma-*, *Alpha-*, and

Betaproteobacteria, *Bacteroidetes*, *Firmicutes*, *Actinomycetales*, *Planctomycetes*, and *Chlorobi* were found to be associated with soft coral *Alcyonium antarcticum* [37]. Five new actinobacterial genera of *Cellulomonas*, *Dermacoccus*, *Gordonia*, *Serinicoccus*, and *Candidatus Microthrix* along with 19 common actinobacterial genera were reported from soft coral *Alcyonium gracllimum* and stony coral *Tubastraea coccinea* in the East China Sea [38].

Culture enrichment aided in the isolation of higher ascomycetes and basidiomycetes fungal taxa from the coral skeletons [39]. Cultured fungi belonging to genera of *Aspergillus*, *Penicillium*, *Cladosporium*, *Fusarium*, *Microsphaeropsis*, *Paecilomyces*, *Phoma*, *Tilletiopsis*, *Gibberella*, *Isaria*, *Acremonium*, *Debaryomyces*, *Myrmecridium*, and *Nigrospora* were isolated from six species of gorgonians from the South China Sea [45]. Fungi associated with coral *Porites pukoensis* have been isolated, with *Aspergillus* being predominant, and the others consisted of *Penicillium*, *Cochliobolus*, *Acremonium*, *Rigidoporus*, *Gibberella*, *Eutypella*, *Didymellaceae*, and *Curvularia* [46]. To date, fungal spatial and functional relationship with corals is still poorly understood, and very few researchers have broadly explored the fungi associated with soft corals to isolate novel biologically active compounds [47].

17.3 Natural Products from Microbes Derived from Sponges and Corals

The discovery of microbes associated with marine sponges and corals has led to their intense exploitation for an untapped resource of the novel bioactive compounds, for example, polyketides, terpenoids, alkaloids, and non-ribosomal peptides [48–50], which might be ample candidates for the invention of new drug leads for cancer, infectious diseases, and lipid metabolic disorders or as immunosuppressants. Marine *Actinobacteria*, e.g., *Streptomyces*, *Micromonospora*, *Microthrix parvicella*, and *Acidimicrobium*, and particularly obligate marine actinomycetes, *Salinispora tropica* and *Salinispora arenicola*, are the producers of bioactive microbial metabolites [51–53]. Marine sponge-derived fungi, especially endophytic, produce the most of marine natural products among the marine fungi [54]. Some metabolites isolated from microorganisms associated with sponges and corals are summarized in Table 17.1.

Table 17.1 Some metabolites isolated from the sponge- and coral-associated microorganisms [28, 55]

Host	Microorganism	Family	Compounds	Reference
Sponge				
Bacteria:				
Petrosia ficiformis	*Streptomyces* sp. SBT348	*Actinobacteria*	Petrocidin A	[56]
			2,3-Dihydroxybenzoic acid	
			2,3-Dihydroxybenzamide	
			Maltol	
Unidentified	*Brevibacillus* sp.		Ulbactins F and G	[57]
Dysidea tupha	*Streptomyces* sp. RV15	*Actinobacteria*	Naphthacene glycoside SF2446A2	[58]
Haliclona sp.	*Pseudomonas fluorescens* H40, H41	*Proteobacteria*	Diketopiperazine	[59]
	Pseudomonas aeruginosa H51			
Unidentified	*Nocardiopsis* sp. 13–33-15 & 13–12-13	*Actinobacteria*	1,6-Dihydroxyphenazine	[60]
			1,6-Dimethoxyphenazine	
Spheciospongia vagabunda	*Micrococcus* sp. EG45	*Actinobacteria*	Microluside A	[61]
Spongia officinalis	*Streptomyces* sp. MAPS15	*Actinobacteria*	2-Pyrrolidone	[62]
Xestospongia testudinaria	*Serratia marcescens* IBRL USM 84	*Proteobacteria*	Prodigiosin	[63]
Callyspongia spp.	*Pseudomonas* spp. RHLB 12	*Proteobacteria*	Chromophore compound	[64]
Haliclona oculata	*Bacillus licheniformis* T6-1	*Firmicutes*	Fluorophore compound	[64]
Halichondria panicea	*Streptomyces* sp. HB202	*Actinobacteria*	Streptophenazines G and K	[65]
Polymastia boletiformis, Axinella dissimilis, and Haliclona simulans	*Pseudovibrio* sp. W64, W69, W89, W74	*Proteobacteria*	Tropodithietic acid	[66]
Dysidea avara	*Nocardiopsis* sp. RV163	*Actinobacteria*	1,6-Dihydroxyphenazine (produced from co-culture)	[67]

(continued)

Table 17.1 (continued)

Host	Microorganism	Family	Compounds	Reference
Acanthostrongylophora ingens	*Micromonospora* sp. M42	*Actinobacteria*	Manzamine A	[68]
Haliclona simulans	*Streptomyces* sp. SM8	*Actinobacteria*	Kitamycins A and B	[69]
			Antimycins A2, A3, A7, A8, A11, and A17	
Spheciospongia vagabunda	*Actinokinespora* sp. EG49	*Actinobacteria*	Actinosporin A	[70]
Unidentified	*Kocuria palustris* F-276,310;	*Actinobacteria*	Kocurin	[15]
	Kocuria marina F-276,345			
	Micrococcus yunnanensis F-256, 446			
Haliclona simulans	*Bacillus subtilis* MMA7	*Firmicutes*	Subtilomycin	[71]
Aplysina aerophoba	*Micromonospora* sp. RV115	*Actinobacteria*	Diazepinomicin	[72]
Axinella polypoides	*Streptomyces axinellae* Pol001T	*Actinobacteria*	Tetromycins 1, 2, 3, 4, and B	[73]
Halichondria sp.	*Bacillus licheniformis* SAB1	*Firmicutes*	Indole	[74]
			3-Phenylpropionic acid	
			4,41-Oxybis(3-phenylpropionic acid)	
Aplysina polypoides	*Streptomyces* sp. 34	*Actinobacteria*	Valinomycin	[75]
Axinella aerophoba	*Streptomyces* sp. 22	*Actinobacteria*	Valinomycin	[75]
Tedania sp.	*Streptomyces* sp. 11	*Actinobacteria*	Staurosporine	[75]
Tethya sp.	*Streptomyces* sp. T03	*Actinobacteria*	Butenolide	[75]
Aplysina fistularis	*Streptomyces* sp. Hedaya48	*Actinobacteria*	Saadamycin	[76]
			5,7-Dimethoxy-4-pmethoxylphenylcoumarin	
Halichondria panicea	*Streptomyces* sp. HB202	*Actinobacteria*	Mayamycin	[77]
Dysidea arenaria	*Streptomyces rochei* MB037	*Actinobacteria*	Borrelidin, BC194	[78]

Fungi:				
Halichondria okadai	*Trichoderma harzianum* OUPS-111D-4	*Ascomycota*	Tandyukisins B–D	[79]
Hymeniacidon perleve	*Aspergillus versicolor* MF359	*Ascomycota*	5-Methoxydihydrosterigmatocystin	[80]
Unidentified	*Aspergillus sydowii* ZSDS1-F6	*Ascomycota*	(Z)-5-(Hydroxymethyl)-2-(60)- methylhept-20-en-20-yl)-phenol	[81]
Melophus sp.	*Penicillium* sp. FF001	*Ascomycota*	Citrinin	[82]
Axinella corrugata	*Penicillium* sp.	*Ascomycota*	Dipeptide cis-cyclo(leucyl-tyrosyl)	[83]
Xestospongia testudinaria	*Stachybotrys chartarum* MXH-X73	*Ascomycota*	Stachybotrin D	[84]
Xestospongia testudinaria	*Aspergillus* sp.	*Ascomycota*	(Z)-5-(Hydroxymethyl)- 2-(61 -methylhept-21 – en-21 -yl)phenol	[85]
			Aspergiterpenoid A	
			(−)-5-(Hydroxymethyl)- 2-(21,61, 61 -trimethyltetrahydro2H-pyran-2-yl)phenol	
			(−)-Sydonic acid	
Callyspongia sp.	*Epicoccum* sp. JJY40	*Ascomycota*	Pyronepolyene C-glucoside iso-D8646-2-6	[86]
Psammocinia sp.	*Aspergillus insuetus*	*Ascomycota*	Insuetolides A	[87]
			Strobilactone A	
			(E,E)-6-(60,70-Dihydroxy20,40-octadienoyl)-strobilactone A	
Unidentified	*Aspergillus clavatus* MFD15	*Ascomycota*	1H-1,2,4-Triazole-3-carboxaldehyde 5-methyl	[88]
Petrosia sp.	*Aspergillus versicolor*	*Ascomycota*	Averantin	[89]
			Nidurufin	
Unidentified	*Trichoderma* sp. 05FI48	*Ascomycota*	Trichoderins A, A1, and B	[90]

(continued)

Table 17.1 (continued)

Host	Microorganism	Family	Compounds	Reference
Phakellia fusca	*Pestalotiopsis maculans* 16F-12		Xylariterpenoids H, I, J, and K	[91]
Corals				
Muricella abnormaliz	*Aspergillus* sp.	*Ascomycota*	Penilumamides B–D	[92]
			22- *O* -(*N* -me-L-valyl)afl aquinolone B	
			22- *O* -(*N* me-L-valyl)-21-epi- afl aquinolone B	
			Afl aquinolones A and D	[93]
Sarcophyton sp.	*Eurotium rubrum*	*Ascomycota*	Eurothiocins A and B	[94]
Dichotella gemmacea	*Aspergillus* sp.	*Ascomycota*	Aspergilones A and B	[95]
Sarcophyton tortuosum	*Chondrostereum* sp.	*Ascomycota*	Chondrosterins F–H	[96]
Dichotella gemmacea	*Aspergillus* sp.	*Ascomycota*	17-Epinotoamides Q and M	[97]
			Cordyols D and E	
Sarcophyton sp.	*Pestalotiopsis* sp.	*Ascomycota*	(±)-Pestalachlorides C and D	[98]
Sarcophyton sp.	*Acrogenotheca elegans*		Phenylalanine derivative 4′-OMeasperphenamate	[99]
			Aspochalasin A1	
			Cytochalasin Z24	
Sarcophyton sp.	*Alternaria* sp.	*Ascomycota*	Tetrahydroaltersolanols C–F	[100]
			Dihydroaltersolanol A	
			Alterporriols N–R	

S. tortuosum	*Chondrostereum* sp.	*Ascomycota*	Chondrosterins A–E	[101]
Scleronephthya sp.	*Micromonospora* sp.		Jadomycin B	[102]
Cladiella sp.	*Aspergillus versicolor*	*Ascomycota*	Cottoquinazoline D	[103]
D. gemmacea	*Curvularia lunatus*	*Ascomycota*	Cochliomycins A–C	[104]
Annella sp.	*Aspergillus sydowii*	*Ascomycota*	Aspergillusenes A and B	[105]
			(+)-(7S)-7-O-Methylsydonic acid	
			Aspergillusones A and B	
D. gemmacea	*Aspergillus* sp.	*Ascomycota*	(+)-Methyl sydowate	[106]
			7-Deoxy-7,14-didehydrosydonic acid	
			7-Deoxy-7,8-didehydrosydonic acid	

17.4 Mass Production of Natural Products from Cultured Microbes Derived from Sponges and Corals

The microorganisms are able to synthesize a vast number of primary and secondary metabolites. However the quantities produced are very low for the industrial scale in the view of the industrial biotechnologists [107]; hence, the mass production efficiency of the microbial bioactive metabolites needs to be improved.

17.4.1 Fermentation Optimization

The optimization of fermentation condition depends on the type of microbial strain and target metabolite [56–58, 62, 82–85, 92], since the standard conditions may not favor the expression of a majority of microbial biosynthetic pathways [70, 108, 109]. The fermentation optimization includes fermentation method and production medium (carbon and nitrogen sources), along with the physical-chemical factors which include salt concentration, pH, temperature, agitation, aeration, incubation time, and competition/interaction between microorganisms [110–114].The solid substrates are widely used for mass production of fungal metabolites, but not much preferred for actinomycetes and bacteria [92, 115–117].

The traditional method of one parameter each a time for factorial optimization might not produce accurate results, so the statistical methods are helpful in this aspect. The widely used statistical tools for the optimization of critical factors of mass production culture conditions are Plackett-Burman (PB) design and response surface methodology [62, 112, 113]. The PB design method is useful to select the critical control factors through the evaluation of the relative importance of bioprocess culture conditions and nutrients on the biomass and metabolite yield in liquid culture. The variables include the medium components, e.g., carbon and nitrogen sources, pH, temperature, incubation time, inoculum concentration, agitation, and aeration [111–114]. Response surface methodology (RSM) is useful to elucidate the interaction of selected critical variables of the bioprocess medium and selection of optimized conditions for the enhanced production of biomass and metabolite yield.

Different factors may hinder or induce the rate of biosynthesis of a novel or known marine microbial natural product or biomass during the mass production. The production medium, physicochemical factors, fermentation conditions, and carbon and nitrogen sources influence the efficient mass production and recovery of microbial natural products [70, 111–113, 117, 118]. The ideal conditions for growth and biosynthesis of secondary metabolites are not indeed the same, and even each organism obliges contrarily. The physiological and chemical regulators vary with diverse microorganisms and different metabolic pathways. Therefore, the individual optimal zones are required to improve the qualitative and quantitative secondary metabolite production. For instance, effective yield of antitrypanosomal active metabolite was observed from ISP2 medium with calcium alginate beads [70]. The sponge-associated fungus *Aspergillus carneus* was able to produce 3 new and 14 known compounds in the rice medium without sea salt than the rice medium with sea salt and modified Czapek medium [116]. Higher yield of (+)-terrein was achieved from the optimized mass production of *Aspergillus terreus* strain PF26 derived from a marine sponge than the un-optimized culture conditions [112]. Two new and one known lumazine peptides, along with a new cyclic pentapeptide, were isolated from the static, submerged fermentation of gorgonian-derived fungus *Aspergillus* sp. XS-20090B15. Further, L-methionine induced the isolation of new penilumamide B in comparison with traditional culture [93].

17.4.2 Efficient Finding and Preparation

Microorganisms are ubiquitous, and they thrive under different environmental conditions. Diverse habitats will influence different class of bioactive metabolites. Moreover, the production of microbial bioactive compounds will be affected by microbial strain selection, production mediums, fermentation conditions, microbial or chemical elicitors or inducers used, and the balance between biosynthesis and biotransformation during the mass production [70, 111–119].

A conventional method of natural product discovery depends on bioassay or chemotypes. Natural product discovery programs through traditional way are not supportive, time-consuming, laborious, and need more resources. Recent technologic advances have simplified the screening and efficient production of microbial bioactive natural products in addition to proposing the unique opportunity for re-establishment of microbial natural products as a more significant source of drug leads. The bacterial and fungal genome sequence information show the link between known natural products and the genes encoding their biosynthesis as analyzed by various software tools, such as antiSMASH, SMURF, CLUSEAN, ClustScan, and so on. Moreover, gene clusters and chemistry of the compounds progressively exploit to classify known natural products to discover new ones. Further, biosynthetic pathways responsible for the production of specific natural products enable a better understanding of mechanisms or interactions during the metabolite production under culture condition [120, 121].

The biosynthetic potential-based strain prioritization may help for natural product discovery, through pathway-specific probes [120] and high-throughput real-time PCR [121]. Moreover, the optimized mass production methods [94, 95, 112–117] and analytical approach of collective LC-MS and UV profile of each active extract help the systematic analysis, early de-replication, and screening with an LC-MS library to known or novel compounds [63–66, 122–125]. Comparative study has showed the utility of standard solvent partitioning (SSP) and accelerated solvent extraction methods (ASE) related to overall yields, solvent consumption, processing time, and chemical stability of both fractions [121]. In the past two decades, the excellent applications of combinatorial chemistry and high-throughput screening (HTS) technologies, genome sequencing, proteomics, metabolomics, and other methods have changed the entire scenario of finding natural products and the ways of harnessing its intricacies [126].

17.4.3 Activating Silent Gene Cluster

There are an increased number of cryptic or orphan pathways discovered; they are new sources to mine novel bioactive natural products. The developments in our understanding of microbial genome sequence, cluster arrangements, and metabolic pathways, and growth conditions, help to improve the natural product yield. Complete genome sequencing and mining are an alternate approach for the exploration of known or novel microbial species to analyze their metabolic potential [127]; however, these biosynthetic pathways are sometimes silent [128] or rarely expressed under standard laboratory conditions [129].

The traditional screening method incudes the selection of indigenous strain, followed by strain improvements through a series of mutational selection for the enhanced growth and metabolite yield [130]. It was suggested that new environments led to discover new microbial species to isolate novel bioactive natural products [131]. Thus, cryptic biosynthetic gene clusters could be activated by changing the cultural conditions. The OSMAC (one strain many compounds) principle is to mine and discover the new bioactive compounds through different approaches [132, 133].

17.4.3.1 Microbial Co-culture

Microorganisms show an active interspecies interaction with each other for available nutrients, space, and other resources for their existence in natural environments. Besides, the interaction may be beneficial or detrimental; the coexistence may incur production of novel bioactive secondary metabolites [134]. Therefore, microbial coexistence under laboratory conditions may induce activation of cryptic biosynthetic gene clusters which led to the innovative prospects. The co-culture strategy helps us to study the interspecies interactions responsible for the production of novel compounds with diverse structure and distinct bioactivities, such as antimicrobial and anticancer compounds [135]. Besides, this strategy has other benefits in comparison with pure cultures, such as in finding novel compounds or enhancing the yield of biological molecule, increase in the growth rate, and better utilization of mixed substrates. For example, based on the investigations of interspecies metabolic diversity of sponge-derived *S. arenicola* and *S. pacifica*, the *S. pacifica* induced the production of new rifamycins O and W from *S. arenicola* and known rifamycins and saliniketals [136]. Three new and ten known compounds isolated from sponge-derived *Actinokineospora* sp. EG49 and *Nocardiopsis* sp. RV163 were the results of co-culture induced biosynthesis [67, 137]. A novel keyicin, a poly-nitroglycosylated anthracycline, was produced by the co-culture of marine ascidian-associated *Micromonospora* sp. Strain WMMB235 and marine sponge-associated *Rhodococcus* sp. Strain WMMA185. The biosynthetic gene cluster analysis of both strains and sequencing results of keyicin BGC confirm that the compound is from the *Micromonospora* sp. [126]. Though many researchers have conducted experiments on co-culture and synergistic microbial interactions, via coax between two or more than two microorganisms, but in reality, the challenges and questions related to the methods are still unanswered [138, 139].

17.4.3.2 Epigenetic Regulators

Putative biosynthetic gene regulators for the production of bioactive secondary metabolites of particular interest have been proved to be unique in different ways from previously understood models of gene regulation. The epigenetic regulators act as a signaling molecule by the regulation of putative biosynthetic genes and induce a variety of responses in microbes, for example, N-acetyl-D-glucosamine (GlcNAc), suberoylanilide hydroxamic acid (SAHA), DNA methyltransferase inhibitor (5-AZA), proteasome inhibitor (Bortezomib), and sodium citrate. The N-acetyl-D-glucosamine-mediated elicitation toward three sponge-derived actinomycetes led to the induced production of 3-formylindole and guaymasol in *Micromonospora* sp. RV43, the siderophore bacillibactin, and surfactin antibiotic in *Rhodococcus* sp. RV157 and improved the production of minor metabolites, actinosporins E–H in *Actinokineospora* sp. EG49 [140].

The influence of SAHA on *Aspergillus terreus* strain PF26 associated with a marine sponge in the biosynthesis of (+)-terrein was investigated. The epigenetic modifier shows the higher impact on (+)-terrein production than the control by stimulating the biosynthesis of the precursor, 6-hydroxymellein [141]. Optimized precursor-directed mutasynthesis has produced higher yield of BC194, a derivative of borrelidin from the *Streptomyces rochei* MB037 derived from the marine sponge *Dysidea arenaria* [78]. Bortezomib, a protease inhibitor, has induced the production of new bergamotene derivatives (xylariterpenoids H–K) from *Pestalotiopsis maculans* 16F-12 derived from marine sponge [91].

17.4.3.3 Gene Engineering

Majority of microbial natural product biosynthetic gene clusters (BGCs), relatively under standard laboratory conditions, are either transcriptional silent or expressed at deficient level, so these are the significant challenges for the discovery of novel natural products [142]. Analysis of microbial genes responsible for the biosynthesis of secondary metabolites usually depends on gene knockout and heterologous expression. Hence, the BGC identification and manipulation are accessible from the complete genome sequencing [128, 143]. For this purpose, some sponge- and coral-associated microorganisms are yet to be cultivated to study their true biosynthetic potential for microbial natural product discovery [130, 144].

The actinomycetes, especially the genus *Streptomyces*, harbor dozens of BGCs per genome [145]. Recently, advanced activation of cryptic or silent BGCs was carried out through the genetic approaches, such as either to unlock the suppression of BGC gene expression in the native hosts [146] or directly bypass the regulatory system by refactoring and reconstructing controlling elements in BGCs in the heterologous hosts [147–149]. Heterologous microbial hosts are an unusual choice, to bypass the task of removing introns and stitching genes by PCR to ensure the correct expression in the model hosts, such as *E. coli*, yeasts, and filamentous fungi [150]. Eukaryotic microorganisms have large and complex gene networks. The complexity and lack of understanding of the physiology of filamentous fungi, compared to bacteria, have delayed rapid development of these organisms as highly efficient hosts for homologous or heterologous gene expression [151].The fungal biosynthetic gene clusters mRNA processing will be complicated for heterologous gene expression.

17.5 Summary and Future Perspectives

The microorganisms associated with marine sponges and corals are the primary sources of marine bioactive natural products, which are least studied and under exploration for the discovery of novel drug leads. Marine microbial bioactive natural products, which are majorly from *Streptomyces* and filamentous fungi, include terpenoids, polyketides, alkaloids, non-ribosomal peptides, phenazines, indolocarbazoles, sterols, butenolides, and cytochalasins. Optimized mass production studies are helpful to achieve high yield of microbial bioactive compounds. Lack of sufficient yield of the pure natural compounds hinders the analysis, structural elucidation, biological activity assays, and further drug developments. So, to achieve a higher yield of the compounds, further developments are required for mass production studies as well as to reduce the labor and other requirements. These aspects are helpful for the upcoming researchers to take up further challenges to produce the novel bioactive marine microbial natural products with pharmaceutical development potentials, such as antimicrobials, antituberculosis, and anticancer compounds.

Acknowledgments We gratefully acknowledge financial supports provided from the Natural Science Foundation of China (NSFC) (31861143020, 41776138, 41742002, U1301131, 41176127, 41076077), and High-Tech Research and Development Program of China (2013AA092901, 2011AA090702, 2007AA09Z447, 2004AA628060, 2002AA608080).

References

1. Montaser R, Luesch H. Marine natural products: a new wave of drugs? Future Med Chem. 2011;3:1475–89.
2. Thomas T, Moitinho-Silva L, Lurgi M, Björk JR, Easson C, Astudillo-García C, et al. Diversity, structure and convergent evolution of the global sponge microbiome. Nat Commun. 2016;7:11870.
3. Moitinho-Silva L, Nielsen S, Amir A, Gonzalez A, Ackermann GL, Cerrano C, et al. The sponge microbiome. GigaScience. 2017;6:1–7.
4. Blunt JW, Carroll AR, Copp BR, Davis RA, Keyzers RA, Prinsep MR. Marine natural products. Nat Prod Rep. 2018;35:8–53.
5. Hooper JNA, van Soest RWM. Systema Porifera: a guide to the classification of sponges. New York: Kluwer Academic/Plenum Publishers; 2002.
6. Hentschel U, Hopke J, Horn M, Friedrich AB, Wagner M, Hacker J, et al. Molecular evidence for a uniform microbial community in sponges from different oceans. Appl Environ Microbiol. 2002;68:4431–40.
7. Lafi FF, Fuerst JA, Fieseler L, Engels C, Goh WWL, Hentschel U. Widespread distribution of poribacteria in demospongiae. Appl Environ Microbiol. 2009;75:5695–9.
8. Webster NS, Taylor MW. Marine sponges and their microbial symbionts: love and other relationships. Environ Microbiol. 2012;14:335–46.
9. Palomo S, González I, de la Cruz M, Martín J, Tormo JR, Anderson M, et al. Sponge-Derived *Kocuria* and *Micrococcus* spp. as Sources of the new Thiazolyl Peptide Antibiotic Kocurin. Mar Drugs. 2013;**11**:1071–86.
10. Taylor MW, Thacker RW, Hentschel U. Genetics. Evolutionary insights from sponges. Science. 2007;316:1854–5.
11. Schmitt S, Tsai P, Bell J, Fromont J, Ilan M, Lindquist N, et al. Assessing the complex sponge microbiota: core, variable and species-specific bacterial communities in marine sponges. ISME J. 2012;6:564–76.
12. Sun W, Zhang F, He L, Karthik L, Li Z. Actinomycetes from the South China Sea sponges: isolation, diversity, and potential for aromatic polyketides discovery. Front Microbiol. 2015;6:1048.
13. Vacelet J. Electron microscope study of the association between bacteria and sponges of the genus *Verongia* (Dictyoceratida). J Microsc Biol Cell. 1975;23:271–88.
14. Vacelet J, Donadey C. Electron microscope study of the association between some sponges and bacteria. J Exp Mar Biol Ecol. 1977;30:301–14.
15. Friedrich AB, Fischer I, Proksch P, Hacker J, Hentschel U. Temporal variation of the microbial community associated with the Mediterranean sponge *Aplysina aerophoba*. FEMS Microbiol Ecol. 2001;38:105–13.
16. Xi L, Ruan J, Huang Y. Diversity and biosynthetic potential of Culturable Actinomycetes associated with marine sponges in the China seas. Int J Mol Sci. 2012;13:5917–32.
17. Jiang S, Sun W, Chen M, Dai S, Zhang L, Liu Y, et al. Diversity of culturable actinobacteria isolated from marine sponge *Haliclona* sp. Antonie Van Leeuwenhoek. 2007;92:405–16.
18. Abdelmohsen UR, Pimentel-Elardo SM, Hanora A, Radwan M, Abou-El-Ela SH, Ahmed S, et al. Isolation, phylogenetic analysis and anti-infective activity screening of marine sponge-associated actinomycetes. Mar Drugs. 2010;8:399–412.
19. Vicente J, Stewart A, Song B, Hill RT, Wright JL. Biodiversity of Actinomycetes associated with Caribbean sponges and their potential for natural product discovery. Mar Biotechnol. 2013;15:413–24.
20. Su P, Wang DX, Ding SX, Zhao J. Isolation and diversity of natural product biosynthetic genes of cultivable bacteria associated with marine sponge *Mycale* sp. from the coast of Fujian, China. Can J Microbiol. 2014;60:217–25.

21. Versluis D, McPherson K, van Passel MWJ, Smidt H, Sipkema D. Recovery of previously uncultured bacterial genera from three Mediterranean sponges. Mar Biotechnol. 2017;19:454–68.
22. Caballero-George C, Bolanos J, De Leon LF, Ochoa E, Darias J, D'Croz L, et al. Fungal diversity in marine sponges from highly diverse areas in the Isthmus of Panama. In: Lang MA, Sayer MDJ, editors. Proceedings of the 2013 AAUS/ESDP Joint International Scientific Diving Symposium Curacao; 2013. pp. 23–30.
23. Wang G, Li Q, Zhu P. Phylogenetic diversity of culturable fungi associated with the Hawaiian sponges *Suberites zeteki* and *Gelliodes fibrosa*. Antonie Leeuwenhoek. 2008;93:163–74.
24. Wiese J, Ohlendorf B, Blumel M, Schmaljohann R, Imhoff JF. Phylogenetic identification of fungi isolated from the marine sponge *Tethya aurantium* and identification of their secondary metabolites. Mar Drugs. 2011;9:561–85.
25. Holler U, Wright AD, Matthee GF, Konig GM, Draeger S, Aust HJ, et al. Fungi from marine sponges: diversity, biological activity and secondary metabolites. Mycol Res. 2000;104:1354–65.
26. Yu Z, Zhang B, Sun W, Zhang F, Li Z. Phylogenetically diverse endozoic fungi in the South China Sea sponges and their potential in synthesizing bioactive natural products suggested by PKS gene and cytotoxic activity analysis. Fungal Divers. 2013;58:127–41.
27. Zhou K, Zhang X, Zhang F, Li Z. Phylogenetically diverse cultivable fungal community and polyketide synthase (PKS), non-ribosomal peptide synthase (NRPS) genes associated with the South China Sea sponges. Microb Ecol. 2011;62:644–54.
28. Sun W, Anbuchezhian R, Li Z. Association of Coral-Microbes, and the ecological roles of microbial symbionts in corals. In: Goffredo S, Dubinsky Z, editors. Medusa and her sisters: the Cnidaria, past, present and future. Springer Press. pp. 347–357
29. Ainsworth TD, Wasmund K, Ukani L, Seneca F, Yellowlees D, Miller D, et al. Defining the tipping point: a complex cellular life/death balance in corals in response to stress. Sci Rep. 2011;1:160.
30. Thompson JR, Rivera HE, Closek CJ, Medina M. Microbes in the coral holobiont: partners through evolution, development, and ecological interactions. Front Cell Infect Microbiol. 2015;**176**:1–20.
31. McDevitt-Irwin JM, Baum JK, Garren M, Vega Thurber RL. Responses of coral-associated bacterial communities to local and global stressors. Front Mar Sci. 2017;4:262.
32. Rohwer F, Seguritan V, Azam F, Knowlton N. Diversity and distribution of coral-associated bacteria. Mar Ecol Prog Ser. 2002;243:1–10.
33. Bourne DG, Morrow KM, Webster NS. Insights into the coral microbiome: underpinning the health and resilience of reef ecosystems. Annu Rev Microbiol. 2016;70:317–40.
34. Frias-Lopez J, Zerkle AL, Bonheyo GT, Fouke BW. Partitioning of bacterial communities between seawater and healthy, black band diseased, and dead coral surfaces. Appl Environ Microbiol. 2002;68:2214–28.
35. Galkiewicz JP, Stellick SH, Gray MA, Kellogg CA. Cultured fungal associates from the deep-sea coral *Lophelia pertusa*. Deep-Sea Res I. 2012;67:12–20.
36. Harder T, Lau SCK, Dobretsov S, Fang TK, Qian PY. A distinctive epibiotic bacterial community on the soft coral *Dendronephthya* sp. and antibacterial activity of coral tissue extracts suggest a chemical mechanism against bacterial epibiosis. FEMS Microbiol Ecol. 2003;43:337–47.
37. Webster NS, Bourne D. Bacterial community structure associated with the Antarctic soft coral, *Alcyonium antarcticum*. FEMS Microbiol Ecol. 2007;59:81–94.
38. Yang S, Sun W, Tang C, Jin L, Zhang F, Li Z. Phylogenetic diversity of Actinobacteria associated with soft coral *Alcyonium gracllimum* and stony coral *Tubastraea coccinea* in the East China Sea. Microb Ecol. 2013;66:189–99.
39. Kendrick B, Risk MJ, Michaelides J, Bergman K. Amphibious microborers: bioeroding fungi isolated from live and dead corals. Bull Mar Sci. 1982;32:862–7.

40. Bourne DG, Munn CB. Diversity of bacteria associated with the coral *Pocillopora damicornis* from the Great Barrier Reef. Environ Microbiol. 2005;7:1162–74.
41. Yakimov MM, Cappello S, Crisafi E, Tursi A, Savini A, Corselli C, et al. Phylogenetic survey of metabolically active microbial communities associated with the deep-sea coral *Lophelia pertusa* from the Apulian plateau, Central Mediterranean Sea. Deep-Sea Res. 2006;53:62–75.
42. Wegley L, Edwards RA, Rodriguez-Brito B, Liu H, Rohwer F. Metagenomic analysis of the microbial community associated with the coral Porites astreoides. Environ Microbiol. 2007;9:2707–19.
43. Hong MJ, Yu YT, Chen CA, Chiang PW, Tang SL. Influence of species specificity and other factors on bacteria associated with the coral *Stylophora pistillata* in Taiwan. Appl Environ Microbiol. 2009;75:7797–806.
44. Amend AS, Barshis DJ, Oliver TA. Coral-associated marine fungi form novel lineages and heterogeneous assemblages. ISME J. 2012;6:1291–301.
45. Zhang XY, Bao J, Wang GH, He F, Xu XY, Qi SH. Diversity and antimicrobial activity of culturable fungi isolated from six species of the South China Sea gorgonians. Microb Ecol. 2012;64:617–27.
46. Li J, Zhong M, Lei X, Xiao S, Li Z. Diversity and antibacterial activities of culturable fungi associated with coral *Porites pukoensis*. World J Microbiol Biotechnol. 2014;30:2551–8.
47. Putria DA, Radjasab OK, Pringgeniesc D. Effectiveness of marine fungal symbiont isolated from soft coral *Sinularia* sp from Panjang Island as antifungal. Prog Environ Sci. 2015;23:351–7.
48. Gunatilaka AAL, Wijeratne EMK. Natural products from bacteria and fungi, in Phytochemistry and Pharmacognosy. UNESCO-Encyclopedia of Life Support Systems (EOLSS). http://www.eolss.net/eolss_sitemap.aspx
49. Karuppiah V, Sun W, Li Z. Natural products of Actinobacteria derived from marine organisms. In studies in natural products chemistry, ed. Atta-ur-Rahman. 2016;48:417–41.
50. Hoffmeister D, Keller NP. Natural products of filamentous fungi: enzymes, genes, and their regulation. Nat Prod Rep. 2007;24:393–416.
51. Montalvo NF, Mohamed NM, Enticknap JJ, Hill RT. Novel actinobacteria from marine sponges. Antonie Van Leeuwenhoek. 2005;87:29–36.
52. Berdy J. Bioactive microbial metabolites. J Antibiot. 2005;58:1–26.
53. Mincer TJ, Jensen PR, Kauffman CA, Fenical W. Widespread and persistent populations of a major new marine actinomycete taxon in ocean sediments. Appl Environ Microbiol. 2002;68:5005–11.
54. Bugni TS, Ireland CM. Marine-derived fungi: a chemically and biologically diverse group of microorganisms. Nat Prod Rep. 2004;21:143–63.
55. Indraningrat AAG, Smidt H, Sipkema D. Bioprospecting sponge-associated microbes for antimicrobial compounds. Mar Drugs. 2016;14:87.
56. Cheng C, Othman EM, Stopper H, Edrada-Ebel R, Hentschel U, Abdelmohsen UR. Isolation of Petrocidin a, a new cytotoxic cyclic dipeptide from the marine sponge-derived bacterium *Streptomyces* sp. SBT348. Mar Drugs. 2017;15:383.
57. Igarashi Y, Asano D, Sawamura M, In Y, Ishida T, Imoto M. Ulbactins F and G, polycyclic thiazoline derivatives with tumor cell migration inhibitory activity from *Brevibacillus* sp. Org Lett. 2016;18:1658–61.
58. Reimer A, Blohm A, Quack T, Grevelding CG, Kozjak-Pavlovic V, Rudel T, et al. Inhibitory activities of the marine streptomycete-derived compound SF2446A2 against chlamydia trachomatis and *Schistosoma mansoni*. J Antibiot. 2015;68:674–9.
59. Santos OCS, Soares AR, Machado FLS, Romanos MTV, Muricy G, Giambiagi-deMarval M, et al. Investigation of biotechnological potential of sponge-associated bacteria collected in Brazilian coast. Lett Appl Microbiol. 2015;60:140–7.
60. Karuppiah V, Li Y, Sun W, Feng G, Li Z. Functional gene-based discovery of phenazines from the actinobacteria associated with marine sponges in the South China Sea. Appl Microbiol Biotechnol. 2015;99:5939–50.

61. Eltamany EE, Abdelmohsen UR, Ibrahim AK, Hassanean HA, Hentschel U, Ahmed SA. New antibacterial xanthone from the marine sponge-derived *Micrococcus* sp. EG45. Bioorg Med Chem Lett. 2014;24:4939–42.
62. Sathiyanarayanan G, Gandhimathi R, Sabarathnam B, Seghal Kiran G, Selvin J. Optimization and production of pyrrolidone antimicrobial agent from marine sponge-associated *Streptomyces* sp. MAPS15. Bioprocess Biosyst Eng. 2014;37:561–73.
63. Ibrahim D, Nazari TF, Kassim J, Lim S-H. Prodigiosin—an antibacterial red pigment produced by *Serratia marcescens* IBRL USM 84 associated with a marine sponge *Xestospongia testudinaria*. J Appl Pharm Sci. 2014;4:1–6.
64. Skariyachan S, Rao AG, Patil MR, Saikia B, Bharadwaj Kn V, Rao Gs J. Antimicrobial potential of metabolites extracted from bacterial symbionts associated with marine sponges in coastal area of gulf of Mannar biosphere, India. Lett Appl Microbiol. 2014;58:231–41.
65. Kunz AL, Labes A, Wiese J, Bruhn T, Bringmann G, Imhoff JF. Nature's lab for derivatization: new and revised structures of a variety of Streptophenazines produced by a sponge-derived *Streptomyces* strain. Mar Drugs. 2014;12:1699–714.
66. Harrington C, Reen F, Mooij M, Stewart F, Chabot J-B, Guerra A, et al. Characterisation of non-autoinducing Tropodithietic acid (TDA) production from marine sponge *Pseudovibrio* species. Mar Drugs. 2014;12:5960–78.
67. Dashti Y, Grkovic T, Abdelmohsen UR, Hentschel U, Quinn RJ. Production of induced secondary metabolites by a co-culture of sponge-associated Actinomycetes, *Actinokineospora* sp. EG49 and *Nocardiopsis* sp. RV163. Mar Drugs. 2014;12:3046–59.
68. Waters AL, Peraud O, Kasanah N, Sims J, Kothalawala N, Anderson MA, et al. An analysis of the sponge *Acanthostrongylophora igens'* microbiome yields an actinomycete that produces the natural product manzamine a. Front Mar Sci. 2014;1:54.
69. Viegelmann C, Margassery LM, Kennedy J, Zhang T, O'Brien C, O'Gara F, et al. Metabolomic profiling and genomic study of a marine sponge-associated *Streptomyces* sp. Mar Drugs. 2014;12:3323–51.
70. Abdelmohsen UR, Cheng C, Viegelmann C, Zhang T, Grkovic T, Ahmed S, Quinn RJ, Hentschel U, Edrada-Ebel R. Dereplication strategies for targeted isolation of new Antitrypanosomal Actinosporins a and B from a marine sponge associated-*Actinokineospora* sp. EG49. Mar Drugs. 2014;12:1220–44.
71. Phelan RW, Barret M, Cotter PD, O'Connor PM, Chen R, Morrissey JP, et al. Subtilomycin: a new Lantibiotic from *Bacillus subtilis* strain MMA7 isolated from the marine sponge *Haliclona simulans*. Mar Drugs. 2013;11:1878–98.
72. Abdelmohsen UR, Szesny M, Othman EM, Schirmeister T, Grond S, Stopper H, et al. Antioxidant and anti-protease activities of Diazepinomicin from the sponge-associated *Micromonospora* strain RV115. Mar Drugs. 2012;10:2208–21.
73. Pimentel-Elardo SM, Buback V, Gulder TAM, Bugni TS, Reppart J, Bringmann G, et al. New Tetromycin derivatives with anti-Trypanosomal and protease inhibitory activities. Mar Drugs. 2011;9:1682–97.
74. Devi P, Wahidullah S, Rodrigues C, Souza LD. The sponge-associated bacterium *Bacillus licheniformis* SAB1: a source of antimicrobial compounds. Mar Drugs. 2010;8:1203–12.
75. Pimentel-Elardo SM, Kozytska S, Bugni TS, Ireland CM, Moll H, Hentschel U. Anti-parasitic compounds from *Streptomyces* sp. strains isolated from Mediterranean sponges. Mar Drugs. 2010;8:373–80.
76. El-Gendy MA, El-Bondkly AA. Production and genetic improvement of a novel antimycotic agent, Saadamycin, against dermatophytes and other clinical fungi from endophytic *Streptomyces* sp. Hedaya48. J Ind Microbiol Biotechnol. 2010;37:831–41.
77. Schneemann I, Kajahn I, Ohlendorf B, Zinecker H, Erhard A, Nagel K, et al. Mayamycin, a cytotoxic polyketide from a *Streptomyces* strain isolated from the marine sponge *Halichondria panicea*. J Nat Prod. 2010;73:1309–12.

78. Li Y, Zhang F, Banakar S, Li Z. Comprehensive optimization of precursor-directed production of BC194 by Streptomyces rochei MB037 derived from the marine sponge Dysidea arenaria. Appl Microbiol Biotechnol. 2018;102:7865–75.
79. Yamada T, Umebayashi Y, Kawashima M, Sugiura Y, Kikuchi T, Tanaka R. Determination of the chemical structures of Tandyukisins B–D, isolated from a marine sponge-derived fungus. Mar Drugs. 2015;13:3231–40.
80. Song FH, Ren B, Chen CX, Yu K, Liu XR, Zhang YH, et al. Three new sterigmatocystin analogues from marine-derived fungus *Aspergillus versicolor* MF359. Appl Microbiol Biotechnol. 2014;98:3753–8.
81. Wang JF, Lin XP, Qin C, Liao SR, Wan JT, Zhang TY, et al. Antimicrobial and antiviral sesquiterpenoids from sponge-associated fungus, *Aspergillus sydowii* zsds1-f6. J Antibiot. 2014;67:581–3.
82. Subramani R, Kumar R, Prasad P, Aalbersberg W. Cytotoxic and antibacterial substances against multi-drug resistant pathogens from marine sponge symbiont: Citrinin, a secondary metabolite of *Penicillium* sp. Asian Pac J Trop Biomed. 2013;3:291–6.
83. Scopel M, Abraham W-R, Henriques AT, Macedo AJ. Dipeptide cis-cyclo(Leucyl-Tyrosyl) produced by sponge associated *Penicillium* sp. F37 inhibits biofilm formation of the pathogenic *Staphylococcus epidermidis*. Bioorg Med Chem Lett. 2013;23:624–6.
84. Ma XH, Lo LT, Zhu TJ, Ba MY, Li GQ, Gu QQ, et al. Phenylspirodrimanes with anti-HIV activity from the sponge-derived fungus *Stachybotrys chartarum* MXH-X73. J Nat Prod. 2013;76:2298–306.
85. Li D, Xu Y, Shao C-L, Yang R-Y, Zheng C-J, Chen Y-Y, et al. Antibacterial Bisabolane-type Sesquiterpenoids from the sponge-derived fungus *Aspergillus* sp. Mar Drugs. 2012;10:234–41.
86. Peng JX, Jiao JY, Li J, Wang W, Gu QQ, Zhu TJ, et al. Pyronepolyene C-glucosides with NF-kappa B inhibitory and anti-influenza a viral (H1N1) activities from the sponge-associated fungus *Epicoccum* sp. JJY40. Bioorg Med Chem Lett. 2012;22:3188–90.
87. Cohen E, Koch L, Thu KM, Rahamim Y, Aluma Y, Ilan M, et al. Novel terpenoids of the fungus *Aspergillus insuetus* isolated from the Mediterranean sponge *Psammocinia* sp. collected along the coast of Israel. Bioorg Med Chem. 2011;19:6587–93.
88. Manilal A, Sabarathnam B, Kiran GS, Sujith S, Shakir C, Selvin J. Antagonistic potentials of marine sponge associated fungi *Aspergillus clavatus* MFD15. Asian J Med Sci. 2010;2:195–200.
89. Lee Y, Li H, Hong J, Cho H, Bae K, Kim M, et al. Bioactive metabolites from the sponge-derived fungus *Aspergillus versicolor*. Arch Pharm Res. 2010;33:231–5.
90. Pruksakorn P, Arai M, Kotoku N, Vilchèze C, Baughn AD, Moodley P, et al. Trichoderins, novel aminolipopeptides from a marine sponge-derived *Trichoderma* sp., are active against dormant mycobacteria. Bioorg Med Chem Lett. 2010;20:3658–63.
91. Li Y, Zhang F, Banakar S, Li Z. Bortezomib-induced new bergamotene derivatives xylariterpenoids H–K from sponge-derived fungus *Pestalotiopsis maculans* 16F-12. RSC Adv. 2019;9:599–608.
92. Chen M, Shao CL, Meng H, She Z-G, Wang C-Y. Anti-respiratory syncytial virus prenylated dihydroquinolone derivatives from the gorgonian-derived fungus *Aspergillus* sp. XS-20090B15. J Nat Prod. 2014;77:2720–4.
93. Chen M, Shao CL, Fu XM, Kong C-J, She Z-G, Wang C-Y. Lumazine peptides penilumamides B-D and the cyclic pentapeptide asperpeptide-a from a gorgonian-derived *Aspergillus* sp. fungus. J Nat Prod. 2014;77:1601–6.
94. Liu Z, Xia G, Chen S, Liu Y, Li H, She Z. Eurothiocin A and B, sulfur- containing benzofurans from a soft coral-derived fungus *Eurotium rubrum* SH-823. Mar Drugs. 2014;12:3669–80.
95. Shao CL, Wang CY, Wei MY, Gu Y-C, She Z-G, Qian P-Y, et al. Aspergilones A and B, two benzylazaphilones with an unprecedented carbon skeleton from the gorgonian-derived fungus *Aspergillus* sp. Bioorg Med Chem Lett. 2013;21:690–3.

96. Li HJ, Chen T, Xie YL, Chen WD, Zhu XF, Lan WJ. Isolation and structural elucidation of chondrosterins F-H from the marine fungus *Chondrostereum* sp. Mar Drugs. 2013;11:551–8.
97. Chen M, Shao CL, Fu XM, Xu R-F, Zheng J-J, Zhao D-L, et al. Bioactive indole alkaloids and phenyl ether derivatives from a marine-derived *Aspergillus* sp. fungus. J Nat Prod. 2013;76:547–53.
98. Wei MY, Li D, Shao CL, Deng D-S, Wang C-Y. (±)-Pestalachloride D, an antibacterial racemate of chlorinated benzophenone derivative from a soft coral-derived fungus *Pestalotiopsis* sp. Mar Drugs. 2013;11:1050–60.
99. Zheng CJ, Shao CL, Wu LY, Chen M, Wang K-L, Zhao D-L, et al. Bioactive phenylalanine derivatives and cytochalasins from the soft coral-derived fungus, *Aspergillus elegans*. Mar Drugs. 2013;11:2054–68.
100. Zheng CJ, Shao CL, Guo ZY, Chen J-F, Deng D-S, Yang K-L, et al. Bioactive hydroanthraquinones and anthraquinone dimers from a soft coral-derived *Alternaria* sp. fungus. J Nat Prod. 2012;75:189–97.
101. Li HJ, Xie YL, Xie ZL, Chen Y, Lam C-K, Lan W-J. Chondrosterins A-E, triquinane-type sesquiterpenoids from soft coral-associated fungus *Chondrostereum* sp. Mar Drugs. 2012;10:627–38.
102. Sun W, Peng C, Zhao Y, Li Z. Functional gene-guided discovery of type II polyketides from culturable actinomycetes associated with soft coral *Scleronephthya* sp. PLoS One. 2012;7:e42847.
103. Zhuang Y, Teng X, Wang Y, Liu P, Li G, Zhu W. New quinazolinone alkaloids within rare amino acid residue from coral-associated fungus, *Aspergillus versicolor* LCJ-5-4. Org Lett. 2011;13:1130–3.
104. Shao CL, Wu HX, Wang CY, Liu Q-A, Xu Y, Wei M-Y, et al. Potent antifouling resorcylic acid lactones from the gorgonian-derived fungus *Cochliobolus lunatus*. J Nat Prod. 2011;74:629–33.
105. Trisuwan K, Rukachaisirikul V, Kaewpet M, Phongpaichit S, Hutadilok-Towatana N, Preedanon S, et al. Sesquiterpene and xanthone derivatives from the sea fan-derived fungus *Aspergillus sydowii* PSU-F154. J Nat Prod. 2011;74:1663–7.
106. Wei MY, Wang CY, Liu QA, Shao C-L, She Z-G, Lin Y-C. Five sesquiterpenoids from a marine-derived fungus *Aspergillus* sp. isolated from a gorgonian *Dichotella gemmacea*. Mar Drugs. 2010;8:941–9.
107. Amagata T, Minoura K, Numata A. Gymnastatins F–H, cytostatic metabolites from the sponge-derived fungus *Gymnascella dankaliensis*. J Nat Prod. 2006;69:1384–8.
108. Uzair B, Ahmed N, Ahmad VU, Kousar F. A new antibacterial compound produced by an indigenous marine bacteria—fermentation, isolation, and biological activity. Nat Prod Res. 2006;20:1326–31.
109. Finore I, Di Donato P, Mastascusa V, Nicolaus B, Poli A. Fermentation technologies for the optimization of marine microbial exopolysaccharide production. Mar Drugs. 2014;12:3005–24.
110. Brinkmann CM, Marker A, Ipek Kurtböke D. An overview on marine sponge-symbiotic bacteria as unexhausted sources for natural product discovery. Diversity. 2017;9:40.
111. Yang LH, Miao L, Lee OO, Li X, Xiong H, Pang KL, et al. Effect of culture conditions on antifouling compound production of a sponge-associated fungus. Appl Microbiol Biotechnol. 2007;74:1221–31.
112. Yin Y, Gao Q, Zhang F, Li Z. Medium optimization for the high yield production of single (+)-terrein by *Aspergillus terreus* strain PF26 derived from marine sponge *Phakellia fusca*. Process Biochem. 2012;47:887–91.
113. Bashir ZA, Ahmad A, Md-Nor S, Usup G. Factors affecting bioactivity of secondary metabolites produced by *Streptomyces* sp. PT1 using Plackett-Burman design. Adv Environ Biol. 2012;6:3043–51.
114. Nagai K, Kamigiri K, Arao N, Suzumura K, Kawano Y, Yamaoka M, et al. YM-266183 and YM-266184, novel thiopeptide antibiotics produced by *Bacillus cereus* isolated from a

marine sponge—I. taxonomy, fermentation, isolation, physico-chemical properties and biological properties. J Antibiot. 2003;56:123–8.
115. Tian YQ, Lin XP, Wang Z, Zhou XF, Qin XC, Kaliyaperumal K, et al. Asteltoxins with antiviral activities from the marine sponge-derived fungus *Aspergillus* sp. SCSIO XWS02F40. Molecules. 2016;21:34.
116. Özkaya FC, Ebrahim W, El-Neketi M, Tanrikul TT, Kalscheuer R, Müller WEG, et al. Induction of new metabolites from sponge-associated fungus *Aspergillus carneus* by OSMAC approach. Fitoterapia. 2018;131:9–14.
117. Grkovic T, Abdelmohsen UR, Othman EM, Stopper H, Edrada-Ebel R, Hentschel U, Quinn RJ. Two new antioxidant actinosporin analogues from the calcium alginate beads culture of sponge-associated *Actinokineospora* sp. strain EG49. Bioorg Med Chem Lett. 2014;24:5089–92.
118. Gao Y, Yu L, Peng C, Li Z, Guo Y. Diketopiperazines from two strains of South China Sea sponge-associated microorganisms. Biochem Syst Ecol. 2010;38:931–4.
119. Yin Y, Xu B, Li Z, Zhang B. Enhanced production of (+)-terrein in fed-batch cultivation of *Aspergillus terreus* strain PF26 with sodium citrate. World J Microbiol Biotechnol. 2013;29:441–6.
120. Xie P, Ma M, Rateb ME, Shaaban KA, Yu Z, Huang S-X, et al. Biosynthetic potential-based strain prioritization for natural product discovery: a showcase for Diterpenoid-producing Actinomycetes. J Nat Prod. 2014;77:377–87.
121. Johnson TA, Morgan MVC, Aratow NA, Estee SA, Sashidhara KV, Loveridge ST, et al. Assessing pressurized liquid extraction for the high-throughput extraction of marine-sponge-derived natural products. J Nat Prod. 2010;73:359–64.
122. Cremen PA, Zeng L. High-throughput analysis of natural product compound libraries by parallel LC–MS evaporative light scattering detection. Anal Chem. 2002;74:5492–500.
123. Tormo JR, Garcia JB, DeAntonio M, Feliz J, Mira A, Diez MT, et al. A method for the selection of production media for actinomycete strains based on their metabolite HPLC profiles. J Ind Microbiol Biotechnol. 2003;30:582–8.
124. Lang G, Mayhudin NA, Maya I, Mitova MI, Sun L, Sun L, et al. Evolving trends in the dereplication of natural product extracts: new methodology for rapid, small-scale investigation of natural product extracts. J Nat Prod. 2008;71:1595–9.
125. Genilloud O, Gonzalez I, Salazar O, Martın J, Tormo JR, Vicente F. Current approaches to exploit actinomycetes as a source of novel natural products. J Ind Microbiol Biotechnol. 2011;38:375–89.
126. Adnani N, Chevrette MG, Adibhatla SN, Zhang F, Yu Q, Braun DR, et al. Co-culture of marine invertebrate-associated bacteria and interdisciplinary technologies enable biosynthesis and discovery of a new antibiotic, Keyicin. ACS Chem Biol. 2017;12:3093–102.
127. Zerikly M, Challis GL. Strategies for the discovery of new natural products by genome mining. Chembiochem. 2009;10:625–33.
128. Scherlach K, Hertweck C. Triggering cryptic natural product biosynthesis in microorganisms. Org Biomol Chem. 2009;7:1753–60.
129. Hertweck C. Hidden biosynthetic treasures brought to light. Nat Chem Biol. 2009;5:450–2.
130. Parekh S. Strain improvement. In: Schaechter M, editor. The desk encyclopedia of microbiology. San Diego: Elsevier/Academic; 2004. p. 960–73.
131. Keller M, Zengler K. Tapping into microbial diversity. Nat Rev Microbiol. 2004;2:141–50.
132. Zahner H, Drautz H, Weber W. Novel approaches to metabolite screening. In: Bu'lock JD, Nisbet LJ, Winstanley DJ, editors. Bioactive microbial products: search and discovery. New York: Academic; 1982. p. 51–70.
133. Bode HB, Bethe B, Hofs R, Zeeck A. Big effects from small changes: possible ways to explore nature's chemical diversity. Chembiochem. 2002;3:619–27.
134. Fredrickson AG. Behavior of mixed cultures of microorganisms. Annu Rev Microbiol. 1977;31:63–87.

135. Rutledge PJ, Challis GL. Discovery of microbial natural products by activation of silent biosynthetic gene clusters. Nat Rev Microbiol. 2015;13:509–23.
136. Bose U, Hewavitharana AK, Vidgen ME, Ng YK, Shaw PN, Fuerst JA, et al. Discovering the recondite secondary metabolome Spectrum of *Salinispora* species: a study of inter-species diversity. PLoS One. 2014;9:e91488.
137. Cueto M, Jensen PR, Kauffman C, Fenical W, Lobkovsky E, Clardy J. Pestalone, a new antibiotic produced by a marine fungus in response to bacterial challenge. J Nat Prod. 2001;64:1444–6.
138. Hesseltine CW. Microbiology of oriental fermented foods. Annu Rev Microbiol. 1983;37:575–601.
139. Adnani N, Vazquez-Rivera E, Adibhatla SN, Ellis GA, Braun DR, Bugni TS. Investigation of interspecies interactions within marine *Micromonosporaceae* using an improved co-culture approach. Mar Drugs. 2015;13:6082–98.
140. Dashti Y, Grkovic T, Abdelmohsen UR, Hentschel U, Quinn RJ. Actinomycete metabolome induction/suppression with N-Acetylglucosamine. J Nat Prod. 2017;80:828–36.
141. Xiao L, Yin Y, Sun W, Zhang F, Li Z. Enhanced production of (+)-terrein by *Aspergillus terreus* strain PF26 with epigenetic modifier suberoylanilide hydroxamic acid. Process Biochem. 2013;48:1635–9.
142. Ren H, Wang B, Zhao H. Breaking the silence: new strategies for discovering novel natural products. Curr Opin Biotechnol. 2017;48:21–7.
143. Fox EM, Howlett BJ. Secondary metabolism: regulation and role in fungal biology. Curr Opin Microbiol. 2008;11:481–7.
144. Gross H. Genomic mining-a concept for the discovery of new bioactive natural products. Curr Opin Drug Discov Devel. 2009;12:207–19.
145. Liu G, Chater KF, Chandra G, Niu G, Tan H. Molecular regulation of antibiotic biosynthesis in *Streptomyces*. Microbiol Mol Biol Rev. 2013;77:112–43.
146. Li SR, Li YY, Lu CH, Zhang JL, Zhu J, Wang HX, et al. Activating a cryptic ansamycin biosynthetic gene cluster to produce three new naphthalenic octaketide ansamycins with n-pentyl and n-butyl side chains. Org Lett. 2015;17:3706–9.
147. Bachmann BO, Van Lanen SG, Baltz RH. Microbial genome mining for accelerated natural products discovery: is a renaissance in the making? J Ind Microbiol Biotechnol. 2014;41:175–84.
148. Li YX, Li ZR, Yamanaka K, Xu Y, Zhang WP, Vlamakis H, et al. Directed natural product biosynthesis gene cluster capture and expression in the model bacterium *Bacillus subtilis*. Sci Rep. 2015;5:9383.
149. Ross AC, Gulland LES, Dorrestein PC, Moore BS. Targeted capture and heterologous expression of the *Pseudoalteromonas alterochromide* gene cluster in *Escherichia coli* represents a promising natural product exploratory platform. ACS Synth Biol. 2015;4:414–20.
150. Lubertozzi D, Keasling JD. Developing *Aspergillus* as a host for heterologous expression. Biotechnol Adv. 2009;27:53–75.
151. Palmer JM, Keller NP. Secondary metabolism in fungi: does chromosomal location matter? Curr Opin Microbiol. 2010;13:431–6.

Chapter 18
Marine Enzymes from Microbial Symbionts of Sponges and Corals

Loganathan Karthik and Zhiyong Li

Contents

Abstract Sponges and corals are great pool of diverse microbes because they are closely associated with microorganisms that occur either extracellularly or intracellularly. The recent studies have revealed new microbial communities and novel compounds from sponges and corals. Many reports are found toward the availability of antibiotics from the sponge-/coral-associated microorganisms; very few reports are available for the enzymes, but there are no scientific research reports with the potentiality in medical and biotechnological applications. Therefore, there is an urgent need of exploration of marine enzymes from sponges and corals for human and environmental perspectives. This chapter will focus on current and future prospects of marine enzymes from microbial symbionts especially sponges/corals.

Keywords Sponges · Corals · Symbionts · Enzymes

L. Karthik · Z. Li (✉)
Marine Biotechnology Laboratory, State Key Laboratory of Microbial Metabolism,
School of Life Sciences and Biotechnology, Shanghai Jiao Tong University,
Shanghai, People's Republic of China
e-mail: zyli@sjtu.edu.cn

Z. Li (ed.), *Symbiotic Microbiomes of Coral Reefs Sponges and Corals*,
https://doi.org/10.1007/978-94-024-1612-1_18

18.1 Introduction

Marine microorganisms have a specific association with marine invertebrates which include sponges, and corals, etc. The microorganisms from the maritime-based environment have attracted considerable attraction toward them because of huge source of natural products for biotechnological application (Fig. 18.1).

The interrelationships between the marine sponges and microorganisms which might be considered as food either permanently or temporarily were found to be incredibly complex and far being understood at the present [1, 2]. Sponges (phylum, Porifera), that are on the evolution-based scale can be termed as ancient metazoans, originated 700–800 million years ago. They are grown especially in the tropical, temperate, and also freshwater niches [3, 4]. Particularly marine sponges are populated from intertidal zones to thousands of meters deep in the ocean [5].

The sponges consist of three major lineages, which include *Calcarea* (including 5 orders and 24 families), *Demospongiae* (15 orders and 92 families), and *Hexactinellida* (which include the 6 orders and 20 families). The numbers of sponge species raise to 15,000, which might be higher [6].

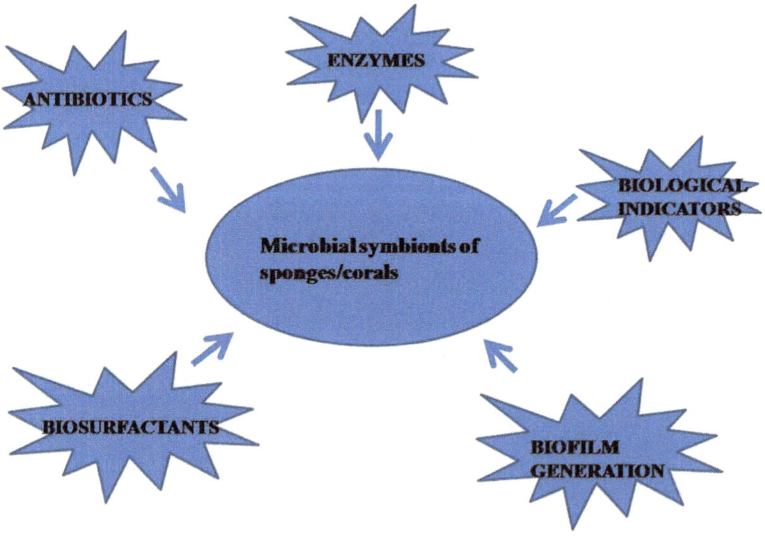

Fig. 18.1 Biotechnology potential of sponge- and coral-associated microorganisms

18.2 Sponge Microbial Association: Abundance, Diversity, and Specificity

The marine sponge mesohyl (extracellular matrix) contains a huge number of microorganisms, found in abundance surrounding the choanocyte chambers, rings of the flagellated cells that form foundation for the sponge aquiferous systems. Howbeit, the bacteriocytes of sponge species also grow, sponge species also host the microorganisms from the endosymbionts, and the pinacoderm cell also consists of increased densities of cyanobacterial species [7]. The microbial densities of the sponge are 10^9 cells/cm^3 of sponge tissue. Some sponges have high microbial abundance (HMA), whereas some species were found to have low-microbial-abundance (LMA) sponges (10^5–10^6 bacteria/cm^3 of tissue) [8].

Some genera show considerable variations which link to the total microbial abundance in certain hosts, e.g., *Chloroflexi*, *Actinobacteria*, *Cyanobacteria*, and *Poribacteria* [9–12]. The microbiomes of HMA and LMA sponges shared some functional features [13, 14].

When compared to human, sponge microbiome contains increased species diversity, whereas there is low phylum-level diversification [15]. There are about 52 different microbial phyla from which the candidate phyla were found to be associated with sponges [16–18]. The density of symbiont communities was found to be different in species ranging from a few distinct operational taxonomic units (OTU) to thousands of genetically distinct symbionts per host taxon [16, 18, 19], most of which are considered metabolically active [20]. The taxa *Gammaproteobacteria* and *Alphaproteobacteria*, *Actinobacteria*, *Chloroflexi*, *Nitrospirae*, *Cyanobacteria*, the candidate phylum *Poribacteria*, and *Thaumarchaea* were dominant in sponge [21, 22].

The marine fungal species when compared to the marine bacteria associated with the sponges were not fully explored but were found to have omnipresent relationship with sponges as reported by Namikoshi et al. [23] and Thakur and Muller [24]. There are twenty one orders from the *Ascomycota* which include the *Boliniales*, *Botryosphaeriales*, *Capnodiales*, *Chaetosphaeriales*, *Claromycetales*, *Diaporthales*, *Dothideales*, *Eurotiales*, *Helotiales*, *Hypocreales*, *Microascales*, *Moniliales*, *Mucorales*, *Onygenales*, *Phyllachorales*, *Pleosporales*, *Polyporales*, *Saccharomycetales*, *Sordariales*, *Trichosphaeriales*, and *Xylariales*, and also there are 8 orders from the *Basidiomycota* including *Agaricales*, *Agaricostilbales*, *Corticiales*, *Malasseziales*, *Polyporales*, *Sporidiobolales*, *Tremellales*, and *Wallemiales* that were reported from the marine sponges by Höller et al. [25], O'Brien et al. [26], Wang [27], Baker et al. [28], Li and Wang [29], Liu et al. [30], Paz et al. [31], Ding et al. [32], Rozas et al. [33], Wiese et al. [34], Zhou et al. [35], Suryanarayanan [36], Thirunavukkarasu et al. [37], and Yu et al. [38]. The orders *Saccharomycetales* and *Malasseziales* are not filamentous fungi. The sponge-derived fungal species have been found to be containing huge deposits of natural products, which are dealing with the enzymes isolated from the sponge-associated fungi that are very uncommon.

18.3 Enzymes from Sponge-Associated Microbes

The sponges are microbial living organisms which belong to the phylum *Porifera*. It is a multicellular organism, with pore-covered bodies and channels for circulation of water, food, and oxygen and removal of wastes. It released three times more oxygen and more organic matter than they consume when it is associated with photosynthesizing endosymbionts. Due to this, sponges have unique ecological niches. Large volumes of the seawater are consumed consisting of the organic particles by the sponge species; hence the scientific people have gathered their attention toward the microbes in sponges and the sponges itself which produce hydrolytic enzymes to convert the present organic matters into nutrients. For instance the hydrolysis of the agar can be achieved by the bacterial genus *Cytophaga* which was isolated from the sponge *Halichondria panicea* [39]. Currently more research is going on bioactive compounds from sponge-associated microbes, but only few reports are available on enzymes from sponge-associated microbes.

Marine sponges are one of the major sources producing halogenated organic compounds (bromoindoles, bromophenols, and bromopyrroles) in marine environment. The bacteria are harboring 40% of the biomass of *Aplysina aerophoba* sponges. *Desulfovibrio* bacteria (anaerobic sulfate reducers) are predominant in this sponge. Ahn et al. [40] discovered that the dehalogenation activity of *Desulfovibrio* spp. with different haloaromatics, for example, 2-bromophenol, 3-bromophenol, 4-bromophenol, 2,6-dibromophenol, 2,4,6-tribromophenol, and 3,5-dibromo-4-hydroxybenzoate, was reduced by debromination under methane-generating and sulfidogenic conditions with no activity observed in the presence of the nitrate. The increased salt tolerance of two novel esterases, namely, EstB1 and EstB2, were isolated and characterized from the genomic library *Bacillus* sp. associated with the marine sponge *Aplysina aerophoba*. The stability of the enzymes was stable in DMSO till 50% (v/v) followed by methanol, ethanol, and 2-propanol [41].

There are about 56 bacterial species which were isolated from the 6 marine sponges, namely, *Spirastrella* sp., *Phyllospongia* sp., *Ircinia* sp., *Aaptos* sp., *Azorica* sp., and *Axinella* sp. These strains are characterized and screened for amylase, carboxymethylcellulase, and proteases as reported earlier by Mohapatra et al. [42]. The screening resulted in the 44 strains to be positive for amylase, 46 for carboxymethylcellulase, and 34 for protease production, respectively. The genera belonging to *Alcaligenes*, *Alteromonas*, *Bacillus*, and *Micrococcus* were able to produce increased activity toward the amylase > 0.5 IU/ml. The genera, namely, *Alcaligenes*, *Alteromonas*, *Arthrobacter*, and *Planococcus*, and one unidentified bacterium were found to be 100% for carboxymethylcellulase activity. The protease activity was observed from the isolated bacteria. The bacterial species belonging to the genera *Alcaligenes*, *Alteromonas*, *Bacillus*, *Corynebacterium*, *Flavobacterium*, *Micrococcus*, and *Vibrio* and one of the unidentified bacterium showed protease activity ranging from 0.050 to 0.971 IU/ml. These results confirm that the enzymes producing strains are widely distributed in the bacterial genera with particular emphasis on sponge-associated strains. Especially 10–35% strains of *Bacillus* and 50–100% strains belonging to the genera of *Alcaligenes* are able to produce large amounts of enzymes like amylase, carboxymethylcellulase, and protease.

In addition, the research carried out by Mohapatra and Bapuji [43] states that the *Arthrobacter ilicis* bacterial species and the *Mucor* sp. characterized from the sponge *Spirastrella* sp. are capable of producing acetylcholinesterase and amylase enzymes, respectively. Also, these enzymes are heat tolerant and are not affected by the cations present in the seawater which include Na^+, Ca^{2+}, and Mg^{2+} at the higher concentrations [44].

By using the metagenomic technology, novel enzymes with a potential in environmental biotechnology were identified like oxidoreductase and halogenases from marine sponges [27].

The bacterium, *Micrococcus* sp., showed 80% urethenase activity which was isolated from the marine sponge, *Spirastrella* sp. The EDTA did not affect the bacterial species and other major cations found in the seawater and also found to be resistant toward 20% ethanol (v/v). Therefore, the removal of urethane from the alcoholic beverages could be the possible usage [45].

The presence of the collagenolytic activity in the bacterial species isolated from the marine sponge *Cymbastela concentrica* was confirmed by polyphasic approach [46]. It was also found that the other bacterial species have genes encoding collagenolytic enzymes, while there was a low abundance of microbes isolated which included *Zobellia* sp. KMM 3665; AB084262, *Vibrio crassostreae*; AJ582809, *Bacillus pumilus*; EU236743, *Shewanella* sp. MJ5323; and DQ531951 possessing collagenolytic activities.

The thermotolerant protease was reported from the marine endosymbiotic species, *Roseobacter* sp. (MMD040), which was isolated and characterized from the marine sponge *Fasciospongia cavernosa*. In pH 9.0, it showed 92.5% activity and 89% activity at 50°C. These results indicate that it can be developed for industrial application [47]. It was also reported that the increased yields of the enzyme amylase-producing strain, *Halobacterium salinarum* MMD047, were isolated from marine sponge *Fasciospongia cavernosa* [48]. A biotechnologically important enzyme, esterase, was isolated from the bacterial species from the marine sponge *Hyrtios erecta*, that was identified by metagenomic approach. It showed activity against acetate (5.6 U/mg), butylate (5.1 U/mg), and caproate (2.8 U/mg) substrates, with thermal stability and salt tolerance property [49].

The alkaline lipase was reported from an endosymbiotic *Pseudomonas* sp. (MSI057) from the marine sponge *Dendrilla nigra* [50]. In the same sponge, extracellular cellulolytic enzyme from *Marinobacter* sp. (MSI032) was reported [51].

From the metagenomic library of the marine sponge, *Haliclona simulans*, the 58 lipase-positive isolates were reported. The sequence analysis revealed the putative lipase gene lpc53E1, which encodes 387 amino acids with a molecular mass of 41.87 kDa. The optimal activity was observed with p-nitrophenyl palmitate (C16) at 40 °C, in the presence of 5 M NaCl, pH 7. When the p-nitrophenyl palmitate (10mM) was added as the substrate, it showed increased lipase activity of 2700 U/mg at the temperature of 40°C with the presence of 5 mM Ca^{2+} and 5 M NaCl [52].

Feby and Nair [53] screened the amylase, protease, gelatinous, lipase, deoxyribonucleic, phosphatase, and urease from bacterial species associated with two demosponges, *Dysidea granulosa* and *Sigmadocia fibulata*. *Gammaproteobacteria*, *Firmicutes*, and *Actinobacteria* were isolated from *Dysidea granulosa*, and in the case of *S. fibulata*, the *Betaproteobacteria* were isolated from the sponges. In the said scenario, sponge-associated bacteria express multiple enzymatic activities greater than four, and also they reported that *Vibrionales* was the main source for multiple enzyme production.

Dupont et al. [54] studied the prokaryotic community associated with the sponge *Asbestopluma hypogea* and its antimicrobial and antioxidant and chitinolytic activity. They found that 16 % of the bacterial isolates were positive for chitinolytic activity and they also suggested the involvement of the microbial species in the digestion processes of crustacean prey. In these, *Streptomyces* sp. S1CA strain had a potential in antimicrobial and antioxidant and chitinolytic activity.

In our lab, a significant work was carried out on enzymes from sponge-associated microorganisms.

The culture-independent metagenomic approach was taken for the screening of the complex microbiome species from the marine sponge *Ircinia* species that confirmed the presence of putative lipase gene *lipA*. The optimal activity was present at pH 9.0 with the presence of 5 mM Ca^{2+} and some of the organic solvents, like the methanol, acetone, and isopropanol at 40 °C. The SDS-PAGE confirmed 30 kDa to be the molecular weight for LipA; the main purpose can be in the detergent industry and enzyme-mediated organic synthesis. The study has widened the lipolytic gene and showed that marine sponges can be an important source for the novel enzymes [55].

The actinobacterial isolate *Streptomyces* sp. DA11, isolated from the South China Sea sponge called *Craniella australiensis*, was able to produce the chitinase enzyme which exhibited antifungal activity. The optical activity was observed at pH 8.0 at 50°C and salinity 45 g‰ psu. In comparison to the terrestrial organism-derived enzymes like the chitinase from the marine microbial sources, they have increased pH and salinity tolerance, which would contribute to the special biotechnological applications. The present study has shown the first report on the sponge-associated microbial chitinase [56]. Using Plackett-Burman design and Box-Behnken response surface methodology, the chitinase activity of 1559.2 U/g cell dry weight (36.43 U/mL) and the maximum cell dry weight of 23.3 g/L were reached after incubation of 72 h, which were 39.2-fold and 2.6-fold higher than that of the basic medium [57].

The bacterial isolate *Bacillus pumilus* B106 from the South China Sea sponge *Halichondria rugosa* was found to produce lipase enzyme, and there was an increased tolerance toward salinity, pH, and temperature. The study also extends our understanding of possibility of sponge-associated bacteria in biotransformation of chemical compounds [58].

Currently we are working on urease from *Bacillus atrophaeus* C89 isolated from *Dysidea avara* and its role in heavy metal detection in water pollution (data not shown). No enzymes have been reported from microbes associated with sponge families such as *Agelasidae, Astroscleridae, Calthropellidae, Geodiidae, Pachastrellidae, Thrombidae, Dictyodendrillidae, Acanthochaetetidae, Alectonidae, Hemiasterellidae, Placospongiidae, Polymastiidae, Stylocordylidae, Tethyidae, Timeidae, Trachycladidae, Bubaridae, Dictyonellidae, Heteroxyidae, Halisarcidae, Calcifibrospongiidae, Phloeodictyidae, Lubomirskiidae, Malawispongiidae, Metaniidae, Metschnikowiidae, Palaeospongillidae, Potamolepiidae, Spongillidae, Spongillina incertae sedis, Plakinidae, Azoricidae, Corallistidae, Desmanthidae, Isoraphiniidae, Lithistida incertae sedis, Macandrewiidae, Phymaraphiniidae, Phymatellidae, Pleromidae, Scleritodermidae, Siphonidiidae, Vetulinidae, Latrunculiidae, Microcionidae, Rhabderemiidae, Desmacellidae, Esperiopsidae, Guitarridae, Hamacanthidae, Merliidae, Podospongiidae, Chondropsidae, Coelosphaeridae, Crambeidae, Crellidae, Dendoricellidae, Desmacididae, Hymedesmiidae, Iotrochotidae, Phellodermidae, Tedaniidae, Samidae*, and *Spirasigmidae* of the class *Demospongiae* and *Baeriidae, Lepidoleuconidae, Trichogypsiidae, Achramorphidae, Amphoriscidae, Grantiidae, Heteropiidae, Jenkinidae, Lelapiidae, Leucosoleniidae, Sycanthidae, Sycettidae, Minchinellidae, Petrobionidae, Clathrinida incertae sedis, Clathrinidae, Leucaltidae, Leucascidae, Levinellidae, Soleneiscidae, Lelapiellidae, Murrayonidae, Ancorinidae, Chondrillidae,, Suberitidae, Callyspongiidae, Niphatidae, Petrosiidae, Neopeltidae, Theonellidae, Acarnidae, Raspailiidae, Isodictyidae, Mycalidae, Myxillidae, Plysinidae, Pseudoceratinidae*, and *Paramurrayonidae* of the class *Calcarea*. There are no reports of microbially originated enzymes from the class *Hexactinellida*.

18.4 Ecology and Physiology: Urea Hydrolysis

Many of the microbial populations are able to produce urease enzyme, which hydrolyzes urea to ammonia and carbamate and therefore maintaining the nitrogen level in the atmosphere [59]. The enzyme urease is produced by diverse bacterial species at a very early stage of their growth during early lag phase. These bacterial populations enhanced the protein synthesis in host sponge by the breakage of urea to ammonia. The microorganisms could play a major role in nutritional, physiological, and ecological mode of the host sponge and its environment. Our group reported the first insight on bacterial potential in urea utilization by detecting the transcriptional activity of *ure*C gene as well as the phylogenetic diversity of bacteria with *ure*C gene [60].

18.5 Defensive Enzymes from Endosymbionts

The endosymbionts produce the enzymes which are defensive in nature, along with the industrial enzymes, like the phospholipase, as the first line of defense. In 2009, Selvin [61] reported the production of an extracellular enzyme as phospholipase A2 (PLA2) from the sponge-associated bacterial isolate *Streptomyces dendra* sp. nov. MSI051 that resulted in more or less similar phospholipase A2 activity. It was also reported that the enzyme is responsible in the host against predatory and disturbances in the habitat.

18.6 Enzymes Responsible for Secondary Metabolite Production

Polyketide synthase, non-ribosomal peptide synthetase (NRPS) [62], and FADH-dependent halogenase [63] are multifunctional enzymes which are responsible for several new antibiotics production. Using metagenomic approach, polyketide synthase [64–66], non-ribosomal peptide synthetase gene clusters [67], and functional enzymes [68, 69] were reported from marine sponge-associated microbiota (please see Chap. 12 for details).

18.7 Coral Microbial Association

The living organisms belonging to the phylum Cnidaria and class Anthozoa are termed as corals, which are immobile otherwise known as the sessile marine invertebrates and are further classified into stony corals and soft corals. The oldest recorded coral dates back to around 450–500 million years to the Ordovician Period of the Paleozoic Era as reported by Stanley [70] and Tapanila [71]. They were found to be in interrelationship with a diverse and significantly richer quantification of bacteria, archaea, and fungi, which in turn influence the anatomical and physiological functions of coral hosts as well as its ecosystem and habitat. The coral-associated microorganisms were found to be acting as key player in nutrient cycling, coral health, and coral reef resilience [72, 73].

The coral-associated microorganisms are divided into four main functional groups, namely, (a) mutualistic bacteria with possible roles in coral nutrition; (b) pathogenic bacteria; (c) bacteria which can act as a probiont, aiding the growth of beneficial bacteria but limiting the growth of pathogenic forms; and (d) purely commensal bacteria with no impact on the other three groups [74]. These microorganisms play an important role in the symbiosis relatively which play an unknown role in the coral health and disease. The *Acidobacteria*, *Actinobacteria*, *Alphaproteobacteria*, *Aquificae*, *Bacteroidetes*, *Betaproteobacteria*, *Caldithrix* KSB1, *Chlamydiae*, *Chlorobi*, *Chloroflexi*, *Cyanobacteria*, *Deferribacteres*,

Deinococcus-Thermus, *Deltaproteobacteria*, *Epsilonproteobacteria*, *Firmicutes*, *Fusobacteria*, *Gammaproteobacteria*, *Gemmatimonadetes*, *Lentisphaerae*, *Nitrospirae*, OD1, *Planctomycetes*, *Spirochaetes*, *Tenericutes*, *Thermotogae*, TM7, *Verrucomicrobia*, WS3, *Zetaproteobacteria* from bacteria, *Crenarchaeota*, *Euryarchaeota* from Archaea, and *Ascomycota*, *Basidiomycota*, and *Chytridiomycota* from fungi were reported [75]. The compounds isolated from coral-associated microbes exhibited a broad spectrum of activity against pathogens, including coral pathogens [76, 77], but only few reports are only available in enzymes producing coral-associated microbes.

18.8 Enzyme from Coral-Associated Microbes

The total mean viable count of bacteria isolated from the soft coral *Lobophytum* sp. was 1.6×10^5cfu/g wet wt. Eight bacterial genera were isolated; in these genus *Pseudomonas* is dominant. It was screened for industrial enzymes such as amylase, carboxymethylcellulase, and protease. As a result, coral-associated microbes showed low to high enzyme activity [42].

The protease from *Halomonas meridiana* RA001 associated with *Acropora* sp. coral sample was reported from Palk Bay, southeast coast of India [78]. Anithajothi et al. [79] screened enzymes involved in melanin synthesis pathway (phenoloxidase (PO) and peroxidases (POD)) and free radical scavenging enzymes (super oxide dismutase (SOD), catalase (CAT)) and glutathione peroxidase (Gpx) in selected scleractinian corals such as *Acropora formosa*, *Echinopora lamellosa*, *Favia favus*, *Favites halicora*, *Porites* sp., and *Anacropora forbesi* collected from the southeast coast of India. The phenoloxidase activity was significantly lower than that of zooxanthellae except for *Favia favus*. *Favia favus* followed by *Echinopora lamellose* showed maximum antioxidant defensive enzymes. They concluded that these enzymes can be used as biomarkers for identifying the susceptibility of corals toward coral bleaching induced by pathogen.

The protease activity of three fungal isolates was reported from soft corals collected from Andaman and Nicobar marine water (lack of full information) [80]. *Bacillus aquimaris* MKSC 6.2 was isolated from a soft coral *Sinularia* sp., from Merak Kecil Island, West Java, Indonesia, which showed increased α-amylase activity, and it has the ability to degrade raw corn, rice, sago, cassava, and potato starches with the adsorption percentage in the range of 65–93% [81].

18.9 Conclusion

Sponge-associated symbionts synthesize industrially important enzymes which have been reported so far from geographically different regions such as the Bay of Bengal, India, South China Sea, France Sea, Baltic Sea, etc. (Table 18.1). This book

Table 18.1 Industrially important enzymes from sponge-associated microbes

Family	Sponge	Symbiont	Enzymes	References
Halichondriidae	*H. panacea* (Baltic Sea)	*Cytophaga* sp.	Agarase	[39]
Aplysinidae	*Aplysina aerophoba* (Marine Biological Station, Banyuls sur Mer, France)	*Desulfovibrio* spp.	Halogenase	[40]
Aplysinidae	*Aplysina aerophoba* (source not mentioned)	*Bacillus* sp.	Esterase	[41]
Spirastrellidae *Spongiidae* *Tethyidae* *Siphonidiidae* *Axinellidae*	*Spirastrella* sp., *Phyllospongia* sp., *Ircinia* sp., *Aaptos* sp., *Azorica* sp., and *Axinella* sp. (different location of Bay of Bengal, India)	*Alcaligenes*, *Alteromonas*, *Bacillus*, *Corynebacterium*, *Flavobacterium*, *Micrococcus*, *Vibrio*, and one unidentified bacterium	Amylase, carboxymethylcellulase, and proteases	[42]
Axinellidae	*Cymbastela concentrica* (Bare Island)	*Zobellia* sp. KMM 3665, *Vibrio crassostreae*, *Bacillus pumilus*, *Shewanella* sp.	Collagenase	[46]
Spirastrellidae	*Spirastrella* sp. (intertidal region of Havelock Island, Andaman Sea, India)	*Arthrobacter ilicis*	Acetylcholinesterase	[43]
Spirastrellidae	*Spirastrella* sp. (intertidal region of Havelock Island, Andaman Sea, India)	*Mucor* sp.	Amylase	[48]
Spirastrellidae	*Spirastrella* sp. (intertidal region of Havelock Island, Andaman Sea, India)	*Micrococcus* sp.	Urethanase	[45]

(continued)

Table 18.1 (continued)

Family	Sponge	Symbiont	Enzymes	References
Thorectidae	*Fasciospongia cavernosa* (peninsular coast of India)	*Roseobacter* sp.	Protease	[47]
Thorectidae	*Fasciospongia cavernosa* (peninsular coast of India)	*Halobacterium salinarum*	Amylase	[48]
Thorectidae	*Hyrtios erecta* (coast of Ishigaki Island, Okinawa, Japan)	Metagenome	Esterase	[49]
Darwinellidae	*Dendrilla nigra* (southwest coast of India)	*Pseudomonas* sp.	Lipase	[50]
Darwinellidae	*Dendrilla nigra* (peninsular coast of India)	*Marinobacter* sp.	Cellulose	[51]
Chalinidae	*Haliclona simulans* (Kilkieran Bay, off the west coast of Ireland)	Metagenome	Lipase	Selvin et al. [82]
Dysideidae	*Dysidea granulosa* and *Sigmadocia fibulata* (Kavaratti Island, Lakshadweep, west coast of India)	*Gammaproteobacteria*, *Firmicutes*, and *Actinobacteria*	Amylase, protease, gelatinase, lipase, deoxyribonucleic, phosphatase, and urease	[53]
Chalinidae		*Betaproteobacteria*		
Cladorhizidae	*Asbestopluma hypogea* (coast of La Ciotat, France)	*Streptomyces* sp. S1CA	Chitinase	[54]
Irciniidae	*Ircinia* sp. (Yongxing Island, South China Sea)	Metagenome	Lipase	[55]
Tetillidae	*Craniella australiensis* (South China Sea)	*Streptomyces* sp. DA11	Chitinase	[56]
Halichondriidae	*Halichondria rugosa* (South China Sea)	*Bacillus pumilus* B106	Lipase	[58]

chapter brings out the fact that members of the class *Demospongiae* are not only the richest producer of medicinally important bioactive compounds, but it is also a good producer of industrially important enzymes in association with microbes.

From the 92 families of the class *Demospongiae*, 15 families associated microbes have been isolated and characterized to be industrially important enzymes, which include the families *Halichondriidae, Aplysinidae, Spirastrellidae, Spongiidae, Tethyidae, Siphonidiidae, Axinellidae, Spirastrellidae, Thorectidae, Darwinellidae, Chalinidae, Dysideidae, Cladorhizidae, Irciniidae*, and *Tetillidae*.

Both bacteria and fungi have been isolated from a wide range of marine sponges. The diversity and symbiotic relationship of bacteria have been studied to a greater extent than that of fungi isolated from sponges, and even the enzymes also more reported from bacteria than fungi associated with sponges. Only one report was available for fungi (*Mucor* sp. for amylase). *Axinellidae, Spirastrellidae, Chalinidae, Cladorhizidae, Irciniidae*, and *Halichondriidae* family-associated symbionts were reported for only enzymes but not for antibiotics.

The coral-associated symbionts or isolates which produce the industrially important enzymes are reported from geographically different regions from Palk Bay, southeast coast of India, Bay of Bengal, India, Andaman and Nicobar, and Indonesia. The nine families from the class Anthozoa have been identified to produce industrially important enzymes.

Sponges and corals are a pool of novel microorganisms. When compared to terrestrial microorganisms, it is capable to tolerate different temperatures, at diverse pH and saltiness. It is more suitable for industrial process. The discovery of potent microbial associates producing enzymes has opened up a new era in marine industry. So far, many studies are carried out for antibiotics, but for enzymes only few reports are available. Hence, researchers and investigators should turn their attention toward the production of such valuable industrial and pharmaceutical enzymes from sponge- and coral-associated microorganisms.

Acknowledgments We gratefully acknowledge financial supports from the Natural Science Foundation of China (NSFC) (31861143020, 41776138, 41742002, U1301131, 41176127, 41076077), and High-Tech Research and Development Program of China (2013AA092901, 2011AA090702, 2007AA09Z447, 2004AA628060, 2002AA608080) and the Chinese Post-Doctoral Funding (No.15005188).

References

1. Wilkinson CR. Symbiotic interactions between marine sponges and algae. In: Reisser W, editor. Algae and symbioses: plants, animals, fungi, viruses, interactions explored. Bristol: Biopress Limited; 1992.
2. Steindler L, Huchon D, Avni A, Ilan M. 16S rRNA phylogeny of sponge-associated cyanobacteria. Appl Environ Microbiol. 2005;71:4127–31.
3. Hentschel U, Hopke J, Horn M, Friedrich AB, Wagner M, Hacker J, et al. Molecular evidence for a uniform microbial community in sponges from different oceans. Appl Environ Microbiol. 2002;68:4431–40.

4. Radjasa O, Sabdono K, Junaidi A, Zocchi E. Richness of secondary metabolite producing marine bacteria associated with sponge *Haliclona* sp. Int J Pharm. 2007;3:275–9.
5. Fusetani N, Matsunaga S. Bioactive sponge peptides. Chem Rev. 1993;93:1793–806.
6. Fieseler L, Horn M, Wagner M, Hentschel U. Discovery of the novel candidate Phylum "Poribacteria" in marine sponges. Appl Environ Microbiol. 2004;70:3724–32.
7. Wilkinson CR. Cyanobacteria symbiotic in marine sponges. In: Schwemmler W, Schneck HEA, editors. Endocytobiology, endosymbiosis and cell biology. Berlin: de Gruyter; 1980. p. 553–63.
8. Hentschel U, Usher KM, Taylor MW. Marine sponges as microbial fermenters. FEMS Microbiol Ecol. 2006;55:167–77.
9. Noyer C, Hamilton A, Sacristan-Soriano O, Becerro MA. Quantitative comparison of bacterial communities in two Mediterranean sponges. Symbiosis. 2010;51:239–43.
10. Abdelmohsen UR, Bayer K, Hentschel U. Diversity, abundance and natural products of marine sponge-associated actinomycetes. Nat Prod Rep. 2014;31:381–99.
11. Moitinho-Silva L, Bayer K, Cannistraci CV, Giles EC, Ryu T, Seridi L, et al. Specificity and transcriptional activity of microbiota associated with low and high microbial abundance sponges from the Red Sea. Mol Ecol. 2014;23:1348–63.
12. Bayer K, Moitinho-Silva L, Brümmer F, Cannistraci CV, Ravasi T, Hentschel U. GeoChip-based insights into the microbial functional gene repertoire of marine sponges (high microbial abundance, low microbial abundance) and seawater. FEMS Microbiol Ecol. 2014;90:832–43.
13. Fan L, Reynolds D, Liu M, Stark M, Kjelleberg S, Webster NS, et al. Functional equivalence and evolutionary convergence in complex communities of microbial sponge symbionts. Proc Natl Acad Sci U S A. 2012;109:E1878–87.
14. Bayer K, Kamke J, Hentschel U. Quantification of bacterial and archaeal symbionts in high and low microbial abundance sponges using real-time PCR. FEMS Microbiol Ecol. 2014;89:679–90.
15. Lozupone CA, Stombaugh JI, Gordon JI, Jansson JK, Knight R. Diversity, stability and resilience of the human gut microbiota. Nature. 2012;489:220–30.
16. Webster NS, Taylor MW, Behnam F, Lücker S, Rattei T, Whalan S, et al. Deep sequencing reveals exceptional diversity and modes of transmission for bacterial sponge symbionts. Environ Microbiol. 2010;12:2070–82.
17. Schmitt S. Assessing the complex sponge microbiota: core, variable and species-specific bacterial communities in marine sponges. ISME J. 2012;6:564–76.
18. Reveillaud J, Maignien L, Eren AM, Huber JA, Apprill A, Sogin ML, et al. Host-specificity among abundant and rare taxa in the sponge microbiome. ISME J. 2014;8:1198–209.
19. Lee OO, Wang Y, Yang J, Lafi FF, Al-Suwailem A, Qian P-Y. Pyrosequencing reveals highly diverse and species-specific microbial communities in sponges from the Red Sea. ISME J. 2011;5:650–64.
20. Kamke J, Taylor MW, Schmitt S. Activity profiles for marine sponge-associated bacteria obtained by 16S rRNA vs 16S rRNA gene comparisons. ISME J. 2010;4:498–508.
21. Simister RL, Deines P, Botté ES, Webster NS, Taylor MW. Sponge-specific clusters revisited: a comprehensive phylogeny of sponge-associated microorganisms. Environ Microbiol. 2012;14:517–24.
22. Hentschel U, Piel J, Degnan SM, Taylor MW. Genomic insights into the marine sponge microbiome. Nat Rev Microbiol. 2012;10:641–54.
23. Namikoshi M. Distribution of marine Wlamentous fungi associated with marine sponges in coral reefs of Palau and Bunaken Island, Indonesia. J Tokyo Univ Fish. 2002;88:15–20.
24. Thakur NL, Muller WEG. Biotechnological potential of marine sponges. Curr Sci. 2004;86:1506–12.
25. Höller U, Wright AD, Matthee GF, Konig GM, Draeger S, Aust H-J, et al. Fungi from marine sponges: diversity, biological activity and secondary metabolites. Mycol Res. 2000;104:1354–65.

26. O'Brien HE, Parrent JL, Jackson JA, Moncalvo J-M, Vilgalys R. Fungal community analysis by large-scale sequencing of environmental samples. Appl Environ Microbiol. 2005;71:5544–50.
27. Wang G. Diversity and biotechnological potential of the sponge-associated microbial consortia. J Ind Microbiol Biotechnol. 2006;33:545–51.
28. Baker PW, Kennedy J, Dobson AD, Marchesi JR. Phylogenetic diversity and antimicrobial activities of fungi associated with *Haliclona simulans* isolated from Irish coastal waters. Mar Biotechnol. 2009;11:540–7.
29. Li Q, Wang G. Diversity of fungal isolates from three Hawaiian marine sponges. Microbiol Res. 2009;164:233–41.
30. Liu W, Li C, Zhu P, Yang J, Cheng K. Phylogenetic diversity of culturable fungi associated with two marine sponges: *Haliclona simulans* and *Gelliodes carnosa*, collected from the Hainan Island coastal waters of the South China Sea. Fungal Divers. 2010;42:1–15.
31. Paz Z, Komon-Zelazowska M, Druzhinina IS, Aveskamp MM, Shnaiderman A, Aluma Y, et al. Diversity and potential antifungal properties of fungi associated with a Mediterranean sponge. Fungal Divers. 2010;42:17–26.
32. Ding B, Yin Y, Zhang F, Li Z. Recovery and phylogenetic diversity of culturable fungi associated with marine sponges *Clathrina luteoculcitella* and *Holoxea* sp. in the South China Sea. Mar Biotechnol. 2011;13:713–21.
33. Rozas EE, Albano RM, Lôbo-Hajdu G, Müller WEG, Schröder HC, Custödio MR. Isolation and cultivation of fungal strains from in vitro cell cultures of two marine sponges (Porifera: *Halichondrida* and *Haplosclerida*). Braz J Microbiol. 2011;42:1560–8.
34. Wiese J, Ohlendorf B, Blümel M, Schmaljohann R, Imhoff JF. Phylogenetic identification of fungi isolated from the marine sponge *Tethya aurantium* and identification of their secondary metabolites. Mar Drugs. 2011;9:561–85.
35. Zhou K, Zhang X, Zhang F, Li Z. Phylogenetically diverse cultivable fungal community and polyketide synthase (PKS), non-ribosomal peptide synthase (NRPS) genes associated with the South China Sea sponges. Microb Ecol. 2011;62:644–54.
36. Suryanarayanan TS. The diversity and importance of fungi associated with marine sponges. Bot Mar. 2012;55:553–64.
37. Thirunavukkarasu N, Suryanarayanan TS, Girivasan KP, Venkatachalam A, Geetha V, Ravishankar JP, et al. Fungal symbionts of marine sponges from Rameswaram, southern India: species composition and bioactive metabolites. Fungal Divers. 2012;55:37–46.
38. Yu Z, Zhang B, Sun W, Zhang F, Li Z. Phylogenetically diverse endozoic fungi in the South China Sea sponges and their potential in synthesizing bioactive natural products suggested by PKS gene and cytotoxic activity analysis. Fungal Divers. 2013;58:127–41.
39. Imhoff JF, Stöhr R. Sponge-associated bacteria: general overview and special aspects of the diversity of bacteria associated with *Halichondria panicea*. In: Müller WEG, editor. Sponges (Porifera), Marine molecular biotechnology, vol. 1. New York: Springer; 2003. p. 35–57.
40. Ahn YB, Rhee SK, Fennell DE, Kerkhof LJ, Hentschel U, Haggblom MM. Reductive dehalogenation of brominated phenolic compounds by microorganisms associated with the marine sponge *Aplysina aerophoba*. Appl Environ Microbiol. 2003;69:4159–66.
41. Karpushova A, Brummer F, Barth S, Lange S, Schmid RD. Cloning, recombinant expression and biochemical characterization of novel esterases from *Bacillus* sp associated with the marine sponge *Aplysina aerophoba*. Appl Microbiol Biotechnol. 2005;67:59–69.
42. Mohapatra BR, Bapuji M, Sree A. Production of industrial enzymes (amylase, carboxymethylcellulase and protease) by bacteria isolated from marine sedentary organisms. Acta Biotechnol. 2003;23:75–84.
43. Mohapatra BR, Bapuji M. Characterization of acetylcholinesterase from *Arthrobacter ilicis* associated with the marine sponge (*Spirastrella* sp.). J Appl Microbiol. 1998;84:393–8.
44. Mohapatra BR, Banerjee UC, Bapuji M. Characterization of a fungal amylase from *Mucor* sp. associated with the marine sponge *Spirastrella* sp. J Biotechnol. 1998;60:113–7.
45. Mohapatra BR, Bapuji M. Characterization of urethanase from *Micrococcus* species associated with the marine sponge (*Spirastrella* species). Lett Appl Microbiol. 1997;25:393–6.

46. Yung PY, Kjelleberg S, Thomas T. A polyphasic approach to the exploration of collagenolytic activity in the bacterial community associated with the marine sponge *Cymbastela concentrica*. FEMS Microbiol Lett. 2011;321:24–9.
47. Shanmughapriya S. Optimization of extracellular thermotolerant alkaline protease produced by marine *Roseobacter* sp (MMD040). Bioprocess Biosyst Eng. 2008;31:427–33.
48. Shanmughapriya S, Kiran GS, Selvin J, Gandhimathi R, Baskar TB, Manilal A, et al. Optimization, production and partial characterization of an alkalophilic amylase produced by sponge associated marine bacterium *Halobacterium salinarum* MMD047. Biotechnol Biproc E. 2009;14:67–75.
49. Okamura Y, Kimura T, Yokouchi H, Meneses-Osorio M, Katoh M, Matsunaga T, et al. Isolation and characterization of a GDSL esterase from the metagenome of a marine sponge-associated Bacteria. Mar Biotechnol. 2010;2010(12):395–402.
50. Kiran GS. Optimization of extracellular psychrophilic alkaline lipase produced by marine *Pseudomonas* sp. (MSI057). Bioprocess Biosyst Eng. 2008;31:483–92.
51. Shanmughapriya S, Kiran GS, Selvin J, Thomas TA, Rani C. Optimization, purification, and characterization of extracellular mesophilic alkaline cellulase from sponge-associated *Marinobacter* sp. MSI032. Appl Biochem Biotechnol. 2010;62:625–40.
52. Selvin J, Kennedy J, Lejon DPH, Kiran GS, Dobson ADW. Isolation identification and biochemical characterization of a novel halo-tolerant lipase from the metagenome of the marine sponge *Haliclona simulans*. Microb Cell Factories. 2014;11:72.
53. Feby A, Nair S. Sponge-associated bacteria of Lakshadweep coral reefs, India: resource for extracellular hydrolytic enzymes. Adv Biosci Biotechnol. 2010;1:330–7.
54. Dupont S, Carre-Mlouka A, Domart-Coulon I, Vacelet J, Bourguet-Kondracki M-L. Exploring cultivable Bacteria from the prokaryotic community associated with the carnivorous sponge *Asbestopluma hypogea*. FEMS Microbiol Ecol. 2014;88:160–74.
55. Su J, Zhang F, Sun W, Karuppiah V, Zhang G, Li Z, et al. A new alkaline lipase obtained from the metagenome of marine sponge *Ircinia* sp. World J Microbiol Biotechnol. 2015;31:1093–102.
56. Han Y, Yang B, Zhang F, Miao X, Li Z. Characterization of antifungal chitinase from marine *Streptomyces* sp. DA11 associated with South China Sea sponge *Craniella australiensis*. Mar Biotechnol. 2009;11:132–40.
57. Han Y, Li Z, Miao X, Zhang F. Statistical optimization of medium components to improve the chitinase activity of *Streptomyces* sp. DA11 associated with the South China Sea sponge *Craniella australiensis*. Process Biochem. 2008;43:1088–93.
58. Zhang H, Zhang F, Li Z. Gene analysis, optimized production and property of marine lipase from Bacillus pumilus B106 associated with South China Sea sponge *Halichondria rugosa*. World J Microbiol Biotechnol. 2009;25:1267–74.
59. Collins CM, D'Orazio SE. Bacterial urease: structure, regulation of expression and role in pathogenesis. Mol Microbiol. 1993;9:907–13.
60. Su J, Jin L, Jiang Q, Sun W, Zhang F, Li Z. Phylogenetically diverse *ureC* genes and their expression suggest the urea utilization by bacterial symbionts in marine sponge *Xestospongia testudinaria*. PLoS One. 2013;8:e64848.
61. Selvin J. Exploring the antagonistic producer *Streptomyces* MSI051: implications of polyketide synthase gene type II and a ubiquitous defense enzyme phospholipase A2 in host sponge *Dendrilla nigra*. Curr Microbiol. 2009;58:459–63.
62. Hutchinson CR. Polyketide and non-ribosomal peptide synthases: falling together by coming apart. Proc Natl Acad Sci U S A. 2003;100:3010–2.
63. Bayer K, Scheuermayer M, Fieseler L, Hentschel U. Genomic mining for novel FADH2-dependent halogenases in marine sponge-associated microbial consortia. Mar Biotechnol. 2013;15:63–72.
64. Piel J, Hui D, Wen G, Butzke D, Platzer M, Fusetani N, et al. Antitumor polyketide biosynthesis by an uncultivated bacterial symbiont of the marine sponge *Theonella swinhoei*. Proc Natl Acad Sci U S A. 2004;101:16222–7.

65. Fieseler L. Widespread occurrence and genomic context of unusually small polyketide synthase genes in microbial consortia associated with marine sponges. Appl Environ Microbiol. 2007;73:2144–55.
66. Fisch KM, Gurgui C, Heycke N, Van Der Sar SA, Anderson SA, Webb VL, et al. Polyketide assembly lines of uncultivated sponge symbionts from structure-based gene targeting. J Nat Chem Biol. 2009;5:494–501.
67. Schirmer A, Gadkari R, Reeves CD, Ibrahim F, DeLong EF, Hutchinson CR. Metagenomic analysis reveals diverse polyketide synthase gene clusters in microorganisms associated with the marine sponge *Discodermia dissoluta*. Appl Environ Microbiol. 2005;71:4840–9.
68. Schleper C, Swanson RV, Mathur EJ, DeLong EF. Characterization of a DNA polymerase from the uncultivated psychrophilic archaeon *Cenarchaeum symbiosum*. J Bacteriol. 1997;179:7803–11.
69. Kennedy J. Functional metagenomic strategies for the discovery of novel enzymes and biosurfactants with biotechnological applications from marine ecosystems. J Appl Microbiol. 2011;111:7877–99.
70. Stanley GD. The evolution of modern corals and their early history. Earth-Sci Rev. 2003;60:195–225.
71. Tapanila L. Direct evidence of ancient symbiosis using trace fossils. Paleontol Soc Pap. 2008;14:271–87.
72. Ainsworth TD, Thurber RV, Gates RD. The future of coral reefs: a microbial perspective. Trends Ecol Evol. 2010;25:233–40.
73. Krediet CJ, Ritchie KB, Paul VJ, Teplitski M. Coral-associated micro-organisms and their roles in promoting coral health and thwarting diseases. Proc R Soc B. 2013;280:20122328.
74. Klaus JS, Frias-Lopez J, Bonheyo GT, Heikoop JM, Fouke BW. Bacterial communities inhabiting the healthy tissues of two Caribbean reef corals: interspecific and spatial variation. Coral Reefs. 2005;24:129–37.
75. Sun W, Anbuchezhian R, Li Z. Association of coral-microbes, and the ecological roles of microbial symbionts in corals. In: Goffredo S, Dubinsky Z, editors. Medusa and her sisters The Cnidaria, past, present and future. Cham: Springer Press; 2016. p. 347–57.
76. Ritchie KB. Regulation of microbial populations by coral surface mucus and mucusassociated bacteria. Mar Ecol Prog Ser. 2006;322:1–14.
77. Shnit-Orland M, Kushmaro A. Coral mucus-associated bacteria: a possible first line of defense. FEMS Microbiol Ecol. 2009;67:371–80.
78. Anithajothi R, Duraikannu K, Umagowsalya G, Ramakritinan CM. The presence of biomarker enzymes of selected scleractinian corals of Palk bay, Southeast Coast of India. Biomed Res Int. 2014;2014:684874.
79. Anithajothi R, Nagarani N, Umagowsalya G, Duraikannu K, Ramakritinan CM. Screening, isolation and characterization of protease producing moderately halophilic microorganism *Halomonas meridiana* associated with coral mucus. Toxicol Environ Chem. 2014;96:296–306.
80. Poosarla A, Tulasi CDSLN, Rajan PR. Isolaton of soft corals associated fungi from andaman and nicobar marine water and screening for antimicrobial and protease activity. J Pure Appl Microbiol. 2012;6:221–9.
81. Puspasari F, Nurachman Z, Noer AS, Radjasa OK, van der Maarel MJEC, Natalia D. Characteristics of raw starch degrading α-amylase from *Bacillus aquimaris* MKSC 6.2 associated with soft coral *Sinularia* sp. Starch. 2011;63:461–7.
82. Selvin J, Kennedy J, Lejon DPH, Kiran GS, Dobson ADW. Isolation identification and biochemical characterization of a novel halo-tolerant lipase from the metagenome of the marine sponge *Haliclona simulans*. Microb Cell Fact. 2012;11:72.

Appendices

Z. Li (ed.), *Symbiotic Microbiomes of Coral Reefs Sponges and Corals*,
https://doi.org/10.1007/978-94-024-1612-1

Appendix 1: Pictures of Sponges and Corals in the South China Sea Provided by Mr. Bing Wang

Demospongiae sponge

Demospongiae sponge

Demospongiae sponge

Demospongiae sponge

Demospongiae sponge

Demospongiae sponge

Demospongiae sponge

Demospongiae sponge

Demospongiae sponge

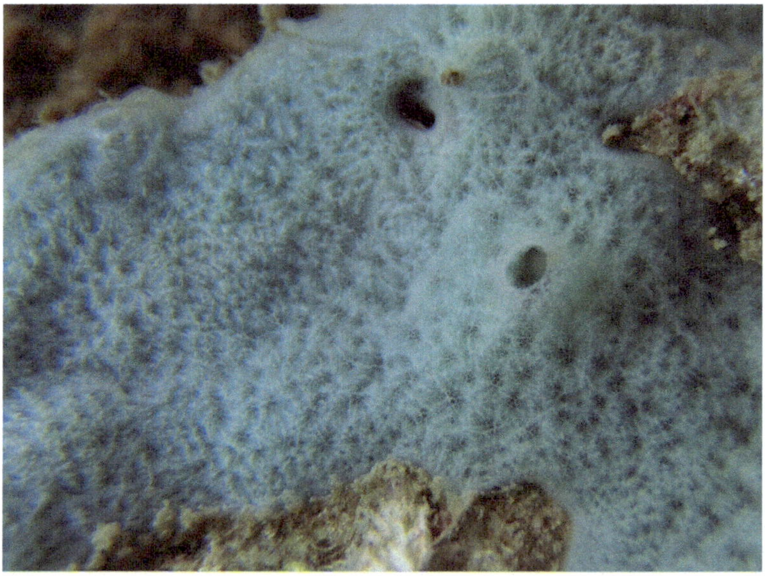

Demospongiae sponge

Demospongiae sponge

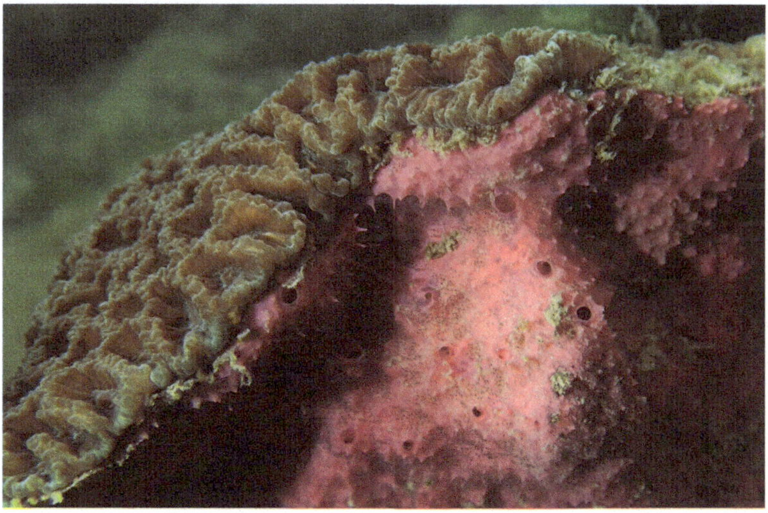

Demospongiae sponge

Demospongiae sponge

Demospongiae sponge

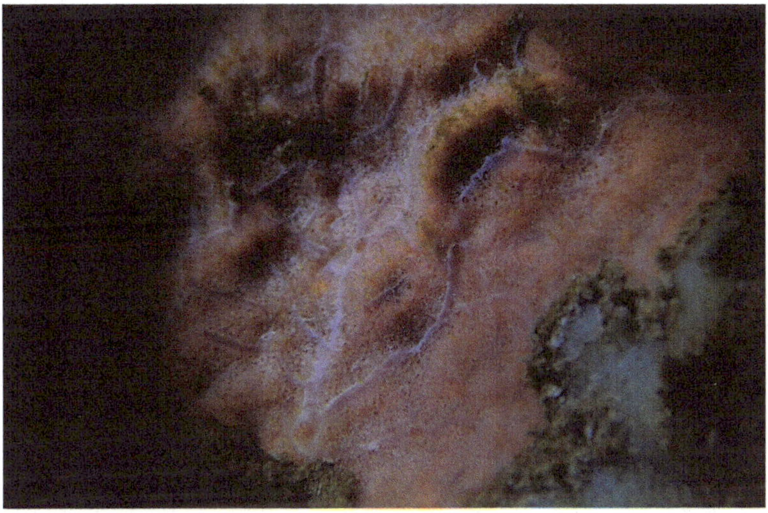

Demospongiae sponge

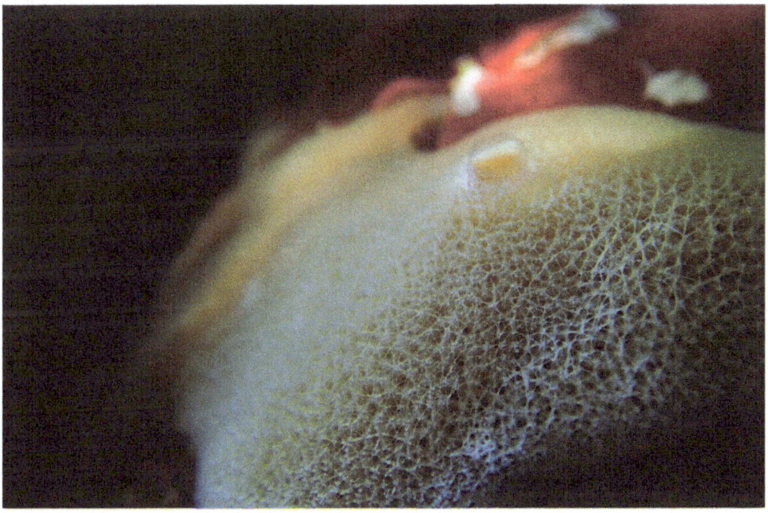

Demospongiae sponge

Demospongiae sponge

Demospongiae sponge

Demospongiae sponge

Demospongiae sponge

Turbinaria peltata

Porites aranetai

Goniopora stutchburyi

Goniopora columna

Pavona decussata

Montastrea curta

Leptastrea pruinosa

Platygyra carnosus

Favia speciosa

Favia maritima

Favia lizardensis

Favites pentagona

Favites acuticollis

Favites abdita

Hydnophora exesa

Acanthastrea echinata

Montipora peltiformis

Acropora solitaryensis

Acropora digitifera

Acropora pruinosa

Appendix 2: Pictures of Sponges and Corals in the South China Sea Provided by Mr. Baolin Liao

Goniopora djiboutiensis

Favites flexuosa

Favia speciosa

Favia maritima

Tubastraea coccinea

Goniopora lobata

Favia lizardensis

Pavona decussata

Acropora pruinosa

Montipora peltiformis

Xestospongia testudinaria

Chalinula sp.

Haliclona cymaeformis

Printed by Printforce, the Netherlands